I0061821

Edwin Lankester

The family medical guide

A complete popular dictionary of medicine and hygiene

Edwin Lankester

The family medical guide
A complete popular dictionary of medicine and hygiene

ISBN/EAN: 9783337201159

Printed in Europe, USA, Canada, Australia, Japan

Cover: Foto ©berggeist007 / pixelio.de

More available books at **www.hansebooks.com**

THE

FAMILY MEDICAL GUIDE:

A COMPLETE POPULAR DICTIONARY

OF

MEDICINE AND HYGIENE.

COMPRISING

ALL POSSIBLE SELF-AIDS IN THE TREATMENT OF DISEASES, ACCIDENTS, EMERGEN-
CIES, ETC., WITH ARTICLES ON GENERAL PHYSIOLOGY; ON DIET AND FOOD;
ON THE DIFFERENT DRUGS, PLANTS, AND MEDICAL PREPARATIONS USED
IN GENERAL PRACTICE; DEFINITIONS OF TECHNICAL TERMS USED
IN MEDICINE; RECIPES FOR THE PREPARATION OF EVERY-
THING USEFUL IN THE DOMESTIC TREATMENT OF
DISEASE, ETC., ETC., ETC.

EDITED BY

EDWIN LANKESTER, M.D., F.R.S., Etc., Etc.,

ASSISTED BY DISTINGUISHED MEMBERS OF THE ROYAL COLLEGES OF PHYSICIANS
AND SURGEONS, LONDON.

American Edition,

*CAREFULLY REVISED, WITH ADDITIONS AND AMENDMENTS,
BY COMPETENT AUTHORITIES.*

NEW YORK:

E. [...]N AND COMPANY,

25 BOND STREET.

1881.

Copyright by
E. R. PELTON & CO.,
1880.

PREFACE.

—•◦•—

THE rapid progress of medical science has rendered antiquated much of that which the public, as distinguished from professional medical men, have been taught by previous books on Domestic Medicine. And it is to be further noted, that no work whatever which deals with the *preservation* of health —that is to say, Hygiene—is now available for the use of intelligent but non-professional men. With a view to the production of a work which should remedy these deficiencies, the various subjects here treated have been referred to the most competent authorities—men specially skilled in the departments of medical science of which they treat. By this means the most recent acquisitions in medicine and surgery have been made available for popular use—it being the aim of the writers, while avoiding all technical phraseology, to expound their subjects in such a fashion as to be intelligible to all, and at the same time to secure the most rigorous scientific accuracy.

By this means, further, a knowledge of that all-important matter, the maintenance of personal and public health, may be generally communicated ; for as day by day the mode in which diseases are spread becomes clearer, so day by day the plans for arresting their diffusion become better understood and more readily applicable. It has, for example, been tolerably clearly made out that overcrowding, bad ventilation, and improper nourishment are the main, if not the sole, causes of Typhus Fever ; that bad drainage and the mingling of sewage with water are the origin of Typhoid Fever, and the chief means of spreading Cholera ; that by the vigorous carrying out of vaccination and re-vaccination Small-pox may be stamped out ; and so on. These matters, of *vital* importance to the public, which have, if not concealed, certainly not been made plain to them hitherto, are here fully discussed.

THE FAMILY MEDICAL GUIDE will be found to contain special references to those accidents which may daily befall any of us, where immediate help makes all the difference between life and death, but where no skilled medical aid is at hand. On such occasions an intelligent man or woman, using the information here furnished, may be of immense service. So also there are times in family life when a knowledge of the signs or *symptoms* of disease may warn an anxious parent in time to save the life of a beloved child, whereas, did no such knowledge exist, the malady might be allowed to drift onward till past all

remedy.   Instances might easily be multiplied—let us be contented with refer-
ring to a ruptured blood-vessel and the inset of croup.

Briefly, then, to diffuse a knowledge of medical matters in a manner intelligible to all, but in matter strictly accurate, is the aim of this book.

It only remains to add, that in preparing the present American edition, the
entire work has been subjected to a most careful, minute, and laborious revision,
the language being rendered clearer and more simple, and the directions more
specific and precise.   Besides this, articles which were applicable only to English
local conditions have been stricken out and others (entirely new) substituted
for them ; some important articles which seemed inadequate have been
expanded ; and American statistics and regulations have been substituted in
those numerous cases where statistics and regulations form a part of the exposi-
tion.   Nevertheless, it should be added, that with innumerable changes in
matters of detail, the original plan and purpose of the MEDICAL GUIDE, and its
characteristic and universally-conceded excellences, have been scrupulously
preserved.

Among these excellences are its great scope and comprehensiveness.   It
contains many times more articles than any other book of household medicine,
and while special attention is given to those common diseases which frequently
occur, the other and rarer "ills that flesh is heir to" are also treated of in a
complete and authoritative manner.

Another feature of great value, on which much pains have been bestowed, is
the articles on the *Medicines, Drugs, Plants, and Preparations used in medical
practice*, each of which has a separate article.   This is of great value, because
there are many occasions when one knows that a certain medicine is good for
a certain disease or state of the system, yet does not know how or in what
doses to administer it, or what effects it ought to produce.   The information
given on all these points is very clear and precise.

Still another feature, which will probably prove more frequently helpful than
any other, is the particular attention given to those *sudden emergencies or accidents*
which are liable to occur at any time and which require immediate action.
These are treated of in such articles as ACCIDENTS, BRUISES, BLEEDING,
BURNS and SCALDS, BITES and STINGS, CHOKING, DROWNING, DISLOCATIONS,
FRACTURES, FROZEN LIMBS, GUNSHOT WOUNDS, RAILWAY ACCIDENTS,
POISONING, SNAKE-BITES, SUNSTROKE, and SPRAINS.

Special attention is also invited to a class of articles which need only be
named in order to show their character and value ; such as those on FOOD,
DIET, DIGESTION, and INDIGESTION ; on AIR and VENTILATION ; on WATER,
ALCOHOL, WINES, etc.; on INSANITY ; on NURSING and FEVERS ; on CLIMATE,
HEALTH RESORTS, and MINERAL WATERS ; on BATHING and GYMNASTICS ;
on EYE, VISION, and OPHTHALMIA ; on EAR and DEAFNESS ; on the HEART,
the LIVER, the KIDNEYS, the BRAIN, the STOMACH, the MUSCLES, the NERVES,
the BLOOD, and other subjects in human physiology.   Some of these articles,
in length and importance, are equivalent to special treatises.

A last and most important consideration is that, as most of the articles were written by London physicians of recognized ability and the highest reputation, *the methods of treating diseases recommended in this book comprise the very latest discoveries and improvements in the science and practice of medicine.* In this respect it is probably twenty years ahead of the average medical practice in this country, outside the few largest cities ; and many, perhaps most, physicians would learn much from it regarding the uses of medicines and the improved modes of treating disease.

NOTE BY DR. LANKESTER.

I SHOULD not have undertaken the editorship of this work had I not been fully assured that the professional gentlemen who have written the chief part of the articles were fully competent to the task. They all possess the highest qualifications, and some of them are attached to public institutions, so that their individual opinions may be regarded as of importance.

Being especially anxious that the public should be instructed more or less on the subject of the laws of health, I have not hesitated to connect my name with this book, for I am deeply convinced that it is for the benefit of the public that they should be instructed in the laws of life. In as far as the prevention of disease could be dealt with in this book, it has been done.

On the whole, I believe the book will be found more up to the science of the time than any previous attempt made to popularize the practice of medicine and surgery. Should reference to this dictionary lead persons to read some of the scientific treatises on physiology, or to interest themselves in introducing into schools the study of this the most important branch of human knowledge, it would be the highest reward and the greatest gratification I could obtain in having connected my name with it.

EDWIN LANKESTER, M.D., F.R.S., ETC., ETC.

HINTS TO THE READER.

1. The arrangement of the articles is strictly alphabetical, as in a dictionary or cyclopædia. The great advantages of this system will be made apparent by even a very slight trial.

2. Diseases are usually treated under their medical names, but the popular name of each is inserted in its proper alphabetical place, and numerous cross-references enable anything wanted to be readily found.

3. In consulting an article, bear in mind that the other articles referred to in it from time to time, or at the end, will often contain important information not found in the article itself.

4. When a technical or unusual medical term is used, an explanation of it will always be found under the word in its proper alphabetical place.

5. In consulting the article on any disease, it is well, after reading what is said about the disease itself, to read also what is said about the medicine that may be recommended. All of the commonly used drugs and medicines are treated of in their proper alphabetical places.

6. The reader is earnestly recommended to glance through the book, page by page. Only in this way can one learn what it really contains; and many useful things will be found which would never be specially looked for.

FAMILY MEDICAL GUIDE.

☞ *The arrangement of topics in this book is strictly alphabetical, as in a dictionary or cyclopædia.*
☞ *Be sure to read "Hints to the Reader" on preceding page.*

A.

ABDOMEN is the name given to that large cavity situated in the lower part of the trunk, and bounded above by the chest (from which it is separated by a muscle called the diaphragm), and below by the pelvis. All around are various muscles and membranes, and outside of all the skin ; these various layers are called the abdominal walls. The cavity is lined by a smooth membrane, the peritoneum, which enables the different viscera to move freely upon one another. The contents are numerous and important, and consist—1. Of those organs which are essential for the absorption of the food which is swallowed by the mouth ; namely, the stomach, intestinal canal, liver, and pancreas. 2. Of the kidneys needful for the removal of different materials from the blood which flow away in the urine. 3. Of various blood-vessels which convey blood to and from a part ; and of the spleen which seems to play a part in the proper formation of the blood. The position of the viscera may be more readily understood by mapping out the front surface of the abdomen. Take two pieces of tape and lay each vertically in the same line as the nipples ; lay two more transversely across the abdomen, the one at the level of the lower ribs where they can be felt to terminate, and the other lower down over the prominent crests of the hip-bones. The region is now divided into nine compartments, thus—

Right Side.	Centre.	Left Side.
Hypochondriac.	Epigastric.	Hypochondriac.
Lumbar.	Umbilical.	Lumbar.
Inguinal.	Hypogastric.	Inguinal.

The Liver occupies the right hypochondriac and the epigastric regions, and in some cases may be so enlarged as to reach into the lumbar and umbilical regions. The stomach is found chiefly in the epigastric region, but its extent varies with the amount of distension of the organ. The Spleen lies in the left hypochondriac region. The Kidneys lie one on either side of the spine, far back in the lumbar regions. The large and small intestines are coiled up in the remainder of the cavity and move freely upon each other. The food after entering the stomach passes down the small intestines and then down the large bowel into the rectum, whence it is discharged. In the stomach the food is acted upon by the gastric juice, and all the albuminoid substances, as meat, the gluten of bread, etc., are here dissolved and prepared for absorption. The liver and pancreas pour their secretions into the upper part of the small intestines, and mix with the food about six inches after it has passed the stomach ; they act upon the fatty or oily part of the food, and enable it to be afterward readily absorbed. The whole of the intestinal canal, which is about 25 feet long, is freely supplied with vessels which absorb the soluble parts of the food and supply the blood with new material, which is essential for the human economy. The walls are partly made of involuntary muscular fibres, which are arranged in a circular and longitudinal manner round the bowel ; when these fibres contract, the food is propelled along the whole length of the intestines by degrees. The Kidneys are the organs by which the urine is excreted from the body ; besides water there are various salts and organic matter which are constantly being removed from the blood, and which, if retained, may cause injurious consequences. The abdomen varies much in size ; in very fat people it may attain very large dimensions from the accumulation of fat in the abdominal walls and around the viscera. The bowels may be full of flatus or gas, and so cause an increase in size. Any tumors in the abdomen (such as ovarian, hepatic, or those formed in the liver), and pregnancy, vastly alter both its shape and size. Cysts or cavities containing fluid, have a similar effect, and then the prominence of any particular part depends upon the situation of the new growth. The abdominal cavity is sometimes filled with fluid, and then the patient is said to be suffering from ascites. In the ordinary process of breathing the diaphragm descends and increases the capacity of the chest ; when from any cause this is interfered with, shortness of breath is the result ; so any tumors or accumulations of fluid, fat, or gas, tend to produce embarrassed breathing by preventing free action of the diaphragm. In some women who have borne many children, the abdominal walls become flabby and weak, and remain distended ; in such, numerous transverse lines over the lower part of the abdomen show the previous distension.

ABLUTION, a washing away, internal or external. See BATHING.

ABNORMAL, a term used by medical men to denote anything irregular, out of order, or not in accordance with health.

ABORTION means the premature expulsion of the fœtus from the womb. Sometimes the word miscarriage, or premature labor, is applied to such an accident, especially during the later months of pregnancy. The causes which lead to abortion are numberless, but in the main they may be grouped into causes affecting the life of the fœtus either directly or indirectly. The condition of the mother has a powerful influence, certain states of constitution having a great tendency to cause expulsion of the immature fœtus. When this constitutional susceptibility exists, and especially when there has been a habit of aborting, very slight causes will bring it on. In such individuals, drawing a tooth, running up and down stairs, severe coughing, or any violent emotion, will suffice to bring on movements of the womb. Hence it is of the utmost importance to avoid any of these exciting causes. Some of these causes operate by separating the attachments of the fœtus from the mother ; but others, as scarlet-fever, smallpox, typhus, etc., which commonly cause abortion, first of all kill the fœtus, and then lead to its expulsion. The death of the fœtus is followed by the expulsion, but by no means always at the same period. Bleeding, too, into the womb, commonly leads to abortion. One of the most certain causes of abortion is syphilis, whether affecting the mother or child, but especially if both are under its influence.

Symptoms.—The signs which announce the onset of abortion are not always very noticeable. They are generally languor, uneasiness, and some pain in the back. After the pains begin they resemble those of ordinary labor. There is some slight discharge of glairy matter often tinged with blood, and pains beginning in the back and loins extend round into the abdomen and downward into the thighs. These recur at intervals until the fœtus is expelled. In patients who have acquired the habit of aborting, the fœtus often comes away with hardly any inconvenience. Sometimes there is much bleeding with abortion. This is technically known as " flooding."

Treatment.—The great thing to know in dealing with abortion is up to what period it can be stopped, and after what period it is to be fostered. Roughly it may be said, that bleeding is the sign that expulsion has become inevitable ; if there are slight pains and glairy discharge only, we may hope to arrest it ; but if the pains begin to come on at regular intervals, and there be bleeding, we can hardly hope to avert the mischief. To stop or prevent an abortion, the first thing is *absolute rest.* The body should only be lightly covered, and all excitement of every kind avoided. The patient should have cooling drinks, and cold should be applied to the abdomen. With these precautions a dose of opium will sometimes arrest the contractions of the womb. Thirty drops of laudanum are required for this purpose, or small doses of tincture of Indian hemp, 12 or 15 drops for a dose, in cold water, to be given every 2 or 3 hours. Should these measures fail, the fœtus must be allowed to come away, due precaution being taken against loss of blood. Plugging, as it is called, is the best precaution, but here the case passes into a stage where skilled attendance is requisite if obtainable.

ABORTION, CRIMINAL.—By this is meant unlawful attempts, successful or unsuccessful, to procure the premature expulsion of the fœtus. Such attempts are by no means unfrequent, with a view to avoid shame and disgrace, or even to avoid inconvenience, but it should be distinctly understood that not only is the act felony, but should death occur to the unfortunate woman the crime rises to murder in the eye of the law. Moreover, it is to be understood that abortion is at all times attended with risk to life, and that abortion unlawfully induced is more dangerous than that arising from natural causes. In all cases it may be said that the risk to the mother equals the risk to the child, and though such a consideration would not always weigh with the unhappy woman who desires to be rid of the proof of her guilt, it will with an accomplice. The plans commonly had recourse to for procuring the expulsion of the fœtus are of two kinds. One is by giving medicines which acts on the womb, either directly or indirectly : these often give rise to inflammation of the interior of the abdomen, and so death. Another plan, still more dangerous in uninstructed hands, is to obtain the death of the fœtus by means of instruments, and subsequently wait for its expulsion. As already said, this often leads to the death of the mother before it does to that of the child.

ABRASION, a rubbing off of the epidermis of the skin leading to the exposure of the true skin below. In treating abrasions from any cause, all dirt or poisonous matter should be carefully washed away with lint and tepid water. The wound may then be dressed in any simple way with lint dipped in tepid water and covered with oil-silk, or a simple dressing of lint and cerate.

ABSCESS is a term commonly applied to a painful and inflamed swelling which after a certain course, in most instances very rapid and acute, in others slow and indolent, terminates in the discharge of a yellowish creamy fluid called pus or *matter.* This affection varies very much in its extent and severity, and in the seat of the swelling. A gumboil, a whitlow, and the large and painful collection of pus frequently formed in the female breast during suckling, are all instances of abscess. Though commonly recognized as an affection of the surface of the body, abscess may attack some important internal organ, as the brain, the liver, or a kidney. Indeed, there is no structure or organ which enjoys immunity from the possible deposit of pus and formation of abscess. In considering the *symptoms* of the affection it is necessary to distinguish those swellings which are much inflamed, very painful and rapid in their course, from those which grow slowly and gently, and with little if any redness of the skin. The symptoms by which we may know the first variety of swelling or *an acute inflammatory abscess* are these : A very hot and painful swelling covered by stretched skin of a bright red hue, most intense at the centre. As the swelling increases in size the pain becomes very severe, and has a characteristic throbbing or pulsating character. In the further course of the affection, the skin and soft parts around the inflamed swelling become puffy and retain for a short time the impression of the finger. As the centre of the abscess becomes more painful and inflamed it loses its hardness and gradually *ripens* or breaks down into pus. The skin at this part becomes thinner, more prominent, and loses its bright red color, presenting the well-known sign of *pointing*, a light yellow or bluish spot. The whole swelling is now soft, and by making gentle pressure alternately with the fingers of each hand, a sensation may be generally felt of a small wave of fluid moved from side to side. The abscess finally bursts and discharges the contained pus through one or more small apertures formed in the thinnest portion of skin. The discharge at first is profuse, and consists of a thick yellowish fluid ; as the cavity of the abscess contracts and closes it becomes clear and thin. During the process of healing the outer layers of skin about the seat of the abscess peel off. The progress of an abscess toward ripening and the discharge of pus is usually accompanied by constitutional symptoms proportional in severity to the size of the swelling and the amount of inflammation. ,These symptoms are : shivering, general uneasiness, feverishness, headache, and wandering pains in the back and joints. Acute abscess is generally the result of debility or a bad state of the blood, and is often met with after fever and during suckling (see BREAST). In persons who have subsisted for some time on bad or insufficient food, any slight injury, as a bruise or cut, may result in inflammation and the formation of pus. Inflammatory diseases of bones and glands in scrofulous subjects are frequent causes of abscess. In the second variety of abscess the symptoms are much less severe. The swelling increases in size very slowly, and with little pain or tenderness. The skin remains for a long time free from inflamma-

tion or puffiness, until the pus has collected in such quantity as to cause its distension and attenuation. There is then a slight blush of redness and the matter or pus is discharged through a small opening, as in the acute abscess. This variety is known by the name of chronic or cold abscess.

Treatment.—In the early stage of a painful inflamed swelling on the surface of the limbs or the trunk an attempt may be made to prevent the formation of pus by applying cold lotions and leeches, and keeping the affected part at perfect rest. If the patient, however, has had much shivering and complains of throbbing pain about the swelling, one should at once carry out such measures as may further the *ripening* and *pointing* of the abscess. Nothing favors the rapid formation of healthy pus so much as nourishing and easily digestible food, as soups, beef-tea, eggs, etc. Ale and small quantities of wine, or some spirit, may be given. The severe and throbbing pain of the abscess will be relieved by the frequently repeated application of hot poultices made of linseed meal, bread, or bran. A prolonged stay in a close room is to be avoided, as fresh pure air is almost as essential for speedy recovery as good diet. When the skin in the centre of the abscess has become thin and changed in color, and there is a distinct feeling of fluid beneath, an incision may be made with a lancet at this part for the purpose of letting out pus. The fluid should be allowed to flow away spontaneously, as forcing it out by pressure not only causes much pain, but increases the inflammation. After the abscess has been opened or has burst spontaneously the application of the poultices should still be continued for some days. When the discharge has become thin and scanty the poultice may be replaced by water-dressing ; that is, by pieces of lint dipped in cold water and covered by some waterproof material, as oil-silk or gutta-percha tissue.

ABSINTHE is a strong liqueur much used in some parts of Europe, especially France. It is flavored with wormwood, which itself is a bitter tonic, but the liqueur used too freely gives rise to symptoms somewhat resembling those of chronic alcoholism (see ALCOHOLISM), but differing in certain minor respects. The word Absinthism has been introduced to designate this state. See WORMWOOD.

ABSORBENTS are medicines which soak up, or in any manner neutralize acid or noxious matter in the stomach and bowels. (See ANTACIDS and CHALK.) Another kind of Absorbents are a set of minute vessels which are distributed over the whole body, and have the power of soaking up the food from the stomach and intestines, and also the effete materials in all parts of the body, and carrying them into the blood.

ABSORPTION is a physiological term applied to that process by which the chyle is taken up from the food in the intestines as well as the removal of the effete materials of the tissues of the body by the vessels called absorbents, or lymphatics. The fluid in these vessels is called lymph, and that in the lacteals chyle. The two great factors of the blood are lymph and chyle. See CHYLE, LYMPHATICS.

ABSTINENCE.—This term is commonly applied to complete or partial deprivation of food by one's own voluntary act. As such, it may be productive of good, or it may be productive of harm. Certain articles of food and drink invariably disagree with certain individuals, and abstinence from these is only an act of wisdom on the part of the person so affected. On the other hand, abstinence is quite capable of doing great harm. A deficient supply of duly nutritious food inevitably leads to disease ; no matter what is the reason for the deficiency. Voluntary abstinence from food enjoined by certain churches during certain periods often does great harm if injudiciously carried out. In any doubtful case it should only be done under proper medical authority.

ACACIA.—The gum Acacia which ordinarily occurs in rounded masses called *tears*, is procured from various species of the Acacia tribe growing in the desert parts of Africa. There it is sometimes used as food. In this country the gums commonly employed for domestic, commercial, and other purposes are called gum Acacia, but in reality are the product of many other trees, cherry gum being largely used. In medicine, gum Acacia is chiefly used to suspend heavy powders when given in liquid, bismuth for example ; and, when swallowed slowly, to allay cough. The gum, in solution, is sometimes given after a corrosive or irritant poison has been swallowed, to protect the coats of the stomach.

ACCIDENTS.—Of all the evils which flesh is heir to, there is nothing in which the benefit of present help in time of need is so welcome as in " an accident." Be it ever so slight or ever so severe, from a small bruise or sprain to an alarming hemorrhage or railway " smash," the assistance, or aid, rendered by some one whose presence of mind and common-sense are equal to the occasion, is invaluable. We intend in this article to point out, in the various forms which an accident may take, *What to do.* In the first place, if possible, before entering into any detail—*Send some one for the nearest physician.* The accidents which most commonly happen are bruises, sprains, burns, scalds, cuts, punctures, foreign bodies, such as splinters, fish-bones, needles, shot, etc., in the various structures of the body, or in its several passages, such as the

nose, ears, throat, eyes, etc. ; broken bones, bones put out of place, or *dislocated,* serious injuries to large blood-vessels, suffocation from drowning or hanging, suspended animation, poisoned wounds, bites from rabid animals or snakes. We shall, therefore, proceed to offer some ready methods of dealing with each of these, and bearing in mind that a little knowledge is a dangerous thing, particularly in Surgery, be as untechnical, plain, and straightforward in our directions as possible.

The following articles should be in every home : Old linen, which may be formed into lint (charpie) by being scraped with a blunt knife on one side ; laths of various lengths ; roller bandages, which may be made from old sheeting, and torn about $2\frac{1}{2}$ or 3 inches wide ; some cotton wool ; a few broad tapes ; some old wide handkerchiefs or neckties ; a pair of good scissors, and a pair of forceps or pliers ; adhesive or diachylon plaster. The simplest materials can be turned to account. Thus, old newspapers, rolled up, make excellent splints ; bandboxes, with the bottom knocked out, are capital makeshifts, if the bedclothing is required to be kept off a broken or wounded limb. The nap of an old hat plucked off and plugged into a cut is often of great service. A very small amount of mechanical skill may be turned to the most valuable account, such as the use of a pocket-knife saw, for the manufacture of extempore splints, which should be well padded with cotton wool, old linen, handkerchiefs, tow, or any handy material, and applied comfortably, but firmly, on both sides of a broken bone which has been previously *reduced*—that is, its ends put in apposition. (See SPLINTS.) Bandages and slings can be made out of old sheeting, towels, or handkerchiefs, and, to facilitate application, should be rolled up. (See BANDAGES.)

Bleeding.—Hemorrhage, or bleeding (see HEMORRHAGE), is of two kinds, namely, arterial or venous : in the former the blood is scarlet, and spurts out of the wound in jets ; in the latter it is dark purple, and rather oozes than gushes out. In the more trivial cuts, the edges of the wound merely require to be brought together with strips of adhesive plaster, *not too close together*, and if the bleeding be obstinate, a pad of lint should be firmly bound over the seat of wound by a roller-bandage. (N.B.—Adhesive plaster should always be taken off when it gets *black*.) A piece of lint steeped in some styptic, or perchloride of iron, is of great service. If a main artery be wounded, pressure should be made immediately with the thumb or fingers, *between* the *wound* and the *heart*, bearing in mind that the arteries carry blood *to* the extremities ; or a *tourniquet*, as it is termed, be extemporized, by tying a knot in a handkerchief, and tying the handkerchief round the limb, so that the knot presses immediately *above* the wound ; a piece of stick thrust between the handkerchief and the limb, and a twist or two given it, will arrest the hemorrhage most effectually. In the case of some large vein being wounded, a stout pad of lint, or linen, graduated, *i.e.*, somewhat conical in form, should be thrust into the wound, with the apex of the pad downward, and retained by means of a well-applied roller-bandage. The great vessels most commonly wounded, because most exposed, are : the great artery of the thigh (femoral), the course of which may be indicated thus—divide the fold of the groin, and in an imaginary line draw from *its centre* to the *middle* of the inside of the knee, the vessel will be felt beating ; the two small arteries at the wrist (radial and ulnar). The course of the great artery of the arm (brachial) may be thus indicated : a line drawn on the inner side of the swell of the great muscle (biceps), commencing on the other side of the armpit, and ending about the middle and rather to the inner side of the bend of the elbow. The great vessels of the neck (carotid artery and jugular vein) are sometimes cut in attempts at suicide, or stabbing, and are very difficult for any one but a surgeon to attempt to treat, and very rapid in the result, although great service may be done by immediate pressure with the thumb in the wound. (See CUT THROAT.)

Bruises.—If an injury be inflicted on the skin by some instrument which does not break it, a bruise or contusion is the result. Bruises vary in degrees of severity, the most simple being a discoloration of the skin accompanied by some amount of swelling and pain, the black or blue color being owing to some of the small superficial blood-vessels which are distributed to the skin having been burst by the blow, and the contained blood *effused*. The discoloration, however, does not usually come on until some little time after the receipt of the injury. In the more severe forms, large vessels may be ruptured and the blood escape into the surrounding textures, or the various structures of a limb, for example, may be entirely crushed, giving rise to conditions which will be more conveniently treated under other headings. (See ANEURISM, AMPUTATION.) *Treatment.*—Ordinary bruises may be treated by the application of lint steeped in cold water and laid upon the part ; if more severe, by the use of some stimulating liniment, such as the common ammonia liniment, or camphor liniment rubbed gently into the injured part when the pain is subsiding ; brandy, spirits of wine, vinegar, or a solution of alum or tannin, frequently do good. These remedies, with *perfect rest* of the seat of injury, and a due attention to the state of the bowels, and a moderate diet, will be found sufficient. If, after the receipt of a severe bruise, the swollen part appears to throb violently, the case is in all probability of a serious nature, and professional aid should be at once sought.

Sprains.—When, through any sudden or violent wrench, a joint, or tendons, or the structures connecting the several parts of the body, become strained, pain, swelling, and ecchymosis (bruise) occur ; and if this happen to any of the larger joints, such as the knee or elbow, the result may be most serious ; especially if the individual to whom the accident happens has been intemperate in drink and is of weak health. *Treatment.*—Perfect rest, and the parts maintained at rest by splints and bandages (see BANDAGES), if necessary ; warm fomentations, the bowels to be kept open, and the living moderate. Should the sprain be very severe, and if there be great heat, swelling, and throbbing, leeches' should be applied to the affected part, cold-water douching, and some stimulating liniment, such as camphor or opodeldoc.

Burns and Scalds.—For the more complete account and method of treatment of these accidents, refer to BURNS. The great thing to be attended to in these cases is that great care is to be taken *not to tear away the clothing from the burned or scalded surface*—supposing, of course, that the seat of the injury is in any part so covered—as by so doing the cuticle or scarf skin is stripped off, and a large, raw, ulcerating surface is the result, and the process of healing greatly complicated. (See BURNS.)

Bites or Stings.—In the case of stings of bees or wasps, as some faintness or depression results, stimulants should be at once administered, such as brandy and water, or sal-volatile ; and the seat of injury should be carefully searched for the sting, which is generally left in, while the wound should be treated with sal-volatile, vinegar, or eau de Cologne. Supposing the throat be stung in drinking, there is, of course, great danger of suffocation, and leeches may be applied outside the throat, and a gargle of hot salt-and-water used immediately, and medical attendance sought at once. In *snake bites*, powerful stimulants, such as hot brandy and water and ammonia, should be given freely. In the case of poisoned wounds of the fingers, etc., the cut or puncture should be immediately sucked, and bathed in warm water ; and no hesitation whatever should be shown in this proceeding, as the danger of the poison lies in the fact of its being absorbed *by the skin*, and not in its being introduced into the system *by the stomach*. The bleeding should be freely encouraged. Cauterizing the wound may be employed ; but it is doubtful if much good results from this proceeding, unless it be done thoroughly and on the instant, by some such agent as nitric acid, caustic, potash, carbolic acid, or even a red-hot iron. (See SNAKES, BITES OF.)

ACETIC ACID may be prepared in various ways, but that commonly used is obtained from the distillation of wood in appropriate retorts, whence it is also called pyroligneous acid. It is used in medicine in two forms—the strong, or glacial, and the dilute. The glacial acetic acid, so called from being normally solid or in the condition of ice, is chiefly used for external applications, the most frequent being for the destruction of warts.

ACHOLIA signifies absence of bile, and this occurs in acute atrophy of the liver, and in some other diseases of that organ. (See JAUNDICE.)

ACIDITY OF THE STOMACH is a frequent symptom of indigestion, and arises from the food taken being converted by decomposition into an organic acid of some kind. The treatment consists in avoiding those articles of diet which produce acidity, as sugar, butter, and starch, and in taking medicines which will correct acidity. One of the best of these is bicarbonate of potash, which may be taken with some tonic, as tincture of orange-peel, in doses of 10 or 15 grains three or four times a day. (See ANTACID, GASTRODYNIA, INDIGESTION.)

ACNE is a term given to the small raised spots so often met with in youth, on the back and over the shoulders and on the chin. It is most common about the age of puberty, and sometimes covers the face so much as to cause great disfiguration. It is met with in both sexes. The surface of the skin is not really level, but there are numerous little pits or pores which open on the surface, and end in small pouches called sebaceous follicles; now, when the channels get blocked up, the contents of these follicles increase, and as there is no outlet a little pimple is produced, with a small black depressed centre, marking the seat of obstruction; sometimes the spots are red from the vessels around becoming congested, and if there is much irritation a little matter may be found. It is very easy to squeeze out the contents by pressing the pimple between the fingers, or it may be done by using a watch-key in a similar way. In some cases they are very troublesome, and resist ordinary treatment. Cold bathing every morning, and frequent friction of the skin with rough towels so as to excite a healthy action, are very useful proceedings. Active exercise should be taken, and a diet not containing too much animal food or stimulants. A mixture of sulphur and alcohol rubbed together, so as to form a smooth paste, is an excellent application for acne; it may be rubbed over the skin every night and washed off in the morning. A popular remedy is to take every morning a teaspoonful of fresh yeast or balm in a glass of beer, and it certainly does seem to do good in some cases. When the spots occur on the chin or

upper lip, and are very numerous, it is best not to shave, as much irritation may be caused by that process. It is a disease generally which may last a long time, as fresh spots often appear as the old ones are healing.

ACONITE.—The common plant, *Aconitum Napellus*, or Monk's-hood, is one of the most deadly poisons known. Every portion of the plant is poisonous, but the root especially so. In medicine, preparations both of the green part and of the root are employed —the former furnishing an extract, the latter a weaker and a stronger tincture, the stronger intended for outward application only, and called a liniment. The tincture is the only preparation which should be used, and then with the greatest caution. When taken into the mouth it causes a curious tingling sensation, by and by followed by numbness. When swallowed, a similar sensation is produced in the hands and feet; but its most important action is on the heart, the number of whose beats is reduced by it, and the force of the pulse considerably lessened. The number of respirations per minute is also diminished. Should the use of the drug be carried too far, great muscular weakness is the result; extreme faintness is produced, and in fatal cases the poisoned individuals seem to die by stoppage of the heart's action. It is chiefly used for two purposes: as a means of relieving pain—that is, as an *anodyne*; and as a means of keeping down inflammation—that is, as an *antiphlogistic*. The latter practice has been seized upon by homœopathics as their own, the chief remedy they use in inflammation being aconite. For relieving pain, the remedy may be applied locally or given internally. It is chiefly used in neuralgia and that special form of the same malady known as sciatica. When given internally, the dose should not exceed *five minims or drops* of the tincture, and this should only be repeated after an interval of four or five hours and with great caution. When used externally, the best plan is to rub the tincture, mixed with a little soap liniment to facilitate the process, into the painful spot by means of a piece of warm flannel or soft glove. Care must be taken to see that there is *no crack or injury of the surface*, which would render its application in this manner most dangerous. It has also been used internally in acute rheumatism, in gout, and in certain forms of heart-disease.

ACONITIN is the alkaloid, as it is called of aconite, that is to say, its active principle, which happens to possess an alkaline reaction and otherwise to behave like an alkali. It is ordinarily obtained from the root, not of the *Aconitum Napellus*, but from that of the *Aconitum ferox*, a native of India, and formerly much used as a poison there. Aconitin is one of the most powerful poisons known, and should not be handled save by

competent persons. Its properties are those
of aconite in an exaggerated degree.

ACRID, a term applied to any substance
which produces irritation, more especially of
the stomach. Thus, poisons that produce inflammation, pain, and heat in the stomach
are called acrid.

ACTEA is the root of *Actæa racemosa,* a
native of North America ; it is commonly
called black snake-root, and a decoction of it
has been much used here in domestic practice for coughs. It has been strongly recommended in rheumatism.

ACUPRESSURE signifies a method of
arresting hemorrhage from an artery,
wounded either by accident or in the slips of
an operation, by passing a needle under it,
and thus pressing it against adjacent structure—just, in fact, as the stalk of a flower is
compressed against the coat when fastened
in with a pin. It is preferred in many instances by surgeons to the ligature. (See
ARTERIES, WOUNDS OF.)

ACUPUNCTURE is a method of lessen-
ing pain, such as neuralgia, by thrusting
needles, some two or three inches long, into
the painful part. It has been suggested that
the relief caused by the proceeding is owing
to the fluid contained in the nerve-sheaths
being thus allowed to escape. It is a favorite
proceeding with the Chinese.

ACUTE HYDROCEPHALUS, a form
of inflammation of the membranes of the
brain, or meningitis, common between two
and five years of age, but occurring occasionally in the adult, and nearly always proving
fatal. (See MENINGITIS.)

ADDISON'S DISEASE is the name
given to a somewhat obscure malady, first described by Dr. Thomas Addison, of Guy's
Hospital, London. It is almost invariably
associated with disease of certain structures
in the body, the functions of which are unknown, but which, being situated immediately above either kidney, are called the
suprarenal capsules. Moreover, the disease is almost invariably of the kind called
tubercular, and is frequently associated with
similar disease elsewhere. The most marked
feature of the disease is a gradual bronzing
of the skin, which goes on until the patient is
of the deepest mulatto tint, or even quite
black in some parts of the body. This the
patient probably only remarks when his attention is called to it ; but in all probability he
complains first of all of great debility and
feebleness, his appetite becomes impaired, and
very likely his stomach becomes irritable and
he habitually vomits his food. Such patients
commonly persist in standing postures of debility, the head and shoulders hang forward,
and there is about them a very peculiar, listless expression. There is, as a rule, no complaint beyond that of weakness, though sometimes vomiting and difficulties in breathing

are present. Unfortunately, as far as we
know, the well-pronounced disease about
which there can be no mistake is fatal.
Nothing seems to do any good. Strengthening remedies of all kinds have been tried,
and life has been prolonged as much as five
years, but finally with termination in death.
Ordinarily the disease lasts a much shorter
time—on an average 18 months. This tendency to a fatal termination, however, renders
it of the greatest importance that we should
bear in mind that there are other circumstances which give rise to skin-bronzing, especially in females. During pregnancy the skin
of females becomes very much darker than
usual, especially on the breast and abdomen,
and even in some an increased darkness of
complexion is noticeable during menstruation. This form of bronzing passes away
with its cause.

ADHESION is a term used to express
the union or ready healing of the divided portions of soft structures after wounds
and ruptures. In healthy subjects and under
favorable hygienic conditions, the edges of a
wound in the soft parts, and the extremities
of a ruptured tendon, will, if they have been
kept in contact, become in a short time bound
together either by the process of healing by
direct or *immediate* union, or, what is more
frequently the case, by the formation of an
intermediate cicatrice or scar. Adhesion
generally results from the pouring out into
the wound of a clear tenacious fluid called
fibrine or *coagulable lymph,* which becomes
organized and subsequently converted into
scar structure. The process of union between divided soft parts, by means of the
early formation of a scar from this coagulable
lymph, is called *healing by primary adhesion.*
When the edges of an old and gaping
wound, the surface of which is raw and discharges pus, can be brought together so as to
unite, the process of adhesion in this instance
is called *healing by secondary adhesion.* This
variety of healing is sometimes observed in
neglected wounds, extending through the
thickness of the lips, and in the union of two
or more fingers after bad burns.

ADIPOSE, a medical term for the fatty
tissue which is more or less prevalent throughout the body ; when it occurs in isolated
nodules under the skin, fatty tumors are said
to have formed. When there is a large accumulation of adipose tissue beneath the skin
all over the body, the individual becomes stout
and unwieldy. This condition seems natural
to some people, but in others it is induced by
excessive drinking and sedentary work. (See
BANTINGISM, AND WEIGHT AND HEIGHT.)

AERATED BREAD. (See BREAD.)

AERATION is the term applied to the
process by which the air taken in during
breathing is absorbed by the blood in the
lungs. (See RESPIRATION.)

AFFUSION, to which the term cold is commonly prefixed, is a mode of treatment sometimes had recourse to in narcotic poisoning. It is also a powerful means of reducing bodily heat, if that be too high, and was at one time employed in fever for that purpose. For this purpose baths of regulated temperature have now in great measure superseded affusion. The patient is seated or placed in a position where the water will do no harm, and four or five buckets of cold water poured over his head and chest from a height of two feet or more. The colder the water and the greater the height, the greater the effect of the remedy. After this the patient is carefully dried and placed in bed.

AFTER-BIRTH. (See PLACENTA.)

AFTER-PAINS is the term applied to those pains which follow on the expulsion of the child and its appendages in a labor otherwise quite natural. Sometimes they are very severe. They are more common in women who have previously borne children. They begin shortly after delivery, and may continue, if unchecked, for four-and-twenty hours. They are commonly due to efforts on the part of the uterus to get rid of clotted blood which may have collected in its interior. If so, these blood-clots should be removed and the pains will commonly cease. They may also be due to distension of the bladder or bowel. These should therefore be seen to. After all causes are removed a sedative may be given, such as a small dose of opium, or warmth applied to the abdomen.

AGUE, a disease characterized by paroxysms of fever occurring at intervals, brought on by a malarious poison. Each attack has a cold, a hot, and a sweating stage, and is followed by a period of complete cessation of fever. (See INTERMITTENT FEVER.)

AIR.—The air is a gaseous envelope which surrounds the earth, and which is commonly called the atmosphere. It is composed of nearly four parts of nitrogen and one part of oxygen ; the oxygen is the most important constituent, as it is essential to the support of animal and vegetable life, and hence was called by the older chemists vital air ; the nitrogen serves chiefly to dilute the oxygen, and prevents oxidating processes taking place too rapidly. The atmosphere is a mere mechanical mixture of these two gases, and not a chemical compound like water, in which the constituent parts, hydrogen and oxygen, are much more intimately united. The air is kept of an uniform density in consequence of a principle known to chemists as the diffusion of gases, by virtue of which there is a thorough intermixture of the two elements. This is due to the absence of cohesion among the particles of which gases and vapors consist. When mercury and water are shaken up together they soon separate on standing, and the denser fluid sinks to the bottom of the vessel ; but however much gases may differ from each other in density, they will soon mix thoroughly if free communication is allowed between them. Oxygen has a specific gravity of 1.1056, and nitrogen of .972, air being taken as the unit ; but by means of diffusion the heavier gas does not accumulate near the surface of the earth, but is uniformly distributed through the air. Chemical actions, too, on the face of the earth are constantly taking place, and oxygen is being removed from, while carbonic acid and other gases are added to, the atmosphere ; these do not sink to the lower level of the air, although heavier than either oxygen or nitrogen, but rapidly mix and become equally diffused. And this is a very important process, as without it life could not well be maintained in the vicinity of manufactories or in large towns, owing to the rapid accumulation of impurities. Carbonic acid and water are the most common impurities in common air ; the former is derived from the process of respiration in animals and vegetables, from the combustion of coal and gas, and from other chemical processes. It may be recognized by placing some lime-water in a saucer in a room where several people have been sitting ; a pellicle will soon form over the surface, owing to the carbonic acid having combined with the lime and formed chalk. Carbonic acid is very prejudicial to life, and therefore close rooms should be avoided, and a proper supply of pure air should be constantly passing through the room ; ventilation is easily procured by opening the door or window occasionally ; the fire, also, by creating a draught up the chimney, causes a current of air to enter the room. The amount of water varies with the state of the atmosphere. If a glass of cold water be brought into a warm room the outside soon becomes bedewed with moisture, owing to the cold glass condensing the aqueous vapor of the air into visible drops. The dew on the grass in the early morning is a common example of the presence of moisture ; when the sun rises the earth becomes warmer and the dew escapes into the air as invisible vapor. Evaporation from rivers, lakes, and seas is the source of the moisture. Ammonia exists in small quantities, about one part in a million of air ; it is mainly from this source that vegetables obtain the nitrogen which they require to form their seeds and fruit, for they do not seem able to assimilate the free nitrogen of the atmosphere. Ozone is present in fresh air, but not in the close air of towns, as it is decomposed by the organic matter. It is supposed to be formed by the discharge of electricity, and to be an active kind of oxygen. In large towns carburetted hydrogen, sulphuretted hydrogen, and sulphurous acid may exist in minute traces. Many substances which occur in small quantities, such as dust or the minutely divided particles of inorganic bodies, may be looked upon as accidental impurities. More important than these are volatile organic impurities, which probably in a great measure influence the healthiness of a locality. The low fevers and agues met with in marshy districts are caused by the presence of some organic impurity, and any one walking along a river-bank of an evening when the water is low must have recognized at various times a very disagreeable odor. Any one who passes from the fresh air into a crowded room becomes aware of the existence of organic impurities. The poor, who herd together in small, close rooms, seem to prefer the warmth to ventilation, although in doing so they act in a way very injurious to health. Many fevers are conveyed by means of the air from one locality to another ; in this way scarlet-fever or measles may spread ; this, too, will account for the rapid extension of cholera from one country to another. Air may be inodorous and yet not healthy, for there may be no smell, although particles emanating from a fever patient may be floating in it ; it may be odorous and yet healthy, as in the vicinity of gas-works, tan-yards, and tallow-melting ; these smells, though disagreeable, are not injurious to most people. In some localities, as in the vicinity of copper or iron works, the air becomes loaded with the impure gases emanating from the furnaces ; in some parts this is so injurious as to prevent trees or plants growing for some miles around.

Since air is essential for the continuance of life, it is most important to breathe it as pure as possible. Every room should be well ventilated, as otherwise carbonic acid is apt to accumulate and produce a feeling of drowsiness and languor with headache. When this gas accumulates to more than four parts in ten thousand of air it is injurious to health. Organic bodies in a state of putrescence should be destroyed or buried so as to prevent any noxious particles spreading into the atmosphere. Excreta should be removed, and the cause of any bad smells arising from water-closets or cesspools should at once be seen to. By breathing pure air and by removing everything which tends to contaminate it, a great deal may be done to prevent the onset of disease. (See FEVERS, VENTILATION.)

AIR-PASSAGES form the channel by which the air can enter the lungs ; different names have been given to each part of the tube. At the back part of the mouth, and just in front of the œsophagus, or gullet, is a chamber called the larynx ; this cavity communicates above with the mouth and nose, and can be closed by a valvular lid called the epiglottis ; below, it is continuous with the trochea, or windpipe, a capacious, circular tube, lying in front of the neck ; at the level of the top of the sternum or breastbone this

tube divides into two branches, called bronchi, one of which goes to either lung ; these again, on arriving at the lungs, break up and subdivide into a number of very fine branches, and at last end in dilated extremities with very fine walls, which are called the air-cells. In the ordinary state, the epiglottis is open, and air can go in and out the larynx with ease, and enter or pass away through the mouth and nose, for both these cavities communicate freely with each other behind, but when the act of swallowing takes place the epiglottis falls over the upper opening of the larynx and prevents any food going that way. Anything which prevents the entrance of air into the lungs by obstructing the air-passages, as a coin or marble in the windpipe, spasm of the epiglottis, inflammation within the larynx or a growth in that tube, or pressure from without as in cases of hanging and strangulation, will cause great distress in proportion to the amount of the obstruction, and, if very great, will cause death by apnœa.

ALBINO is a name to an individual whose hair, skin, eyes, etc., are deprived of all coloring matter ; in other respects they may have perfect health. They are generally short sighted, and the pupils of the eye have a pink color ; the hair is thin and of a silvery-white color. Nothing, of course, can be done for this singular condition.

ALBUMEN is a chemical compound found in the tissues of both plants and animals. The best example of it in the animal kingdom is the white of the egg. It is also found in all animal blood and the nerves and brain. It is found in the juices of many kinds of vegetable food, as in cabbages, asparagus, and potatoes. It is in all these instances in solution in water. It is easily discovered by the facility with which it coagulates by heat. It is thus that the white of the egg becomes solid by boiling. Blood and the nervous organs of animals are also coagulated and thickened by exposure to heat. It belongs to that class of alimentary substances which are called proteinaceous, nitrogenous, or flesh-forming. As a part of our food, it supplies the waste of the nerves and muscles of the body. It sometimes appears in the urine and constitutes the disease called albuminuria. (See ALBUMINURIA, BRIGHT'S DISEASE.)

ALBUMINURIA, or the presence of albumen in the urine, is a symptom met with in many cases of disease of the urinary organs. It is known by boiling the urine in a glass tube over a spirit-lamp, when a white, flocculent precipitate is thrown down, and this deposit is not dissolved on adding a little nitric acid. In all cases of Bright's Disease this substance is present in greater or less quantity in the urine ; in many cases of heart-disease, when associated with dropsy of other parts ; in many febrile disorders, as typhus and typhoid fevers, diphtheria, etc.; in all cases where blood is also present in the urine (see HÆMATURIA), and whenever there is pus in the urine, as when a stone is present in the bladder or when there is inflammation of the lining membrane of that organ. Albuminuria itself requires no special treatment; it is only an important symptom of grave mischief elsewhere, and so the treatment that should be adopted must have reference to the particular disease which is the cause of the albuminuria. (See BRIGHT'S DISEASE.)

ALCOHOL is the active principle of wines, spirits, beers, and other fermented beverages. It is formed during the process of fermentation from fruit sugar. When fruit sugar is exposed to the action of a ferment it loses carbonic acid gas, and is converted into alcohol. This substance has a special power of acting upon the nervous system; it is on this account that it has been taken largely by mankind in fermented beverages. Alcohol acts upon the nervous system, producing first a pleasant stimulation, then great excitement, and finally a state in which the person who takes it is more or less unconscious. This state is called drunkenness. Habitually taken in small quantities it produces great disturbance of the nervous system (ALCOHOLISM). When taken in large quantities it causes complete derangement of the brain and nervous system, which often terminates fatally, and always leaves it more or less permanently deranged (see DELIRIUM TREMENS). The nervous is not the only system affected by the constant abuse of alcoholic drinks. The mucous membrane of the stomach becomes inflamed, and indigestion is produced. The liver is subject to a peculiar disease from its action (see CIRRHOSIS). The blood is deranged, and the nourishment of the various organs is interfered with. A series of changes take place in the heart, the liver, and the kidneys, which are known by the name of Fatty Degeneration (DEGENERATION, FATTY). Such an agent should be carefully watched on account of its very general use. The inquiry is often made as to how much alcohol can be taken in a day with impunity. It is quite safe to say that children and young healthy adults can have perfect health without taking it at all. Where small quantities are required it may be stated that from one to two ounces of pure alcohol is all that can be safely taken from day to day.

The question is often discussed as to whether alcohol is in any way consumed in the system or thrown off the body in the same state in which it was taken. Whilst there is abundant evidence that alcohol when taken in large quantities is thrown off by the various excretory organs, the latest researches show that a considerable quantity of alcohol taken is decomposed and thrown off the system in other forms. In some forms of disease alcohol is the only substance acting as a nutrient that can be absorbed into the blood, and in these cases it acts as a stimulant to the heart and brain, and also supplies aliment to the body. In the low forms of fevers and other exhaustive diseases of the body it is the sheet-anchor of the physician, and life is imperilled where its agency is disregarded. Alcohol is used in medicine of various degrees of strength; chiefly for dissolving out the active ingredients of various remedies. Such solutions are called tinctures. Its properties are, as is well known, stimulant; as such it is used in the form of wine and spirits.

We subjoin a table of the quantities of alcohol contained in various common forms of beverages:

Quantity of Water, Alcohol, Sugar, and Acid contained in one Imperial Pint of various Fermented Beverages.

Name of Beverage.	Water.	Alcohol.	Sugar.		Acid.
	Ozs.	Ozs.	Ozs.	Grns.	Grains.
Pale Ale........	17½	2½	0	240	40
Mild Ale........	18¾	1¼	0	280	38
Strong Ale......	18	2	2	136	54
Cider...........	19	1	0	100	150
Port............	16	4	1	2	80
Brown Sherry..	15½	4½	0	360	90
Pale Sherry.....	16	4	0	80	170
Claret..........	18	2	0	0	161
Burgundy.......	17½	2½	0	0	160
Hock...........	17¾	2¼	0	0	127
Moselle.........	18¼	1¾	0	0	140
Champagne.....	17	3	1	133	90
Madeira........	16	4	0	400	100
Brandy.........	9½	10½	0	80	100
Rum	5	15	0	0	100
Gin (best)......	12	8	0	0	100
Gin (retail).....	16	4	¼	0	100
Whiskey........	10½	9½	0	0	100

ALCOHOLISM, the result of a long continued abuse of intoxicating liquors, causes a serious change in the blood, and then of the various tissues of the body. The liver may become fatty or cirrhosed, the heart weak and flabby, and unable to perform its functions so well as before; the kidneys are liable to waste, the lungs to become emphysematous, and the patient short of breath. The brain also shares in the general mischief, and many of the nerve-cells waste through being badly nourished; the mind in consequence becomes affected; there is loss of memory, giddiness at times, disagreeable dreams, and restlessness at night, occasionally flashes of light appear before the eyes, and the patient wakes up in the morning with no appetite for breakfast, and a feeling of sickness. His nervous system, too, is weakened; any excitement or trouble affects him; in advanced cases the tongue and hand tremble, and weeping comes on with a very slight cause. If the heart be affected, he may be troubled with fainting, which at length may prove fat.l. Many of these cases may go on for years without seeming to h.ve their health impaired by the quantity of spirit they take, but sooner or later it tells upon them, and any acute illness will quickly carry them off ; such a course of life induces premature old age. Very little can be done for the habitual drunkard, who may cease for a short time from drinking while he may be suffering from some ailment caused by his habits of life ; a sedative may be given at bed-time to enable him to sleep better, and for the dyspepsia or indigestion which accompanies this disease some bitter tonic, as gentian or quassia, may be given with nitric or hydrochloric acid two or three times a day ; small doses of strychnine and arsenious acid have been taken with advantage, but these are very poisonous remedies, and.must only be taken under medical advice.

ALE. (See BEER.)

ALIMENT. (See FOOD, DIET.)

ALKALIES are the oxides of certain metals a d their salts. Ammonia, which is not the base of a metal, acts as the alkalies, and is called the volatile alkali. (See POTASS, SODA, AMMONIA, LIME, MAGNESIA, LITHIA.)

ALKALOID, a term applied to those vegetable principles which act chemically like alkalies, such as quinine, morphine, strychnine, etc.

ALLOPATHY is a term applied to the practice of medicine as carried on by the great mass of medical practitioners. It is opposed to the term Homœopathy, in which diseases are supposed to be cured by remedies which produce the same effect on the system as the disease ; hence the axiom, *similia similibus curantur.* On the other hand, Allopathy is supposed to cure by remedies which produce effects different from those they are given to cure ; and a contrary axiom is assumed, *contraria contrariis curantur.*

ALLSPICE is the fruit of a tree belonging to the same family as the clove. It contains an agreeable volatile oil, which is used for giving flavor to bread-sauce and other articles of food. (See PIMENTO.)

ALMONDS are the seeds of a species of *Amygdalus,* and are of two kinds, sweet and bitter. The sweet almonds are brought to table, and in countries where they grow form an important article of diet. They contain starch, oil, and albumen. The oil is often expressed and used as salad oil. The bitter almond contains, in addition to the fixed oil, a peculiar oil known by the name of oil of bitter almonds. The almond itself and this volatile oil are used in cookery for the purpose of giving flavor to custards, cakes, puddings, etc. It is also used in perfumery. The smell of the oil is imitated by an artificial compound, nitro-benzol, obtained from coal-tar, and it is often sold in the shops for oil of bitter almonds. In the shops two kinds of

oil of bitter almonds are sold, the pure and impure. The impure contains hydrocyanic acid, and is very poisonous, while the pure contains no poisonous principle. The taking impure oil of bitter almonds by mistake or design is a frequent cause of death.

ALOES is the thickened juice of various species of plants, called aloes, growing in many parts of the world. That used in medicine is chiefly brought from Africa and the West Indies. Few remedies are more useful than aloes, and it enters into many preparations used in medicine. Its most prominent properties are purgative. When given as such it had better not be given in large doses ; half of an ordinary pill, that is, two or three grains, taken regularly for a time, being generally quite enough to insure regular action of the bowels. It seems to act as a tonic as well as a purgative. In larger doses it is said to produce piles, and should not be employed where these exist, neither should large doses be given during pregnancy.

ALTERATIVES is a term applied to medicines which are supposed to alter the condition of the blood and tissues without exciting any sensible action of the excretory organs. Thus, small doses of the mercurial preparations are regarded as alteratives.

ALUM is a compound crystalline body having as its essentials alumina and sulphuric acid with potass or ammonia. It is an *astringent* substance, and hence is used to lessen discharges of many kinds. It is also used as a gargle in sore throat. When heated it melts and becomes powdery ; this, which is called burnt alum, is often used for ulcers when they become flabby. Along with decoction of oak bark, in the strength of 2 or 3 grains to the ounce of decoction, it constitutes one of the safest local applications for ordinary discharges.

AMAUROSIS is used to express imperfect vision or total blindness due to some unhealthy changes in the back of the eye, in the optic nerve or nerve of sight, or in the brain. It also includes various nervous affections of the eye in which there is no apparent change of structure to account for the failure of vision. Since the introduction of the ophthalmoscope, an instrument by means of which the interior of the eye-ball can be illuminated and examined, a great number of well-marked affections of the visual apparatus have been ranged under this heading. The chief *causes* of amaurosis are the following : Diseases of the brain, as apoplexy, inflammation, tumors, abscess ; affections of the nerve of sight ; tumors growing within the eye-socket and disease of the soft parts surrounding the eyeball ; inflammation of the retina and choroid, two membranes of the eye ; certain changes in these membranes associated with Bright's disease of the kidneys, with syphilis, and with diabetes. It is believed by some authorities that excess in tobacco smoking is a not infrequent cause of blindness. Amaurosis occasionally results from debility and during convalescence from fever, diarrhœa, and profuse hemorrhage, and may occur in pregnant and hysterical women, and in children affected with intestinal worms. The most common and important cause of impaired vision, however, is debility and congestion of the interior of the eyes due to the prolonged use of these organs under certain conditions. Use of the eyes, however long continued, can do very little harm so long as the light is not very intense and the regarded objects are diversified in form, size, and color, but when minute objects are closely watched for a long time under a bright light, especially when one eye only is used, as in microscopical examinations, injury of the retina or visual membrane is likely to result. Long-continued exercise of the eyes in very hot and badly-ventilated rooms with glaring lights is another frequent cause of amaurosis. Watchmakers, draughtsmen, compositors, and needlewomen are peculiarly exposed by their occupations to amaurosis. The sudden exposure to bright light of a person who has previously remained for a long time in dusk or total darkness may give rise to impairment of vision, a fact which should be remembered in the treatment of convalescents from long and severe illness. There are many different forms of amaurosis. It may affect one or both eyes. In some cases it consists in total blindness, in others, in slight weakness of vision. It may be permanent, temporary, or intermittent. Sometimes it comes on suddenly, but in most instances slowly, and at first almost imperceptibly. It may be attended with severe local and general symptoms, as intense pain in the eyeball, headache, vomiting, giddiness, convulsions, and palsy ; or, on the other hand, it may cause no uneasiness to the patient except what arises from the failure of a most important sense. The following are the chief *symptoms* of the affection, especially of the slow form, that results from long-continued abuse of vision : Difficulty in reading print or writing, the letters being doubled, or halved, or distorted, obscured, or discolored ; the appearance of small black specks, like particles of soot, floating before the eye ; the appearance of larger fixed specks ; a dense mist before the eye, varying in color at different times ; flashes of bright yellow or blue light appearing when the eyelids are closed ; distortion of objects, especially of flame ; an iridescent and rainbow-like halo around flame and strongly illuminated objects ; in some cases intolerance of light, in others a desire for light ; the pupil is generally large and dilated and the iris immovable, but this is not a constant symptom ; pain and a sense of fulness in the eyeball ; with these symptoms

is associated a gradual failure of vision, until the power of appreciating the shape and color of external objects is quite lost.

Treatment.— In the treatment of amaurosis one must seek for the probable cause of the disease. In the common instances of impairment of vision coming on in debilitated subjects who have been engaged in occupations necessitating a long-continued examination of minute and brightly-illuminated objects or hard study during the night, much good may be done by a discontinuance of such pursuits, by strengthening the system. Quinine and steel drops may be taken with advantage, but more beneficial than any medicinal agent will be found good living, fresh air, and a change of scene and occupation. The bowels should be freely relieved and kept open by blue pills, Epsom salts, or a frequently-repeated black draught. Wine and beer ought to be taken. A bright light must be avoided, and the patient, when taking exercise in the open air, should wear spectacles with glasses of a light blue tint. This treatment, it must be remembered, is applicable only to amaurosis brought about by such avoidable circumstances as want of fresh air and good diet and an incautious use of the eyes. In other forms of the affection, a proper use of remedial means is to be based upon a recognition of its true cause, which may exist either in a general affection of the body or in disease of some distant organ.

AMBER is a hard, semi-transparent substance of a yellow color. It has the character of a resin, and is supposed to be of vegetable origin. An oil is obtained from it called *Oleum succini*, which is extensively used as an embrocation in rheumatism and whooping-cough.

AMBLYOPIA. (See VISION, DOUBLE.)

AMBULANCES.—Numerous conveyances have been designed, and are in use, for the carriage of the disabled, and as it would be almost impossible to refer to them all in detail, some few practical suggestions, which would refer equally to civil as to military practice, on the proper position of wounded men during the act of transportation, with some remarks on their removal by bearers when no conveyances are at hand, will be of value. Ambulance conveyances, however borne, are constructed for carrying patients, either lying at full length or sitting. The recumbent position is undoubtedly the best in the case of severe wounds, and in cases of shock or faintness from hemorrhage, as it is the position in which the several parts of the body are subjected to the least amount of concussion. The sitting posture is, as a general rule, only adapted for those whose injuries are of a comparatively slight nature. The semi-recumbent position is where the trunk of the patient is raised, and supported at a certain angle with the lower part of the body and lower extrem'ties ; the knees are also raised, and the thighs bent and supported at an angle with the legs. This position is very desirable in wounds of the chest, owing to the feeling of oppression in breathing, preventing the re- cumbent position and the jolting of the sit- ting. If no conveyance be at hand, the as- sistance of bearers must be resorted to, and it will be convenient to mention some method of affording help when only *one* attendant or bearer is at hand. If the wound be in the head, neck, or upper part of the trunk, the patient should partly support himself, with a stick in one hand (or musket), while his other hand and arm lean upon the upper part of the back and distant shoulder of the attendant who walks by his side. At the same time the attendant should place his near arm across the neck of the wounded man, reaching round and partly encircling his body with the fore- arm and hand, so as to support the trunk. If more than one attendant is available, a regular litter is at hand. The first method they may adopt is that of carrying the patient by the two bearers joining hands beneath the thighs, while their arms which are not thus occupied are passed round his loins. A second and better method of joining two hands for the semi recumbent support of a patient is as follows : The ad- vanced right and left hands of the two bear- ers are closely locked together, and the wrists brought into contact ; at the same time their other hands are made to rest upon an l, in a certain degree, grasp each other's shoulders on the same sides respectively. One of the best methods is that of a four-handed seat with crossed arms, known commonly as the "sedan chairs," and to effect it readily the following directions are useful : The bearers, A. and B., stand thus : A. on the left side, B. on the right. A grasps with his right hand the left arm of B.; B. grasps with his left hand his own right arm. B. then grasps with his right hand the left arm of A. ; A. with his left hand his own right arm.

AMENORRHŒA means an absence of the usual flow which generally occurs at regular periods in women from the time of puberty until middle life. Menstruation usu- ally begins between the age of twelve and fourteen'; in some cases it is very much later, in a few it never appears at all, and in many the "periods" are very irregular. Amenor- rhœa may result either from retention or sup- pression of the flow of blood ; in the first case it depends upon some malformation of the organs of generation, and a simple surgi- cal operation is necessary for its cure ; it may be suspected in a young girl who has all the usual symptoms and discomfort attending that period, but from whom there is no ex- ternal flow. Suppression of the flow is far more common ; it occurs naturally in every case of pregnancy ; it is often brought about

by exposure to cold, by sitting on the damp grass, or by getting wet in the feet ; it en- sues in the course of many exhausting dis- eases, as consumption, kidney disease, can- cer, etc. ; it is met with for a time after the patient has passed through a severe illness, as typhus or typhoid fever, and finally it may result from disease of the uterus and ova- ries. In many cases of anæmia, amenorrhœa is generally found to exist : in fact, a condi- tion which brings about one will generally cause the other, so that patients suffering from this complaint are usually extremely pale ; the lips and inside of the eyelids lose their usual color ; the patient is very liable to headache, palpitation of the heart, faintness, and lassitude. When the cause is due to preg- nancy, of course nothing need be done, as the symptom is quite natural ; in other cases tonics are of the greatest service, and espe- cially those preparations which contain iron. Moderate exercise in the open air, without tiring the patient too much, is necessary ; a generous and wholesome diet, avoidance of late hours and close rooms, early rising and ffesh, bracing air, with cold bathing, will do much good ; but some of these remedies can- not be adopted when the patient is suffering from other diseases which really are the cause of the amenorrhœa. Stimulants are not needed, but a glass of beer or a glass or two of light wine, may be taken at dinner. The bowels should be kept open by aperients, and a warm bath should be given at bed-time just before the time when the " period" should recur. These remedies are most valuable when the amenorrhœa is dependent upon anæmia, on overwork, exposure to cold, etc.; but in cases of consumption, cancer, uterine, ovarian, and kidney diseases, etc., there is no occasion to treat the amenorrhœa, but at- tention must be directed to the more impor- tant malady.

AMMONIA is used, both by itself and combined with other chemical agents, for a variety of purposes. It is nowadays got from gas-house refuse, but used to be obtained by burning hartshorn, whence it got the same name. Its application as burnt feathers is familiar to many ; to more it is known in the form of smelling salts. Pure ammonia, or, as it is called by chemists, caustic ammonia, is rarely used. When ammonia is given in- ternally for its stimulant virtues, its carbonate is used—sal volatile, or smelling salts. It is used as aromatic spirit of ammonia in doses of from a few drops on sugar to a teaspoonful or more (m. v. to 3j.) to relieve flatulence, to remove the feeling of sinking, and to get rid of acidity and heartburn. It is also useful in some forms of headache. It frequently does much good in the chronic bronchitis of old people, when their winter cough is compli- cated with copious tenacious expectoration. Ordinary smelling salts may be used as an emetic in certain forms of poisoning ; about a teaspoonful should be given dissolved in lukewarm water. Liquid ammonia has been of late used with much success as a remedy for snake-bite. It has been given internally in considerable quantity along with brandy or whiskey. In Australia it has been repeatedly injected into veins for snake-bite, and the treatment has proved there quite successful, but has failed in India. The old remedy, hartshorn and oil (*freshly prepared*), will be found very useful for the stings of bees and wasps, the bites of mosquitoes, gnats, and such like.

Acetate of Ammonia has been a good deal used in practice as liquor of acetate of ammonia, commonly called Mindererus' Spirit. Its dose is from ten drops to a teaspoonful (m. x. to 3j.), and it has commonly been given to relieve feverishness, as in ordinary fevers, colds, etc. In such like forms of disease it is supposed to cool the skin by promoting perspiration, whence it is called a diaphoretic. It may be given along with sweet spirits of nitre and a few drops (2 or 3) of antimony wine, when the skin is hot and dry and the pulse quick.

Hydrochlorate of Ammonia, also known as Chloride of Ammoniacum, more commonly as Sal Ammoniac, is a remedy of exceeding great value in certain forms of headache. It is not easy to say which form it is exactly suited for, but it often succeeds with headache and tic-douloureux when everything else has failed. Twenty or thirty grains should be taken for a dose. It is rather nauseous, and may be given in beer. It is also useful in certain female complaints, especially when the periods have been irregular or have prematurely ceased.

AMMONIACUM is what is called a gum resin—that is to say, it is both gummy and resinous in character. It comes from the north-east of India. A plaster of it, combined with mercury, has been a good deal used for getting rid of old swellings. It is of most use in the chronic bronchitis of old age, and is given in doses of from five to twenty grains.

AMNESIA, or loss of memory, is met with in some cases of apoplexy. It varies in amount. At times the patient will lose all memory of recent events, while there is a clear recollection of the past ; at other times the converse may exist, while generally there is more or less forgetfulness of everything.

ANÆMIA is a condition in which there is an impoverished state of the blood, and where the patient is very pale and in a state of general debility. Anæmia may exist alone or in conjunction with other exhausting diseases, as consumption or cancer. In the first variety the patients are generally young women who are employed in close workshops and confined places from morning to night ; or they are women who have lived badly, and

having had several children, are suffering from over-lactation. In such people the whole surface of the skin is paler than usual, and the lips and lining membrane of the eyelids, instead of being rosy, are of a pale pink color. There is also a feeling of general debility and an inability for much exertion. Palpitation of the heart, headache, pain in the back and in the left side, are commonly met with. This disease arises chiefly from want of pure air and light, and from living badly. The *treatment* consists, therefore, of moderate exercise every day in the fresh air, and working or living in rooms with plenty of light and means for ventilating the apartment whenever the air becomes impure. The diet should be light and nourishing ; a moderate amount of animal food should be taken, but anything which is heavy or which causes indigestion should be avoided. Stimulants should not be resorted to, but a glass of beer with a meal may be beneficial. A cold bath in the morning is often of great service, and for those who can afford it, change of air and scene is very useful ; such persons may take horseback exercise, and they should go to the sea-side or some place where the air is bracing and refreshing. Tonic medicines are of great value, and for this purpose preparations containing iron are most beneficial ; a patient may thus return to the bloom of health in a very few weeks. When women are at the same time suffering from over-lactation it may be advisable to wean the child. Anæmia is often associated with a temporary cessation of the menstrual function, but this is usually restored with the improvement of the general health. The habits of town-life predispose to this disease in a great measure, and in all cases country air is most beneficial. In young girls tight lacing is often most injurious, as it prevents due expansion of the chest and prevents the free entrance of air into the lungs, a process which is most important for the various changes which are constantly going on in the blood. Persons who are anæmic are very often nervous and hysterical, and all sources of mental worry or anxiety should be avoided as far as possible. It is a disease which too often arises from neglect of the common principles of health and from the artificial life of modern society ; but it is one which, unless treated early, may lay the seeds of future and more serious diseases.

ANÆSTHETICS are remedial agents which take away the sensibility from a part or the whole of the system. Those substances which, when externally applied to any particular organ, take away its power of sensation are called local anæsthetics, whilst those which are taken internally and act through the blood are called general anæsthetics. The same substances are generally capable of acting in both ways. Thus chloroform and ether, both of which are general anæsthetics when applied locally, especially by means of a spray, are capable of producing on a particular part an entire want of sensation. The agents which produce this impression on the nervous system are very numerous. The properties of these substances may be consulted in the articles ACONITE, BELLADONNA, CHLOROFORM, ETHER, OPIUM.

ANASARCA is a state in which there is a general swelling of the body and extremities, caused by an effusion of the serum of the blood into the loose cellular tissue under the skin. The skin in such cases is generally very pale, and when the finger is pressed upon it, a little pit or depression is formed as the effused fluid is by the pressure squeezed away. It is generally associated with kidney disease, and often occurs after scarlet-fever, when that organ is also affected. It is generally noticed first in the face and genitals, where the skin is looser than elsewhere, and the patient's face has then a puffy look. In more chronic cases, when the kidneys are much diseased, the whole surface of the body may become puffy and swollen; the lips especially are liable to suffer, and they are worse after they have been in a dependent position, as the fluid gravitates to the lowest parts, and thus it often happens that the face is most swollen in a morning. When only one limb is thus affected, it is said to be œdematous, but when the whole surface is puffy, the patient is said to have anasarca. Both conditions, however, arise from the same causes. In anæmic girls, where the blood is much impoverished, and thinner than usual, œdema of the feet not infrequently occurs, but this variety may be cured by treatment. Swelling of the legs and other parts of the body may thus arise from various states in which the blood is altered in quality, and these cases are in a great measure amenable to treatment. Other cases, however, arise, where the blood is altered in quantity, such as those where there is some mechanical obstruction to its flow. When the return of blood to the heart is prevented by obstruction of the veins, the parts behind the obstruction become so full of blood that the serum exudes from the distended vessels, and soaks the tissues around. The swelling of the arm and hand, which sometimes is caused by falling asleep in a chair while the arm is hung over the back, is due to the prevention of the proper return of the blood current; in like manner, a tight bandage will produce similar effects; but these are easily remedied. A more serious class of cases are those in which the obstruction is seated in the heart or lungs, and arises from disease of those organs. It is not uncommon in persons who have bad winter cough and shortness of breath for some time to find a swelling of the legs; this arises from the obstruction to the flow of blood through the lungs. The *treatment* of such

cases will depend upon the nature of the cause, and for this purpose a careful inquiry into all the symptoms of the case is required. Relief may usually be obtained by preventing the affected part from remaining in too dependent a position; thus the legs may be considerably reduced in size by placing them in a horizontal position. Pricking the extremities with a needle so as to allow the effused fluid to escape is often of great advantage; but here great care must be taken, and the parts should be wrapped in flann ls wrung out in hot water, and gradual oozing may be permitted for some days. This plan is more successful in cases resulting from kidney disease than in those in which the heart is affected.

ANCHYLOSIS is a term used in surgery, signifying a fusion or welding together, either partially or totally, of the end of bones at the joints, as a result of injury or disease. It is commonly divided into three varieties: *false*, that is, where the bag containing the joint oil becomes thickened and thus impedes motion. *Ligamentous*, where two opposed articular surfaces become unnaturally united. *Bony*, when the last-named form of anchylosis becomes ossified or converted into bone tissue. With regard to treatment, the only form in which a non-professional man can be of assistance is in the first-named, when, after the inflammation following a severe wrench or sprain, where stiffness threatens, *passive* motion, *i.e.*, for the bending or extension of the joint, is performed by another person, assisted by shampooing, warm baths, rubbing or emollient applications. The second or third stages come under the surgeon's care.

ANEURISM is a pulsating tumor communicating, either directly or indirectly, with the calibre of an artery (see ARTERY). If its *sac* is composed of the arterial coats, it is called a *true* aneurism; if formed by surrounding tissues, owing to a wound being made in the vessel, it is termed *false*. True aneurism is the result of disease of the arterial coats, and its mode of formation varies either by these coats evenly dilating into a pouch, in the walls of which atheroma (see ATHEROMA) is apparent, or by the giving way of the internal and middle coats owing to disintegration, the external one forming the wall of the cyst, or by rupture of the internal and middle coats during some violent muscular effort, in which the arterial tunics are stretched beyond what they are able to bear. The contents of the sac are blood, fluid or coagulated, and layers of fibrin. There are several kinds of true aneurisms; thus, surgeons speak of the *tubular*, *i.e.*, when the sac is uniformly dilated; the *sacculated*, when the sac is unequally dilated; the *dissecting*, when the blood gets between the coats of the artery. Tumors situated over an artery may have a pulsation communicated to them by the underlying vessels, and simulate aneurism;

but such a tumor can be generally told from an aneurism, from the fact that by pressure the latter can be emptied, and refills immediately the pressure is removed ; and on listening to it closely with a stethoscope (see STETH-OSCOPE), a peculiar thrill, or rush, can be heard, caused by the blood passing through it.

Treatment.—The chief means adopted for the cure of aneurism are the ligature, pressure, and flexion. That by ligature consists in passing a stout hempen thread round the artery, between the aneurism and the heart, so cutting off the current of blood through the main trunk, the circulation being gradually re-established by what is termed *collateral* means ; that is to say, the work of the main vessel is thrown upon its smaller branches, and by their dilatation the blood finds its way into the limb beyond the point of ligature. Pressure may be exerted upon an aneurism either by mechanical means, such as tourniquets, or by the fingers—*digital* compression. The treatment by flexion consists in flexing or bending the limb, such as the leg upon the thigh, or the forearm upon the arm, where the aneurism is situated, as it frequently is in that portion of an artery immediately opposite to the point of flexure of the limb in which it lies, such as in the ham (popliteal aneurism), or bend of elbow. In the case of false aneurism, resulting from rupture or puncture of an artery, pressure should be immediately applied between the heart and the supposed point of escape of blood, until surgical aid arrives. There are several other methods of treatment, which, however, need not be discussed here.

Internal Aneurism.—Though in a work like this it is necessary to refer to internal aneurism for the sake of completeness, not much need be said with regard to it. The forms of aneurism, as above defined, which most frequently occur internally, are aneurisms of the great vessel of the body, the aorta, or of its branches, or aneurisms occurring within the head. Cerebral aneurisms—aortic aneurisms—may be situate either in the cavity of the chest or in the abdomen. In either situation a certain number of them admit of surgical operation, which is sometimes successful ; while some of them yield to remedial treatment, especially if accompanied by absolute rest in bed. In all cases where such a thing is suspected, the earliest opportunity should be taken of consulting a skilled practitioner on the subject

ANGELICA root is produced by the plant known as *Gaudea Angelica*, the *Angelica Archangelica* of the botanist. It contains a pleasant volatile oil, and is used as a stimulant and carminative in medicine. The stem of the same plant is preserved in sugar, and used as a sweetmeat.

ANGINA PECTORIS or **BREAST PANG** is fortunately not a disease of fre-

quent occurrence. It comes on in paroxysms, in which there is a struggling for breath, intense pains about the region of the heart, and a terrible sense of impending death. The anguish is extreme while it lasts, but it passes off and leaves the patient apparently tolerably well till the next attack. The face is pale, the body covered with sweat, and the sufferer perfectly sensible. The attack does not last long, ordinarily only a minute or two, though sometimes longer. It always recurs, but at no fixed interval ; it may be absent weeks or months, but the intervals tend to become shorter the longer the disease lasts. They come at any time—night or day—whether the patient be walking about or lying down. As to the *cause* of these attacks, that is obscure ; generally there is some malady of the heart itself. It may be fatty, or its own particular vessels may be diseased and the circulation through them obstructed, or both may exist. These are the diseased conditions most frequently found. After one seizure, as already pointed out, another is to be dreaded as likely, and accordingly, during the interval, everything ought to be done that will conduce to the patient's health. During the paroxysm stimulants like brandy, aromatic spirit of ammonia, ether, and spirit of chloroform ought to be given. Recently a drug, nitrate of amyl, has been found to control the attack, but this is too dangerous for common use without medical advice.

ANGOSTURA BARK is obtained from South America. It is not much used in medicine, but its name has been given to a kind of " bitters" a good deal employed. It is tonic in its properties, and is also said to be of some use in ague and similar tropical fevers. A simple infusion of the bark might be used with advantage in domestic practice for loss of appetite and weak digestion.

ANGULAR CURVATURE is a disease of the spine often met with in scrofulous or rickety children, and resulting from the erect posture being assumed when the spinal column is too weak to bear the weight of the head and upper part of the body. The spine is curved so that the convexity looks backward ; the ribs often bulge out more on one side than on the other ; and the chest is much encroached upon, so that there is less room than usual for the lungs to expand. It is a sign generally of serious constitutional import, but its nature and treatment will be treated of in the article on RICKETS.

ANIMAL HEAT. (See HEAT, ANIMAL.)

ANISEED furnishes a volatile oil with stimulant properties. A drop or two may be given on sugar to allay windy spasms.

ANKLES (WEAK) is an affection depending upon weakness of the flexing and extending muscles of the ankle-joint, or on a rickety condition of the bones of the leg. To remedy this condition, high-heeled boots should be worn, with the *inner* edge of the heel thicker than the outer or a stout webbing bandage should be applied carried round the ankle *from* the *inner* side of the foot (see BANDAGES). Cold water douching, and some astringent lotions, such as arnica lotion, are sometimes of value.

ANODYNES. (See NARCOTICS.)

ANTACIDS are any kind of medicine which counteract the formation of acids in the system. The alkalies and alkaline earths are the best antacids. The carbonate and bicarbonate of these substances are better than the pure alkalies and earths, which act as caustics, and unless largely diluted, injure the coats of the stomach. The best forms of antacid medicines are the bicarbonates of soda and potash, and the carbonates of lime and magnesia. (See CHALK.)

ANTHELMINTICS are medicines which are given to expel worms from the intestinal canal; they are generally given in conjunction with a purgative, so as to favor their action and to get rid more effectually of the parasite. The anthelmintics in most common use are only three or four in number. 1. The extract of male-fern, which is given in cases of tapeworm, and which should be taken on an empty stomach after fasting. 2. Santonin, a crystalline white neutral principle, turning yellow on exposure to the light; it should be given when a round worm or lumbricus is present in the intestines; five grains of this substance with an equal quantity of compound jalap powder will prove effectual in a child from six to ten years of age. 3. Kousso is part of a plant growing in Abyssinia; it may be given for tape-worm, but it is not in common use. 4. Kamela is an orange-red powder which purges freely, and is used in India for tape-worms; it is a powerful anthelmintic, but is seldom administered in this country. In addition to the above anthelmintics, there are many purgatives, as rhubarb or jalap, which will bring away worms, but they have no special character beyond their purgative action. Injections or enemata of salt and water, or solution of the perchloride of iron and infusion of quassia, are very useful in the treatment of thread-worms in children. (See ENTOZOA and PARASITES.)

ANTHRAX is the technical name for carbuncle. (See CARBUNCLE.)

ANTIDOTE is the name given to any remedy which, whether used externally or internally, is capable of counteracting the effect of a poisonous agent. (See POISONING; also see each particular poison.)

ANTIMONY is most commonly employed in combination with cream of tartar, when it is called tartarated antimony, or more commonly, tartar emetic. It is a powerful medicine, and must be cautiously used. In small doses not exceeding half a grain, it promotes

perspiration, in larger doses producing nausea, and in still larger doses vomiting; for the latter purpose two or three grains suffice. It is best given as antimony wine, from 10 to 30 drops, to produce perspiration; a teaspoonful or more to produce sickness. As a remedy it used to be much employed in internal inflammation, especially those of the lungs, on account of its power of controlling the circulation, but is now seldom resorted to. So, also, it was used in croup, and by many is still so employed. It is useful in promoting expectoration in the earlier stages of bronchitis, when the chest is sore and the cough dry. It is very useful in feverish colds, promoting perspiration and relieving the aching pains then often experienced. It had better be combined with a few grains of Dover's Powder. The dose is from 3 to 5 grains. Antimony has occasionally been used as a slow or secret poison. The symptoms it produces are sickness, tendency to vomit, complete loss of appetite, and extreme debility. Its detection is easy.

ANTIPHLOGISTICS are remedies which are supposed to oppose inflammation in any part of the body, and act as antagonists to any excitement or stimulation going on in the body from disease. Bleeding is one of the most powerful antiphlogistic remedies. The salts of antimony and mercury are also antiphlogistic remedies.

ANTISCORBUTICS are medicines and articles of diet that counteract the effects of sea-scurvy, or any tendency to that disease. The most efficient antiscorbutics are uncooked vegetables, and lemon-, lime-, and orange-juice. (See SCURVY.)

ANTISEPTICS are agents that counteract the effect of putrescency in the living or dead organism, as carbolic acid, charcoal, common salt, vinegar. (See DEODORANTS, DISINFECTANTS.)

ANTISPASMODICS are those medicines which overcome pain, cramp, or spasm in the human body, as ether, opium, assafœtida, and brandy.

ANUS (ARTIFICIAL) is an unnatural opening in some part of the walls of the abdomen, communicating either directly or indirectly with an orifice in the intestinal canal. Through this opening, when it is placed in the lower portion of the gut, the whole of the excrement is passed, and a mixture of excrementitious matter and partially digested food, when it is placed higher up and near the stomach. The most frequent causes of the affection are penetrating wounds of the abdomen, neglected strangulation, and mortification of a rupture, and the ulceration set up by the presence of a foreign body in the intestine; but an artificial anus is sometimes formed intentionally by the surgeon in cases of obstruction of the bowels, or to relieve the severe pains caused by the flow of excrement.

over a cancerous growth in the rectum. In some instances, instead of a large opening into the intestine, there exist one, two, or more minute orifices through which but a small quantity of excrement, and that in a liquid state, is passed. To this latter condition the name of *fæcal fistula* is given by surgeons. Artificial anus is a most annoying affection, and, when placed near the stomach, terminates sooner or later in death, owing to the debility caused by the discharge of partially digested food. In consequence of the frequent and involuntary flow of excrement from the orifice, the patient complains of uncleanliness, and suffers from pain and irritation in the skin about the opening. Severe colic is also a frequent affection in cases of this kind. Occasionally there is a prolapse or protrusion of a considerable portion of bowel through the artificial opening, a very painful condition, and necessitating immediate relief and return of the displaced tube. The palliative treatment of artificial anus consists in frequently repeated cleansing of the skin around the opening and the application of lead lotion, zinc ointment, or a lotion of zinc and tannin. For the purpose of hindering the constant discharge of excrement, a plug of metal or wood must be worn. In some instances a metal case has been worn for the purpose of catching the discharge. These appliances vary much in form and construction.

AORTA is the name given to the large vessel which arises from the left ventricle of the heart, and thence conveys the arterial blood by numerous branches to the various parts of the body. It is an elastic tube about three inches in circumference at first, but afterward becomes considerably narrower. In the first part of its course it is nearly vertical for about two inches and just behind the sternum; it then forms an arch, and, curving from right to left, and from before backward, it descends by the side of the vertebral column, through the diaphragm into the abdominal cavity, and there divides into two terminal branches called the iliac arteries. From the upper portion of the arch arise three great trunks, the innominate, left carotid, and left subclavian; these vessels send the blood to the head and neck and upper extremities. From each side of the descending aorta are sent off numerous branches which supply the lungs, and the thoracic and abdominal walls, while from the anterior aspect in the lower part of its course vessels are given off which convey the blood to the stomach, liver, kidney, pancreas, spleen, and intestines. This vessel is often liable to disease. As people advance in age or when their blood becomes impoverished by disease, the walls of this artery are liable to decay through receiving insufficient nourishment, and degeneration of the coats takes place in consequence. As a

result of this the vessel becomes more rigid, and there is difficulty in the conveyance of the blood to the various organs; sometimes an uniform dilatation of the aorta occurs, at others a bulging of the wall takes place at one spot, and gives rise to an aneurism, a state attended with great danger. People who are subject to gout, or who indulge in drink, or those who have kidney disease, are liable to have degeneration of the coats of the vessel; it is rarely met with in children, and is altogether a chronic condition arising from a vitiated state of the blood, and often produced by neglect of the ordinary principles of health. Great exertion, such as is brought on by rowing or by athletic sports, tend to cause dilatation of this vessel, and it is said to occur among soldiers as a result of wearing too tight stocks, which impede the passage of the blood to the head and neck. It seems likely that amongst men who train severely, harm is done by suddenly returning to a liberal diet after living for weeks in rigid discipline. Malformation of this vessel sometimes occurs in fœtal life; it is in rare cases given off from the right ventricle instead of the left; for such cases no treatment can be of any avail, and death generally takes place in early life.

APERIENTS are medicines which act on the bowels and enable them to expel their contents. They are also called purgative medicines. They act for the most part in making the muscular coat of the bowel contract more vigorously than usual, and some set up an irritation of the lining membrane; some simply empty the bowel, others drain away a certain quantity of fluid from the blood. Aperients are divided into different classes according to their nature and action. 1. There are the *simple* aperients, as senna, castor-oil, and rhubarb; prunes, figs, tamarinds, and sulphur also belong to this class; they simply empty the bowel, do not cause much griping pain, and are useful in many cases of disordered stomach; when administered, some warm stomachic, as ginger or peppermint, is generally given with the dose. 2. The *saline* aperients, as Epsom salts or sulphate of magnesia, Rochelle salt, citrate of magnesia, sulphate of potash, etc. These may be taken in an effervescing form, and are useful when there is any fever present with the constipation. 3. *Drastic* aperients, like colocynth and jalap, cause much purging and drain the blood of fluid also; they act as direct irritants to the intestinal canal. 4. *Hydragogue* aperients, as elaterium, scammony, and gamboge, cause very watery evacuations, and hence are used in cases of dropsy arising either from disease of the heart or kidneys, so as to diminish the quantity of fluid which is effused into the different tissues. 5. *Cholagogue* aperients, or those which are supposed to act more especially on the liver, as mercury, tart avacum, and podophyllin. 6. *Emmenagogue* aperients, or those which act more especially on the womb, as aloes, etc. The reader must refer to each drug for any further description of its action.

APHASIA, or loss of the faculty of speaking, occurs in certain cases of hemiplegia of the right arm and leg ; this must be distinguished from aphonia or loss of voice : in the former the faculty is lost, in the latter the mechanism is interfered with. The persons so affected will probably understand what is going on around, but are unable to ask for anything, and if they speak at all, will limit themselves to the use of monosyllables. Even if asked to spell their own name they will fail to do so, and in reply to any question they generally reiterate the same expression. The handwriting is affected, too, in most cases, and although they know what particular letter to write, they are unable to put it in writing. Often in the course of a few weeks or months recovery slowly takes place, and every day they will learn a few fresh words until they acquire a tolerably large vocabulary. Nothing can be done specially for this singular symptom, except daily educating the patient, beginning with simple words and short phrases.

APHONIA implies loss of voice. It is very frequent in cases of common cold, or catarrh, and then the patient can hardly speak above a whisper, and there is frequently more or less pain or feeling of soreness in the throat and chest. The best thing to do is to wrap some warm, dry flannel round the throat, and inhale steam by the mouth. This may easily be done by pouring boiling water into a jug and then draw in air by the mouth while the face is held over the vessel ; still better, the individual may use an inhaler made for the purpose and of various kinds ; such can easily be obtained at a druggist's. It is necessary, as far as possible, to avoid going out at night, or even during the day when the air is cold and raw, or when there is much fog. This variety is very curable, although it is much more common in some people than in others, and more so in women than in men. Another kind is met with in some cases of hysteria, and chiefly in highly nervous young women. It occurs quite suddenly, and even while conversing as usual, and often without any marked cause. Generally, however, there is some emotional cause which has brought on the complaint, as fright, mental worry, loss of a relation, or trouble in pecuniary affairs. When it happens for the first time it is apt to alarm the patient, but there is really no danger in it. Frequently the voice returns as quickly as it went, and to the great surprise and delight of the patient and her friends, but the complaint is very liable to come back again. There is no actual disease, but the muscles which are brought into action in the mechanism of speech are not affected by the will, and hence will not act. As a rule, the voice will return in a month, if the throat be galvanized, even though the patient has not spoken above a whisper for a year ; sometimes the voice will return immediately after the first application of the galvanic current ; in a very few cases the disease is more obstinate and may last for a long time. Change of air and horse exercise, easy circumstances, and the removal, if possible, of the exciting cause, will often effect a cure. Sudden shocks have been known to bring back the voice at once. In the last stage of consumption, aphonia is often met with, and here, too, there is a roughening or ulceration of the vocal cords which are mainly concerned in the production of speech. In other cases there may be warty growths on the vocal cords which interfere with speech. These may be seen by means of the laryngoscope, an instrument provided with a mirror, by means of which the inside of the upper art of the air-passages may readily be seen ; and indeed this instrument is very useful in making out the cause in all the varieties of aphonia. The treatment for these growths consists in their removal. Foreign bodies in the larynx or upper part of the air-passages, as coins or marbles, etc., are obvious causes of loss of voice ; they should be removed without delay.

APHTHÆ (or *Thrush*) are small round ulcers covered with whitish patches of a sloughy appearance which occur in the mouths of children and not unfrequently extend downward into the stomach. They constitute the disease called thrush in infants. In adults they rarely occur except in the worst stages of certain fevers or allied conditions, where the bodily powers are at the lowest possible ebb. In children they often begin as small white blebs or blisters on the tongue and insides of the cheeks. These burst, and the little ulcers remain behind. Sometimes a number of these grow together, and form a single mass covered with the whitish or yellowish leathery-looking substance. Beneath the membrane the sores are red and angry looking. They occasion great discomfort to the poor infant, and frequently interfere with its powers of taking food. Not unfrequently the condition follows or accompanies measles. In dealing with such a condition the first thing is to keep up the child's strength by careful feeding. Raw meat well pulped may be given if necessary. As the bowels are generally disordered they should be attended to, a little gray powder being perhaps the best opening medicine. This may be required, even supposing the child has diarrhœa. Lime-water is a valuable internal remedy, and should be given always with milk. For the lips and mouth borax and honey or glycerine, well smeared on, is perhaps the best application. Or a wash consisting of 60 grains of sulphate of soda to the ounce of water may be freely applied by a feather or brush.

APNŒA is the name given to the mode of death which results from not allowing the entrance of air into the lungs. Death does not take place directly, but may occupy three or four minutes ; a simple experiment may be made upon a dog : thus, if a string is passed round the trachea and suddenly tightened so as to prevent any air ent ring, the animal will struggle very severely, and will die in from four to five minutes. In cases of suffocation, where the head or a pillow is placed over the nostrils, death is produced in a similar way ; and also in those cases where a cord is drawn round the neck, as in the process of strangulation. When hanging takes place death may be produced by fracture of the spine, or by apoplexy ; yet sometimes it is produced by apnœa when the drop is not very great, and then death is not so quick as where injury to the spinal cord is met with. In drowning, also, death is produced by apnœa ; for a time the person may be enabled to keep on the surface and inhale air, but as exhaustion comes on he sinks, and the water above him effectually prevents the entrance of air, and death must ensue very shortly. It will be seen, therefore, that death can hardly take place in less than four minutes, even when the strangulation is very complete, and of course it takes much longer in the majority of cases. Every means should therefore be taken to restore the respiration as soon as possible, so long as the body is warm, and by continuing to do so for a long time many cases have been successfully restored to health, although quite insensible and apparently dead. Any person found hanging should be at once cut down, and all pressure removed from the neck ; the patient should be placed in the open air, and artificial respiration should be performed at once. Similar treatment ought to be adopted in cases of strangulation, suffocation, or drowning. (See DROWNING and HANGING.)

APOPLEXY is a state in which a person falls down suddenly and lies without sense or motion, while the breathing is often labored and noisy, and the pulse beats often with unnatural force. To this condition the name of *coma* has been applied. A person thus attacked is unable to think or to feel, or to make any voluntary movement, but the functions of the respiratory and circulatory organs still continue, although their action is more or less interfered with. The attack does not always come on in the same way. In some cases the person falls down in a deep sleep, with a flushed face and labored breathing, and convulsions may ensue, or rigidity and contraction of the muscles of the arm or leg. In other cases the coma is not the earliest symptom : there may at first be sharp and sudden pain in the head, then faintness

and pallid skin and vomiting ; after a l pse of time, which may vary from a few minutes to several hours, the patient becomes heavy and stupid, and sinks into a state of coma. And these cases are nearly always fatal, and more dangerous than those which commence suddenly. There is yet a third set of cases, in which the apoplectic state is not so well marked, and in which the patient may become paralyzed without actually losing consciousness. There is sudden loss of power on one side of the body, and to this kind of paralysis the name *hemiplegia* is applie l (see HEMIPLEGIA) ; the patient may be sensible, and able to answer questions and give an account of the attack, but very frequently speech is affected, and there has been some transient giddiness. These cases are less formidable than the others above-mentioned. Sometimes the patient soon gets well, and the paralysis passes away completely ; or he may recover to a certain point and be able to walk about, but only partially regain the power of moving his leg and drag it after him in walking ; or the leg may improve and his arm remain weak. Occasionally no improvement takes place, and the person becomes bedridden and perhaps unable to walk, while he is still more or less sensible, and after a lapse of some weeks or months, he finally dies of exhaustion. In the first method, then, there is coma with paralysis ; in the second, paralysis followed by coma ; and in the third paralysis without coma. When a person falls down in a fit of apoplexy he is quite unconscious of anybody or anything around him ; if pinched or pricked with a pin he will not feel it ; if spoken to loudly he will not take any notice ; if a limb be raised by a bystander it will fall helplessly down again. The breathing may be heavy and noisy, or irregular, and when he takes a breath it is attended by a snoring noise, and his cheeks puff out when he empties his chest on expiration. The face is sometimes flushed or of a dusky appearance ; the eyes are generally closed, and the pupils smaller than natural. Often, too, one side of the face is palsied and the mouth is awry, because it is drawn over to the healthy side , when this occurs, the patient cannot masticate his food well, because it lodges between the gum and cheek of the affected side ; when he tries to whistle the paralyzed cheeks puff out in a helpless manner ; occasionally, also, he is unable to close the eye on the palsied side, and care should be taken to keep it covered, as otherwise the wind or dust, or some foreign matter, may get into it and cause considerable irritation. Sometimes convulsions occur, or one limb may be rigid. The bowels are often sluggish, and a motion is passed in the bed unconsciously ; the urine, too, will flow or dribble away without the patient being aware of it. This state does not, as a rule, last long, and death may take place in a few hours ; in other cases the coma may still continue, and the patient may linger on for several days, and even for a month ; but generally, if death does not take place in twenty-four hours, there is considerable hope of recovery. The deep sleep by degrees passes off, the patient becomes partially sensible of persons or objects around him, and is able to swallow some nourishing liquid ; but although consciousness may thus return, the memory is often much affected, and the patient is low-spirited, and ready to weep on any occasion, or he may remain more or less imbecile for the rest of his life. When the coma has passed away, the hemiplegia, or palsy of one side of the body, may yet remain for some weeks or months ; in some cases complete recovery may take place ; others are bedridden for life, while many obtain a partial use of the palsied arm or leg. As a rule, the face is the first to recover, then the leg, and lastly the arm. The speech is sometimes thick for some weeks, but it generally improves at an early period: and there are certain cases in which although the patient knows clearly what is said, yet he cannot answer properly, and keeps to one or two monosyllables on all occasions, and so makes absurd replies ; in such cases the ordinary process of writing cannot be performed. Apoplexy attacks people of all ages : but it is far more common after fifty years of age ; it is met with in both sexes ; it is found not only in full-blooded people with a red face, short thick neck, and stout frame, but also in thin and spare people. Some persons are more or less liable than others in consequence of some hereditary taint of constitution. A patient generally has some warnings before a fit comes on : headache, sickness, and giddiness coming on in advanced life are threatening symptoms ; or there may be double vision or squint, or numbness of a limb, and the familiar sensation of pins and needles ; in other cases loss of memory and a mistaken use of words are signs of the coming attack. Anything which makes the heart beat faster or fills the head with blood may excite an attack ; much bodily exercise, as galloping on horseback or hastening to catch a train, or running upstairs ; any violent mental shock or fright, or any kind of excitement—may bring on a fit in those who are liable to the disease, and therefore should be avoided. Straining at stool and any stooping position should be guarded against. And it is important to remember the danger that may arise from neglecting these precautions, as an attack may with due care be warded off.

Trea'ment.—When a person is in a fit, his neckcloth or any tight part of the dress should be loosened : he should be kept in a horizontal position and placed on a bed or couch, with his head slightly raised. A piece of linen, dipped in vinegar and water, may be laid across the forehead, and hot bottles should be applied to the feet if they are cold. Perfect quiet should be kept, and the blinds may be drawn down, so as not to let too much light into the room, and only one or two people should be allowed by the bedside.

APPENDIX VERMIFORMIS is a small portion of the cæcum (see INTESTINES) which hangs down in a worm-like shape in the centre of the abdomen. It is remarkable for no known use being assignable to it. At the same time, it is often a source of disease and death. In the passage of the food through the intestines it often happens that a cherry-stone, lemon or orange seed, or other such hard substance, drops into the cavity of the appendix and produces inflammation, ulceration, and often death. This catastrophe may be suspected when intense pain occurs in the abdomen over the seat of the cæcum.

APPETITE, as used in medical language, means a healthy desire for food. Loss of appetite, technically called anorexia, is one of the most common symptoms of disease, as the return of appetite is one of the most certain signs of returning health. This returning appetite is one of those things which require somewhat careful management. It is sometimes not a bad rule to make a patient hungry and keep him hungry ; certainly anything like overtasking the stomach or its powers is to be carefully avoided. At first, as the powers of the stomach have been greatly weakened, small quantities of food, carefully prepared, ought to be given and repeated frequently if necessary. Disordered appetite is frequently a symptom in females of hysteria ; it may also occur in early pregnancy.

APPLE is the fruit of a species of *Pyrus*, and one of the most popular of fruits. Apples contain malic acid, which gives them their acid flavor, and a varying quantity of sugar, pectin, cellulose, and salts. Practically, they are divided into eating and cooking apples, the latter having more cellulose and acid, and less sugar. When eaten it is advisable to take off the peel and remove the core, as those parts are less digestible. In cooking, the peel should be removed and sugar added. If roasted, the peel should be allowed to remain on. Apples, like other vegetable products, contain saline matters which act beneficially on the system. One or two apples a day, according to their size, may be taken advantageously as an anti-scorbutic.

AQUA FORTIS is an old Latin name, meaning strong water, for nitric acid. (See NITRIC ACID.)

ARACHNITIS is a name sometimes given to inflammation of the membranes of the brain. (See MENINGITIS.)

ARCUS SENILIS is a term applied to the narrow opaque zone which may be observed near the margin of the cornea of many aged persons. It usually affects both eyes, and varies in tint, according to its period of duration, from a pale gray to a dense chalky white. It does not extend to the circumference of the cornea, but leaves a narrow ring of clear corneal tissue between its sharply defined outer margin and the commencement of the white external tunic of the eyeball, the sclerotic. It commences as an indistinct semi-opaque crescent seated near the upper or lower margin of the cornea, which crescent gradually extends around the whole circumference of this portion of the eye and increases in width and opacity. This appearance is due to fatty changes in the circumferential parts of the cornea, and is, according to many authorities, an indication of fatty degeneration of the heart and other internal organs of the body. This so-called arcus senilis, though most frequently met with in old subjects, often makes its appearance on persons of thirty or forty, especially in those who in consequence of physical or mental exhaustion, or from hereditary predisposition, become prematurely aged.

ARDENT SPIRITS. (See ALCOHOL.)

ARISTOLOCHIA, a genus of plants, so called from its being thought a remedy promoting recovery after child-birth. There is a species called the Virginian snake root, having the doubtful reputation of being a remedy against the bites of serpents. It is used in medicine, and is a stimulant and tonic, and given in cases of debility and ague.

ARM, BROKEN. (See FRACTURES.)

ARNICA, the root of the *Arnica montana*, as a remedy, is mostly used in homœopathic practice. Its tincture is most commonly used as an outward application in sprains and bruises, and for these it is excellent.

AROMATICS are drugs which have a pleasant smell, agreeable flavor, and slightly stimulating properties. Most of the essential oils belong to these groups of substances.

ARRACK is a kind of distilled spirit, much used in the East, and is obtained from fermented rice, betel nuts, and the sap and fruit of palms. It is prepared with less care than European distilled spirits, and contains pure oil and other substances, which produce headache and other disturbances of the nervous system.

ARROWROOT is the name given to an alimentary substance obtained from the tubers and roots of various plants. Genuine arrowroot is, however, obtained from the root stock of various species of *Maranta*. Arrowroot is a white powder, consisting entirely of the granules of starch. It is sold in shops under the names of West Indian, East Indian, and Bermuda arrow-roots. From whatever part of the world they come, they have the same properties. They fetch different prices, according to their whiteness, absence of flavor, and other properties. Like all mylaceous food, arrowroot becomes thickened by boiling in water, and can be made the recipient of other substances, as wine, brandy, sugar, spices, etc. It is on this account extensively used in the sick-room. It should, however, be recollected that it has no higher dietetical value than other forms of starch. (See CORN FLOUR, SAGO, TAPIOCA.)

ARSENIC is the common term for what is more strictly called arsenious acid, or white arsenic. It is both a dangerous poison and a powerful remedy. The quantity given should therefore be exceedingly small, never exceeding the twelfth part of a grain. It is best given in the solution known as Liquor Arsenicalis or Fowler's Solution, of which three, four, or five drops may be given in water immediately after a meal. Notwithstanding the disastrous consequences of large doses, given in the small ones described it is very valuable in certain complaints. Even in tropical fevers of the same class in which quinine has been given and failed, arsenic will sometimes succeed. While, on account of its cheapness and small bulk, arsenic may in certain cases take the place of quinine, it may almost always be superadded to the latter with advantage, or better still, be continued when the quinine has been left off. There are some kinds of headache, especially one called brow-ague, which seems to be allied to true ague: in these arsenic does good, as it also does in others more distinctly neuralgic in character. It has also been given for some forms of nervous disorder. Of all remedies arsenic seems to be that which is of most use in skin diseases, especially those of a scaly or scurfy kind; where much purulent matter is produced it seldom does much good. Its effects are sometimes marvellous. Small doses should be given very regularly in the way indicated above, and if any smarting of the eyes comes on, it should be discontinued for a time, and again resumed in smaller quantity. In cancerous affections arsenic has sometimes been used locally for its destructive effects, but this is dangerous, and lives have been lost by the practice. In *poisoning by arsenic* the contents of the stomach should be promptly evacuated, and as an antidote the hydrate peroxide of iron given. The antidote is made by taking a druggist's stock-bottle of tincture of muriate of iron, adding to it the contents of the Liquor Ammoniæ bottle, and pouring off the fluid at the top. The precipitate at the bottom is the antidote.

ARTERIES are elastic tubes conveying blood *from* the heart, to which, after having nourished the various structures to which it is distributed by means of the *capillaries*, or ultimate branches, it is carried back by the veins. For the general reader it is sufficient to state that an artery consists of three coats or coverings : (1) An outer one, composed of elastic fibrous tissue ; (2) A middle, composed of muscular fibre in a great measure ; and (3) An internal composed of *epithelium*, of which there are several subdivisions, which we need not specify here. The elasticity of the coats of an artery serve to assist in the propulsion of the blood throughout the system. The arterial system is divided into two main parts. one springing from the left ventricle of the heart, and carrying the blood by means of the *aorta*, the great artery of the body, and its branches to the head, trunk, and limbs ; and a secondary system (*pulmonary*) upon which the former depends, namely, that arising from the right ventricle of the heart, which throws the spent blood, already sent back to the heart by the veins, to the lungs to be converted (*arterialized*) into fit and proper blood for distribution by the first-named system. (See HEART.)

ARTHRITIS properly signifies *any* inflammation of a joint, but in surgery the term is most frequently associated with rheumatism : thus we hear of *Chronic rheumatic arthritis*, a disease characterized by an alteration of all the structures composing a joint. It afflicts rheumatic and gouty patients, and its symptoms are, a racking, gnawing, wearing pain in any joint, generally dependent on weather, accompanied by a grating feeling when the joint is used, and an audible evidence of friction of the opposed surface of the articulation. Opiate embrocations, warm douches, Turkish baths, seem to be good local means of alleviation.

ARTIFICIAL MINERAL WATERS. (See MINERAL WATERS, ARTIFICIAL.)

ARTIFICIAL RESPIRATION is used in cases of drowning, or after an overdose of chloroform has been given, or whenever death by apnœa has ensued, and there is a chance of saving the patient. For the various methods to be adopted, see DROWNING.

ASCARIDES are commonly called *thread-worms*. They look to the naked eye like short bits of white thread ; they are of two kinds, male and female, the latter being about half an inch in length, and longer and larger than the former. They live chiefly in the lower part of the bowel, and may accumulate there in vast numbers ; sometimes they crawl out or pass out when the bowels are relieved. Great itching is caused in the surrounding parts when they are present. When recently expelled they are seen to move about like little maggots. This worm is chiefly found in the stage of infancy and childhood, and only rarely attacks the adult. The symptoms by which it may be known to be present are, picking of the nostrils, fœtid breath, distension of the stomach, and irritation about the anus and genitals as well as the actual passage of the worms. In female children a discharge from the genitals is not uncommon in consequence of the irritation caused by the worms. The precise cause is not clear ; it is found among the children of the poor and those of dirty habits, and is said to be caused by eating blackberries, apples, etc., and it seems most common also in those who are fond of sugar or sweet articles of diet. It is certain that the ova producing the worms are taken in with the food. It is not difficult to get rid of this parasite ; an occasional purge, with an injection every morning up the bowel of a solution of common salt, and careful attention to the diet, which should be plain and wholesome, will usually suffice.

ASCITES signifies dropsy of the abdomen. In such cases there is an accumulation of fluid in the cavity of the abdomen ; this is sometimes so great as to cause extreme distension, and then the patient can only breathe with difficulty. Its most common causes are diseases of the liver, heart, or kidneys ; in cases arising from mischief in the two latter organs, dropsy of other parts, and especially of the legs, is liable to ensue, but not so when the liver is affected. Cirrhosis of the liver arising from drink causes an obstruction to the passage of the blood, and so the abdominal cavity becomes full of fluid ; the patient is then of a sallow or yellow color, loses flesh, and also his appetite. The abdomen is round and swollen, and blue lines may be seen running over the surface in consequence of the fulness of the veins. It is a disease which is gradual in its course, and may last some months or years. Hot fomentations may be applied over the surface of the abdomen to relieve any pain which may exist, and sometimes tapping the swelling and letting out the fluid is very beneficial. Chronic inflammation of the peritoneum or scrofulous disease of the mesenteric glands in children is a frequent cause of ascites, and here the general health must be attended to, for the local disease depends in those cases on the constitutional taint. Cancer of the liver or other abdominal organ may cause ascites in the course of its progress, but in such cases no remedies are of much avail except in so far as they relieve the patient from suffering.

ASIATIC CHOLERA. (See CHOLERA.)

ASPHYXIA is the name given to the mode of death which occurs in drowning, suffocation, strangulation, and in some cases of hanging. The term *apnœa*, is however, a more correct designation. (See APNŒA.)

ASSAFŒTIDA is a foul-smelling gum resin, brought from the East, where it is more appreciated than with us. It is, perhaps, best given in the form of a pill, but in any shape it renders the recipient disagreeable. It is much used in the treatment of hysteria, and is supposed by many to be highly beneficial. Combined with aloes, it is useful in certain forms of flatulent colic, especially in hysterical women.

ASSES' MILK. (See MILK.)

ASTHENIA, a medical expression used to indicate a want of power or strength in the system.

ASTHMA is a peculiar but familiar disease of the breathing organs which is characterized by a painful gasping for breath, coming on suddenly, and passing away without necessarily leaving injury to the lung behind it. We may speak of at least three varieties of the disease : 1st. The first spoken of, which is also called Spasmodic Asthma ; 2d. Asthma occurring as a sign of other disease, i.e., Symptomatic Asthma ; and 3d. Certain peculiar varieties of the disease, of which hay-fever is the most important. **Spasmodic Asthma** is so called because it is supposed to be due to spasm or violent contraction of the air tubes, whereby air is prevented from reaching the interior of the lung. Most frequently the disease comes on without any warning, and commonly occurs an hour or two after midnight, the patient being suddenly roused from his sleep by an attack. There is a feeling first of all of constriction, which grows till there is a fearful struggle for breath. The patient most frequently has recourse to the open window, and there holding firm with his hands, so as to enable him to use the powerful muscles of the upper arm for breathing, he may remain for hours gasping for breath. Over the chest various kinds of unusual sounds are heard, the skin becomes cold, and the temperature falls sometimes many degrees. Subsequently this gives way to sweating from fatigue. By and by relief comes, the patient begins to cough, expels some pellets of mucus, and before long falls asleep. Next day, most probably, the sides ache all over from the fatigue undergone. During the intervals of attack the patient may be tolerably well, but as a rule is not strong, and practically may be said to be a valetudinarian— one constantly looking after his health. The disease is often most capricious, attacking the individual in apparently the most healthy situations, and leaving him alone in smoky, apparently unhealthy quarters. Not unfrequently asthma is hereditary, and commonly sets in about middle life. The immediate cause in bringing on the fit is very variable, but of all causes the most certain and most frequent is eating a heavy meal late in the day. Suppers, especially of an indigestible kind, are prominent causes of an attack. Asthma seldom directly destroys life, however bad the patient may seem in the fit. Many who are subjects of asthma live a good long life, the reason probably being that they are forced to take care of themselves. Generally, however, the disease induces other conditions, especially of heart and lung, which indirectly prove fatal. **Symptomatic asthma,** as far as the paroxysm is concerned, resembles the other form, only being connected with disease of the lung or heart, its conditions are not the same, and the fits are not subject to the same laws. The conditions of lung most commonly associated with asthmatic attacks, are chronic bronchitis and emphysema. Sometimes the order is reversed, and the asthma gives rise to these. The third variety of asthma is that which is induced by certain peculiar causes, hay-fever (which see) being among them.

Treatment.—Treatment can do much and yet little for asthma. The grand rule to be observed is to avoid everything likely to set up the attack, particularly unwholesome articles of food. Abstinence from food must not be carried too far, as we sometimes have seen it done, and so the onset of other diseases favored. If the attack has been brought on by an injudicious meal, by all means let the stomach be emptied. But in the paroxysm other remedies are to be thought of. Foremost among these we should place a pipe of tobacco or stramonium, a few whiffs of which will frequently act like a charm. Stramonium inhaled is also a powerful remedy. For many chloroform or ether is best, but requires careful management, while in others a draught of hot brandy and water, or strong coffee, is best. In the interval the health of the patient should be carefully attended to, iodide of potassium and arsenic being among the most approved remedies to be then given.

ASTIGMATISM is a term signifying irregular refraction ; that is to say, that light, as it passes through the transparent portions of the eye to the retina, is acted upon differently by sections of these portions, thus producing a blurring of the object ; or, while one portion of the viewed object appears distinct, the one next to it seems smudgy. It is dependent upon several causes, such as some original defect in the eye, the results of wounds of the eyeball, or displacement of the lens. Spectacles recommended by some competent oculist are the means of relief. (See EYE.)

ASTRINGENTS are drugs which act by causing a shrinking or puckering of the tissue to which they are applied if strong enough ; they coagulate albumen, and check the flow of blood from a part. Some act locally, and may be applied to a wounded surface ; others are absorbed into the blood, and check hemorrhage from a distant part. Tannic and gallic acids are found in gall-nuts, and are most powerful astringents; iron, zinc, some salts of lead, and more especially the acetate or sugar of lead, are v ry efficacious ; catechu, logwood, chalk, and kino are also astringents. Matico may be applied locally to a wound to check bleeding. Turpentine is a very useful astringent in some cases, and more especially when there is hemorrhage from the lungs, when it can be inhaled with steam. In cases of diarrhœa, chalk, opium, and catechu or sugar of lead can be given, while if there is hemorrhage from the womb or kidney, iron and mixtures containing tannin or gallic acid are necessary. Iron and tannin or gallic acid must not be given together, as they form a disagreeable and nauseous inky compound. The reader must refer to the different drugs mentioned for an account of them.

ATAXY, a peculiar affection of the spinal cord, in which the patient loses control over the movements of his limbs. (See PROGRESSIVE LOCOMOTOR ATAXY.)

ATHEROMA, a degeneration which is very liable to occur in old age, as a natural result of senile decay ; it occurs in earlier life in those who have led fast and intemperate lives, and is one of the many changes which result in bringing the sufferer to an early death. (See DEGENERATION.)

ATMOSPHERE. (See AIR.)

ATROPHY or **WASTING** is a term used generally and specifically. Thus, if the arms and legs waste in any disease, they

are said to be atrophied, or their muscles are said to be atrophied. The same often happens in a paralyzed part. There are, however, certain specific forms of disease to which the term atrophy applies. Such are acute yellow atrophy of the liver (see LIVER), progressive muscular atrophy (see PROGRESSIVE.)

ATROPIA, the active principle of Belladonna. (See BELLADONNA.)

AURA is the name given to certain peculiar sensations which sometimes give warning of the immediate onset of an attack of epilepsy. These feelings are of very various kinds. Perhaps the most common are a feeling of a stream of water or air—cold or hot —gradually creeping up from an extremity toward the head. This feeling reaches a certain point, and then the patient becomes unconscious. It directly ushers in the attack. The period during which this lasts is short, generally exceedingly so ; but its importance rests on this, that occasionally, if it can be stopped, the fits do not occur. For this purpose machines are sometimes worn which, being touched, suddenly and powerfully grasp the arm or leg, certainly in some cases working the desired result.

AURICLE is the name given to two of the cavities of the heart from their resemblance to an ear. (See HEART.)

AUSCULTATION is the art of ascertaining the condition of the internal organs of the body, especially the lungs, by the aid of the ear. (See PERCUSSION, STETHOSCOPE.)

AXILLA is the anatomical name for the arm-pit.

AZOTE is a name for nitrogen gas. Substances, such as certain foods, are called azotized, on account of their containing nitrogen. (See FOOD.)

B.

BAKERS' ITCH is a form of skin disease produced on the hands of bakers by the irritation of the yeast used in making bread. (See PSORIASIS.)

BALDNESS.—Although this term implies complete removal of the hairs either over the whole scalp or over portions of it, it is proposed to describe under this head the causes and nature of the different varieties of thinning and atrophy, as well as of loss of the hair. Baldness, or alopecia as it is technically called, may be partial or general, temporary or permanent. It is best known in the form of senile baldness : that disappearance of the hair from the crown of the scalp which is one of the changes indicating general structural decay and advancing age. In some individuals the head becomes bald during middle life, and in others it is well covered by hairs even at a

very old age. These differences depend upon two influences : that of general health and strength of constitution, and that of hereditary peculiarities. In this form of baldness, whether due to senile or premature decay, the hairs first become gray and then white ; they no longer present their usual appearance, but are short, split, and very dry and crisp. The scalp at the same time becomes thin and tense. At last the white hairs are shed, and no others are formed : complete baldness is then produced, and the thin scalp becomes smooth and shining. These changes always begin on and are very often limited to the vertex of the head ; they are due to senile shrinking of the tissues of the scalp and obliteration of the hair follicles—those small depressions in the skin in which hair originates. Of temporary baldness there are several varieties. The most common form, perhaps, is that general thinning which is caused by exhausting diseases, as for instance, fevers, by bodily decay, and by great mental emotion. Sometimes extensive thinning, or even total loss of hair, may be seen in children and young adults, apparently strong and in good health, and without any affection of the scalp to account for this serious condition. It has been suggested that this early loss of hair may be due to failure of nervous power, or to cessation of the natural reproducing function of the hair-bulbs and hair-forming apparatus. Temporary baldness is also very frequently produced by parasitic diseases of the scalp, such as **favus** and the different forms of **tinea.** According to the nature of the disease it is general or partial : in favus the whole scalp is affected, and in tinea decalvans there is complete baldness only over small patches. Thinning of the hair is a symptom of venereal disease ; in cases of specific sore-throat, associated with a pimply or scaly eruption on the limbs and chest, the hair is very loose, and comes out in large quantities when it is combed or brushed ; in some instances the patient becomes quite bald. This affection, however, is usually temporary, and the hair grows again after the course of the general disease has been averted by suitable remedies. The congenital and senile varieties of baldness do not yield either to local or general treatment. In the former class of cases, one must wait patiently until the formative organs of the hair are well developed, and in the latter class the loss of hair is to be regarded as an inevitable result of advancing age. The application of stimulating washes only irritates the skin, and may do much mischief. In temporary baldness, occurring during convalescence from fever or other exhausting maladies, the hair usually grows again as the patient becomes restored to the previous condition of health and strength. In cases where the hair becomes thin and loose in consequence of debility or want of tone, local stimulant is the best treatment ; in slight forms of the affection, cold water should be poured over the head every morning, and the scalp then well rubbed with a rough towel ; at night, a wash made up of equal parts of glycerine and sal volatile should be rubbed into the scalp at the roots of the hairs. In more advanced forms of baldness from debility, tincture of iodine may be painted over the most denuded portions of the scalp two or three times in the course of the week. Shaving the scalp also does good in bad cases. The best treatment for the excessive thinning of the hair which sometimes occurs in girls who are growing quickly consists in the cold douche, friction with a rough towel, and the local ap-

plication of glycerine and sal volatile or glycerine and lime-water. The hair when it begins to grow again is soft and downy, but in course of time resumes its natural appearance. While it is in this state care should be taken not to apply any oil or pomatum to the scalp. In baldness from constitutional debility or disorder attention should be paid to the important organs of the body. The stomach is very often at fault, and sometimes the nervous system is affected. In cases of this kind the baldness must be attacked indirectly with anti-dyspeptic remedies and nervous tonics.

BALM OF GILEAD is the name given to a juice which exudes from the branches and leaves of various species of Balsam-odendron. It eventually becomes solid. Its properties, like those of other so-called balms and balsams, have been much exaggerated.

BALSAM OF COPAIBA. (See COPAIBA.)

BALSAM, FRIAR'S. (See BENZOIN.)

BALSAM OF PERU is a thick, treacly looking substance with a peculiar odor called "balsamic." It was at one time much used in the treatment of wounds ; now it is sometimes applied to bed sores, but seldom used. Occasionally it is given in chronic bronchitis of the aged. It is rarely used.

BALSAM OF TOLU. (See TOLU BALSAM.)

BANDAGES consist of strips of linen, calico, or flannel, of various breadth and of any length. The best material is stout unbleached calico ; but a strip of sheeting, or strips of an old petticoat or dress, are very serviceable. They should be rolled up firmly for use, as they are applied by unrolling them over the part to be bandaged. There are many forms of bandages in use among surgeons which have received various names, either from their shape or inventor ; but it will be found that as they all depend for their efficacy upon the dictates of "common-sense," the more complicated forms may be omitted. There are, however, some few plain rules for guidance in the application even of the simplest bandage which can be used ; as the *manner* in which it is bound round the limb makes all the difference to the comfort of the patient. It will be found most convenient to hold the roller on the inner side of the limb (if it be a limb) to be bandaged, so that in the case of the *right side* being operated on, the bandage is held in the operator's *right* hand, and *vice versâ ;* and for quickness in application the portion which is still unwound should be *underneath* that which is being wound round the limb—in fact, the bandage should form a sort of continuous figure of 8. On first starting off, rather more than the circumference of the limb should be unwound and cast around the part, and the hand not employed in holding the bandages made to tuck the free end under the first complete turn. If this slight manœuvre be dexterously done, the bandage will never slip, unless purposely unwound. It is then lightly but firmly wound round the limb by a series of turns as far as required. Now it is evident that in the case of a well-shaped, muscular limb, this winding cannot be made evenly, as it will not lie flatly ; the simple device of "reversing" is then employed : it consists of taking a "turn" in its application, and bending it upon itself by changing the surface of the roller which is applied to the skin by making an acute angle or reverse at each turn, and giving it a sharp "twitch" at each. In bandaging the arm or leg, it is best to commence with a few turns round the hand or foot first, whether it be for the retention of splints or dressings. Bandages should always be applied with an equable pressure throughout, and not too tightly. Any person possessed of the slightest ingenuity or neatness of hand would, after a few hints from a good hospital nurse, or an intelligent surgeon, learn the essentials of bandaging in a very short time. Bandages, such as the above, may be rendered hard and strong by smearing their successive turns with gum, plaster of Paris, glue, paste, or white of egg, which speedily sets, serving the double purpose of bandage and splint. (See SLINGS, SPLINTS.)

BANTINGISM is a term applied to a system of diet by which it is proposed to make fat people thin, and which succeeded in the case of a Mr. Banting, who wrote a pamphlet on the subject. The great principle recognized in the system is the withdrawal from the diet of those articles of food, such as bread, potatoes, sugar, fat, and butter, which are known, when taken in excess, to produce obesity. There is no doubt, when the accumulation of fat in the body interferes with exercise and produces shortness of breath and palpitations, that means should be taken to reduce the fat-making articles of diet. At the same time this may be done too violently, and persons have been thus greatly injured. If stout persons wish to reduce themselves, they should diminish the quantity of bread, sugar, fat, and butter in their diet, but not suddenly leave off anything to which they have been habituated through a long life. (See WEIGHT AND HEIGHT.)

BARBADOES LEG. (See ELEPHANTIASIS.)

BARBADOES TAR is a species of naphtha, found naturally in the island of Barbadoes in great abundance. It is a thick bituminous substance, like molasses, having a pungent odor resembling tar. It is only used in this country as a horse medi-

cine, but in the West Indies has a reputation in bronchial and pulmonary diseases.

BARBERRY, the common name of the *Berberis vulgaris*. The fruits are of a red color, and contain a sweetish acid juice, which is reckoned febrifuge. An active principle is also obtained from this plant, called Berberia.

BARK. (See CINCHONA.)

BARLEY is a well-known grain, valuable as an article of diet, both as an addition to soups and broths, and, when ground in the form of meal, as a nutritious food. The medicinal drink known as barley-water is made from the pearl or Scotch barley, and if carefully made is a pleasant and soothing drink in diseases of the throat and chest.

BARRENNESS. (See STERILITY.)

BARYTA is a mineral product, a preparation of which is used in medicine under the name of solution of the muriate of baryta, or chloride of barium. It is prescribed in scrofulous affections, glandular enlargements, and cutaneous diseases, and care must be taken in its administration, as it is very powerful.

BASILICON is known as Royal Ointment, and is an old-fashioned remedy for ulcers, wounds, and abrasions. It is of three kinds : the black, made with pitch ; the green, in which the flowers of melilot form a part and the yellow, made of wax, rosin, and lard, and is the only one now in use.

BATHING (SEA) is a remedy potent for good or evil. There are few constitutions so delicate that they will not bear sea-bathing if the process of preparation is carefully gone through, but that is all-important. It is true, no preparation is needed for a strong man, but then he does not so much need the benefit to be acquired from the baths. To a delicate female sea bathing is often like the renewal of life, but it must be carefully gone about. If the patient has been accustomed to a cold bath in the morning, the only change required first of all is the substitution of salt water for fresh ; if she has not, she must use the sea-water tepid first of all, gradually accustoming herself to the water the temperature of the sea. Next, a sunny day having been selected, she may try, when the sun has been well out, a bath from the beach, but only remaining in the water long enough to be completely wetted from head to foot, then well rubbed dry, and a gentle walk along the shore. This should not be attempted within less than three hours after breakfast, but by degrees the time may be lessened, breakfast being made less and less of a meal till it consists merely of a cup of milk, which it is better to take in all cases before proceeding to bathe. By and by the morning hours may be used for bathing, these are the best ; the bath should not last more than ten minutes, and a smart-ish walk should follow. *Never bathe with a full stomach or when feeling cold before entering the water.* It is not advisable to bathe more than once a day at the sea-side, except a morning bath at home be taken and subsequently one in the sea in the forenoon.

BATHS, whether regarded as means of preserving or recovering health, are of very great importance. The baths employed by a very considerable number of our countrymen daily for the purposes of cleanliness and the preservation of health are cold baths—that is to say, their average temperature is under 60°. If intended more for cleanliness, water of a higher temperature, from 60° to 100°, is employed. They differ too in respect of mode of use, for whereas the cold bath is administered variously, as shower, plunge, shallow, or sponge baths, the warm bath is almost entirely restricted to what is called the shallow bath. The douche is perhaps the most powerful mode of administering the cold bath, but is commonly used as an appendage to the so-called *Turkish* or *Roman bath.* This form of bath consists merely of heated air, and though habitually used by many healthy persons, belongs rather to the category of baths used as remedial measures, than to those in daily use. As remedial measures, hot-air baths are sometimes very useful, especially perhaps in cases of sub-acute rheumatism, colds, and the like. They are, however, to be used with caution, as to many individuals they are dangerous, producing unpleasant sensations in the head. This is especially the case if high temperatures are in use. As a rule, it may be said that 140° F. is quite high enough for all useful purposes, and the time of remaining in the bath should be regulated rather by the effect produced in bringing out perspiration than by other considerations. As a rule such baths terminate either with a cold plunge or a douche, which to many is the pleasantest part of the whole. *Medicated baths* are in use in this country; some prepared so as to resemble mineral waters abroad, others constituted on a different principle, and used mainly in the treatment of skin disease. One variety, viz., alkaline baths, have been found of great use in chronic or sub-acute rheumatism. Iodine baths have been used for the same complaint, and for advanced syphilis. Sulphur baths have been used in lead-poisoning, as well as in itch. The most important of these baths, if indeed it deserves the name of bath at all, is the so-called mercurial vapor bath. This is of undoubted value as a remedy in syphilis. About 30 grains of calomel is placed in a kind of saucer surrounded by hot water, the whole kept heated by a spirit-lamp, The patient sits on a cane-bottomed chair. The apparatus is placed below, and the whole covered with a blanket, so as to completely envelop patient and all. In about 20 minutes both water and calomel will have disappeared, after which the patient is put to bed. This valuable plan of treatment should never be had recourse to without advice. The mustard foot-bath is a favorite remedy among females, who ascribe to it considerable powers of bringing on the menstrual flow. (See HYDROPATHY and SITZ BATH.)

BAY is the name given to the *Laurus nobilis.* It is the true laurel of the ancients, the sweet bay of the English. The *berries* of the bay are aromatic, and are used as spices in food. They are also employed in medicine, and act as carminatives and stimulants. The *leaves* also contain an aromatic oil, and smell very pleasantly, and are used in the same way as the berries.

BAY CHERRY, or Bay Laurel, is the name given to the common *Prunus Laurocerasus.* This shrub goes by the name of the Laurel in our gardens. Its leaves and fruit contain oil of bitter almonds and hydrocyanic acid. They are used in the same manner as oil of bitter almonds. (See ALMONDS.)

BAY-SALT is a name given to a form of common salt, chloride of sodium, which is prepared from sea-water by evaporation in the sun. It is chiefly manufactured in the Mediterranean, and has nothing to recommend it beyond common salt, except that it contains some of the other saline constituents of sea-water.

BED-SORES are large unhealthy ulcers formed over the hips, buttocks, and the lower parts of the back of bedridden persons. They are due to long-continued pressure on these parts, to a vitiated state of blood, and to general debility, and are met with in the subjects of fever, paralysis, broken back, and in very old people who have been in bed for a long time. In cases of palsy of the lower half of the body, bed-sores are very large and deep, and spread with rapidity. In these cases ulceration is favored by loss of sensibility and power of motion, and by the contact of fæces and urine which are passed without the knowledge of the patient. A bed-sore commences as a dusky-red patch on the skin, which becomes excoriated. After the separation of the cuticle the surrounding soft parts become swollen, and the inflamed integument is converted into a gray or black slough, from the under surface of which there is a discharge of thin matter. This sloughing process extends both superficially and deeply until a large cavity is formed, which in some instances exposes bone. In old or very debilitated subjects death is frequently the result of this affection. Except in cases of palsy and broken back, the existence of a bed-sore bears witness to the incompetence or carelessness of the nurse. In cases of long-continued illness and confinement to bed, injurious pressure on the back and hips may be prevented by the use of soft pillows and air and water cushions, and by a constant attention to cleanliness. Draw sheets should be placed over the lower half of the bed, and be frequently renewed, and the buttocks and back ought to be washed twice in the day with a weak lead lotion or spirits of wine, and afterward carefully dried. When a red patch makes its appearance on the skin, collodion should be applied and the inflamed part protected from further pressure by means of a circular air cushion perforated in the centre. When the skin is broken, resin ointment will be found a good dressing. The treatment of large sloughing sores consists in the use of poultices sprinkled with charcoal or chloride of lime, and in supporting the strength of the patient by good diet and alcoholic stimulants. In the first stage of bedsore when the skin is simply reddened, the contact for ten minutes of a bladder containing ice, followed by the application of a linseed-meal poultice, will often prevent further mischief.

BEEF-TEA.—This valuable preparation is of such constant use in sickness, that it may almost rank as a medicine, and it is very important that all those who have the care of invalids or delicate persons should fully understand its value and composition. The great object of the cook should be to extract every particle of nutriment from the beef she uses ; and in order to secure really good nutritious beef-tea, there must be no stint in the quantity of material used. A pound of gravy beef will not make above a pint of really strong beef-tea ; but there are many degrees of strength required, according to the purpose for which it is consumed. Where it is necessary to feed a patient with spoonfuls of beef-tea, and to get as much nourishment taken as possible to assist recovery, an excellent and delicious extract or essence of meat can be made by cutting up about a pound of gravy beef and placing it in a jar, with alternate slices of a nice large turnip and a little salt. Add no water, but cover the jar tightly and let it stew in the oven for six or eight hours. When taken out a most fragrant and nutritious cupful of extract of beef will be there, which will contain all the life-giving constituents of the meat. Very different is the action of such beef-tea as this on the system from that of the extract of meat sold as Liebig's extract, which, though excellent as an addition to other food, and most valuable as a digestive, will never support life or failing strength alone. (See MEAT, EXTRACT OF.)

BEER is a form of alcoholic beverage made from the fermentation of roasted germinating grain. When a seed begins to germinate its starch is converted into sugar. By roasting the process of germination is arrested, and the dried grain, under the name of malt, is used for making beer. Beers are not, however, alone fermented malt. In this country and now nearly all over Europe the flower of the hop is added before the fermentation is commenced, and a bitter taste and tonic quality is given to beer which is not possessed by wines or spirits. Hence medicinally beers act as stimulants and tonics. Beer is sold according to the way it is made, under various names. Thus we have ales, porter, and stout. Ales are mild, strong, and bitter. Mild ales contain from half an ounce to an ounce of alcohol in the pint. Those which contain the least amount of alcohol are most to be recommended as ordinary articles of diet. Strong ales contain from one ounce to an ounce and a half of alcohol in the pint, and ought only to be used when the stimulant effects of alcohol are required. Bitter, pale, or India ales contain from one to two ounces of alcohol, and have a larger quantity of hops than either mild or strong ales. In some cases of disease where tonics are necessary to enable the stomach to digest its food, they are to be greatly commended. Porter and stout are brewed to suit a prejudice. They are brewed with over-roasted or blackened malts, and thus get a dark color. London porter contains from three quarters of an ounce to an ounce of alcohol in the pint, while stout contains an ounce and a half. They are supposed to be more nutritious than ales, but as the blackened malt of which they are made contains no nutritious property, this is evidently a mistake. (See TABLE in ALCOHOL.) All these beers are bottled for sale. The only difference bottling makes is that the carbonic acid gas liberated during fermentation is kept in the bottle and passes out mixed with the beer. In some cases this carbonic acid has apparently the power of assisting digestion, and as a matter of experience is preferred to draught beer. In some states of the system beer is a most objectionable article of diet. The unfermented saccharine matter undergoes changes in the stomach which communicate certain properties to the blood, favorable to the generation of such diseases as rheumatism and gout. When these diseases are not produced there is a general condition of the system brought about in which attacks of serious disease are rendered much more liable to a fatal termination than they otherwise would be.

BEESWAX. (See WAX.)

BELCHING. (See ERUCTATION.)

BELLADONNA, technically known as *Atropa Belladonna*, is a native of Great Britain, but is easily cultivated here. All parts of the plant are active, but those chiefly used are the leaves and the root. From the leaves are prepared a tincture, and an extract from the root, the alkaloid atropia, and a liniment. The effects of belladonna are very striking, especially in allaying pain and arresting muscular spasm. In large doses it is poisonous, and its attractive berries not unfrequently prove fatal to children. When taken internally the drug produces a dryness of the throat, and sometimes an eruption on the skin. These have been said to resemble scarlet fever, and hence this substance has been used by homœopaths in the treatment of scarlet fever, and it has been tried, but vainly, as a preventive of that disease. One of its most notable effects is dilatation of the pupil, so that the iris or colored part of the eye almost disappears, and sight is impaired. This occurs whether the medicine is given internally or applied externally, and is taken advantage of by oculists in dealing with eye-diseases. In many spasmodic or convulsive diseases belladonna is of use. Thus, it has been used· in asthma, in whooping-cough, in epilepsy and neuralgia. In inflammation of the eye, when there is danger of the pupil becoming closed permanently, belladonna is of the greatest possible use by removing the edges of the iris as far as possible from each other. It is also of use in the incontinence of urine in children. Briefly, it may be said that wherever there is much local pain, especially with stiffness depending on ·muscles, belladonna does good. Here it may be applied locally, either as a plaster made from the extract, or as a liniment containing a good deal of a strong tincture. In palpitation of the heart depending on disease of that organ, a plaster worn constantly over the part gives very great relief. In chronic rheumatism the liniment well rubbed in is of great value. The tincture of belladonna should not be given in greater doses than 10 drops, nor the extract in more than half a grain. The liniment is not intended for internal use. The extract is often given in purgative pills to prevent griping.

ATROPIA is the alkaloid or active principle of belladonna; combined with sulphuric acid it is used by oculists to dilate the pupil of the eye. Atropia should not be given internally, but may be applied under the skin.

BELLY. (See ABDOMEN.)

BELLY-ACHE. (See COLIC.)

BENZOIN or **GUM BENJAMIN** is a resinous exudation from a plant growing in the Eastern Archipelago. Combined with aloes, storax, and balsam of Tolu dissolved in spirit, it used to have a great reputation as a vulnerary or application to cut surfaces. This compound was known as Friar's Balsam. It is rarely given internally.

BERBERRY. (See BARBERRY.) ·

BERRIES, poisonous. Children often eat poisonous berries, and show symptoms of illness, before it can be found out exactly what they have eaten. When such a suspicion exists, the safest way is at once to administer an emetic of the easiest and quickest kind. Mustard and water is nearly always obtainable, or salt and water; and tickling the inside of the throat with a feather till vomiting comes on, repeating the operation till all substances seem to be expelled ; then give vinegar and water or milk to the patient to neutralize the effect of the poison in the stomach.

BETEL is used in the East Indies as a masticatory. It is the fruit of a species of palm called *Areca Catechu.* This fruit contains tannic acid, and it is on account of the astringent properties of this substance that it is used. When chewed the nut is cut up and placed in a leaf of the *Piper Betel* and mixed with a small quantity of lime. Cloves, cinnamon, and other spices are sometimes added.

BILE or **GALL**, the name of the secretion formed by the liver, which is emptied into the gall-bladder, from whence it flows into the intestines, where it mingles with the food. It is of a green color, and intensely bitter taste, hence the term "bitter as gall " (see LIVER). When the action of the bowels and stomach are reversed, and vomiting occurs, the bile comes into the mouth, and is readily recognized by its bitter taste. It is not, however, all bitter tastes in the mouth that arise from bile. This arises more often from butyric acid (see BUTTER). The gall of animals, more especially that of the ox, is used in medicine as a tonic, and in cases of deficient biliary secretion. It has been employed as an application to bruises and sores, but its utility in these cases is very questionable, and it is therefore not to be recommended. (See OX-GALL.)

BILIOUS HEADACHE. (See HEADACHE.)

BILIOUS FEVER. (See REMITTENT FEVER.)

BIRDS AS FOOD.—Next to the flesh of mammalia, that of birds is most consumed as food by man. Several species are domesticated in this country, and used as food, whilst a large number of wild birds are consumed. Upwards of 170 species have been recorded as eaten by man in various parts of the world. The flesh of birds has not been so carefully analyzed as that of the mammalia. It contains, generally, more of the principle creatin, and this is especially the case with wild birds. Young birds contain albumen and gelatine, whilst older birds contain fibrine. The flesh of birds contains but little fat; this is more especially the case in wild birds. Domestic fowls are fattened, more especially in the form of the capon. The goose and duck become fat by abundant feeding in domestication. The flesh of birds presents a greater variety of flavor than that of any other class of animals. These flavors are dependent on the presence in the flesh of compounds which are not well understood. As a rule the flesh of carnivorous birds have a stronger flavor than those which are herbivorous or graminivorous.

BIRTH. (See LABOR.)

BIRTH-MARK. (See NAEVUS.)

BISMUTH is used in medicine in two forms, the sub-nitrate and carbonate. The former is the more commonly employed. It is exceedingly useful in certain kinds of irritation of the stomach, especially when food gives rise to pains, but should be given in good large doses of 20 or 30 grains. As it is quite insoluble it must be given in something which will suspend it; gruel will do. Gum Arabic is commonly used for the purpose. A liquid form of the remedy has been prepared by a chemist named Schacht, whence it is known as Schacht's Solution of Bismuth. It is a good and useful preparation. Some people prefer the carbonate to the nitrate; its effects are similar.

BISTOURY.—A surgical instrument, usually carried in the pocket-case. It is a long, narrow-bladed knife, and may be straight or curved.

BITES. (See INSECTS and SNAKES.)

BITTER ALMONDS. (See ALMONDS.)

BITTER-SWEET. (See DULCAMARA.)

BLACK DRAUGHT is a popular name given to an infusion of senna, with Epsom salts or sulphate of magnesia. (See SENNA.)

BLACK DROP is a solution of opium in verjuice, the juice of the crab apple. It is sold in the shops as a patent medicine. One drop is equal to 12 drops of laudanum. It is supposed to disturb the system less than other liquid preparations of opium. (See OPIUM.)

BLACK SNAKE ROOT. (See ACTEA.)

BLACK VOMIT is a term applied to the dark-colored fluid that is thrown up in many fevers. It consists mainly of decomposed blood. It is often seen in yellow fever, and is considered one of the most disastrous symptoms of that disease. (See FEVER.)

BLACK WASH is made by adding calomel to lime-water, and is used as an external application for venereal and other sores.

BLADDER.—This organ is situated in the pelvis, in front of the rectum in the male, and of the womb in the female. It is a hollow cavity, made up chiefly of muscular fibres which enable it to contract, and lined within by a smooth coat of epithelium. It has three openings; two small ones on its posterior aspect, where are the ureters, these being the small tubes which convey the urine from the kidney on each side into the blad-

der; in front there is also the opening into the urethra, or canal which allows the passage of the urine out of the body. The bladder, like other organs, is liable to disease; it may be inflamed and cause intense pain (see CYSTITIS). It may become dilated from being too full of urine, or its walls may become paralyzed, as in some cases of disease of the spine (see PARAPLEGIA). A calculus, or stone, may become deposited or form in this cavity, requiring for its removal the operation of lithotomy or lithotrity. Or the prostate, a gland which is situated at the neck of the bladder, may become enlarged, as in old people; or a tumor, either cancerous or simple in its nature, may be developed.

BLEBS, or **BULLÆ**, are large vesicles, like little blisters, which form on the surface of the skin in some diseases, and very frequently in the later stages of erysipelas of the face.

BLEEDING is a procedure not often adopted now, except in cases of heart or lung disease, where there is great obstruction to the circulation. Formerly, nearly every one was bled as a matter of course in the spring and in the autumn of the year. (See ACCIDENTS, HEMORRHAGE, AND VENESECTION.)

BLINDNESS. (See AMAUROSIS, COLOR BLINDNESS, and EYE).

BLISTER.—Any substance which, applied to the skin, raises the outer cuticle or scarf-skin, and fills the space between that and the true skin with water or scrum is called a blister. There are many substances used for raising blisters. The most commonly applied blister is made from Spanish fly, or cantharides; besides which, mustard, croton oil, nitric acid, and many other substances, are sometimes used. Blisters are considered by many physicians to be most valuable, as they are most powerful remedies. They frequently produce a desirable depletion of the system, and do away with the necessity for bleeding. The ordinary blistering plaster, is composed of lard, suet, rosin, wax, and Spanish flies, a piece of which mixture is spread on adhesive plaster, cut to the proper shape and size. All blisters should have a margin of at least half an inch. The plaster must be spread with the thumb, smoothly and evenly, and not less than the thickness of the thumb-nail in depth. The time a blister takes to rise depends, first, on the part where it is placed, and, secondly, on the temperament of the patient; the period is, however, usually between eight and eighteen hours. It is best to apply a blister before going to bed, as the patient then generally sleeps through the early stage of the process. As soon as the blister has been formed the plaster should be gently taken off, and the bag of fluid carefully nicked with a sharp pair of scissors at the lower part, so as to insure the escape of all the serum, which should be carefully prevented from running on to the skin. Care must be taken not to remove any of the outer skin. A warm bread poultice, inclosed in a piece of muslin, should then be applied, and kept on for an hour. When this is removed, the blistered surface should be dusted with violet powder, and covered from the air, a little fresh powder being added from time to time. This method of dressing a blister generally causes it to heal in a few hours, and prevents the cracking, smarting and stiffness that often follows the application of ointment, or washing the part. Blisters are always liable to affect the kidneys, and, in some constitutions, produce very painful results. To prevent this, the patient should drink freely of barley water, with about a scruple of powdered nitre in each quart, during the time the blister is on, and for a while after its removal. A mustard blister is seldom used, unless very severe counter-irritation is required, and is a severe and painful remedy. Tartar emetic, rubbed into the skin, is another counter-irritant, and produces a crop of vesicles, which fill with serum, like the chicken-pox.

BLOOD is that fluid which is formed from the food of animals, out of which all the organs of the body are developed. The blood of men, when drawn and looked at with the naked eye, is a red liquid. When allowed to stand a few minutes it coagulates, and is separated into two parts : a solid part, called *clot*, and a liquid part in which the clot floats, and which is called *serum*. If a drop of blood is placed under a microscope before it coagulates, it is found to consist of two parts—a liquid called *liquor sanguinis*, and a number of small flattened globules or cells, which are called *blood-globules*. The latter are of two kinds, red and white. The white globules are rounder, rougher, and larger than the red ones. They differ further in containing a more solid central part or kernel, and do not form more than one in a hundred of the blood-globules altogether, though the lymph is full of them. These kernels or nuclei are supposed to escape and become red globules. The red globules are very minute, and when measured are found to be $\frac{1}{3500}$th of an inch in diameter—that is, 3500 of these minute bodies could be placed side by side on a single linear inch ; these red globules are depressed both above and below, and in man and those animals which suckle their young, have no kernel , but in birds, reptiles and fishes they have one. The red globules are composed of a red substance, called hæmoglobin, which can be, separated into an albumen called globulin, and a coloring matter, which gives the thin red color, and is called hæmatin. The size and shape of the blood-globules

vary in different animals. In sheep, oxen, and deer, they are smaller than in man, and are much larger in reptiles ; they are oval in birds and fishes. A knowledge of the forms of the blood-globules has sometimes led to the detection of crime, by revealing the exact nature of blood-stains found upon clothes after the commission of crime. The liquor sanguinis consists of water, albumen and saline matters. When blood coagulates, an albuminous body, which has been called blood-fibrin, is formed, and separates, entangling the blood-globules, and constitutes the clot. The serum which is left holds in solution most of the albumen and saline matters. The serum also contains various other matter, such as coloring and odoriferous principles, with dissolved fatty matters. It also contains oxygen and nitrogen gases, carbonic acid, and a little ammonia. Thus constituted, it is carried by means of the heart and arteries to all parts of the body. On coming in contact with the delicate structures of the body it supplies them with new materials, by which they perform their various functions, and carries away those particles which have done their duty in the work of life. In its course through the body it is carried to various glands, which separate from it those compounds which are to be thrown off from the body. In the liver it gets rid of certain products which form the bile, and which appear to be again taken up into the blood in the bowels. In the kidneys it gets rid of a substance called urea, which, like the albumens, is formed of all four of the organic elements, and represents these substances in a changed and effete form, by which they are thrown out from the body. (See HEART).

Blood might be said, with some degree of truth, to be the most important constituent of the body. It is the means whereby every structure which is worn out for the time being is renewed, and the means whereby its débris is washed away. It is clear, therefore, that any serious impairment of its properties in either of these respects, must speedily lead to alterations of the greatest importance throughout the system. But, besides such apparent changes, there are supposed to be others whereby the whole system becomes altered in such a fashion through the agency of the blood, that no part · can be said to be in a condition of health. Such diseases have been called general diseases or blood diseases. But as it is by no means clear that in these the blood is affected more than any other part, we shall, in dealing with morbid states of the blood, confine ourselves to those conditions where that fluid is tangibly altered in some respect. Now, blood consists of two parts, a solid and a fluid ; the former consisting of what are called blood-corpuscles, red and white, which float in the fluid part of the blood. Either of these may exist in a morbid state, and so we shall try to consider the diseased conditions of each separately. When the blood is poor in quality—that is to say, when its red corpuscles are deficient—whatever other change may have taken place, the patient is said to suffer from Anæmia (which see). These corpuscles may also be imperfectly colored ; at all events the patient is pale, the gums white, and sometimes there is a greenish-white tinge all over the body. This last condition exists in what is called chlorosis. There is, however, another condition, perhaps allied to these, in which not only are the red corpuscles imperfectly constituted, but the white ones greatly exceed their usual proportion. This condition is described as those of white-celled blood, Leucæmia or Leucocythæmia. In it the spleen is generally greatly enlarged, and other organs may be so also. There is a condition where the blood is infected from a suppurating wound, in which pus is supposed to be found, though this has been by no means found. This condition is characterized by the formation of abscesses in all parts of the body, especially the internal organs. It is common in war after injuries to bone. It goes by the name of pyæmia, i.e., pus in blood. In all of these conditions, the solids of the blood are concerned ; in those which follow it is the fluid part which is at fault. First among these, a very rare form of disease, one where the fluid seems to contain an excess of fat in it. To this the name of piarhæmia or lipæmia has been given. More common is that condition which gives rise to Diabetes, where there is an excess of sugar in the blood. The source of this sugar, is immaterial, though it is commonly supposed to be the liver ; but as excess of sugar in urine is called glycosuria, so excess of the same substance in blood is called glycæmia or glycohæmia. Wherever sugar is found in the body, it is thence removed by the blood, and if in excess is removed by the urine. Now this is the constant rule with another substance, the removal of which is essential to life. This substance is urea, and the condition characterized by excess of urea in the blood is called uræmia. This condition supervenes in diseases of the kidney, which interfere with their function, and is the common mode in which these prove fatal (see URÆMIA). There are yet other morbid states of the blood, in which the fluid portion is altered. One of these is characterized by a general yellow tint of the body. Bile is circulating with the blood, and the body turns yellow in consequence. This is jaundice, or, if the bile has not been formed, and its unformed materials are circulating, there will be no jaundice ; this is

acholia or suppression of bile. There are three other conditions in which the blood is altered, but in a way as yet unknown to us. In all of them there is an unusual tendency for the blood to break out of its vessels, though in other respects they are widely different. There is in some individuals an uncontrollable tendency to bleeding, from their birth upwards ; this is spoken of as a hemorrhagic diathesis or hemophilia. Again, there are two conditions acquired by insufficient food and exposure, the causes of which we know, though the changes in the blood are unknown. These are sea scurvy and land scurvy, or purpura (see SCURVY). In certain epidemic diseases, the blood undergoes a change which favors bleedings under the skin, from the gums, etc. Such maladies are then said to assume a hemorrhagic type.

BLUE DISEASE. See CYANOSIS.

BLUE OINTMENT, the popular name of the mercurial ointment of the Pharmacopœia. It is made with fresh lard and pure quicksilver. (See MERCURY.)

BLUE PILL, the popular name of the mercurial pill of the Pharmacopœia. (See MERCURY.) It is made with conserve of roses and pure quicksilver.

BLUE STONE. (See COPPER.)

BOIL. It is hardly necessary to define what a boil is to unprofessional readers. As they frequently depend upon the state of health, constitutional treatment is necessary. Locally, if soft, red, and painful, a hot linseed-meal or bread poultice should be applied, and changed when cold (see POULTICE), and a clean cut made well into it with a sharp penknife or lancet. If *indolent* (or slow in coming to a head), a mixture of equal parts of glycerine, extract of opium, and belladonna, with about twenty times its bulk of resin ointment, is a most excellent application, or iodine paint in obstinately indolent cases. The constitutional treatment consists in the administration of tonics, such as iron, quinine, and ammonia ; the bowels should be kept open, but not purged.

BONE is the hard parts of the vertebrate animals which form their skeleton. Bones are divided into two sorts, according to their composition, and are called cartilaginous and osseous. The former are characterized by the absence of phosphate of lime, while the latter consist of from 40 to 60 per cent of that material. In the living human body the bones contain a considerable quantity of water ; when dried, they are found to consist of about one third of organic matter and two thirds earthy matter. The organic matter consists of fat and gelatine. The teeth are composed of the same materials as the bones, but they contain less organic and more mineral matter. The enamel of teeth contains only two per cent of animal matter. In some animals, as the so-called cartilaginous fishes, comprising sharks, rays, and many of the larger fishes, there is little or no mineral matter in the bones at all. In some parts of the human skeleton, as the cartilages of the ribs and joints, there is very little mineral matter. Although the bones are very hard, like all other tissues of the body they are developed from cells. Originally the bony skeleton in the young of the higher animals is composed almost entirely of cartilage. Gradually bony matter is deposited in the cartilage, and the osseous takes the place of cartilaginous tissue. It is some years after birth that the cartilaginous skeleton of the infant becomes fully converted into bone. Bony matter is, however, formed after birth independent of cartilage, as is seen in the union of bones after a fracture, or in the formation of new bones in cases of necrosis. (See NECROSIS.) When a very thin slice of bone is examined under the microscope, it is found to consist of fibrous, hard material, in which are a series of radiating bodies—black spots with lines running in all directions—looking like minute insects. These are really little cavities, and are called "bone-lacunæ;" they are the active agents in the growth of the bone, and maintain the bones in their integrity. These cavities radiate around certain centres or tubes, which are called the Haversian canals, and which serve as passages for the minute blood-vessels and capillaries which nourish and cause the bone to live. The cartilages present much simpler cells than those of bone, and between them are deposited much larger quantities of intercellular matter than in bone. The cartilages also possess fewer blood-vessels than the bones. The teeth resemble bone in their ultimate structure. On the outside of all teeth is the enamel, which contains very little animal matter, and a great deal of mineral matter. The outside of the fangs of the teeth are covered with bony matter, while the mass of the tooth is made of a substance called dentine, which stands between the bony matter and enamel in the quality of hardness, and is full of very little tubes, which meet in the middle of the pulp. It is this substance which is so largely developed in the tusks of the elephant, and produces ivory. The bones of animals are sometimes recommended as articles of diet. Boiled with their meat on they yield gelatine and fat, and portions of the phosphate of lime are supposed to be yielded in the solution. They should never be thrown away, but used for making soup. Bone dust is used for making jellies, and is a better way of introducing phosphate of lime into the system. Ivory dust is obtained from the refuse of the ivory-turner's lathe and saw, and contains more phosphate of lime than bone dust. In rickets and softening of the bones and scrofula these things may be used as articles of diet with advantage. The marrow of bones

is principally fat, and, after boiling the bones, it may be taken by invalids where a fatty diet is required. Bones when damp are liable to decompose, and when used for cooking purposes should be employed fresh and well crushed before they are cooked.

BONE-FELON. (See WHITLOW.)

BORAX, known to chemists as biborate of soda, is chiefly used as a domestic remedy for children whose mouths are sore with thrush. It is mixed with honey, and smeared all over the inside of the mouth.

BOUGIE is a long and smooth cylindrical instrument used in the treatment of stricture of the urethra, rectum, or any other canal leading to the interior of the body. There is great variety in the size and composition of bougies, some consisting merely of threads of spun glass, or of horsehair, others forming thick cylinders made of gutta-percha, of caoutchouc, or of linen hardened by wax or gum-copal. Instruments similar in form and purpose are also made of metal, but to these the name of sounds is usually given. The treatment by bougies consists in passing the instrument through a stricture, and allowing it to remain for a time, in order to produce by pressure gradual relaxation of the contracted portion of the canal. The parts of the body into which a bougie is introduced in disease are the urethra, the gullet, the rectum, the entrance to the womb, and the Eustachian tube or canal leading from the back of the throat to the internal ear. When a rapid dilatation of the contracted passage is required, recourse is had to bougies formed of some material which will readily expand when moistened. Instruments of this kind are usually composed of catgut, of pieces of compressed sponge, or of the stem of the sea-tangle (Laminaria digitata).

BOWELS. (See ABDOMEN, ENTERITIS, and MELAENA.)

BRAIN.—The brain is a complicated structure formed of nerve-tissue, and constituting a most important part of the nervous system of man. It is enclosed in a bony cavity called the skull, and is thereby protected in a great measure from external injury; it has also three special membranes covering it : the dura mater, a fibrous texture lining the skull ; the arachnoid, a fine delicate membrane lining the dura mater, and covering also the brain ; and finally, the pia mater, a tissue rich in vessels, which here become of very minute size, and, running into the brain, supply that organ with blood. The brain may be looked upon as a prolongation of the spinal cord, which, when it enters the cranial cavity, spreads out in a fan-like manner and forms those large masses of nerve substance known as the cerebrum and cerebellum. The brain is formed in two nearly, if not quite, symmetrical halves, which are partially joined together, so that a close communication exists between each division. Each part is composed of a vast number of white fibres, which form a great proportion of the bulk, while externally there is a shell of gray matter where the nerve-cells are met with, and where the active functions of the brain in great measure are developed. This shell, or superficial layer of gray matter, is about $\frac{1}{10}$ to $\frac{1}{4}$ of an inch in thickness, and is in man and the higher animals very much convoluted, so as to increase the superficial area ; the convex surface of the brain is marked with a number of sulci, or grooves, into which the vessels of the *pia mater* dip and supply the nerve-cells with nutriment. The white fibres merely convey impressions, while in the gray matter reside the functions of the mind. A fair, but not accurate, analogy may be made by comparing the gray matter to an electric battery, and the white fibres to the telegraphic wires : the one generates force, the other only conducts it ; yet both are liable to deceive ; there may be changes in the gray matter, as in cases of insanity, or the fibres may be broken and paralysis ensue. Those functions of the brain which are called the intellect, emotion, and will, and which together make up the mind, have their seat in the outer gray shell, which is made up of layers of delicate nerve-cells, freely supplied with blood ; but each nerve-cell communicates with other nerve-cells, and with distant parts by means of fibres, and these fibres pass down through the spinal cord and ramify all over the body under the name of nerves. For instance, we desire to move a hand ; through the influence of the will, or volition, an impression is sent by means of these fibres from the surface of the brain down to the right muscle or muscles of the arm which have to perform the movement ; the direction of the current here is from the centre to the periphery or circumference. Again, when a finger is pricked, the sensation is really felt in the brain and conveyed there by another set of fibres ; it is not until the brain receives the message, not until it knows what has taken place, that we feel the sensation called pain. The direction of the current is here from the periphery to the centre : the first set of fibres are called motor or excito-motor fibres, and the second set sensory or excito-sensory fibres. Thus there is a complete circle formed by the brain and nerve-fibres, and by this means most of our movements are performed, our sensations felt, and our knowledge of the external world acquired. Besides this large circuit there are smaller ones, and some of these do not include the surface of the brain, and therefore their movements are made unconsciously ; such movements are termed reflex movements, and the will has no power over them. Besides this active gray matter and the white fibres in the brain, there are numerous local centres in each half, termed nuclei ; they also are formed of nerve-cells, and from them proceed various nerves which have special duties to perform. These nuclei are seated in the lower part of the brain, and near its middle line, and the fibres from them form nerves which, emerging from the base of the brain, pass through various holes in the skull, called foramina, and then they supply the parts for which they are destined. These nerves are twelve in number on each side, and they have very different functions ; some are for ordinary motion, as the third, fourth, and sixth pairs, etc. ; some for common or ordinary sensation, as the fifth pair ; some for special senses, as the first pair for smelling, the second for sight, the eighth for hearing, and part of the ninth for taste. The brain is not a solid body, but hollowed within into various cavities, called ventricles, which are lined by a fine epithelial membrane, and contain a little serous fluid. The lateral ventricles are the largest, and are found one in each hemisphere ; the third ventricle lies below them, while the fourth is smaller and more posterior : it is here that many important nerves arise, and any injury to this spot will cause rapidly fatal results ; the fifth ventricle is very small and unimportant. The internal carotid and the vertebral arteries on each side are the chief vessels which give the brain its blood ; on entering the skull, they divide and form very free communication with each other at the base ; then, more minutely dividing, they ramify all over the surface of the brain and enter its substances ; the blood returns by passages hollowed out in the dura mater, called sinuses. The various diseases of the brain are described under the names by which they are commonly known, and reference must therefore be made to those articles for information ; the following is a list of the subjects to be referred to : Inflammation of the membranes, or Meningitis ; Apoplexy ; Hemorrhage, Cerebral ; Cerebral Softening, Malformation of the Brain, Hydrocephalus, Paralysis, Hemiplegia, Palsy, Fits, Epilepsy, and Coma.

BRAIN FEVER. (See MENINGITIS).

BRAN.—When wheat is ground in the mill it is separated into two portions, the flour and the bran. The flour is the inner portion of the grain, while the bran is the outside. A grain of wheat is really a fruit, and the bran is the outside epicarp. It is separated on account of its coloring the flour and making it look coarser. It is, however, frequently retained and mixed with the flour made into bread. Such bread is called *whole meal* or *brown bread.* Although the bran is so often rejected it nevertheless contains constituents which render it a useful article of diet ; indeed it contains more flesh-forming matter and mineral matter containing phosphate of lime than the flour does. Where persons can digest brown bread it is undoubtedly more economical than white bread. Bran, however, in its coarse condition acts upon the bowels, and while it forms a very excellent diet where the bowels are confined, it is on that account to be avoided where the bowels act too freely. When the bran is reduced to a flour with the rest of the wheat, the bran is not so likely to act upon the bowels.

BRANDY, a form of distilled spirits. It is usually distilled from some form of wine, and peach kernels are added to it while being distilled, which gives it its characteristic flavor. Brandy usually contains more alcohol than other distilled spirits, and on this account is more frequently used as a stimulant in disease. There is nothing in brandy to make its action in any way peculiar. The very small quantity of oil of bitter almonds or hydrocyanic acid afforded by the peach kernels could not in any way affect its action. (See ALCOHOL).

BREAD.—All food is called by this name which is made from the flour of grains or seeds and made into a dough and baked. At the present day the most common form of bread is that made from the flour of wheat. Other flours are used, as those of rye, barley, maize and millet, but the flour of these grains is unsusceptible of fermentation, thus this kind of bread is heavier than that whch is fermented. By the process of fermentation bread is made *vesicular,* hence we divide bread into vesiculated and unvesiculated, or into unleavened and leavened bread. All flour made into loaves, cakes, biscuits, etc., is unleavened bread, whilst the flour which has the dough mixed with yeast, in order to start fermentation, is called leavened bread. There is another way of producing vesiculation in bread, and that is by what is called aëration, and bread thus made is called aërated bread. This bread is prepared by adding carbonate of soda to the flour and an acid, and as the acid expels the carbonic acid gas from the soda the bread becomes aërated. Another way is by injecting carbonic acid into the dough, which on being expelled vesiculates the bread without interfering with its composition. Bread contains the same substances as the flour from which it is made. In the process, however, of preparation, a quantity of water mixes with the bread, which reduces the quantity of alimentary matter it contains. The principal constituents of bread are starch and gluten, which exist in the proportions of about four to one. This is the proportion in which the system requires the two classes of food (see FOOD) which are represented by starch and gluten. It is on this account that bread is regarded as a typical food. Unleavened bread is less digestible than leavened bread, and hence in none of its forms is used so extensively. Bread leavened with yeast is sometimes found to disagree with weak stomachs, and in these cases the aërated bread is to

be preferred. The addition of butter to bread appears to increase its digestible property, and adds to the alimentary properties represented by the starch.

BREAST.—At the age of puberty the breasts both of boys and girls are subject to swelling and tenderness, which is perfectly natural, and subsides of itself after a short period of inconvenience. SORE NIPPLES.—These painful cracks, or excoriations, are best treated by painting them over with collodion, and the nipple should be protected from the child's mouth or from the woman's clothing by means of a metallic or caoutchouc shield (not vulcanized) sold by druggists for that purpose. Washing the nipples several times daily with a solution of alum is of great service (see NIPPLES). The breast is peculiarly subject to tumors, the characters and diagnostic features of which may be found in any work on scientific surgery. There is, however, one form of disease, of which it seems important to say a few words, and that is cancer (see CANCER). It first commences as a swelling, attracting notice by its presence, is hard, with a tendency to increase in breadth rather than prominence, seems adherent to the structures above it instead of rolling from under them ; causes the nipple after a while to contract or pucker in, and when pain comes on it is characterized by a severe, shooting pain. After a while the neighboring structures become involved, and the skin ulcerates, and the cancer spreads, the *glands* in the armpit become hardened or indurated, and the health and strength rapidly decrease. The question of removal or of palliation must be referred to a surgeon. The most frequent disease of the female breast which the public are acquainted with, perhaps, is *abscess.* Now, this abscess may arise from several causes : lactation, blows, cold, neglect in suckling, sore nipples, etc., and is attended with swellings, great pain, tenderness, fever, and shivering. The breast should be fomented or poulticed, "slung" with a handkerchief or bandage, and the bowels kept open by a mild purgative : as soon as the matter " points," *i.e.,* comes to the surface, an incision into the abscess should be made with a sharp lancet, taking care that the incision be *vertical* and not *across,* as, in the latter case, the matter will " pocket," or burrow, and fresh abscesses form. Poultices should be applied, and tonics of iron, quinine, bark and ammonia, port wine, etc., given.

BREAST-BONE, a common name for the sternum, a bone which runs down the front of the chest, and to which the cartilages of the ribs are attached.

BREAST-PANG. (See ANGINA PECTORIS.)

BREATH, Shortness of. (See DYSPNŒA.)

BRIGHT'S DISEASE is a name applied to several affections of the kidney which are dependent upon an altered condition of the blood, and generally associated with dropsy and with albumen in the urine. Nephritis is the scientific term applicable to this affection ; the disease may be either *acute* or *chronic*, so that acute nephritis is synonymous with acute Bright's disease, and chronic nephritis with chronic Bright's disease. 1. *Acute Bright's disease* may occur from a cold, from a blow, from taking substances, like turpentine or cantharis, which irritate the kidney ; but more usually it follows some acute febrile disturbance, and more expecially it is associated with scarlet fever. About the second or third week after the commencement of scarlet fever, the patient may find his urine of a dark, porter color, and rather diminished in quantity ; at the same time he will feel lassitude, probably slight pain across the loins, and there may be puffiness of the eyelids and loose parts of the skin ; if kept in bed, the urine in a few days becomes paler, but still looks very cloudy and deposits a copious sediment on standing ; when boiled, a flocculent precipitate is thrown down, because the albumen which is present becomes coagulated. At times convulsions occur, because the blood gets altered in consequence of the kidney disease, and these may be very numerous and end fatally ; at the same time less water is passed. At other times the glands of the neck become large, swollen and painful, and, if they soften and burst, they add seriously to the danger of the case. As a rule, acute Bright's disease has a tendency to recover in two or three weeks ; by degrees, the urine resumes its normal color, more is passed, and finally the albumen disappears ; the dropsy also subsides, and the patient merely feels weak and looks pale.

Treatment.—Since the kidneys cannot do their work, the skin, lungs, and bowels have to take an increased activity ; hot baths do good by causing sweating, and giving free action to the excretory power of the skin ; they may be given at bed-time and repeated every night ; the water should be about 95° Fahr. to 98° Fahr., and the patient may remain in it from five to ten minutes, then quickly dried and put to bed at once. Purgatives should be given so as to freely purge the patient, and for this purpose compound jalap or compound scammony powder are excellent. Rest in bed in a warm room is most important, nor ought the patient to think of leaving his room until all the dropsy and acute symptoms have subsided. The appetite is generally very fair, and light, nourishing food may be taken, as bread and milk, beef-tea, broth, a little mutton, rice pudding, arrowroot, and gruel. During convalescence, great care must be exercised in not sending the patient out too soon, especially when the weather is cold, nor should cold bathing be too soon resorted to ; flannel should be worn next the

skin, moderate exercise may be taken, and a nourishing diet ; no stimulants are required in this disease, but after recovery a pint of beer or two glasses of sherry or claret may be taken every day with benefit ; tonics containing iron and quinine will relieve the debility and the anæmia.

2. *Chronic Bright's disease* occurs in three forms : (*a*) a large fatty kidney ; (*b*) a large waxy kidney ; (*c*) a small contracted kidney. (*a*) The fatty kidney occurs in scrofulous or consumptive people and in those who drink sometimes ; the course of the disease is very long, and may last for years ; it comes on gradually, and the first thing the patient may notice is that he passes less water and that his legs swell ; this swelling is caused by dropsy of the lower extremities, and is worse at night than in the morning ; the skin is pale, and pits readily on pressure (see ŒDEMA). The urine is small in quantity, often darker than usual, throws down albumen on boiling, and there is more or less deposit when it stands. In many cases much relief may be obtained if the case is treated in time, but the dropsy then may come back and spread upwards so that the abdomen becomes distended with fluid (see ASCITES). The breathing is then impaired, and the more so if any hydrothorax is present, as the lungs are encroached upon and there is less space for breathing. The heart has more work to do, and becomes hypertrophied and thicker and larger than usual ; there is often nausea or vomiting. headache, and now and then bleeding from the nose. The face becomes pale or sallow, and the skin all over the body may become œdematous. There is no fever nor pain ; the appetite is often very fair, and the chief distress arises from the dropsy which is met with in the various tissues.

Treatment. — The treatment consists of rest in bed, hot-air or hot-water baths to encourage the action of the skin, and purgatives which shall cause watery motions ; for this purpose jalap, scammony and cream of tartar are to be recommended. If the dropsy in the legs be very great, these limbs may be pricked with a needle or a lancet, in several places, so as to let the fluid out ; the legs should then be wrapped in hot flannel, and a mackintosh placed on the bed underneath ; in this way many quarts of fluid may escape in a few days and give great relief to the patient ; he will feel much easier, and will pass more water and breathe freely now and then ; however, inflammation of the legs will come on after the puncture. Although relief may thus be given for a time, the patient will ultimately die worn out by the constant drain of albumen from the blood, or suppression of urine may come on and cause convulsions, coma, and death.

(*b*) The waxy kidney is so called because its appearance has some resemblance to wax ; it is associated with a different set of symptoms. It occurs in those who have suffered from diseased bone, scrofulous abscesses, or from syphilis, or who have been exhausted by wasting diseases. Nearly always the loin and spleen share in the general mischief, and become much larger than usual. Such patients have very little, if any, dropsy during the whole of the illness, but they pass a large quantity of pale-colored urine, which contains plenty of albumen, but deposits hardly any sediment. The course of this disease is also very chronic, and may go on for years ; it occurs in children as well as in adults ; as in the last case, there is no fever, nor is the appetite impaired particularly. Death will eventually take place under similar conditions to those mentioned under the fatty kidney. Purgatives are less called for here, as there is seldom dropsy ; a nourishing light diet should be given, and tonics containing iron and quinine ; warm clothing must be worn, and moderate exercise may be taken when the weather is fine and mild.

(c) The small, contracted kidney occurs chiefly in gouty people and in those who drink much. The disease comes on very insidiously, and attains considerable progress before it is often found out ; dropsy is rarely present, but much less water is passed than in the last case, although there is often more than usual ; it is pale in color, deposits very little sediment, and contains only a little albumen ; the urine should always be carefully tested in gouty people to see if they have any kidney disease. Debility, headache, a sallow expression, occasionally nausea and bleeding at the nose, are symptoms met with in this form of disease. With it is often associated disease of other organs, as the cause which sets up mischief in the kidneys will also cause changes in other tissues ; the vessels often have fat or saline matter deposited in their walls, and are then said to become atheromatous and calcareous ; the tissues supplied by these vessels are therefore badly nourished, and suffer in consequence ; the lungs become emphysematous, and the patient short of breath ; the brain also is frequently involved, and hemorrhage into its substances or white softening may result, and cause apoplexy (see APOPLEXY). There will then be premonitory or warning symptoms, of giddiness, headache, and loss of memory, followed by a stroke or fit, and hence those who suffer from this disease are always in a serious condition. The heart may become diseased, and inflammation of the pericardium or endocardium may ensue and add to the danger (see PERICARDITIS and HEART). Death may, therefore, take place by one of these complications, carrying the patient off, or the urine may become suppressed and cause convulsions, followed by coma and death.

Treatment.—The treatment is similar to

that mentioned in the last variety. In all cases of Bright's disease the eyes are apt to become affected from changes taking place in the retina, which cause dimness of vision and even blindness. The origin of Bright's disease is always in the blood, and that fluid in turn becomes still further altered by becoming contaminated with materials which ought to be passed off by the kidneys, but which are retained in the system ; the blood also becomes poor in quality by being daily drained of albumen—one of its most important constituents. Pallor, debility, loss of flesh, and defect in the general nutrition arise from this cause.

BRIMSTONE. (See SULPHUR.)

BROMIDE OF POTASSIUM. (See POTASS.)

BROKEN BONES. (See FRACTURES.)

BROMINE, an elementary substance found in sea water, in company with chlorine and iodine. It is found in combination with sodium, and is used extensively in medicine.

BRONCHI, the name given to the air-passages which pass from the windpipe, and are distributed to the whole of the lungs (see LUNGS). The air-passages are subject to inflammation and an increase of the mucus, constituting a very common form of disease. (See BRONCHITIS.)

BRONCHITIS is an inflammatory disease of the lining membrane of the bronchial tubes. This disease may have an acute and a chronic form. 1. *Acute Bronchitis.*—This complaint is a very common one, and is very liable to attack persons in the winter and at times when the east or the northeast winds are prevalent. It commences with the symptoms of a common cold ; there is first a feeling of chilliness and aching pains in the limbs ; the patient is thirsty and feverish, with languor and headache, loss of appetite and restlessness. There is an uneasy feeling of soreness behind the sternum or breast-bone, and this is increased on taking a deep inspiration or in going out into the cold air. At first there is a dry, hacking cough, and very little phlegm is brought up ; in two or three days the cough becomes looser, and the expectoration is more abundant ; the latter is frothy, viscid, and shortly becomes of a greenish-yellow color ; this is attended with relief to the patient, and the feeling of soreness and constriction in the chest then goes away. Wheezing sounds are heard in the air-passages, and may be felt when the hand is placed on the chest or back ; these will partly disappear after the phlegm has been coughed up, and then recur again to be coughed away in return. The sounds are due to the air passing over the viscid mucus, which more or less fills the bronchial tubes. With proper care and rest this disease can be checked and recovery soon take place.

Treatment.—As soon as the patient feels

ill he should go to bed, and keep there until he is well again ; in this way an attack may be checked in a few hours ; the air should be warm, and for this purpose a fire may be lighted and the temperature kept up between 60° Fahr. and 63°Fahr. There should also be a certain amount of moisture in the air, and to effect this a kettle of boiling water may be placed on the fire, and the steam allowed to pass into the apartment ; this may be done two or three times a day, and for about ten or fifteen minutes at a time. A warm bath before going to bed is also a most useful remedy, but the patient should be well dried and put to bed directly afterwards, so as to encourage free perspiration. Some are in the habit of taking a Turkish bath when they have an attack coming on ; the only inconvenience in this procedure is the return home through the cold air afterwards. Whichever plan be adopted, the great object in the treatment is to promote the action of the skin, and cause a moderate amount of sweating. A warm glass of whiskey and water, or port-wine negus, may be taken at bed-time, and this with much comfort to the patient : there is, however, no occasion to give any large amount of stimulants for this complaint ; many domestic remedies are as useful as any medicine. A little prepared barley boiled in half a pint of milk, to which is added a wine-glassful of whiskey, some grated nutmeg with sugar, and lemon-juice, added to taste, will be found a very agreeable potion at night-time. The patient will not care to eat any solid food at first ; bread-and-milk, rice pudding, or one made of arrow-root, tapioca, or ground rice, may be given; broth or beef-tea or chicken-broth ; jellies may also be taken ; as a rule, hot, bland fluids are most enjoyed. A hot linseed-meal poultice may be placed on the chest, and renewed when it becomes cold (see POULTICE). The patient should wear a flannel shirt next the skin, as it will absorb the perspiration as well as a linen shirt, and has the advantage over the latter of not causing that damp and chilly feeling which the linen produces if the bed-clothes are removed by chance and the wet shirt is cooled. Cotton wool laid on the chest is often as good as a poultice, and never forms a damp, heavy mass, which the latter does, if not properly made. Hot local applications assisted by warm and moist air, and a few domestic remedies, will generally suffice to cure an attack of acute bronchitis ; if, however, the disease be neglected in its early stage, and the patient be exposed to draughts and cold air, serious symptoms may arise. Such symptoms would be indicated by the lips becoming of a purplish color, while the cheeks would be pale and livid ; the expression becomes more and more anxious, while the entrance of air into the chest is more difficult, and the patient makes painful efforts to

breathe. Delirium may come on, and rapid sinking ; in these cases the patient dies from apnœa, and the bronchial tubes are choked up with the viscid secretion which he had not strength to expectorate. It is here, when the obstruction to the flow of blood through the lungs is so great, that bleeding may at times be advisable, and afford great relief. Cupping-glasses may be applied to the chest with much benefit, while stimulant expectorants, as ammonia and ether, may be given internally. Although the cough may be very troublesome, and the patient may complain much of want of rest, yet it is often very dangerous to give opium or any preparation of that drug, as it will add to the congestion and may hasten a fatal termination.

Acute Bronchitis in Children is of much graver importance than in the adult, and a great deal of the mortality in childhood arises from this disease. In children the mischief is very apt to spread down the bronchial tubes even to the smallest branches, while in the adult the main branches are, as a rule, the seat of the disorder ; and it is in proportion to this downward extension that the relative danger lies ; for the more the smaller tubes are affected, the less can the blood become properly aërated, and death may take place from suffocation. This disease begins with the symptoms of an ordinary cold, and for some days perhaps nothing more serious appears ; but by degrees there is more fever and restlessness ; the heat of the skin, as shown by the thermometer, is much above the average, the pulse rapid, the breathing quick and wheezing ; the cough is more frequent and painful, and then the veins of the forehead and neck stand out, and the face is flushed. The child feels as if the chest were stuffed, and wheezing sounds may be felt on both sides when the hand is placed over the back or front of the chest. At bed-time the fever and cough are generally worse, and the child is more restless ; then it will often sleep for several hours and awake with a fresh accumulation of mucus and phlegm in the chest, which causes it to make vigorous efforts to expel by coughing ; vomiting may come on, and this may give relief by freeing the tubes of mucus and allowing of easier respiration. The tongue is moist throughout, the appetite bad, while there is more or less thirst. If the little patient become worse, the face may be pale while lips are rather livid ; the nostrils dilate with each inspiration, and the breathing is more hurried and difficult. Convulsions often precede a fatal termination ; generally death takes place without much suffering, as the child passes gradually into a sleepy and unconscious state.

Treatment.—The treatment should begin as soon as possible, as any delay is dangerous. The child should be placed at once in bed, wrapped in a flannel blanket or shirt, and the temperature of the room kept between 60° Fahr. and 65 Fahr. A warm bath may be given at the outset, so as to encourage the action of the skin, and the child should then be quickly dried before a fire and placed in bed. Cotton wool should be laid on the chest, or a large piece of spongiopiline wrung out of hot water ; a hot linseed-meal poultice may also be similarly used. These applications may be renewed when they become cool, and the same process may be repeated for twenty-four or thirty-six hours ; occasionally a little mustard may be mixed with the linseed-meal. The bowels may be opened by a dose of castor oil, and much relief may be experienced in doing so, as, in children, when inflammation of the lungs occurs, there is often diarrhœa or disturbance in the intestinal canal. An emetic of ipecacuanha wine may be given if there is much wheezing and stuffing of the chest, and this will often relieve much discomfort. Expectorant medicines should be given, so as to enable the patient to expel the mucus which is being poured out into the bronchial tubes. At the same time, the strength must be carefully supported ; milk should be given freely, and, if necessary, a few doses of brandy may be put in also. Beef-tea or veal-broth may be given alternately ; solid food is not to be given, nor will the child care for it, as long as there is much fever ; the diet should then be liquid, while it is also light and nourishing. When all the severe symptoms have subsided the patient may return gradually to its usual diet, and the exhibition of some steel wine, or other tonic, will expedite the recovery. Yet for some time care should be taken that the child is not exposed to cold, or to any influence which might bring on a recurrence of the attack.

2. *Chronic Bronchitis.*—This is a very common disease, and is very prevalent during the winter months, causing a great deal of mortality. It may occur at any age, but is most usually met with in middle-aged or old people, and in those who suffer from emphysema. Cough, shortness of breath, and expectoration are the three most constant symptoms of chronic bronchitis ; the patient is pretty free all the summer and during warm weather, but as the winter comes on he takes cold very readily, and the above symptoms come on and prove more or less troublesome all the winter ; they will disappear again in the summer, only to recur when the cold weather returns. Each attack is, in fact, an acute bronchitis, and the symptoms are the same as those which have been detailed above. Those who are liable to a cough every winter generally become sooner or later emphysematous. They are unable to undergo any great exertion because they are so short of breath ; the chest does not expand so much as usual, and they often require support when the breathing is more difficult than usual ;

they cannot lie down at night, but prefer a reclining posture in bed ; the lips are livid and congested, and the eyes bright and watery. Palpitation of the heart is common, and a feeling of fullness at the bottom of the sternum. The circulation of the blood through the lungs being obstructed, the large veins become distended, and dropsy of the legs is very common in those who have suffered long with this complaint. The sleep is often disturbed at night from attacks of difficulty of breathing, and this is much aggravated on a foggy night. The appetite is injured, and any indigestible food makes the patient worse, by causing distention of the stomach, and thereby encroaching on the thoracic space by pushing up the diaphragm.

Treatment. — The best treatment for chronic bronchitis in a variable climate like ours, is change of air and passing the winter in the south, or in some place where the variations of temperature are slight, and where there is an absence of fogs and east wind. To the majority of people this is, of course, impossible, and the treatment must be directed to avoiding as much as possible any exposure to cold, or to any of the exciting causes of this disease. To those who are engaged in out-door occupations, and exposed to all the inclemency of the weather, but little can be done except to alleviate any distressing symptoms that may arise ; thick boots should be worn, so as to prevent damps and cold feet, for the circulation in such people is sluggish, and they cannot take active exercise to keep themselves warm. Flannel should always be worn next the skin, and warm baths may be occasionally taken, so as to keep the functions of the skin in good order. The diet should be nourishing, but easily digestible : meat may be taken at least once a-day, and a pint or two of beer, but starchy food, as potatoes, bread, etc., should only be taken in moderate quantities. Such people should go out after sunset as seldom as possible, and they should not talk in the open air on a cold day, but breathe through the nose, as in this way the air is somewhat warmed before it passes down the bronchial tubes ; a respirator is often of great service. For those who are not obliged to work, much benefit will be found by only going out on fine and mild days, and by avoidance of night air. Some people are obliged to keep the house nearly all the winter, otherwise they get an acute attack of bronchitis which may prove serious. A mild winter is a great boon to those who suffer from this disease. By using these precautionary measures an attack may be warded off ; but when such an attack does come on, the patient should at once place himself under treatment and go to bed ; the treatment then will be the same as that mentioned above for acute bronchitis. (See ASTHMA and EMPHYSEMA.)

BRONCHOCELE. (See Derbyshire Neck.)

BRONZED SKIN is a peculiar discoloration occurring in Addison's disease; very little is known as yet about its nature, and there does not seem to be any means of remedying the color. (See Addison's Disease.)

BROOM, the common name of the *Spartium scoparius.* The tops or ends of the branches are employed in medicine. They are emetic and purgative, and in small doses they act as diuretics. Hence an infusion of broom-tops is used as a remedy for dropsy. A juice of broom-top is prepared and sold.

BROW AGUE is a form of headache in which the pain recurs regularly at a fixed hour. It has not necessarily anything to do with malaria, such as induces true ague, but headache in an individual who has been exposed to such influences, is apt to assume an intermittent type. (See Headache and Intermittent Fever.)

BRUISE, or ecchymosis, is a painful and livid swelling at or near the surface of the body, which is caused by external violence, as a fall or a blow inflicted by some blunt object. It is met with in most cases of contusion, and also with fractures and dislocations, and is caused by the rupture of blood-vessels and the pouring out into the subcutaneous soft-tissues of blood or blood-stained fluid. Bruises vary very much in extent, color, size, and situation. In the slightest form there is a small and superficial patch of a light or dark red color, and attended with very little swelling. In the most severe cases a soft swelling is formed as large as a child's head, or the whole of a limb is swollen and of a black or dark blue color. In some cases the bruise is formed among the muscles and at some distance from the skin, and at others it involves only the surface of the limb over a greater or less extent. The rapidity with which a bruise is formed also varies according to the situation of the injured part. Where the skin is in close proximity to subjacent bone, and is bound down by unyielding tissue, the blood is effused slowly, but in a blow upon the eyelids or upon the breast a large livid swelling is rapidly formed. It has been stated that a bruise when slowly formed presents a deep blue, and when rapidly formed a red or livid red color. In fractures of the bones of the leg and fore-arm there is often extensive bruising which is associated with the formation of large blebs or bulbs on the surface of the skin, which are distended by black or purple fluid. In contusions of the scalp in children a large collection of blood is often formed under the skin; this is usually soft at the centre and very hard at its margin, and feels very much like a depression in the skull. A large bruise when fully developed is of a purple color, mottled with yellow and greenish yellow patches. As the blood becomes absorbed and the bruise fades the purple gives way to changing shades of brownish-red, green, and light yellow. These changes commence at the margins of the bruise. The rapidity with which the disappearance takes place varies according to the age and general condition of the patient: the process is slower in old than in young persons, and in strong healthy men than in persons who are feeble and ill-nourished. The effused blood, even in very extensive bruises is usually wholly removed by absorption, but it now and then occurs that a collection of fluid blood caused by an injury to an unhealthy individual, instead of becoming absorbed, sets up inflammation in the surrounding tissues and forms an abscess which bursts and discharges unhealthy ill-smelling matter or pus mixed with soft clots of blood.

Treatment.—In the treatment of recent bruise the first object is to check further effusion of blood. This may be best done by applying cold, and by elevating, if possible, the injured part above the level of the body in order to retard the circulation; layers of lint dipped in cold water and frequently renewed, or a bladder containing ice, are the best cooling applications. If the bruised parts be very tense and painful some leeches may be applied near the margins of the dark blue patch. After the acute stage of pain and heat has passed off, and the indications remain of any fresh effusion of blood, the treatment should be directed so as to favor absorption of the fluids and to remove the swelling; for this purpose the most useful agents are the tincture of arnica, a lotion composed of two ounces of spirits of wine to twelve ounces of water, or a solution of sulphurous acid. The large transparent blebs which form over very extensively bruised surfaces should be pricked with a sharp needle and then covered with cotton wool, which will absorb the dark-colored fluid which is thus allowed to trickle away. Bruises, however large, soft and permanent they may be, ought never to be opened; since when once the clotted blood which they contain is exposed to the external air, decomposition takes place and results in inflammation and the formation of large abscesses. (See Accidents.)

BRYONY.—There are two plants called by this name, and both of them used occasionally in medicine. One is the *Bryonia dioica*, and belongs to the order *Cucurbitaceæ*. It is called red bryony on account of the color of its berries. The black bryony is the *Tamus communis*, and belongs to the same order as the sarsaparilla. It has no active properties.

BUBO is an inflammation of a lymphatic gland, usually situated in the groin, and having as a cause some venereal affection. Abscesses in the groin, however, may be caused by some injury to the leg or foot, and may be the result of ulcers of the legs, or they may come on after hard walking, riding, or over-exercise. The treatment is that of acute abscess. (See ABSCESS.)

BUCHU is the leaf of a plant growing in South Africa. It is not much used now, but Sir Benjamin Brodie had a great belief in its efficacy in irritation or inflammation of the bladder connected with disease of the kidneys. It is best used as infusion, and should be employed in good large doses, almost as a drink.

BULLÆ, a name given to the blisters or vesicles which appear on the surface of the body in some forms of skin diseases.

BUNION consists in a subcutaneous swelling seated on the inner side of the ball of the great toe. In its early stage it is a thin-walled sac, filled with clear fluid, and then causes very little uneasiness, but subsequently, in consequence of constant pressure and friction, becomes hard and tender. Sometimes, particularly after active exercise, the swelling becomes very painful and inflamed, and forms an abscess. The develop ment of a bunion is caused, in most instances, by a distortion of the great toe, and is much accelerated by the use of tight boots, and by much walking. When the bunion is young, and exists as an indolent movable swelling, with a thin wall and fluid contents, firm pressure with the fingers, or a sharp tap with a heavy object may cause it to burst, and bring about a cure. In those cases where the swelling has existed for some time, and become hard and painful, very little can be done except to wear boots made large and roomy over the toes, and with the sole thicker at the outer than at the inner edge, so that the foot in walking may be thrown more upon the outer part. When the bunion becomes very tender, and the skin covering it red and inflamed, the treatment should be immediate, and consist in rest, and the application of one or two leeches and warm fomentations.

BURGUNDY (See WINES)

BURNS AND SCALDS.—By the term *burns* is meant in surgery the result of the application of excessive heat to the surface of the body, by means of some heated solid body, or as flame. A scald implies the contact of some hot or boiling fluid with the body. As these two kinds of injury present the same appearances, are attended by similar constitutional symptoms, and require the same treatment, they will be described together in this article. Burns and scalds are very serious accidents. Even in their slightest forms they are very painful, and when severe are attended by bodily prostration and congestion in the internal organs. When a considerable portion of the integument of a limb has been destroyed, the patient during the whole course of a long-continued treatment is threatened by various fatal maladies, as visceral inflam-

mation, perforation of the intestine, lock-jaw, and pyæmia. Even after the wound has closed there is generally danger of distortion and hideous disfigurement from the contraction of the resulting scar. The simplest and most convenient classification of burns and scalds with regard to their results on the surface of the body is that which arranges these injuries in the three following classes : That in which the injury causes inflammation and nothing more ; that in which inflammation is followed by destruction and sloughing ; and, finally, that in which sudden charring or complete destruction is produced in that part to which heat is applied. It should be stated that in all extensive burns the two former re sults, and in very severe burns, all these re sults, may be observed on the same patient Around a patch of charred tissue there will be a zone of hopelessly injured but yet living and sensitive skin, and around this again a zone of inflammation, the bright red tint of which passes gradually into the pale color of healthy skin. Simple and transient inflammation is generally the result of a scald. In mild injuries of this kind the skin is merely reddened. This is by far the most trivial result of the application of heat, so long as the surface injured is not very extensive. A scald of this kind produced on the chest or abdomen of an infant, by boiling water, or over the whole surface of the body of an adult in consequence of submersion in a vat of hot fluid may, however, prove rapidly fatal. The next variety of local injury is vesication or blistering ; this is a very common result, and takes place after both scalds and burns. The reddened skin is covered by blisters or blebs, varying in size, and containing a clear yellowish fluid. This condition is well-marked in severe scalds of the hand and fore-arm ; immense bladder-like swellings suddenly appear about the back of the wrist, and sometimes the epidermis of the fingers and the whole hand is separated in the shape of a glove from the parts beneath. Healthy and well-nourished subjects soon recover from the effects of burns which do not pass beyond the stages of in flammation and vesication, but for some hours after the infliction of the injury, suffer much more acute local pain than those whose skin has been deeply destroyed by the action of heat. In the second class of burns the skin is more or less disorganized. At the seat of the injury may be seen soft and elevated patches of a dark-gray color, each surrounded by reddened skin and blisters. These patches in the course of time separate from the surface of the body, leaving large sores, from which there is a free discharge of pus or matter. In some cases the whole thickness of the skin, in others, but its upper layer, is thus disorganized and thrown off. In cases of sudden complete destruction of the surface of the body the burnt skin is hard, dry, and tough, like parchment. It is quite insensible, although pressure upon it may act upon the nerves of deep-seated parts and so cause pain. The color of the destroyed patch varies ; it is sometimes yellowish-brown, at others deep black. The thickness of the burnt part varies according to the intensity of the injury and the duration of the application of heat to the surface of the body. The integument only may be involved, or skin, muscle, and all the soft parts composing a limb may be thoroughly charred.

The chief dangers of burns and scalds lie in the severe general disturbance to which they frequently give rise, especially in young children. In the first forty-eight hours the shock may be fatal, or the patient may be speedily carried off in consequence of congestion of the brain, lungs, or abdominal organs. Immediately after a severe burn, the surface of the body is cold, the pulse weak or almost imperceptible, the lips blue, and the eyes fixed and glazed. When with these symptoms are associated delirium and convulsions, and the patient complains of no pain, death is generally close at hand. The intensity of this state of shock is proportionate, not so much to the depth of the burn, as to its superficial extent, and the age of the patient. Complete destruction of the hand or foot of a strong and healthy adult will be attended with less prostration and collapse than a simple scald reaching over the front of the chest and abdomen of an infant. From the third to the fifteenth or sixteenth day the chief sources of danger are fever, diarrhœa, inflammation of the stomach and intestines, lungs, and brain. During this period death sometimes occurs from a giving way of a certain point of the coats of the *duodenum*, or that part of the small intestine into which the stomach opens. From the commencement of the third week until the period when the wounds are completely closed, the patient is exposed to the risks of pyæmia, tetanus, and hectic fever. If the burn has been extensive there is generally a profuse and exhausting discharge of very fœtid *matter* from these wounds.

Treatment.—In the treatment of severe burns the first and most important point is to endeavor to bring the patient out of the state of shock. When the extremities are cold, and the intense pain of the injury is expressed only by a feeble cry, the body should be wrapped in warmed blankets, and brandy and hot water be administered, care being taken that no more brandy be given after the patient has commenced to revive. In the next two weeks in complicated cases the diet must be light, and saline draughts and frequent purgatives should be prescribed. When the patches of burnt and disorganized skin have been thrown off, and large ulcers are left, from which there is a copious discharge of

matter, it is necessary to support the health of the patient by good diet, a free supply of wine or spirits, and by medicinal tonics. The local treatment of burns and scalds which do not proceed beyond superficial inflammation or blistering of the surface of the skin, consists in the application of such agents as may serve the threefold purpose of reducing inflammation, relieving pain, and preserving the injured part from the air. The following are some of the very many methods that are used in ordinary surgical practice : to dredge the burnt part with flour or starch, so as to form a thick crust or paste ; to apply a thick layer of soft cotton wool and to fix this by a loose bandage ; to lay on strips of lint or cotton rags steeped in a mixture of equal parts of linseed oil and lime-water ; to use as a varnish a mixture of collodion and castor oil, two parts of the former to one of the latter ; sweet oil, white paint, vinegar and whitening are all useful applications. Cotton wool or flour may be recommended as the most suitable agent in all cases in which the skin has not been injured to any great depth. The dressing, when once applied must not be disturbed for several days, so that the surface of the skin may be protected from cold and irritation until the inflammation has ceased. Fresh cotton wool is to be applied over the former dressing of this material, and fresh flour laid on until the old and new layers form together a thick crust. After the dressing has been detached from the burnt surface by the discharge from the ruptured blisters, and the smell from the wound is very offensive, the whole should be carefully removed after soaking in warm water. The best application for the raw surface thus exposed is the ordinary chalk ointment or lead lotion. When there is much blistering and the blebs are very large, a small prick should be made into each before the dressing is applied, in order to allow the contained fluid to drain away slowly. In treating very superficial but widely extended scalds on the chest or abdomen of infants, great caution must be taken not to apply cold water or cooling lotions of any kind ; the best agent in cases of this kind is warm cotton wool. In the treatment of more severe burns which produce sloughing and destruction of skin, stimulating applications are the rule. Of these the most approved are spirits of turpentine, spirits of wine, a mixture of lime-water and linseed oil, a liniment composed of one ounce of resin ointment and half an ounce of turpentine, an ointment of carbolic acid, a mixture of carbolic acid and boiled linseed oil —one part to ten. The dressing should be covered over with layers of cotton wool or well-carded oakum which can be removed daily. After the separation of the burnt portions of skin the raw and ruddy wound may be dressed with lotions consisting of a weak solution of sulphate of zinc or of copper. This part of the treatment of burns demands the greatest care and judgment on the part of the physician. On the one hand he has to guard against general exhaustion and other results of a profuse discharge of matter, and on the other, against the too rapid formation of a sore which may be thick and contractile, and a cause of great subsequent distortion. The deformities so frequently observed after burns in front of the neck in the bend of the elbow, and in the hand are produced in the following way : An extension wound left after the separation of destroyed skin is allowed to scar over rapidly, while the movable parts on its neighborhood are, for the sake of ease, retained by the patient too closely to each other. Thus in a burn in front of the neck the head is raised on a pillow above the level of the body, and the chin depressed toward the chest. In burns of the upper extremity the arm is bent and the fingers closed upon the palm. The dense scar formed over the wound naturally tends to keep the bent parts in their acquired position and by the contractile properties of its tissue, subsequently increases the distortion and deformity. It is necessary, therefore, in treating burns, to keep the edges of the wound apart and to let the surface heal slowly. Of course with extensive wounds nothing can prevent the formation of a contractile scar, which will drag upon the skin of neighboring parts , but much may be done to prevent deformity by careful dressing and by keeping the injured part of the body in a good position.

In concluding this article a few hints may be given as to the best means of immediate action in burns and severe scalds. In most instances where the clothes have caught fire the burnt person loses presence of mind, and rushes out of the room with a draught by which the flame is fanned and its ravages increased. By standers, or those summoned by the loud cries of the patient, should at once do all in their power to prevent action of this kind. If the burning clothes cannot at once be torn away, the person should be laid upon the ground and covered by a rug, a piece of carpet or a table-cover, and turned over quickly from side to side. In order to prevent any draught, the doors and windows ought at once to be closed ; water, when close at hand, should, of course, be thrown over the flame. On removing the clothes from a badly burnt person, great care must be taken not to tear away at the same time any portion of skin or to rupture any blisters. The body-clothing over the seat of injury ought to be cut in small pieces, each of which is to be raised gently ; if any part of this should adhere to the burnt surface, it had better be left until the arrival of the physician. The stockings when tight should be well soaked with oil before removal. In scalds of

the upper or lower extremities the injured parts should be immersed in tepid water before the clothes are taken off.

BURSÆ are closed bags of thin membrane, containing an oily fluid ; they exist in great numbers near the surface of the body, and are placed between bone and skin, and between bone and muscle or tendon, so as to favor the free and ready movement of the latter structures. The tendons moving the fingers and toes also pass through large bursal sacs seated in front of the wrist and at the sides of the ankle. The most important of the superficial bursæ, or those seated between skin and bone, are, one in front of the knee-cap or *patella ;* one at the back of the elbow ; one over the bony projection at the outer part of the hip. In addition to these structures, which exist as constant and normal parts of the human body, there may be other bursæ which are accidental in character, and formed in abnormal situations, in consequence of the application of unwonted pressure and friction. These accidental bursæ are developed in club feet, and over the prominence formed by curvature of the spine. Bursæ in their healthy condition contain but just the amount of fluid necessary for lubricating purposes, and are scarcely to be recognized during life. In consequence of continued pressure, however, or of injury, these membranous sacs become distended from increased collection of fluid, and form prominent swellings, which, with further pressure or injury, become very painful and inflamed. Any superficial bursæ may be thus affected, but the most common and the best known instances are the bunion, the housemaid's knee, and the student's elbow, in which the bursæ at the inner side of the great toe, that over the knee-cap and that at the back of the elbow are respectively enlarged. Another frequent situation of distended bursæ is the back of the wrist, where it is met with as a smooth, painless, and very movable swelling, to which surgeons apply the name of ganglion. When bursæ are subjected to constant or frequently repeated pressure, they often inflame, and gradually become harder in consequence of an increase in the thickness of their walls, at the expense of the contained fluid. They are sometimes converted into abscesses, and then constitute a source of great danger, as the neighboring joint may become affected, or the bones exposed and diseased. Much good may be done by *treatment* in the early stage of an enlarged bursa ; rest of the affected member, and a removal of irritating and compressing agents, must be insisted upon if there be any pain or uneasiness in the swelling. When the parts are quiescent, some means may be taken for removing the disease ; firm pressure with the hands, or a sharp tap with a large volume or some heavy object, will cause rupture of the distended bag, and effusion of the contained fluid ; this, however, is a rough and very painful proceeding. When the walls of the bursa have become much thickened, nothing short of a surgical operation for the removal of the swelling can give relief. In cases of inflammation, and where the formation of an abscess is threatened, leeches, warm fomentations or poultices, and rest in the recumbent position, are essential means of treatment.

BUTTER is a common name given to a variety of fatty substances used as food, whether obtained from the vegetable or animal kingdoms. Thus a fat extracted from a plant in Africa is called Shea Butter. The solid oil obtained from nutmegs is called butter of nutmegs. The term is, however, more generally employed to designate the fatty matter found in the milk of animals. The most common source of butter is the milk of the cow. The quantity of butter contained in different kinds of milk varies. Thus there is 3 per cent in human milk, 1½ per cent in ass's milk, and 3½ per cent in cow's milk. Butter is sold in the markets as fresh butter

and salt butter. The latter has a certain quantity of salt added to prevent its becoming rancid. Butter by keeping is very liable to become rancid, a very disagreeable acid being formed in it called butyric acid. Butter often undergoes this decomposition after it has been eaten, and the bitter taste often felt in the throat and attributed to bile is in reality butyric acid. This disagreeable effect of butter is easily got rid of by taking twenty or thirty grains of bicarbonate of soda or pot ash. Butter is the chief of our fatty foods. It is composed of carbon, hydrogen, and a little oxygen. It is consequently more capable of maintaining animal heat and force than sugar or starch. One pound of butter will go as far in that respect as two pounds of either starch or sugar. Butter is not so easily digestible as the last, and frequently disagrees with those who have a feeble digestion. When butter is eaten in larger quantities than is necessary for the purposes of the system it leads to the deposition of fat in the tissues, and persons become corpulent. This can only be remedied by taking less butter or leaving it off altogether.

C.

CÆCUM, a little sac formed in the course of the intestines. (See INTESTINES.)

CÆSAREAN SECTION is an operation which has been adopted in very extreme cases to save the life of the child during a confinement ; in some cases, as in great deformity of the pelvis, where the child cannot pass down in the ordinary way, and the medical man fails to extract it by any other method, this operation may be done to save the life of the mother or child, or both. It has been done in some cases of rupture of the womb during child-birth. The operation consists in making an incision in the lower part of the abdominal wall large enough to introduce the hand, and then an opening can be made in the womb and the child can be extracted. This procedure is in itself very fatal, and is only justifiable as a last resource and after mature deliberation.

CACHEXIA is the term applied to that condition of profound dishealth which accompanies certain morbid states of the system. Thus with cancer in the latter stages of the malady the patient becomes thin, the color becomes sallow, or of a dusky yellow hue, the features are sharpened, and there is a general expression about the patient of hopelessness and care which is very striking. This condition is described as a cachexia. So, again, among those who have long been exposed to malaria or the poison of ague : the skin assuming a dirty whitish hue. All work is done with an effort ; there may be no distinct paroxysm of fever, but there is a feverishness

which is very striking, this too is spoken of as a cachexia. It is plain, therefore, that though the word is generally modified by an adjective, it means the peculiar state of constitution induced by such diseases as those referred to.

CAFFEINE. (See COFFEE.)

CAJEPUT OIL is an oil of a beautiful bluish green tint, obtained by distilling the leaves of a tree growing in the Moluccas. It is powerfully stimulant in character, and is used both externally and internally. Externally, when mixed with olive oil, it may be rubbed into a part as a liniment in certain forms of chronic rheumatism. In some it does very much good, in others little or none. Internally, a drop or two given on a lump of sugar acts as a powerful stimulant and antispasmodic, and is therefore of much use in certain forms of flatulent indigestion and colic. A drop on cotton-wool applied to a decayed tooth will often relieve toothache.

CALABAR BEAN is the product of an unknown plant, and has the remarkable property of causing a contraction of the pupil of the eye when externally applied. It is used in certain diseases of the eye when the pupil is morbidly dilated. (See PHYSOSTYGMA.)

CALCAREOUS DEGENERATION. (See DEGENERATION.)

CALCIFICATION. (See DEGENERATION.)

CALCULUS is a concretion in any gland or organ, whether it be the bladder (urinary calc.); prostate gland (prostatic calc.);

breast (lacteal calc.) ; kidney—salivary glands (salivary calc.). The term, however, in its most common signification, is applied to the bladder, stone in the bladder. (See CONCRE. TIONS, GRAVEL, and STONE.)

CALOMEL. (See MERCURY.)

CALVES-FOOT JELLY. (See GELA. TINE.)

CALUMBA. (See COLUMBA.)

CAMOMILE. (See CHAMOMILE.)

CAMPHOR, notwithstanding its appearance, is a volatile oil, which, however, normally remains solid at ordinary temperatures. It is obtained from the *Laurus Camphora* by the process of sublimation, a kind of rough distillation. It is mostly obtained from China and Japan. In medicine its uses are not very well defined, but it is largely used for flavoring. To this end some lumps of the substance are placed in a bottle of water and allowed to remain for a time. The water acquires the camphor flavor, and is used as a vehicle for other remedies. It is nevertheless highly esteemed by some authorities, especially abroad, as a stimulant in exhausting diseases, such as fevers of the continued variety. It has also been given in insanity, in asthma and a great variety of other affections, being, in effect, one of those remedies which have acquired a reputation in cures whose natural course is not known. It has, however, been of undoubted service in certain of the complaints of women, especially in alleviating pain. Outwardly, camphor is often used in liniments applied to tender surfaces ; and camphorated spirits of wine is a very good application for chilblains. Homeopaths have used it in cholera. The dose of the substance itself varies from two to twenty grains. It is best given suspended in mucilage or made into pills.

CANADA BALSAM is hardly used in medicine, although much employed by microscopists. Being a kind of turpentine, a mixture of oil and resin, its properties are allied to the better known oil of turpentine.

CANCER.—All that modern surgery has hithero done with regard to cancer is a definition of it, its structure, kinds, and history. It is a disease of itself, as ascertained by pathologists, and one of the class of new growths. It is unlike all other tumors, being an infiltration among the natural tissues of the body, and its peculiar structure is only to be discerned by the microscope. It is ineradicable : if cut out it returns, if not at the place of operation, in some other part or tissue. There are, however, cases where a cancerous tumor has been removed at its first appearance, and the patient has enjoyed an immunity from any return for ten, fifteen, or even twenty years. Hitherto nothing satisfactory has been proved as to its causes, neither individuality, locality, mode of life, or condition. Domestic animals and cattle

are equally subject to it, and pathologists have but slight grounds for suggesting its predisposing causes, such as its being in a small percentage apparently hereditary, its existence more frequently in the female sex, and in the aged. Cancer possesses all the characteristics of malignancy, so defined by surgeons, namely, constitutional origin, rapid growth, constant increase, pain, returning if cut out, infiltrating every tissue in its vicinity, and invading the lymphatic glands, is liable to be diffused over the body from *secondary deposits*, is attended with cachexia (CACHEXIA), "resists all treatment, softens inwardly, ulcerates outwardly, resembles no tissue naturally found in the body, and ultimately proves fatal." The several forms of cancer are called from their special features, *a.* Scirrhus ; *b.* Medullary ; *c.* Melanotic ; *d.* Epithelial ; *e.* Osteoid ; *f.* Colloid ; *g.* Villous. The last two named, however, are by some scarcely held as cancer. *a. Scirrhus.* This form of cancer is probably the most frequent, at least in this country, and most commonly affects the female breast, though it is also found in the rectum, eye, testicle, womb, skin, bones, and salivary glands. Its chief feature is its stony hardness, which is due to an abundance of fibrous tissue ; it is nodulated, becomes adherent to the overlying skin, and it has the singular property, not possessed in a like degree by any other tumor, of drawing into it adjoining structures, is subject to severe stabbing or lancinating pain, and to ultimate ulceration. *b. Medullary,* or brain-like cancer, so called from its resemblance to the substance of the brain, or stiff blanc-mange. Encephaloid, or soft cancer, are its synonymous terms. This form of cancer differs form the preceding in possessing none of that peculiar characteristic of drawing toward itself of neighboring structures, but rather that of a disposition to distend and thrust them aside by the *rapidity* of its growth, and by the great accumulation of cancer material in its bulk. It produces greater constitutional cachexia, and is more speedily fatal. It is most frequent in the limbs and breast. It is excessively vascular, and as it nears the surface throws out huge fungating bleeding masses termed hæmatoides, from ulceration. *c. Melanotic Cancer.* The main distinctive feature of this form of cancer is the presence of pigment or coloring cells, which give it a black or dark appearance. The most frequent situation for melanosis is the skin of the eye, and is more common in the horse or dog than in man. It derives its color, as a general rule, from the structures in which pigment naturally exists. *d. Epithelial Cancer,* termed also epithelioma or cancroid, so termed because the microscopic cells found in it differ less from the cells of the part in or near which they grow, than those of the foregoing kinds, and from their analogy to the *natural* epithelial structures. Its chief situation is in the skin, in or near a mucous orifice, *e.g.*, lip, nose, anus, prepuce, scrotum, or tongue. When it exists on the scrotum it forms the so-called chimney-sweeper's cancer. *f. Osteoid Cancer*, a form of cancer occurring usually in bones, and more commonly in the lower end of the femur apparently than elsewhere. It is very rapid and painful in its growth. In the form of cancer the *stroma* is converted into a very dense fibrous tissue, and then into a peculiar bone, which is rough and porous, and very brittle, readily reducible to a chalky powder after maceration of specimen. It seems to be singularly interchangeable with encephaloid. *f. Colloid Cancer*, a form of the disease not regarded by some writers as includable under the term cancer In appearance it is jelly-like, about the consistence of thin glue or tapioca pudding, of rapid growth, and frequently attaining enormous bulk. It is most frequently found in the intestinal canal. *g. Villous Cancer*, not now considered cancerous ; the name has been associated with a vascular growth composed of delicate papillæ, each containing a vascular loop, generally in connection with cancer or epithelium.

Treatment.—With regard to the *Treatment* of cancer, all that can be done is to check the disease as far as possible, and thus endeavor to alleviate suffering and prolong life. In an article of this nature and compass, we cannot possibly enter into the ethics of the treatment of cancerous growths, and, indeed, they cannot but tend to one point, and that is, removal in all cases on their earliest detection. In advanced stages of the disease, palliative treatment, both local and constitutional, must be had recourse to, but death will sooner or later supervene, and all that can be done is to make the end as easy as possible.

CANCRUM ORIS is an ulcerative affection of the gums and cheeks, most commonly met with in children between the ages of fifteen months and five years, and seldom observed in adults. There are two well-marked forms of this disease. The milder form occurs in ill-fed and squalid children during the cutting of one or more teeth. The anterior surface of the gums becomes red and swollen, and bleeds at the least touch. The roots of the teeth are laid bare. The swelling then extends to the inner surface of the cheek, and sometimes to the lips, and in these two situations several small but deep ulcers are formed, from which there is a fœtid discharge. In the course of a week the swollen gum becomes gangrenous and separates from the surface of the jaw, exposing blackened and rough bone. The teeth become loose and fall out. During these changes the tongue is white and furred, the breath fœtid, and the flow of saliva from the mouth almost constant. The glands in the neck are some-

times swollen and painful. There is much difficulty in masticating, and also in swallowing. It should be remembered that in this affection the swelling is generally found on one side only of the mouth, and that it never causes ulceration of the tongue or palate. In these respects it differs from the diffused soreness of the mouth, produced through the administration of large quantities of mercury. The subjects of this affection should be supported by strong beef-tea and broth, together with wine. Castor-oil may be administered every other morning in order to keep the bowels freely relieved. For the purpose of removing the offensive discharge, the mouth should be frequently washed with a lotion containing two teaspoonfuls of carbolic acid to a pint of water. Undiluted claret also forms a good wash, from its astringent action on the swollen gums. Pieces of alum should be given to the patient to suck, and the gums painted twice in the day with a lotion made up of half an ounce of dilute muriatic acid and half a pint of water. Chlorate of potash and bark are the usual internal remedies in this, as in most other ulcerative affections of the mouth.

The second or more severe form of cancrum oris is that commonly known by the name of water-canker. It attacks children who have just recovered from measles, and in large schools or places where many young patients are crowded together sometimes spreads rapidly after an epidemic of this eruptive fever. A similar affection, and one occurring under similar conditions, is sometimes observed in the external genitals of young girls. This is called noma. The affection of the mouth first shows itself as a large and doughy swelling on one cheek ; the skin over this swelling is tense and shining, and at its prominent part presents a patch of a violet or dark-blue color. On the inner surface of the swollen cheek will be found a deep, foul-looking ulcer, the edges of which are generally swollen and irregular. The gums become gangrenous and there is a very fœtid discharge of dark-red or black fluid from the mouth. In the further progress of the disease, the cheek is perforated by the ulcer, and the orifice thus formed is subsequently enlarged by rapid sloughing. The little patients, as a rule, do not seem to suffer much pain from this affection, but about the sixth or seventh day become heavy and sleepy, and finally pass into a state of coma. This form of cancrum oris is usually fatal. The local treatment, to be of any service, demands energy and skill. The best agents for arresting the progress of the mortification are strong nitric acid, and the actual cautery or red-hot iron. The patient should be allowed plenty of fresh air, and wine and good nourishment. It is necessary to wash out the mouth frequently with weak solutions of chloride of zinc or lime, or of carbolic acid.

CANELLA is the bark of a tree growing in the West Indies. It has a hot cinnamon-like taste and slightly tonic properties. It is not often used by itself, but is contained in rhubarb wine. A favorite domestic remedy, called *Hierapiera*, or more familiarly *Hiccory-Piccory*, is composed of equal parts of canella-bark in powder and aloes.

CANKER OF MOUTH. (See CANCRUM ORIS.)

CANTHARIDES, or SPANISH FLIES, are mostly collected in Hungary. More than one species are in use. They are beetles, and their wing cases, which are long, have a fine green color. They are collected by brushing the trees, killed by boiling vinegar, and dried. Before being used they are reduced to powder, from which may easily be obtained a crystalline substance called *Cantharidine*. This powder constitutes the active ingredient in that well-known remedy, a fly plaster. The other ingredients are wax, rum, and fat. This form of the remedy, though the most common, is not the best. Blistering solutions may now be obtained which, when painted on to any part of the body, and a warm poultice applied, more effectually, more speedily, and more painlessly produce vesication. Elegant little blistering plasters prepared in France may also be obtained in boxes ready for application at a moment's notice. Cantharides are rarely given internally ; even externally when long used they may produce troublesome symptoms referable to the bladder and kidneys. They should never therefore be given internally without medical advice. (See BLISTERS.) It constitutes the basis of most stimulant hair-washes, and the application of a blister to the bald surface of the head will frequently have the effect of producing a growth of hair ; usually a few drops of the tincture added to some salad oil will suffice.

CAOUTCHOUC. (See INDIA RUBBER.)

CAPSICUM, or red pepper, employed but rarely in medicine, is the pod or fruit of a plant now frequently cultivated in this country. The pod, which is bright red, is dried and reduced to powder—the well-known Cayenne pepper. This substance is a powerful stimulant, and is chiefly employed as a condiment ; but there is an authorized tincture. Occasionally this is used mixed with water as a gargle. The powder is mostly used to excite appetite by stimulating the stomach. It has been given as a stimulant in delirium tremens.

CARAMEL is a term applied to burnt sugar, and is principally employed for coloring wines and brandies.

CARAWAY.—The fruits of the ordinary caraway are like many others belonging to the same natural order, endowed with a vola-

tile oil, which gives them their peculiar odor and taste. This oil may be obtained in a separate form, and administered in drop doses for flatulence, gripes, etc., but the flavor is very objectionable to many.

CARBOLIC ACID, or PHENIC ACID, is a substance contained in the very complex body we call coal-tar, whence it is obtained by careful distillation. It is met with ordinarily in two shapes, either in crystals or fluid. It is not very soluble in water, but sufficient may be taken up to endow the fluid with valuable properties. Applied without dilution to the skin, it produces something like a burn. It is, however, possessed of powerful antiseptic properties, being one of the most powerful agents known as a means of preventing putrefaction. To this end it is very largely employed for preventing bad smells, for keeping wounds sweet, and for preventing the spread of infectious diseases. Carbolic acid may be obtained for disinfecting purposes, either as crystals, which may be dissolved in water, or in the form of disinfecting powder. A solution as useful as any that can be bought in the shops may be made by purchasing an ounce of the *pure* carbolic acid and mixing it with water, in the proportion of half an ounce to a quart of water. It is an invaluable agent, but on account of its smell, which is far from agreeable, other disinfectants are sometimes preferred to it. Its most convenient strength as a lotion for the prevention of smell and arresting discharge, is one pint to forty of water ; but in the hands of surgeons it is used of much greater strength. Internally, carbolic acid is of much value in indigestion and flatulence. pain and vomiting of fermented half-digested food. A drop or two of the deliquesced acid when the acid becomes fluid by exposure to the atmosphere may be given in mucilage some hours after taking food. It has been used in the form of ointment, or dissolved in glycerine, for the cure of itch with success. And a strong solution is a capital means of getting rid of vermin in the hair of men or animals. It should be well rubbed in, and in a quarter of an hour washed out again with soap and water. (See DEODORANTS and DISINFECTANTS.)

CARBON. (See CHARCOAL.)

CARBONIC ACID POISONING is produced whenever there is an accumulation of carbonic acid in a room or building. It is probable that death is due to the individual being deprived of oxygen as well as to the noxious presence of carbonic acid gas. Carbonic acid is always present in the atmosphere in a small quantity, and it is partly an accumulation of this gas which makes a small, badly-ventilated room smell close and stuffy ; languor and headache then come on, and unless fresh air is admitted injurious consequences may supervene. From this cause, overcrowding is injurious, and these who work together in a room, as dressmakers, etc., are often liable to headache, anæmia, and general debility. Fresh air is of course the simple remedy for this condition of things, while in serious cases of poisoning by the gas, and when the patient is found insensible, removal of the individual into the open air is the first thing to be done, and then artificial modes of respiration must be resorted to. For a description of the methods to be adopted for producing artificial respiration, the reader must refer to the article on DROWNING. Carbonic acid is a gas, and in its pure state is poisonous when retained in the blood. By itself it is not greatly used in medicine, but in the form of soda-water it is often exceedingly useful. It is this agent contained in effervescing liquids which gives them their sparkling character, and which enables them to be retained on the stomach, which would otherwise reject them in fevers and such like disorders. It is partly due to the carbonic acid in it that champagne is frequently retained when nothing else is. As a vehicle for many remedies and many stimulants there is no better agent than a solution containing an excess of carbonic water, the simplest being the soda-water aforesaid. The gas itself has been employed to relieve the pain of cancerous, especially uterine, affections.

CARBUNCLE is an exaggerated boil (see BOIL) most frequently situated where the tissues underlying the skin are of a dense fibrous character, such as the nape of the neck. Carbuncles vary in size, sometimes being as large as an orange. They are very hard, brawny, dreadfully painful, discharging matter from several apertures ; and usually attended with considerable constitutional disturbance, such as fever, hectic, etc., the condition being indicative of blood poisoning. With regard to treatment, it must be both constitutional and local. The strength must be kept up by brandy, wine, and ammonia and bark ; hot fomentations, opiate poultices, and free incision must be made through the thickened implicated tissues.

CARDAMOMS belong to the group of remedies called stimulants and carminatives. The compound tincture of caradamoms is one of those nice preparations which belong to what might be called the confectionery of medicine. It is used to give coloring and pleasant flavor to more powerful remedies.

CARDIALGIA signifies pain in the heart, or over the region of the heart, and may arise from various causes. (See HEARTBURN.)

CARIES is an unhealthy inflammation of bone, causing it to absorb or ulcerate. The most common example of this condition is a decaying tooth. Any bone, however, may be affected, and it is generally caused by some constitutional disorder, such as scrofula or syphilis. It bears somewhat the same relation to bone that ulceration does to the soft tissues of the body. Its treatment is, to rectify the constitutional condition, and locally to remove the diseased bone.

CARMINATIVES mean remedies which are slightly stimulant in their character— which relieve flatulence by expelling gases, and alleviate colicky pains. They almost all contain a volatile oil, on which their properties depend. Favorite examples are ginger, mustard, horse-radish, the different kinds of pepper, cinnamon, cloves, anise, coriander, peppermint, etc. (See STIMULANTS.)

CAROTID ARTERY.— The pulsations felt on each side of the neck are due to arteries which pass from the heart to the brain, and are called by this name. They were first called carotid in the supposition that if pressed upon coma would be produced.

CARPUS, a technical name for the wrist.

CARRAGEEN MOSS is the name given to a sea weed known to botanists as the *Chondrus crispus*. When boiled it produces a decoction which becomes glutinous on cooling. It contains little or no nutritive matter, consisting principally of the starch known as lichenin.

CARRON OIL.—At one time this mixture was supposed to be the best remedy for burns and scalds. It is an oleaginous mixture of lime-water and linseed oil, which looks like the broken yolk of an egg. Its name was given from the fact that it was at the great Carron Ironworks, situated in Scotland, on the banks of the Carron River, that it was first made, about sixty years ago, and its reputation was so great that it was sent from thence all over the country for the cure of burns and scalds. Since that time, science has entirely superseded this remedy by the introduction of others more effectual and less disagreeable.

CARTILAGE, vulgarly called gristle, is a firm, flexible, and highly elastic substance, of a pearly white color. It is met with in joints, and takes part in the composition of certain important organs of the body, hence the division into *articular* and *non-articular* cartilage. The articular variety covers the joint-ends of bones, favoring by its smooth surface easy movement of the joint, and by its great elasticity, freedom from shock and concussion during active exercise of the extremities. Non-articular cartilage is met with in the wind-pipe, the external ear, the nose, and the eyelids. The ribs terminate anteriorly in long pieces of cartilage which pass inward and are united to the *sternum* or breast-bone. Cartilage varies in appearance at different periods of life. In fœtal life and early infancy it is soft and semi-transparent ; in youth and adult life, it presents the well-known bluish white opacity ; and in old age. it becomes hard and yellow, and in some localities, more especially in front of the chest, is converted into bone.

CASCARILLA BARK is the product of a shrub growing in the Bahamas. It has a spicy odor, and a bitter aromatic taste. Its properties correspond, being tonic and stomachic. It is, however, by some esteemed as a remedy in chronic bronchitis, and for the expectoration of consumption. The tincture or infusion may be used in teaspoonful and two tablespoonful doses respectively.

CASEINE is the name given to a product found in plants and animals, and so-called because it constitutes the basis of cheese (see CHEESE). It is one of the flesh-producing constituents of plants, and is found in large quantities in peas, beans, and lentils. It is not used in a separate form.

CASSIA.—Many medicinal articles are known by this general name. They are the product of plants belonging to the natural order *Leguminosæ.* We may give as instances of such drugs, the *Cassia cinnamonium* (cinnamon), *Cassia fistula* (pipe or purging cassia), and *Cassia senna.*

CASSIA PULP is the pulp of a long pod or legume, which has purgative properties. Given by itself it gripes. It is contained in the confection of senna, a very useful preparation.

CASTOR OIL is the produce of the seeds of a plant called *Ricinus communis.* The oil is obtained from the seeds by pressure, which is applied with or without heat. That which is cold drawn keeps longest, and is preferred for use. Castor oil is a mild laxative, producing in its action little or no pain, and leaving, after its effects have gone off, no tendency to constipation. It is on that account strongly to be commended where only a slight action of the bowels is required. Like all oils it is nauseous to the taste, and that is its great objection. Several ways have been devised of administering it. The easiest and quickest way is to administer it in water with a few drops of brandy, peppermint, or other agent to take away its taste. It may also be made into an emulsion with an egg if preferred. According to age, from a teaspoonful to a tablespoonful may be administered at a dose, and repeated in three or four hours if it fail to act.

CATALEPSY, or TRANCE, is a condition so very rare that few men have seen it. It seems half mental, half bodily. In catalepsy the patient, very frequently, but not always, an excitable hysterical female, suddenly seems to lose all consciousness of surrounding objects. They remain fixed in one position ; but, if that is altered by any one standing by, the new position is maintained instead of the former. In this state the individual remains for a time, varying from minutes to hours, or days, and then suddenly re-

covers, knowing nothing of what has passed in the interval. Very frequently it is hard to say whether this condition of things is real or assumed, and the fact that it might possibly be real, that the individual might be alive, and yet seem dead, and so be interred alive, has given rise to much uneasiness in the minds of many. Stories have gone abroad of individuals turned in their coffins, but that has plainly been due to putrefaction changes. Besides, in those conditions which are likely to be most deceptive, the period which elapses between death and burial is generally sufficient to establish marked signs of putrefaction, such as are altogether incompatible with life. Allied to catalepsy is ecstasy, where the individual seems buried in contemplation of some curious object. This, too, mostly occurs in women, and is mainly hysterical.

CATAMENIA. (See MENSTRUATION.)

CATAPLASM. (See POULTICE.)

CATARACT may be defined as impairment or loss of vision, due to opacity of the crystalline lens. This affection is occasionally met with in new-born infants, but occurs most frequently in old people as a result of certain senile changes within the eyeball. In about eighty per cent of the cases of cataract the patients are above the age of fifty. It may, however, occur at any age as a consequence of wound of the lens, or of simple concussion of the eye-ball by a severe blow. It has been stated that diabetic patients, and females who are suckling, are specially liable to become affected with cataract. It generally affects both eyes, commencing in one before the other. In the cataract of old people the pupils will be found opaque, and of a pale amber or grayish white color. The opacity is most marked in the centre, and fades away toward the circumference into gray cloudy specks. This amber-colored or grayish opacity is due to contraction and hardening of the central portion of the hitherto transparent lens. This condition usually comes on slowly, and the sight gets gradually worse for months or even years, until there is almost complete blindness. During the development of the affection the patient is much troubled by a mist or haze surrounding all white or pale objects. This mist is not removed on regarding the objects through spectacles. Black spots are often observed as if floating before the affected eye. Vision is improved by a subdued light, while there is great intolerance of bright or strong light. There is occasionally much dazzling and agitation of regarded objects both in the day time and by candle light. The flame of a lamp or light of any kind is surrounded by a broad misty halo. Objects are increased in number and distorted. In daylight the patient sees objects more readily when the back is turned to the window and the eyes are shaded by the hand. Pain in the eyeball is rarely complained of ; and the patient is not troubled, as in cases of amaurosis and disease of the retina, by prismatic or rainbow-like halos around flame. The movements of the iris remain free during the progress of the disease. Before the sight has been much impaired, considerable temporary improvement may be produced by putting atrophine drops, or a small disc of atrophized gelatine, into the eye; the pupil is dilated and thus permits rays of light to pass through the peripheral and less darkened portion of the lens. THE CATOP-TRIC TEST OF CATARACT.—When a lighted candle is placed at a short distance from the front of the healthy eye, three reflected images of the flame are distinctly seen arranged from before backward. The first and third of these images are erect, and when the position of the candle is altered move in the same direction. The middle image is inverted, and when the candle is moved upward or downward, moves in the opposite direction. This middle image is reflected from the posterior surface of the lens, and the third image from the anterior surface of this body. When the lens is rendered opaque by cataract, the inverted or middle image is much obscured and in advance of the cataract altogether absent. This is called the catoptric test, by which cataract may be distinguished from amaurosis or blindness due to disease of the retina. Cataracts may be roughly divided into hard, soft, and fluid. Hard cataract is the most frequent form, and is rarely met with in persons under thirty-five years of age. The opacity varies in color from pale gray to mahogany. It most frequently presents an amber tint. The diseased lens is harder and most opaque at its centre than at its circumference. When extracted it has the consistence and somewhat the appearance of soft bees-wax. In soft cataract the lens is swollen and of a milky or bluish white color. When fluid the cataract has an uniform grayish white color, and looks like a small mass of thick gruel. Soft and fluid cataract may be present at any period of life, but in most instances during infancy.

Treatment.—The treatment of cataract is almost exclusively operative. During the early stages of the affection the failing sight may be temporarily improved by dropping in a solution of atrophine, or by smearing the upper lid with extract of belladonna in order to produce enlargement of the pupil. When both eyes are affected, and the patient strong and healthy and free from gout, the lens may be removed with considerable chance of a successful result. There are three chief methods of operating on cataract : *Extraction*, in which the lens in removed through an incision made in the cornea ; *absorption or solution* of the lens by breaking up its substance with a needle, and allowing it to become saturated in the fluid of the anterior chamber of the eyeball, which acts as a solvent ; thirdly,

the old operation of *couching*, which consists in moving the lens backward into the interior of the eyeball, and, at the same time, depressing it below the level of the pupil. Of these operations, *extraction* is the one most frequently performed.

CATARRH means simply a running, such as we have from the nose in a cold ; the name of the symptom has, however, been transferred to the condition which produces it, and so one may hear of catarrh of the stomach, bowels, bladder, etc. (See COLD and INDIGESTION.)

CATECHU is of two kinds, the puce and the black. The latter is not now contained in the Pharmacopœia ; it is obtained from the wood of a species of acacia. The puce catechu is extracted from the leaves and young shoots of plants growing in Siam and other parts of the eastern hemisphere. It occurs in irregular pieces, generally approaching to cubes. It contains a kind of tannic acid to which it owes its properties. The substance is a powerful astringent, and may be used in a variety of complaints, but is most frequently given for diarrhœa when there is no inflammation present, and when it does not depend on bilious derangement. For this purpose, *i.e.*, arresting the diarrhœa, the compound catchu powder, consisting of catechu, kino, rhatany, cinnamon, and nutmeg, may be given in doses of 20 or 30 grains. For relaxed sore throat, such as singers and public speakers often suffer from, catechu, in the form of a lozenge, is very beneficial. These may be obtained from most druggists, and should be allowed to melt gradually in the mouth. The infusion may be used as a gargle, and combined with charcoal, finely powdered, catechu forms an excellent dentifrice when the gums are spongy and expose the base of the teeth.

CATHARTICS are a class of medicines that act on the bowels strongly as purgatives ; such as senna, castor-oil, Epsom salts, gamboge, colocynth, etc.

CATHETER.—A catheter is a curved tube, made either of silver, india-rubber, or gum-elastic, for the purpose of drawing the urine from the bladder in cases of stricture of the urethra. The size, or calibre, varies from a tube the size of a bodkin, to that of a large lead pencil.

CATNIP (also called *catmint*) is a plant that grows wild in the fields throughout the United States. The leaves, which are much used as a domestic remedy, are aromatic, pungent, and somewhat bitter to the taste. Cats are very fond of them, and are said to use them medicinally. The leaves may be preserved by drying in the sun and keeping them in a dry place. They should be plucked when the plant is in bloom. *Catnip Tea*, the form in which catnip is administered, is an infusion made by pouring boiling water on the leaves and allowing them to steep. It acts as a tonic and to some extent as an antispasmodic ; in a much diluted form, with a little sugar in it, it is often given to very young infants to sooth them and to expel the wind from their stomachs.

CAUL is a term applied, first, to the membrane which covers the bowels, also called omentum. Second, it is applied to the membranes enveloping a new-born babe, when they extend over the head. There is a superstition that persons possessing the caul of a new-born babe will not be drowned at sea, and at the present day it is not an uncommon thing in England, to see a caul advertised for sale. The origin of this superstition is obscure.

CAUSTICS are a number of very different substances which possess the common property of burning and destroying any part of the living body with which they may be in actual contact. The best known of these agents, perhaps, are sulphuric acid, and the nitrate of silver, or lunar caustic, so much used in surgical practice. They vary much in form and composition, and in their modes of action. The majority act either as oxidizing agents or by withdrawing water from the living tissues. They all cause great pain, and may produce much mischief when applied by those unacquainted with the special indications for treatment by caustics. The chief objects for which these agents are used in surgery are the following : To set up counter-irritation ; to keep down overgrown granulations or "*proud flesh ;*" to arrest the progress of ulceration ; to destroy cancerous growths and ulcers ; to open abscesses ; to stay absorption in poisoned wounds. A large abscess just on the point of bursting and covered by thin integument can be opened by the application to the softest and most inflamed part of its surface, of a thin stick of hydrate of potash or *potassa fusa*, care being taken to prevent the deliquescent caustic from flowing over the surface beyond the seat of its application. A small black slough is formed, which separates in the course of eighteen or twenty-four hours, and leaves a free opening into the interior of the abscess. This method, however, though less terrifying, is much more painful than the use of the lancet in the hands of a skilled surgeon. In dog-bites it is advisable to cauterize the wound or wounds with some mineral acid applied on a glass tube or some non-vegetable conductor. Nitric acid is generally used for this purpose, but hydrochloric acid and the oil of vitriol answer equally well.

CAUTERY is an agent employed for applying intense heat to superficial parts of the body. There are three kinds of cauteries, namely, *potential*, *actual*, and *galvanic*. The first term was applied by old surgeons to the various forms of caustic applications described under the heading of *Caustics*. The actual cautery consists in a rod or knob of iron heated to incandescence, and the galvanic cautery is formed of wires heated by a galvanic battery. Surgeons use the *actual* cautery with the following objects : To produce counter-irritation ; to arrest bleeding ; to destroy cancerous and other tumors on the surface of the body ; to stop the progress of hospital gangrene, and to destroy the edges of large fistulous openings. The *galvanic* cautery is chiefly used for destroying the wells of the long sinuous passages occasionally left after the discharge and contraction of an abscess.

CAYENNE PEPPER. (See CAPSICUM.)

CELLS, minute parts of the structure of animals and plants, always microscopic, but assuming various sizes, forms, and conditions. The essential of every living cell is a particle of matter called protoplasm, containing the four elements—nitrogen, carbon, oxygen, and hydrogen. From such particles, both in animals and plants, the cell-walls are formed. The cell-walls in plants are composed of cellulose, and in animals of gelatine. According to their age and functions, cells are solid or contain water with various contents, floating or dissolved. All growth takes place by the multiplication of cells, and diseases are produced by the cells acting in an abnormal way.

CELLULITIS. (See PELVIC CELLULITIS.)

CELLULOSE is an insoluble substance which composes the cell-wall of plants. It is the basis of all wood and timber. It is very hard in the stones of plums, apricots, etc., and very soft in oranges, pears, and other fruits. It is constantly taken into the stomach with unprepared vegetable food. It is not digested in the stomach of man, although it seems to supply food material to many of the lower animals. In estimating the quantity of alimentary material in any article of diet the quantity of cellulose should always be deducted.

CEPHALALGIA. (See HEADACHE.)

CERATE is a term applied to an unguent in which wax forms an ingredient. The white cerate of the druggists' shops is formed of white wax and pork lard. Yellow cerate consists of yellow wax and resin.

CEREBELLUM. (See BRAIN.)

CEREBRAL SOFTENING is a result of disease of the vessels in the brain or of changes taking place in that organ consequent upon previous mischief. It is often associated with cerebral haemorrhage, and it is one of the conditions which produce apoplexy. Hemiplegia, or paralysis of the arm and leg of one side of the body, is then of common occurrence, and this comes on with insensibility or coma. It is a very serious change, indicating long standing disease in the constitution. It occurs in those who have had gout or chronic Bright's disease, and in those who have been intemperate : sometimes, also, as a natural result of old age and great mental exertion. The mind before an attack is often impaired ; the patient has loss of memory, giddiness, vertigo, occasional attacks of faintness, lowness of spirits, and irritability ; the countenance is often sallow, the person thin and shrunken, the eyes marked with an arcus senilis, and the general appearance that of premature decay. (See APOPLEXY, HEMIPLEGIA, and DEGENERATION.)

CEREBRO-SPINAL FEVER is an acute, epidemic disease, characterized by profound disturbance of the central nervous system, and marked by shivering, giddiness, intense headache, delirium, and spasms of various muscles ; there is great prostration, and occasionally a purple eruption appears on the skin. This disease is also known as epidemic cerebro-spinal meningitis, petechial fever, purpuric fever, etc. Age does not seem to have much influence upon this disease ; children, young people, and adults all suffer during an epidemic ; males, however, seem more liable to it than females. Season has a remarkable effect upon this malady ; it occurs especially during the cold months ; of 216 local outbreaks in France and the United States, 166 prevailed between December and May, and by far the majority of attacks have occurred during the winter months. Fatigue, cold, overcrowding, foul air, and dirty dwellings have been put down as exciting causes of this disease, but at present very little is known for certain on these points.

Symptoms.—In most cases the patient feels indisposed before the onset of the disease ; there are pains in the head and various muscles, loss of appetite, and slight shivering ; in some the onset is quite sudden. Acute shivering is followed by severe headache and giddiness, then by profuse vomiting without nausea ; with the sickness there is often neuralgic pain in the abdomen ; after the lapse of a short time, often only a few hours, the mind becomes confused, and the patient grows restless ; there is muttering delirium with occasional cries, or the patient falls into a state of apathy and stupor, or he may be violently delirious. With this mental disturbance there is pain along the spine and limbs, and chiefly in the muscles at the back of the neck and along the spinal column ; the head is drawn backward in consequence of the pain and spasm, and this retraction of the head is a marked and common symptom. As the disease advances, other groups of muscles may be affected in a similar manner, the trunk often being curved backward, and the legs bent upon the thighs ; the sensibility of the skin is also increased. The expression shows acute pain, or is distorted by spasm ; the eyes are suffused and the face pale, with occasional flushings. The temperature of the body is higher than usual, and the pulse weak ; the

tongue is sometimes clean, sometimes foul, and the bowels are either costive or loose ; these two latter signs vary in different epidemics. Purpuric spots appear on the skin, and do not disappear on pressure of the finger ; at first purple and circumscribed, they soon become black and extend their margins, so as to form dark blotches; this is a bad sign, and then the patient may be in a state of collapse and insensibility, and death may take place very rapidly; in other cases hemorrhage takes place from the mucous membrane of the mouth, nose, and intestines. If the disease tends to a fatal result the spasms increase, coma comes on, and death may ensue in from twelve hours to seven or eight days ; if life exist longer, inflammation of the eyes and ears may be set up, as ulceration of the cornea and deafness ; or paralysis of one side of one limb may ensue, or there may be an inflammatory state of the joints. If the disease go on favorably, recovery may take place in three or four weeks, but if the progress is interrupted by any complication, convalescence is much retarded. Inflammation of the lungs, pleura, and heart, swelling of the glands of the neck, and under the ear, disease of the eyes and ears, bed-sores, and joint affections, are met with as complications in this malady. The rate of mortality varies in different epidemics ; the minimumrate is twenty per cent, and it has been as high eighty per cent. Cerebro-spinal fever may be mistaken for typhus fever, but the history, rash, and progress of the disease will clear up doubt ; spinal meningitis and cerebral meningitis may much resemble this malady, but the onset of cerebro-spinal fever is so much more rapid, and the fact of its coming as an epidemic will help to solve any difficulty ; besides, no rash is met with in the last two cases.

Treatment.—1. *Preventive.*—Since so little is known as to the causes of this malady, all preventive efforts must be limited to those sanitary measures which are applicable to all epidemic disorders for the purification of houses and localities. 2. *Curative.*—The treatment of cerebro-spinal fever by remedies is very unsatisfactory ; it is doubtful if the administration of any medicine has been beneficial in doing more than relieving symptoms ; for this purpose opium or morphia has been given to allay pain and spasm. Sulphate of quinine in large doses, and given early, appears to have benefited some cases ; bleeding and mercurial preparations are of no value, and may do harm. The diet must be generous and nourishing, and consist of milk, beef-tea, and soup, etc., but often the patient can only take food with difficulty ; the object must be to sustain the vital powers during the great prostration which marks the acute stage of this disorder. During convalescence the usual principles of diet must be adopted which are detailed under the head of FEVER.

CEREBRO-SPINAL MENINGITIS. (See MENINGITIS.)

CEREBRUM. (See BRAIN.)

CERVIX, a neck, is applied in anatomy to bones, as cervix femoris, the neck of the thigh-bone, and cervix uteri, the neck of the womb.

CEVADILLA. (See SABADILLA.)

CHAFING. (See INTERTRIGO.)

CHALK, or impure carbonate of lime, is only used in medicine when it has been thoroughly washed and purified, when it is called prepared chalk. It acts partly by absorbing and partly by what might be called its drying properties. It is mostly used in summer diarrhœa in the form of *Chalk-mixture*, consisting of chalk, gum acacia, syrup, and cinnamon water. Of these chalk is almost the only active ingredient, and should be given in doses of twenty or thirty grains, or more. It is not desirable to continue its use for too long a time, as if the diarrhœa does not speedily yield to it, other and more powerful remedies should be given. Chalk is also used as the basis of most tooth-powders, either as prepared chalk or as precipitated chalk, the latter being a finer powder. If the gums are at all spongy, a little rhatany powder may be added, or a little powder of cinchona bark and some flavoring agent ; orrisroot is perhaps that most used. Chalk is also used with advantage as an application to raw and weeping surfaces, such as occur in the folds of fat infants ; but on the whole oxide of zinc is preferable.

CHALK STONES.—A white, insoluble substance, deposited in the textures of the bones, joints, or areolar tissue, of gouty persons ; generally in the feet or hands. Its chemical composition is lithate of soda. The swellings produced are very painful if inflamed, and discharge freely ; simple soothing dressings should be applied, and suitable constitutional treatment be adopted. Their removal is rarely admissible.

CHALYBEATE. — Anything containing iron. (See IRON, MINERAL WATERS.)

CHAMOMILE is the flower of the *Anthemis nobilis*, a plant somewhat resembling a daisy. The single flowers, *i.e.*, those having most yellow in the centre, are the best. It has long been a favorite in domestic practice, given as infusion or tea for a variety of complaints. In reality it is not to be despised as a tonic, although there are other and more powerful remedies. The tea should not be given warm, or it will probably cause sickness and vomiting.

CHAMPAGNE. (See WINES.)

CHANCRE. (See VENEREAL DISEASES.)

CHANGE OF LIFE. (See MENSTRUATION.)

CHAPS.—Usually, the disagreeable condition of the skin known as chapping is produced by insufficiently drying it after washing, and exposing it in a wet or damp state to the influence of the weather and the winds. Persons affected with chapped hands should be very careful not to wash them too frequently, and to dry them very carefully, rubbing a little glycerine over them before taking them out of the water, or dusting a little powder over after drying, to remove any moisture that may remain. Sometimes chaps are the result of a scorbutic state of the body, in which case general treatment is necessary, as in the case of persistently chapped lips, which are sometimes very painful. Smoking a pipe will sometimes produce a very painful crack in the lip which is very difficult to heal.

CHARCOAL, or carbon, occurs in nature, as black lead or plumbago, but is ordinarily made artificially from animal or vegetable substances. Wood charcoal is mostly employed externally, and that most frequently in the form of poultice, combined with linseed meal and bread. This poultice is of very great value when sores are fœtid and parts are sloughing away, keeping them moist and warm, while preventing smell. The powder may be used with similar intent. It is sometimes given internally, when patients are suffering from organic disease of the stomach and intestines, accompanied with the formation of foul-smelling gases and acrid fluids. For this latter purpose, however, animal charcoal or purified bone-black is most frequently given, and it is also recommended as a temporary antidote for certain organic poisons, as aconite and strychnine. In either case a tablespoonful should be given suspended in water. It is one of the most valuable disinfectants for household purposes, and should always be employed when dead bodies are deposited in coffins.

CHARPIE, a French name given to a coarse kind of lint, or tow, which is prepared from coarser materials than are employed for the manufacture of lint.

CHEESE, an article of diet made from the milk of various animals. The milk of all animals contains water, saline matters, sugar, butter, and caseine. Cheese consists of a mixture of the two latter substances. When milk is allowed to stand, and acid or fermentable substances are added to it the caseine and butter separate in the form of what is called curd. This curd on being strained is converted into cheese. Cheese is always made in this country from cow's milk. It has a very different appearance and taste according to the way in which it is prepared, the pressure applied to it, the time it is kept, and the substances added to it. When curdled speedily and floating in the water and sugar, which is called the serum or whey of the milk, it is called curds and whey. The butter when removed from milk is called cream, and when milk is curdled quickly by heat and the cream taken off, it is called clotted cream.

When the curd is removed with the butter from the milk, and gently pressed, the product is called a cream or soft cheese. All these cheeses contain the caseine or curd in a form in which it can be easily digested, and in all cases where a fatty and nutritious diet is desirable they are to be highly commended. They may be often used on account of the butter they contain, with great advantage as substitutes for cod-liver oil. When cheeses are made hard they are allowed to stand longer, and then submitted to pressure for varying periods. During this process they undergo various changes. In some a sweet substance is separated from the cheese which collects in little vesicles in the cheese, such as are characteristic of Gruyère and Dutch cheeses. In others a process of moulding sets in which alters very much the flavor. Frequently a portion of the butter is converted into butyric acid which gives a strong flavor. Cheeses vary according to the quantity of butter they contain, and are valued and high-priced as this substance prevails. In some cases, as in double Gloucester and Stilton cheeses, the cream of one milking is added to another milking, thus doubling the quantity of butter. In Suffolk a contrary practice prevails, the cheese being made after the cream has been taken off and made into butter. Caseine when once dried becomes very hard and indigestible, hence the bad favor of Suffolk cheese. Cheeses differ in color according to the quality of the food eaten by the cow, and as colored cheeses have been erroneously supposed to be rich cheeses, in Cheshire they adopt the practice of adding annotta to the cheese. In some countries they add flavoring substances to cheese. In Germany a favorite cheese, called Schabzeigar, is flavored with the common melilot. Cheese is a very nutritious article of diet, on account of the large quantity of caseine it contains. The indigestibility of the caseine, however, sets a limit to its use. The caseine is, however, rendered more digestible by the butter it contains, and the richer cheeses are therefore the best for food. Mixed with other food cheese has a tendency to promote digestion. This is, perhaps, better effected by decaying than by fresh cheese : hence the practice of taking decayed cheese, especially at the end of a meal. (See MILK.)

CHEST.—The chest is one of the three large cavities of the body, and contains not only the heart and lungs, but the great vessels which convey the blood to various parts of the system, the œsophagus, which carries the food to the stomach, and other smaller but important structures. In the skeleton, the chest, or thorax, is seen to be bounded behind by the spinal column, and in front by the sternum or breast-bone, while its lateral boundaries are formed by the ribs which are affixed behind to the spine by movable joints and in front join the breast-bone by their cartilaginous prolongations. Thus a cage or box is formed which is capable of movement in three directions, viz., upward, outward, and forward. Above, the chest is much diminished in area, and is bounded by the structures which form the neck ; below, the diaphragm closes the thorax and separates its contents from those of the abdomen. In the living subject, the chest is lined by a thin, smooth membrane, called the pleura on which the lungs can glide with ease, while, externally, the ribs are covered by the skin ; between these two coverings are numerous muscles, called the external and internal intercostal muscles, by which many of the movements of the chest are performed and respiration is enabled to be carried on ; they are so called because they lie between the costæ or ribs ; the action of the external set is to raise, and that of the internal set is to depress, the ribs. The diaphragm is the most important muscle of respiration ; it is convex toward the chest, while its hollow or concave surface looks toward the abdomen : it is perforated in a few places so as to allow vessels to pass from the chest to the abdomen and *vice versâ :* thus the descending aorta and œsophagus pass through it close together in front of the spine, the one to send the blood to the lower part of the trunk and legs, and the other to communicate immediately afterward with the stomach. Another opening allows the inferior vena cava to pass through to the heart ; this is the large vein which brings the return column of blood from the lower extremities and abdominal organs ; besides these structures, a few nerves and lymphatics pass through. The diaphragm, like all muscles, has the power of contraction, and its fibres are so arranged that during inspiration it descends and allows more air to enter the chest, while, in doing so, it compresses the abdominal contents, while during expiration it rises, and so lessens the area of the chest. The contents of this cavity are numerous and important ; at the back part, and therefore just in front of the spine, is the gullet or œsophagus, a hollow, muscular tube, which allows food to pass directly down from the mouth to the stomach ; close to and in front of this tube is another shorter one, called the wind-pipe or trachea, which, soon after entering the thorax, divides into two branches called bronchi, and these, entering the lungs, break up and subdivide into a vast number of smaller branches, which end in small, dilated, closed extremities called air-cells or air-vesicles ; as they become smaller and smaller, so the wall of the tube becomes thinner and thinner, until at last it is of extreme tenuity ; and this is important, because the air can then readily interchange gases with the blood through this delicate membrane, for the blood runs outside the air-cells in vessels with extremely fine walls also. In this way respiration is carried on, for during inspiration oxygen is carried from the external air down the wind pipe and bronchi into the ultimate air-cells, and then through their thin wall into the blood; while carbonic acid, which is not wanted in the organism, is carried away in the opposite direction during expiration. The greater portion of the cavity of the chest is filled up by the lungs. (See LUNGS.) The *heart* is seated in front of the chest and at its lower part, just between the lungs; it is made up in four compartments, the two right being quite distinct from the two left cavities in health. On the right side the heart receives the venous blood from the inferior vena cavæ, as before mentioned, and also from the superior vena cavæ, which, in a similar way, brings the venous blood from the head and neck and upper extremities; it then sends it on to the lungs through the pulmonary arteries to be aerated. On the left side the heart receives the blood from the lungs and sends it on into the aorta, a large vessel which, after ascending about two inches, curves backward and then passes straight down to the abdomen by the side of the œsophagus; in its course in the chest it has the shape of a syphon-tube. Close by this large vessel runs a very small one, the thoracic duct, which comes from the abdomen, and, entering the blood at the root of the left side of the neck, supplies that fluid with important elements. There are many minor structures which do not require notice here.

Such being briefly a statement of the contents, it is now needful to show where to find the position of the heart and lungs during life; for this purpose the back and front of the chest may be mapped out into districts. On each side are twelve ribs, which can easily be counted in a thin person; above, and in front of the chest, is a clavicle or collar-bone, while behind are the shoulder-blades or scapulæ; these are points which are easily recognized by any one. Take two pieces of tape, and placing one end at the junction of each collar-bone with the sternum or breast-bone, let the other end hang vertically downward, there will then be a narrow central space and a wider one on each side; next lay a piece of tape horizontally across the chest just above the nipples, and another piece parallel to it about three inches below or on a level with the seventh rib; the lower margin of the ribs is the lower boundary of the thorax. In this way nine spaces are marked out, the side ones being of equal size, but the central ones narrower; the lateral ones are named on each side from above downward—sub-clavian or infra-clavicular, from being below the clavicle, mammary or the breast region, and infra-mammary or the region below the breast; the central spaces are termed upper, middle, and lower sternal regions. Continue the horizontal tapes round into the axillary region, and then each is divided into three spaces termed the axillary, lateral, and lower lateral regions from above downward. In a similar way the back may be mapped out; the space over each shoulder-blade is called the scapular region, while that between the scapula and the spine on each side is termed the inter-scapular or vertebral; below these are the dorsal regions, which correspond to the lower portions of the lungs. Further, the part above each clavicle is called supra-clavicular, and the space over the shoulder-blade is called the supra-spinous fossa. All these divisions are, of course, quite arbitrary, and are only of use for easy reference in describing the seat of the disease that may be present.

CHERRY-LAUREL, the *Prunus lauro-cerasus*, is a well-known shrub, the leaves of which, when distilled, yield small quantities of prussic acid. As the quantity varies much, the old-fashioned cherry-laurel water, which owed its efficacy to the prussic acid it contained, is now rarely if ever used, this acid being such a dangerous poison. It should not be used in domestic medicine.

CHICKEN-POX, or *Varicella*, is a contagious but harmless disease of childhood, unattended by any constitutional disturbance, as a rule, and after running its course for a few days ends in complete recovery. Often, several children of the same family have it one after the other, and it seems to occur in an epidemic form at some seasons of the year; it affects both sexes and all classes indiscriminately. After a period of incubation, the length of which is doubtful, a number of little red points suddenly appear on the skin, and in the course of twenty-four hours each has become a small blister, or vesicle, raised above the surface and surrounded by a pink areola or zone. Each vesicle is seen to be formed of a delicate membrane, and within is a little clear, watery fluid. The next day, more red spots appear, which also form blisters, and so on for about three or four days, fresh crops appear, as the previous ones attain a maturer stage. The eruption is most abundant on the back and front of the body, much less so on the face and head, while very few spots appear on the arms and legs; the small blisters are convex, unless they have burst, which, from their delicate structure, is no infrequent occurrence, but even they are different from the central depression seen in the pustules of small-pox, and a careful examination will always show one or two convex vesicles which are still entire. In about a week the vesicles begin to wither and dry up, and in a week or ten days longer the scabs fall off, leaving, as a rule, no scar, but if they are picked or irritated, a small white depression may be left in the skin. Eczema may be produced by rubbing the vesicles, therefore the child should be kept from picking the scab. If the child be in bad health previously, the scabs may fall off and leave circular ulcers which may take some time longer to heal, and then scars are sure to result, but this is a rare occurrence. As a rule, the child need only be kept in the nursery, and not in bed all day long ; occasionally, the little patient is restless and feverish, but, in most cases, it will play about as cheerfully as usual, and appear to have nothing the matter with it. The temperature is generally no higher than in health, the tongue clean, and the appetite good. For a few days the child may be kept in-doors, and the diet should be plain and simple. This disease is by many called the *glass-peck.* Vaccination will not prevent this disease occurring, nor does it seem to have any connection with small-pox. One disease will not prevent the other from attacking the same individual. The absence of pain in the back, fever, and sickness, which are so common in small-pox ; the rapid development of the rash, which is mature in 24 or 36 hours, which comes out in crops, and is more common on the body than on the face, will help to distinguish this mild complaint from small-pox. However, when small-pox occurs in a modified form, it is occasionally difficult to know one from the other.

CHICORY. (See COFFEE.)

CHILBLAIN is a very common and troublesome affection in this country during the winter months. It commences as a mild and superficial inflammation of the skin in certain parts of the body, by preference the feet, hands, ears, and nose, and afterward, when neglected, or badly treated, is converted into a blister or a deep sore. The subjects of this complaint are most frequently young females who have a soft and tender skin, and who perspire much, and its exciting cause is a sudden transition from extreme cold to warmth, or warmth to cold ; generally the former. The combination of moisture with cold, the friction of coarse stockings, and the pressure of tight and badly-made boots are also to be regarded as active conditions in the causation of chilblain. The most common form of chilblain is a small red patch on the skin, to which the patient's attention is first called by a sensation of heat and itching. If this has been neglected or scratched, and no pains taken to regulate the temperature of the affected part, the inflamed patch becomes darker in color and vesicates or blisters, this change being attended by intolerable itching. If the chilblain be still allowed to go on without treatment, it passes into the stage of excoriation, and finally into that of ulceration. The skin becomes raw and discharges matter, and a deep open sore is formed, which in unhealthy subjects extends below the skin and becomes very rebellious to treatment. In badly-nourished and neglected children of the poorer classes, the affection in some instances

rapidly progresses to a state of gangrene. Much may be done to *prevent* chilblain. At the approach of winter those who are liable to become thus affected should endeavor to harden the skin of the extremities, and to accustom the feet to pressure and friction. The feet, as well as the hands, should be frequently washed with cold water, or, if this cannot be used on the lower extremities without danger or discomfort, a solution of alum, or spirits of wine should be rubbed into the skin night and morning. The socks or stockings should be thick and warm, and the boots loose and furnished with good soles. Sudden exposure to cold and wet should be carefully avoided, and the temptation guarded against of suddenly bringing a benumbed hand or foot into close proximity to a fire. Wet feet should be well washed with fresh cold water and then scrupulously dried.

Treatment.—The chief point in the treatment of chilblain is to avoid any fresh exposure to the exciting causes of the affection. The painful and inflamed part must be guarded against extremes of temperature, and against friction and pressure. In the mildest form of the complaint, when there is a faint inflammatory blush, and some slight itching, the inflamed part should be well washed with spirits of wine and cold water, carefully dried and then painted over with a lotion composed of one drachm of carbolic acid, one ounce of glycerine, and four ounces of water. This lotion should be applied every hour by means of an ordinary camel's hair brush. When the chilblain is more inflamed and painful, either of the following lotions will prove serviceable : spirits of turpentine, two drachms ; castor oil three drachms ; collodion, two ounces ; or, camphor, one drachm ; eau de Cologne, three ounces. When there is much blistering, simple starch powder, or zinc ointment will be found the best applications. An ulcerated chilblain requires poulticing, and, when there is much discharge and the sore is covered by large granulations or *proud flesh,* the application of Peruvian balsam, carbolic acid ointment, or wine of aloes. In cases where the sore is deep and obstinate under treatment, and the skin around inflamed, attention should be paid to the general health of the patient, and wine and good nourishment liberally supplied. When the discharge from the sore is very foetid, one may combine with the above-mentioned dressing a solution of chloralum, or one of permanganate of potash, or apply a layer of cotton wool, or carded oakum powdered with charcoal.

CHILDREN. (See DIET.)

CHILD-CROWING. (See LARYNGISMUS STRIDULUS.)

CHILLS are always one symptom of mischief to the system which should not be neglected. Sudden shiverings, known by this name, are often premonitory of an attack of fever of some kind, and it is wise to take early notice of the warning, and endeavor either to ward off the enemy or to mitigate its violence by the administration of a hot bath and good rubbing, so as to restore action to the skin, a mild purgative, a basin of gruel, and a warm bed. These precautions may frequently break the chain of morbid actions in the system, and prevent the threatened disease making its appearance.

CHILLS AND FEVER. (See INTER-MITTENT FEVER.)

CHIMAPHILA, the name given to the American Winter Green, *Pyrola umbellata*. This plant has bitter properties, and is celebrated as a tonic and anthelmintic.

CHIN-COUGH. (See WHOOPING-COUGH.)

CHIRETTA is the entire plant of the *Agathotes Chirayta*, growing in Northern India. It is imported in bundles with flowers and roots attached. The plant is very bitter, with a somewhat peculiar twang about it, and is one of the safest tonics which could be recommended for domestic use. Combined with an acid like the dilute nitro-muriatic acid, it may be given with great advantage in cases of defective appetite where there is no organic mischief. The dose varies from one to two ounces of the infusion.

CHLORAL is formed by the action of chlorine on alcohol, and is a dense colorless liquid, with a caustic taste and a suffocating odor. It has a very soothing effect upon the nerves, and is now extensively used in the treatment of nervous diseases, and as a narcotic. The dose is from 10 to 15 grains in a little water; and more than this should never be taken except as prescribed by a physician, as it is a very dangerous remedy. The signs of an overdose are slow, noisy breathing, with excessive drowsiness or unconsciousness. The *Treatment* of one suffering from an overdose is artificial respiration, as described in the article on DROWNING.

CHLORIC ETHER, now known as spirits of chloroform, consists of a mixture of chloroform and alcohol, and is the form in which chloroform is most frequently given by the mouth. Its properties are, of course, similar to those of chloroform, but as the latter is chiefly administered by the lungs, that is to say, by inhalation, a word may be said as to the uses of chloric ether. As a remedy it is of great value where there is much sickness and disturbance of stomach, and it is a useful adjunct to nauseous medicines, frequently enabling them to be kept down when without it they would inevitably be ejected. It is also of great use in certain spasmodic affections, as cough and prolonged hiccough, asthma, and the like. For a dose, 20 or 30 drops may be given in sugar, or along with some other substance of the same kind.

CHLORIDE OF LIME. (See CHLORINE.)

CHLORINE and *Chlorinated Lime* may best be considered together, seeing that the latter is but a convenient form whereby the other is rendered portable. Chlorinated lime, commonly called chloride of lime, is prepared by pouring gaseous chlorine over quick-lime, which absorbs the gas and acquires certain of its properties. It is mostly used as a disinfecting agent. It acts by virtue of the chlorine, which it slowly evolves when exposed to the atmosphere, or gives it more speedily when spread in a saucer or other flat vessel (non-metallic), and a little acid, such as hydrochloric acid, is added. Chlorine acts as a disinfectant by virtue of its power of oxidizing decaying bodies, and being volatile it readily attacks those in the air as well as on more solid objects exposed to its fumes. Unfortunately these fumes are rather irritating to the throat and the eyes, and frequently cannot be borne by the sick; but for ordinary purposes, such as cleansing close rooms, there is nothing more effectual than chlorine generated from chlorinated lime. But as it attacks many things, being a powerful bleaching agent, and readily destroys colors, and affects metal-work, including gilding, it is somewhat at a discount; but in privies, workhouses, ships, etc., it is of much value. The mode in which it should be used has been already hinted at. To keep down ordinary smells the powder need only be exposed in a flat dish, and renewed from time to time. For the purposes of active disinfection it is better to add to the lime some acid, and shut up the doors, windows, and chimneys for a time, taking care that the place has been well ventilated before re-entering. Chlorine is also used as a gargle, especially when there is much fœtor from the throat, as in scarlet fever, diphtheria, and the like. Here it is exceedingly useful.

CHLORODYNE is a patented remedy for neuralgia, recently introduced, and supposed to consist principally of chloroform, morphine, Indian hemp, and hydrocyanic acid. The dose for adults is 20 to 30 drops.

CHLOROFORM is produced in the crude state by distilling rectified spirit from off chlorinated lime. In appearance, it is a clear, transparent, watery-looking liquid, not, however, mixing with water, and if poured on its surface, forming a layer at the bottom of the vessel. Its odor is also peculiar. As is well known, the mode in which chloroform is commonly administered is by inhalation, and, as is also well known, its use is not unattended with danger. Its administration is therefore, a matter of importance, not to be rashly undertaken save by skilled persons. No exact quantity can be assigned as proper to be given, as some people speedily become insensible under its influence, others again requiring a very much longer time, and a much larger quantity of the drug. Ordinarily the first stage of chloroform inhalation gives rise to pleasurable sensations, unless the chloroform be allowed to irritate the nostrils or interior of the mouth, or the patient is frightened and restless. There is no loss of consciousness, but often a sense of relief caused by comparative freedom from pain. This is followed by another stage, where there is often a good deal of excitability. The patient may laugh and talk at random, and will sometimes fight and struggle. This is most marked in powerful individuals. There is also a loss of sensibility, but hardly of consciousness, though sensations of pain are greatly lessened. There is still power of motion, and, as has been said, struggling is common. Beyond this stage the use of chloroform in midwifery practice should not in ordinary cases be carried, otherwise the labor will be retarded. Of course, if operative interference becomes necessary, anæsthesia must be carried to complete insensibility and motionlessness. If we divide the ordinary results of inhaling chloroform into three groups, the total insensibility and complete loss of voluntary motion constitutes the third; but in strong young men there often intervenes a condition apparently of complete rigidity, sometimes accompanied with trembling, and the patient will become dark in the face. To those unacquainted with the use of chloroform this condition is sufficiently alarming, but it speedily passes away, the limbs become relaxed and totally devoid of power, and respiration is resumed, though often stertorously for a time. Now is the time for operation; there is neither sense, feeling, nor power of motion. Beyond this stage the inhalation of chloroform should never be carried. Even this may be dangerous, for sometimes suddenly the tongue falls back, breathing ceases, and the pulse stops, and death may result before efficient aid can be rendered. Should this come to pass, artificial respiration (see DROWNING) should be used; the patient turned on his face and rolled from side to side; water should be thrown on the face, and smelling-salts applied to the nostrils. The application of electricity to the diaphragm and side of the neck is also of great service, but *prompt measures are always the best.* Often, as the patient begins to recover sensation, there is a fit of vomiting, which may troublesomely be repeated. To avoid this the purest chloroform alone should be used; the patient should have eaten nothing for at least four or five hours, and the last meal should have been light. Ice, or iced champagne, or brandy and iced water are the most grateful remedies for the sickness.

Besides being used to produce insensibility in operations, chloroform is largely used to enable the practitioner to make a correct diagnosis when there is much pain and spasm. It has also been given in asthma (though

here, and used in this way, its use is at least liable to abuse), and in tetanus. In this last named disease, commonly known as lockjaw, the remedy promised to be of use, and patients have been partially kept under its influence for long periods. Recovery has in many instances followed; but as it has done so with multitudes of other remedies, chloroform cannot be relied on as having been the curative agent. In spasmodic and irritable strictures of the urinary passages, chloroform often proves of very great service in enabling an instrument to be passed. Tic douloureux and other forms of neuralgia often yield to its influence, which should not, however, in these cases be carried beyond the first stage. So in toothache arising from a decayed tooth, chloroform often acts a double part; to the part itself it acts as an irritant, and so in many cases does good, while the sedative effect which follows is none the less beneficial. Next, perhaps, after its use in cutting operations, chloroform has been of most signal service in enabling the surgeon to reduce dislocation and strangulated hernia or ruptures. In such cases the patient should be brought thoroughly under the influence of the vapor before anything is done.

CHLOROSIS is a disease in which the patient becomes of a yellow or sallow tint, and in which there is a diminution of the coloring matter of the blood. It occurs chiefly in young women who lead sedentary lives, or sit in close workshops, or in any place where light and ventilation are deficient; such people have a dark border under the eyes, pale complexion, a feeling of lassitude and weakness, headache, pain in the back and amenorrhœa. The treatment is the same as for anæmia, and, in fact, chlorosis is but a variety of the same complaint. *Anæmia* is a term used when the patient becomes pale from other diseases, as cancer, consumption, and kidney disease; *chlorosis* is used when the alteration in the blood is the primary change, and not dependent upon mischief elsewhere. (See ANÆMIA.)

CHOCOLATE. (See COCOA.)

CHOKE DAMP, a name given by miners to all irrespirable gases; but it more especially applies to carburetted hydrogen. Carbonic acid gas is called fire-damp.

CHOKING is an accident liable to happen to hungry persons eating hastily, or to children, and requires the greatest self-control and presence of mind on the part of those who are present. The situation is one of extreme terror, and without some previous instruction, time is apt to be lost in sending for aid which may come too late. The substance which causes the choking may either be at the top of the throat, at the entrance to the gullet, or lower down. If at the upper part of the throat, prompt action will often remove it, either by thrusting the finger and thumb into the mouth and pulling the obstruction away ; or, if it cannot be reached so as to pull it away, a piece of whalebone, a quill, or even a pen-holder—anything at hand—should be seized and pushed down as a probang, so as to force the substance down the gullet. If it be impossible to do this, tickling the back of the mouth with a feather, so as to produce sudden retching, will sometimes dislodge it, a sharp blow on the back will perhaps displace it or a sudden splashing of cold water in the face, causing involuntary gasping. Even with every willingness and dispatch, the patient may become insensible before relief can be afforded, but it must not be assumed for certain that death has taken place ; for it may only be a condition of suspended animation, and the usual remedies in such a case—such as dashing cold water in the face and on the chest, applying ammonia to the nostrils, and inflating the lungs by bellows, etc., should be continued till medical aid arrives.

CHOLERA is of two kinds, what is known as cholera morbus, a disease bad enough but not particularly fatal, and that terribly fatal disorder Asiatic, malignant, or epidemic cholera. In this country this last disease has almost invariably prevailed in its worst form in poor crowded dwellings, among those whose food supply was bad ; whose hygienic conditions were otherwise unfavorable, but especially among those who had a tainted supply of water. Very frequently when cholera prevails diarrhœa also does so, and an epidemic of cholera is frequently ushered in by an unusual prevalence of the other malady. In point of fact in a case of ordinary, though by no means of maximum intensity the disease is ushered in by an attack of diarrhœa. This may last a longer or a shorter period, but speedily the matters passed by the bowel alter in character. They assume a peculiar flocculent or rice-water character. Vomiting, too, comes on, the fluid being thin and colorless. Then follow severe cramps, especially of the abdominal muscles and legs, which become like rigid cord. The flow of urine ceases, breathing and circulation are so much impaired that the body becomes icy cold on the surface, the tongue is cold, and so even is the breath. The lips are blue and shrivelled, the face pinched, the voice is hardly audible, the very eye-balls are flattened. This is called the cold or algid stage of the disease. The condition may go on getting worse till the heart stops, the patient being quite conscious to the end. Frequently it is impossible to tell whether the patient is to live or die, when suddenly the sickness lessens, the body begins to get warm, the face flushes, and restlessness subsides. The patient seems on the very verge of getting well. But sometimes the urine does not flow, or there may be congestion of the lungs or

brain, and so, though reaction has set in, the man may yet perish. Thus in an ordinary mild case of cholera a man will pass through three stages : First, that of premonitory diarrhœa ; secondly, that of collapse ; and, thirdly, that of reaction probably in about forty-eight hours. It is no part of our duty to enter into any discussion of the immediate *cause* of these extraordinary symptoms. There can be no doubt of this, that the disease is produced by some particular poison. This poison may be transmitted through the air by water, or communicated by one individual to another. There can be no doubt that the discharges are one main source of this poison, and hence should be most carefully disinfected.

Treatment.—With regard to treatment, much is in dispute. But about this there can be none : viz., that prevention is better than cure. Accordingly he who would avoid cholera during a cholera season, ought to live by rule and method. First see that his water-closets are in good order, and that every precaution is taken in cleansing and disinfecting them. Carbolic acid is best for this purpose. See that the house is clean, sweet, and airy ; let not foul or decaying matters of whatever kind remain upon the premises. See that the water supply is pure ; if necessary, boil it and pass it through a charcoal filter before use for drinking. Let no stale meat or vegetables, no sausages, game, or substances likely to create digestive disturbance be used ; especially avoid unripe fruit, prolonged abstinence from food, excessive fatigue. Should diarrhœa come on at such a time it had better be stopped, except it be due to some irritant matters in the bowels, when a small dose of rhubarb or grey powder may be given to remove these, and an astringent subsequently administered to keep it from acting too powerfully. Above all avoid strong aperient medicines of every kind. The astringents to be used sould not be powerful ; chalk mixture, sulphuric acid, lemonade, or these with a little opium added, are best. Except in the eyes of a skilled practitioner, nothing more must be done. *No diarrhœa in cholera time is to be neglected.* This is a standing rule. Should collapse come on not much is to be done. It is quite useless to give medicines internally, they only accumulate there, for they are not absorbed. The only thing is to try to keep up the heat in every way which will not *disturb* or *fatigue* the patient ; that is material. The patient is consumed with thirst, and there is no reason under the sun for refusing him drink, if of a wholesome kind. The thing is, wait for the reaction. Should that occur, the life or death of the patient is, to a very great extent, in the hands of his attendants. Above all he must be kept quiet. If his head trouble him and his face is flushed, apply cold to it. If there is much sickness let him have a little ice or ice-water. If his lungs get gorged, warm poultices or turpentine stupes will be best. But the great anxiety is the kidneys, if they go well all may be well, but if not certainly not. If they do not act warmth must be tried, perhaps as a warm bath, but this is a delicate matter and requires caution. If they are acting well and the patient requires a stimulant, let him have some sal volatile. The greatest care must be taken for a few days in fear of a relapse. The food given is of special importance ; broths, soups, and jellies may be given, but certainly not meat. Small quantities, too, at a time must be given and repeated as frequently as necessary. In the collapse a choleraic patient may be said to be in the hands of God ; after, to a very great extent, he is in the hands of his attendants.

CHOLERA INFANTUM is a disease very common during the summer months in children under two years old. It is caused by improper food, or too much food, impure air, and hot weather ; never by teething alone, though this aggravates it. Its first *symptom* is usually a slight feverishness, which is followed by a diarrhœa with thin watery discharges, and a little later by vomiting. Sometimes the vomiting and diarrhœa begin at the same time and these are the worst cases. Within a few days emaciation begins, the hands and feet become cold, the head and surface of the abdomen hot, the face pale and shrunken, the eyes dull and heavy, and the pulse irregular and quick ; by degrees the child becomes sleepy, and finally sinks into a state of insensibility.

Treatment.—At the first symptom of cholera infantum, put the child into a warm bath, or apply flannels dipped in hot water to the bowels, and keep the child in an ordinary posture. Dissolve a teaspoonful of gum-arabic in an ounce of peppermint water, and give a teaspoonful every half hour. A milder astringent may be made by mixing together three ounces of chalk mixture, half an ounce of tincture of kino (or catechu), and half an ounce of compound tincture of cardamoms ; of this mixture the dose for a child eighteen months old is one teaspoonful every two hours if the discharges are frequent, and at longer intervals if not—care being taken to shake the bottle well before pouring out the medicine. If there is much thirst, give 6 or 10 drops of brandy in a teaspoonful of water or mucilage, every 15 or 20 minutes. It is of the greatest importance that the stomach of the child be at rest ; hence, no food or drink should be given for some hours. Further than this nothing can be done without the advice of a physician, and this should be had at the earliest possible moment.

CHOLERA MORBUS is frequent enough in the summer months. It may almost invariably be traced to some injudicious arti-

cle of food, some unhealthy occupation, or some distinct cause. The symptoms are vomiting and purging ; the vomit being bilious and utterly unlike the colorless vomit of true Asiatic cholera. The question to be decided is usually this. Has the vomiting and purging lasted long enough to expel the offending matters ? if so, they had better be stopped ; if not, something should be given to favor their expulsion. Rhubarb and gray powder are perhaps the best things to give, if laxative medicine is necessary ; if not, a few drops (ten or so) of laudanum and sulphuric acid, or five grains of Dover's powder, or ten or fifteen grains of compound chalk and opium powder, will be best to stop the purging. Ice or an effervescing drink will be best to allay sickness, should these be required. The disease may prove fatal to children, but rarely to adults.

CHOLAGOGUES are medicines which act on the liver and increase the flow of bile. Among the most powerful of these, notwithstanding recent assurances to the contrary, we must still place the various preparations of mercury, such as blue pill—mercury with chalk and calomel. Podophyllin is undoubtedly of use in the same way. When the liver is unloaded, such remedies as purgative salts likewise increase the flow of bile, and otherwise do good.

CHONDRUS CRISPUS. (See CARRAGEEN MOSS.)

CHOREA is a disease generally met with in children between seven and twelve years of age, but occasionally it occurs in adult life. This disorder is often dependent on a fright, as the sight of a fire, the bite of a dog, or falling into a river, and then the disease will appear in the course of two or three days by twitchings of the arm and leg and the muscles of the face ; generally one side is more affected than the other, and sometimes the choreic twitchings are confined to one side. The twitchings of the muscles are worse when any one is observing them, or when the child is excited ; they are better when the patient rests in bed, and they always cease during sleep. Since the child cannot control its movements, it is unable to write properly, or walk across a room, or take hold securely of any object ; the tongue is protruded and drawn back with a jerk, and the head is never steady, because the muscles of the neck jerk it about constantly. In most cases the appetite is not affected, nor does the general health seem much impaired ; there is, as a rule, no fever, and generally a recovery may be looked for ; in a few cases, which are of somewhat rare occurrence, the twitchings are so violent that the elbows, knees, and hands become sore and abraded by constantly striking surrounding objects ; there may be fever, delirium, and sleeplessness ; these are signs of grave importance, and are usually associ-

ated with heart disease, or follow on an attack of rheumatic fever.

Treatment.—Chorea, or, as it is popularly
called, St. Vitus's dance, is a very curable disease in ordinary cases ; sometimes it gets
well of itself ; more often it persists for some
weeks or even months, and makes the patient
weak and pale. Early treatment is always
advisable, and it should consist of taking a
cold bath every morning, to be followed by
friction with a rough towel ; tonics are of
great service, and more especially those containing iron. A simple but nourishing diet
should be given, and a certain amount daily
of out door exercise. For those cases in
which fever and delirium are present, very
little hope can exist as to a favorable result,
and a different treatment must be adopted ;
for such, rest in bed, sedatives to allay the
sleeplessness, and a fever diet are desirable.
Chorea occurs in nervous excitable children,
and in those whose parents or other members
of the family have suffered from nervous diseases, as paralysis, epilepsy, and neuralgia.

CHRONIC HYDROCEPHALUS. (See
HYDROCEPHALUS.)

CHYLE is the fluid found in the lacteals
of the small intestine ; it is absorbed by these
small vessels as the food passes down the intestinal canal, and is of a milky, opalescent
appearance, from containing a large quantity
of finely-divided oily matter ; after passing
through the mesenteric glands, where it undergoes certain changes, it enters the *receptaculum chyli*, and then goes on to join the
blood-current by ascending the thoracic duct.

CHYME is the name given to the partially digested food after it has left the stomach
and while passing down the intestinal canal ;
during its passage various soluble substances
are absorbed by the vessels of the small intestine, and the remainder, which is called
fæces, leaves the rectum as excreta.

CICATRIX.—A wound or sore invariably leaves after healing, a distinct mark upon
the skin. This mark is called a cicatrix or
scar. It is of a pale pink or white color, is
made up of firm gristly material, and is covered by a transparent, smooth, and shining
layer of cuticle. These are the characters of
the ordinary cicatrix observed in the seat of
a superficial burn, an incised wound, or a
simple ulcer upon the leg. But the appearances of cicatrices vary immensely according
to the part of the body on which they are
seated, and the nature and depth of wound or
ulcer from which they have been developed.
The scars from burns, from scrofulous sores,
and from malignant or lupoid ulceration, have
each their special characters by which the
nature of the previous disease can be determined after the lapse of several years. After
a sore has healed, the resulting cicatrix always undergoes certain changes. It loses its
ruddy or pink color and becomes whiter ; as

it gets older it glides more freely over the tissues beneath, and at the same time becomes
smaller in superficial extent, drawing upon
the neighboring soft parts so as to produce in
some instances, especially after deep burns,
much distortion and deformity. The principal
change accomplished in scar-tissue in course
of time is a gradual softening and assimilation both in appearance and function to
healthy skin. Scars frequently become diseased. They are sometimes the seat of severe
pain, which is more intense in damp, cold
weather. Obstinate ulceration and cancer are
frequently met with in the seats of old
wounds. The deformities produced by old
scars have been more fully described under
the article BURNS.

CIDER is a fermented beverage made
from the juice of apples. Perry is made in
the same way from pears. The quantity of
alcohol contained in cider and perry varies
according to the amount of sugar in the juice,
and the completeness of the fermentation.
The lower priced cider contains about half an
ounce in the pint, while the higher priced and
that which is bottled contains from one ounce
to two ounces of alcohol in the pint. The sugar
in cider varies. That which is called hard
cider contains only a small quantity, while
sweet cider contains from 100 to 150 grains
in the pint. Cider and perry contain malic
acid. The quantity of this acid contained in
a pint is about 150 grains. It acts on the system in the same way as tartaric and citric
acids. (See CITRIC ACID.) There is nothing
special in the action of cider on the system.
It may be frequently taken with advantage
instead of beer, especially the hard cider,
which contains less sugar, and saline constituents which differ from those contained in
beer. Perry has more flavor than cider, and
more care is taken in its preservation.

CINCHONA, better known *par excellence*
as *bark*, is the product of different species of
trees growing naturally in the Andes of South
America, but now cultivated in Java and Jamaica, and with great success in India. Three
varieties of bark are recognized in our Pharmacopœia, but many more in commerce.
The officinal varieties are the yellow, the pale,
and the red. In these are found two chief
alkaloids, viz., quinine and cinchonine ; most
quinine being found in yellow bark, most cinchonine in pale bark, while the red bark
yields both. It is chiefly to these substances
that the barks owe their great virtue, although
there are others also contained in the different kinds of cinchona of some value. Quinine is used in medicine as a sulphate, cinchonine as a hydrochlorate, the alkaloids themselves being but little soluble. Of the two,
quinine is the more powerful, and the more
employed, but cinchonine is very much
cheaper, and has the properties of quinine
only in a smaller degree. It is, however, apt

to upset the stomach. The discovery of the properties of bark was made in a curious fashion. As the story goes, an Indian sick unto death of the fever of the country, dragged himself from the spot in the forest where he had fallen exhausted, to the nearest pool to quench his thirst. In this pool a tree had fallen and its waters were strangly bitter, but their effect was magical, the fever fit left, and as the man returned to drink again and again he was speedily cured. This led to the discovery of the effects of, as it was then called, Peruvian bark, and on the occasion of the illness of the viceroy's lady, it was recommended. The effects were so satisfactory that, returning to Europe, the viceroy carried with him some of the bark, whose praises were soon spread abroad, and to it his own name of Chinchon, slightly modified, was given. From that date to the present day the tropical fevers which used to be so deadly have been comparatively kept under hand. These fevers are of the intermittent or remittent type, coming on or growing worse at certain definite times, and it is just before these that quinine should be given, and that too in a good large dose of not less than 5 grains. Smaller doses should be given during the whole of the interval. Quinine does not appear to have any power of arresting what are called *continued* fevers, such as typhus and typhoid, but it is of undoubted service during the period of convalescence, and it is now considered to have the power of reducing high temperature even in these diseases. It is perhaps for this reason that some are fond of giving it in rheumatic fever, where, however, it as a rule does no great good. In neuralgia and tic douloureux, especially when they come on at stated periods, quinine is often of immense service, as indeed it is in all affections coming on periodically. Brow ague is one of these. As is well known, quinine is the great remedy in most forms of debilitating disease, especially for the sake of giving an appetite, when that is defective, and so improving the nutrition of the body. For this purpose, a grain or two grains should be given for a dose, a few drops of dilute sulphuric acid being added to enable it to dissolve more speedily. It is however to be noted that a solution so prepared does not keep well, as a kind of fungus forms and grows on it. When given in very large doses, quinine produces a singing in the ears and throbbing in the temples which is far from pleasant. These are signals for a diminution in the dose given. Quinine is now usually prescribed instead of bark, except in certain instances. To give bark in the form of powder, as was formerly the practice, is to administer large quantities of utterly indigestible woody matter, but in certain forms of disease accompanied with great debility, nothing better can be given than decoction of bark, with carbonate of ammonia in large and repeated doses. A draught of this kind will also frequently relieve headache and give appetite. There is an old rule, that quinine should never be given with a furred tongue, but that is nowadays frequently disregarded.

CINNABAR is the red sulphuret of mercury found native, and is one of the chief sources of that metal.

CINNAMON BARK, as used in medicine and confectionery, is the inner bark of the young shoots of a tree growing mostly in Ceylon. The trees are generally pollarded so as constantly to supply a series of young shoots. From it is distilled an oil called oil of cinnamon, which has got the fragrance of the bark as well as its pungency. In medicine it is mostly used as cinnamon water, which may be made from the bark or oil. This with many is a favorite behide of unpleasant remedies. The powder is also used as an adjunct to relieve flatulence or prevent griping. The dose of the bark is immaterial ; that of the oil is from one to five drops.

CIRCULATION OF BLOOD. (See HEART.)

CIRCUMCISION, the operation of removing the foreskin. In the practice of the surgeon this proceeding is frequently necessary, either from preternatural length of the foreskin, or for disease. The operation has been practised by the Jews from the earliest times ; doubtless on account of the resulting cleanliness and probable immunity from infection of venereal disorders.

CIRRHOSIS OF THE LIVER is the name given to a disease in which that organ becomes smaller and firmer than usual. It is known more commonly as the "hobnailed" or "gin-drinker's liver." The first named is derived from the fact that the surface of the liver in these cases is rough and uneven, and the second name denotes the too frequent cause of the disease. Cirrhosis occurs but rarely in children, but is not uncommon in adult life. Among the many evils caused by drunkenness, this gradual wasting of an important organ, together with an increase in the fibrous tissue of the viscus, must take its place ; in other cases it seems brought about by syphilis, or in other diseases where there has been going on for some time a deterioration of the blood. This disease is always chronic in its course and begins somewhat insidiously. Loss of appetite, sour risings from the stomach, a feeling of sickness in the morning, and fœtid breath are often the earliest signs ; these are, in fact, the usual signs of dyspepsia, and the skin, in addition, may have a slightly jaundiced color. The patient may go on for many weeks or months without being much distressed, but he will notice that he is gradually losing flesh and that his strength is failing. After a time, the abdomen will become distended, because the circulation through the liver is obstructed, and the serum in the over-full vessels behind passes through their walls and causes an accumulation of fluid in the abdominal cavity ; the patient is then said to have "dropsy in the abdomen," and the term "ascites" has been applied to this by medical men. (See ASCITES.) The veins, also, over the surface of the abdomen become very full, and the skin is marbled over with blue ramifying lines. The patient is usually emaciated, and the skin has a sallow, yellow color : the pinched expression of face and the absence of fat under the skin contrasts markedly with the distended abdomen. The patient feels weak, and cannot undergo any exertion ; his appetite is bad, and the tongue dry and red ; the presence of the dropsy prevents him breathing with ease, and the tightly-stretched skin gives him pain ; generally, too, the intestines are very full of gas, and so flatulence adds to his discomfort.

Treatment.—Much good may often be done by treatment if the case is taken in time ; in the early stage, before dropsy comes on, the patient must abstain from any excess in stimulants, and take them only in moderation ; since he is also troubled with dyspepsia, the food he takes must be light and nutrient ; cold milk for breakfast is generally well borne on the stomach, with some dry toast or biscuit or bread soaked in it. Mutton and roast beef may be taken, but pork, salt beef, cheese, pastry, and vegetables are not good. If a chop or steak or a piece from a joint cannot be taken, strong beef-tea or broth, or chicken, etc., may be given ; the object being to supply the individual with food which is nourishing in quality and capable of easy digestion. Coffee is better than tea, and cocoa with milk forms a pleasant beverage. Rich food and made-up dishes should be avoided. The mineral acid (as hydrochloric or nitric acid) may be given with some bitter infusion twice a day, and sometimes this checks the tendency to drink and improves the appetite. An occasional aperient should be given, and for this purpose a rhubarb mixture may be prescribed. Should, however, the patient be so far advanced in the disease as to have dropsy, means must be taken to relieve him of the fluid, while at the same time his general health is supported. When there is much ascites, the urine passed in a day is high-colored, diminished in quantity, and deposits a pink sand on standing ; this, as well as the pain in the abdomen and the difficulty of breathing arise from the pressure of the fluid ; hence the object must be to remove the fluid as far as possible ; for this aperients may be given which produce copious watery evacuations ; such drugs are called hydragogues, and among them may be named jalap, scammony, colocynth, and elaterium. As the patient will probably be thirsty, a drink made of lemon-juice, cream of tartar, sugar, and boiling water is very grateful, and, while relieving thirst, it aids also the action of a purgative. If the dropsy be so great that drugs seem to be of insufficient avail, recourse may be had to tapping the abdomen and letting out the fluid. For a time the person will recover in a great measure, but at some future period the fluid may again be effused, and after every tapping there is a diminished chance of ultimately doing much good ; yet in many cases careful diet and judicious treatment may prolong life for many years.

CITRATES are favorite forms of giving many remedies, as they can ordinarily be given in an effervescent state, enabling them to be more readily retained by the stomach. The most common form is technically known as Citro-tartrate of Soda. To this many remedies may be added : it is itself a gentle laxative, and is often used as such in doses of about sixty grains or more. Citrate of Magnesia is a popular preparation.

CITRIC ACID is the substance which gives the pleasant acid flavor to oranges, lemons, and most of our fruits. It is found pure in the lemon, citron, and other fruits of the natural order *Aurantiaceæ*. It can be obtained pure and in a crystalline form from the juice of any of these fruits. Its action on the system is like that of other vegetable acids. It is cooling and refreshing to the taste and is decomposed in the system, acting, probably, in the same manner as sugar. When taken in the form of lemon-juice, lime-juice, or in the fruits of the orange family of plants, it is eminently antiscorbutic. On this account ships going a voyage of more than six weeks are compelled to take a supply of lemon or lime-juice and let the sailors take at least half an ounce a day. The crystallized citric acid does not appear to act as an antiscorbutic. Citric acid and the juices which contain it are employed for making effervescing draughts, when mixed with alkalies. Fourteen grains of citric acid, or half an ounce of lemon-juice, mixed with twenty grains of bicarbonate of potash, makes, with one ounce of water, an excellent effervescing draught.

CITRON is the fruit of a variety of the *Citrus medica*, or common lemon. Its juice contains citric acid, and its peel, or external covering of the fruit, is preserved in sugar, and called candied citron peel. The peel of all the fruits of the orange family of plants contains a volatile oil, which gives it a pleasant flavor, and renders it slightly stimulant.

CLARET. (See WINES.)

CLAVICLE. (See COLLAR-BONE.)

CLAVUS, the peculiar choking feeling felt in the throat in hysteria and in allied nervous affections. (See HYSTERIA.)

CLERGYMAN'S SORE THROAT. (See SORE THROAT.)

CLIMACTERIC is a term derived from ancient writers on disease. It was suggested that there was a tendency in the human body to certain diseases at the end of definite periods. This period was usually fixed at every seven years, when peculiar diseases occurred. The ninth period, or sixty-three years of age, was the time of the commencement of what was called the " grand climacteric." In some instances teeth appear between the sixty-third and eighty-first years of age, after the disappearance of the second or permanent set of teeth. This is called " climacteric teething." For the " climacteric period " of women, see MENSTRUATION.

CLIMATE is used to express a multitude of conditions, some known, some unknown ; but on the whole the idea of heat or temperature is uppermost. Next to that comes humidity or dryness, so that we speak of a hot climate or a cold climate, a moist or a dry one. But to know whether a given place possesses a good climate, that is to say, on the whole a healthy one, we must know something more than is told by the thermometer and rain gauge. Then, as regards cold, it is quite possible, by shutting up an invalid for the winter and by exercising due care, to keep the surrounding atmosphere at any temperature we like and maintain it evenly so. If we send him abroad we seek to send him to a part of the world where not only will the temperature be tolerably high all the year round, but where it will be so even that he will be able to spend a considerable part of his time out of doors. But it is quite plain that if we send a sick man to a place where one day there is a high temperature and next day a low one, though the mean temperature be high, he will be unable to go out when the temperature is low, and will be fatigued by the heat when it is high. Such a climate will be to him worse than useless, he had far better have remained at home. It is not, therefore, the mean temperature of a place we have to study, but its extremes, its highest and its lowest points, and we must judge of its eligibility by these. But this is not all. Still air, whether hot or cold, is much more bearable than is moving air or wind. So we may have a windy place with a tolerable temperature in the shade altogether unsuited to the wants of the invalid. Moreover, the situation of the proposed residence must be considered with regard to the prevailing winds ; even in windy regions some sheltered nooks are to be found which will possess a vegetation characteristic of regions possessed of a much higher temperature. So, too, with regard to humidity, the rain gauge is no great criterion. In certain regions a vast quantity of rain pours down, fills up the rain gauge for two or three inches, and then passes away not to return for many weeks. In other regions it rains more or less every day, and the whole atmosphere is impregnated with moisture. Here there may, by the rain gauge, be a smaller rain-fall than in the other instance, but the climate will be as different as may be. Nor is it only the mode in which the rain falls which has to be taken into consideration ; the mode in which it disappears after falling is of the first importance. Here soil comes in. Suppose the soil a clay soil up to the surface, the rain-water cannot percolate through this, and so if the land slopes it runs off at once, if the land is flat it lies on its surface. If instead of a clay soil we have a sandy one, the rain-water will sink through the porous earth almost as soon as it has fallen, but its ultimate destiny depends on other circumstances. If the sandy soil slopes, the water will speedily run off, but if it forms a valley, let us say, and has beneath it a bed of clay, the ground water will only sink as far as the clay and remain there concealed to view, indeed, but as actually present as if it were above the ground instead of below it. In certain districts, instead of the surface being made up of sand or gravel, it is composed of vegetable débris, resting on a subsoil of mud. This constitutes a marsh, and such an association of things gives rise, under the influence of the sun's rays, to a power of unknown nature, but well known by its effects, what we call malaria. The exact import of ground water with regard to health is not yet clearly known ; but enough is known to render it an important element in forming an estimate of climate. Certainly, for invalids suffering from diseases like consumption or chronic rheumatism, localities characterized by excess of ground water are to be avoided. Exposure to the sun is another thing to be taken into account, though not, perhaps, quite so much as exposure to the prevalent winds. Finally, in forming an estimate of the value of a given place as a health resort, we must not forget to take into account the conditions which man himself imposes on a locality. The very great importance of drainage in adding to the healthiness of a locality is daily becoming more and more appreciated. It has been as clearly demonstrated as a thing can be, that bad drainage brings in its train diseases of the class called Zymotic. Better far a situation which does not promise so much in point of climate than a town, however eligible in other respects, which is totally deficient in hygienic necessities. Taking the two great factors in climate as our guide, we might classify climates by them, especially if we take into consideration one other alluded to, viz., wind. Where the atmosphere is moist and the temperature tolerably and uniformly high, we have a climate which is commonly called *relaxing*. Supposing, now, the temperature is high and the air excessively dry, we have a kind of climate of which Nice affords a good example. It is customary to call such climates *exciting*. Taking now the element wind into consideration, we find certain climates characterized by stillness of atmosphere and tolerable dryness, without this being excessive. Finally, we have climates, tolerably common in this country, where they are ordinarily associated with sea air. These are characterized by a temperate atmosphere, neither too hot nor too cold, fresh breezes, and the absence of all oppressiveness or humidity. Mountain air, too, belongs to this class, which is denominated *bracing*. Bracing climates are, perhaps, more used by tired people than by invalids. At the present day, when competition is so keen and work is so continuous, it has come to be recognized by many that the only conditions on which the pace of the world can be kept up is by taking a long annual holiday. Such men, wearied out by their exertions, seek a climate instinctively which would be unsuitable for an invalid. As to the others, it is hardly possible to lay down rules, for this is daily becoming more apparent, that the selection of a climate in a given case is, to say the very least of it, quite as important as the selection of a medicinal remedy, and a physician should always be consulted in regard to it. The kind of climate best adapted for an invalid having been settled on, it is desirable that he should undertake his journey with due care and attention. The grand rule is to avoid fatigue as far as possible. It is of much less importance to arrive at the destination rapidly than to arrive at it in a condition to be benefited by the change. Short journeys and long rests should be the rule ; but the application of the rule must, it is plain, be regulated by circumstances and the condition of the patient. Flannels next the skin should always be worn on such journeys, and frequently also after the destination has been reached. A small medicine chest is often found very useful, except the patient be going to regular invalid resorts where the English pharmacy is always obtainable. The maladies most likely to be improved by change of climate are diseases of the lungs or chronic bronchitis, consumption, inflammation of the air-tubes and larynx, asthma, neuralgia, gout and rheumatism, derangements of the functions of the womb, and certain diseases of the kidneys. Delicate strumous children are often very greatly benefited by change of air, especially to the seaside, if an appropriate place can be found. Above all, change is beneficial in convalescence from acute disease. For this purpose no special climate need be chosen—change is the great thing, and so it is when intended to prevent serious illness, especially of a mental kind. (See HEALTH RESORTS.)

CLOTHING is, perhaps, too much a matter of fashion to be touched on with great advantage in a work like this. Nevertheless, certain sound rules may be laid down. In civilized lands certain materials and shapes are supposed to be incumbent on every one, whereas abroad every one endeavors to suit his dress to the climate. The great thing is to avoid extremes of heat and cold, and if we are unable to change our clothes to suit the altered conditions, we must endeavor to wear such as will suit either emergency. There is only one thing which will enable everybody to do this, and that is flannel. It by no means follows that the flannel need be thick, it may be as thin as possible, yet from its power as a non-conductor of heat it will enable any one to step from a strongly heated room into the cold air with impunity. In winter it is desirable that we should not wear the same clothing in-doors as we do out, hence the general use of great coats. One of the most important articles of clothing for health's sake is shoes. It is of the utmost importance, especially to delicate people, to keep the feet dry. For this purpose nothing is so serviceable as India-rubber goloshes. Indeed, that material is now being used for similar purposes in a variety of ways. The end is the same in all to keep the body dry. If it cannot be kept dry the great rule is, change as speedily as possible ; but when men in full health have for a long time been exposed to the elements, they may do much as they please.

CLOVE-HITCH is a knot in use among surgeons for the purpose of reducing dislocations ; on account of the very firm hold it has upon the limb to which it is applied. It may be made either with a cord or a jack-towel. It is made thus : a good-sized loop is taken, held by both hands, that portion held in the right hand is then twisted *under* that held in the left, again looped and twisted *upon itself*, and the resulting loop placed *upon* the former one ; the limb is then placed in the double loop thus formed, and the two ends drawn up. The knot thus formed cannot slip.

CLOVES are the unexpanded flower-buds of a tree growing in the East Indian Islands. They yield an oil, generally when we see it, brown, but at first much lighter in color. This has the hot burning taste of the cloves, which, indeed, chiefly owe their properties to it. Their virtues are described as stimulant, aromatic, and carminative ; they are useful for relieving flatulence and the distension it causes. The oil is sometimes given to prevent purgatives griping, but most frequently is used to allay the pain of hollow teeth. A drop poured on a bit of cotton wool is introduced into the orifice to alleviate the pain.

CLUB-FOOT, or talipes, signifies a deformity of the feet, caused by the contraction of the muscles or tendons of the leg ; mostly congenital. The different forms are: 1. *Talipes equinus*, that is, when the heel is raised and the individual walks on the ball of the foot. 2. *T. varus*, when the patient walks on the outer edge of the foot. 3. *T. valgus*, when the inner edge is trodden on ; and 4, *T. calcaneus*, when the patient walks on the heel. The treatment of these cases requires surgical and mechanical interference, such as the cutting of the tendons which interfere with the proper direction of the foot, and the application of suitable apparatus.

CLYSTER. (See ENEMA.)

COCCULUS INDICUS is the fruit or berry of a climbing shrub growing in India and the East Indian Islands. It contains an active principle called Picrotoxine. In medicine cocculus berries are only used reduced to a powder, and made into an ointment for the destruction of vermin on the heads of children or the coats of domestic animals ; and for this but rarely. Nevertheless large quantities are imported, and are used by brewers to adulterate beer, thereby adding to its intoxicating qualities.

COCHINEAL can hardly be said to be a medicine. It is, however, used to color certain tinctures. It is the product of a certain insect found on the cactus growing in Mexico. Carmine is prepared cochineal.

COCOA is prepared from the seeds of a plant known to botanists as the *Theobroma Cacao*, and is a native of South America. The seeds, several of which are produced in a long pod, are roasted before being used. When thus prepared they are called cocoa nibs, and are sometimes boiled, and made into a decoction in this form. More frequently, however, the seeds are crushed and made into a paste, and sold as "cocoa paste." When flavoring matters, such as vanilla, are added, the preparation is called "chocolate." Cocoa and chocolate are the same substance, but the latter is flavored. Sugar is sometimes added to cocoa, and always to chocolate when sold in the form of a sweetmeat, of which there are great varieties. While cocoa differs greatly from tea and coffee in its composition, it nevertheless resembles them in possessing an alkaloid called theobromine, which acts in the same manner on the system as theine (tea). In addition to this substance cocoa contains a half part by weight of a fixed oil or butter, which gives a very decided character to its action on the system. Cocoa also contains in every pound three ounces of flesh-forming matter, so that it constitutes a food at once heat-giving and flesh-forming, and admirably adapted to all the wants of the system. Cocoa as an article of diet, is to be recommended in all cases where a nourishing and heat-giving diet is required, and with milk and sugar forms a very healthful food for breakfast or supper. The fat does not agree with some persons, and it is in these cases that an infusion or decoction of the "cocoa nibs" is recommended. The prepared cocoas which are advertised, generally contain an addition of starch in the form of arrow-root, sago, or tapioca, and are to be recommended no farther than as they diminish the original quantity of fat in the cocoa.

COCOA NUTS are the produce of a species of palm which grows in the East Indies and Ceylon. The seed is large with a thin shell. The shell is lined with a white flesh, and the interior contains a fluid which is called the milk of cocoa nut. The flesh contains oil and albumen, and is used extensively as an article of diet in the countries where it grows. It is eaten in this country to a limited extent, and made into puddings, cakes, etc. Like the kernel of all nuts, it is not very digestible, but is not injurious.

CODEIA is one of the alkaloids contained in opium (see OPIUM). It has recently been recommended in diabetes.

COD-LIVER OIL is one of the most valuable remedies we possess, and yet it should be looked upon rather as a food than as a medicine. It is prepared from the liver of the cod-fish, which at certain seasons of the year is richer in oil than at others. Other kinds of livers are employed, but those of the cod yields oil most abundantly. It is essential to understand that about cod-liver oil properly prepared there is nothing disagreeable either to the smell or the taste. The great virtue of cod-liver oil probably lies in its easy digestibility. It has been supposed to derive its useful character from the small quantity of iodine and bromine it contains, but this is out of the question. Probably in those cases where it does so much good, it re-establishes the balance of nutrition, enabling other substances to be made use of in the animal economy which were before rejected. Under its use patients sometimes marvellously increase in weight and improve in general appearance, their blood becoming richer, and their complexions ruddier, even though they are taking no iron. Its use is thus indicated in a great variety of exhausting diseases, especially those where there is chronic wasting, with gradual emaciation. Individuals, too, with swollen glands, which themselves interfere with nutrition, are almost invariably benefited by its administration, and one of the most intractable complaints known—chronic rheumatism and rheumatic gout—frequently yield to its influence. Where there is scrofula, and a tendency to phthisis, cod-liver oil is invaluable ; it has, indeed, enabled us to combat these diseases with no slight hope of success. Even in advanced pulmonary consumption, its effects are wonderful. Nor is it of less value in treating the diseases of bones and joints occurring in scrofulous persons. In the wasting diseases of childhood, there is no remedy to be compared with cod-liver oil. In rickets and chronic hydrocephalus, *i.e.*, water in the head, its effects are very marked, but perhaps even more so in the disease known as tabes mesenterica, where nutrition is impaired by disease of the lymphatic glands in the abdomen. It is a common assertion on the part of patients, that they cannot take the oil because it makes them sick. The first thing to be done under such circumstances is to secure the purest and most palatable oil possible; it should be perfectly clear, and have not the slightest trace of rancidity about it. Good oil has a slight fishy smell and a slightly fishy taste, that is all. Next, it should be given in small doses; let the patient begin with a teaspoonful, or even less, a few drops, if necessary, only begin without making him sick. It may be given in anything the patient fancies, water, milk, orange wine, curaçoa, etc., and it is best given within half an hour after a meal; some like it best immediately after. Even this may upset the stomach, and then it must be tried the last thing at night, after the patient has lain down in bed. The patient once accustomed to it, the dose must be gradually and carefully increased, taking care not to overload the stomach, until he can take a tablespoonful or more three or four times a day. When everything fails, it is to be rubbed into the stomach and bowels with the hand or a warm flannel. This plan is especially adapted for children. Supposing, however, that cod-liver oil cannot be borne, and that the smell produced by its inunction is too objectionable, what is to be done? Then two other oils may be tried, viz., almond oil and the best salad oil; in no case should secondrate materials be tried. Or again, that oldfashioned remedy, rum-and-milk, may be recommended. Either so or in a more aristocratic mixture of champagne and cream.

COFFEE is the name given to the seeds of a plant known to botanists as the *Caffea Arabica*, and belonging to the order *Cinchonaceæ*. The seeds are contained in a berry, which when ripe is of a red color. Each berry contains two seeds, which is covered over with a tough membrane called the "parchment." The seeds vary in size, and the best are obtained from Yemen, which yield the best Mocha coffee. They are brought into this country in their green state, and subjected to a process of roasting. They are then ground and boiled, and made into the decoction which we call coffee. The coffee beans or seeds contain an active principle, called caffeine, which is identical with theine which is contained in tea. This substance acts powerfully on the nervous system, and is undoubtedly the agent which makes coffee an acceptable article of diet. During the roasting of the coffee bean, the constituents of the seed are converted into certain empyreumatic oils, one of which, called caffeic acid, gives a peculiar odor to the roasted seeds. These oils are stimulant, and give the flavor to coffee. The action of coffee is sedative and stimulant,

and when taken with milk and sugar it is one of the most popular and beneficial of beverages. In cases of narcotic poisoning, after the stomach has been emptied of the poison by emetics or the stomach pump, the administration of strong coffee without milk or sugar is an excellent remedy. Coffee is often adulterated with chicory. There is no objection to chicory on the score of its action on the system. It, however, contains no caffeine, and only diminishes the taste of the caffeic acid of the coffee. Some persons prefer to add a little chicory to their coffee, and there is no objection to the practice, but they had better be bought separate.

COLCHICUM, as employed in medicine, is either the bulbous underground portion (more correctly termed a corm) or the seeds of the meadow saffron. From the corm are prepared an extract and a wine of colchicum, and from the seed a tincture. Whatever preparation is used, colchicum seems to have the effect of increasing the flow of bile, of diminishing the force and rapidity of the heart's action, and if large doses are used causes vomiting and purging. The specific action of colchicum is, however, on the pain of the gouty paroxysm, which it relieves in a marvellous manner. Nor is the mode in which this is brought about at all well known, for small doses will have this effect, even without any purging, which is ordinarily beneficial to the disease. Its use is nevertheless followed by some prostration, and a tendency to faintness, which is far from agreeable, even though the pain has gone. It is sometimes used in acute rheumatism, but does not produce the same wonderful effects as in gout, what good it does being probably due to its reducing action on the heart. There is, however, one fact which is of vital interest to those who suffer from gout, they may kill the pain with colchicum, but they do not cure the disease, and in all probability this will return sooner and more violently after being choked off with colchicum, than had it been allowed to exhaust itself, or other remedies, as alkalies and alkaline purgatives, employed. Of the ordinary extract, about a grain should be given for a dose, of the wine and tincture 10 or 15 drops every four hours.

COLD, of which catarrh is the most prominent symptom, is perhaps the most frequent malady in this country. Its causes are manifold, and the consequences of catching cold are infinitely various, from merely a slight temporary inconvenience to speedy death. A fatal result is not, however, likely to happen except damaged organs be attacked, but then a very slight cause may suffice to bring about fatal results. Colds are frequently felt to date from some particular period, but frequently their onset is not appreciated for a time. Very likely there is some shivering and sneezing, with lassitude, pains in the back, loins, and limbs, with tightness of the forehead, and an unnaturally dry state of the lips and nostrils. These speedily give way to excessive discharge from the nostrils, first watery and acrid, later mucus or mucous and purulent matter. There is hoarseness and slight sore throat, the eyes water, feverishness, loss of appetite, thirst, and quick pulse. Sometimes small vesicles, called herpes, appear on the lips or about the nose. These symptoms do not last long ; they either pass away, or become aggravated if the inflammation passes onward into the interior of the lung. If the latter, there will probably be some difficulty in swallowing from the pain of the inflamed parts, and there may be loss of voice —temporary merely—and some difficulty in breathing. Not much treatment is required for such cases, though most people do something for them ; as already pointed out, in a great majority of cases they pass away in due course, though they may recur again, and if the individual is subject to them, not only must greater care be taken to escape catching cold, but every endeavor must be used to get rid of them speedily, lest they affect the lungs. Various plans are used with various success ; perhaps the best is for the patient to put his feet in hot water, take ten grains of Dover's powder, with a good glass of " something hot," and get straight into bed. If the throat is bad, a water compress had better be used. Most likely the patient feels better in the morning, whether from the attack passing away, or the result of the remedies. A seidlitz powder, or some such slight laxative had better be taken, and the patient may return to his usual routine. (See INFLUENZA.)

COLIC is a form of disease characterized by a severe twisting pain in the bowels, especially in the region of the navel, and occurring in paroxysms. This pain, though severe for the time being, and alarming, is not really of a dangerous character. It indicates no inflammatory mischief, and whereas the pain of colic is relieved by firm pressure, that from inflammation in the same region is greatly aggravated thereby. The complaint is accompanied by constipation, and frequently by vomiting, but there is no fever, and no quick pulse as in inflammation. Such attacks commonly arise from some indigestible article of food, or some portion of the food has decayed in the bowel, and given rise to foul gases, which distend the bowel and give rise to pain. As a rule such painful conditions are signs of intestinal disturbance which necessitate some opening medicine, of which for this purpose castor oil is best. An ounce may be given with a few drops of laudanum, not more than ten—or a little spirit of chloroform—or yet again it may be followed by a glass of hot spirit and water. Such means speedily give relief. (See FLATULENCE.) Besides this simple form of colic, there are others associated with the introduction of mineral poisons into the body. Chief among these is lead colic or painter's colic, to which also other names have been given. Among painters, especially those engaged in working white or red lead, and who are not exceedingly cleanly in their habits, the disease is of frequent occurrence. (See LEAD-POISONING.)

COLLAPSE. (See SHOCK.)

COLLAR-BONE is the bone which on either side is situated between the sternum and the shoulder-joint. (See FRACTURE.)

COLLODION is the product of gun cotton dissolved in ether and spirit. When exposed to the air, the ether speedily evaporates, and leaves the dissolved gun cotton in a thin film on the surface to which it had been applied. This film is air-tight, and has a considerable contractile power, hence it is useful for cuts or other trifling injuries about the face, and, indeed, may be applied to any surface from which it is desired to exclude the atmospheric air. Thus herpes, erysipelas, burns, sore nipples, chapped hands, bleeding from leech-bites, etc., may be successfully treated. A preparation called flexible collodion, made by adding Canada balsam and castor oil to ordinary collodion, is, however, much more useful in many instances than ordinary collodion, as it does not crack on being bent or stretched.

COLLYRIUM is a Greek word, signifying anything that will stop a running or disease of the eyes. Collyriums are of two kinds ; one to subdue inflammation in the ball or coats of the eye, and the other of a stimulating nature, such as is used in chronic affections of the organ, to excite the vessels to a healthier action. A very usful eye-water for inflamed eyes is made of rose-water, sulphate of zinc, and sugar of lead ; should it cause pain, it is too strong, and may be reduced by the addition of water. Lotions for weak eyes that require stimulating are made by mixing about 6 grains of sulphate of copper with 6 ounces of water or elder-flower water. Nitrate of silver in solution, very weak, is also a valuable application, but must be carefully used.

COLOCYNTH is the pulp of a kind of gourd growing along the southern and eastern shores of the Mediterranean. The fruit itself is shaped something like an orange, and the pulp is exceedingly tough and felt-like. It is only used in the form of pill, but there is an extract for combining with other similar substances, and a pill containing hyoscyamus as well as colocynth. Colocynth itself is a powerful purgative, giving rise to much griping and plentiful watery evacuations, but it is seldom if ever given alone, being ordinarily combined with other purgatives and some carminative. It is mostly used when a speedy and effectual opening of the bowels is desirable. It is too irritating to be used as an habitual purgative, but on occasions there are few more effectual than three or four grains of compound extract of colocynth, with one or two grains of calomel, or, better still, two pills may be made, each containing four grains of the pill of colocynth and hyoscyamus, with one grain of calomel. This is a very desirable compound when the liver is deranged from overloading of its portal vessels, or when the bowels have been long confined and their action is irregular and torpid.

COLOGNE. (See EAU DE COLOGNE.)

COLON. (See ABDOMEN.)

COLOR - BLINDNESS is a subject which, but recently well understood, promises to become one of the most important in the science of optics. The structure and mode of action of the eye are explained in the article on VISION (which see). People are color-blind when their retina will not perceive some of the rays of light. Light is made up of three primary colors, which when overlapping in the spectrum are known as the colors of the rainbow. Each color is produced by the different velocity with which the undulations of the imponderable particles that fill all space beat upon the eyeball. The eye can only see those colors which represent a certain velocity of the undulations, and in ordinary vision these colors are red, blue, and violet, with their compounds, green (blue and yellow), orange (red and yellow), and purple or violet (red and blue). According to what is known as the Young-Helmholtz theory of vision, there are in the eye "three kinds of nerve-fibres : stimulation of the first produces the sensation of red ; of the second, that of green ; of the third, that of violet." All three kinds of fibres are excited, however, by each color of the spectrum, as there is no spectral color "into which but one primitive color enters exclusively." The most frequent forms of color-blindness are those of insensibility to red or green—violet-blindness being extremely rare. The insensibility is due to the absence of the nerve-fibres requisite to the perception of the given fundamental color ; and it is a structural defect which when congenital, no amount of training or practice can cure, though there are means of alleviating the resulting inconveniences. The effect of red-blindness (also called *Daltonism*) is that green and red do not appear to be different colors, but simply different shades of the same color. Not only so, but "the whole chromatic system will be upset," though of many colors, as of yellow and of violet, the perception of the color-blind will not greatly differ from that of the normal vision. "All colors containing their defective one will be grayish, and this in proportion to their individual amount of defect. The red-blind, for instance, will see only such colors as can be produced by the

combination of green and violet. Spectral red, which feebly excites the perceptive organs of green and scarcely at all those of violet, must consequently appear to the red-blind a 'saturated' green of a feeble intensity. Feebly luminous red does not excite the perceptive organs of green, and it therefore seems to them black."

Color-blindness is of much more frequent occurrence than could have been supposed until adequate tests for it were devised. As the result of over ten thousand testings among New England teachers and scholars, Dr. B. Joy Jeffries estimates that in the United States from three to four in every hundred males are color-blind ; and this agrees with Professor Holmgren's investigations in France and elsewhere. Among females the percentage is very much smaller.

Tests for Color-Blindness.—The simplest and most effective test is that devised by Professor Holmgren, of Upsala, Sweden. It consists in requiring the person examined to match a sample color with shades from a large number of differently colored worsteds placed before him together. All mention of the *names* of colors should be avoided ; for the nomenclature is so imperfect and so imperfectly or differently understood by most people that it is a sure source of confusion and mistakes. It should never be forgotten that a person may be perfectly able to recognize colors and shades of colors, and yet not be able to *name* them when shown him. To *match* a piece of red or green worsted placed in his hand, without naming its color, is the sure test ; and it cannot be evaded by any device to which a color-blind person will be apt to resort in order to conceal his defect.

It will be seen, of course, that color-blindness is of practical importance as well as physiological interest. No color-blind person should be employed in railway or marine services where the safety of hundreds may depend upon the correct interpretation of a signal which he is utterly unable to distinguish.

COLOSTUM. (See MILK.)

COLOTOMY, an operation devised for opening the bowel in the left loin, in cases where there is an obstruction in the lower part of the intestines. (See OBSTRUCTIONS.)

COLUMBA is the product of a plant growing in Zanzibar and Eastern Africa. The part used in medicine is the root, which is shaped not unlike a carrot. This is sliced and dried and is imported. It contains a large quantity of starch, and hence certain of its preparations are apt to spoil. In itself it is an excellent tonic, very mild and unirritating to the stomach, and as it does not blacken with iron constitutes a remarkably good vehicle for that remedy. When the stomach persistently rejects other and more powerful tonics this will often be retained, and the best mode of taking it is in an effervescing draught, a teaspoonful of the tincture for a dose. The infusion is also much used—it should be freshly prepared. The powder, combined with carbonate of soda and rhubarb, is an excellent domestic remedy for irritative dyspepsia—15 or 20 grains might thus be given.

COMA is a state of deep sleep or insensibility in which the patient lies perfectly unconscious of what is going on around. It is caused by a great many conditions, and these will here be shortly enumerated ; it is not a disease, but a symptom of disease, and therefore the reader must refer to the diseases mentioned if he wish to have a more detailed account. 1. Coma may proceed from drunkenness, because the blood is for a time poisoned, by the amount of alcohol taken, and the brain suffers in consequence ; this state usually passes off in a few hours ; there is a history of the patient having taken too much ; the breath will smell of spirit or beer, the face is flushed, the breathing is noisy, accompanied with puffing of the cheeks, and the man will be in a helpless stupid state. It is often difficult to distinguish this state from apoplexy or from fracture of the skull, but in the latter case there will be a history of a fall, and in both it will be very difficult to rouse the patient, while a drunken man can be roused if he be galvanized, or put under a stream of cold water, or made to vomit. Yet in all cases of doubt, the patient should be placed quietly on a bed or mattress, and not taken about from a hospital to a police cell, or from one place to another ; a few hours will soon suffice to make sure of the nature of the complaint, as by that time the drunkard will have recovered. 2. Coma may result from apoplexy, or in other words from a clot of blood in the brain, or white softening of that organ, or from a plug in the vessels supplying it. The patient will then fall down insensible and there will be paralysis of the arm and leg of one side (HEMIPLEGIA), or the mischief may be so great that he lies in a passive state, and all the limbs are paralyzed and drop by their own weight when they are raised. (See APOPLEXY and HEMIPLEGIA.) 3. Coma comes on in cases of poisoning by opium or carbonic acid ; the one may be induced by taking morphia or laudanum, the other by exposure to the gas, as when persons have been suffocated by burning charcoal in a non-ventilated room. In cases of opium poisoning every means should be used to rouse the patient by walking him about, slapping him with towels, giving hot and strong coffee, and applying mustard poultices to the calves or hot bottles to the feet ; where carbonic acid is the poisonous agent, the person affected should at once be taken into the open air, the mouth opened, and the tongue drawn forward, and artificial respiration must be re-

sorted to. (See DROWNING.) 4. Coma is often met with in the last stage of kidney disease ; the urine is diminished in quantity and perhaps hardly any is passed ; the patient complains of headache, sickness, convulsions, and in three or four days he may pass through a state of stupor into coma and death. Purgatives, cupping over the loins, and diaphoretics may be used, but not with much hope to the patient. 5. Coma ensues from a blow on the head with or without a fracture of the skull ; in such cases blood is generally poured out between the dura mater and the skull, and this, pressing on the brain, will cause the insensibility. Little can be done beside keeping the person perfectly quiet in bed, and applying ice to the head. (See FRACTURES.) 6. All the forms of *Meningitis* are accompanied by coma before death, and this condition is also met with in many cases of fever, as typhus, typhoid, and scarlet fevers, pyæmia, etc. (See MENINGITIS, etc.) 7. After an epileptic fit the individual is insensible and in a comatose state for a short time varying from a minute or two to twenty minutes or half an hour ; but the coma soon passes off, and no special treatment need be adopted for that symptom. (See EPILEPSY.) During an hysterical attack the patient may go off into an insensible state, but the timely administration of cold water, or a galvanic current will usually have a most beneficial effect. Finally, some persons may feign this condition for a short time from a morbid desire to create sympathy ; any of the usual methods of causing pain will expose the imposture, and prevent perhaps the recurrence of the malady.

COMBUSTION, SPONTANEOUS, rests upon somewhat doubtful authority. At various times it has been reported that individuals have taken fire and been consumed to ashes. Such a fate is described in one of a popular novelist's works, but if ever such a case did occur no one certainly has heard of one recently. The fact, however, that the bodies of living persons, may, under exceptional circumstances, attain to an extraordinary combustibility, rests on somewhat better authority. Such a condition, however, though often confounded with the former, implies something totally different. The individuals who have been supposed to attain to this superior combustibility have ordinarily been inordinate spirit drinkers. Cases have been recorded where such individuals were partially consumed—the burning commencing from comparatively slight causes.

COMPOSITION OF THE HUMAN BODY.—The human body is composed of the same elements as are found entering into the composition of the mineral substances found on the earth's surface. The following is a list of the quantities of the various elements found in a human body weighing 154 pounds :

	lbs.	ozs.	grs.
Oxygen	111	0	0
Hydrogen	15	0	0
Carbon	21	0	0
Nitrogen	3	9	0
Phosphorus	1	12	190
Sulphur	0	2	217
Calcium	2	0	0
Fluorine	0	2	0
Chlorine	0	2	382
Sodium	0	2	116
Iron	0	0	100
Potassium	0	0	290
Magnesium	0	0	12
Silicon	0	0	2
Total	154	0	0

It will be seen that the four first elements are oxygen, carbon, hydrogen, and nitrogen. These are non-metallic elements, and enter largely into the composition of all organic beings. It is by the presence and abundance of these elements that plants and animals are distinguished from minerals. No organic being can be developed without these four elements, hence they have been called organic elements. The next elements of importance are undoubtedly phosphorus and sulphur. They are both found associated with the four organic elements in such compounds as caseine, fibrine, and albumen, hence they may be very properly called pseudo-organic elements. Chlorine, fluorine, and silicon are non-metallic elements. The rest are metals. Of these sodium is most abundant, and iron and silicon are least so. Nevertheless, they are necessary. Thus iron, although in small quantity, produces disease when it is absent, and there is no set of diseases more easily detected and none so easily remedied as those dependent on the absence of iron in the body. Even the absence of the small quantity of silicon is accompanied by diseased conditions. Thus the enamel of the teeth in a healthy condition contains silicon, but if this is absent the enamel is not properly formed, and the teeth quickly wear away and become comparatively useless. The elements, however, are not found in the body in their pure state, but are mixed together, forming the following compounds :

	lbs.	ozs.	grs.
Water	1·1	0	0
Gelatine	15	0	0
Fat	12	0	0
Albumen	4	3	0
Fibrin	4	4	0
Phosphate of Lime	5	13	0
Carbonate of Lime	1	0	0
Fluoride of Calcium	0	3	0
Chloride of Sodium	0	3	376
Chloride of Potassium	0	0	10
Sulphate of Soda	0	1	170
Carbonate of Soda	0	1	72
Phosphate of Soda	0	0	400
Sulphate of Potash	0	0	400
Peroxide of Iron	0	0	153
Phosphate of Potash	0	0	100
Phosphate of Magnesia	0	0	75
Silica	0	0	3
Total	154	0	0

There are many other substances belonging to both the organic and inorganic series of compounds which have been described by chemists, but the above are the principal permanent compounds found in a human body. *Water* is composed of oxygen and hydrogen gases, and constitutes four parts of the bulk, and sometimes even more, of the whole organic kingdoms. Water is the great agent by which the food is dissolved and taken up and discharged from the body. *Gelatin* is composed of the four inorganic elements, and is found in the cell-walls of the animal tissues. It is especially abundant in the bone-cells and the skin. It is soluble in hot water and is identical in composition in the human body and that of the lower animals. *Fat* is a compound of carbon, hydrogen, and oxygen. It is distributed over the body in the adipose tissue, and its increase is the principal cause of obesity. It is also found in the marrow of the bones, in the joints, and other parts. *Albumen* contains the four organic elements. It is found in the blood, and the principal substance entering into the composition of the nerves. *Fibrin* differs but very slightly from albumen. It is not found dissolved in the blood like that substance, but is suspended in it and coagulates when blood is allowed to stand. It enters into the composition of the muscular tissue. *Phosphate of lime* is found in the bones. A half part, by weight, of the human skeleton is composed of phosphate of lime. *Carbonate of lime* is found also in bones, in the proportion of about ten per cent. It constitutes almost wholly the skeleton of corals and shell-fishes, where but little phosphate is found. *Fluoride of calcium* is also found in the bones of human beings. It is often found in large quantities in fossil bones, and the quantity of fluoride is said to be a guide to the age of the bones in which it is found. *Chloride of sodium* is found in the blood. It is necessary to the life of human beings as well as all other animals. This salt, which is found in sea-water, determines the existence of those forms of plants and animals which live in the sea. Its use is universal among mankind, and they suffer from disease when deprived of it. *Carbonate, sulphate, and phosphate of soda*, are other forms of sodium which are found in the blood and tissues of the human body. *Chloride of potassium*, and the same salts, as of sodium, are also found in the body, but are not so abundant as the latter. Nevertheless, the small quantity of potassium is necessary to the proper nutrition of the capillaries and vessels of the body. When persons are kept long on salt meat it is well known that they get scurvy, and this seems to arise from the absence of potass salts in the food. A still smaller quantity of *magnesia* than of the alkaline metals is constantly found as a constituent of the human body. (See Food.)

COMPRESSION is a term used by surgeons to imply pressure upon the brain caused through severe injury. The symptoms indicating this condition are : total insensibility and loss of motion, slow, noisy, and deep respiration, a slow and laboring pulse, partial or general palsy, one side of the body being usually paralyzed, involuntary discharge of the contents of the bowels and retention of urine, dilatation of the pupils, and closing of the eyelids. These symptoms are sometimes associated with delirium, restlessness, convulsions and vomiting. The causes of this state are various : it may be due to fractured skull and depression of bone upon the surface of the brain, to effusions of blood within the skull, to internal suppuration or to the presence of some foreign body, as a bullet or piece of exploded gun barrel. The condition is a very serious one, and in most instances terminates in death.

CONCRETIONS may occur in many internal organs, but the most important are those which occur in the intestine, gall-bladder, and kidney and urinary bladder. Intestinal concretions, or calculi, are of rare occurrence in the human being, but in ruminant animals they are not uncommon. In man they occur in the great gut most frequently. They consist for the most part of imperfectly crystallized salts and indigestible fibrous or other matters arranged round a nucleus which may be a gall-stone, the stone of a fruit, or any such foreign body. Some concretions consist entirely of hardened fæcal matter, or if chalk and magnesia have been largely swallowed they may form something of the kind. Hair, cotton, and paper may be found in mass, having been swallowed owing to a depraved appetite. In animals balls composed of hair which has been removed by licking, are perhaps the most common form of concretion. These may occur either in the stomach or in the intestines. Most frequently these masses are only discovered after death, but it is quite possible that they may gradually increase to such a size as to obstruct the bowel and so give rise to signs of intestinal obstruction which may prove fatal. Occasionally these concretions are passed by the bowel, or if they get very low down they may be broken up, but they are at all times dangerous. (See GALL-STONES and STONES.)

CONCUSSION is a term used by surgeons to express a severe shattering of some internal organ in consequence of a fall or heavy blow. It is a condition that gives rise to certain definite symptoms, varying according to the particular organ injured, which though severe and alarming are less grave than those resulting from actual rupture and laceration of tissue. It is probable, however, that the symptoms of concussion are always due to some local injury, and not merely to an excessive vibration of the minute elements of the tissues. In concussion of bone, for instance, there is frequently some separation of the external membrane or *periosteum*, and in concussion of the brain rupture of the small blood-vessels and effusion of blood. The best-known form of this injury is *concussion of the brain*. This condition varies very much in intensity, and may manifest itself either as a simple stunning or by complete bodily prostration and loss of consciousness. For the purposes of description it may prove convenient to give two typical forms of concussion, although it must be remembered that in most instances the symptoms of one form will be found to coexist with those of another. In all instances the symptoms of concussion follow an injury either from direct or indirect violence to the head. In the first form the patient experiences a sudden weakness and muscular trembling in the limbs, especially the lower, and cannot walk without staggering ; at the same time there is a ringing sound in the ears, and dimness of sight. These symptoms soon pass away after the patient has rested for a time in a darkened room. In the second form of concussion, the patient becomes deadly pale, and is at once deprived of consciousness, of hearing, and sight, and of the power of motion. The skin is cold and the pulse weak. The eyelids are closed, and the arms and legs bent upon the body. The breathing is slow and regular, and the patient, when spoken to loudly and called by his name, will open his eyelids or give some other sign of recognition. It is important to bear in mind these two facts, as they serve to distinguish concussion from compression and other more severe injuries of the brain. In some cases there is slight and transient shivering. This state lasts in the majority of instances but a short time after the injury, generally about one hour, when the patient wakes up for a time and then passes into a lethargic condition, which varies in duration according to the age and constitution of the patient and the severity of the injury. Recovery from the early and intense symptoms of concussion is indicated by increased temperature of skin, and by movement of the limbs, but chiefly by vomiting, which should be looked upon as one of the most favorable symptoms of this injury. *Uncomplicated concussion* is by no means that fatal disease one would be inclined to conclude from newspaper reports. It is doubtful whether it ever causes death directly. It often leaves in its train, however, a set of symptoms and certain chronic affections which may render the patient permanently disabled, or even bring on early death. The following are occasional sequelæ of simple concussion : irritability of the brain marked by a hasty violent temper, or by speedy excitement after drinking spirits or wine ; loss of memory ; temporary or permanent insanity ; loss of sight and of hearing ; muscular weakness and general trembling ; loss of flesh and general debility ; in flammation of the membranes of the brain. These affections are due in many cases to indiscretion on the part of the patient, who as soon as the symptoms of concussion have passed away, but while the brain is still sluggish and enfeebled, returns to his former habits and probably undertakes the duties of some responsible post.

Treatment.—The following are the chief points to be remembered in the treatment of concussion : to place the patient upon a bed or couch in a darkened room ; to free the neck and chest from all articles of daily clothing ; to keep the head raised : and to apply cold wet cloths over the forehead. In cases where there is intense prostration, and the surface of the body is cold, the patient should be placed in bed, between blankets, and hot water bottles be placed near the feet and armpits. Friction with the hand may also be used to keep up the circulation. Spirits and other stimulants *must not be given*. As soon as the patient has become sensible, some hot broth or beef-tea, may be administered. The after-treatment of concussion consists in perfect rest, both of mind and body, free purgation, and a mildly nutritious diet. Alcoholic stimulants are still to be avoided. If severe headache come on, or slight impairment of the mental faculties be observed, a blister or strong mustard poultice should be applied to the back of the neck, and the bowels be freely opened. In the treatment of concussion, as of other severe injuries of the head, there is no urgent necessity for removing all the hair. Cold may be readily applied to the head by means of ice or cold compresses over the forehead and, in the female, considerable relief may be given by allowing the long hair to keep moist by constant immersion in a vessel of cold water. In concluding these remarks upon concussion of the brain, it is necessary to state that in very many instances this affection is complicated with or followed by certain symptoms that indicate serious injury to the skull or its contents. Hence the popular dread in cases of this kind, and the frequent reports of death from this cause, which is due not to concussion merely, but to concussion *plus* compression or laceration of the brain. Even *stunning* may be followed by fatal brain mischief. In all cases of supposed concussion, therefore, the treatment should be directed with great caution, judgment, and skill.

Concussion of the Spinal Cord.—Of this affection there are two forms ; one in which several well-marked symptoms immediately follow a severe blow upon the spine or a fall upon the buttocks or back ; and the other in which the injury, generally a violent shaking of the whole body, gradually results in the

course of months in paralysis of the lower extremities and other grave disorders. The latter affection will be described in the article on RAILWAY INJURIES. The first or acute form of spinal concussion is marked by the following symptoms : pain in the back at the seat of injury, general bodily prostration, weakness of the lower limbs, and difficulty in walking, numbness in the feet, and diminished sensation of the skin of the lower extremities, difficulty in making water, swelling of the abdomen due to distension of the intestines with gas. These symptoms usually subside in the course of two or three weeks, and the patient makes a good recovery. In some instances, however, concussion of the spinal cord terminates in permanent weakness or even complete palsy of the lower limbs with retention of urine. The treatment of this injury consists in keeping the patient in bed, and in cupping the back or loins or applying leeches, and afterward by giving tonics and nourishing food.

CONDIMENTS. — Those substances which are added to food with which salt is taken are called by this name. They are mostly derived from the vegetable kingdom, and contain peculiar vegetable oils. These oils have not only their own flavor, but they act as stimulants in the system. They act beneficially by their effect upon the nerves and secretions of the stomach. The principal vegetable condiments are, pepper, Cayenne pepper, mustard, horse-radish, onions, garlic, peppermint, thyme, caraways, anise, dill, fennel, samphire, and others, most of which are separately noticed in this volume.

CONFECTIONS are preparations of medicines ordinarily semi-solid, and containing sugar or honey. They are chiefly used for making pills, and some of them have no active power ; confection of roses, for example. Confection of senna is a useful purgative.

CONFINEMENT. (See LABOR.)

CONFLUENT SMALL-POX is said to occur when the pustules run together and form large and unsightly scabs. (See SMALL-POX.)

CONGESTION implies a fulness of blood and a retarded circulation in a region or organ of the body. This condition is called by pathologists *local hyperæmia*, while a tendency to general fulness of blood—an excess, as it were, of this fluid in all parts of the body—is called *plethora*. In local congestion the following changes take place : The amount of blood circulating through the affected part is much increased ; the temperature is raised, and there is usually pain and a sense of heaviness ; the veins and the minute tubes between these vessels and the arteries are overdistended with blood, which is frequently poured out through rents into the tissues around ; the blood is of a darker color. In advanced local congestion there is complete arrest of the circulation in many of the minute vessels. The principal cause of local congestion is obstruction to the return of blood by the veins from the affected part of the body. The obstructing or compressing agent may be present either internally or upon the surface of the integument. In addition to this, debility from fever or some other severe illness, mal-nutrition,and senile decay,are frequent causes of local congestions. Internal piles from obstruction to the circulation through the liver and the veins of the abdomen, swelling of the feet during a prolonged convalescence, and the inflammed and ulcerated legs so frequently observed in old persons are well-marked instances of congestion due to the above causes. The *treatment* of painful congestion consists in the removal of any cause of obstruction to the blood-flow, in application of leeches or cupping glasses to the affected part, and in attention to the general health and condition of the patient. The details of treatment will be treated more fully in articles on HÆMORRHOIDS, ULCERS, and VARICOSE VEINS.

CONIA is the active principle of *Conium maculatum*, or Hemlock. (See HEMLOCK.)

CONJUNCTIVA is the anatomical name for the thin and sensitive membrane that covers the front of the eyeball, and is reflected above and below along the posterior surfaces of the lids. At the inner junction of the eyelids this membrane forms a small red fold called the semilunar fold, which represents in man the large nictitating membrane or third eyelid found in birds. In children the conjunctiva is quite transparent, but as age advances it becomes dusky and yellow, and is rendered more and more opaque by the presence of large blood-vessels.

CONSERVES are preparations in many respects analogous to confections, and used like them.

CONSTIPATION is a symptom which is often attended with much discomfort and pain to the patient, and one which may be due to disease of the bowels, or to an imperfect performance of their function. In the natural course, the food, after digestion by the stomach, passes down into the intestines, and by the contraction of their muscular coats, it is propelled onward, to be discharged once or twice a day from the rectum as excreta or faeces. Any disease, as ulceration or cancer of the bowel, which obstructs the passage of the food, will therefore cause constipation, and any condition which produces a paralyzed or sluggish state of the muscular walls of the bowel will likewise cause constipation by removing or interfering with the propelling power. Remedies which increase the contractile power and cause a more rapid flow of the excreta are termed purgatives. 1. Habitual constipation is not unusual in women after a confinement, in people of a nervous temperament, and in those who lead a sedentary life ; those, also, who are in the habit of frequently taking opening medicine, pills, etc., are liable to it, and nothing is to be more deprecated than the custom adopted by some of constantly taking some aperient. In such cases an altered diet will nearly always suffice, and cause no after ill effects. A glass of cold spring water taken first thing the morning has a most beneficial effect on some ; brown bread has a marked laxative action, and should be eaten instead of white bread ; roast apples, figs, prunes, and stewed fruit are valuable auxiliaries. With these simple remedies should be combined a sharp walk every day, and when advisable, a cold-water batn should be used every morning ; some adopt the habit of smoking a pipe after breakfast for the purpose, and this, no doubt, is serviceable ; but probably the benefit is due to the regularity which it induces, and habit is a most important element in preventing constipation ; no one should postpone the process, and in health the performance of the function ought to occur regularly about the same hour every day. An occasional aperient may be required, and then a mixture containing Epsom salts or some similar preparation can be ordered ; the effervescent citrate of magnesia is often given, and better still, for those who can afford it, a wine-glass or two of Pullna water may be taken first thing in the morning with great benefit. By the use of these means, habitual constipation may nearly always be cured, if it has not lasted too long. Some advise the use of injections or enemata daily ; but this process is disagreeable and not generally adopted in this country. In children a similar treatment may be adopted, while in infants an altered diet, and a little magnesia occasionally mixed with the milk, will suffice for a cure. Many boys have been dosed with brimstone and treacle with the object of "purifying the blood," but adoption of the above plan is far better. 2. Constipation may come on from some growth or ulceration in the intestines which prevents the progress of the excreta ; there will be then more or less vomiting, which will for a time relieve the distension, pain over the seat of mischief, swelling of the abdomen, loss of flesh, and frequent sickness will accompany the constipation. If the obstruction be high up, as in the stomach, nothing can be done for the constipation ; if low down and in the rectum, means may be taken to make an artificial opening in the loins to let out the faeces, but this can only be done in extreme cases. (See COLOTOMY.) 3. Constipation may be only occasional, and due to taking indigestible food, as nuts, unripe fruit, etc., or to taking too large a quantity at once ; the tongue will then be foul and white or yellowish, the abdomen full and painful, and a feeling of sickness may ensue. If the pain is very intense, so as to make one suspect *enteritis*, an opiate should be given to allay the urgent symptoms, and the constipation may be left alone for two or three days ; then a small dose of castor oil or some mild purgative may be given. In cases of peritonitis, some liver diseases, emphysema, and other chronic affections, this symptom may prevail, but the treatment must then vary with the special cause. A regular action of the bowels should always take place in health every day, and for his purpose, fresh air, light, active exercise, and a wholesome diet are the best provocatives. No doubt, occasionally a purgative is required, but they should not be taken indiscriminately, as they tend to induce after a time the very condition for which in the first place they are given.

CONSUMPTION is the disease to which technically the name Phthisis, or wasting, is applied. By it is meant that form of lung disease where first of all there is a deposit of new material in the substance of the lung. After a time this softens and breaks down. It is expectorated and leaves behind cavities. This process is accompanied by fever of a peculiar kind, and general wasting of the body ; whence the name. Such is the pathologic history of the disease, deposit and subsequent softening ; but the deposit is by no means always of the same kind, though that was taken for granted up to a very recent period. The processses which lead to this deposit are two in number—one is inflammation of the lung substance, and the other is a deposit of a new growth called tubercle. Most frequently the two processes are associated, for the deposit of the new growth sets up inflammation and its consequences. When the substance of the lung becomes inflamed, we have to deal with a very different set of phenomena than when the air passages alone are so affected. The disease may assume a very acute form, such as cannot be mistaken, or it may steal on insidiously, especially if it spreads from the air tubes to the lung substance. The consequence of such an inflammation is the choking up of the little cavities of which the lung consists in a portion of its substance, and the material thus deposited may either remain there for a length of time, or at once proceed to soften and break down. In this process the damaged material of the lung too may take part. It may soften as well as the newly-deposited substance, and breaking down and being expectorated, leave behind a cavity in the substance of the lung. This process may go on quickly or slowly, sometimes very slowly, especially if other changes go on at the same time, such as indurate the texture of the lung, as what has been called fibroid phthisis, a very slow form of the malady. But again there may be a deposit of new substance, the process being by no

means inflammatory, and this new growth which is laid down in the substance of the lung is called tubercle. Once deposited, its history is the same, or nearly so, as that of the inflammatory material laid down in the ,lung cavities. It softens and breaks down, the injured texture of the lung doing so also, and so a cavity is formed. There is yet another mode and kind of deposit. That due to syphilis. This is, perhaps, if a diagnosis can be made, the most hopeful variety of the disease. To both the former varieties of disease there may be a strong hereditary proclivity ; if so, this is a circumstance which tells most unfavorably on behalf of the patient, for there are few diseases in which a hereditary character is more prominent than that which commonly goes by the name of pulmonary consumption. It is of the very first importance that this disease should be diagnosed in the earliest stages, for it is then that certain of its forms may be treated with tolerable confidence of success, and all can be dealt with to most advantage. That form which promises most by timely treatment is the inflammatory form, especially that which comes on in a patient who has long been in depressed health from whatever cause. It commonly begins with a slight cough, which, however, persists, and will not go away, and the patient gets gradually thinner. The respiration indicates feebleness, being wavy in character, or even jerking. Besides this, there are certain sounds only to be appreciated by a skilled ear. If with all this, there is a bad family history, the case is one demanding prompt action. This may be taken with good hope of success. Fever, heat of body, best indicated by the thermometer, is a most important element in such cases. If it keeps high, the chances do not improve ; if it gradually diminishes and totally disappears, the patient may be said to have regained his health.

Take now a case of tubercular consumption. It may arise from the former, or it may be developed from the products of some long standing disease of other organs, or one lung may infect the other. This form is not so common as the other. Its origin is very insidious ; but having begun it goes on. There is considerable uneasiness. At night the temperature is high ; and there are troublesome night sweats. There is a persistent cough, and very likely pain in one side. The appetite is very capricious, and very likely there is diarrhœa. With such cases, too, a huskiness, or even loss of voice, is by no means uncommon. This rarely occurs in any other variety of consumption, and so may be looked upon as proof positive of the existence of this form, if any consumption be present. For the public, however, it is of less consequence to know what variety of consumption the patient labors under, than to know that he has got some form of the disease, or is likely to have it, if the malady he labors under is not arrested. Accordingly, an abstract of the commonest signs are now given ; for we desire to impress the public very earnestly that it is in its earliest stages that consumption is remediable. The earliest symptoms are very probably connected with digestion ; the appetite becomes capricious ; there are pains in the chest, with some cough often dry and hacking, with a small quantity of frothy expectoration. There is debility, flushing of the face on the slightest exertion ; at other times the countenance is pale, except there be a hectic patch of red in the middle of the cheek. The eyes look unusually white and pearly ; there is some fever at night, and a tendency to night sweats. Very likely there is some spitting of blood. This occurs in a very considerable proportion of cases, and is often the earliest symptom calling for attention. Provided it is clear that the blood does not come from the gums or throat, any bleeding by the mouth, especially in a young person, demands attention. As the disease advances emaciation advances, so that the joints become enlarged by shrinking of the limbs, and the fingers commonly become clubbed at their points. The night sweats and diarrhœa are the great means of reducing the bodily strength and substance ; but in some instances, excessive expectoration aids materially in this untoward process. At the same time the capricious appetite and the imperfect digestion leave the bodily supply very deficient. During all this time the spirits of the patient are good, and it is often distressing to see one doomed to an early death talking of the future in a tone of assured confidence. A very troublesome complication often seen is fistulæ in the lower bowel, which, if not relieved taxes the patient's strength sadly. On the other hand, there is always a risk that if an operation be attempted the wound will not heal, and so the latter risk is worse than the first. Usually, if the disease be not arrested, the patient dies of exhaustion ; sometimes he is suffocated or bleeds to death, consciousness continuing to the last. But this result is by no means necessary ; and the dread of the disease as being universally and unerringly fatal, which was wont to prevail, has been shown to be without just foundation. Undoubtedly, if a patient with a bad family history is seen for the first time when the disease is well advanced we have little ground for hope. True, also, that the tubercular form of the disorder is less amenable to treatment than is the inflammatory. Yet due care being exercised, there are few cases which cannot be benefited ; a goodly number can be cured completely, or, at all events, the lungs so healed that each may be enabled to lead a good long life in moderate comfort.

Treatment.—The first and greatest point of all is the selection of the conditions under which the patient is to live. Unfortunately, in too many instances, this is not possible ; but where it is possible, and the disease is in an early stage, much may be done. On the continent of Europe, there may be found in different health resorts, people who have all their lives had bad chests, but who by wandering from health resort to health resort, according to the season of the year, are able to maintain life comfortably. If such a thing is not possible, we must try next to select the most favorable conditions possible. The first great point in selecting an abode is the avoidance of damp ; it should be situated on a dry and porous soil. This is even of greater importance than temperature, though that too is important, inasmuch as fresh air is a necessity, and daily exercise in the open air even in winter is a thing very greatly to be desired. Such patients must take the greatest possible care of themselves—no risks must be run. They must live plainly ; but their food must be nutritious. They must avoid excitement, but cheerful society is of the greatest possible value. They must not fatigue themselves, but daily exercise is incumbent. They must not be exposed to too great heat ; but cold is even more to be dreaded. Hence the rule, which is a good one, always to keep indoors between sunset and sunrise. They must try to keep the skin open ; but they must avoid perspirations. Hence baths must be regulated in temperature for the individual—tepid, cool, or cold, as the case may be. The bowels must be kept open, but if they are loose, the diarrhœa must be checked. Finally, such patients are on no account to go without flannels ; while the outer clothing should be changed, if desirable, to suit the different periods of the day and year. At all times it must be warm, so as to avoid risks from cold. For those in the very early stage of phthisis nothing perhaps does so much good as a sea voyage in a mild climate—to the West Indies, for instance, though many prefer the longer voyage to Australia or New Zealand ; often these do great good, but they must be undertaken early, or the result will be the reverse of favorable. Change of climate in females is apt to provoke derangement of the menstrual function. This should be seen to, as any excessive flow would be very weakening. This, moreover, has to be borne in mind, that in consumption this function almost entirely ceases, and generally does so altogether. (See CLIMATE and HEALTH RESORTS.)

Patients, the subjects of consumption, have often, early in the disease, a rooted objection to fat as an article of food. This is the more important, as of all substances it is to them the most necessary. If therefore, they refuse to take fat as food, we must endeavor to give it as medicine. The form of fat which is most easily digested is cod-liver oil. If this be given, it need have no disagreeable taste, beyond a slight fishy flavor which to many is agreeable. It is to be given to the patient cautiously. Often one will say, " I cannot take cod-liver oil ; it always makes me sick. I have tried again and again always with the same results." You inquire, and find they have been endeavoring to take one or two tablespoonfuls at a time. As a matter of course they get upset ; but if they reduce the quantity to a teaspoonful, or even a few drops, they gradually get accustomed to it, and can take almost any quantity. Cod-liver oil is, however, food, rather than medicine, and the best time of taking it is just after a meal. The fish oil used in this way should be entirely devoid of color ; *every trace of color is an impurity.* Next to cod liver oil as a remedy comes iron. This, too, is best dealt with as a food ; that is to say, given along with the meals. The best preparation is the reduced iron, which can be taken in soup. If this is not attainable, the freshly prepared carbonate should be given. Next to these remedies, to be taken along with or just after food, to add to the value of food, comes anything which will aid digestion. Pepsine, as procured from the pig's stomach, is for this purpose exceedingly useful, enabling the food to be digested with ease and comfort, when otherwise it would only pass into the intestines, there to putrefy and ferment, and so set up diarrhœa. Four or five grains may be taken for a dose just after a meat meal. If that does not suit, meat digested beforehand might be tried. When cod liver oil cannot be taken, other kinds of oil may be tried. Of these the best are cream and salad oil. When no oil can be taken rubbing it into the skin does good ; but it creates a horrid smell, which is very trying to the poor patient. •Syrup of the iodide of iron may be given along with the oil, and often does good. Iodide of potassium seldom does, except the disease be syphilitic in its origin. ·If prescribed at all, it had better be given in decoction of bark. But of bark, the best preparations are the compound tincture and the liquid extract, given in doses of a drachm or so three or four times a day. It is often useful to combine some acid with the bark ; the best is the dilute nitro-muriatic acid, in doses not exceeding twenty drops. If the perspirations be very troublesome, it is customary to give dilute sulphuric acid ; but any acid does good. On the other hand, it is very frequently good to give alkalies instead of acids. These certainly, combined with bitters, very greatly strengthen the appetite and aid digestion. Liquor potassæ is commonly given in doses of 5, 10, or 20 drops, sometimes with bark, sometimes with gentian, or other bitter. Certain remedies called hypophosphites have been highly extolled in the earlier stages of the disease.

Their value is uncertain, though they often *seem* to do good. Counter-irritation is a means of treatment which has been unjustly decried and unduly lauded ; undoubtedly it does good if employed judiciously, viz., so as not to weaken the patient. It is of most benefit when the pleura is concerned, and the patient cannot lie in certain positions on account of pain. It must, however be employed cautiously ; best by some liniment as croton oil and turpentine mixed. Flying blisters, kept on for only a few hours ; these speedily create counter-irritation, while they do not cause it to such an extent as to be injurious.

Of the complications to be dealt with one or two yet remain to be noticed. First comes bleeding. This it must ever be remembered is a very serious matter, for it may be the cause of death. When it comes on, absolute rest must be enjoined, cold applied to the chest, ice taken internally, and gallic acid with sulphuric acid freely imbibed. Oil of turpentine is also of service, though perhaps less directly. Night sweats have been alluded to ; mineral acids, if not otherwise forbidden, are best for them. Diarrhœa must be dealt with carefully. It must never, however, be allowed to weaken the patient ; chalk, opium, and acids are the best remedies. If the throat is bad, nitrate of silver is the best application. For the cough a little opium, or hyoscyamus, or belladonna may be given ; but it is better treated on general principles.

CONTAGION, a name applied to the poison which is supposed to be the cause of many fevers, and also to the mode in which it spreads, viz., by contact with the infected person. Scarlet fever, measles, typhus fever, etc., are thus said to be *contagious,* because the disease will spread from one person to another who comes in close proximity to the one affected. (See FEVERS.)

CONTINUED FEVERS is a name applied to a group of febrile disorders in which the duration of the feverish period is prolonged for several days or weeks. The group includes typhus, typhoid, and relapsing fevers. (See FEVERS.)

CONTUSIONS.—By this term is generally understood a form of injury in which there is more or less laceration of the soft parts near the surface of the body, while the skin remains unbroken. In fact, the term may be applied to any injury caused by heavy pressure, or a sharp blow, so long as there is no external wound. When, in connection with much crushing and tearing of the soft parts, the skin is broken, the injury is then called a *contused wound.* With fracture of bones, dislocation, sprains, and other injuries from violence, there is always some amount of contusion. There is pain over the seat of injury, increased by pressure or movement of the limb; there is also considerable swelling, with more or less discoloration, due to rupture of blood-vessels, and accumulation of poured-out blood. Contusions vary much in extent and severity, from simple bruising of the skin to complete disorganization of the whole thickness of a limb. The prospects of speedy recovery from a severe contusion depend in the first place upon the amount of laceration in the subcutaneous soft tissues, and in the second upon the age and general condition of the patient. In children and healthy persons, very large collections of effused blood are absorbed with rapidity, provided that there be no communication with the external air through a wound in the skin. In persons with a tendency to so-called rheumatic pains in the limbs and back, contusions are frequently followed by persistent stiffness of the injured part, and a dull heavy pain, which is more severe during wet weather, or with an easterly wind. When the system has been weakened by chronic alcoholism, bad or insufficient nourishment, or by some chronic disease, the contused parts become inflamed, and the seat of a large diffused abscess. In the most severe cases of contusion, where all the soft parts of a limb are crushed and thoroughly disorganized, and the large blood-vessels torn, gangrene is an inevitable and often fatal result. The *treatment* of contusions and contused wounds is described fully in the articles on BRUISES, DISLOCATIONS, FRACTURES, and RAILWAY INJURIES. It consists chiefly in rest of the injured limb in an elevated position, in the application of cold lotions or iced water, and tincture of arnica. In cases of superficial bruising, the last named agent is of great service.

CONVALESCENCE means the period of recovery from an acute or chronic disease.

CONVULSIONS may be said to mean violent and involuntary contractions of the muscles of certain parts of the body or of the whole of it, lasting for a longer or a shorter time, and very frequently returning in paroxysms. Sometimes the convulsions or muscular contractions last a considerable length of time without relaxation. To such a spasm or convulsion the term *tonic* has been applied, and is such as characterizes the disease called lockjaw or tetanus. Common cramp is an affection of the same kind, but of more limited duration, and affecting only a small part of the body. Again, if the spasms alternate with relaxations, as occurs in what are sometimes called the *jumps*, the spasms are described as *clonic*. Such spasms or convulsions associated with complete insensibility constitute an epileptic or epileptiform seizure. The causes of convulsions are manifold, but it would seem as if there is a certain amount of evidence to support the notion that all act by suddenly depriving the motive part

of the brain of a due supply of blood. This arrest of blood-flow may be brought about in many ways, by plugging of the vessels, by powerful contraction of their muscular coats, etc. Sometimes poisonous matters in the blood as in the condition known as uræmia, give rise to convulsions, especially in pregnant women. In children, too, where the nervous system is readily excited, irritation in a remote part of the body may be reflected in the brain and cause convulsions. Such is the explanation of convulsions from teething, worms, etc. In most instances, convulsions are accompanied by loss of consciousness. This is the case in those forms commonly called fits, but in hysterical fits or convulsions there is no loss of consciousness, and the tongue is not bitten. In dealing with convulsions in adults, perhaps the best plan is to wait quietly till the convulsion is over, and then try to prevent its recurrence. As a rule, interference with the individual during convulsions is mischievous. He should never be held or further controlled than is necessary to prevent him from hurting himself. His dress should be loosened, plenty of fresh air allowed to circulate around him, and none save those engaged in looking after him should be allowed to come near him. As he begins to revive, a little cold water to swallow may do good, but as soon as possible he ought to be got to bed and undressed—if not previously so—and left to himself. Very likely he will fall into a quiet slumber, and when he awakes there should be an urgent inquiry into the cause of the convulsions; among which albuminuria should never be forgotten. Should he not recover consciousness, but pass from convulsions to coma, as it is called, very probably the cause of the attack has been the rupture of a vessel and the effusion of blood into the brain substance. Nevertheless, it is quite true that uræmia may take the same course and terminate, too, in coma, or complete insensibility, with dilatation of the pupils of the eyes. Salaam or nodding convulsions are very rare forms of the malady, peculiar to children. See INFANTILE CONVULSIONS; PREGNANCY; PUERPERAL CONVULSIONS, and PUERPERAL FEVER.

COOLING MEDICINES. (See REFRIGERANTS.)

COPAIBA is a mixture of oil and resin obtained from various species of trees growing in South America. It is of a thickish consistence and is yellow in color; its odor is disagreeable. By distillation the oil may be separated from the resin; this oil is allied to turpentine both in character and properties. Copaiba acts as a stimulant, especially to mucous membranes; and as it is discharged from the body chiefly by the lungs and urinary organs, it acts chiefly on the mucous membranes of these. Hence it is of use in the bronchitis of elderly people who want stimulation, and in discharges from the urinary passages. It is chiefly used in the treatment of the latter, but is apt to disorder the stomach, so that local applications to the part affected are generally preferable. In large doses it produces a peculiar rash on the skin. Sometimes it is given for threadworms. The dose of the balsam, as it is called, is about 30 drops ; of the oil, 10 drops. The balsam is exceedingly disagreeable to the taste, and is apt to return for a time. It is best given in capsules. It communicates its smell to the breath.

COPPER itself is not used in medicine, but as copper vessels are much used in cooking and are liable to be attacked by their contents, so as to produce a poisonous compound, it is of some importance. The compound so formed is verdigris, an impure acetate ; it gives rise to vomiting and purging. This salt is formed by introducing anything containing vinegar into the copper, or allowing its contents to ferment. Sometimes the acids of fats separate, and in like manner attack the containing vessel ; for this reason nothing should be allowed to stand in the copper, and it should be carefully cleaned after each time of using. The oxalic acid sometimes used for cleaning the outsides of kettles is a dangerous poison, and should not be used. If anything of that kind be employed, exceedingly weak nitric acid is best, and the vessel should be carefully rinsed out with water immediately. *Sulphate of copper,* better, known, perhaps, as bluestone, is the most important preparation of copper used in medicine. It occurs in somewhat irregular crystals, and is sometimes known by the name of blue vitriol. It is sometimes given internally in small doses as an astringent in obstinate cases of diarrhœa. Given in large doses it acts as a speedy emetic, and is used for this purpose in narcotic poisoning. Externally it is used as a kind of stimulant application to sores, which may be lightly stroked with a crystal when indolent and indisposed to heal. It is also used in lotion for some discharges, or as an application to flabby ulcers. The dose as an astringent is about half a grain ; as an emetic about five grains. The strength of the lotion should ordinarily be about a grain or two grains to the ounce of water. In poisoning with copper, vomiting should be promoted by copious draughts of warm water and a solution containing tannic acid prepared, such as tincture of galls, tannin itself, or oak bark, which should be given as an antidote to any of the substances not expelled by the vomiting which itself has produced, for it may be said at all times to act as an irritant to the stomach.

COPPERAS. (See DISINFECTANTS.)

CORIANDER is the fruit of an umbelliferous plant ; its properties are similar to those of caraway and a variety of other seeds and fruits, etc. It is stimulant and carminative.

CORNEA. (See EYE.)

CORN-FLOUR is a preparation of starch, and so called because it was originally prepared from maize or Indian corn. The term has also been applied to starch prepared from rice and other grains. In the preparation of the maize and other grains, in order to make corn-flour, the husk and gluten of the seed are separated by grinding and the action of water. In the use of corn-flour as an article of diet, it should always be recollected that it contains little else than the granules of starch (see STARCH). Starch acts on the system as a heat and force giver, but not as a flesh former. Corn-flour should, therefore, always be given with milk. The excellence of corn-flour is to be tested by its freedom from flavor, as that which affords least flavor most readily takes flavor from other things.

CORNS.—If a portion of the cuticle or scarf skin becomes greatly thickened, and penetrates into the true skin, causing great pain and annoyance, it is called a corn. Corns are commonly spoken of as *hard* and *soft :* the hard are those situated on the more exposed surfaces of the foot, where the cuticle gets dry and hard, and the soft where the cuticle is moist—generally between the toes. With regard to treatment, the first thing to be attended to is to have the boots or shoes made accurately to fit the feet, of soft leather ; the feet should be washed often, and the inside of the socks just over the corns rubbed with slightly moistened soap. Glacial acetic acid, applied to the surface frequently, will generally be found to disperse them. Care must be taken that the acid is applied to the corn only, and not to the surrounding tissue. The parts around the corn should be covered with oil or soap. In soft and painful corns seated between the toes nitrate of silver is a useful application.

CORONER'S COURT is one of the most ancient institutions, and took its rise at the time of Alfred the Great. It was originally instituted for the purpose of inquiring into the cause of the death of those who had suddenly or unaccountably died, or had been found dead, or were known to have been slain by others. In all cases where persons have suddenly and unexpectedly died, and there has been no medical attendance, or the medical man has been called in so late that he cannot give a certificate of the cause of death, an inquest should be held. All medical certificates to be regular must be given by a "legally qualified practitioner," and only such a practitioner can be called on by the coroner to make a *post-mortem* examination of the body, although a *post-mortem* examination is most frequently made in cases of inquests on persons suddenly dying or found dead. This need not be done, should the coroner or the jury not demand it. The Coroner's Court is called into action in various ways. The police are enjoined to give notice to the coroner of any sudden death, or of persons found dead or dying, or any kind of violent death by accident or design. Any person can inform the coroner of the death of an individual, if they suspect in any manner they have improperly come to their death. The coroner has power to hold inquests in cases of natural deaths, if there is reason to believe the disease could have been prevented, or was aided and abetted by neglect or malice, or the breaking of a law. Thus landlords are liable for the death of their tenants by fever, if it can be proved the fever broke out as the result of bad drains or other bad arrangements, which might have been removed. In all cases of accident, however slight, which lead to death, the Coroner's Court must be put in action. Medical men and other persons certifying to death without mentioning the accident, may be proceeded against for misdemeanor, and the coroner may order the exhumation of the body for the purpose of holding an inquiry should he deem it necessary. In all cases where persons are in doubt about the necessity of an inquest, the coroner should be appealed to. The proceedings in the Coroner's Court are as follows : On receiving notice of a death, the coroner forwards the notice to a constable, who then proceeds to inquire into the circumstances of the death, and forwards all the information he can obtain to the coroner. If the coroner deems the case one for inquiry, he sends his precept to the constable requiring him to call a jury at a certain time and place. When the jury is assembled, their names are called over, they choose a foreman, and an oath is administered to them to the effect that they shall deliver a " true and impartial verdict." They then proceed with the coroner to view the body, which is the most necessary part of the proceedings, as the inquiry can only proceed upon view of the body (" super visum corporis"), and the dead body is a part of the evidence. At the view, the jury are expected to examine the body to see if there be any marks of violence upon it, or any other indications of how the person may have come to his death. On the reassembling of the jury after the view, evidence is taken on oath of all parties who know anything of the cause of the death under investigation. The coroner determines if a post-mortem examination be necessary, and gives a special order for that purpose. Post-mortem examinations are not necessary when persons have been seen to be killed or drowned, except in cases where it will be thought to throw light on the cause of suicide, as revealing a diseased condition of the brain. After the examination of the witnesses, the coroner sums up, and the jury deliver their verdict,

which is entered upon an inquisition and signed first by the coroner, then by the foreman, and at least eleven of the jury. The coroner then issues his warrant for the burial of the body, which up to that time is in his custody, and cannot be removed without his sanction. The coroner has the same power in his court as a judge, and can commit persons to prison for contempt of court, and can lock up a jury until they unanimously agree upon a verdict. If upon the inquest any one is found guilty of homicide, he is to be committed to prison by the coroner, and the inquisition with the evidence is turned over to the proper court.

CORPULENCE. (See BANTINGISM and DIET.)

CORROSIVE SUBLIMATE, known to chemists as perchloride of mercury, at once partakes of the nature of mercury (which see), and possesses distinctive features of its own. Apart, therefore, from its property as a mercurial, corrosive sublimate, as its name implies, is a powerful irritant or corrosive, and is consequently a dangerous poison. By virtue of its irritant character, it may be used as a wash or lotion to indolent ulcers, and here its mercurial character coming in, it is of much value in those of a syphilitic nature. It is also used as a collyrium or eyewash, and a gargle in sore throats. A weak solution (one grain or two grains to the ounce) is an excellent remedy for scabies and that condition of lousiness so troublesome in old people. This lotion is also the best remedy for crab lice. In a case of poisoning with corrosive sublimate vomiting is generally prompt, and should be encouraged, but the great remedy to be promptly got ready is *white of egg*, an antidote with which it promptly combines to form an insoluble compound. But there is no time to be lost, as death may follow in a few hours. It seems to affect the kidneys as well as the stomach, for in cases of poisoning by it the urine is scanty and often bloody.

CORYZA. (See NOSE, DISEASES OF.)

COSTIVENESS. (See CONSTIPATION.)

COTTON-WOOL is of use, not only as a means of applying remedies directly to some one spot as, for example, a carious tooth or a suppurating ear, but is also valuable for protecting exposed surfaces. In this way it is often used for burns, but as the products of suppuration accumulate below it, and are there confined, some surgeons object to its use in this way. It is of much benefit to the sufferer when plentifully applied to a joint the subject of rheumatism or gout. In such situations it ought further to be covered with oiled silk or gutta-percha tissue ; so as to form a kind of steam bath. For these purposes carded cotton rather than cotton wadding should be used.

COUCHING. (See CATARACT.)

COUGH is an exceedingly troublesome symptom of very various diseases. It may arise from irritaton of the air passages or of the lungs, of any kind, from the slightest inflammation to the most severe. It may arise from aneurism or heart disease, or it may be connected with indigestion, or merely hysterical, and due to no cause in particular. Sometimes, as in whooping-cough, it constitutes the main part of the disease. At all times it is a troublesome and sometimes a most distressing symptom. Cough is mainly due to a kind of reflex or reflected irritation. The source of this irritation may be in the lungs or out of it, though most frequently in it. This, conducted by some nerve or other to the breathing nerve centre, sets up violent expiratory efforts while at the same time the chink between the throat and windpipe is nearly closed. These efforts are renewed until all the available air is exhausted in the chest and the patient is forced to desist until he can draw a breath. If the tendency to irritation is great, the inrush of cold air is quite enough to set him coughing again, and so on to complete exhaustion. To a worn patient, as one in an advanced stage of consumption, such attacks of coughing are inexpressibly wearing and to be avoided at all hazards. True, they may be avoided by deadening the sensibility of the part by opium, but this deadens it too to the healthy stimulus of the secretions which need removal. On the whole, therefore, the best thing is a regulated temperature, as far as it can be kept even, and inhalation of steam if advisable. These do great good, but they do not and cannot prevent all attacks, especially in the morning, when the accumulated secretions of the night have to be got rid of. When the coughing is aggravated by a consistant titillation in the throat, relief may be had by dissolving a bit of hoarhound or lemon candy in the mouth, and swallowing the saliva. A pinch of salt dissolved on the tongue is also good. For a child nothing is better than the old-fashioned " Brown Mixture" (obtained at any drug-store). When a child has a cough as accompaniment of a cold, small doses (10 drops) of syrup of ipecacuanha may be given with benefit. For adults, when a cough accompanies a cold, the following is a good remedy: Mix 8 teaspoonfuls of molasses, 40 of vinegar, 2 of antimonial wine, and 4 drops of laudanum ; take two teaspoonfuls at night and one in the morning.

COUGHING BLOOD. (See HÆMOPTYSIS.)

COUNTER-IRRITATION.—A system of remedies intended to relieve internal inflammations by the pain and action excited in the skin immediately above the part affected. Mustard plaisters, blisters, and embrocations are examples of counter-irritants.

COUP-DE-SOLEIL (sun-stroke). (See HEAT STROKE.)

COWHAGE, or *Cowitch,* consists of minute hairs from the pod or fruit of a plant growing in the West Indies. The sharp prickles getting inserted into any part of the body give rise to intolerable itching. It used to be given to get rid of worms, on the principle, it has been suggested, of tickling them to death. It is not now officinal.

COW-POX, or *Vaccinia,* is a disease which is met with on the teats and udder of the cow ; it forms an eruption made up of numerous little blisters with watery contents, and this fluid when introduced into the system of man will produce a similar affection, and such persons are not liable to suffer from small-pox. It was an important observation made by Jenner, nearly a century ago, that those who had cow-pox from milking cows affected with this disease were free from the danger of small-pox, and this great discovery led him to adopt vaccination as a means for preventing man suffering from that dangerous disease. (See VACCINATION.)

CRABS, the common name for the *Pediculus pubis,* so called from its resemblance to the crustacean, is a loathsome insect, which gets into the roots of the hairs, about the pubes, and is characterized by an intolerable itching ; it is readily got rid of by rubbing in any mercurial ointment, the best being the white-precipitate (ammonio-chloride), and taking a hot bath soon after. Prompt measures should be taken to get rid of these creatures, and in fact all parasites, as they are exceedingly prolific.

CRAMP means violent and involuntary muscular contraction. It is, perhaps, most readily induced by cold, especially after prolonged exercise. This, perhaps, is the reason why it so often occurs in swimmers, and is supposed to account for a good many of the lives lost in water. The best remedy is rubbing, especially with some stimulant application, as spirit, but it is the rubbing which is the most valuable part of the process. (See CONVULSIONS.)

CRAZINESS. (See INSANITY.)

CREAM is the name given to the butter of milk, when cow's or other milk is allowed to stand, so that the butter floats. It consists principally of butter, and when placed under the microscope is found to consist of little globules, which, from their peculiar action on light, give the white appearance to milk. Cream is more digestible than butter, and where it can be procured fresh is a very desirable article of diet where fatty food is required for health. It may be taken with advantage in cases where cod-liver oil is needed but cannot be taken. (See MILK.)

CREAM OF TARTAR, or *Bitartrate of Potassa,* is deposited in an impure condition (Argol) in wine casks when the wine has been allowed to stand for a time. The sediment is purified by washing, etc., and as cream of tartar is employed in medicine in various ways. In small doses it is cooling and tends to increase the flow of urine ; in larger doses it is a purgative, producing copious watery stools. For the latter purpose it is commonly combined with jalap (as compound powder of jalap) or scammony. Such a combination is largely used in certain forms of dropsy, especially such as depend on acute inflammation of the kidney, as after scarlet fever. As a refrigerant and diuretic, cream of tartar is best given as a habitual drink. An ounce of the substance may be added to a quart of boiling water, sugar added to taste, and a few slices of lemon allowed to float in the mixture ; a wineglass full or more to be taken now and again. The use of such a simple remedy has sometimes a wonderful effect in cures of dropsy, the more especially as it may be used as an auxiliary to more powerful means.

CREASOTE is one of the numerous substances produced in the destructive distillation of wood for the purpose of obtaining acetic acid. It is or ought to be a colorless, transparent liquid with a peculiar odor and burning taste. It was at one time extensively employed in medicine, but since the introduction of carbolic acid the purer substance has greatly superseded the use of the less pure creasote of whose composition we know little. It might be expected that the virtues of creasote would resemble those of carbolic acid in preserving substances from putrefaction, and, indeed, it is largely used for rendering wood less liable to decay. In medicine too its properties are analogous, thus a drop or two of creasote have been found most useful for arresting vomiting connected with fermentative changes in the food. Again, it has been found of use in arresting certain changes within the system, as the excessive formation of sugar, characteristic of saccharine diabetes. Diarrhœa depending on fermentative or putrefactive changes in half-digested food may be arrested in like fashion. In hæmorrhage from the stomach creasote is often of use, provided, of course, it does not arise from liver disease. Its vapor, mixed with that of hot water, has been highly commended in chronic bronchitis and phthisis, with excessive or fœtid expectoration. Carbolic acid may be used in the same way. As an application to wounds and sores, creasote, like carbolic acid when properly diluted, is very valuable. As before said, it is a powerful remedy, one or two drops is enough for a dose. For a lotion or gargle half a drachm may be added to a pint of water, with which, however, it does not mix readily and does not at all dissolve. Acetic acid aids the combination. A drop of creasote on a bit of cotton placed in the hollow of a decayed

and aching tooth almost always affords relief.

CREATINE is a substance composed of carbon, hydrogen, nitrogen, and oxygen, and is found in the juice of the flesh of all animals. A pound of flesh yields upon an average about five grains. The quantity varies in different animals. The flesh of fowls yields the largest quantity. The flesh of fish contains it in larger quantities than beef or mutton. Creatine is obtained in colorless transparent crystals, and, dissolved in water, it has a slightly bitter taste. It unites with the various acids forming salts. When creatine is boiled with alkalies, a new alkaloid is produced called sarkosine. This substance is also found in the flesh of animals. If creatine is boiled with hydrochloric acid it produces *creatinine*. This substance also forms salts with the various acids, and is found normally in flesh. These alkaloids are probably the result of the decomposition of the flesh of animals. They are found with the extract of meat, but whether they exert any power on the system is not known. It is, however, an interesting fact to know that they belong to the same class of substances as quinine and theine, both of which exert a great influence on the nervous system and are used in medicine and food. (See TEA, AND MEAT, EXTRACT OF.)

CRIMINAL ABORTION. (See ABORTION.)

CROCUS. (See SAFFRON.)

CROTON OIL is the oil expressed from the seed of a plant growing in the East Indies. It is of a lighter or deeper yellow in tint, varying with exposure. Its odor is unpleasant, and its taste exceedingly acrid. This oil is extremely irritant, and as such is employed both externally and internally. Thus a drop is frequently added to five grains of compound colocynth pill as a remedy in obstinate constipation, or again in seizures of an apoplectic nature ; a drop or two is let fall on sugar, and deposited in the back of the throat to be swallowed. In either case a copious action of the bowels ordinarily follows. Another very good plan is to add a drop of croton oil to a tablespoonful of castor oil, also with a view to increasing its activity where the bowels are obstinately confined. It is, however, too powerful to be used as an ordinary purgative. Externally, croton oil rapidly brings out a crop of small pustules, and acts as a counter-irritant, but is rather painful. It is therefore better to add a small quantity of the oil to some liniment, and rub in the mixture. Thus a drachm may be added to an ounce of soap liniment or turpentine liniment. Such a combination is of great use in certain stages of consumption, when there is pain in the walls of the chest, or again in certain forms of neuralgia or muscular rheumatism. It should be used night and morning till a crop of pustules make their appearance, and after that, at more distant intervals for a somewhat longer period.

CROUP is an inflammatory disease of the larynx or upper part of the wind-pipe, and is very common in children between two and five years of age. It is attended by very noisy breathing, and this, which is a marked symptom, is due to spasm of the glottis, which by being thus narrowed, prevents the free entrance of air into the lungs. The child feels as if it were going to be choked, and it makes violent efforts with the muscles of the chest, so as to increase the supply of air within. Some children are very liable to croup, and have it several times over ; in such it is generally a spasmodic affection, and due to some nervous irritation. The symptoms, although alarming at first, will often pass off in a few hours. It is seen also in children who have had measles followed by some bronchial or laryngeal affection. Croup must be distinguished from diphtheria, in the course of which disease a similar noise is sometimes heard during inspiration, but that disease is catching, is far more fatal, and is attended with the formation of a false membrane in the air-passages, and also in the throat and nose ; such cases, however, are too often called croup.

Treatment.—Croup is a disease in which no delay should take place in treatment, as imminent danger may ensue from suffocation. Hot sponges wrung out of hot water should be at once applied to the throat, and the patient should be made to inhale steam by putting before him a jug of boiling water. A small mustard poultice may be applied to the front of the throat, and when the urgent symptoms have subsided, a large hot linseed-meal poultice should be wrapped round the throat, and renewed, until the breathing is all right again. Vomiting is a source of great relief in many cases, and for this purpose syrup of ipecacuanha may be given in doses of a teaspoonful repeated every 15 minutes until vomiting occurs. Medical advice should be obtained as early as possible, as domestic treatment must not be relied upon in this disease. When a child is attacked more than once with this complaint, less fear need be entertained, as it is probably spasmodic, and will go off again if treated early. After an attack care should be taken not to expose the child to draughts ; flannel should be worn next the skin, and a comforter wrapped around the throat ; regular diet should be given, and the bowels should be kept open once a day, and any cause likely to set up nervous irritation should be removed. For *Spasmodic Croup*, see LARYNGISMUS STRIDULUS.

CUBEBS are a kind of pepper cultivated in Java. Unlike most other kinds of pepper the berries retain their tails. They have a taste something like pepper, and a disagreea-

ble odor. They are almost exclusively employed for arresting discharges from the urinary passages, however produced ; but on account of the unpleasant and unmistakable odor they give the breath, are not nowadays very much employed. Cubebs have also been used for chronic inflammation of the bladder, and for the relief of internal piles. The dose varies from 15 grains to 2 drachms. An oil is obtained from the fruit, which has similar properties. Its dose is about 10 drops.

CUMIN is the fruit of an umbelliferous plant, having properties like Carraway.

CUPPING is a method of local blood-letting, practised for the relief of inflammation and congestion in internal organs. The instruments used are bell-shaped glasses, varying in size, and a *scarificator*, which is a brass case containing ten or a dozen lancets, the edges of which can be made to start out by touching a spring. The operation is performed in the following manner : After the skin over the affected part has been well washed with a sponge dipped in hot water, it is covered by one or more cupping glasses, the air within which has just been rarefied by the flame of a small spirit lamp, or of pieces of blotting paper steeped in spirits of wine, and then ignited. Up to this point, the proceeding is called *dry cupping ;* but if it be desired to draw blood, each glass is removed, the scarificator applied and discharged, and the glass again heated and placed over the small lined wounds formed by the lancets. As the rarified air within the cupping glass cools and becomes condensed, the skin rises up as a dome-shaped swelling, and blood is sucked out from the numerous lancet-wounds. In this way, from 15 or 16 ounces of blood may be drawn from the surface of the body. As the scars formed after scarification are permanent, and very distinct in consequence of their number and arrangement, it is necessary to avoid cupping exposed parts of the body.

CURCUMA. (See TURMERIC.)

CURRY POWDER is a compound of condiments and spices introduced from the East Indies, and is employed to give flavor to stewed meats, which are usually mixed with rice. Genuine curry consists of turmeric, cardamoms, ginger, allspice, cloves, black pepper, coriander, cayenne, fenugreek, and cumin. Of course the flavor of the curry will vary according to the preponderance of one or more of these ingredients. Although the constituents of curry powder are unobjectionable, it may be taken in such quantities, or the more stimulant substances may be introduced in such quantities, as to become injurious to the stomach and render it powerless to digest without the curry. Such curry also predisposes to the drinking of alcoholic beverages, and is often taken to produce such a thirst. Curry powder, when used in small quantities, is an agreeable aromatic, and certainly acts beneficially in hot climates by recalling to the stomach the circulation, otherwise exclusively excited by the action of the sun on the skin. Curry powder is much preferable to alcohol as a stimulant of the stomach in hot countries, as it does not affect generally the nervous system, nor act destructively on the secreting surfaces of the stomach or bowels.

CUSSO, or *Kousso.*, are the flowers of a plant growing in Abyssinia, where they are largely used as a remedy for the tapeworm so prevalent there. The remedy has not been very much used in this country, partly on account of its price, partly on account of its adulteration. Half an ounce of the flowers is to be infused in half a pint of water, and swallowed, flowers and all. Half an hour after, a dose of castor oil (half an ounce) should be taken.

CUT-THROAT is generally the result of an attempt at suicide or murder with a knife or razor. In such cases, supposing that the escape of blood has not caused immediate death, the first object is to prevent further bleeding. The wound must be cleansed with a sponge and warm water, and the bleeding vessels secured by ligatures. Next the edges of the wound or wounds should be brought together by stitches, care being taken to keep the patient's head forward by means of suitable bandages, so that the cut edges may be approximated. If the air-tube (trachea) be divided, care must be taken that matter from the gullet does not interfere with respiration, and supposing the gullet itself be wounded, sufficient nourishment must be allowed to pass downward, and possibly a small tube, leading from the mouth, or from the wound to that part of the gullet below the injury, may be required for a time.

CUTS. (See ACCIDENTS.)

CYANIDES are salts of metals and the compound radical cyanogen. Prussic acid is a compound of hydrogen and cyanogen. (See PRUSSIC ACID.) The cyanides of the metals, when placed in contact with organic substances containing hydrogen and oxygen, are accompanied by prussic acid and an oxide of the metals are formed. Cyanide of potassium acts in this way and is often used as a poison. It is employed by photographers for the purpose of washing their plates, and is now a common article of commerce. It is in this way it has come to be used as a poison.

CYANOGEN. (See CYANIDES.)

CYANOSIS is a term applied to the blueness or lividity of the skin which is so often observable in children who are born with malformation of the heart. The blueness is most marked in the hands and feet, in the ears, tip of the nose, and, in fact, in those parts where the circulation is slowest and

most languid. It is due to the veins and small vessels being too full of blood as a consequence of the obstruction to the circulation through the heart. The malformations are of various kinds, and will hereafter be considered. (See HEART.) Such children rarely live many weeks, if the obstruction be very great; in other cases, they may go on for some months, or even years, but they are very liable to die, if attacked by any acute disease. Cyanosis is seldom noticed until a month or two after birth, and often it is associated with convulsions; at the same time the child suffers from shortness of breath, which is worse on coughing or making any exertion. This disease also hinders the development of the child, and hence the tissues become badly nourished. There is generally some œdema or swelling of the extremities, because the serum of the blood oozes through the coats of the distended vessels into the loose tissue under the skin.

Treatment.—Very little can be done in the way of treatment, as cyanosis depends upon a condition of things which is incurable; yet life may be prolonged, and certainly distress may be alleviated, by taking care that the child is not exposed to whooping-cough or to cold, so as to catch bronchitis or pneumonia, or to the contagion of measles and scarlet fever. The child should be fed in the usual way, and may be taken out in the open air on fine and warm days. The extremities should be kept nice and warm by thick woollen gloves and socks, and friction with the hand may be daily used on those parts in order to encourage the flow of blood through them. The object of the treatment is to prevent the child taking any disease which might yet farther embarrass the circulation, while the flow of blood through the tissues is kept up by increasing the ordinary heat of the body by warmer clothing. Persons who have suffered from emphysema and bronchitis, who are short of breath, and have had a cough every winter for many years, become blue about the lips and ears, and often have swelling of the legs. Here again, these people are suffering also from an obstructed circulation, and they also are really cyanotic; but usually the term cyanosis is confined to children who are laboring under malformation of the heart.

CYSTITIS is the technical term for inflammation of the bladder. The symptoms are great pain in the region behind the scrotum or purse, in the groins, and lower part of the back, tenderness over the bladder. Very frequent desire to pass water, attended with great efforts to do so, and a whitish ropy mucus deposited in the urine, accompanied with feverish symptoms. *Treatment:* Hot baths, and hot fomentations. the administration of calomel, and castor oil, to relieve the abdominal circulation, the pain to be allayed by opium or morphia, either internally, or as a suppository, and copious draughts of bicarbonate of potassa and lemon juice. Infusion of buchu has also been used with benefit.

CYSTS are tumors consisting of a limiting bag, or sac, which contains either solid, semisolid, or fluid matters. (See GANGLION and HYDATIDS.)

D.

DALBY'S CARMINATIVE is a popular carminative. It contains carbonate of magnesia, tincture of assafœtida, tincture of opium (laudanum), and the oils of anise, peppermint and other volatile oils. (See CARMINATIVES.)

DALTONISM is a condition of the eye in which the individual is not able to distinguish one color from another. Dr. Dalton, the celebrated chemist, labored under this defect, hence the name. (See COLOR-BLINDNESS.)

DANDELION is the root of the common dandelion of our fields gathered during the winter months. It yields when cut a bitter milky juice, to which some rectified spirits may be added to make it keep (one pint to three of juice), and the whole used as a medicine. The value of this remedy, which is commonly called taraxacum, is exceedingly doubtful; some attribute to it great virtues; others deny them altogether. Certain it is that in the hands of most regular practitioners it does not now obtain such a reputation as formerly. It is ordinarily given in cases where the liver is supposed to be out of order, but generally with other remedies of a more powerful character. To give it a fair trial, the juice above referred to should be given in teaspoonful doses three or four times a day.

DANDRUFF, a disease of the scalp, attended with the production of scales on the skin. (See PITYRIASIS.)

DATURA. (See STRAMONIUM.)

DEADLY NIGHTSHADE. (See BELLADONNA.)

DEAFNESS is the one common symptom of a great majority of diseases of the ears meaning thereby not absolute, or as it is sometimes called "stone" deafness, but merely defect in the power of hearing. Thus it may arise from obstruction of the outer ear, from perforation of the membranes of the tympanum or drum of the ear, from inflammation of the middle and inner ear, from paralysis of the nerve of hearing, or from obstruction of the Eustachian tube from whatever cause. A word of explanation is required as to the structure of the ear to understand this aright. The ear consists of three portions besides those which we see externally. The opening we see which leads into the skull is the outer ear (or Auricle), and it reaches to a certain depth. There it is terminated in a membrane which completely blocks up the passage, and which is called the membrane of the tympanum or drum of the ear. Beyond this comes the middle ear, or tympanum, which contains air, admitted to this portion of the ear by a tube which reaches down and opens into the back of the throat. This tube is called the Eustachian tube. Between the drum of the ear and the wall of the innermost cavities of the ear are four little bones, which are jointed and movable, so that being attached to the drum of the ear and to a corresponding membrane between the middle and inner ear, any movement of the one is immediately communicated by them to the other. In the inner ear, or labyrinth, which is filled with fluid, are expanded fine membranes in which the nerve of hearing terminates, so that any movement in the fluid is at once communicated to the nerve and from the nerve to the brain. Vibrations in the atmosphere being the cause of most sounds appreciated by us, these vibrations or waves act upon the drum of the ear, then through the small bones on the membrane between the middle and inner ear; its vibrations set the fluid in motion, whereby the nerve is affected and the sound appreciated. Anything which interferes with this process will cause deafness. Hence we may divide the causes into those affecting the outer, middle, or inner ear. Children often put foreign bodies, as peas, glass beads, slate pencil, etc., into their ears. These may, by obstructing the passages, interfere with hearing. They should not be rashly interfered with if they cannot be removed by syringing with water.

One of the most common causes of deafness is the accumulation of the substance we call wax in the ear. This sometimes becomes very hard and gives rise to noises in the head, deafness, etc. It can generally be removed by syringing with soapsuds. If not, put in a drop or two of glycerine and a bit of cotton wool for a day or two till it soften, and then try again. The injected fluid should be warm, and should not be sent in violently but gently. Sometimes after fevers there remains behind a discharge from the ears in children which is apt to give rise to deafness. The ear may be inflamed and give rise to much pus, and the tympanum may be perforated. Sometimes the cuticle becomes thickened and the epidermis accumulates in masses very hard to remove, unless previously softened by glycerine or warm oil and liquor potassæ. In ear cases syringing is the only method applicable to removing obstructions which can be recommended to the public. Another affection of the outer ear is a fleshy growth called a polypus. These always require surgical treatment, and often special skill, so as to remove them without injury to the parts concerned. It has already been pointed out that the drum of the ear or membrane of the tympanum may inflame, if so, in all probability, the middle ear will become affected in the same way, as the membrane which lines the middle ear lines also the drum on its inner side. Sudden and intense pain is characteristic of this lesion, which in all probability goes on to destruction of the drum of the ear and complete deafness. There may also be a chronic inflammation of the inner cavity leading surely to permanent deafness. Leeches to the back of the ear generally do good in these affections, but other remedies are generally required. In the acute affection it may save the drum to perforate it and allow the purulent fluid in the middle ear to escape. Rupture of this membrane, may be accidental, from blows on the ear, loud noise, etc. Such rupture is shown by the fact, that if the patient hold his nose and breathe out with all his might the air will escape by the Eustachian tube and outer ear. Such an injury can in great measure be repaired by introducing a little cotton wool into the ear, quite down to the perforated drum. This is quite as good as much more expensive and complicated arrangements. Should the discharge persist, the organ must be washed out by a very dilute solution of carbolic acid.

Another form of deafness is produced by the growing of the little bone which terminates the series in the middle ear, and which is called the *stapes*, to its attachments by bony union. This sadly interferes with its mobility, and consequently with its powers of inducing movements in the fluid of the inner ear. The Eustachian tube, already spoken of is not unfrequently blocked up and causes deafness. This happens when we have a cold. Sometimes this exceeds the amount of deafness ordinarily produced, for in inflammation of the back of the throat the disease may spread upward and affect the middle ear. These maladies are to be dealt with by healing the throat in the first instance, when probably the ear will get well. There are, however, a certain number of cases which require other treatment, but that must be of a skilled kind. Certain varieties of deafness, the origin of which is not well known, are called nervous. These so-called nervous deafnesses are, however, most probably, due to some cause which we cannot make out. Thus the whole group of chronic inflammations of the inner ear, now well known and duly appreciated, if not always readily cured, were at one time supposed to be due to nervous causes. In old persons, in whom nervous deafness is supposed to be most common,

there is usually thickening adhesion or other changes in the bony structure to account for the dull hearing. Finally, we may have complete deafness from destruction of the nerve. That sometimes follows blows on the head, or fracture rupturing the nerve ; or disease of the brain, affecting the part where the nerve comes or whence it arises. Little or nothing is, however, known of diseases affecting the innermost ear of all. We have purposely said little as to the treatment of maladies giving rise to deafness, for if a little syringing won't do, measures must be taken which imply great skill, and the risk is great, for inflammation may readily spread from the ear to the brain, and life be forfeited.

DEATH-RATTLE. (See RATTLE.)

DEATH, SUDDEN. (See SUDDEN DEATH.)

DEBILITY, or *Weakness*, generally is one of the innumerable signs of imperfect nutrition or actual disease. Frequently the onset of a serious disease is marked by unusual debility, while recovery from it is always accompanied by the same. Thus in fever an individual naturally strong and robust begins to feel out of sorts, has headache and what he fancies is indigestion. Often he thinks to shake off these uneasy feelings by a good long walk ; he sets out in good spirits, but soon finds to his astonishment that what used to be the easiest of tasks has become an insurmountable difficulty. He returns home and takes to bed, very probably not to rise till he has passed through a dangerous encounter with death. When he begins to recover and the fever leaves him, his most marked symptom is debility ; but whereas formerly this was the prelude to disease and was irremediable till the disease was over and gone, his debility may now be combated with great success. Nourishing food and good wine will do wonders, but the food must be very nourishing and very easily digested—perhaps the best to begin with is essence (not extract) of meat. Of wine, perhaps the best to begin with is old Madeira, but this is scarce. A dry sherry like Manzanella will do. Spirits are not good, as the stay in good wine is far preferable ; but rather give good sound spirit and water than bad wine. These two forms of debility are typical of many others, and have therefore been chosen. The debility of childhood and old age differ from these in this—they require considerable warmth as well as appropriate food, and that is all-essential.

DECLINE. (See CONSUMPTION.)

DECOCTIONS are preparations of remedies which have been prepared by boiling the substance in water for a longer or shorter period. The length of time required for preparing a decoction should partly at least depend on the solubility of the substance to be extracted from the drug. This and the quantity of the drug used are generally so adapted the one to the other as to make the dose of a decoction about two tablespoonfuls. Of course in domestic as in ordinary practice, no drug should be converted into a decoction whose properties depend on a volatile substance, as this of course would be expelled by boiling. Most decoctions should be strained while hot, otherwise on cooling they deposit a sediment. In the pharmacopœia there are but two compound decoctions, that is, decoctions which contain more than one ingredient. These are, the compound decoction of aloes and the compound decoction of sarsaparilla.

DEFECATION is a term applied in pharmacy to the removal of the lees or sediment of any liquid. It is also used to express the act of discharging the fæces from the bowels.

DEFERVESCENCE, a name used to signify the fall in the temperature which occurs when convalescing from acute disease.

DEFORMITIES.—For a description of the various conditions which may be included under this term the reader is referred to articles under the following headings : ANCHY-LOSIS, CLUB-FOOT, FLAT-FOOT, HARE-LIP, KNOCK-KNEE, MALFORMATION, SPINA-BIFI-DA, RICKETS, WRY-NECK.

DEGENERATION.—During the development of the fœtus, active changes go on, and the simple elements which are found in the very early stages of the embryo are developed into the more complex structures which are met with in childhood and adult life. But as time goes on, these various structures undergo different changes and in process of time waste and decay, so that in old age the various organisms of the body are much less vital, and naturally decay ; old age is, in fact, a period when vital changes take place very slowly, and the tissues of the human frame are more or less in a state of degeneration. But it frequently happens that degeneration takes place much earlier in life, either from hereditary causes, from bad living, or from disease acquired in adult life, as syphilis, excessive drinking, etc. Thus an individual may be subject to (1) fibrous degeneration ; (2) fatty degeneration ; (3) waxy degeneration ; (4) melanotic degeneration.

(1) *Fibrous* changes take place naturally in old age, but they occur in those who are the subjects of syphilis, and in those who drink much : such people are in a state of premature decay ; they often suffer from gout, are sallow and shrunken in appearance and of weak constitution. This form of degeneration is met with in various parts of the body. The liver becomes firmer than usual, is often of smaller size than natural, and passes into the condition known as cirrhosis (see CIRRHOSIS). The kidneys waste and become uneven and granular on the surface ; the urine is light in color, and there may be found albumen in testing for it ; such a state is known

as chronic inflammation of the kidney (see BRIGHT'S DISEASE). The heart is generally larger than usual from having more work to do, and there is often disease of the outer or inner coat ; sometimes the valves are diseased and death ensues from the serious mischief engendered. The brain does not escape ; in some the membranes become opaque and thickened and the organ itself shrinks ; there is loss of memory, giddiness, pains in the head, sleeplessness, and flashes of light before the eyes. The lungs become adherent to the chest wall and the respiratory power is diminished. These changes do not take place equally in all organs, and the amount of disease varies much in different individuals ; in some one organ is more particularly affected, in others, general disease is present. The arteries, too, which convey the blood to the different parts of the body share in the degeneration ; their inner coats become roughened and brittle, and yellowish opaque patches are seen ; they are then said to have become *atheromatous*. Often, too, there is a deposit of lime salts in the coats of the vessels, and then the walls become brittle and may rupture, or the wall may dilate at one point and form an aneurism or dilated sac on one side of the artery. Such vessels are said to have undergone *calcification* or *calcareous degeneration*. Whenever the vessels which supply blood to a part become in this way the seat of serious disease, the tissues nourished by them undergo degeneration, in consequence of not being properly nourished ; in this way *cerebral hæmorrhage* may occur from the rupture of a diseased artery in the brain, or *white softening* of that organ may come on from the nerve-tissue being badly nourished ; or, again, a thrombus or clot may form in one of the vessels from fibrine being deposited on the roughened internal surface, and then also softening may occur. All these changes generally come on in advanced life and set up the condition known as apoplexy ; if recovery take place, the patient may still be incapacitated from work by the consequent paralysis of one side or other, and he is then said to be suffering from hemiplegia.

(2) *Fatty Degeneration* is of very common occurrence, and is often found in parts which having done their duty in the economy are no longer wanted ; thus it occurs in a normal way in the arteries or womb after a confinement, for in this manner the muscular elements which were developed in that organ while the fœtus or embryo was present become broken down and absorbed by the blood and lymphatics. Fatty change also is met with in the liver-cells after a meal, and to a less extent in the kidney and even in the blood, but these conditions occur in a healthy state. Fatty degeneration is generally found in scrofulous or consumptive people, in those who drink much, and in those who are the

subjects of any urinary disorders. The liver is very often the organ chiefly affected ; it becomes larger than usual, of a pale fawn color and greasy to the feel. The changes go on very slowly, and at first are not noticed by the patient ; there is no pain, no jaundice, and, in fact, very little inconvenience ; but after a time there is dyspepsia, loss of strength, pallor of skin, and dropsy of the abdomen or skin may occur ; often, too, the kidneys share in the mischief and become much larger than usual and the tubes of those organs are full of oil ; less water is passed than usual, and the urine, beside being scanty, is dark in color, of high specific gravity, and deposits albumen on boiling and a sediment on standing. This change aids in causing the onset of dropsy and hastens the progress toward death. Just as the fibrous degeneration brings on a cirrhosed liver (see CIRRHOSIS) and the contracted kidney of Bright's disease (see BRIGHT'S DISEASE) so the fatty form of degeneration brings on a fatty liver (see LIVER) and the fatty form of Bright's disease. But it seldom happens that this change attacks one organ only, and in general the heart, muscular system and vessels suffer, as well as the liver and kidney, and sometimes it is associated with the fibrous degeneration. The heart becomes flabby, and weaker than usual, and often dilates in consequence of being unable to bear up against the pressure of the strain ; such people are in general of middle or advanced life, and are subject to fainting fits of an alarming nature, to palpitation on exertion and general distress of breathing. This, too, is a common cause of sudden death, for the left ventricle may become over-distended and then is unable to contract, so that the patient dies of syncope. The muscles of the body in these cases are flabby and badly nourished, so that there is loss of strength and inability for exertion. Yet the patient often looks stout, because there is plenty of fat under the skin, and so wasting does not go on in that tissue. The minute vessels in the different organs undergo, likewise, a fatty change, and may produce symptoms and diseases like those mentioned under the head of ATHEROMA, above.

(3) *Waxy degeneration* is much less common than the above two varieties. It is met with in rickety or scrofulous children, in those who suffer from disease in the joints, as hip-disease, etc., and in those who are the subjects of syphilis, inherited or acquired, or in those who have had ague, cancer, or some chronic wasting disorder. This form generally attacks the liver, spleen, kidneys, and intestinal canal. The liver is much larger than usual, and is translucent in appearance ; it may become so large as to fill up a great part of the abdomen, and cause a feeling of weight in the right side ; pain is seldom present, and never severe ; jaundice does not

occur, and there is seldom any dropsy. There is a feeling of fatigue and general debility ; the appetite is impaired, and the patient may suffer from dyspepsia ; the skin is often of a deadly pallor, but there is not much loss of flesh ; when the blood is examined under the microscope or chemically, it is found to be deficient in red corpuscles, and more watery than usual. When the kidneys are affected, the patient passes a very large quantity of water, free from deposit or nearly so, very pale in color, and containing albumen ; often at first there may be some blood present, but this does not last long ; it is the daily drain of albumen from the system in these cases which is so exhausting to the patient. The result of the intestinal canal being affected is shown by a troublesome diarrhœa sometimes accompanied by bleeding from the bowels, and this may easily be excited by any error of diet. The spleen becomes large, but does not cause pain, nor, in fact, any symptoms easily noticeable by the patient ; but it aids in causing those changes in the blood which are met with in this disease. The reader must refer to the articles on the LIVER, KIDNEY, and SPLEEN, for further information ; the waxy change in the kidney forms one of the varieties of Bright's disease. The term waxy is used, not to signify the actual change that takes place, but the common appearance presented by various organs when so affected ; some have adopted the term of *amyloid* or *lardaceous* degeneration for the same affection. These three forms of degeneration are all chronic in their nature, and may last for years before they cause death ; but in the mean time, they have caused serious changes in various tissues of the body, and made the individual very liable to acute disease. Often such patients seem in very good health, and men wonder why in the midst of such apparent vitality, death may so suddenly in some cases take place ; but it is not really the acute change or disease which kills, but the slowly-proceeding course of events, which, beginning years previously, finally carry off the patient in an unexpected manner. These changes are often the result of follies in early life, and proceed from an excess in eating or drinking, or in the too eager pursuit of pleasure ; and while for a long time the system does not seem to be affected by such a course of living, yet in the end disease is sure to supervene.

Treatment.—The physician cannot give back health, he cannot restore those tissues to a normal state ; but he may do much to arrest its course, and to place the patient under more favorable conditions for preserving life. The diet should be carefully regulated and no excess of any kind be allowed. Each meal should be light and nourishing, and easily digested, consisting of meat once, and if possible, twice a day, light puddings, fruit, toast or bread, milk, cocoa or tea and coffee ; salt meat is not so good, nor are vegetables of much service. A rich dish should be avoided, and the cooking should be made as simple as possible. A pint of beer a day, or two or three glasses of sherry or claret, or some light wine, may be taken ; but no excess should be taken. Open air exercise is very valuable, and a daily walk or ride should be ordered. Early rising, and a cold bath every morning, is of much service, if the patient be well enough to take one, while late hours must be avoided, and also hot overcrowded rooms. Medicines may be taken if there is much debility, and for this purpose iron and quinine are the most valuable, or a mixture containing hydrochloric or nitric acid, with some bitter infusion, may be taken before a meal for the purpose of inducing an appetite.

A much less important form of degeneration, and one that calls for no treatment, is the deposit of pigment in various parts of the body, and it is known as *melanosis* or *melanotic degeneration*. (4.) *Melanosis* is the name given to a condition in which coloring matter, or pigment is found in various parts of the body. In infant life very little pigment is met with, so that the lungs are of a pink color, and the iris is blue in tint ; but as years roll on, the iris becomes colored of various tints, and the lungs become of an iron-gray color, or even black. This is quite a natural process, and is a result of the various tissue changes which occur in advanced life, so that in old age there is much more pigment than there is in a child. Nevertheless coloring matter may be deposited as a result of disease. Thus in those who work much in dust, as coal miners, knife grinders, etc., much pigment is deposited in the lungs. In cancer of different organs, and more especially in the liver, pigment is found in large quantities and gives a black appearance to the disease. Very little is known as to the cause of this change, nor are any symptoms caused by it which call for any special treatment. Sometimes the brain and spinal cord are the subject of this peculiar change ; it may depend in some cases on an alteration in the coloring matter of the blood, in others it seems to be influenced by the inhalation of injurious materials by the air-passages as in knife-grinders' disease. Under the term of degeneration some include *atrophy*, but this has already been considered (see ATROPHY). By this term should be meant simple wasting, without any disease being present in the tissue itself, just as when the muscles of a blacksmith or athlete may diminish by disease, or when a leg or an arm wastes from paralysis. Atrophy occurs in almost all forms of degeneration ; but this is itself part of the disease, and one condition cannot be dissociated from the other.

DEGLUTITION is the act of swallowing food after mastication, and is performed by the joint action of the muscles of the cheek and tongue assisted by the throat : so that when the food has been softened in the mouth and incorporated with the saliva, it is carried down the gullet by the act of deglutition.

DELIQUESCENCE is the condition in which certain substances become damp and absorb moisture from the atmosphere. Many preparations of potash are liable to this condition, unless well protected from the air.

DELIRIUM means that accompaniment of acute disease wherein the mind wanders, and incoherent talk is the result. It is common in many acute disorders, especially fevers, and is more common in the young than the old. We speak of two kinds of delirium, active and passive. The active being something merely indicative of mental derangement, without any tendency to action, but in the fierce delirium of some diseases there is violent exertion to get out of bed, shouts, and rage depicted in the countenance. This is sometimes the case in certain inflammations of the brain. The other form of delirium, in which the mind seems to be wandering is generally seen in exhaustive fevers, such as typhus. These patients will often, however, if desired, answer questions correctly. The delirium seems due to the circulation of poisoned blood or of imperfectly aerated blood in the brain. If, therefore, the circulation be relieved in any way it will probably pass off at least for a time. Delirium is most frequent in the night and is of very various omen. In acute disease it may merely indicate a sharp attack of the malady ; but in other instances, as in inflammation of the lungs, the onset of delirium is a very bad sign, showing, as it does, that the lungs are no longer capable of purifying the blood sent to them.

DELIRIUM TREMENS, or *Delirium a Potu*, is one of the consequences of chronic alcoholism. There may be said to be two varieties of the malady : one of spontaneous origin after prolonged drinking ; the other coming on as the result of an accident to those who are habitual drinkers without being drunkards. The malady consists of hallucinations, and trembling of different parts of the body, which last may be taken as a sign of weakness for the patient has in all probability eaten little or nothing for some time past, having lived mainly on intoxicating drinks. The chief symptoms are sleeplessness and restlessness, with delirium generally busy, but not very often violent, except driven to violence through mortal terror, for the patient is a terrible coward. The objects seen are very often loathsome creeping things, rats and serpents, and the like, in the existence of which the patient fully believes. When hearing is affected, and that is generally the case, he hears people calling him names, taking away his character, and so on. He will not rest in bed, and is constantly getting up, but will lie down again quietly if told to do so. As to his physical condition, the face is usually pale and wild-looking, the skin moist or clammy, the tongue coated and tremulous, and the pulse quick and soft. There is complete loss of appetite, and the bowels are generally confined. In delirium tremens there is no great risk of the individual hurting himself or others, except driven to it by mortal terror, as is sometimes the case, and then he will stick at nothing. During the night he is worse than during the day. This does not last long ; if spirits are abstained from the condition ordinarily ends in eight-and-forty or two-and-seventy hours, in profound sleep, from which the patient wakes weak, but in his right mind, and ordinarily very repentant. Occasionally the condition terminates fatally ; if so the temperature rises and he gets no sleep ; feebleness increases till the delirium is muttering merely. Death commonly comes in weak convulsions. Broadly, it may be said such cases should not occur. The cause is almost invariably excessive spirit drinking, though other intoxicating liquors may have a similar effect. As a rule, such attacks come on when the stimulus is withdrawn, and nothing given to take its place. The other kind of delirium tremens is not unfrequently seen in brewers' draymen, who might be looked on as splendid specimens of men, yet if they meet with any accident they are very liable to a kind of delirium which adds considerably to the risk from their injury.

Treatment.—The treatment of delirium tremens is a simple thing, and yet it requires the greatest possible judgment. The strength of the patient is the great thing to look to : if that is good, all should go well. If the attack is an acute attack, that is to say, coming on after a bouse of six or eight days, and if the patient has eaten nothing, or little or nothing during all that time, as is very often the case, the first thing is to get him to take food. Here, too, perhaps, it is not advisable to cut off stimulants entirely, but to substitute others for those to which the individual has been accustomed. The aromatic spirit of ammonia and spirit of chloroform will be found best. With these precautions, giving too a little brandy, if necessary, the patient will do well enough for eight-and-forty hours, very miserable, it may be, but in no risk whatever. At the end of that time, his bowels having been well opened in the interval, let him have a good dose of morphia subcutaneously, half a grain or so, and be put quietly to bed. There is little use of trying to keep him in bed up to that time : very likely he will go to sleep and wake a new man. Tipplers are more easily or with more difficulty managed.

These, if they have had a debauch after a long course of habitual tippling, make perhaps the worst patients. Their appetite and digestive powers are completely gone, and they have no reserve of strength ; very likely they have had one or two attacks previously, but disregarded them. Such must be handled with care ; their strength must be kept up, and stimulants of any or every kind given, if necessary, and the critical period must be watched for with care. If it comes, then opium in full dose ; or if it does not come, opium or chloral must still be given to try to bring it on. It is in such cases that the injection of morphia under the skin is of most manifest benefit. Given by the stomach it would very probably not be absorbed ; given by the arm it must be (see HYPODERMIC INJECTION). In these cases, too, the temperature must be carefully watched. If it go up, cold must be applied, best perhaps in the form of the tub pack ; but if the head is very hot and the face flushed, the shower-bath may be tried, or perhaps what is better, ice applied while the rest of the body is packed. But nourishment must be given hour by hour, or even at shorter intervals, or the patient may sink. The other variety is comparatively mild, a little extra drink may have induced an attack of the " horrors." For this, perhaps, the best remedy is a good emetic and purge, or a good long walk. Here, too, the strength must be the guide as to the treatment.

DELITESCENCE, a term sometimes used to signify the sudden termination of an inflammation.

DELIVERY. (See LABOR.)

DELUSION. (See INSANITY.)

DEMENTIA is that form of insanity where the mind gradually fades away or becomes a perfect blank. It is a common finale to other kinds of insanity, and a kind of it occurs in a considerable majority of very old people. The other kind of insanity allied to it is imbecility or idiocy—but by idiocy we imply a total absence of reasoning powers from the birth, and by imbecility we mean a marked absence throughout life of such powers as are possessed by the average of mankind. In dementia there is marked apathy to everything and everybody, though sometimes the patients are subject to fits of restlessness. Their manner is what we would call childish or silly, and they repeat words without attaching any meaning to them. Dementia constitutes the last stage of many forms of madness, and is altogether beyond hope of recovery. In old people, however, there may be a brightening up before death. (See INSANITY.)

DEMODEX FOLLICULORUM, the name given to a little animal which is found in the follicles of the skin. The function of these follicles is to secrete oil, with which the skin

is naturally lubricated. In these follicles, especially those on the sides of the nose, the minute animal in question is found to reside. It
belongs to the family of mites, and has eight
imperfect legs and is about the tenth of an
inch in length. It is perfectly harmless, only
occasioning a slight inflammation, which produces a minute pustule and destroys the animal. It must not be confounded with the
itch mite, which is much smaller and quickly
spreads over the whole body. (See ITCH.)

DEMULCENTS are a class of medicines
composed of bland unirritating substances,
most of which form with water a viscid solution. They are given in cases where the alimentary canal is irritated or inflamed, and
are supposed to have an effect on even the
respiratory passages when taken by the
mouth. The substances employed contain
sugar, gum, or starch, and sometimes all three
of these are combined. Pearl barley, Iceland
moss, liquorice, marsh-mallow, oatmeal, linseed, gum acacia, and tragacanth, are a few
examples of this class of substances. Decoctions of these things may be made useful vehicles for the administration of other and
more powerful agents.

DENTIFRICE. (See TEETH.)

DENTITION, or the process of teething
in children, begins generally at the sixth or
seventh month of infant life, and is not usually completed until the twenty-fourth or
thirtieth month. The first indication of teething is shown by the increased flow of saliva
into the mouth; at first the salivary glands
seem to have very little to do, and the mouth
has a tendency to become dry, but about the
fourth or fifth month the mouth will be seen
to be constantly filled with saliva, and the
child continually drivelling. Teeth have come
through at the fourth, or even as early as the
third month, but in these cases, when the first
are so forward, the rest come out at the usual
time, and there is a longer interval between
each appearance. The two middle incisors
of the lower jaw generally appear first, then
the corresponding teeth in the upper jaw, and
next the lateral incisors of the lower jaw.
There is no definite order about the cutting of
the remainder, but usually the four anterior
molars next appear, then the four canine
teeth, and lastly the four posterior molars. Children, therefore, have only twenty
teeth, and these are called "deciduous or
milk" teeth, because, during childhood they
gradually fall out and make way for the "permanent" teeth, which are more numerous.
The arrangement of teeth in each jaw of an
infant is, therefore, the following: 1 posterior molar, 1 anterior molar, 1 canine, 1
lateral incisor, 2 middle incisors, 1 lateral incisor, 1 canine, 1 anterior molar, 1 posterior
molar. The "permanent" teeth are thirty-
two in number; each "milk" tooth is replaced by a "permanent one, and in addition

three molars are developed in each jaw on
each side, so that in adult life there are two
rows, each row containing sixteen teeth instead of ten. The process of development in
the infant as regards dentition does not go
on regularly, and it often happens that parents undergo considerable anxiety by seeing
several teeth appear, and then observing
none appear for some time; there is seldom,
however, any real ground for anxiety, for the
process is one natural to every child, and no
attempts on the part of the medical man or
nurse can hurry the course of nature. Three
or four months may elapse between the appearance of the lateral incisors and the anterior molars, while a still longer interval may
intervene between the latter and the posterior
molars. Dentition is generally attended with
more or less suffering and constitutional disturbance; but, on the other hand, there can
be no doubt that a great many infantile disorders are put down as due to this cause,
when perhaps there is no relation between
the two. When teeth appear, mothers not
only suckle their children, but give them thickened food, or even solid food, under the idea
that such food is more strengthening for them
than milk or liquid diet; and herein arises a
great mistake, in consequence of which many
infants suffer; it is not the teething in those
cases that does the harm, but it is because the
delicate stomach is overloaded with food
which it cannot digest; an alteration of the
diet soon gives great relief. Yet dentition is
attended with disorders of its own. Many a
child cuts a tooth without any more discomfort than an increased flow of saliva and dribbling from the mouth; at other times the
gum becomes tense and shining, while the
mouth is hot, and the child restless and feverish; the position of the new tooth may be
seen by the prominence of the gum over it;
often there are small ulcerations on the
tongue, gums, or inside the lips. The child
may be fretful and cry out as if in pain, and
it may sleep badly at night, and perhaps have
a convulsion. In yet more severe cases, inflammation of the gum may occur, attended
by considerable fever and disturbance of the
digestive functions. Small unhealthy ulcerations may occur on the gum just where the
tooth has pierced, and these give much pain
to the child; such cases, however, are very
rare, and in general terms it may be said
that, with due care to diet the process of
dentition is very simple, attended with but
very slight constitutional effects, and very
seldom fatal in its result. Dentition is but
one of the many important changes which
take place in the system in the development
from infancy to childhood, and in a delicate
and complicated framework like that of an infant, very slight external causes will set up
local uneasiness and general derangement;
errors of diet or a common cold will at once

give rise to serious symptoms ; thus it is that if they are teething at the same time, any disorders are put down to that process, when it would be nearer the truth to look on dentition as only a part of the general development which is then rapidly going on, and treat the case accordingly. It was formerly a common, but a barbarous custom to lance a child's gum frequently, under the idea that the convulsions, or the fever, or any other derangement that might be present, arose from the mechanical pressure of the tooth in its effort to pierce the gum ; this is quite an erroneous theory, and the foolish practice gives the infant much pain, and, in the great majority of cases, no ease at all. Undoubtedly there are times, when the gum is much swollen and red, and the child is evidently in pain about the mouth, that lancing the gum at the affected spot will give relief, but such cases only rarely occur ; prominence of the gum over the place where the tooth is about to come, attended with a paleness of the part, is quite a natural appearance, and not one to call for any operative interference. When a child has a convulsion from teething then "lancing" is useful, but then it must be clearly shown that the little patient is really "cutting a tooth" at the time, and that the "fit" is due to that cause.

As a rule, medical interference is rarely wanted during dentition. The diet must be carefully regulated, and the mouth may be moistened with barley-water, if it is hot and painful ; if any ulcerations are present, a lotion containing chlorate of potash is very useful ; a small piece of soft linen may be dipped in the lotion and then the inside of the mouth can be dabbed with it. For the same purpose borax and water are useful ; many use borax and honey, but the latter is sweet and apt to become sour and undergo fermentation, and when this takes place it only aggravates the ulceration. If the child be suckling it may be kept at the breast, and no other diet need be given if the mother has sufficient milk ; if, however, the baby has been weaned, more care will be required, as it will not digest its ordinary food ; thin arrowroot with a third of milk may be given, or milk and water sweetened according to taste ; once a day a little chicken broth or veal broth may be given. The bowels are frequently disordered in these cases, and at the outset a rhubarb powder may be given with a little soda and gray powder ; one dose will be enough, and it need not be continued. Should there be much vomiting and inability to retain food on the stomach, lime-water may be mixed with the food, in the proportion of one part of lime-water to four parts of milk and water. It allays vomiting and diarrhœa by neutralizing any acidity in the stomach, which is apt to be superabundant at these times. Solid food should not be given,

as it is painful to a child to eat, and therefore it swallows its food hastily without digesting it, and thereby adds to the intestinal derangement. In addition to the local ulceration and affection of the gums in teething, and besides the derangement of the digestive functions, convulsions may be added as a not uncommon occurrence, and they seem to occur in consequence of the greater susceptibility to any irritant cause of the infant's nervous system ; indigestion and diarrhœa would help to bring about a similar state of things ; careful dieting and a warm bath during the fit are the best measures. A slight purgative may be given if required, and now and then lancing the gums may be necessary. During dentition the temperature of the body may suddenly rise several degrees, and this usually occurs at bed-time while in the morning the fever may be much less, or perhaps absent ; the very suddenness of the rise of temperature would negative the idea of any fever coming on, and point to some reflex source of irritation. Lastly, there are some skin disorders which are liable to appear during the process of teething ; just as sometimes occurs after vaccination, eczema may appear on the skin from the constitutional irritation ; so, during teething, from a similar cause, eczema and impetigo may occur. The former may appear in scabs on the scalp or behind the ears, and in the flexures of joints ; when the scabs come off, a moist red surface is left which will soon again become encrusted. Impetigo appears on the chin and cheeks generally as angry, red spots, with a little pustule in the centre about as large as a pin's point ; these rapidly spread by scratching, and give great annoyance to the child. The nature and treatment of each will be described under separate heads. (See ECZEMA and IMPETIGO.) It is not always wise to cure these rashes at the time of dentition, as serious symptoms have resulted ; the local mischief may be kept in check, and during the intervals of dentition the skin disease may be cured.

The *second dentition* consists in the replacement of the "milk" or "deciduous" teeth by others which succeed them ; this important change takes place in childhood, and commences about the seventh or eighth year of life ; some time before this however, the germs of the new teeth begin to develop. The sixteen teeth in each jaw are arranged as follows : 3 molars, 2 bicuspids, 1 canine, 1 lateral incisor, 2 middle incisors, 1 lateral incisor, 1 canine, 2 bicuspids, 3 molars. In the replacement of the "milk" teeth, the development takes place with considerable regularity. First the middle incisors fall out and are renewed ; and then a similar process takes place with the lateral incisors ; the anterior temporary molars are followed by the anterior bicuspid teeth ; then the posterior temporary molars are replaced in a like way by the bicuspid teeth ; this latter change occurs about a year later than the former : the canine teeth are the last to be exchanged ; finally in the succeeding year the second pair of true molars appear, while the third pair, or the " wisdom teeth" may not appear for three or four years, or even longer ; now and then they cause considerable pain and distress during their development.

DEOBSTRUENTS are a class of medicines supposed to remove obstructions from any part of the body, especially chronic enlargements, tumors, etc. Plasters, iodine, turpentine, and other local stimulants are thus called, and also the stimulus of friction either with a brush or the hand, when applied to the skin.

DEODORANTS are substances which purify the air and remove noxious vapors or gases which may be injurious to human life ; they also check the growth of fungoid or infusorial organisms. The diffusion of gases in the atmosphere, and their consequent dilution, the currents of air caused by the variations in the temperature, oxidation and the fall of rain, are the main processes by which nature is constantly carrying on a purifying action. Yet, in the large centres of population, chemical agents are required to destroy the various poisonous elements which would otherwise accumulate to a dangerous extent. Carbonic acid, ammonia, sulphuretted hydrogen, and various organic substances, some odorous, others not, are the chief impurities met with. Air purifiers or deodorants may be in the form of solids or liquids and absorb the substances from the air, or they may be gaseous, and passing into the atmosphere act on the various impurities.

1. *Solid Deodorants.* Charcoal is the most effectual, and has the remarkable power of separating gases and vapors from the atmosphere, and oxidizes rapidly a great variety of substances. Animal charcoal is better than any other variety. It should be exposed to the air in bags or saucers ; its effect is very marked with sewage gases ; it absorbs sulphuretted hydrogen and purifies the air from the organic emanations of disease ; it is very useful, therefore, in the wards of a hospital or in a sick room ; after death some should be placed in the coffin, or laid about the room for a few days. Quicklime absorbs carbonic acid and may be employed for that purpose ; the carbolates of lime and magnesia and a mixture of lime and coal tar are useful, but not so effectual nor so easy to obtain as charcoal.

2. *Liquid Deodorants.* Solutions of potassium permanganate, zinc chloride, and lead nitrate are often used ; they should be exposed in thin layers in flat dishes, or cloths may be dipped in the solution and hung about the room. Not being volatile they act only on the air which comes in contact with them, but

even then may do a great deal of good. Chlorides of lime and soda, and solutions of sulphurous acid act chiefly by the gases which they evolve.

3. *Gaseous Deodorants.* These air purifiers act as a powerful means of freeing the air from impurities. The principal are ozone, chlorine, iodine, nitrous and sulphuric acids, carbolic acid, tar fumes, acetic acid, and ammonia. *Ozone* is supposed to be a modified form of oxygen, and may be produced by the action of electricity ; it is found in the air in increased quantity after any electric phenomena ; it may be evolved by partially immersing a stick of phosphorus in water in a wide-mouthed bottle, or by heating a platinum wire by an electric current. It destroys organic matter, and acts as an oxidizing agent. *Chlorine*, when given off in large quantities, is very irritating to the air passages, but in small quantities it is very valuable. Chloride of lime or soda may be moistened with water and placed about the room in shallow vessels ; the gas is then slowly given off ; if a quicker effect is desired a little weak sulphuric acid may be added, which will liberate the chlorine more rapidly. Chlorine decomposes sulphuretted hydrogen and sulphide of ammonium very soon ; it no doubt destroys organic matter in the air, and its powerful effect may be seen by its property of bleaching organic colors and destroying odors; it abstracts hydrogen from the compound and indirectly oxidizes it. This gas has an unpleasant odor, although its action is very effectual. Iodine will decompose sulphuretted hydrogen and arrest putrefaction, but it is not so diffusive as chlorine, and therefore less useful ; it is also very irritating ; a small quantity may be placed on a hot plate, when the gaseous fumes will be given off. *Nitrous acid* is made by placing clean copper in nitric acid and water ; a colorless gas (nitrogen dioxide) is given off, and this, combining with the air, forms the red fumes of nitrous acid. It is a powerful agent for oxidizing organic matter, and it has this property by very readily parting with its oxygen and reforming the nitrogen dioxide, which in its turn will again combine with the oxygen of the air and make nitrous acid ; in this way the oxidizing action may be continued for a long time. It rapidly removes the smell of the dead-house, but is extremely irritating and offensive ; hence the room should be cleared of people while the deodorizing process is going on. *Sulphurous acid* is easily made by burning sulphur ; it decomposes sulphuretted hydrogen, and acts powerfully on organic matter ; this gas is extremely useful ; a small quantity burned in the morning in a hall, or on a staircase, will purify a house very readily and no disagreeable smell will remain : it is thus useful in a children's hospital or in a nursery wherever there is close air. *Carbolic acid* is

now much used for deodorizing air; when weak its smell is rather pleasant. It is prepared from coal tar. The solid acid may be placed in a saucer about the room, or, still better, some of the solution may be sprinkled about. It conceals all odor and arrests putrefactive changes, and seems to have the power of stopping the growth of fungi. Tar fumes, vinegar, or acetic acid, and ammonia, are old remedies for a similar purpose, but they are not very effectual.

It will thus be seen that some deodorants decompose sulphuretted hydrogen and sulphuphide of ammonium, as chlorine, iodine, and sulphurous acid; others destroy the organic molecules which float about in the atmosphere, as ozone and carbolic acid. It is, however, most important to remember that none of these agents, valuable as they are, can take the place of ventilation and free currents of air. In a sick-room a small fire should be kept up, and the door or window opened for a short time three or four times a day, but not so as to place the patient in a draught; it is important to have the air thoroughly renewed. Charcoal should be placed about the room, or some carbolic acid should be sprinkled about. Deodorants are not only of much service in purifying the air in the above-mentioned cases, but they are equally valuable in destroying the noxious emanations from sewage; for this purpose numerous measures have been suggested. Charcoal may be employed, but it is not so useful here as in purifying the air. Dry earth has a good effect; it is used in earth-closets and has been found very valuable in large institutions and in camps; the excreta are at once covered over and no effluvia escape; in this way diarrhœa and typhoid fever appear to have been prevented. Quicklime and water may be added to the sewage until a deposit occurs. The lime forms insoluble salts, and decomposes the sulphuretted hydrogen; it delays but does not prevent the decomposition of animal and vegetable matters. The salts of alumina mixed with charcoal are very useful. The alumina when it is thrown down forms a very bulky precipitate, which entangles the suspended matters, and this is aided by the albumen of the blood, which coagulates during the process. A deposit soon forms and an odorless fluid is left, free from any poisonous influences; the deposit is valuable as a manure, and being concentrated can be easily removed. Perchloride of iron is also useful: it decomposes sulphuretted hydrogen and carbonate of ammonia, which is so often met with in sewage. A solution of chloride of zinc (half a pound to a gallon of water) may be used: it will destroy ammoniacal compounds and organic matter: it delays decomposition for some time. Permanganate of potash must be used in very large quantities to have much effect on sewage; it is useful in deodorizing excreta, and may be poured on the stools of patients suffering from cholera or typhoid fever. The preparations from coal tar, as creasote, carbolic acid, and cresylic acid, are very valuable agents in purifying sewage; they may be obtained as powders or crystals or liquids; the latter are the most useful, as they mix readily with sewage; one part of the liquid carbolic acid, if good, may be mixed with eighty or one hundred parts of water and poured into a cesspool, or on a dung-heap, or used in a water-closet. In seasons when cholera is about, or any infectious epidemic is raging, or in hot weather, when the effluvia from drains is very abundant and offensive, several gallons of weak carbolic acid should be poured into the drains; the water also which sluices the water-closets may have some of the acid poured into it, and in a similar way all latrines or urinals in public places should be treated; in this way much good may be done and noxious smells in a great measure prevented. It does not follow that because air smells badly it is therefore impure in proportion; gas works or tan works may be disagreeable, but they are not injurious; again, a cesspool or drain may not smell much, but the exhalations may be most dangerous. A noxious smell is like a Davy lamp to a miner, a warning of danger, but it is not itself the danger. By keeping in mind the evils arising from impure air, close rooms, noxious emanations, and sewage contaminations, and by using every means to procure ventilation and to remove the impurities by chemical means, a vast deal of good must result, and many diseases may be prevented.

DEPILATORIES are substances used for removing superfluous hairs. A variety of them is offered in the drug-stores, but they usually have either an arsenical or a caustic basis, and are consequently highly injurious to the skin. Plucking out the hair by the roots is by far the best way of removing it. In cases where this will not answer, the following depilatory is less objectionable than any that can be bought already prepared: Take a half-pound of slaked lime, and a half-ounce of orpiment in powder; mix and keep in a well-corked bottle. To apply, mix a small portion with water to the consistence of cream, spread it upon the objectionable hair, and let it remain about five minutes, or till it begins to burn the skin; then remove with an ivory or bone paper-knife, wash the part with water, and apply a little cold cream.

DERBYSHIRE NECK, called also *goître*, signifies a swelling in the neck, owing to enlargement of the *thyroid* body. It is a common disease in some of the cantons of Switzerland, especially the Tyrol and valley of the Rhone (cretins). Its cause is somewhat obscure, being assigned usually to (in

Switzerland) the use of snow water, and confinement in damp, close valleys, inactivities, and want of occupation ; it is frequently associated with idiocy.

DETERMINATION OF BLOOD is a phrase erroneously applied to the feeling of a rush of blood to the head in those who are liable to apoplexy and some other nervous diseases.

DEW-POINT is the point at which the mercury stands in the thermometer when aqueous vapor is deposited from the atmosphere on a cold object in the shape of minute globules of water. The atmosphere always contains aqueous vapor in greater or less quantity, and when a cold substance is brought into a warm room, this vapor is deposited on the cold object in the shape of minute globules of water ; this may be easily shown by bringing a glass of cold water into a warm room ; in a few minutes the outside of the glass is quite moistened with dew, which will again become vapor when the glass has become warmed to the temperature of the room. And a similar process takes place on the earth's surface. During the day the earth is warmed by the sun's rays, at night this heat is given off by radiation, but not equally from all objects ; thus metals have very little radiating power, especially when polished, while plants, grass, sand, and the ground readily radiate their heat ; they thus become much cooler than the surrounding atmosphere ; in consequence the aqueous vapor in the air is condensed on their surfaces in the shape of dew or minute globules of water. In hot countries and in hot seasons this is most noticed, as the earth being so much heated during the day gives off the heat very rapidly during the night. The state of the sky also exercises a marked influence on the dew-point. If the sky is cloudless, the earth radiates heat very considerably, and therefore becoming very much chilled, there is an abundant deposit of dew. But if there are clouds, these radiate toward the earth, and so as less chilling occurs, there is only a slight deposit of dew. Wind also affects the quantity of vapor deposited. If feeble, it increases it, because it renews the air ; if strong, it diminishes it, because it heats the bodies by contact. The formation of dew is greater in proportion to the moisture present in the air. The dew-point will vary, therefore, under many conditions, and to obviate any sources of error very ingenious methods have at various times been invented ; instruments used for finding out the amount of aqueous vapor in the air at any given time are called hygrometers. (See Hygrometer.)

DEXTRINE is a compound of carbon, hydrogen, and oxygen, and is one of that series of bodies which is called the dextrine series. When the seeds of a plant germinate,

the starch they contain is converted into sugar. This is well seen in the process of malting, in which the grain is made to germinate, and when the sugar is formed, the process of growth is arrested by roasting, and malt is formed. The sugar of the malt is then fermented and made into beer. If the process of germination is arrested after the starch has begun to change, and before the sugar is formed, dextrine is obtained. Dextrine differs from starch in that it is soluble in water, and not colored blue by iodine, and from sugar in the absence of sweetness, and the capability of fermentation. Its name is given by the property it possesses of producing right-handed rotation upon a ray of polarized light. It has some of the properties of gum, such as forming an adhesive liquid with water, and is used for gumming the backs of stamps, and other purposes. Dextrine is used for making moulds, and has been employed as a splint in surgery. Dietetically, it resembles gum, and does not appear to be absorbed into the blood, so that it is valueless as an article of diet. Gum may be said to be stereotyped dextrine. Liquorice, or Spanish juice, is a kind of dextrine ; so also is pectine. It is, in fact, because a number of substances closely resembling dextrine, of which starch stands below, sugar at the top, and dextrine in the middle, that they have been called the dextrine series. To these belong lichenin, or lichen starch— the starch of sea-weeds and lichens—manna, and various kinds of sugar, none of which are of any value as articles of diet, only as they can be converted into sugar, dissolved in water, and taken up into the blood.

DIABETES is a disease in which a very large quantity of saccharine water is passed daily by the patient, accompanied by great thirst and general debility. It may attack people at any age, but it is far more fatal, and runs a much more rapid course in children and young people than in adults, or in those of advanced age ; among the latter, a small quantity of sugar in the urine may be present at one time and absent at another, and these cases seldom need cause much anxiety. It is, at present, a moot question whether the liver or the blood is at fault in this disease, nor is it yet settled what part the nervous system may play in this affection, but it is an ascertained fact, that irritation in certain parts of the brain will produce sugar in the urine. The kidneys are not the seat of mischief ; they merely allow the sugary urine to pass, and, in doing so, suffer more or less in the process.

Symptoms.—Great thirst, dryness of skin, and passing an immense quantity of urine, are the most marked features in this disease. The thirst is so great, that the patient is always wanting some liquid, and will drink as much as four or five gallons of water a day in some cases ; two or three gallons is a very common amount. Since they pass so much urine, the other tissues of the body are drier than usual ; the skin feels dry and harsh, and an eczematous eruption is liable to break out, and sometimes boils form ; the nails are dry, and frequently chip in consequence. The bowels are confined and the motions are generally firm and dry. The urine is light in color, but much heavier than usual, from the great quantity of sugar present in it ; the usual specific gravity of urine varies from 1015 to 1020 ; in this disease it may rise to 1030 or 1040. Sugar may be detected in the urine in several ways : 1. Take equal amounts of urine and liquor potassæ in a test-tube and boil ; the solution will become, first yellow, then orange, and, finally, of a deep horse-chestnut color. 2. On adding to the urine a drop or two of a strong solution of sulphate of copper (blue vitriol) and then three or four drops of liquor potassæ, the precipitate first formed will dissolve, and the solution will have a dirty bluish-green color ; then boil, and an orange precipitate will come down. 3. To a large test-tube full of the urine add a little yeast ; invert the test-tube in a dish which also contains urine, and let it stand in a warm place for twenty-four hours ; gas will then have formed in the upper part of the tube, which will be due to the carbonic acid evolved, for fermentation has taken place, and the sugar has been converted into alcohol and carbonic-acid gas. 4. A more delicate test is obtained by allowing the passage of a ray of polarized light to pass through the urine : when sugar is present, the ray is turned to the right. The appetite is generally good, and even excessive, in some cases ; the tongue often dry and red, and the tempera-ture rather lower than usual ; the patient sleeps well, and the general health may go on for a long time without being seriously im-paired. In the course of time, there is more or less wasting of the body, and a liability to disease of the lungs. Cataract is, also, by no means an uncommon complication. In young people and children, the disease often runs a very rapid course, and may cause a fatal result in six weeks from the commence-ment of the symptoms ; much more common-ly it lasts for two or three years, and, in old people, sugar may occasionally appear with-out any harm resulting. Death often takes place by suppression of urine, followed by stupor, coma, and, perhaps, convulsions ; or, it may occur, through general exhaustion, or from disease of the lungs.

Treatment.—The treatment of diabetes gen-erally consists in placing the patient upon a diet from which all starchy or saccharine arti-cles of food are, as far as possible, excluded. Brown bread, bran biscuits, meat, green veg-etables, and milk, etc., may be allowed, but ordinary bread, sugar, rice, potatoes, etc.,

are prohibited. (See DIET.) Persistence in this plan is often followed by much relief to the patient, causing him to gain strength, les-sening his thirst and the quantity of urine, but it will not avail in curing him ; nor, at present, is any remedy known which can erad-icate this disease ; nevertheless, much may be done to alleviate the urgent symptoms and prolong life in some cases. Since there is so much thirst, a large allowance of water must be given, as much, in fact, as the patient likes. Raw meat has been found to be bene-ficial in some cases. Various preparations, as diabetic bread and diabetic biscuit, have been recommended, but few can continue their use long, as they eventually tire of keep-ing to a restricted diet.

There are certain cases in which persons pass a very large quantity of water without having any sugar in the urine : they are then said to be suffering from *polyuria ;* this dis-ease is of much less importance.

DIACHYLON is the name given to ad-hesive plaster, both spread and unspread ; though in the Pharmacopœia the name is con-fined to the litharge plaster (*Emplastrum plumbi*). (See LEAD.)

DIAPHORETICS are a class of medi-cines which exercise an almot exclusive action on the skin, producing perspiration and there-by reducing fever. Such drugs as ipecacu-anha, antimony, squills, camphor, and opium are of this sort.

DIAPHRAGM.—The diaphragm is the chief muscle of inspiration ; it divides the chest from the abdominal cavity, and is per-forated in several places so as to allow of the passage of various vessels and nerves ; above it, are the lungs and heart while immediately below it are the stomach, liver, spleen, and intestines. During inspiration it descends, so as to increase the cavity of the chest and to give the lungs room to expand, while it as-cends during expiration. When there is any distension of the abdomen, as from the pres-ence of a large tumor, pregnancy, ascites, or flatulence, the action of the diaphragm is im-paired and respiration is made more difficult.

DIARRHŒA can hardly be considered as a disease in itself, but rather as a symp-tom of other diseases. It is characterized by frequent loose evacuations from the bowels, due to functional or organic derangement of the small intestines, and produced by either local or constitutional causes. The evacua-tions vary in consistence and quality as well as in quantity and frequency ; they may be fluid or semi-fluid ; sometimes they are watery and serous, at other times they are mixed with mucus and occasionally a little blood. In children diarrhœa is readily pro-duced and is often very troublesome to cure ; it may be acute or chronic, and in both forms is dangerous to life. Under five years of age the mortality from this cause is greater than at any other period of life, and the greatest liability is shown during the period of teething, from six months to eighteen months or two years of age. As regards the time of year, diarrhœa is far more prevalent in the summer and autumn than in spring or winter.

Causes.—The causes may be divided into the *local* and *general*. Among the former may be classed indigestible food, bad air and water, parasites in the bowels, and any irritating matters poured into the intestines. Among the latter may be enumerated several diseases which produce cachexia, and induce a diminution of the vital powers. Consumption, diseases of the liver and kidneys, and many cases of blood-poisoning are accompanied by diarrhœa ; it is also associated with many fevers at their onset, and it is a prominent symptom in typhoid or enteric fever. Chronic malarious complaints are often complicated by diarrhœa, and it may occur in many cases where there is a degeneration of tissue in any of the abdominal viscera. In these cases diarrhœa appears as a symptom, while in other cases the diarrhœa appears to be produced more directly and to be due to the operation of bad water or a vitiated atmosphere acting on the blood or through the nervous system. Diarrhœa may be beneficial and due to an effort of the irritated bowel to throw off its noxious contents, as in cases where bad meat or putrid fish or unripe fruit is taken ; in other cases it aids to bring about a fatal result by exhausting the patient ; hence to properly treat a patient, it is important to make out thoroughly the cause ; this is sometimes difficult, as two or three causes may be conspiring to produce the same effect. 1. In summer diarrhœa prevails epidemically during the hot months ; it is never absent during that time of the year, but varies very much in its spread. The disease is probably due to the heat causing a general relaxation of the tissues, but its exact nature is not yet made out ; checked secretion of the skin, and miasmatic influences generated by the heat may aid in producing it. When genuine summer diarrhœa is very severe, it is commonly called English cholera, or choleraic diarrhœa, but it is advisable not to use such terms, as they lead to the impression that it is either related to cholera, or a mild form of it, whereas there is no proof at all of the connection, and it causes a panic in the minds of timid people, which is often very injurious. 2. *Impure* air is a common cause. Those who are engaged in any occupation which exposes them to the influence of decaying animal or vegetable matter, decomposing manure, or emanations from drains or sewers are very liable to this complaint. A leaky waste-pipe from a water-closet may allow the foul air to enter into a house and cause diarrhœa among all the in-

mates. 3. *Indigestible* food will cause diarrhœa ; it may do so by the direct irritation on the bowel or by producing a vitiated state of health. Prisoners in jails or all large collections of people suffer sometimes in this way. The excessive mortality of infants is often due to this cause, and it is very common just after they have been weaned, and when they are put too soon upon an altered diet. In children brought up by hand, and in those in foundling hospitals, this disease is often met with. Starvation frequently causes diarrhœa, and in the chronic wasting which attends many lingering diseases, the badly nourished tissues supply the blood with materials which rapidly decompose and cause colliquative diarrhœa. 4. Impure water is another common cause of this complaint. Spring water, which contains much saline matter, will cause a looseness of the bowels, especially in those who are not used to it. Any water which contains decomposing animal matter or sewage is a fruitful source of diarrhœa, and it may cause it either by direct irritation of the bowels, or by introducing poisonous matter into the blood. Water from wells into which drainage can enter is always hurtful to life, and should be carefully avoided. 5. Any irritants introduced into the stomach may produce diarrhœa ; eating a large quantity of unripe fruit will frequently bring on purging, vomiting, and griping pains in the stomach ; fruit is by many parents forbidden to their children during the summer months ; this is a mistake in the opposite direction ; a moderate quantity of ripe fruit is most beneficial to health, and when cooked and made into puddings or tarts it will do no harm. Fish which is not perfectly fresh may cause considerable distress with sickness and purging ; so also will mussels and periwinkles sometimes ; in some people oysters have a similar effect. Tainted meat, and the preserved meat which turns sour on keeping the tins open too long in which it is kept, may produce this complaint. Purgative medicines, the mineral acids and caustic alkalies, many common berries which grow in our hedge-rows, etc., will cause diarrhœa by irritation of the mucous lining of the alimentary canal. 6. Malaria, or the damp, faint-smelling emanations from a marshy district, are very injurious ; in summer-time, walking along the banks of a river, where the stream is sluggish and rank vegetation abundant, it may be often noticed that, as evening comes on, a faint damp smell is present, which is very injurious to those subjected to it. 7. Bad dwellings and cold damp houses, especially those which are situated low and badly drained, will cause this disease ; and so also insufficient clothing will aid it.

Symptoms.—In simple irritative diarrhœa, the patient will feel, a few hours after a meal, some flatulence and pain in the bowels, followed by loose evacuations ; this purging may and generally does relieve the pain ; the motions are feculent, and consist of a brown fluid containing small lumps of solid fæces ; if the purging continues, the motions become more liquid, and contain mucus. Generally the diarrhœa will cease of itself as the noxious cause is removed by the purging. The pulse does not rise, and there is no increase in the temperature ; the tongue may be slightly coated with a moist white fur. Eating too much and too rapidly, and taking purgative medicines, will bring on these symptoms. If the cause should be diseased or putrid food and water, then the diarrhœa will be more severe and exhausting ; there will be considerable constitutional disturbance ; the pulse may become feeble, and the surface of the body is colder than usual : in some respects these cases resemble the milder forms of cholera, but they are not often fatal. *In children* simple irritative diarrhœa is common in the summer and autumn months. If the attack comes on in previously healthy children, it generally attacks, and is attended with vomiting of the contents of the stomach; at first the excreta are natural, then they assume a yellow color, which changes to green on exposure to the air, or they are slimy and mixed with mucus ; sometimes white particles of undigested milk are found in the discharges ; the purging often comes on just when the child is suckling or taking food, and this gives rise to the popular notion that the "food runs through the child as fast as it takes it ;" it is hardly necessary to say that this notion is quite untrue, and the excreta in those cases are the remains of food previously taken, and which happen to be evacuated at the same time that fresh food is given. As the child returns to health the fæces become less watery, and resume their yellow color, or they may remain for some time green and slimy ; the disorder usually goes away in four or five days, although there is a tendency to diarrhœa on very slight irritation. Such cases are not accompanied by fever or much constitutional disturbance in the majority of cases ; the appetite is much impaired, and the child will cry for water, as it is more or less thirsty. The tongue is moist, but not much coated, while the papillæ may appear more prominent than usual, and look like bright red spots on the surface of the tongue ; there is very little pain or tenderness over the abdomen, and if there is any it is relieved by the purging. Great loss of flesh rapidly ensues, and in two or three days a fat, healthy child will lose greatly in weight and its flesh become loose and flabby ; the face is pale, and the eyes appear sunken, while the child sleeps badly and is fretful and languid. In the diarrhœa produced by *teething* the symptoms do not come on so suddenly, and they are slower in their course,

while they generally disappear when the tooth is cut. In whooping-cough, and after recovery from measles and other febrile disorders, diarrhœa may come on and assume a chronic form, which may kill the child by exhausting its strength. Diarrhœa is very obstinate in some children just after they are weaned, and is due to the altered diet ; more or less pain over the stomach, constant vomiting and loss of flesh, accompany the purging. Sometimes this simple diarrhœa passes gradually into an inflammatory form, and in these cases the symptoms are graver and there is much more constitutional disturbance ; vomiting and purging come on with great frequency, the stools are like green water, and consist chiefly of mucus from the bowel mixed with some feculent matter ; at times a little blood passes, and often there is "prolapsus ani," or a protrusion of the lower part of the bowel. The skin is dry and hot, the pulse quick, the child heavy and peevish ; at first restless, it soon passes into a half-drowsy state, and likes to lie quietly on its mother's lap ; very little suffering occurs, although there is generally a little abdominal pain before each evacuation ; the tongue becomes red, and the child is very thirsty ; there is loss of appetite, and the stomach rejects what is given ; great loss of flesh ensues, and in a few hours a plump child may become emaciated. Yet a fatal result is not very common, and a marked improvement generally takes place in two or three days ; a danger exists in these cases becoming chronic, which they are apt to do, and wear the child out in the course of a few weeks. *Parasites* in the bowels will bring on diarrhœa in children and adults. In such cases the worms may be passed, and the nature of the case is then clear, in other cases the breath is fœtid, the abdomen larger than usual, the appetite increased and difficult to satisfy, and there is often grinding of the teeth at night and picking of the nose ; the round worms and the thread-worms are most likely to cause this complaint. Inflammatory diarrhœa may occur in the adult as well as in the child, and will produce much the same symptoms ; in some of these cases it is difficult to distinguish it from dysentery ; but in the latter there is ulceration of the large bowel, which does not take place in the former disease.

Summer diarrhœa sets in suddenly ; there is copious vomiting and purging, and the stools are copious and liquid ; generally there is much pain and cramp of the abdominal muscles or of the muscles of the calf of the leg. The tongue is dry, and the patient thirsty ; great exhaustion will supervene, and in very severe cases the pulse will become feeble, the voice not raised above a whisper, and the general surface of the body lowered in temperature. The majority of cases are amenable to treatment, and are well in two or three days; the mortality is very small. Diarrhœa appears in the course of typhoid fever, and many chronic wasting disorders; but the symptoms and treatment in such cases will come under notice in describing the separate diseases.

Treatment.—The treatment of diarrhœa must vary with the cause; and when that is ascertained then only can any good be done. In many cases it is not advisable to check the purging, especially where there is some irritant present in the bowel, which is keeping up the flux. Often a change of diet, which should be light and nourishing, is of great benefit; and any food that is taken should be given in small quantities at a time. If the purging be due to unripe fruit or indigestible food, a dose of castor-oil at bed-time, followed by greater care in the diet, will suffice for a cure; but if the purging continue, and should prove at all exhausting to the patient, then it should be stopped. In the case of infants who have been recently weaned, and are suffering from this complaint, cold milk and water should be given, with a little broth or beef-tea, while solid food should be avoided; no opium should be given, as it is a very dangerous remedy at that age; an aromatic mixture of chalk, flavored with peppermint or anise, will be useful to keep the purging in check; if it depend on teething, it will probably cease as soon as the tooth has been cut, and a simple diet, with saline medicine, will promote a cure. *Summer diarrhœa* is generally curable in two or three days, without the use of medicine; but if it persist, and the tongue be clean, a mixture of chlorodyne with camphor-water and tincture of ginger will check it; should the tongue be much coated and the abdomen distended and painful, a gentle purgative should be first given; a dose of Gregory's powder or castor-oil will be sufficient. If the symptoms are severe, and there is collapse and cramp in the legs, then medical aid should at once be called in; in the mean time, iced water or milk may be given, and mustard-plasters applied to the calves of the legs; stimulants are not of much avail, and should be given in very small quantities. If impure air is the cause, removal to another place is naturally the most efficient remedy; should it break out in a camp, the troops should change their position. If it occur from the bad air of a drain or sewer, means should be taken at once to flush the pipe, see that there is nothing blocking it up, and take care that the foul air should be carried out into the open, and not enter the house; often the air accumulates in the waste-pipe of a water-closet, and when the pan is raised, the poisonous gas rushes up before the water from the cistern can come down; this can be remedied by letting the waste-pipe communicate directly with the open air, so as to prevent the stagnation of the noxious air in the pipe and each closet should be provided with a small cistern of water immediately above it. Should the cause be indigestible food, or if the latter be insufficient in quantity as well as in quality, a more generous diet will afford relief; and this subject is one of much importance, as the comfort of many individuals depends upon it. Any food which has been found to disagree with a person should be avoided; the meals should be taken at regular intervals, and not too much should be taken at a time. If impure water is causing diarrhœa, the remedy clearly is to improve the supply; the cisterns supplying a house should be examined, to see if they are clean and sweet; no cistern supplying a privy or water-closet should also supply the drinking-water, although this simple hygienic condition is rarely observed. If a well receive any surface drainage, it must be condemned at once, and no one should be allowed to use it; this is most important when the diarrhœa is very prevalent, as a whole village may be attacked at once from this cause. No stagnant water should ever be drunk, and every care should be taken that the source of the water should be kept free from decomposing animal or vegetable matter. If the purging follow any irritant food, as putrid fish or meat or poisonous berries, a mild purgative may be given, so as to remove speedily the cause; and this should be followed by a light and simple diet until recovery take place. Malarious influences must be avoided if they cause this complaint, and frequently change of residence is the only cure. In short, to cure diarrhœa, the food, the water, and the air should be looked to, and every means taken to insure their purity; and as towns increase in size, so, unless careful sanitary means are taken, the danger to human life will be continually on the increase.

In chronic diarrhœa, where the cause is some serious constitutional disorder, no rules can here be laid down, as they must be treated according to the special requirements of the case. A few general rules may here be mentioned for the treatment of diarrhœa, and they may be adopted in addition to removing the cause. 1. *Diet.*—This should not consist of too much farinaceous food, as arrowroot, tapioca, etc. Iced milk is most refreshing, and can be readily borne when nothing else can be retained on the stomach; where necessary, a little brandy may be added. Beef tea and broth carefully made are very good; but the latter should contain no vegetables, and the former must be made by stewing the meat first and then raising it to the boiling point; in this way the most nourishment is gained, while much is lost if the meat be first put into boiling water. An egg may be beaten up in the milk or in tea, and prove beneficial; but hard-boiled eggs must be avoided. Sago, rice, and such like foods often produce flatulence ; the lean of a mutton chop or a piece of well-boiled mutton may be given, but potatoes and most vegetables are inadmissible. Chicken and veal are good ; but pork, bacon, or salt beef are too indigestible. In the case of children, lime-water mixed with the milk proves of great service. A large quantity of fluid should not be given at a time ; but if the patient be very thirsty, the mouth may be moistened frequently with iced drinks. Cooked fruit may be given, and the pulp of grapes or ripe orange-juice ; unripe fruit must be carefully avoided. 2. *Residence and clothing*.—A damp, low situation is injurious, and removal from malarious influence is essential. Flannel should be worn next the skin. 3. *Change of climate*.—Change is of great benefit to many who suffer from this complaint ; but the place selected must vary with the nature of the case and the state of the patient. A long journey must not be recommended, unless the invalid is strong enough to bear it. 4. *Medicines*.—In malarious cases, quinine is of great value, or dilute nitric acid may be given, with infusion of gentian or calumba ; as a rule, these cases require some tonic to promote a cure : in addition, iron, strychnia, and salicine may be useful. Opiates are rarely, if ever, called for in the case of children ; they are of much value when given to the adult in certain cases, but, on the other hand, they are frequently injurious : where they cause flatulence and abdominal distension, followed by discomfort, vomiting, and pain, they must be avoided. They are chiefly of use in the inflammatory forms of this complaint, and they may often be given combined with a purgative, so that the bowel is soothed, while the astringent action is counteracted. Astringents may be given when the diarrhœa is not due to an irritant, or, when arising from that cause, the purging has gone on without being checked. Chalk mixed with mucilage is commonly given, and to this may be added some tincture of catechu and peppermint-water ; in some long-standing cases tannin, gallic acid, and kino are beneficial ; in others the sulphates of iron or copper are resorted to ; ipecacuanha in certain cases is resorted to with benefit. No rule, however, can be given as to when these drugs are to be given. Where simple remedies fail, medical advice must be sought.

Rest in bed or on a sofa is advisable in all cases of diarrhœa. When a patient has been suffering from diarrhœa in a chronic form, he is very liable to relapses, and any error of diet, or exposure to an exciting cause, may be most deleterious. In all cases, therefore, not only should care be taken during convalescence, but for some time after it ; and this is the more advisable in those persons who have suffered abroad, and, after a cure effected by absence, return to the place where they previously suffered.

DIASTOLE signifies the dilatation of the cavities of the heart, which follows immediately after their contraction. (See HEART.)

DIATHESIS implies a peculiar state or constitution of the individual, which renders him specially liable to a disease or group of diseases ; in this country there are well-marked diatheses, as the gouty, the nervous, the tuberculous, the bilious, etc.

DIET.—The question of diet is one of much importance, and it is one which is somewhat difficult to determine accurately, as it must vary much with the taste and condition of the individual. Men require more food per day than women, and those engaged in hard manual exercise require more than those employed in sedentary work. Different periods of life make a difference in the quantity as well as in the quality of the food taken, so that this subject naturally divides itself into the diet required for (1) adults, (2) children, and (3) infants. But, in the first place, must be given a short account of the different kinds of food. Physiologists have for a long time divided food into five classes, viz., the starchy or saccharine, the oleaginous or fatty, the mineral or saline, the albuminous, and the aqueous. 1. The *starchy* or saccharine food forms a large element in the composition of wheaten bread, rice, arrowroot, potatoes, sago, etc. Starch is a complex chemical compound of carbon, hydrogen, and oxygen ; such bodies are termed hydro-carbons ; when burnt off in the human economy, they form water and carbonic acid by entering into combination with oxygen ; the fat or adipose tissue in each individual is stored up in consequence of the amount of starchy food taken ; infants are much fatter than adults in proportion, because they are accustomed to live on so much farinaceous food, and such a diet contains a great deal of starch. Starch becomes converted into sugar when it mixes with the saliva ; in this way, when a piece of bread or potato is well masticated a large amount of sugar is formed, and this new compound is much more soluble in the stomach than starch ; the secretion from the pancreas which flows into the intestine has a similar action to that of the saliva. It has been said that starchy or saccharine diet forms hydrocarbons when taken into the system ; now these hydro-carbons build up the fatty tissues, and just in the same way as a person who drinks much beer will get too stout, so one who lives on too much starchy food will become corpulent ; hence those who go into training avoid this kind of diet, and take more meat, which gives them muscle and strength. 2. The *oleaginous* or fatty kind of food is commonly known ; all butter, lard, suet, the fat part of meat, and rich greasy foods consist of this variety ; like the last kind, they consist of carbon, hydrogen, and oxygen, and form, when used up in the human

economy, water and carbonic acid ; they differ in not being acted on by the saliva or
pancreatic juice, but they are made into an
emulsion by the secretion from the liver, and
thus being divided into extremely minute particles or globules, they are in a fit state for
absorption by the lacteals, which are so numerous in the small intestine. This kind of
diet tends to make persons fat, and a common
example is seen in the case of those who take
cod-liver oil ; those who naturally are too
corpulent should avoid taking saccharine or
fatty substances as far as possible. 3. The
mineral or saline variety of food is found in
nearly every article of diet : common salt is a
familiar example ; in ordinary drinking
water, in milk, in bread, and, in fact, in every
animal and vegetable product, there is more
or less saline matter ; it is one of the most
important constituents for the formation of
tissues, and during fœtal life the child is
nourished by a fluid which contains a good
deal of common salt : wherever vital changes
go on rapidly, saline matters are essential ;
without them the health fails, and many diseases have arisen from the want of salt during
long sieges. 4. The *albuminous foods.*—
These consist of hydrogen, nitrogen, carbon,
and oxygen, and are also called azotized substances from their containing azote or nitrogen. The lean of all kinds of meat, the white
of an egg, the caseine of milk, the gluten of
bread, are common examples of this kind of
food. While saline substances are readily
absorbed by the stomach without undergoing
any change, albuminous bodies have first to
be digested in the stomach, and, when thoroughly acted on by the gastric juice, they are
taken up by the vessels and introduced into
the economy ; they help to build up muscular
tissues, and hence are much required by those
who lead active lives, and who undergo much
exertion. 5. The *aqueous* or watery portion
of our food is too well known to require
notice ; no substances that we eat are so absolutely dry as to contain no water, and we
drink not only to quench thirst but also to
dissolve solids and render them more easy of
absorption. Finally, there are certain condiments, as mustard, pepper, pickles, etc.,
which are not essential as foods, but which
tickle the palate and cause an increased secretion of saliva and gastric juice, and by
doing so help to promote digestion.
From this brief survey it will be seen how
valuable milk is during the growth of children, for it contains in a liquid and soluble
form all the elements necessary for the growth
of the body ; the cream contains oily matter,
while the remaining liquid portion consists
chiefly of water holding in solution saline,
saccharine, and albuminous matters. Bread,
again, is a valuable article of diet, inasmuch
as it contains water, salt, starchy and albuminous materials. Thus two cheap and common articles of diet contain all that is essential for human life. An ordinary joint of
meat contains albuminous matter, as shown
by the amount of lean, associated with more
or less of fat and aqueous material, and in
addition a small quantity of salt. An egg is
a good example of a mixture of the different
kinds of food ; the yolk contains much fatty
matter, while the white of an egg is made up
chiefly of albumen, water, and salt. Beer
contains saline and saccharine matter in combination with alcohol and water ; the hops
only aid in giving a bitter flavor, and promoting the appetite ; the strength of the beer
depends on the malt it contains ; this fluid, it
will be seen, tends to promote the growth of
fatty tissues, but is not of much use in the
formation of muscle. Tea, coffee, and cocoa
contain chiefly water when taken as beverages, and, in addition, sedative and tonic
properties ; cocoa contains a good deal of
fatty and saccharine matter, and is very
nourishing. With this brief survey, rules
will now be laid down which may be employed in the diet of infants, children, and
adults ; but in each case circumstances must
modify the amounts given, and various modes
of dressing the food by devices in cookery,
may, of course, in any case be adopted.

Diet of Infants.—Too much care cannot
be taken in the bringing up of children ; a
great many of their complaints are due to
errors in feeding, and a large mortality annually results from this cause. An immense
amount of ignorance still remains on this
point : often solid food is given at far too
early an age, when the delicate stomach is
unable to digest it, and flatulence, pain, and
sickness frequently occur. Attention to the
following simple rules will be found of the
greatest benefit ; they have lately been drawn
up by a committee of the London Obstetrical
Society :

Suckling.—Nature provides breast-milk as
the proper food for infants, and suckling is
by far the best way of feeding a child. Provided the mother or wet nurse has plenty of
milk, and is in good health, the child requires
and should have no other food but the breast-
milk until about the sixth month. The milk
itself, for the first few days, acts as a laxative,
and no other aperient is necessary. Should
the formation of the milk be delayed, a little
cow's milk, diluted with an equal quantity of
warm water, and slightly sweetened, may be
given until the mother is ready to nurse. The
child should, for the first six weeks, be put to
the breast at regular intervals of two hours
during the day. During the night it requires
to be fed less often. As the child gets older,
it does not require to be fed so frequently.
A child soon learns regular habits as to feeding. It is a great mistake, and bad both for
the mother and child, to give the breast whenever the child cries, or to let it be always

sucking, particularly at night. This is a common cause of wind, colics, and indigestion.

How a nursing mother or wet nurse should be fed.—A nursing woman ought to live generously and well, but not grossly. It is a common mistake for wet nurses to live too well, and this often causes deranged digestion in the child. Should a nursing woman suffer from dizziness, dimness of sight, much palpitation and shortness of breath, or frequent nightsweats, it is a sign that suckling disagrees with her, and that she should cease to nurse.

Mixed feeding, when the mother has not enough milk.—When the mother has not enough milk to nourish the child, other food may be given, especially during the night. This should consist of the best milk, with one third the quantity of warm water added. This plan of combining breast-feeding with bottle-feeding is better than bringing up the child by hand alone.

Weaning.—The child should not be weaned suddenly but by degrees, and, as a rule, it should not be allowed to have the breast after the ninth month. After the child has cut its front teeth it should have one or two meals a day of some light food, such as bread and milk or nursery biscuits, and these may be gradually increased until the child is weaned. When the child is about from seven to ten months old, according to its strength, it may have one meal a day of broth or beef tea, with crumb of bread soaked in it, or it may have the yolk of an egg lightly boiled. When it is about a year and a half old, it may have one meal a day of finely minced meat ; but even then milk should form a large proportion of its diet.

The food of grown-up people bad for children.—Meat, potatoes, and food, such as grown-up people eat, are often given to young infants. This *kind of food*, and *all stimulants*, are entirely unsuitable, and are common causes of diarrhœa and other troubles.

Hand-feeding.—If the child must be brought up by hand, the chief rule to remember is, that the food should resemble as closely as possible the milk provided for it by nature. Milk, and milk only, should be used for this purpose. Cow's milk is generally used, but ass's or goat's milk is good. Two thirds pure and fresh milk, with one third the quantity of hot water added to it, the whole being slightly sweetened, should be used. A tablespoonful of lime-water may often, with great advantage, be added to the milk, instead of an equal quantity of the warm water. The milk should be given from a feeding-bottle, which should be emptied and rinsed out after every meal, and the tube and cork or teats kept in water when not in use. Perfect cleanliness is most important ; otherwise the milk may turn sour and disagree with the

child. The child should be fed regularly, just as if it were suckled ; it is a bad habit to give it the bottle merely to keep it quiet. The milk diet only should, as a rule, be given until the child begins to cut its teeth, when other food may be gradually commenced, as recommended under the head of " Suckling." When milk is found to disagree, other food should be given under medical advice. Most of the mortality from hand-feeding arises from the use of arrowroot, corn flour, and other unsuitable kinds of food, which consist of starch alone, contain no proper nourishment, and should not be used as substitutes for milk.

Diet of Children.—It is difficult to lay down accurate quantities of the different kinds of food suitable for children in health ; some require and can digest more than others ; some take also more active exercise than others. To make his subject clear, the following tables are given as a guide for parents :

1. From eighteen months to two years old. Breakfast at 7.30 A.M. A large cup of new milk, with a good slice of stale bread ; or half a pint of hot bread and milk. If breakfast is taken early, a cup of milk may be given in about three hours' time. Dinner should be taken about 1.30 P.M. It may consist of some good beef tea or broth, in which some bread crumbs or a well-mashed potato may be mixed. A cup of milk and water may also be given. At 6 P.M. a large cup of good milk may be given, with a slice of bread and butter. No other meal need be given, as the child, when healthy, ought to sleep all night, and it is bad to accustom it to wake in the night and cry for food. Yet, if it should do so, a little milk and water may be given. Farinaceous food should not be given at an early age to any extent, as the stomach is overloaded by that means and fails to digest it properly.

2. From two years to three years old. Breakfast at 7.30 A.M. A large cupful of milk, with a slice of bread and butter, and now and then the lightly boiled yolk of an egg. At 11 A.M. a cup of milk may be given. For dinner, a large cup of beef tea or broth, or a little finely cut-up roast mutton ; or three or four tablespoonfuls of gravy, in which bread crumbs, or a mashed potato may be mixed. A small quantity of rice pudding, with plenty of milk, or a piece of custard pudding. At 6 P.M. some milk may be given, or a little tea with plenty of milk in it, together with some bread and butter or some toast and butter. Now and then a little stewed fruit may be given, or occasionally a little jam. Rich things should be avoided, and no child should be allowed to eat too much. A diet can always be modified from day to day to please the palate, but milk should form a main ingredient in the morning and evening meals.

3. From three to ten years of age. The amount of food given will vary with the age and appetite of the child. Breakfast at 7.30 A.M. A basin of bread and milk, with some thick slices of bread and butter. Occasionally a lightly boiled egg may be given. At 11 o'clock A.M., a small slice of bread and butter may be given, if required, with a little water or milk and water. Dinner at 1.30 P.M. Some lightly boiled mutton, or a slice of roast beef or mutton with plenty of gravy ; bread should be eaten with it, or a mashed potato. Occasionally some other vegetable may be given, as well-boiled cauliflower, or asparagus heads or some turnip and carrots well done. A light pudding may be given, as rice, custard, ground rice, etc. At times a fruit pudding, well cooked, may be tried, or well stewed fruit is beneficial. Considerable variety may be adopted at this age, provided that too much is not given, and that it is digested well. Broth or soup may be substituted once or twice a week for the meat. Boiled salt beef, pork, and veal are not so easily digested as fresh beef and mutton. Cheese is not advisable. Prunes, figs, almonds, and raisins, and such like fruits, may be given now and then with advantage ; but any excess should be avoided carefully. Biscuits, nuts, preserved foreign fruit, walnuts, and dates are less digestible. A roasted apple, well sugared, or stewed pears are very nice, and suitable for children ; and occasionally some jam, as raspberry, strawberry, or currant preserve, may be given with bread at tea-time. At 6 P.M. milk and water, or tea with plenty of milk, may be given ; also bread and butter. Plain seed cake, or a slice of an ordinary home-made plum cake, may occasionally be substituted, or a sponge cake. No stimulants need be given at all at this age. There is seldom any occasion for supper ; if required, a thin slice of bread and butter may be given.

4. From ten to fifteen years old. The same diet as No. 3, only now more may be given in proportion to the age ; boys, too, often require more than girls, as they undergo more active exercise. A good meat meal should be given at mid-day, but, it is not required oftener. For breakfast, cocoa and milk is very nutritious, or a basin of oatmeal porridge with fresh milk may be substituted for the bread and milk. These diets presuppose that the child is in good health, and that active exercise is taken ; but if disease be present some modification may be required, and for this medical advice should be sought. In the treatment of children's diseases more than half the success is due to the careful arrangement of the diet. Many diseases are simply due to the engorgement of the stomach produced by over-eating, and it is owing often to the fond but foolish mother, who, to please the child, gives it anything which it fancies,

without considering whether or not it is hurtful. A little firmness in refusing a child too much food, and a little more knowledge on the part of the parents of the chief articles of diet which a child should have, would save their offspring from many an illness and much discomfort.

Diet of Adults.—It is difficult to lay down any strict rule as to the amount of food to be taken in twenty-four hours for grown-up people : men require more animal food than women, and those engaged in active exercise require much more than those who live a sedentary life. Laborers can get through much more work in a day when well fed than when living on a meagre diet. The different kinds of food should be well apportioned ; it is equally bad to live on a purely farinaceous diet, as it would be to take only fat or meat ; what is required for a state of health is to take a fair proportion of each. Hunger is the best test as to when a man should eat ; but it is often very unwise to go on eating until a feeling of fulness and engorgement ensues ; it is important also that meals should be taken with regularity, as it a is very bad plan to allow intervals of varying length between meals. A great many people eat too much, and this has its evils, just as much as excess of drinking. It has been estimated that the food required every twenty-four hours by a man in full health, and taking free exercise, is, of meat 16 oz., bread 19 oz., fat 3½ oz., and of water 52 fluid ounces ; that is, about 2½ lb of solid food and about 3 pints of fluid. The amount of fluid taken varies with the work and with the season of the year ; those perspiring a great deal require the most, and in summer time many take more than three pints in the twenty-four hours ; the fluid here includes any liquid taken. Of the different kinds of meat, mutton and roast beef are the most digestible ; salt beef, bacon, pork, and veal would rank next in order ; some sorts of fish are digestible, but there is not so much nutriment in them as in a corresponding quantity of meat. Bread is taken at most meals, either as bread or toast ; or the same elements of food are taken, as biscuit, cake, etc., and it forms an important element of diet ; not only is it cheap, but it contains four out of the five kinds of food ; life can be sustained for a long time on bread and water. Pastry is an abomination, and is very rarely well made ; it is heavy, greasy, and indigestible. Fruits or preserves made into puddings or tarts are very excellent articles of diet, and so are light puddings made of rice, arrowroot, tapioca, etc. Salt should be taken with food, and generally it is present in a greater or less degree in most kinds of food. Sugar seems necessary in early life, but the desire for it is lessened as we grow older. Both the saccharine and fatty foods should be avoided by those who are too corpulent, as they both help to build up fatty tissues. The system of Banting rested on this principle, and there is no manner of doubt that fat people can become greatly reduced in weight, and keep themselves down, by attending to this rule ; the change should not be made too suddenly, but no danger need be feared on that score. The following substances should be avoided by a fat man, or at least taken only in moderation : Fat of meat, bacon, pork, etc. ; white bread, potatoes ; starchy food, as tapioca, rice, arrowroot, sugar, beer, and heavy wines or spirits. The following articles may be taken without fear of forming too much fat : Brown bread, toast, biscuits, rusks, lean of any kind of meat, fish, fowl or game, green vegetables, as cauliflower, asparagus, and lettuce, celery, fruit, either cooked or fresh, jams in moderation, and light wines. Of course, no one can strictly adhere to this plan, but attention to its leading principles will soon make a vast improvement in the ease and condition of the individual. For people who are thin, a converse plan may be in part adopted. Climate makes a great difference in the appetite, as has been shown by the large amount eaten by sailors who have served in arctic expeditions. Some people have been known to eat 8 lb. of meat a day, and others from 12 lb. to 20 lb. per day, in those northern regions. (See BANTINGISM, COMPOSITION OF BODY, and FOOD.)

DIGESTION.—The alimentary canal is the great channel whereby new material is introduced into the blood, and it is in this canal that the important function of digestion takes place. A man swallows daily a certain amount of meat, bread, butter, water, vegetables, etc., and it has been computed that the amount of chemically dry solid matter taken daily by a man of average size and weight amounts to about 8000 grains ; he also absorbs by his lungs about 10,000 grains of oxygen every twenty-four hours, making a total of 18,000 grains (or nearly two pounds and three quarters avoirdupois) of daily gain of dry solid and gaseous matter. Of this quantity about 800 grains, or one tenth part of solid matter, leaves the body daily as excreta, and as no solid matter in any quantity leaves the body in any other way, it follows that in addition to the quantity of oxygen absorbed by the lungs, about 7200 grains of solid matter must pass out of the body in gaseous or liquid secretions, supposing the man to keep the same weight. The urine, the perspiration, and the expired air from the lungs, carry off nearly all this quantity in their secretions. All the substances used as food are classified and discussed in the articles on DIET and FOOD. When these different foods are swallowed various changes take place. Starchy compounds are very insoluble, but the saliva converts these during mastication into sugar, and this passing down

into the stomach is easily soluble ; hence arises the necessity for well masticating bread, biscuits, potatoes, toast, rice, and arrowroot, etc., so that all the starch may be thoroughly converted, or else indigestion may ensue. Albuminous compounds, as the lean of meat, etc., should be well masticated so as to tear up each portion into minute pieces and enable it to be easily acted upon by the gastric juice when it gets into the stomach ; no chemical change takes place in the mouth with regard to this group, nor with the next two groups either ; the only change is a mechanical one, and by this means the food is well mixed together and finely divided. The œsophagus is merely a tube to convey the food from the mouth to the stomach, and takes no part in digestion. The stomach is a dilated chamber where the food remains for a time to be digested and to be acted upon by the gastric juice. This important secretion, poured out from the walls of the stomach in great quantity during digestion, renders soluble all the proteids or albuminous compounds ; and the more finely divided these bodies are, the easier does the process go on. When meat is swallowed hurriedly, or when tough, fibrous, and indigestible food is taken, the action of the gastric juice is lessened and indigestion results. Thus, in the course of three or four hours after a meal, the stomach contains all the proteids, amyloids, and minerals, in a state of solution, for water in some form is always taken with food ; only the fatty matters as yet are unaffected. Passing down into the small intestine the food is now called *chyme*, but it does not go far when it meets with the bile and the pancreatic juice, which, acting on the fatty matters, form an emulsion, whereby the oily particles are so minutely divided as to render them capable of being absorbed by the lacteals and vessels of the small intestines. Thus, either in the mouth, stomach, or intestinal canal, the various kinds of food are so acted upon as to render them capable of absorption, and this process goes on not only in the stomach but all the way down the intestines, so that the blood is supplied after every meal with a fresh stock of food to make up for the losses which are continually going on in other parts of the body. There is, however, always a residue of indigestible matters in the food, so that all the chyme is not absorbed, but the remainder is excreted daily and known as fæces. As the coat of the intestines is in part made of muscle, it is constantly contracting in waves and gently pushing the chyme forward so as to bring it in contact with different parts of the canal, and finally to expel the indigestible remainder. If this process from any cause go on too rapidly then diarrhœa will result, and if it continue the patient will lose flesh, because those substances escape which ought to be absorbed by the blood ; or again, if there be disease of the mesenteric glands or walls of the intestines, as in some cases of wasting disease in children, in cancer of the bowels, etc., then absorption will not go on properly and emaciation will be the consequence. Foods vary much in their degree of solubility, and hence arises the importance of careful diet in those who have a weak digestion, or in those who are convalescing from an illness. By bearing in mind the importance of mastication and digestion in early life, much suffering may be avoided in after years, and many of those who are confirmed invalids and martyrs to indigestion might have been free from disease had they paid more attention to diet. Not only should the food be easily digestible, but it should not be swallowed too hastily ; it should always be taken at regular intervals, and rest after a meal for a short time is advisable ; also too much should not be taken at once, so as to make the individual feel distended and uncomfortable. (See DIET, FOOD, and INDIGESTION.)

DIGITALIS, or *Foxglove*, though a common plant, is one of our most valuable remedies. The leaves of the plant are used, and from them may be extracted an active principle, not an alkaloid, called digitaline. Its two officinal preparations are an infusion and a tincture. The leaves are taken when about two thirds of the flowers are expanded, and, as a rule, the second year's leaves are the best. Digitalis acts as a sedative on certain important organs, especially on the heart, and that, too, through one special nerve called the pneumogastric or vagus. This nerve serves as a kind of fly-wheel to the heart. Stimulation of it in any shape diminishes the rapidity, while it increases the force of the heart's action. Paralysis of it, on the other hand, increases the rapidity, leaving the force *pari passu*. Now digitalis stimulates this nerve, and therefore steadies the heart. Under its influence the heart no longer beats frequently and imperfectly expels its contents, it acts more slowly and more perfectly. Hence, had dropsy arisen from the heart's imperfect pumping, the fluid is again taken up by the vessels, and carried through the kidneys, whose functions of course in this fashion increased, so that the medicine acts as a diuretic. Of course, if the use of the drug be carried too far, a totally opposite result will follow—the heart's action will be reduced too far, and may even cease. When digitalis has been too freely given, therefore, there is considerable danger of paralysis of the heart—it may stop, and so death ensue. This is most likely to be the case if the patient attempts any unusual exertion, or even sits up in bed. It is, however, to be noted that in patients the subject of heart disease, this remedy may be given for months with only good effects. It is chiefly used as a remedy in heart disease, where it is most valuable if the proper cases are selected.

It has also been used in delirium tremens in large doses, but this treatment has not been generally accepted. A more valuable application perhaps is to the treatment of acute mania. Digitaline is sometimes used in the same malady, being injected under the skin if there is any difficulty in getting the maniac to take it. At all times digitalis should be used with caution, and is one of those remedies which in appropriate cases do much good, but in badly selected cases may kill. The infusion of it will keep as the best preparation ; the dose should always be less than one tablespoonful.

DILATATION occurs in various organs of the body. 1. In the heart, in many cases in which there is disease of the valves, or where the wall of that organ is fatty and weak. 2. In the air-cells of the lung, forming the condition known as emphysema. 3. In the bronchial tubes, in persons who have long been subject to winter-cough. 4. In the bladder, when the patient has suffered from stricture for some time. 5. In the kidney, if the ureter be blocked up by a stone, so that this organ may be distended into a large cyst and become quite useless. 6. In the ventricles of the brain, as in some cases of meningitis. Thus there is hardly any cavity in the body which may not become dilated as a result of some form or other of disease, but the treatment will consist in removing the cause for the dilatation, if possible, as little or nothing can be done for the organ when it has once been well dilated.

DILL is the fruit of *Anethum graveolens*, a plant belonging to the hemlock family. It contains an oil which may be distilled from it. To this oil it owes its property, and from it, or from the fruit, is prepared dill-water, the form in which the substance is commonly used. It is almost entirely employed in the maladies of children, accompanied by flatulences. It is much used as a domestic remedy, and is the more to be commended for this, inasmuch as though useful it is harmless. A teaspoonful of the water may be given for a dose, or a drop of the oil let fall on sugar.

DILUENTS are a class of remedies made use of to quench thirst, or to make the blood, heated and thickened by fever, thinner and cooler. Toast and water, barley-water, lemonade, and such like beverages, are of this class.

DIORESIS implies an excessive flow of urine.

DIPHTHERIA is a specific contagious disease, occurring generally in an epidemic form, and characterized by a peculiar inflammation of the mucous or lining membrane of the fauces, pharynx, and upper part of the air-passages ; sometimes the disease spreads to other parts of the mucous membranes ; there is also generally some affection of the spleen and kidney, together with much general prostration. This disease seems to have been known for the last two thousand years, and under various names it has prevailed with great severity in different countries. It has often been confounded with croup and scarlet fever, and it was not until recent epidemics that this disease has been clearly and generally recognized. Children and young people are more liable to it than adults, and more girls suffer from it than boys ; in like manner women are more liable than men, and the weakly of either sex are more prone to the disease than the strong and healthy. Climate and season do not seem to exercise any influence on the disorder ; it occurs with equal severity in the winter as in the summer months, and in its symptoms and mortality it is the same in hot as in cold countries ; yet various epidemics differ in severity and in extent. It is quite clear that this complaint is contagious, but in what way is not so manifest ; at one time an isolated case will appear in a country village and not spread widely, while on another occasion a whole district will suffer severely ; if one inmate of a house be attacked, most of the other members of the family will suffer, too, if they come in contact with the patient. The infectious matter is capable of diffusion into the air and carried to distant parts, but it is more common for people who inhale the patient's breath or who come in close contact with the sufferer to suffer most. No atmospheric condition is known which tends to favor the spread or check the progress of the disease ; it is very doubtful if the disease can be taken from one house to another by an unaffected person, but the presence of one sick person in a house is sufficient for its communication to another, although the two may be kept as separate as possible. As in most epidemics the mortality is greatest at the outset, and this is probably due to the most susceptible and most susceptible being attacked first. Although every care be taken to cleanse and purify an apartment in which a patient has suffered from this disorder, yet the infection will sometimes cling to it with remarkable tenacity. " In a country house in Scotland a visitor suffered from this disease while occupying a chamber in which a case of diphtheria had occurred eleven months before." The time between exposure to the disease and the first appearance of the disorder varies very much ; in some cases the period has only been thirty hours, in others several days elapse. The infection may be disseminated for some time after convalescence has been established ; thus in one case, a little boy, who had been absent from his family for three weeks, went to see his sisters who were recovering from the disorder, and caught it ; other instances are recorded where, five or six weeks after the commencement of the illness other persons have been attacked. There seems to be a predisposition on the part of some people to take this disease more readily than others ; those who are highly nervous or have undergone much mental activity, and those who have suffered from exhaustion or bodily fatigue are more liable than others. The disease seems to attack indifferently all classes of society.

Symptoms.—The onset of an attack is marked by lassitude and prostration, aching in the back and legs, pallor of the skin, and pain in the throat ; in children there may be diarrhœa headache, giddiness, and a stupid condition. The pulse becomes quick and may beat 120 or 140 times a minute, but the respirations are not particularly increased . The tongue is moist and slightly coated ; the appetite is impaired, and there is more or less thirst. The urine is pale and generally contains a little albumen ; this may be known on heating a small quantity in a test-tube ; if a hazy cloud appear, which does not go away on adding a little nitric acid, it shows that albumen is present. More marked are the local symptoms. The throat is sore, and it is difficult and even painful to swallow, and this pain extends often to the ears, and there is a feeling of stiffness in the muscles of the neck. On looking inside the mouth there will be found some swelling and redness of the soft palate and tonsils and the back part of the throat ; if the inflammation extend upward into the nasal passages there may be a glairy discharge from the nose, or, if it spread downward into the larynx, symtoms similar to those met with in croup will appear. There will then be hoarseness and weakness of the voice, with cough and crowing inspiration, and if the obstruction be very great there will be imperfect expansion of the chest, pallor of the face, and lividity of the lips.

When the inflammation extends into the larynx the mortality, especially in children, is very great, and it has been estimated that one half of the fatal cases die from this cause ; in adults this extension of the disease proves less dangerous, and they are often able to expectorate large pieces of the false membrane. Sometimes the mischief is confined entirely to the larynx, but more generally the fauces will be found affected also. The most characteristic appearance in diphtheria is the presence of a membrane which covers more or less the parts about the upper and back part of the mouth ; this membrane is soft and of an ashy-gray color, and when removed leaves behind a red and raw surface, and then it rapidly reforms again ; the swelling of the mucous membrane and the amount of false membrane may be so great as to prevent swallowing, and so endanger life by preventing enough air from entering the lungs. This membrane, too, may appear on any abraded surface, on a mucous membrane, or on the skin ; if there be an open wound anywhere the surface will cease to heal and become covered with this unhealthy membrane. Hemorrhage occasionally takes place on attempting to remove the deposit from any affected surface, so that much care must be taken whenever this is attempted. The inability to swallow is sometimes very great, and when fluids are taken they are often apt to come back through the nose ; at the same time there is a loss of sensibility in the fauces and soft palate ; complete inability to swallow seldom comes on before the third or fourth week of the disease, and it arises from paralysis of the muscles of deglutition ; this condition is a very serious one and adds much to the danger of the case ; the pulse may become weak and slow, and death may occur suddenly from fainting on any undue exertion. The paralysis may extend to other parts of the body, and these become affected at a later period ; in this way the legs or arms become useless for a time, and the muscles of the neck may be so paralyzed that the patient is unable to move his head. Loss of power and irregular action of the muscles of the pharynx is the earliest and most common form of nervous affection in this disease, and it may disappear rapidly and leave no mischief behind, but sometimes it lasts for many weeks or months and retards convalescence. The patient is often unable to articulate clearly from imperfect movement of the tongue, and tingling sensations are often felt in the tongue and lips. Every case of diphtheria must be regarded with anxiety, as it is attended with considerable danger ; any extension of the deposit in the fauces, the onset of a hoarse voice, or croupy breathing, and the occurrence of hemorrhages, are serious symptoms. The mortality varies in different epidemics, and some of them have been marked by an immense percentage of deaths ; occasionally all the members of a family have been rapidly carried off in succession, but the average of deaths ranges from one in three at the height of the epidemic to one in seven or ten at its close.

Treatment.—This consists in general means and local measures. There is no drug which can be looked upon as a specific, nor are there any means in our power to eliminate the disease when once it has attacked an individual, yet a great deal may be done at the outset if the disease is recognized sufficiently early. As a local remedy a solution of nitrate of silver should be thoroughly applied to the diseased surface of the throat, but not forcibly so as to rub off the membrane and cause bleeding to follow. Hydrochloric acid and honey have been used for a similar purpose, but in all these cases medical advice must at once be sought, as it is dangerous to employ merely popular remedies. The patient must be put in a well ventilated room, but free from draughts, and if the weather

be cold there should be a fire in the room ; the air should be between 60° and 65° Fahrenheit, and it may be kept moist by boiling water in a kettle on the fire and letting the steam pass into the room occasionally. Complete rest must be obtained, as there is always great prostration, and any exercise or movement on the part of he patient should be avoided so as to store up all his strength. Milk may be given to the extent of three or four pints a day, and brandy can be mixed with it, if it is necessary. Beef-tea, chicken-broth, and eggs may also be given ; it is of no use giving solid food as the patient will not care for it, and it will create pain in swallowing. The general treatment in fact is similar to that which has been laid down in the article on Fevers ; in this disease, however, great care must be taken that the food is given in small quantities at a time, and slowly, because in consequence of the paralysis of the muscles of deglutition which often ensues, the act of swallowing is rendered dangerous. Where there is much obstruction in the larynx the operation of tracheotomy may be resorted to, but this proceeding is attended with a very small amount of success, and is nearly always followed by a fatal result in very young children. When convalescence begins, the return to solid diet must be slow and gradual ; for many weeks the nutriment should be light and wholesome, and not too much should be taken at a time. As soon as the patient can be removed with safety, and without carrying infection to others, removal to country air or the seaside is most beneficial, and it is the more needful in these cases as there is so much prostration and anæmia for many weeks afterward ; yet, even in bad cases, the health will in time be thoroughly restored. Cold bathing, tonic medicines, moderate exercise, or even a sea voyage, are very valuable aids in restoring the health. In cold weather a bath is not advisable unless the chill is taken off the water, but in summer time it is most refreshing and strengthening ; carriage exercise may at first be taken, or a short walk during the fine part of a day, but no great exertion should be made, and the patient should rest as soon as a tired feeling comes on. Although this disease is not so communicable by the clothes as scarlet fever and some other disorders, yet it is always advisable that any articles of clothing should be thoroughly disinfected before being worn again, and for this purpose they may be placed in an oven and exposed to a high temperature ; a similar remark will apply to the bedding, curtains, sheets, etc, of the room in which the patient has lain. When possible, the house in which the disease has broken out should be well cleansed and fumigated ; it may be kept empty for this purpose, for a week or ten days, and chloride of lime may be sprink'ed about the rooms on the floors.

The great danger in diphtheria seems to be from inhaling the breath of an affected person, therefore communication with other people should be avoided as far as possible, and this is most important in the case of children, who are very susceptible to this complaint. Nothing is more foolish than for friends to come and see those who are affected with some contagious or infectious disorder, unless they are either doctors or nurses, or are of some use to the patient; by giving way to a morbid sympathy or curiosity, they run the risk of catching the disease itself or of communicating it to others, and in this case, as in many others, prevention is better than cure.

DIPHTHERITIC PARALYSIS. (See Paralysis.)

DIPSOMANIA is the name given to that horrid craving for drink which is either developed or is innate in some men—more rarely in women. Whatever be its cause, whether brought on by a man's own doings, or, as some would have us believe, hereditary, the man who becomes the subject of dipsomania is no longer a free agent, and he ought to be dealt with as such. Undoubtedly such can be reclaimed, but only by stopping all supplies of liquor until they have regained their power of self-government, which they have utterly lost. It is terrible to see a man who has been brought up well, and whom you have known as a true gentleman, become from whatever cause a dipsomaniac. The man who was once the soul of honor becomes a liar, whose word you cannot for a moment trust. He was honorable in his dealings, he becomes everything that is the reverse, and will not hesitate to steal to gratify his horrid appetite. This, indeed, is a consummation much to be desired, for if he does he can be locked up and cured, for if he is not locked up he will not be cured. If you take his money from him he will pawn his clothes, if you search the house every night his cunning will defy you; he will get drink unless you lock him up, and if you do so you do it at your peril. An action for false imprisonment would lie, and you might be mulcted in ruinous damages. Yet, while you talk to them and are with them these people are manageable; it is only when they escape from your sight that they straightway go wrong.

DISINFECTANTS are substances which are used to purify the air of those noxious products which emanate from persons in certain states of disease. It is supposed that when persons are suffering from various contagious disorders, as small-pox, scarlet fever, measles, typhus fever, diphtheria, certain particles emanate from them, and passing into the air, carry the disease to other people. These germs, which are floating about in the atmosphere, may be easily carried to other localities, and set up fresh centres of disease. It is therefore very important to use such measures as may be possible to destroy these germs, and so diminish the propagation of the disorder. In small-pox and the above-mentioned fevers it seems likely that the morbid products are given off from the skin or in the expired air, while in cholera and typhoid fever the evacuations from the bowels are looked upon as the chief source of danger. Patients, as a rule, are more dangerous to others when the malady is subsiding, or during the convalescent period; thus in scarlet fever when a child's skin is peeling the disease is very liable to be propagated. _Chlorine_ is one of the best disinfectants, and if the air of the room could be thoroughly charged with this gas, all the poisonous particles would most likely be destroyed; this cannot be well done because the gas is so irritating, and the same remark will apply to the use of nitrous and sulphurous acids; nevertheless, when the room is empty and requires fumigating, these gases can be used freely and with great benefit. In any case of fever chloride of lime can be laid in saucers or shallow dishes about the room, and then enough chlorine will be given off to produce a faint smell of the gas in the apartment. _Carbolic acid_ is perhaps as effectual, and is not so disagreeable; solutions of it diluted with water may be sprinkled about the room, or cloths dipped in the fluid may be hung up; in other cases powdered _carbolate of lime_ may be placed near the patient. All these disinfectants should be also placed outside the apartment in the hall or on the staircase, and the hands of the attendants should be washed in weak carbolic acid (one part to two hundred of water) before they leave the sick-room.

Dr. Elisha Harris gives (in one of the "Hampton Tracts for the People") the following "Simple Rules for Disinfection:"

(1.) **Quicklime.**—To absorb moisture and putrid fluids, use fresh stone lime finely broken; sprinkle it on the place to be dried, and in damp rooms place a number of plates or pans filled with the lime powder; whitewash with pure quicklime, freshly slacked in water, and of a creamy thickness. If the apartment has been contaminated by foul gases or by contagious disease, add a pint of the best fluid carbolic acid to every gallon of fluid whitewash.

(2.) **Charcoal Powder.**—To absorb the putrid gases, the coal must be _dry and fresh_, and should be combined with _Lime;_ this compound is the _Calx Powder_, as sold in shops. Any person can prepare it.

(3.) **Chloride of Lime.**—To give off _chlorine_, to destroy putrid effluvia and to stop putrefaction, use it as lime is used, and if in cellars or closets the _chlorine gas_ is wanted, pour strong *vinegar* or diluted *sulphuric acid* upon plates of Chloride of Lime occasionally, and add more of the Chloride.

(4.) **Carbolic Acid.**—This may be diluted at the rate of from *ten* to *forty* parts of water to *one* of the best fluid acid. Use this solution for the same purpose as copperas is used ; also to sprinkle upon any kind of garbage or decaying matter, and on foul surfaces of any kind. For sprinkling foul or unhealthy streets and alleys, gutters and foul places, mix 1 to 3 parts of this crude carbolic acid with 100 parts of the copperas solution, and sprinkle the entire surface every four or five days during the summer and autumn. It is chiefly valuable to prevent putrefaction.

(5.) **Sulphate of Iron (Copperas) and Carbolic Acid.**—To disinfect privies, cesspools, drains, and sewers, and especially the vessels, grounds, or places in which the discharge from the sick with any fever or diarrhœal diseases are evacuated or cast away : Dissolve eight or ten pounds of sulphate of iron in three or four gallons of water, and add a pint of best fluid carbolic acid (if it is desired, and a good quality is at hand), stir or agitate it briskly to make a complete solution. Use this disinfectant as follows :

To keep privies and water-closets from becoming infected or offensive, pour a pint of this solution into every water-closet pan or privy seat, every evening.

To disinfect masses of filth, privies, sewers, or drains, gradually pour in the solution, hour by hour, until every part of the mass or foul surface has been thoroughly disinfected.

To disinfect the discharges from the sick, let a small quantity of this solution be constantly kept in all vessels into which the discharges are voided from the body, and let every privy and every place where the discharges are cast away be thoroughly saturated with the disinfecting solution. Wherever cholera or dysentery, and any of the fevers are present in a house or neighborhood, and wherever persons are arriving from infected places, the daily use of this disinfecting fluid should be maintained. *Bed-pans and chamber-vessels* are disinfected with this strong solution, using a gill or more at a time, and emptying and thoroughly cleansing as soon as used by the patient.

(6.) **Sulphate of Zinc and Common Salt.** —Dissolved together in water in the proportions of 4 ounces of the zinc sulphate to 2 ounces of salt to the gallon of water, is one of the best disinfectants for cloth'ng as soon as taken from the sick, or their beds. All infected clothing should be packed in such a disinfecting fluid until it is boiled in the wash. This is the disinfectant for clothing recommended by the National Board of Health. (See DEODORANTS and FUMIGATION.)

DISLOCATION means the displacement of one or more bones, and a separation, either partial or wide and complete, of those surfaces which are covered by cartilage or gristle, and in their natural condition remain in close contact forming a joint or articulation ; thus, when the shoulder or elbow is said to have been " put out," there is a dislocation or separation of the upper extremity of the humerus from the shoulder-blade, and of the bones of the fore-arm from the lower end of the humerus or long single bone of the arm. Dislocation is in most instances caused by external violence, generally a fall. Now and then a joint which had previously been severely injured, undergoes what is called spontaneous dislocation, the bones being suddenly displaced by rapid or excessive use of their muscles. This form of dislocation is occasionally met with in the shoulder and lower-jaw. In most cases of dislocation caused by external violence, the ligaments or sinews which fasten together the ends of bones forming joints are torn and stretched, and in many instances the skin is bruised and the soft parts about the joint are swollen and very painful. The most severe form of injury is that called by surgeons a compound dislocation ; here a deep wound is present, which extends from the surface of the injured region to the dislocated bone and leads directly into the joint. Simple dislocation at one of the joints of the arm or leg when *promptly* and properly treated, is by no means so serious an accident as fracture in a similar situation, but when overlooked and not reduced shortly after the accident, becomes a source of great and long-continued annoyance to the patient. If surgical assistance be put off for three or four days, the nature of the injury will be obscured by swelling of the soft parts about the joint, and any attempt to move the damaged limb will then cause great pain. The difficulty and pain attending the reduction of a dislocation, bears a direct proportion to the duration of the injury. In the great majority of cases of dislocation at a large joint, it is impossible after an interval of twelve hours to replace the separated bones without administering some anæsthetic. Dislocation is by no means so frequent an accident as fracture. According to the statistics of the French surgeon Malgaigne, the proportion is as one to eight. The joint which is most prone to dislocation is the shoulder ; next in frequency are the hip, elbow, and ankle. The number of cases of dislocated shoulder far exceeds those of dislocations of all other joints put together.

The signs which indicate a recent dislocation are : inability to move that part of the limb immediately below the seat of injury, the displaced extremity of the bone being fixed in its unnatural position ; pain in the injured joint ; this, though usually not so acute as the pain attending a broken bone, varies in intensity in different cases ; sometimes the displaced bone presses upon one or more large nerves and then causes much suffering. Inability to move the joint, and pain, are in most cases associated with some swelling of the soft parts near the seat of dislocation, and with bruising of the skin ; the latter however, are not constant symptoms. Finally, but more important than any other sign, there is deformity in the joint ; this can usually be recognized at the first glance ; sometimes there is flattening as in dislocation at the shoulder, and at other times well-marked unnatural prominence of one or more bones, as in the elbow and ankle. It should be borne in mind that all these symptoms are common to dislocation and to fracture near the joint-end of a bone. It is very often an extremely difficult point with surgeons to decide whether a certain injury in the region of a joint be a fracture, a dislocation, or a combination of these two kinds of injury. In fracture the segment of bone below the injury is generally very movable, and one may generally detect on moving the broken parts a peculiar grating noise, called by surgeons *crepitus*, which is felt rather than heard both by patient and medical attendant. In pure dislocation, this crepitus is absent. Another distinction between dislocation and fracture consists in this ; when once the deformity attending the former injury has been removed by surgical manipulation, or setting of the joint as it is called, it does not return, as the head of the displaced bone when brought back into its socket remains there ; with fracture, on the other hand, there is a constant tendency for the fragments of bone to become displaced, until they are joined together by young bone in the course of the treatment. In consequence of the resistance of the muscles which move the dislocated bone, the treatment of this injury, even when quite recent, is by no means easy, and necessitates, particularly in large joints like the shoulder and hip, the application of a considerable amount of force. If the dislocation has been left unreduced for a week or ten days, this difficulty is much increased, and it becomes necessary to place the patient thoroughly under the influence of chloroform, for the purpose of relaxing the tense muscles which prevent the return of the displaced bone. In dislocation of six weeks' or two months' standing, the chances of a reduction will depend upon the situation of the injury. If the shoulder be affected, well directed attempts under the influence of an anæsthetic will probably replace the head of the bone. With the hip, however, the chances of recovery are not so good at the end of a month after the receipt of the injury. After an interval of six months the case is generally hopeless ; the limb, however, does not remain immovable and quite useless ; the head of the dislocated bone by its pressure forms for itself a fresh socket in that portion

of the adjacent bone on which it was thrown at the time of the accident ; fresh sinews are formed, the muscles adapt themselves to the altered state of things, and a new joint is formed which allows the patient considerable, though far from perfect, use of his arm or leg. This favorable process is carried on only in strong and healthy individuals.

The surgical treatment of dislocation consists : in pulling at the part below the injured joint—extension ; in keeping the parts immediately above the joint fixed—counter-extension ; and in endeavoring with the hand to elevate or replace the dislocated bone—manipulation. In recent cases of dislocation of the shoulder and elbow and of most smaller joints, a sufficient amount of extension may be obtained by the unaided efforts of the surgeon and one or two assistants ; but in recent displacement of the upper extremity of the thigh bone, and in most old dislocations of other joints, the pulleys are required. Counter-extension is generally kept up by means of a napkin, jack-towel or folded table cloth. The reduction of the bone is indicated by a sudden snap, the form of the joint and its functions are at once restored, and the pain is very much relieved, the whole limb also recovers its natural length and position in relation to the rest of the body. The subsequent treatment consists in confining the limb for a period varying with the size and situation of the injured joint. In dislocation of the shoulder, the arm is bandaged to the side of the body for about two weeks, and in dislocation of the hip it is necessary for the patient to remain in bed for some time. By keeping the joint at rest, the pain and irritation caused by the injury are removed, and the torn sinews are repaired. The muscular power and general tone of the limb are restored by shampooing and friction, affusion of cold water, and the use of liniments. Premature use of an injured limb prevents the complete repair of the structures of the joint, and favors a return of the dislocation on the application of slight violence. Recovery can never be complete, as the dislocated joint always remains weaker than any of the other sound joints. In a healthy and young or middle-aged subject, whose injury has been properly treated, this difference is scarcely appreciable, but in old people and those who are rheumatic or gouty the joint remains more or less stiff and painful, and is much affected during climatic changes and after exposure to cold and wet. The immediate and provisional treatment of a patient who has received a severe injury producing dislocation of some important joint—the hip, for instance —should be the same as that carried out in a recent case of fracture. (See FRACTURE.) In the following paragraphs a short sketch will be given of the symptoms and management of the most common forms of dislocation :

Lower-Jaw.—This bone when dislocated is carried forward, and its front part is depressed so that the lower row of teeth projects beyond the middle teeth of the jaw above. The jaw is fixed in this position, and the mouth cannot be closed, the cheeks, are flattened, the tongue is slightly protruded, and the saliva flows over the lower lip and chin. There is generally severe pain below the ears. This dislocation may be produced by opening the mouth very widely, as in yawning or attempting to masticate large pieces of food, or by a blow or fall upon the chin when the mouth is open. So long as the jaw is in this unnatural position the patient is unable to speak plainly or to swallow. Sometimes only one head of the bone is put out of place, and then the teeth are displaced laterally away from the seat of dislocation. This accident, as has been stated above, may in many cases be easily produced by slight causes. When surgical aid cannot be obtained, an attempt may be made to reduce this dislocation by placing the thumbs, protected by pieces of linen, over the last teeth on either side of the lower jaw, and, while depressing these teeth by raising the chin with the fingers, the jaw will then probably return with a sudden snap. When the dislocation occurs only on one side, one thumb only is to be placed on the corresponding molar teeth. The jaw should then be kept in its place by a handkerchief or bandage, and the patient must for many weeks restrict as far as possible the movements of the bone, for the sake of avoiding a recurrence of the dislocation.

Shoulder-Joint.—Dislocation occurs more frequently in this than in any other joint of the body. The hemispherical head of the arm-bone moves on a very shallow socket on the outer margin of the shoulder-blade, and is retained there by a loose membranous sac called the capsular ligament; this arrangement admits of great mobility and free latitude of motion to the arm-bone, which on the application of great force, either directly upon the shoulder or indirectly through the arm, may be readily displaced. The usual causes of dislocation in this region are blows upon the shoulder or falls upon the elbow or hand. Dislocations of the shoulder vary very much in the direction taken by the head of the displaced arm-bone. It is most frequently forced downward into the arm pit, but may be carried inward, forward, or backward. The following are the signs of a downward dislocation of the arm-bone: pain and loss of motion in the joint, slight elongation of the upper extremity and tilting outward of the elbow, which cannot be brought in contact with the side of the body without causing the patient great pain; the patient leans over to the side of the dislocation, and supports the elbow of this extremity with the opposite

hand; the hollow of the arm-pit is occupied by a hard swelling which moves with the rest of the arm, and can be felt distinctly to be the displaced extremity of the bone. In the absence of surgical aid, an attempt may be made to reduce this dislocation by raising the injured arm from the side of the body, the patient being seated in a chair, and placing the knee in the arm-pit, and then gently depressing the arm over this, which should serve as a fulcrum. Another method, which however, is not so safe, is to place the heel, the boot having been removed, in the arm-pit and to pull down the whole arm, the grasp being taken either at the wrist or just above the elbow. The patient must lie at full length on a mattress. If any grating should be felt or heard on moving the injured limb, these attempts ought not to be continued. Dislocation at the shoulder-joint is very liable to return, unless care be taken to confine the injured arm in bandages for some weeks after the reduction.

Elbow.—Dislocations at this joint are common in children. The most frequent form is the displacement of the bones of the fore-arm backward; they are sometimes displaced to one or other side, and, in some rare instances, forward. The backward dislocation is usually caused by a fall on the palm of the hand. The following are the symptoms of this injury: The whole of the upper extremity appears to be shorter than its fellow; the fore-arm is half bent and the thumb and outer surface of the wrist is turned forward; at the back of the elbow there is a considerable hard projection formed by the dislocated upper extremity of the cubit; in front of the elbow the lower extremity of the arm-bone is unnaturally prominent. Every attempt to bend or straighten the fore-arm causes acute pain. In lateral dislocation, which is always partial, the nature of the injury is indicated by increased width of the elbow, particularly in front, and by unnatural prominence of one or other of the lateral projections at the lower extremity of the arm-bone. The lateral dislocation more frequently occurs outward. The reduction of a recent dislocation at the elbow can generally be effected without much difficulty; the patient having been placed in a chair, the surgeon, resting his foot on the seat, applies his knee to the front of the elbow, he then grasps the wrist and bends the forearm around his knee, taking care at the same time to press backward the upper extremities of the bones of the fore-arm, in order to free them from the lower part of the single bone of the arm. If this method should not succeed, an attempt may be made to reduce the bones by forcibly extending the fore-arm, while an assistant keeps up counter-extension by grasping the arm firmly above the elbow and holding it in position.

Thumb.—The first, or metacarpal bone, which extends from the wrist toward the web of the digits, is sometimes dislocated either backward or forward at its upper extremity ; this, however, is a rare accident. The second and third bones are frequently dislocated backward over the heads of the bones above, in consequence of falls upon the end or contracted surface of the thumb. Forward dislocations also occur, though very rarely. In dislocation of the second from the head of the first bone, the reduction in many instances may be readily performed, either by pulling out the end of the thumb, or by forcibly bending the displaced portion backward, and pressing forward the dislocated extremity of the second bone. When the injured thumb is short, or a firm grasp it cannot be obtained for the fingers, a finger bandage or piece of broad tape may be tied round the thumb in a *clove-hitch*, and extension made with this. In some cases of the above dislocation, and in almost all cases of dislocation of the third bone of the thumb, reduction is extremely difficult, and before it can be effected it is often necessary to administer chloroform, and even to perform a cutting operation. Dislocations of the fingers are not so frequent as those of the thumb, and are treated in same manner.

Hip-Joint.—Dislocation occurs more frequently at this than at any other joint of the lower extremity, although the rounded head of the thigh bone works in a deep socket, is retained there by strong ligaments, and is protected in all directions by thick and powerful muscles. There are several varieties of dislocation at the hip-joint. The most common form is displacement of the head of the thigh bone, backward and upward, upon the back of the large hip-bone. This injury is usually caused by the individual falling while bearing on his shoulders a heavy load, or by a fall of some large and heavy mass upon his shoulders while the body is bent forward. The following are the symptoms : The injured limb is shortened to the extent of one inch and a half or two inches ; the knee is turned inward and bent forward ; the foot is also inverted and the toes rest upon the surface of the sound foot : the head of the thigh-bone forms an unnatural projection above and behind the situation of the hip-joint, and the natural roundness of this joint has disappeared ; the lower extremity cannot be moved by the patient without suffering. The reduction of this, as of all other dislocations of the hip, whether they be old or quite recent, requires the application of much force, and it is necessary in most cases to put the patient under the influence of chloroform. There are two methods of restoring the head of the thigh-bone to its socket ; it may be done by forcible traction or by what is called manipulation. If the former method is to be

practised, the patient is laid on his back upon a mattress, and a jack-towel is passed between the dislocated hip and the perineum, and fixed behind the patient's head, either to a bed-post or to a hook firmly screwed into the wall ; while counter-extension is kept up by this, traction is made on the thigh-bone by cords and pulleys fixed at one end to a staple in the wall in front of the patient, and the other to a padded leather belt or long towel fastened around the lower part of the thigh, the knee being turned inward and the whole thigh brought over the opposite limb. The surgeon sometimes endeavors to raise the displaced head of the thigh-bone by means of a second jack-towel passed under the upper part of the thigh, and around the back of his neck and shoulders. In reduction by manipulation no pulleys or apparatus of any kind are required. The patient having been put under the influence of chloroform, the surgeon bends the knee and hip of the injured extremity to the utmost extent, then rolls the thigh-bone outward, and finally abducts the thigh or forces it directly outward and away from the median line. During these manœuvres the head of the thigh-bone generally slips with an audible noise into its socket. *When no surgical assistance is at hand*, the patient should be laid on his back, and his legs be fastened together above the knees by a broad bandage. In cases where reduction has not been effected, there is much lameness and pain in the injured joint when the patient gets up for the first time, but in the course of a few months a new joint is formed, and the range of movement is much increased. The limb is of course shortened and patient must always limp, but with the aid of a thick sole to the boot worn on the injured side, these inconveniences may be very much diminished. The next in frequency to the above is dislocation forward and downward. Here the limb is lengthened to the extent of two inches or more ; the knee is bent forward and the whole limb is widely separated from its fellow ; the body is bent forward and slightly directed to the injured side ; the foot is pointed forward and downward. This dislocation may be reduced by the pulleys, counter-extension being made as in the former injury by a jack-towel passed around the crutch. Extension of the limb having been made downward and outward, the head of the bone should be pulled outward by means of a towel applied round the upper part of the thigh. Dislocation forward and upward is a very rare injury. According to Sir A. Cooper, " it happens when a person while walking puts his foot into some unexpected hollow on the ground ; and his body, being at the moment bent backward, the head of the bone is thrown forward upon the pubic bone. In this injury the foot is turned outward and the whole

limb separated from its fellow, rolled outward and shortened. In the groin can be felt a
large hard mass which is the displaced head
of the thigh-bone. In the reduction of this
dislocation the surgeon makes extension of
the limb downward and in a line behind the
axis of the body, so that the thigh-bone may
be dragged backwards ; to effect this the surgeon places the patient near the edge of the
bed or couch, so that the injured limb can
hang down.''

Knee-joint.—The leg may be displaced forward, backward, or to either side. The
dislocation is nearly always partial, as might
be expected from the extent of the opposed
surfaces of the thigh-bone and the tibia, or
larger bone of the leg. In lateral displacement the nature of the injury is apparent at
first sight, in consequence of the unnatural
projection of the inner or outer condyle of the
thigh-bone. The leg is generally twisted
upon its axis. These injuries are usually
caused by violent and sudden twists of the
knee or by heavy blows. Dislocations forward or backward are most serious injuries,
and are associated with much tearing of the
ligaments and soft parts around the articulation ; in the backward dislocation, the lower
end of the thigh-bone projects in front and
the hollow at the back of the joint is occupied by the displaced head of the leg-bone.
The dislocation forward, which is a rare injury, is associated with rupture of all the important ligaments of the knee and of some of
the hamstring tendons. Reduction may be
effected by extending the leg while the thigh
is fixed by counter-extension. After reduction, cold should be applied to the injured
knee, and the patient be kept in bed for about
three weeks.

Knee-cap.—This bone may be displaced
inward, outward, or upward. In some
cases, it is half twisted upon its axis, so that
its outer or inner edge rests upon the front of
the lower extremity of the thigh-bone. The
most frequent injury is dislocation outward ;
this, like the other varieties of dislocation, is
generally caused either by a blow on the side
of the knee, or by a sudden lateral movement of the body made by the patient, in
order to avoid being knocked down by any
passing object, or being run over. The bone
may be either partially or completely displaced. In the dislocation edgeways, the
knee-cap forms a very distinct unnatural
prominence in front of the knee ; and its
sharp edge, in most cases the outer one, can
be felt immediately under the tense skin. A
laterally displaced knee-cap may generally
be restored to its proper position without
difficulty by raising the limb high above the
level of the body, and then depressing the
prominent edge of the dislocated bone.
Sometimes the bone may be taken between
the thumb and finger, and lifted into its prop-

er place. A twisted knee-cap cannot be replaced so easily, and sometimes remains im-
movably fixed. Reduction of this form of
dislocation may be produced by bending the
knee forcibly and suddenly ; or, if this does
not succeed, by making the whole limb
straight, and then pressing down the prominent edge of the bone. The subsequent pain
and inflammation in the joint should be
treated by keeping the knee at rest, and by
applying ice or frequently renewed cold compresses. A person while walking sometimes
strikes the everted foot against some hard projecting object, and immediately feels an acute
pain in the knee, which prevents him from
walking. The leg, though stiff and painful when
the patient is erect, can be readily moved when
he lies down. After the accident the knee
begins to swell. Similar symptoms are sometimes produced through a sudden inward
twist of the leg upon the thigh. In cases of
this kind, there is dislocation of the semilunar cartilages, two flat gristly structures of
a horse-shoe shape, which are fixed to the
margins of the upper surface of the leg-bone.
One or both of these become detached and
slip between the opposed surfaces of the
bones of the thigh and leg, and so give rise
to the severe sickening pain which is so characteristic of this injury. Reduction may be
readily effected by first extending the leg
upon the thigh, and then suddenly bending it
backward until the heel touches the corresponding buttock ; the other hand of the surgeon being placed at the same time upon the
front of the knee. This dislocation is very
apt to return, and for this reason the patient
should, when he takes exercise, wear a bandage or tight knee-cap.

Ankle-joint.—Dislocation at this joint is
generally associated with fracture of one or
both bones of the leg. The uncomplicated
dislocations are those in which the foot is
moved forward, or its upper part driven upward, between the two bones of the leg.
Both these injuries are extremely rare. In
those dislocations which are complicated with
fracture, the foot may be dislocated outward, inward, or backward. In the first
and third class of cases, the slender outer
bone of the leg is generally broken ; and in
the second class both bones across their
lower ends. In simple dislocation of the foot
forward or upward an attempt may be
made to bring about reduction by grasping
the instep of the injured foot with one hand
and the heel with the other, and making extension, while the leg is kept steady by an assistant. In the complicated dislocations, the
chief object of treatment will be to reduce
the fracture, and to retain the foot in its proper position by splints until the bones are set.
The astragalus, an irregularly shaped bone
which articulates with the lower surfaces of
the bones of the leg, is sometimes thrown

forward upon the upper surface of the foot, and forms there a hard prominent tumor. This, if it cannot be reduced by forcibly extending the foot while the patient is under chloroform, may give rise to much local mischief, and the formation of abscesses, which will necessitate its removal by a surgical operation.

DISPENSARIES are institutions founded and kept up by charitable people for the relief of the poor. They differ from hospitals in not having beds for in-patients, and in the fact that out-patients are visited at their homes by the physicians and surgeons, or by the resident medical officer. Out-patients attend also for advice at the dispensary at certain fixed times, and for this they must, in the first place, obtain a letter of admission from a governor or subscriber.

DISSECTION WOUNDS.—Under this heading we may conveniently class not only such poisoned wounds as are encountered by those professionally engaged in the examination of dead bodies, but such as are frequently met with in individuals, who may be in any way exposed to contact of decaying or putrescent animal matter introduced into the system by some local wound or abrasion. During the decomposition of animal matter, substances are formed which have a most deleterious effect if introduced into the blood of a living animal. When the patient suffers from the effects of inhalation of such poison only, he exhibits symptoms of sickness, diarrhœa, or dyspepsia, and the poison is quickly eliminated by change of air, stimulants, or aperients. But in the case of such poison being inoculated, the symptoms are severe and are frequently fatal. The most dangerous cases are those where the body dissected has died recently of erysipelas, puerperal fever, or pyæmia ; much more so than from the body in a mere state of advanced putrefaction ; and the more recently after death the greater the danger, as it seems that the virus has a tendency to neutralize itself after a short time. In dissecting-rooms, after a puncture or scratch, there is rarely any great danger of more than a local sore, as the modern antiseptic processes of injecting the vessels for preserving the subjects for the lengthened period necessary for dissection, have rendered them all but innocuous, unless, of course, the operator is in a low state of general health (in which instance the atmosphere of dissecting-rooms and hospital wards should be most studiously avoided). It is not usually the *severe* wound which infects, as the individual recognizes it at once, and sucking it, or the immediate application of a caustic, or poulticing, generally reduces the effects to mere locality, but it is from some insidious scratch or abrasion, which has existed perhaps some time, and has been so small as to pass unheeded, that the more serious cases

arise, and at some period, perhaps six or eighteen hours later, the patient begins to feel unwell, depressed, sick, shivery, with severe headache, and a sharp, rapid pulse. Supposing a finger to be the seat of inoculation ; pain and tenderness in the shoulder, perhaps, at first directs his attention to his case ; afterward there is severe pain and swelling in the arm-pit, and upon examination there will be seen red, regular lines, along the fore and upper arms, proceeding from the seat of inoculation toward the arm-pit, indicating the course of the lymphatics. Abscess forms after a while, perhaps, accompanied with diffuse suppuration of the surrounding areolar tissue. There is always intense constitutional disturbance. Sometimes the influence of the morbid poison is so virulent that the patient dies of the precursory fever before sufficient time has elapsed for any local disease to appear, and there are cases on record of patients dying within forty-eight hours after the receipt of the wound. Sometimes diffuse cellular abscesses occur in remote parts, such as the knee or hip. Sometimes diffuse inflammation commences at the seat of injury and extends up the arm, accompanied by cutaneous erysipelas.

Treatment.—With regard to the treatment, the indications are to endeavor to eliminate the poison from the blood, to support the strength and relieve pain, and promote the discharge of pus. In the first instance, by purging, diaphoretics, and diuretics, these to be maintained till elimination seems to be complete ; afterward tonics, fresh air, and exercise. In more urgent cases, calomel and opium are frequently beneficial, and after suppuration has been freely established, iron, bark and ammonia are indicated—all abscesses should be opened at once. Thirst should be quenched by effervescent drinks ; beef-tea, wine, or brandy should be given to support the pulse. In very severe cases doses of quinine and mineral acids are of great service. Locally, hot fomentations, poultices to the inflamed and swollen axilla, or elbow, and free incision as soon as any decided swelling with softening be detected, and it is mainly upon prompt attention to the evacuation of pus or serum that success in treatment depends.

DISTILLED SPIRITS are made by distilling alcohol from some of the various forms of fermented liquor in which it exists. Distilled spirit is not, however, pure alcohol ; but contains varying quantities of water. A spirit having a density of 920, water being 1000, is called proof spirit in this country, and when distilled spirits contain more or less alcohol than this, they are said to be under or above proof. The most common forms of distilled spirits which are used in this country are brandy, gin, rum, and whisky. *Brandy* is distilled from wine, and its peculiar flavor is produced by the addition of peach kernels to the liquid while distilling. It also contains œnanthic and acetic ethers. *Gin* is obtained from fermented grain, to which the berries of the juniper are added to give a flavor. Other flavoring substances are used, such as cinnamon, cloves, etc., and also sugar for making what is called " Cordial Gin." *Rum* is procured from fermented sugar and molasses in the West Indies. Its peculiar odor depends on butyric ether, and a flavor is sometimes given by adding pineapples. *Whisky* is distilled from fermented grain. Many other distilled spirits are drunk in various parts of the world. *Arrack* is made in the East from rice or from betel nuts, or the sap of various species of palm. *Liqueurs* are also alcohol distilled with various substances to give it a flavor, and large quantities of sugar are also added. The favorite liquor of the French is *Absinthe*, which is a spirit distilled from wormwood. Spirits are made from all fruits containing sugar, as apples, oranges, pears, artichokes, maize, and other things. Honey is capable of vinous fermentation, and a beverage called *Mead* is made from it. Proof-spirit is used for making tinctures. These consist principally of solutions of vegetable substances directly in the proof-spirit. Sometimes vegetable substances are added to proof-spirit and then distilled. Such preparations are called Spirits in the Pharmacopœias.

DISTILLED WATER. (See WATER.)

DIURETICS are medicines which increase the flow of urine, whether directly or indirectly ; as a class, it is an extremely unsatisfactory one, though doubtless many substances have the power of increasing this secretion. The flow of urine may be increased in various ways : thus, in disease of the heart, by strengthening the action of that organ, as is done by digitalis. But there are certain substances which seem to act directly on the kidney, and to stimulate it in such a way as to give rise to a free flow of urine. Such are cantharides and turpentine. But they must be given cautiously, or they will create irritation and really diminish the flow. Juniper, too, acts in this way. We may also foster the flow of urine by introducing certain salts into the system ; these have to be got rid of somehow, and are usually expelled by the kidney, notably increasing the flow of water along with them. Such are citrate and acetate of potassa, cream of tartar, etc. One of the most efficient and most readily obtainable diuretics is broom. The tops are boiled and the fluid used. This, combined with cream of tartar and juniper or gin, which is flavored by juniper, will often be found to be an efficient and harmless remedy in cases of dropsy, for it is in that disease that diuretics are mostly required. Alcohol itself is diuretic, and a glass of beer will often produce a copious flow of urine. Alkaline diuretics are frequently of use in getting rid of refuse material which, if not expelled, would give rise to maladies such as gout.

DIZZINESS. (See VERTIGO.)

DONOVAN'S SOLUTION was the name given to a valuable combination of arsenic, iodine, and mercury. It was found to be of special value in the treatment of skin diseases connected with syphilis, but not limited to these. It has fallen greatly out of practice, and is not now contained in the Pharmacopœia. Ten to twenty drops were given for a dose, but it had to be given with all the precautions attaching to other preparations of arsenic.

DOUCHE signifies a stream of hot or cold water which is poured over the body ; it is used in the ordinary process of shampooing, and sometimes it is ordered as a remedial agent, as in cases of chorea and hysteria.

DOVER'S POWDER, so called from its inventor, is known in the Pharmacopœia as Compound Ipecacuanha Powder. It contains ipecacuanha and opium, a grain of each in ten of the compound powder, and so it must be prescribed with a due regard to the quantity of opium it contains. Ten grains is the usual full dose. It was originally a secret preparation, and then contained some saltpetre along with the two other ingredients ; now instead it contains sulphate of potass. It is a powerful diaphoretic—that is to say, it promotes free perspiration, and is consequently of great use in many maladies. It does not agree with everybody, and at all times it is advisable to take precautions against cold after its use. In the feverish stage of a common cold, this remedy is particularly valuable, and frequently cuts short the malady. When the patient is cold and shivery, but the skin hot and the nose stuffed, ten grains of the powder at bed-time, putting the feet in hot water at the same time, and promptly getting covered over with the bedclothes, will commonly induce a profuse sweat, and will probably greatly benefit the patient. A cold sponge is advisable next morning, and the bowels must be seen to if confined. In the dropsy which follows scarlet fever this remedy is frequently used, but must be used with great caution ; other diaphoretics, especially those which contain no opium, should have the preference.

DRAGON'S BLOOD is the common name of an Indian plant (*Pterocarpus draco*), from which exudes a red-colored resin—once used in medicine as an astringent—but now only employed to color tooth-powder, or by French polishers to give a deep color to their wood.

DRAINAGE. (See HOUSE.)

DRAIN-FEVER. (See TYPHOID FEVER.)

DRASTICS is a name given to purgatives whose action is somewhat violent; such are elaterium, gamboge, jalap, scammony and the like. Put it down as a broad rule that they should never be used except by medical advice. More constipation is caused by the use of such remedies than perhaps arises from any other cause whatever.

DRESS. (See CLOTHING.)

DROPPED WRIST is an affection met with among painters, and others who work much with lead : it consists in a paralysis and wasting of the muscles of the arm ; the right arm is generally more affected than the left, and the muscles on the outside of, more than those on the inside of the arm ; the result of the wasting is that the patient cannot raise the hand when the palm is turned downward. (See LEAD-POISONING.)

DROPSY is a term applied to any accumulation or effusion of fluid under the skin or in a cavity of the body, occurring in diseases of the heart, liver, lungs, or kidneys. Sometimes the legs only are swollen, and this may proceed simply from enlarged veins in the upper part of the leg without any other serious disorder. At other times the abdomen becomes swollen to a very great size, and when the breathing becomes impaired the operation of tapping and drawing off the fluid may be had recourse to. Dropsy affects the most dependent parts, and hence the legs are more swollen at night after walking about. When the dropsy is all over the body, as in some diseases of the kidney, it is called *Anasarca* ; when limited to one part, it is spoken of as *Œdema* of that part ; when in the abdominal cavity it is called *Ascites ;* in the cavity of the chest, *Hydrothorax ;* when in the cranial cavity the name *Hydrocephalus* is applied. Under these headings more detailed accounts are given.

DROWNING is a frequent form of violent death. In a case of pure drowning the individual at first sinks to a certain depth and then ascends to the surface of the water, where, if he be not a good swimmer, he struggles to clear his lungs and mouth, and to obtain fresh air. As water is generally taken in with the inspired air the patient sinks again for a short distance, and then by his exertions again succeeds in reaching the surface. These struggles are repeated until the lungs and stomach are filled with water, and the general specific gravity of the body is increased. The body then sinks to the bottom of the water. The duration of this contest for life will vary according to the sex, age, strength, and general condition of the individual. Fat persons float more readily than those who have large bones without any unusual amount of adipose tissue. Women and children float longer than adult males, their skeletons being smaller and the fat more abundant. The loose flowing clothes of the female

tend to buoy up the body, in consequence of their spreading out and floating upon the surface of the water ; the tight-fitting dress of the male on the other hand, when thoroughly saturated with water, has a tendency to sink the body more deeply. After the final submersion the dying individual still endeavors to breathe, and the remaining portion of air is forced out from the lungs by the entrance of more water, and rises in bubbles. Death is preceded by convulsive movements of the extremities, the patient having by this time become unconscious and insensible. According to Dr. Taylor, who accidentally experienced all the phenomena of drowning up to this point, "there is not the least sensation of pain, and as in other cases of asphyxia, if the individual recover, there is a total unconsciousness of suffering during the period when the access of air was cut off from the lungs." The question as to how long a human being may be submerged, and yet be recoverable, has not yet been clearly settled. According to the officers of the Humane Society, persons who have been under water for more than four or five minutes do not generally recover. But on the other hand cases have been reported in which recovery took place after submersion lasting for fourteen minutes, and even half an hour. According to Dr. Taylor, however, the recorded cases of restoration after submersion of half an hour and upward are to be regarded as "extravagant fables." Dr. George Harley, who was a member of a committee appointed by the Royal Medical and Chirurgical Society to consider the subject of apnœa, believes "that if a person be completely submerged, and the entrance of water to, and the exit of air from, the lungs not prevented, recovery would be impossible after two minutes. On the other hand if the air passages were closed against the entrance of water, and the chest kept full of air, we see no reason for thinking that a human being would perish either more slowly or more quickly than a dog placed under similar circumstances—namely, in from four to five minutes." (Holmes's System of Surgery, article Apnœa.) It is held by many that those individuals who fell into a state of syncope or fainting at the time of submersion may survive for a longer period under water than those in whom the heart's action is maintained until the last, and who die from pure apnœa. The following are the appearances generally presented by a body which has been recovered shortly after death by drowning : The surface of the body cold and of a white color, mottled here and there by large patches of lividity, the face also pallid ; the jaws closed and the lips and nostrils covered by a frothy foam ; the tongue swollen, but not protruded ; the eyes half open and the upper lids livid and somewhat swollen ; the knees and elbows bent ; the hand clinched, and mud, or sand, and sometimes portions of weed, found included in their grasp, the skin of the fingers is sometimes excoriated, and mud or sand is found underneath the nails. The stomach and air-passages and sometimes the lungs contain much water. The vessels of the lungs are engorged with black fluid blood. All the important internal organs are much congested. The blood on the left side of the heart has been found to contain more water than the blood left in the cavities of the right side. The right side of the heart contains much more blood than the left side. In a body that has been in the water for a long time, general putrefaction has taken place. The skin where not covered by clothes is of green or blue color, and the face much swollen and distorted. The gases formed by putrefaction and decomposition of the tissues collect and render the body lighter than its bulk of water, so that it rises to the surface and floats there. The period at which the drowned body rises varies according to the depth of the water, the character of the water—whether it be salt or fresh, and its temperature. In inquiries as to how a body found in water came to its death : whether in the first place it was due or not to drowning, and next whether in the former case the drowning was accidental, suicidal, or homicidal, great importance is attached to the presence or absence of the following post-mortem appearances : excoriations of the fingers ; sand or mud under the nails ; portions of water-plants, or mud grasped in the hand ; a rough or contracted skin—the so-called goose-skin ; water in the stomach, especially when this contains plants, duckweed, and other substances resembling those which exist in the water from which the body has been taken ; froth on the mouth and nostrils ; mucous froth containing mud or sand in the air-passages ; water in the lungs. On no one of these signs ought a positive assertion to be based, but in the presence of all or of several of them the probability that death occurred from drowning is very great. The circumstances attending the death cannot very readily be determined, and the questions as to whether it was accidental or intentional, and whether it was the result of suicide or homicide, are extremely difficult to answer. When there are no marks of violence upon the surface of the body, this point cannot be considered by a medical man, and must be decided upon other evidence. When marks of injury are present, it has to be considered whether these might not have been caused by the fall of the individual against some hard substance at the time of immersion, or by the rubbing of the body against sharp and hard obstacles after death, or if the marks be such as to indicate intentional infliction before immersion, whether these were such as would be inflicted by one intending suicide.

Treatment of the apparently Drowned. —The following very useful directions have been published by the Royal National Life-boat Institution. The leading principles of these are founded on those of the late Dr. Marshall Hall, combined with those of Dr. H. R. Silvester, and are the result of extensive inquiries which were made by the institution in 1863–64 among medical men, medical bodies, and coroners throughout Great Britain. These directions are in use in Her Majesty's fleet, in the Coast-guard service, and at all the stations of the British army at home and abroad:

"I. Send immediately for medical assistance, blankets, and dry clothing, but proceed to treat the patient *instantly* on the spot, in the open air, with the face downward, whether on shore or afloat, exposing the face, neck, and chest to the wind, except in severe weather, and removing all tight clothing from the neck and chest, especially the braces. The points to be aimed at are—first and *immediately*, the restoration of breathing ; and secondly, after breathing is restored, the promotion of warmth and circulation. The efforts to *restore breathing* must be commenced immediately and energetically, and persevered in for one or two hours, or until a medical man has pronounced that life is extinct. Efforts to promote *warmth and circulation*, beyond removing. the wet clothes and drying the skin, must not be made until the first appearance of natural breathing ; for if circulation of the blood be induced before breathing has recommenced, the restoration to life will be endangered.

"II. To Restore Breathing.— *To clear the throat.* Place the patient on the floor or ground with the face downward, and one of the arms under the forehead, in which position all fluids will more readily escape by the mouth, and the tongue itself will fall forward, leaving the entrance into the windpipe free. Assist this operation by wiping and cleansing the mouth. If satisfactory breathing commences, use the treatment described below to promote warmth. If there be only slight breathing, or no breathing, or if the breathing fail, then turn the patient well and instantly on the side, supporting the head, and excite the nostrils with snuff, hartshorn, and smelling salts, or tickle the throat with a feather, etc., if they are at hand. Rub the chest and face warm, and dash cold water, or cold and hot water alternately on them. If there be no success, lose not a moment, but instantly *imitate breathing.* To imitate breathing, replace the patient on the face, raising and supporting the chest well on a folded coat or other article of dress. Turn the body very gently on the side, and a little beyond, and then briskly on the face, back again, repeating these measures cautiously, efficiently, and perseveringly, about fifteen times in the minute, or once every four or five sec.

onds, occasionally varying the side. (*By placing the patient on the chest, the weight of the body forces the air out ; when turned on the side this pressure is removed, and air enters the chest.*) On each occasion that the body is replaced on its face make uniform but efficient pressure with brisk movement on the back between and below the shoulder-blades or bones on each side, removing the pressure immediately before turning the body on the side. During the whole of the operations let one person attend solely to the movements of the head and of the arm placed under it. The result is *respiration* or *natural breathing*, and if not too late, *life*. While the above operations are being proceeded with, dry the hands and feet, and as soon as dry blankets or clothing can be procured, strip the body, and cover or gradually reclothe it, but taking care not to interfere with the efforts to restore breathing.

"III. Should these efforts not prove successful in the course of from two to five minutes, proceed to imitate breathing by Dr. Silvester's method, as follows : Place the body on the back on a flat surface inclined a little upward from the feet ; raise and support the head and shoulders on a small firm cushion or folded article of dress placed under the shoulder-blades. Draw forward the patient's tongue, and keep it projecting beyond the lips—an elastic band over the tongue and under the chin will answer the purpose, or a piece of string or tape may be tied around them, or by raising the lower jaw the teeth may be made to retain the tongue in that position—and remove all tight clothing from about the neck and chest, especially the braces. *To imitate the movements of breathing :* Standing at the patient's head, grasp the arms just above the elbows, and draw the arms gently and steadily upward above the head, and *keep them stretched* upward for two seconds. (By this means air is drawn into the lungs.) Then turn down the patient's arms, and press them gently and firmly for two seconds against the sides of the chest. (By this means air is pressed out of the lungs.) Repeat these measures alternately, deliberately, and perseveringly, about fifteen times in a minute, until a spontaneous effort to respire is perceived, immediately upon which cease to imitate the movements of breathing, and proceed to *induce circulation and warmth.*

"IV. Treatment after Natural Breathing has been Restored.—*To promote warmth and circulation*, commence rubbing the limbs upward, with firm grasping pressure and energy, using handkerchiefs, flannels, etc. (By this measure the blood is propelled along the veins toward the heart). The friction must be continued under the blanket or over the dry clothing. Promote the warmth of the body by the application of hot flannels, bot-

tles, or bladders of hot water, heated bricks, etc., to the pit of the stomach, the armpits, between the thighs, and to the soles of the feet. If the patient has been carried to a house after respiration has been restored, be careful to let the air play freely about the room. On the restoration of life, a teaspoonful of warm water should be given, and then, if the power of swallowing have returned, small quantities of wine, warm brandy-and-water, or coffee should be administered. The patient should be kept in bed, and a disposition to sleep encouraged.

"**General Observations.**—The above treatment should be persevered in for some hours, as it is an erroneous opinion that persons are irrecoverable because life does not soon make its appearance, persons having been restored after persevering for many hours.

"**Cautions.**—Prevent unnecessary crowding of persons round the body, especially if in an apartment. Avoid rough usage, and do not allow the body to remain on the back unless the tongue is secured. Under no circumstances hold the body up by the feet. On no account place the body in a warm bath unless under medical direction, and even then it should only be employed as a momentary excitant."

Dr. Bain's Method of Performing Artificial Respiration.—This method has been thus described by Dr. Bain himself : " The patient being laid on his back on a table, if convenient, the mouth and nostrils are to be wiped dry, the clothes from the upper part of the body, at least, having been removed. The operator stands at the head of the patient, placing the fingers of each hand in the armpits, in their front aspect, with the thumbs on the collar-bones, and pulls the shoulders horizontally toward him with a certain degree of power. Upon relaxing his pull the shoulders and chest return to their original state."

DROWSINESS is a symptom which naturally precedes sleep ; it is often the forerunner of serious mischief in those who are the subjects of Bright's disease, and it is then accompanied by a scanty flow of urine, and such persons may afterward pass into a fatal state of coma or insensibility. It occurs also as a result of living in an overcrowded or badly-ventilated room, in consequence of an accumulation of carbonic acid gas. It precedes the fatal stupor of those who are frozen to death in the snow.

DRUNKENNESS.—Alcohol, when swallowed, is speedily absorbed by the veins of the stomach and mixed with the blood, and then, by its poisonous action on the brain, spinal cord, and nerve trunks, produces the symptoms of acute alcoholism or drunkenness. In mild cases the pulse becomes rapid, the face hot and flushed, and the eyes bloodshot ; if more drink be taken, there is confusion of intellect and partial paralysis of the voluntary muscles, and the drinker feels giddy, reels, and experiences more or less difficulty in articulating properly, as the muscles of the tongue become paralyzed ; he becomes maudlin and afterward noisy and delirious, and finally sinks gradually in a state of deep stupor. On the following day there is general prostration, with nausea and occasional vomiting. In fatal cases of poisoning by alcohol, the state of stupor passes into one of true coma ; the drinker becomes quite unconscious and insensible, and cannot move ; respiration ceases, and finally the action of the heart is arrested. Some few instances have been recorded, in which speedy death had been caused by a very large quantity of spirit. Orfila mentioned an instance in which a soldier died instantly after drinking for a wager eight pints of brandy ; and Professor Christison, in his work on Poisons, refers to a case of a London cabman, who for a bribe of five shillings drank at a draught a whole bottle of gin, and in a few minutes dropped down dead. In cases of death from large quantities of alcohol, the patient speedily passes into a state of marked coma, which is sometimes accompanied with convulsions. The intensity and character of the symptoms of alcoholic poisoning, and the rapidity with which they come on, vary much in different persons, even when about the same quantity of alcohol has been taken by each.   The more concentrated the spirit, the more rapidly is drunkenness produced. The speedy absorption of alcohol into the blood is favored by an absence of food from the stomach. When much alcohol is taken on a full stomach the ordinary symptoms of drunkenness are associated with excessive vomiting. The mental symptoms, such as noisy talk, sentimental and maudlin utterances, and delirium, vary according to the character of the individual.   When excessive drinking is combined with the consumption of strong tobacco or cigars, drunkenness comes on quickly, and is indicated by much reeling, much mental confusion, and vomiting. During the state of drunkenness, alcohol is present in the urine and sweat, and the odor of the spirit or wine which has been taken is very perceptible in the breath. It has been stated that about one fourth of the quantity of alcohol is eliminated unchanged in the course of forty-eight hours. The action of alcohol on the nervous system is indicated by the double vision, the difficulty of articulation, the partial palsy of the muscles of the lower extremities, and the mental condition of the individual. Drunkenness may be produced by inspiring the concentrated vapor of alcohol. Persons employed in large wine-cellars, and who have been occupied for many hours in bottling spirits, and also anatomists, who

have been engaged in the dissection of specimens preserved in strong alcohol, may be readily intoxicated by the spirituous vapor.

Consideration of the predisposing causes of alcoholism will render evident the hopelessness of all attempts by mild or ordinary legislative means to prevent the consumption of stimulants so widely prevalent in this country. General education and increased wages to the laboring classes, with an amelioration in their moral and hygienic conditions, will no doubt produce vast improvement in this portion of the community, in respect to the diminution of drunkenness ; but still the occupations, both mental and bodily, of those who are the typical representatives of a highly civilized and commercial nation, necessitating, as they do, excessive energy and intense mental excitement and mental tension, which are invariably followed by nervous exhaustion and depression, must induce, in many instances, a craving for stimulants. Poverty, serious disappointments in life, and pecuniary embarrassments, are all predisposing causes of alcoholism. Monotony of occupation is also another frequent predisposing cause. Dr. Anstie, in an article on Alcoholism in Dr. Reynolds's System of Medicine, states that "amongst monotonous occupations, especially when combined with much confinement in close rooms, there are none which have furnished him with so many and such serious cases of alcoholism as the trades of shoemaker and barber. The want of active out-door exercise, of course, represses elimination and much increases the evil." Dr. Anstie has seen "few more desperate cases than some which have occurred in barbers, who have been habitually confined in miserably small shops, and at the same time have received enough money to pay for a great deal of drink." Finally, we meet with those unfortunate persons whose tendency to indulge in alcohol has been caused by an inherited morbid condition of the nervous system.

The diagnosis of advanced alcoholic intoxication is a question of great difficulty and importance, and has been much discussed of late in consequence of the increasing number of those unfortunate instances in which persons have been confined in police-cells while in a state of insensibility due to some other and perhaps fatal condition. A person is found lying in the streets in a state of complete insensibility, and it is then at once assumed by those who find him that he is dead drunk, whereas it is just as probable that his unfortunate condition may be due to one of the following causes : cerebral apoplexy, concussion from an injury to the head, compression of the brain from fracture of cranium or traumatic intra-cranial hemorrhage, opium-poisoning, or uræmic poisoning from disease of the kidney. A medical man, when asked to give an opinion on a case of this kind, has to pass over in his mind the characteristic symptoms of each of the above affections. If the face be flushed, and the conjunctivæ red and swollen, if the breath smell strongly of liquor, and if the man, when aroused, supposing it is possible to do so, talks maudlin or sentimental nonsense, the case is clearly one of drunkenness. If the face be pale, the surface of the body cold, the pupils contracted, and if the patient, when aroused, speaks but a few words, and then relapses into a state of unconsciousness, the case is considered to be one of concussion. If the breathing be stertorous, the face drawn on one side, the pupils dilated, one or more limbs paralyzed, and the patient in a state of confirmed coma, perfectly unconscious and insensible, the case will probably be regarded as one of cerebral compression, due either to apoplexy or to injury. In a doubtful case the medical man would endeavor to draw off some urine from the bladder by means of a catheter, and then if he found, on boiling this urine, or on adding to it a few drops of strong nitric acid, that there was a dense white and cloudy deposit, he would probably assume, in the absence of any other cause for the state of insensibility, that the patient was suffering from the effects of uræmic poisoning. It should be remembered, however, and this is the chief cause of all the difficulty and annoyance attending these cases, that although the insensible person may have been drinking freely, and that a strong odor of alcohol in the breath is most unmistakable, the insensibility may not be the direct effect of the drunkenness. He may have had a fall and injured his head ; fatal injuries to the brain may occur without any external signs, save a slight graze or bruise of the scalp. Apoplexy may have occurred while the man was in a state of intoxication. Again, a person not very intoxicated may be rendered insensible by exposure to cold and wet. And then again, it must be remembered that the symptoms of cerebral concussion or compression may be marked by the peculiar symptoms of alcoholic intoxication. So great is the difficulty attending the diagnosis of one of these so called police cases. Indeed, it may be said to be impossible, in a large majority of instances ; and therefore it ought, we think, to be laid down as a rule, that every person found lying insensible in the streets, or in any place where he is alone and away from his own residence, should at once be placed under medical treatment—if possible, in a hospital—until the real nature of his case has been cleared up by the further course of the symptoms. (See DELIRIUM TREMENS and DIPSOMANIA.)

DRY CUPPING. (See CUPPING.)

DULCAMARA, better known perhaps as bitter-sweet (*Solanum Dulcamara*) is a remedy of very doubtful value. It has a certain reputation in family practice, but when tested in public practice it has been found wanting. It has been commended for certain forms of skin disease, especially those of a scaly nature, but most probably it has no real influence over them.

DULCAMARINE is an extract, not a true proximate principle obtained from the twigs of bitter-sweet.

DUMB AGUE. (See INTERMITTENT FEVER.)

DUMBNESS is usually associated with deafness, and but few instances are met with where it is not so. Occasionally, however, there is some congenital malformation of the organs of speech, which prevents the power of articulation, and in rare instances it is recorded as arising from the entire neglect in childhood of exercising the function. There has recently been a system introduced by which the dumbness consequent on deafness is overcome, and the deaf person is made to articulate sounds by aid of sight—that is to say, after seeing a word formed and pronounced by the lips of another person many times, the power of imitation enables him to do the same, and produce an articulate sound, so that he can no longer be said to be dumb.

DURA MATER, a thick fibrous membrane, which lines the skull and spinal column, and forms a covering for the brain and spinal cord.

DYSENTERY, which is an inflammatory affection of the great gut, giving rise to ulceration, mucous, and blood stools, straining and much pain, is well known and very fatal. The causes of dysentery are not quite clear, but it generally appears among soldiers after long exposure to wet in low districts with insufficient food. It always tends to make its appearance in marshy districts where malaria prevails. It generally begins with some uneasiness and griping pains in the abdomen, and there is much desire to go to stool. At first this gives relief, but by and by no relief follows, so then the patient seems to desire to sit on the stool constantly. What comes away consists at first of badly formed motions, but by and by they become more scanty, then mucous and even bloody, sometimes mixed up with small hard masses called *scybalæ*. The desire continues to increase, the attempt to gratify the desire increases the pain, the stools alter more and more, becoming bloody, fœtid, and with shreds of membrane in them sometimes too ; there is purulent matter. The urine is frequently voided and is generally high colored and scalding. There is at the same time more or less fever, and there is great restlessness and sometimes cramps. The tongue is furred and dry. The pulse small and quick. Great thirst and complete loss of appetite. Perhaps these gradually abate, the purging and straining become less frequent, and the rest in the intervals is more complete, gradually the patient gets better, but his bowels remain in a troubled state for a long time to come. Sometimes, on the other hand, and this is especially the case where the malady is epidemic, the patient gets worse, the bowels become inflated and the abdomen tender, the tongue becomes dry and glazed or aphthæ form on it, and the insides of the cheeks. The evacuations are exceedingly offensive and passed under the patient, the whole body has a corpse-like odor, coma comes on, and death soon follows. Very frequently in warm countries or in epidemics, dysentery is complicated by ulcers of the liver. Sometimes the ulcerations perforate the gut and set up peritonitis, or the gut may mortify. In warm countries, too, the disease may become chronic, and the nutrition of the body is so sadly interfered with that the patient withers away. The bowels continue during this time very irregular, and the discharges most offensive.

Treatment.—Very much may be done by treatment, especially when the disease is not epidemic. The diet should be scanty but nutritious, hot poultices or cold compresses applied to the abdomen, and strict rest enjoined. It is desirable to remove all hardened fæces which may set up irritation, and for this purpose nothing suits so well, or gives the bowels so much relief, as copious injections of warm and very thin gruel. These having been removed, a totally different plan must be adopted : no more copious injections, but injections of an ounce or two of starch, containing 30 drops of the liquid extract of opium. This may be repeated if necessary. At the same time it is desirable to give internally full doses of ipecacuanha, consisting of not less than from 30 to 60 grains, in any form which may be deemed desirable. It may be repeated in 6 hours if necessary. If the patient gets over this, another kind of treatment must begin. Tonics must be given carefully, the bowels attended to, and every sign of relapse closely watched. The diet must then be nourishing, but not bulky. Remedies may be given to prevent the contents of the bowels from putrefying, which often causes much discomfort. Sulphate or hyposulphate of soda, or sulphocarbolate of soda, or carbolic acid, may be given for this purpose. If dysentery become chronic, change of climate is important, a mild and agreeable atmosphere doing great good.

Dysentery in Children.—This is much more frequent in this country than the dysentery of adults, and is caused by indigestible food, unripe or decayed fruit or vegetables, the breathing of impure air, exposure to cold, strong cathartic medicines, and bad or impure water. Its *symptoms* are a constant desire to

go to stool, resulting in small discharges of bloody mucus ; there is considerable fever, griping pain usually near the lower portion of the intestines, causing the child to scream at times as if in fright, with cold shivers, and more or less delirium. *Treatment.*— Domestic treatment of dysentery cannot be relied on, but in the absence of a physician proceed thus : If there has been any constipation during the previous day or two, give a moderate dose of castor oil ; when the bowels have moved, as a result of this, dissolve a teaspoonful of gum arabic in an ounce of peppermint water, and give a teaspoonful every half hour. A better mode of treatment, however, is this : Make a little thin starch, and to one tablespoonful of this add one drop of laudanum ; inject it into the child's bowels with a small syringe, and keep it there as long as possible. This should be repeated every four hours until the disease is arrested. Increase the quantity of laudanum by one drop for each year up to five. From the first the child should be kept as quiet as possible, as rest and warmth and a recumbent posture are essential to comfort and recovery.

Dysentery is infectious by evacuations, and therefore all bed-pans or other vessels used by the patient should be scalded each time with boiling water. The privy-vaults and water-closets should also be disinfected with sulphate of iron or carbolic acid.

DYSMENORRHŒA, or difficult menstruation, is a disease which is very harassing to the patient, and often very obstinate to cure. It affects more especially women who are nervous, or of a rheumatic and gouty tendency. It may occur at any time in the child-bearing period of life, and affects both the married and the single. The pain is felt in the lower part of the abdomen on each side, just above the groin and in the region of the ovaries ; pain is also felt in the back and in the womb itself ; it is generally most severe a day or two before the " period" comes on, and is relieved when the flow takes place. Between the different times the patient may enjoy good health. In some the pain is due to neuralgia of the ovaries or uterus, and in such cases medicines containing quinine are useful, and bromide of potassium is a valuable sedative. In others the pain is due to the vessels of the part being too full of blood, and when the congestion is relieved great benefit ensues ; a hot hip bath and leeches applied to the neck of the womb or to the abdomen over the seat of the pain will give relief. In all cases belladonna may be applied locally with much benefit in allaying the pain. During the interval the general health should be looked to : if able to bear it, moderate exercise every day in the open air should be ordered ; if too feeble, a carriage drive may be taken. Avoidance of late hours, of overwork in close and confined rooms, a generous and wholesome diet, with an occasional aperient, will aid in curing this disease.

DYSPEPSIA. (See INDIGESTION.)

DYSPNŒA, or *shortness of breath*, is a symptom often met with in many diseases. It occurs naturally after running fast, and it is due to the alteration, for the time being, in the quantity of blood passing through the lungs, and the amount of air entering the chest. In emphysema, chronic bronchitis, and in almost all diseases of the lungs, larynx and trachea, difficulty of breathing occurs as a symptom ; in many cases of kidney and heart disease it is almost always met with, and such invalids are thereby prevented from running up-stairs or hurrying themselves ; in fact, any condition which alters the amount of the blood which usually flows through the lungs per minute, as in affections of the heart, or which alters the amount of air entering the lungs, as in affections of the air-passages, will result in producing this symptom. It comes on when the patient makes any exertion, or on exposure to cold air, as going out on a raw foggy morning, and sometimes it will make the patient wake suddenly from his sleep, and forms then what is commonly called an asthmatic attack. Rest in bed, in a room with moist air of the temperature of 65° to 70° Fahrenheit, is to be recommended when the patient is very much troubled in this way ; avoidance of cold air must be insured ; in some cases a respirator is advisable. During an attack much relief may ensue from taking an expectorant mixture containing ether and ammonia and squills. Generally by these means an attack may pass off, but at times, and during the last few days of life, in those who are dying of heart disease or bronchitis or dropsy, the dyspnœa is very distressing ; the face then shows great distress, the lips are blue or livid, the veins of the neck stand out like blue cords, and there is but little movement of the chest. In some cases bleeding from the veins of the arm will for a time relieve the congestion ; in others, pricking the legs with a needle, when they are very dropsical, will give much ease by withdrawing some of the excess of fluid. When there is hydrothorax or œdema of the lung, brought about either by kidney or heart disease, less air can enter the lungs and dyspnœa must be more or less present, and the space in the chest is further encroached upon by the increased size of the heart usually found in these cases. When the dyspnœa comes on in consequence of a foreign body, as a marble or a coin getting into the larynx or windpipe, surgical means must at once be resorted to, and when it is due to inflammation of that tube, as in cases of croup or diphtheria, special remedies must be used, which will be detailed when treating of those diseases. Finally, an aneurism of the aorta may cause dyspnœa by pressing on the windpipe or setting up a spasm of the epiglottis by pressing on the nerve supplying that part of the windpipe ; however, but little can be done in such cases.

DYSURIA, pain or difficulty in passing urine, may arise from a great variety of causes. Some connected with disease of the organs concerned, others arising from altered conditions of the fluid itself ; stone ; stricture ; inflammation of the bladder and urinary passages, are all important causes, and to these the reader is referred. Alterations in the urine giving rise to dysuria, are commonly excessive acidity or the presence of calculi, either of which may give rise to great irritation, and a tendency to pass water without, however, bringing any away. Albuminuria also gives rise to something of the same kind, the patient frequently emptying his bladder. Sometimes there is a kind of neuralgic pain connected with urination, which gives rise to much discomfort. If the pain arises from acidity, alkalies will speedily relieve it ; if due to spasms, a pipe of tobacco is perhaps the best remedy.

E.

EAR.—The structure of the ear is explained in the articles on DEAFNESS and HEARING. The diseases or affections of the ear may be referred to those three chief portions of which it consists, thus—(1) The affections of the External Ear, or Auricle. (2) Those of the Middle Ear, or Tympanum. (3) Those of the Internal Ear, or Labyrinth.

(1) Affections of the Auricle.—The auricle is subject to severe cutaneous affections, the most important being—*chronic erysipelas* and *chronic eczema*. In *chronic erysipelas* the ear becomes greatly swollen, its skin is dry, red, and covered with epithelial scales, with derangement of general health. The treatment consists in cleanliness and free exposure to the air. If the inflammation is considerable, poultices are of use, and an astringent lotion should be applied. Glycerine is a valuable application. In *chronic eczema* the auricle is considerably swollen, and covered with yellow crusts, exuding fluid. The itching and irritation are considerable. The meatus, or passage, must be well syringed out with warm water, to prevent the accumulation of discharge, and the ear itself bathed with some astringent solution, or with glycerine. The scabs are to be removed by poulticing. Excoriations of the auricle in children are readily cured by attention to cleanliness, or by the use of astringent lotions.

Gout affects the external ear, and is a common cause of deafness. The burning heat and gnawing pain commence about midnight, gouty deposits are found in the auricle, and there is congestion of the cartilage. The treatment is the same as that for gout in other parts of the body. (See GOUT.)

Tumors are frequently met with. For example, enlargement or hypertrophy of the lobes, frequently met with in women who wear heavy earrings, cystic, fibrous, and malignant tumors, all requiring surgical interference.

The external meatus, or *passage*, is liable to *accumulation of the cerumen or the natural wax*, and deafness is frequently due to this condition. This is one of the most frequent affections of the external ear-passage, and when detected should be removed by syringing. The syringe should be fitted with a very small nozzle, and the instrument itself capable of affording a strong and continuous stream. The water injected should be warm, and a small ear-spout placed beneath the ear to catch the return current and its additions. The ear should be plugged with cotton wool after the operation is complete, for a day or two. If the wax be very hard and firm, a few drops of oil introduced for a few nights will facilitate its dislodgement. In syringing the ear, the error is commonly made of introducing the nozzle of the syringe obliquely to the side of the head, instead of placing it at right angles and pressing the *tragus* (the eminence over the opening) forward. As excessive sensitiveness is the after-result, it is best to fill the ear with some fine cotton-wool.

Some *abscesses* or *boils* often form in the meatus, causing intense pain. Hot poultices and fomentations applied to the ear give great relief, and in severe cases, free purges and leeching may be advisable. The constitutional treatment consists in the administration of tonics, of which iron is of the most value. This external meatus is very frequently the seat of inflammation, which may be either acute or chronic. In acute inflammation there is at first a dull aching pain, enlargement of the glands of the neck, and impairment of hearing, followed perhaps by a discharge of mucus or muco-pus, and considerable derangement of the health. Syringing the passage with warm water gives great relief, hot fomentations and poultices and the internal administration of morphia, all exposure to draughts carefully avoided, and the general health attended to. The chronic form is generally a sequel to the foregoing, and is often caused by prolonged bathing and neglect in drying the hair; weak astringent lotions, such as a weak solution of acetate of lead or of nitrate of silver, are of great use.

In children this complaint is invariably associated with derangement of the health, and quinine, cod-liver oil, and iron are indicated.

Polypi may form anywhere in the passage, and not unfrequently on the membrane; one form soft, pulpy, and vascular, and the other firm and fleshy. These growths cause deafness, and set up an offensive discharge. The treatment consists in their removal.

Foreign bodies in the meatus have been separately treated of. (See FOREIGN BODIES IN EAR.)

2. **Affections of the Tympanum.**—*The Membrane of the Ear* (see HEARING) is liable to both injury and disease. *Rupture* may occur from a variety of causes, such as the introduction of foreign bodies, a blow, sudden deafening noises, violent syringing, with improper introduction of the syringe (see above), violent blowing of the nose, vomiting, coughing, etc. The symptoms are slight pain, generally a little bleeding, and perhaps impairment of hearing, although this condition is not necessary. *Inflammation* of the membrana tympani, like all other forms of inflammation, may be acute or chronic, a consequence of cold, gout, scrofula, or syphilis. The symptoms are pain, itching, and slight deafness. Ulceration may take place, and perforation of the membrane ensue. It is to be treated according to circumstances, dependent on its causes; beyond this the treatment is the same as that for inflammation of the external meatus. *Perforation* of this membrane occurs after certain conditions, such as ulceration of its substance from internal inflammation; the diagnosis of this perforation is easy. In the first place it can be detected by a *speculum*; again, the patient, by closing the mouth and nostrils, can blow air (if the Eustachian tube be not obstructed) through it, and the patient is moreover somewhat deaf. If possible, attempts must be made to close the orifice by the application, in slight cases, of lunar caustic to the edges of the wound, or by the introduction of cotton wool, or by the artificial membrana tympani. This artificial membrana tympani consists of a thin circular plate of vulcanized India-rubber, to which a silver wire is attached obliquely as a handle. At first this apparatus should only be worn for an hour or two, and always removed on retiring to rest. Occasionally the hearing becomes re-established after it has been worn for some time. The Eustachian tube, or passage of communication between the middle ear and the pharynx is liable to several forms of disease, obstruction, and a permanently open condition. The cavity of the tympanum is liable to severe inflammation, arising from cold, scrofula, or a sequence of scarlet fever. The usual symptoms of this condition are discomfort in swallowing or blowing the nose, headache, and intense pain in the ear, and more

or less deafness. The constitutional symptoms are severe, and in adults delirium is present, and convulsions in children. These conditions terminate either in resolution or by the formation of abscesses, the matter by bursting through the membrane, or by inflammation of the mastoid cells and mastoid bone. Salines should be given, if due to gout or rheumatism, colchicum, or iodide of potass. Locally, steam, poppy-head fomentations, linseed, onion, or garlic poultice, and small blisters on the mastoid process of the temporal bone—just behind the ear in fact.

(3) **Affections of the Internal Ear.**—The function of hearing may be impaired, or completely destroyed by the results of the severer diseases of the middle ear, when suppuration has followed either of them. The auditory nerve, which is found lying within the labyrinth, is subject to functional diseases, causing what is termed nervous deafness. True neuralgia has been described as occurring occasionally.

Earache.—The *earache*, or *otalgia*, is a neuralgic affection, occurring in fits of excruciating pain, darting over the head and face, generally caused by bad teeth. It may be partly relieved by syringing the ear out with warm water, to which a little laudanum has been added—say, twenty drops to a wineglass of water. Hot fomentations should be applied to the ear, or tincture of aconite or belladonna, painted behind the auricle. The state of the bowels is to be carefully attended to, and free action obtained by purgatives.

EARTH CLOSETS are contrivances recently introduced for superseding water-closets, particularly in country places. In them, instead of pulling the handle and allowing a flood of water to sweep away all matter from the pan into the sewer, the same handle allows, from a hopper, a quantity of dry earth to fall and cover the evacuations. The dry earth completely prevents any smell from arising from them, and apparently prevents all further decomposition. After a time the accumulated matters may be removed, and constitute a valuable manure. No doubt in cities the general application of such a plan would not be easy, but in the country the plan is valuable. In cities drainage is carefully looked after by the proper authorities, and it is a man's own fault if his drains in his own house are bad; but in the country drainage is hardly possible, except into a cesspool, which is a perpetual source of risk to all in the neighborhood. In the country there is no sufficient water power to rinse out the drains—they get choked up, and typhoid fever breaks out. All this is avoided by the use of earth closets. As soon as they get filled they can be emptied; they give rise to no smells, and, as far as we know, to no fevers. Only one or two precautions are necessary to make them work well. Fluid

excretions should, as far as possible, be kept apart from solid excreta, and the earth used should be well dried before use. Imperfectly burned wood ashes, mixed with ordinary loam dried, makes the best kind of earth to use. Sand does not suit well.

EAU DE COLOGNE, a much esteemed perfume, which derives its name from the city where it is so largely manufactured. It is a distillation in alcohol of various sweet-scented substances, and is most refreshing and grateful as an application in cases of headache and exhaustion. A bad nervous headache may often be relieved by diffusing eau de Cologne over the part affected by means of a vaporizer or even by dipping a handkerchief into it mixed with water and then gently fanning the patient till the eau de Cologne evaporates, and removes the heated condition of the head.

ECCHYMOSIS. (See BRUISES.)

ECHINOCOCCUS is the name given to the parasite found in hydatid cysts, and when occurring in man, the echinococci are developed from the tapeworm of the dog. (See HYDATIDS.)

ECLAMPSIA. (See PUERPERAL CONVULSIONS.)

ECRASEUR is an instrument that has been devised for the purpose of removing tumors by a combined process of crushing and tearing. Its use is attended with much less bleeding than that of the surgeon's knife, and for this reason it has been applied with success to cancer of the tongue, internal piles, and other vascular growths.

ECSTASY, a peculiar form of intense nervous and emotional excitement. (See CATALEPSY.)

ECTHYMA is the name given to a skin disease. It consists of large, circular, raised pustules, surrounded by a livid purplish zone. They occur generally on the extremities, and are always isolated ; the fingers and legs are very common seats of the eruption. If the pustule is pricked, an unhealthy, greenish-colored fluid exudes, and a scab forms ; then in about three weeks this scab falls off, and leaves no ulcer beneath, but simply a red scar. As soon as one heals another appears, and so they may be seen in all stages at the same time. It mostly occurs in children, and especially when from any cause they are in a debilitated condition, as after recovery from measles or scarlet fever, or from bad living. Plain but wholesome diet, with fresh air and exercise, will improve the general condition ; while steel wine taken internally, and zinc ointment applied to the spots, will generally complete a cure. An occasional aperient may be required. There is no pain attending this eruption ; if there should be, or if the scabs require removing, a hot linseed-meal poultice may be applied at bed-time. Care should be taken that the child does not knock the pustules or scratch the head off.

ECTOZOA are animal parasites which have their "habitat" on the surface of the human body. The following are the most common varieties which are met with :

1. The *acarus*, or itch-insect. There are two kinds, male and female ; the latter burrows in the epidermis and there deposits the ova. They are generally found among very dirty and poor people, and occur on the outside of the arms and very often between the fingers, also on the trunk ; but rarely above the shoulders or below the knees. By means of the suckers or ambulacria, they have powers of locomotion, while with their mandibles they are enabled to cut through the epidermis, and extract fluid from the tissues. The female seldom leaves her burrow except at night ; when disturbed by scratching, they crawl with great rapidity over the skin, and readily pass from one person to another ; so that the complaint is easily caught. Great itching accompanies the presence of these insects, so that the skin may often be seen covered with scratch marks, and in some parts bleeding from the friction used. The acarus likes tender skin, and hence is more common in youth than in old people. The disease, although extremely troublesome, is easily cured. An ointment, composed of sulphur and lard well mixed together, should be thoroughly rubbed in every night until the skin which is rubbed feels a warm glow, and the next morning the patient should be well washed in hot water with coarse soap and a flesh-brush. This method, if repeated three or four mornings properly, will generally effectually cure. Every night the person should be wrapped in an old shirt which can be destroyed as the disease is cured.

2. *Pediculi*, or lice. Of these there are different kinds. Some are found on the hair of the head, and chiefly at the back part ; they are of a pale drab color, and much longer than they are broad ; they crawl about in the hair, and deposit their ova on the hair by means of a gummy kind of substance. At first these ova are close to the root of the hair ; but as the hair grows they may be found an inch or two off the skin ; but by that time the ova have escaped, and only left the empty sac in which they lay attached to the hair ; these sacs with their contents are commonly called " nits." In dirty children they may be seen in abundance, when the back hair is separated, as little shining particles arranged along the hairs. Great itching is produced by the presence of these parasites and even eczema may be produced by the scratching ; the hair then becomes matted together, and the head gets into a very filthy state. The best treatment is to cut off the hair as close as possible, and rub in every night for three or four nights some white precipitate ointment ; this should be done, however, with care. A solution of carbolic acid

(one part of the acid to sixty parts of water) will kill all the lice, but not the ova. Others are found in the hair of the genitals, to which the names of crab-louse, or "crabs" have been given. (See CRABS.)

3. The harvest-bug, which often attacks people when walking through a stubble-field in the autumn ; it is a small red insect which causes intolerable itching.

4. In the West Indies a most troublesome creature is the chigoe, or *Pulex penetrans*. It penetrates the skin, and there lay its eggs, producing, in consequence, an irritable sore.

ECTROPION. (See EYE.)

ECZEMA is a skin eruption of very common occurrence. It is a non-contagious disease and is characterized by the presence of minute vesicles hardly seen without a lens. These spots may terminate by the fluid in the vesicles being reabsorbed, or excoriations may form which leave a raw red surface from which a watery liquid oozes ; as the liquid dries, it forms dirty scabs on the affected part which present a very loathsome appearance. This common disease may be produced in a great many ways. Heat may cause it, and then it is called *eczema solace* or heat spot ; thus it may be produced when a fair or tender skin is exposed to the sun on a hot summer's day. Contact with irritating substances will produce it, and so it is found among grocers, affecting the hands of those who deal much in sugar ; potboys are, from a similar cause, very liable to it. It sometimes occurs in those engaged in working with quicksilver, or in those who have taken an undue amount of mercury ; this is, however, very unusual, and probably depends on some peculiar state of the constitution on the part of the individual affected. The eruption begins usually on the groins and thighs ; it is commonly produced in the flexures or folds of the skin in fat and dropsical people, and when the folds are raised, a raw, moist, and red surface is seen. The skin at first is red, and is accompanied by heat and tingling ; it is apt to extend very rapidly, but although the surface affected may be large, yet the disease does not go below the skin itself. On this red angry-looking skin, numbers of minute glittering vesicles soon appear, and these vesicles are due to a very small portion of the epidermis being raised up by a little serum or watery fluid beneath ; at first they are clear and almost pellucid, but the contents become opaque, and under favorable circumstances dry up ; more often these little vesicles burst, and the fluid escapes and dries up into gum my masses on the surface ; in doing this it entangles any dust or dirt that may be present, and thus forms large ugly-looking scabs ; this takes place when the part is neglected, for with proper care such scabs should not be allowed to form. This disease is a moist skin eruption, and hence, when the scabs are taken away, the skin looks raw and moist ; great discomfort from this cause is felt by the patient also ; the clothes which he wears next the skin become stiffened and dirty with the discharge, a fœtid smell is often present ; the rash spreads, the vesicles increase in size from the irritation and the acrid juice which is being poured out, the skin swells up, feels hot and uncomfortable, and the poor sufferer is in a state of great worry and distress. The constant irritation makes him feverish ; he sleeps badly at night and is always wanting to scratch himself, and this only tends to make him worse. In children it runs a similar course to what it does in the adult. Owing to their delicate condition, any dis turbance of the constitution is liable to cause the appearance of this rash, especially in those who are at all strumous. In infants, after birth, a red rash often appears, merely from the irritation of the air or clothes on the tender skin ; this is well known under the common name of *red gum ;* it is easily cured by washing the surface with warm water and using zinc ointment. After vaccination, eczema often appears, and it is owing to this in a great measure, that so much prejudice is felt against vaccination : now this operation, simple as it is, and valuable as it is in its results in preserving humanity from small-pox, cannot be done without some slight disturbance of the constitution, and then this disease often appears ; among the ignorant and the filthy, the rash soon spreads and forms dirty, fœtid scabs, which lead people to imagine that their children are suffering from some horrible and dangerous disorder ; no popular prejudice can be more unfounded, as eczema is in nearly all cases curable by a little care and cleanliness, and it is in no degree attended by danger. Teething is another cause, and here again the irritation and febrile disturbance brought about by that process, acts in a similar way to the above. Fat children often have this eruption in the folds of the chin, beneath the knees, in the bend of the elbow, and very often in the nates or round the buttocks ; this is generally due to the irritation caused by the passage of the excretions, and to a want of due cleanliness and proper changing of linen. The head in children is a very common seat of eczema ; it begins on the scalp in the usual way, and comes behind the ears, leaving angry red places from which oozes moisture ; often there is a running from the ears, and lumps appear beneath the chin or at the back of the neck, and these are due to the lymphatic glands becoming enlarged from the affection of the skin in the neighborhood. This disease often comes on after a child is recovering from measles or scarlet fever ; it is met with after an attack of chicken-pock or glasspock, and is often due to the child scratching the vesicles and so irritating them. In the

disease known as itch, eczema appears sometimes ; it is here produced artificially by the scratching of the skin.

Treatment.—In most cases the following treatment will suffice. Smear on the part some simple olive oil, so as to soften the crusts, and then lay on at bed-time a hot linseed meal poultice, so as to cover the part well. In the morning most of the scabs will be removed and a great deal of the dirt, while a moist red surface will be left, and if the process be not repeated, it will soon scab over again ; washing with soap is of no use, as it only further irritates the skin ; let cleanliness be kept up carefully by oiling and washing. If the rash should be in a part where it is difficult to keep on a poultice, it is just as good to wash the part with oatmeal and hot water instead of poulticing. When the surface is in this way cleaned, let zinc ointment be applied all over the sore, and in a short time much improvement will ensue. A child in a filthy state and covered with scabs may thus in a day or two with great care, make great progress. When itch is present, the remedy for that must be used, both diseases cause itching, but the itch does not affect the head, but comes on the body and arms and between the fingers, and is moreover very catching. Children are often much benefited by taking steel wine for a few weeks. It is important to know that while the great majority of cases are thus easily cured, yet the disease is very liable to recur, and in a few obstinate cases seems to defy all treatment. In some scrofulous children the disease will break out in some part or other, the skin at first looking rough, dry, and shiny, as if it were too tight over the part ; then it becomes moist and goes through the usual stages. Relief may be afforded for a time, and the child may grow out of it, but such cases are very troublesome. It is singular that such children very often look well and are playful, and in good condition. The hair should be kept away from the skin where this eruption has broken out. Mere wrapping the part in rags kept constantly wet with cold water will bring about a cure. Fuller's earth is useful ; most of the powders in common use are of no value, as they irritate the skin.

Eczema of the leg is often met with in aged people, and in those who suffer from varicose veins and ulcers. There are two typical forms of eczema in this situation, viz., the acute and the chronic. In the former, the affection comes on quickly and is very painful. The skin is of a bright red color and very tense. Upon this inflamed portion of skin minute blebs are formed, which contain a transparent fluid. As the inflammation subsides, these blebs either dry up and form thin scales, or their contained fluid increases in amount and becomes thick and milky-like pus, causing much irritation to the skin, and finally drying and forming thick yellow or brown scabs. The severe symptoms of acute eczema subside in the course of five or six days, and then the affection either disappears altogether, or, as most frequently happens, it passes into the chronic form. Here the skin is less painful and inflamed, and there is less "weeping," or discharge of thin fluid from the affected surface. The skin, however, and also the parts immediately beneath are swollen and inflamed. Large white and adherent scabs are formed by the drying up of the frequently renewed crops of blebs or vesicles. The chronic form of eczema is a troublesome and obstinate affection, and is generally attended with much itching. Eczema occasionally attacks the nipples. Very small blebs are formed, which break and leave raw surfaces from which there is profuse weeping of thin clear fluid. Like other forms of acute eczema, this is attended with much pain. It attacks women at all periods of life, most frequently girls who have just reached the age of puberty. The chronic form is very troublesome and obstinate. Eczema in the lower extremities of persons troubled with varicose veins may be prevented by the use of bandages or an elastic stocking, cold bathing, and by their avoiding as far as possible much standing or walking. The *treatment* of acute and severe eczema, whether in the legs or on the nipples, consists in administering saline purgatives, and by applying some warmed goulard water mixed with a small quantity of laudanum. For chronic eczema tonic medicine is generally indicated, and also warm baths. The inflamed parts should be frequently washed with simple water or with bran-water. Soap should not be used. The skin around the inflamed patch should also be well washed with weak spirits of wine, and then carefully wiped. The following are some of the lotions most frequently applied to chronic eczema : bicarbonate of soda dissolved in water ; nitrate of silver dissolved in water, with the addition of some sweet spirits of nitre ; a mixture of tannin and glycerine ; bichloride of mercury, proof spirit, and water ; lime-water ; borax and glycerine.

EFFERVESCING DRAUGHTS are often very useful and pleasant in febrile attacks. They can be made from any of the vegetable acids and an alkali. To a tumbler of water the following proportions are sufficient : Bicarbonate of potash, 2 scruples ; tartaric acid, 25 grains ; or, carbonate of soda, ½ drachm ; tartaric acid, ¼ drachm. Add a teaspoonful of capillaire, or any syrup, and you have a pleasant draught.

EFFUSION is the pouring out of any fluid either into a cavity or the cellular tissue of the body. An effusion may be of blood or serum, which is called water. Thus, we have in the first case apoplexy, if the effusion be of blood on the brain ; or water on the brain, if of serum. Likewise on the chest, effusion causes either congestion or water, as the case may be. Effusion also may take place in the joints, or between the skin and muscles.

EGGS are very nutritious articles of food, they contain as much oil and flesh-forming matter as butcher's meat. The white, however, is not so digestible as meat. They enter into the composition of puddings, cakes, buns, and other forms of diet. They are also eaten alone, boiled or fried, and are most digestible when least done. The egg of the domestic fowl is usually eaten, but those of other birds are frequently used. The eggs of the crocodile and other oviparous reptiles are eaten in some parts of the world. The eggs of the woodcock, plover, and other small birds are esteemed a luxury. Those of the duck and goose have a strong flavor, and those of the seafowl are fishy. The eggs of the turkey are rich in flavor, while those of the guinea-hen have a delicate flavor. All birds' eggs may be eaten with impunity. Eggs are most useful and nutritious as articles of diet in the sick-room. They are used for mixing with castor-oil, turpentine, and other strong medicines, to render them more palatable ; also for making mulled brandy and wine. A most nutritious and agreeable drink may be made for invalids, consisting of sherry or brandy beaten up with raw eggs and sweetened with sugar. Eggs may also be given mixed with Liebig's extract of meat.

ELATERIUM is the sediment which falls from the expressed juice of the squirting gourd, or wild cucumber. When ripe it ejects its seeds, hence its name. The juice is set aside after expression, and the sediment is allowed to strain on a linen cloth, after which it is dried on a porous brick. For this reason the thin scales in which elaterium is seen are marked on one side by the impression of the cloth. The drug is an exceedingly powerful one—one eighth of a grain acting as a strong drastic purgative, carrying off much fluid. It accordingly requires to be cautiously given, and should never be employed if there is a tendency to irritation of the bowels. It often causes nausea and sickness, and sometimes gives rise to considerable pain. Its great value is in dropsical accumulations of fluid, as in heart disease. In such cases the fluid may accumulate to an extent that diuretics are unable to cope with it. When that is the case elaterium given in the above-mentioned dose, combined with a few grains of compound extract of colocynth, often does great good, causing an immense drain of water. It should, however, not be used for too long a period continuously. Belladonna is a good thing to give along with it. Elaterium is often bad—it spoils by keeping, and is frequently adulterated, hence it should be procured direct from some reliable druggist.

ELATERINE is the active principle found in elaterium. Its dose is about a quarter that of elaterium.

ELDER FLOWERS.—The water distilled from off these flowers is sometimes used as a vehicle for more powerful medicines. It has no specific properties of its own, but the inner bark of the tree acts as a hydragogue cathartic, and has been used with success in the treatment of dropsies in the form of decoction. Four ounces of this bark may be boiled in a pint of water, and two or three ounces taken as a dose.

ELECTRICITY, or *Galvanism*, is an exceedingly powerful remedial agent, the exact value of which we are only now beginning to appreciate. There are three kinds of electricity in use—first, the so-called static variety, which is obtained by rubbing a glass plate or cylinder, and which may be stored up in Leyden jars. This is also called Franklin's electricity, from its discoverer, Franklin, and is not greatly used in medicine. The most important variety of electricity is that called dynamic, or current electricity, and of this there are two kinds—namely, that which passes in a continuous current from one pole of the battery to the other, and that which is called interrupted, which is a kind of to-and-fro current ; the last is also called an induced current, and as its properties were first investigated by Faraday, it has received his name. It is to these last we must confine our attention, as it is these which are chiefly used in medicine. It would be impossible to enter into full details as to the mode in which these forces are developed ; briefly, the main points are these : Suppose a plate of copper and a plate of zinc are introduced into a vessel containing diluted oil of vitriol ; if now a copper wire be so placed as to touch both of these, active change will be set up in the fluid, and an electric or galvanic current will be set up in the fluid from the zinc to the copper, and out of the fluid by means of the wire from the copper to the zinc. This is the continuous current, and if the pair of plates be multiplied, so will the force increase. Moreover, if two wires are used instead of one, and any portion of the body be introduced between the two, the current will pass through the body so as to go from one wire to another, and its effects on the body will be made manifest. There are many varieties of continuous current batteries, but they are all constructed nearly on the same principle, though the material varies. It is to be noted that the effects are only noticeable when the circuit, as it is called, between the metals which gives rises to the current is complete ; when the circuit is broken, that is to say, when the two wires neither touch directly nor indirectly, there is no current.

But suppose the current to be made to pass through a piece of soft iron, this will be affected like the wires, readily conducting the electricity from one wire to another. If now, however, a coil of thin wire be made to surround the soft metal so as not to touch either it or itself, through the coil of wire will pass a stream of electric force whenever the circuit is opened or shut ; that is to say, whenever either of the wires is made to remove from or to touch the iron centre piece. When the wires touch, the circuit is closed, and the current passes from the copper to the zinc, as usual, but if removed it can no longer pass, and the iron remains unelectrified. The wire coil does not touch the centre, and so the current is said to be induced ; it only passes on opening and closing the circuit, that is, interrupting the current, hence it is also called an interrupted current, but it is plain these interruptions occur in both the original circuit and the induced current, so that in reality both are interrupted, though only one is induced. It would not be possible to give in this short space full details of the kind of cases in which the several forms of electricity may prove or have proved useful. In the first place they may be either applied locally or generally—that is to say, to some one spot or part of the body, or to the whole body. If applied to one definite part, the influence would naturally be to a great extent limited to that part, but if to the whole body, a kind of tonic rather than any distinct or specific effect would be anticipated. Briefly we shall point to two most important sets of cases where electricity does good. First in neuralgia, especially of the face. This most intractable and painful malady sometimes yields in a most surprising manner to the use of electricity, and especially to the continuous current, though cures sometimes do better with the interrupted one. In facial palsy, too, where the muscles are so wasted as to be unable to respond to the interrupted current, prompt contraction follows the application of the continuous one. In ordinary cases of paralysis one great object, at all events, is to keep the muscles properly nourished until the nervous system of what has suffered damage has had time to be repaired. For this purpose, as a rule, the induced current is best, each separate muscle of a limb can be duly exercised and prevented from undergoing degeneration by applying the force to appropriate points. At the same time if the exact nerve centre affected can be made out, the continuous current may perhaps be of service in aiding its recovery. One form of loss of voice connected with hysteria is promptly cured by the interrupted current. In certain forms of paralysis connected with syphilis the continuous current is most useful. These instances might be multiplied indefinitely, but the main principles affecting the use of electricity are above indicated. Of course the kind of electricity to be used, and its mode of application, must be left to a physician.

ELECTRO-BIOLOGY. (See MESMERISM.)

ELECTUARIES are certain forms of remedies into which sugar or honey largely enters. They are of the consistence of a thick paste, and readily admit of being made up into pill-like masses which may be covered over with sugar or any tasteless material. They are much the same as confections, and two of them at least are very valuable remedies. These are confection of sulphur and confection of senna, both sometimes called electuaries ; both are valuable laxatives, the former especially useful to those troubled with piles, the latter as a means of administering a good but nauseous remedy to children.

ELEMI is a kind of resin, with properties allied to turpentine, which is imported from the East. Its ointment is sometimes employed in sluggish sores. It is not given internally

ELEPHANTIASIS is the name given to a condition where limbs swell to enormous proportions from no very definite cause and remain permanently in that elephantine condition. In it the skin and subjacent tissues are greatly thickened and increased in density, but the muscles are destroyed or altered for the worse rather than increased in strength. Most frequently it attacks the lower extremities, sometimes the upper ; less frequently other parts of the body. The skin is the part most affected ; it becomes of a brawny thickness. All kinds of remedies have been tried, even to ligatizing the vessel which supplies the limb, but success has not been great. Some of the cases ligatured have done well, some badly ; bandaging has done good to some. In India such a growth frequently attacks the scrotum, causing it to assume the most portentous proportions. For this, as for the other, removal seems the best remedy, and should not be too long deferred, or the health may suffer irretrievable damage.

ELIXIR is an Arabic word, signifying strength. At one time it was a favorite name for medicines supposed to be particularly efficacious, and where the ingredients were almost entirely dissolved in the menstruum, making it thicker than a tincture. There were all sorts of elixirs sold, but we now only find in the shops elixir of vitriol and paregoric elixir.

ELM BARK is a remedy of very uncertain value ; some thinking it very valuable, and others useless. It is given in the form of decoction as an alterative—whatever that may mean. Certainly we possess far more powerful remedies.

EMACIATION or loss of flesh occurs in many cases, as in cancer, consumption, and starvation, etc., and is due to the tissues not receiving a due supply of nutrition.

EMBALMING is the process of preparing any animal body to resist the decay natural to it. The art of embalming was practised by the Egyptians in perfection, and we have their mummies now to prove their skill. The chief ingredient in all embalming preparations is benzoin, a resin existing in friars' balsam. This, combined with naphtha, is freely used, and bandages soaked in it are bound round the body, after elaborate preparations of spices and resins have been placed within the body itself. At the present day embalming is seldom required even by the rich for their dead, excepting when a long time must necessarily elapse before interment ; and, as in the case of royal personages, where lying in state is practised.

EMBOLISM is a term applied to a condition in which a piece of fibrine in the heart or in a large vessel has become dislodged and carried by the current of the circulation into some distant part. This occurs in some cases of heart disease, and more especially after rheumatic fever has caused disease of that organ ; a clot of fibrine carried into an artery of the extremities would do very little harm, but if carried into an artery supplying the brain, an attack of hemiplegia or paralysis will ensue, and the patient will pass into a state of coma or insensibility. This is a very serious condition and nearly always fatal ; it may be suspected in those who, suffering from heart disease after rheumatic fever, have daily variations of temperature with weakness and pallor, and other well-defined symptoms of fever. An embolon, or plug of fibrine, may be carried from a vein into the heart and cause sudden death by blocking up the pulmonary artery. Such cases are very rare and may come on after a confinement. Death will take place in an hour or two, or may be still more prolonged : the patient will suffer intense agony and distress from a feeling of impending suffocation ; she will toss herself about calling for air although there is plenty entering the chest, and a fatal result will shortly ensue, not because there is not enough air to aerate the blood, but because the blood cannot get to the air to be oxygenated. The symptoms are the same as those met with in death by apnœa, although the cause is reversed in this case. It is doubtful if recovery ever takes place, when once the above symptoms have come on.

EMBRYO is the name given to the earliest appearance of the fœtus when it begins to be developed in the womb.

EMBROCATIONS are forms of remedies intended to be rubbed into a part, whereas a liniment is strictly intended only to be smeared on to it. Nevertheless, the word liniment is now generally used in the widest sense so as to embrace embrocations. (See LINIMENTS.)

EMETICS are medicines or other agents which produce vomiting ; the simplest, and in many cases the most effectual, being a tickling of the back of the throat, especially at the part called the soft palate, with a feather. In medicine several classes of emetics are used ; some cause sickness and faintness ; some by irritating the stomach cause it to get rid of its contents without any great degree of faintness such as accompanies the other. Vomiting itself is a complex act, partly the result of the powerful muscles constituting the walls of the belly, partly the result of contraction of the muscular walls of the stomach itself, that is to say, of the cavity into which the food is received. Of ordinary emetics, ipecacuanha is that most frequently employed ; antimony is also used in the form of antimony wine or tartar emetic, the latter in small doses. These remedies cause much sickness and prostration, and consequently are used chiefly in cases where it is desirable that such a condition should be induced for the arrest or suppression of certain diseases. Thus, in the case of children attacked with croup, or in whom the croupy cough has been brought on by exposure to wet, an emetic of this class is of the greatest possible value, and often succeeds in arresting the disease, especially if accompanied with a warm bath and fostered with lukewarm drinks. Then, again, it must never be forgotten that little children, especially infants, if attacked with cold, cough, and thereby expel from their lungs the matter which has collected there, but this in all probability only reaches the air tubes and gets no farther. The material may it is true, reach the throat and be swallowed, but infants cannot expectorate, and so the tendency is to accumulate phlegm in the chest, whence the rattling noises heard when they have colds. Now it is of vital importance to get rid of this substance, and of all remedies an emetic is the most efficient. Ipecacuanha wine had best be used, and that may be given in repeated teaspoonful doses until the child is sick and the whole is brought up. This may seem harsh practice, but in the end it is safest. In olden times it was the common practice to begin a course of physic as it was called by a vomit and a purge, and it was held that these would often arrest a fever ; now our beliefs are somewhat altered, but we are also inclined to think that the value of emetics is too frequently overlooked. In poisoning, by whatever agent, it is of vital importance to get it expelled from the stomach ; frequently the substance is of a character to induce vomiting itself ; if so, all that requires to be done is to foster this tendency by copious draughts of lukewarm water, or some bland drink ; if not, the stomach must be

emptied, and to this end it must never be forgotten that the readiest means is the best. In most situations common salt, mustard, or smelling-salts are obtainable, and these may be given in the respective doses of a handful of salt, a tablespoonful of mustard, or a teaspoonful of smelling-salts, all freely diluted with water, and to be followed up by copious draughts of lukewarm water. These are especially useful in poisoning with opium or other narcotic agent. For this purpose, too, sulphate of zinc (white vitriol) and sulphate of copper (blue vitriol) are particularly well adapted, but as a rule less readily obtainable. Sulphate of zinc is a very safe emetic, emptying the stomach without giving rise to much nausea. Perhaps the best is a combination of this with ipecacuanha—15 grains of sulphate of zinc and 5 of ipecacuanha, given as usual with much lukewarm water. There are many cases where an individual has partaken of indigestible or unsuitable food, that an emetic, by getting rid of it, does great good, as during the process of vomiting, the liver and gall bladder are compressed, and bile finds its way back into the stomach, thence to be expelled by the mouth. An emetic is often one of the very best plans for getting rid of an accumulation of bile in the liver or its appendages. The process of vomiting commonly also causes a certain amount of perspiration.

EMMENAGOGUES are remedies which are supposed to foster the menstrual flow. They are a most diversified and unsatisfactory group. Very often deficiencies or absence of the flow are due to no local cause, but to bad health generally, especially to the condition known as anæmia. When this is the case it is useless to attempt to restore the local functions until the general mischief is set right. For this reason salts of iron are among the most useful emmenagogues, especially in large towns, and as in these patients the bowels are usually more or less sluggish, especially the lower bowel, it is well to give aloes at the same time. Sometimes the arrest or non-appearance of the flow is due to mechanical obstruction, in which case operative procedure becomes necessary. This is comparatively rare.

EMOLLIENTS are remedies which, when applied locally, soothe the part and diminish irritation. Heat and moisture are perhaps the most important agents ; but these may be used in a variety of ways. Bathing with warm water ; the application of hot poultices, however compounded ; the application of oily or greasy substances, so as to keep the skin lissome and supple, all come within the definition.

EMPHYSEMA is a disease of the lungs, which is attended very often by shortness of breath, cough, and inability to expand the chest thoroughly. It very frequently comes on as a sequel to a winter cough. When a person has been suffering every winter with a recurrence of bronchitis, the air-cells of the lungs become unduly distended, and cannot so well expel their contents ; it thus happens that the lungs become larger and holds more air than usual, but as the air is stagnant in the lungs in a great measure respiration is carried on imperfectly, as the products of combustion are not removed fast enough. For the same reason an emphysematous person cannot take a deep breath, because his lungs being already full of air, he is in a similar state to a man who has taken a deep inspiration, and therefore the further expansion can only be very slight. Such people are generally stout, and have too much fat deposited about them ; this arises from the fact that they are not able, or are indisposed, to take violent exercise, and that the ordinary processes of combustion in the lungs are somewhat impaired. In old people also emphysema occurs, in consequence of the changes which naturally take place in the tissues in old age ; the lungs, like other organs in the body, are less nourished then than usual, and so they are unable to bear the external pressure of the atmosphere ; hence the air-cells dilate, and this is most observed in the upper and front parts of the lungs. This disease is met with also in children, and appears in some cases to be hereditary ; in many cases, however, it follows whooping-cough or some bronchial affection of childhood, and the little patient may be seen with high shoulders, prominent chest, quick but shallow expansion of chest, and rather congested appearance in veins of face. This disease in itself is not dangerous, but it is so often accompanied by bronchitis, and in a great measure induces that disorder, that grave evils may ensue. Besides the shortness of breath and the inability to take fast exercise, there is often palpitation of the heart and pain at the pit of the stomach, because the right side of the heart is full and cannot properly force its contents through the altered lungs. Sometimes the neck and face are swollen, and even the legs may become so too, and then when the finger is pressed on the skin a little pit of depression is formed, and the patient is said to be dropsical. But these results do not occur except in bad cases, and only when the person has been suffering for some years. Bronchitis is the most common affection which coexists with emphysema, and often it is extremely troublesome. Year after year, when the cold weather sets in, the emphysematous person will fear his winter cough, which is almost sure to come on if he is at all exposed in his trade or calling to atmospheric changes. Thus it will be seen that while bronchitis often causes emphysema, the latter is also in its turn a cause for the former, and as each aids in making the other

worse, such a person generally becomes worse after each fresh attack. When a cold comes on the breath is shorter than before, and there is difficulty in breathing ; the patient is wheezy, and feels as if there were a weight lying on his chest ; in a day or two under proper treatment the cough will become looser, and he will find relief by expectorating a good deal of phlegm from the chest. The cough is generally very troublesome first thing in the morning, because the phlegm has been accumulating during the night ; the patient is unable to lie down comfortably, and feels better when propped up in bed. By coughing so much, pain is frequently felt in the lower part of the chest on each side, and this is due to the muscles there being tried with the violent exertion. At times the eyes may be bloodshot, and the veins of the neck stand out during the cough because they are congested. Frequently the patient breathes better when he leans forward, resting on his hands, because then, the shoulders being fixed, expansion of the chest takes place more freely. After many attacks the lips, ears, and nose often are of a livid or purplish tint, owing to long-continued congestion of the vessels of the part. This disease affects all classes, more especially those whose work exposes them to all kinds of weather. A very common cause is indulgence in eating and drinking ; when old gentlemen are seen with large stomachs, shortness of breath, and a wheezy condition of the chest, emphysema is pretty sure to be present. Women, by leading a more domestic life, are less subject to this disease than men.

Treatment.—If a person is predisposed to emphysema by one of his parents having suffered from it, he should avoid exposure to inclement weather as far as possible, and when he has a cold or an attack of bronchitis he should try and get it cured as soon as possible. Avoid excess of eating and drinking, if at all inclined to obesity ; take lean meat rather than fat ; do not eat much bread, or butter, or pastry, or potatoes, but have dry toast, biscuits, or brown bread, and green vegetables ; a little claret or sherry is preferable to beer. Exercise should be taken every day in fine weather, and night air should be avoided. But it generally happens that a person takes no heed of himself until he is actually emphysematous, but even then the observance of the above rules may somewhat prevent further mischief. Removal to a warmer and equable climate is of the greatest service, but this is often beyond the means of most people. When an emphysematous person has taken cold, or has an attack of bronchitis, he should at once to go to bed and keep the room at a moderate temperature of 65° Fahr. or 70° Fahr. ; if too hot, the air is oppressive ; the atmosphere should be moistened by boiling water in a kettle, so that the steam shall pass into the apartment ; moist hot air is what is most grateful to the patient. Avoid any draught of cold air into the room. Place on the chest hot linseed-meal poultices, but care should be taken that they are really hot, and not allowed to lie on until they become a cold damp lump on the chest ; or flannels, wrung out of hot water, may be sprinkled over with a teaspoonful of turpentine, and then placed over the chest or back ; a piece of oiled calico should be laid over the flannel—it not only keeps in the heat, but prevents the clothes becoming wet. The patient should not lie too low in bed, as he will breathe freer if propped up by pillows. Careful attention should be given to the diet ; solid food should be avoided at first, and hot milk or bread and milk may be given, with a lightly boiled egg and beef-tea at intervals. Beer should not be given, and if any stimulant be needed, some port wine negus or a glass of warm whiskey-and-water may be given ; any excess in this direction is bad. Any light farinaceous pudding or some mutton may be tried in a few days, when the appetite returns, but the stomach should not be loaded with food so as to cause distension. The bowels are often confined, and so purgatives may be given occasionally. By adopting these means an attack of bronchitis may disappear in three or four days, but in old people who are in a feeble state it is a very serious complication, and often carries them off very rapidly ; this may be noticed every year at the onset of cold weather. Between the attacks, and during the warm summer weather, the object is to improve the general health, as far as possible by careful diet and tonic medicine. Flannel should be worn next the skin, and warm socks and thick boots. A respirator often gives great relief, as the inspired air is by that means warmer, but it is not so pure. Persons affected in this way should breathe through the nose rather than through the mouth, and they should not talk when out walking in the night air. Great relief is afforded by staying in the house all the winter, so as to avoid being exposed to cold or wet ; but this is a plan which can only be adopted by the few.

EMPYEMA is a disease of the pleura associated with the effusion of pus into the pleural cavity. In many respects this disease presents symptoms closely resembling those met with in pleurisy, but differing in being more intense, and attended with more danger to the patient ; in simple inflammation of the pleura, the products effused have a tendency to become absorbed, and to leave only adhesion of the two surfaces of the membrane, while in empyema adhesions rarely occur, and the matter must be let out by surgical interference ; there is, in fact, in these cases a large abscess in the chest, which, by compressing the lung on that side,

causes great distress and much difficulty in breathing. Persons who suffer from this disease are generally in a bad state of health previously, and are often of a scrofulous constitution. Scarlet fever in children may set up empyema, and it is more common from this cause in early life than among adults. In some who have diseased joints or sinuses in the limbs, with diseased and bare bone, and after amputation of a limb when pyæmia has been set up, secondary deposits in the lungs and empyema are very liable to recur. Those also whose lungs are in a diseased state, as in cases of phthisis and some forms of pneumonia, are liable to this complaint. The bursting of an hydatid cyst into the pleura, the rupture of a tuberculous cavity of the lungs, and the extension of a similar disease in the pericardium, will set up empyema. And, finally, it may come on insidiously without any distinct cause being made out.

Symptoms.—There is at first pain of a sharp and shooting character in the affected side, and this is generally confined to one spot ; the patient cannot cough or take a deep breath without increasing this pain. In a few days, when the fluid is poured out into the pleura, the pain may diminish considerably ; but there is more or less distress of breathing, because, from the pressure of the fluid, air cannot enter the lung on the affected side, and the other lung is called upon to do all the work ; hence the patient lies on his back or diagonally toward the diseased side, so as to give the healthy side of the chest all the room he can to expand. From the first there are the usual signs of fever—a furred tongue, quick pulse, loss of appetite, and much thirst. The temperature, too, of the body rises considerably, and is liable to much daily variation, being high at night and perhaps two or three degrees lower in the morning. When disease is well established, the diseased side of the chest is larger in circumference than the other, and there is bulging of the intercostal spaces ; the veins also are obstructed over the part, and appear as blue lines running over the chest. The dyspnœa is great, and increased on exertion ; each respiration is hurried and shallow ; the countenance is anxious, and sometimes pale or livid. Generally the patient is worse at night, and becomes hotter and more oppressed ; at times a hectic flush appears on the cheeks, at others there is much perspiration over the head and body ; rigors or shivering are very usual in the early stages of the disease, but become less frequent afterward. The rigors, the hectic flush, the daily oscillations of temperature, the increase at night in the intensity of the symptoms, the bulging of the chest-wall, and the distress and prostration of the patient are the indications which show that the case is one of empyema, and not of simple pleurisy ; yet in some cases these symptoms even may be absent, or at least not well marked ; and, again, a person may be attacked with ordinary pleurisy first, and afterward the effused fluid may turn to pus.

Treatment.—The patient must be kept in bed in a warm and well-ventilated room ; the air should be moist, and of a temperature from 60° to 65° Fahr. When there is much pain a few leeches will often give great relief, and then a hot poultice can be applied, or else flannels wrung out of hot water, and covered over with some oiled calico or oiled silk, so as to keep in the heat, and prevent the bedclothes becoming wet. Care should always be taken that, in applying either poultices or hot flannels, the bleeding from the leech-bites (if any have been applied) should have quite stopped, because, if this precaution be not taken, the warmth will encourage the bleeding, which will trickle into the poultice without being seen, and so the patient may lose much more blood than is advisable. Food of a light and nourishing description must be given ; milk, beef-tea, broth, and a moderate amount of stimulant are best borne ; the diet, in fact, is such as may be given in all cases of fever, and will be more fully described under the general head of fever. (See FEVERS.) When there can be no doubt in the mind of the medical man that pus is present, it is certainly advisable to open the chest by a small incision, so as to let it out ; no good can come by delay, as the patient's health will become worse, and no benefit can be expected from leaving the case alone. Yet, should any doubt exist as to the nature of the disease, an exploratory puncture may be made ; this is done by means of a fine trocar and canula, or, in other words, a needle which just fits into a silver tube, and after both have been inserted into the chest-wall, the needle or trocar is withdrawn, while the fluid will escape through the canula or tube ; this is a very simple and harmless operation, and of great use in making sure of the nature of the effused fluid. If pus escape, then there can be no hesitation in tapping the chest, or in performing the operation which is technically known as "*paracentesis thoracis.*" For this purpose an incision, about an inch long, or rather less, is made through the skin, about the sixth or seventh intercostal space, and in the line of the axilla or arm-pit. A trocar and canula, about one fourth or one fifth of an inch in diameter, is then introduced, and when the trocar is withdrawn the pus will run through the tube most readily. Sometimes enormous quantities are removed in this way, and the relief given to the patient is very great in proportion, as now the compressed lung may begin to expand again if the disease has not lasted too long, and the pressure on the other internal organs is removed. The wound should not be allowed to close, but a piece of tubing of gutta-percha

should be kept in, so that any more pus that forms may escape at once, and not accumulate again. Even in very favorable cases pus continues to be secreted and to flow through the tube for days and even weeks after the original puncture. The quantity produced daily gradually diminishes until at length it ceases. All this while the patient will be easier ; he can breathe more comfortably ; there is less fever and hectic ; he will recover his appetite, and rest better at night ; but in all cases that recover convalescence is a very slow process, and tonics, generous diet, cod-liver oil, a visit to the sea-side or country are indispensable aids for regaining health. If the lung cannot expand after the matter has escaped, the chest-wall of the affected side will be pressed in by the external atmosphere, and so be smaller than the other, and in this way such patients often have lateral curvature of the spine afterward. In time the healthy lung becomes much increased in size, and does in a great measure the work of both. The mortality from this disease is considerable, and it is nearly always fatal when arising from pyæmia, or when the patient's health has been worn down by previous disease. In a few cases the pus has made its way through the skin of the chest, and burst externally of its own accord ; but it is best to tap the chest before such a process has taken place.

EMULSIONS are soft, smooth liquids, usually prescribed for coughs, though purgatives can be made into emulsions, as when castor oil is rubbed down with yolk of egg, or milk, or mucilage and syrup. A pleasant cough emulsion is made from almonds, gum-arabic, sugar, water, and a little tolu, paregoric, and sweet spirits of nitre.

ENCEPHALITIS is a technical term for inflammation of the brain.

ENCEPHALOID CANCER. (See CANCER.)

ENDEMIC diseases are those which are peculiar to localities or situations, as goitre to Switzerland, and plica polonica to Poland. Diseases may be endemic and epidemic at the same time.

ENDOCARDITIS means inflammation of the lining membrane of the heart, and is common after an attack of rheumatic fever, and in course of Bright's disease ; it generally is met with in the left ventricle, and sets up a serious affection of the valves. (See HEART.)

ENEMA.—Where, from whatever reason, it is considered not to be advisble to administer food or medicine by the mouth, it is possible, by means of a specially prepared apparatus, to introduce them in a fluid form into the lower bowel. Whatever is so introduced is termed an enema ; formerly a clyster. First, then, an enema may be employed with advantage in cases of prolonged constipation, when it is better and easier to act upon the

hardened mass from below than from above.
Many substances may be employed, but there
is none better than plain soap and water. If
that do not succeed, half an ounce of castor
oil and half an ounce of turpentine may be
beaten up with an egg, and a pint of hot
water added. Usually this will suffice, especially if repeated more than once ; if not,
other means must be taken. It is to be noted
that in making use of enemata for this purpose not less than a pint should be used ; for
the normal stimulus to the bowel to act is
distension. If, on the other hand, it is desired that the enema should be retained instead of being expelled, the smaller the quantity used the better. This is the case when
from disease of the stomach it is impossible
or unadvisable to give food that way, and
small quantities of beef-tea, etc., may be
thrown up the bowel. Then not more than
a couple of ounces should be used at a time.
This too is the case when opium enemata are
prescribed, as they sometimes are for disease
of the lower bowel or neighborhood.

ENERVATION is a term applied to the
weak state met with in cases of nervous debility, and in those who suffer from hysteria
and allied nervous disorders.

ENTERIC FEVER. (See TYPHOID
FEVER.)

ENTERITIS, or INFLAMMATION OF THE
BOWELS, particularly that portion called the
small intestine, is rare as a disease arising of
its own accord. Usually it is the result of
irritants, or comes on in typhoid fever, or is
produced by scrofula. The term, moreover,
is applied only to inflammations beginning in
the inner coat, though it may extend outward and affect all. Inflammation beginning in the outer coat is called peritonitis.
When inflamed, the mucous membrane becomes of a deep red color, almost black, and
very frequently ulcers form on it corresponding with certain glands on its inner wall,
called Peyer's glands and patches. Sometimes these ulcers eat so deep into the gut
that their outer wall is perforated, the contents escape, and peritonitis, a much more
fatal malady, is set up. If the inflammation
be only of a subacute character, the bowel
may be thickened. This sadly interferes with
i.s function, and gives rise to constantly recurring attacks of diarrhœa. The symptoms
of enteritis vary exceedingly in gravity. In
typhoid fever there are often no signs of it
beyond diarrhœa until fatal perforation
occurs, and peritonitis is set up. If all the
coats are affected at once, this is different.
The symptoms then resemble those of strangulation, as it called, there being more the
signs of enteritis. There is intense pain, a
hot skin, quick hard pulse ; the legs are
drawn up to relieve the tenseness of the belly,
and there is nausea and vomiting. The
bowels too, in this form, are obstinately con-

fined, and there may be fœcal vomiting. The
slightest pressure increases the pain, so that
the patient can hardly bear the bed-clothes.
The pulse soon becomes excessively small
and hard, wiry, and imperceptible. If the
mucous membrane alone be affected, the
symptoms are quite different. There is diarrhœa instead of costiveness, and no fœcal
vomiting ; but there is great fever, thirst,
and pain, and the bowels swell with flatus.

Treatment.—In dealing with such a case
we must rely on opium, given both by the
mouth and bowel. If it cannot be retained,
morphia must be given in the anus or over
the bowels. Small doses frequently repeated
are best. Hot fomentations or turpentine
stripes applied to the abdomen give great
relief. Ice to such is both grateful and valuable. Strict quiet is to be maintained in bed,
and no attempt made to open the bowels
until such time as that is urgently needed.
Premature attempts of this kind may cost
the patient his life. The food, too, should be
given in the smallest possible bulk ; as there
is little or no appetite and the body is undergoing no exertion, little food suffices, and so
the bowels may remain locked for a long
time without inconvenience. When it is
judged safe to open them, if they do not act
of their own accord, enemata of soap and
water had better be used. With children
opium must be given with great caution
Lime-water suits them well.

ENTOPHYTA are vegetable parasites
which dwell within the body ; they are found
in some diseases of the mucous membrane of
the mouth and alimentary canal. In the
complaint called thrush, so common in infants, spores of the fungus known as *Oidium
albicans* may be found, also in the false membrane formed in the throat in cases of diphtheria. In certain cases of enlargement and
dilatation of the stomach, fungi are found in
the vomited matters ; the name of *Sarcina
ventriculi* has been given to them ; they form
little square packets of a greenish yellow
color, and are marked by vertical and transverse lines. The yeast plant, or *Tortula
cerevisia*, which is made use of in fermenting
beer or spirituous liquors, is also occasionally found in the stomach and bladder. All
these parasites are of very low organization,
and generally consist merely of bright, round
particles or sporules, or of threads formed by
the union of these spores ; they are only
seen under high powers of the microscope.
They are probably conveyed into the body
from the external atmosphere, and develop
wherever they find a convenient nidus. No
particular treatment need be adopted, as they
are an accompaniment of the disease, and
not the disease itself.

ENTOZOA are animal parasites which are
met with or have their "habitat" within the
human body. They have been divided by

biologists into three classes : 1. *Cœlelmintha*, or hollow worms ; 2. *Sterelmintha*, or solid worms ; 3. *Accidental* parasites, or those having the habits, but not referable to the class, of entozoa.

1. The following are the most common worms met with in the class Cœlelmintha : *a. Ascaris lumbricoides*, or round worm, which is met with in the small intestine, and often passes upward into the stomach. The male measures from four to six inches long ; the female from ten to fourteen. In shape it much resembles the ordinary earth-worm ; it is of a reddish color, round, smooth, and fusiform, tapering gradually at each extremity. They are most common in children between the ages of three and ten years. Their number varies from two or three to twenty or thirty, and they are seldom solitary. *b. Ascaris* or *Oxyuris vermicularis*, commonly known as thread-worms ; they are found in the rectum or lower bowel, and are more frequently met with in children than any other worm. The male measures one-sixth of an inch in length ; the female is from one third to half an inch long. They are not seen when a child is suckling, unless other food be given. (See ASCARIDES.) *c. Tricocephalus dispar*, or the long thread-worm ; it is not very common in this country ; the male measures an inch and a half in length, the female two inches. It is found in the large intestine. These three varieties are sometimes spoken of as nematoid worms, because they belong to the order Nematoda. *d. Trichina spiralis*. A worm rare in this country, but common in Germany. It gets into the system by eating sausages not thoroughly cooked. It is attended by symptoms not unlike those of typhoid fever, in some respects. They are met with in the muscles, where they lie coiled up in little oval cases, which are just visible to the naked eye.

2. The following are included under the class Sterelmintha. *a. Tænia Solium*, or tape-worm. Its length is great, varying from six to ten or twenty feet or more. It is a flat, ribbon-like worm, of a white color, about one third of an inch broad, and made up of segments about an inch long near the tail end, and each fits into the segment preceding. The body is pretty uniform in width, but toward the head, the neck tapers very much, not exceeding often one eighth of an inch, and the segments also are very much shorter. The head is known by four black spots upon it, and these are the suckers by which it clings to the walls of the bowels ; the head is about the size of a pin's head, and is rather wider than the neck. On the front part of the head is a small proboscis, on which is arranged a double row of hooks in a circle. A tape-worm may have several hundred segments ; the ones near the neck are at first immature. The worm increases in length by fresh segments

being produced at the neck, while the fully developed segments near the tail drop off ; each fully matured segment is called a " proglottis ;" when these pass away with the excreta, the patient is known to be suffering from tape-worm. No good is done unless the head is expelled, as yards may come away, but if the head remain fresh growth will take place. This worm is usually solitary, is found in the small intestine, and rarely affects children under three years of age. *b. Tænia mediocanellata* is another kind of tape-worm, and the more common of the two. It resembles the preceding in every respect, except that there is no proboscis on the head and no hooklets. *c. Bothriocephalus latus*, or broad tape-worm ; it is the largest of all, and is often twenty or thirty feet in length and an inch in breadth. The head is blunt and flattened from behind forward ; there are no hooklets ; the anterior segments are narrow at first, but widen gradually, so as to attain their greatest width toward the centre of the body ; toward the tail end, the segments diminish in width, but increase in depth, so that the worm is much thicker in the posterior than in the anterior part, where it is flattened. The total number of joints has been said to be four thousand. These three varieties are the most common, and they are called by some cestoid worms, because they belong to the natural order Cestoda. The mature segment or " proglottis" of these worms contains both male and female orders of reproduction ; when one mature segment has become impregnated with another mature segment by contact with it, eggs are formed. These eggs remain in the " proglottis" until it escapes from the bowel, when the " proglottis" itself bursts from the growth of the eggs within ; when the ova escape in this way, they may be eaten by some animal, or even taken into the stomach by drinking water into which they have got. When the embryo in this way enters into a pig or rabbit, it breaks its shell, and, boring through the intestinal wall, lodges in the tissues ; here it forms a cyst, where it may attain a large size, and develops an animal consisting only of a head and neck. Thus it will be seen that the eggs of a tape-worm in man will not produce a tape-worm in another animal, but a body known as a cysticercus, or an anmial in an intermediate stage ; now when a cysticercus is swallowed by man, the fully developed tænia or tape-worm will be produced. The two stages cannot take place in one animal. There are a great many tape-worms of different kinds, and many animals, as the dog, cat, and rabbit, are liable to them as well as man. We may chance to swallow the ova of the tape-worm in the dog by eating water-cresses, or drinking water in which the embryo has happened to be ; and if this be done, we shall not suffer from tape-worm, but from the intermediate variety, and thus a cyst may form in some organ, and grow so as to cause some inconvenience, and even danger to life. These cysts are often called hydatids, and the liver is the most common seat ; they rarely heal of themselves, but generally form rounded tumors which cause very little pain or disturbance ; they generally contain fluid, and attached to the inner wall of the cyst are those curious bodies known as cysticerci, or the worm in the intermediate stage. These cysticerci, when removed from a cyst alive, may be swallowed by man with impunity, but if given to a dog again, they will develop in its intestine into a mature worm. Tape-worm in man is not caused by swallowing the ova, but by eating meat in which the cysticerci are lying. Pigs and rabbits provide us with the *Tænia Solium*, while oxen may give us the *Tænia mediocanellata*. When the mature worm is developed in us, the ova which escape may in their turn supply these animals with fresh material for forming cysticerci. Hence it is an important thing to burn all portions of worms that are voided. There are some other unimportant varieties ; the first two are the only common ones in this country ; the development of hydatids is very rare. Tape-worm itself is attended with much inconvenience, but very little danger.

3. There are a few *accidental parasites*, as the larva' of the gad-fly, and a few less well known. These resemble the entozoa in dwelling within the body, but they have no anatomical relation to those nematoid and cestoid worms which form the class known under the name of Entozoa. For *treatment* of all these varieties, see PARASITES.

ENTROPION. (See EYE.)

ENURESIS, a technical term for incontinence of urine. (See INCONTINENCE OF URINE.)

EPHEMERA. (See MILK FEVER.)

EPIDEMIC diseases are such as are universally prevalent in a district or country at the same time, and which, having endured for a period, at last disappear or die out. Influenza and cholera are instances of epidemic diseases.

EPIDERMIS is the name given to the epithelial covering of the skin ; the number of layers of epithelial cells or scales varies in different parts of the body, being thinnest on the inside of the arms and legs and on the fingers and toes, while it is very thick on the external surfaces of the body, on the palm and heel, etc. When a blister is applied, it is the epdidermis or scarf-skin which is raised above the true skin, and separated from it by the effused fluid ; so also after an attack of scarlet fever, or some other febrile disorders, it is the epidermis which is shed in various-sized patches.

EPIGLOTTIS, a valve-like membrane which fits accurately over the glottis or upper part of the air-passages, so as to prevent any food going down that way.

EPILEPSY is a diseased condition, the exact cause of which is undetermined, but of which the main features are sudden and total loss of consciousness, and convulsions lasting a longer or shorter period. These attacks have a tendency to recur and ultimately affect the mental powers. Frequently these fits are preceded by a kind of warning (see AURA) ; and if this can be stopped, the attack may be arrested. The attack begins with a sudden pallor of the countenance and a fixed expression of face. Sometimes there is a shriek, and the individual falls to the ground violently convulsed (hence the name *falling sickness*). There is usually foaming at the mouth ; the tongue is thrust forward, and sometimes fearfully lacerated by the teeth. The eyes are often fixed, sometimes rolling and quite insensible. The countenance is diffused, sometimes purple, and the breathing is frequently suspended for a time. The bowels and bladder may discharge their contents. The convulsions may affect any or all parts of the body ; usually one side is worst. Gradually they pass off, and the patient remains quiet and apparently sensible ; this may pass into sound sleep, from which he may recover knowing nothing of what has passed, except from the pain from straining his muscles and the pain from his lacerated tongue. Generally, too, there is headache. The fit may last from a few minutes to half an hour, and may recur sometimes once or twice in one day, often not for very long intervals. There is always some risk to the epileptic from being seized in a situation of danger. They may fall on the face and bruise themselves, or they may fall in the water and drown themselves in a pool a few inches deep. Such are the characters of a severe and well marked fit of what the French call the *grand mal*. The *petit mal* may only mean a slight momentary unconsciousness, instantaneously recovered from, or there may be a faint for a few seconds without any fall or dizziness ; or there may be some twitching of the face or one limb, followed by an absent feeling for a few minutes or moments. The appearance of confirmed epileptics is striking ; they have a stolid, immobile look, are usually very stupid ; and very likely also their moral faculties are obtuse. Epileptic maniacs are an extremely dangerous set. Often, in mania, a fit of violence will take the place of a true epileptic paroxysm, and they are always dangerous before and after the onset of a paroxysm ; it is at these times the homicidal impulse is greatest. Epilepsy is often hereditary, but it may be induced by a variety of causes. Epileptiform convulsions are not, however, to be confounded with true epilepsy. Such often occur as the result of over-mental strain, indigestion, etc., but when the cause is removed they have no tendency to recur ; not so with epilepsy.

Treatment.—The treatment of epilepsy resolves itself practically into what is best done in the intervals of the fits. During the paroxysm, the patient should be let alone, care being taken that he does not hurt himself. The great remedy for epilepsy at the present day is bromide of potassium in full doses. To begin, the patient ought to have at least 10 or 15 grains, three times a day, going up to 30, 40, or even 60 for a dose, if necessary. This does good in a great majority of cases, but in some it does not. In these strycnhine or nux vomica is sometimes given with advantage, but it must be used cautiously, and ought never to be given without a physician's prescription. At the same time every effort must be made to improve the general health.

EPIPHORA means an overflow of tears. (See EYE.)

EPIPHYTA are vegetable parasites met with on the skin or external surface of the human body. There are several varieties, and they form in many cases a very troublesome form of skin affection. They are contagious, because the spores may be taken from one body to another by contact, and they are most frequently met with in children.

1. *Pityriasis versicolor*, or *chloasma*, is due to a parasitic fungus ; the disease is a very common one, and often seen on the chests of poor people, especially of those who wear flannel next to the skin, and who are not very clean. It occurs in fawn-colored or buff-colored patches, so that the surface of the skin is mottled with these discolorations. If a few of the epithelial cells so affected be scraped off, soaked in a solution of potash or soda, and placed under the microscope, a number of small round globules are seen adhering to the scales, and besides may be found filaments formed by the long cells being placed end to end. A solution of sulphurous acid or hyposulphite of soda, applied to the skin after it has been well washed, will often cure this disease ; acetic acid or strong vinegar and iodine paint are effectual, but they are more painful applications. Liver-spots is a name sometimes applied to this affection, while *Microsporon furfur* is the technical name given to the parasite.

2. *Tinea tonsurans* is the name given to the common affection known as ringworm. It occurs in circular patches on the scalp or back of the neck, or on the arms, but it may be found anywhere on the skin. Commencing as a small red patch, it spreads in a circular manner, so that while the centre may be healed, there is an outer ring of a red tint, and covered with a little scurf ; this scurf is due to the scales of epithelium which are being shed ; when these scales are examined, as mentioned above, the spores of the fungus may be seen.

Trycephyton tonsurans is the technical name for the fungus. Although very catching, it may readily be healed by painting the surface affected with a solution of sulphurous acid or vinegar, iodine paint, or solutions of corrosive sublimate ; the first-named is the best, as it is harmless and causes no pain, while fatal results have occurred from the incautious use of the latter. Ink is a domestic remedy, and it may do good sometimes from containing iron and tannin, but it is a dirty application.

3. *Favus* is a troublesome disease of the scalp, which now and then occurs in children. It is due to the presence of a parasitic fungus, *Achorion schonleinii*, which attacks the hair-follicles and the bulbs of the hair itself ; hence the hair becomes brittle and breaks off short. Scales are formed in abundance, and these constitute yellow crusts, whose surface is concave ; they are generally circular in shape, and have a disagreeable mousy odor. The only cure is to shave the head as close as possible, and pull out the hairs at the spot affected, then apply a solution of corrosive sublimate ; this must be done constantly, and whenever a fresh crop appears, but with every care it is very difficult to eradicate.

4. *Microsporon mentagraphytes* is a parasitic fungus met with in the hair-follicles in the disease known as sycosis or mentagra. It occurs in man, and affects the mustachios chiefly, or the hair close around ; the hairs become brittle and break off, while at the roots are little pustules, which break and discharge matter. The hair should, be pulled out, and treated in a similar way to cases of favus.

5. *Microsporon Audouini* is another fungus affecting the scalp in cases of *Tinea decalorus*. This disease may be known by bald, circular patches occurring on the head. While the centre is devoid of hair, the disease spreads at the circumference, and here short broken hairs may be seen. It may be cured by the free application of acetic acid or sulphurous acid. Every bald patch is not due to this parasite ; it is only when the broken-off hairs are found that this is the case.

The spores or sporules in each of these diseases are almost indistinguishable from one another ; they are all contagious, because the fungi may be pretty readily conveyed from one to another. The remedies act by killing the parasites, hence the name parasiticides has been given to them.

EPISPASTICS, that is to say, things that draw, is the term commonly applied to blistering agents, of which the chief are Spanish flies—CANTHARIDES (which see).

EPISTAXIS signifies a bleeding from the nose. The blood supply of the nose is important. The arterial supply is derived from the ethmoidal, spheno-palatine, posterior palatine, and facial, and, as a rule, the veins accompany these arteries ; but some of them, the *emissary*, have no analogy with the above-mentioned arteries, and establish an intricate communication between the nostrils and the cranial veins—a circumstance of some importance, as accounting for the bleeding from the nose in cases of obstinate cephalalgia or headache, and for the " efficacy of derivative abstractions of blood from the nostrils under such circumstances." The causes of epistaxis are idiopathic or traumatic, spontaneous or accidental. Accidental or traumatic is the result of a blow, or by any unusual exertion, sneezing, or violent blowing of the nose. Spontaneous or idiopathic epistaxis has as its causes several different circumstances : thus, capillary hæmorrhage dependent on active or passive congestion, renal and hepatic disease, ulceration, or the presence of polyp. (See POLYPUS.) In young persons of nervous temperament, such symptoms as flushing of the face, buzzing in the ears, and severe headache, are generally relieved by bleeding from the nose. Spontaneous bleeding may also occur in vicarious menstruation, scurvy, fever, or in the hæmorrhagic diathesis. (See HEMORRHAGE.) Epistaxis is frequently a concomitant of declining and advanced life, in which instance it is usually venous.

Treatment.—Simple forms of hemorrhage from the nose, whether accidental or spontaneous, can be readily arrested. Cold applications to the nose and forehead, or snuffing cold water up the nose, a cold key slipped down the back, or cold water dashed to the nape of the neck, or the elevation of the arms as high as possible above the head, are all of great practical use. In some instances the bleeding may be stopped by pressing upon the nostrils with the thumb and finger for some short space of time. The head should be maintained in the erect position, as it is naturally. If these simple methods prove unavailing, a stream of cold water, containing a little perchloride of iron, tannin, or alum, directed through the nostrils, will stop the bleeding. The ultimate resort is the plugging of the nares ; but this must be done by a physician, with an instrument made for the purpose. The plugs should not be kept in for more than forty-eight hours or so, as the confined discharges become very offensive, and indeed dangerous. On removal, the cavities should be thoroughly cleansed with warm-water douches. By the prompt plugging of the nostrils, in severe hemorrhage, many lives have been saved.

EPITHELIUM is a delicate cell membrane, which invests the internal and external surfaces of the body, and which is found lining the various cavities. Over the skin, where there are several superimposed layers, it is known as epidermis ; but it is much thinner over the mouth, nose, lips, and fauces ; it is very thin all the way down the alimentary canal, which it lines throughout. The ureter, bladder, and urethra, the peritoneum, or lining membrane of the abdomen, the pleura, or lining membrane of the thorax, and the ventricles of the brain, all have a thin coating of epithelium. It is found in arteries and veins, and forms a large portion of the liver and kidney ; it occurs in the heart and lungs, and in the various follicles and glands of the skin and mucous membranes. It not only serves as a layer to preserve delicate vessels and nerves from injury, but it takes an active part in the functions of secretion and excretion.

EPSOM SALTS, or *Sulphate of Magnesia,* are one of our most useful and most simple remedies. These salts are now commonly got by acting on dolomite limestone by sulphuric acid. Formerly they were got from wells or sea-water. They occur as fine needles, which are almost identical with those of sulphate of zinc—a somewhat dangerous resemblance. In ordinary doses Epsom salts act as a saline purgative, giving rise to a speedy and free watery evacuation of the bowels. Two drachms or half an ounce would commonly be required. As, however, constipation for a time sometimes follows its use this way, it is perhaps better to give the salt in smaller doses, daily repeated for a time. The addition of a few drops of dilute sulphuric acid renders the salt more palatable. In this way it is best given in the morning, and is an exceedingly valuable remedy for those whose livers are habitually what is called torpid—that is, where there is a tendency to biliousness, with irregular bowels and high-colored urine, such as occurs in men who habitually live too highly.

EPULIS. (See GUMS, DISEASES OF.)

ERGOT is the product of a peculiar fungus which attacks the grains of rye, especially in bad years, and gives rise to a black-looking protuberance from the ear of rye. Hence the name it commonly gets, viz., speared rye. It contains a quantity of oil, and a principle called ergotine. Its best preparation is the liquid extract, but it may also be used as an infusion, if freshly prepared. Ergot exercises its powers, whether directly or indirectly, mainly on unstriped muscular fibre. Hence it acts specially on the minuter blood-vessels, and still more markedly on the womb, especially during pregnancy and at the period of childbirth. Its main use, indeed, in medicine is to stimulate the womb and cause it to contract during delivery. But it must always be used with care, for if given in unsuitable cases it may prove fatal to both mother and child. Of the powder of ergot 20 or 30 grains are given after infusing in water for 20 minutes, grounds and all ; of the liquid extract the corresponding dose would be about 30 drops. This having been given to a woman in childbed—*and the case* *being in every way suitable*—in no long time contractions of the womb, previously suspended from whatever cause, begin ; having begun, they continue almost without intermission until the child is born. These continuous and powerful contractions constitute, indeed, the great danger of ergot, for thereby the child may be destroyed, and if the way be not clear for its expulsion the mother also may be injured. Ergot is frequently of great value in flooding after labor, especially if this arises from imperfect contraction of the womb and the retention of blood clots. From its effects in the smaller arteries, ergot, if long taken, may prove dangerous. It causes such contraction of their calibre that blood is prevented from reaching the extremities in sufficient quantity to keep them alive, and so a kind of mortification, such as sometimes occurs in old men, follows. This only occurs from eating rye bread made of badly prepared grain, containing consequently much ergot.

ERUCTATION is a term applied to the rising of gases into the mouth from the stomach ; it is often a sign of indigestion.

ERYNGO, the Sea Holly (*Eryngium maritimum*). It grows abundantly on almost every sea-coast. A decoction of the root, when made of sufficient strength, is said to act on the kidneys and liver, and is useful in cases of congestion. A confection is made of slices of the root steeped in boiling syrup, which is agreeable and useful in coughs or hoarseness.

ERYSIPELAS of the face is a disease of pretty frequent occurrence, and although causing much discomfort to the patient, is not often attended with much danger. It seems to affect persons of a nervous and excitable temperament, and some seem much more liable to it than others. It is rarely seen in children, but it attacks adults of both sexes, and women are more subject to it than men. It comes on without apparent cause in many causes, but sometimes a blow or exposure to a cold and cutting wind sets up the inflammation. It usually begins at the ear or one side of the nose, and then the redness and swelling extend over that side of the face ; more rarely it crosses over the median line and affects the whole of the upper part the face. Pain and tingling precede the inflammation, and when the latter has reached its height, the eyelid is so swollen that it cannot be opened ; the ear is large, red, and flabby, while the skin adjacent is swollen, red, and painful. Erysipelas is, in fact, an inflammation of the skin, and it is severe, according to the depth to which this tissue is implicated. Sometimes only the upper layer is affected, and then the appearance is like that seen in erythema. There is but slight swelling, and the constitutional symptoms are not severe ; but if the whole thickness of the skin be attacked, and, in addition, the loose cellular tissue beneath, then the inflammation is of graver import, and may spread over a large area. There is, from the first, a high temperature, quick pulse, thirst, often a sore throat, loss of appetite, and a thickly coated tongue. The patient feels very restless, and sleeps badly at night ; in many cases delirium comes on toward evening, and this is mostly observed in those previously addicted to intemperate habits. The bowels are often constipated, and the urine high-colored and containing a little albumen. The mucous membrane of the throat is of a dusky purple color, and swollen in some cases, and when erysipelas attacks this part also, it adds to the danger of the patient by preventing swallowing, and even by causing suffocation. Erysipelas of the face, without any other complication, usually runs a course of six or seven days, when the temperature rapidly runs down, the tongue begins to clean, and all the febrile symptoms disappear, leaving the patient weak and anæmic. But, if the inflammation has affected the deeper layers of the skin, or if the patient has been previously in bad health, matter or pus may form beneath the scalp from extension of the disease upward ; when this occurs, the pus soon burrows about under the scalp, and therefore, when this takes place, an opening must be made to let the matter out at once. The formation of pus may be known by the temperature keeping high, and by the patient having rigors, accompanied at the same time by a doughy, soft swelling on the scalp above where the matter has formed. Large blisters often form over the inflamed skin of the face, and very frequently the hair comes off in large quantities during convalescence, and especially where there has been any inflammation of the scalp.

Treatment.—The patient must be kept in bed, and fed on a light and nourishing diet, in the same way as is described under the general treatment of fevers (see FEVERS). The light should be kept off the patient's eyes, either by placing the person with the back to the window or by having curtains round the bed. The greatest relief is obtained by preventing the access of air to the inflamed skin, and for this purpose flour is commonly dusted over the surface ; a much better plan is to brush or smear the part gently with a mixture of equal parts of castor oil and collodion, or castor oil alone may be used. It effectually keeps off the air and relieves the tightly stretched skin. Some opening medicine may be given at first, if the bowels are confined and the tongue much coated. As a rule, the disease will get well with careful nursing in a few days ; but if the throat be much affected, the case must be watched and means taken to subdue the swelling. Stimulants are not wanted, except where there is much prostration and delirium, and in those cases where matter has formed under the scalp. During convalescence, tonics, containing iron and quinine, may be given, and for some time any exposure to cold winds, etc., should be avoided ; great moderation in the use of intoxicating liquors should also be exercised.

ERYSIPELAS (SURGICAL).—In the great majority of cases of erysipelas, and especially of the more severe forms, the disease has its starting point in a wound, open sore, or large ulcer on the surface of the body. Given a recent contused or lacerated wound on the scalp or the skin of the leg in a badly nourished and debilitated individual living under faulty hygienic conditions, erysipelas will most probably show itself in one of the two following forms : simple or cutaneous erysipelas, resembling in all respects the affection which frequently attacks the face in the absence of any wound or local irritation, and the characters and treatment of which have been described. In some cases of contused wound of the scalp, the redness, swelling, and blistering of the skin of the face are associated with much pain and tenderness over the whole of the head, and a hard brawny condition of the scalp. The patient, after an attack of intense shivering, becomes very hot and feverish, and often loses his senses and raves violently. The tongue becomes brown, and the pulse very rapid. In the course of thirty-six or forty-eight hours, the condition of the scalp undergoes a change ; it is no longer hard and tense, but now very puffy and raised from the surface of the skull by a collection of fluid, which subsequently, if not let out by the surgeon, breaks through at one or more points, and shows itself as thick yellow pus or matter. Along with the profuse discharge from these openings there is a throwing off of foul shreds of a white or yellowish-white color, formed by the death of the soft and tendinous structures between the skin and the surface of the bone. Occasionally considerable portions of the skin are destroyed, and bone is very often laid bare. If the patient should survive the acute stage of this dangerous affection, the erysipelatous redness and swelling disappear, the fever and delirium subside, and the sloughing wounds on the head are replaced by ruddy ulcers, which heal rapidly as the general health improves. In many cases, however, death occurs from one or more of the following causes : the intense general action of the erysipelatous affection which seems to poison the whole mass of blood ; pain and cerebral excitement ; a general affection resembling typhoid fever, which is associated with formation of abscesses in the liver, lungs, and some of the joints ; purging and hectic fever ; exposure and death of a portion of skull, and formation of abscess

between the inner surface of skull and the upper surface of brain. This, which is called the phlegmonous form of erysipelas, may occur after an external injury at any part of the surface of the body or limbs. It is often seen after severe contused wounds or compound fractures of the lower extremities. There is yet another variety of erysipelas, called diffuse cellular inflammation, which may present itself in connection with local irritation or an open and discharging surface, but which is generally due to the introduction into the system of some animal poison, as in dissection-wounds, the bite of a horse, or in snake-bites. Here there is much swelling and hardness of the affected part, intense pain, and rapid sloughing, with formation of spreading abscesses. The skin, however, is not primarily affected, but the cellular tissue beneath, large portions of which are destroyed and become gangrenous. This disease is generally diffused over a wide surface, and spreads rapidly. It does not in all cases appear in the neighborhood of the wound or sore ; but is separated from it by a tolerably wide extent of perfectly sound skin. The constitutional symptoms are very severe, and death generally takes place on the seventh or eighth day, and sometimes earlier. The essential *cause* of erysipelas, though as yet not well determined, seems to be a poison engendered from putrid animal matter. The predisposing causes are to be sought for in the affected individual, and in the condition as to ventilation, living, and the like, under which he is placed. The general state of disorder produced by habitual intemperance and irregular living, disease of the liver and kidneys, indigestion, exhaustion from fever and from profuse discharges, render patients with wounds liable to an attack of erysipelas. Exposure to cold, fatigue, and indiscretion in diet are also predisposing causes. There are some individuals who seem to have a peculiar constitutional predisposition to the disease. It has been asserted that women are more liable to contract it than men. Of all the predisposing causes of erysipelas, deficient ventilation is probably the chief.

Treatment.—In the treatment of wounded individuals, great care should be taken remove all sources of foul and unwholesome exhalations, and to keep up a constantly renewed supply of fresh air. Unremitting attention should also be paid to the cleanliness of the patient and everything about him. The bed-linen ought to be frequently changed, and not be allowed to remain when soiled by discharge. The motions should be at once removed, and a solution of carbolic acid, chloride of lime, or some other antiseptic be poured into the bed-pan. The wound or raw surface should not be wiped with a sponge, but with tow or cotton-wool, which must immediately be thrown away or destroyed.

The patient's bedroom should be emptied of all but indispensable articles of furniture, and bed-curtains be at once removed. The treatment of phlegmonous eryispelas and diffuse cellular inflammation consists in supporting the strength of the patient by alcoholic stimulants and by tonics, the most effectual of which are quinine and the tincture of perchloride of iron. In no other disease is brandy or wine more needed than in erysipelas. The bowels should be freely relieved from time to time. Bleeding and the application of leeches are now but rarely resorted to, and then only in cases of threatening inflammation of the brain in strong and full-blooded patients. Ammonia is a valuable medicinal agent in bad cases of diffuse cellular inflammation from snake-bites and animal poisoning. In the local treatment of the severe forms of erysipelas various agents have been used. Of these perchloride of iron, sulphate of iron, tincture of iodine, and nitrate of silver, or lunar caustic, have proved the most useful. When the swelling is soft and boggy, incisions must be made in order to let out the purulent fluid and shreds of gangrenous subcutaneous tissue. In the absence of medical aid the simplest and best local treatment would be the application around the inflamed parts of flannels dipped into boiling water, and then well wrung, or of linseed-meal poultices, to which when there is a profuse and ill-smelling discharge of pus, charcoal, carbolic acid powder, or chloralum should be added.

ERYTHEMA is the name given to an eruption of the skin which is attended by a diffuse redness over a larger or smaller tract of skin. When the finger is pressed upon an affected portion, it becomes pale for a moment, paler as the blood is pressed from the vessels, but the redness almost directly returns. This disease is something like a mild attack of erysipelas, and in some cases may shade into it, but it is much less severe in character, and although troublesome, is not dangerous. This disease, unlike erysipelas, is not confined to the face and head particularly ; it is not attended with inflammation of 'the true skin, nor with any marked pain or fever. When the skin is dry, as in old people, and when it has somewhat lost its elasticity, it is very apt to become erythematous : the face and neck may become in this condition from walking out in a cold north east wind ; the friction of clothes will also do it, and it may be produced by colored articles of wearing apparel, from which the dye comes off and irritates the skin. These simple cases may be treated by resting the affected part, keeping it covered up from the air, and bathing it with tepid water several times a day. Another kind which is more important, but still very curable, has been styled "*erythema nodosum.*" It is generally seen in children,

and is found on the shins and arms ; the extremities are affected more readily than the body. Dirty purplish patches are to be seen in front of the shins, and these are raised above the surface, and are painful on pressure ; they are worse after walking about. This state is due to blood and serum being effused under the skin, and it is thus different from the other variety. The child is generally pale and in bad health at the time. Rest in bed is a good thing, and let the patient be fed well with plenty of nourishing diet, viz., milk, meat, or strong beef-tea or broth. With the aid of a little medicine of a tonic character, a cure soon takes place. This form is sometimes met with in cases of rheumatic fever. It more frequently affects young women and girls than the male sex ; yet it is met with in feeble boys. There is slight fever with it and a feeling of languor and discomfort. Red, elevated spots then come out in a few days, nearly always on the legs, and they are generally situated along the length of the limb or in a vertical direction. The patches are oval, and may be several inches in length ; they generally assume a more or less rounded appearance. The lumps in a short time become purple, as if they were cold, and this in time dies away, leaving no mark behind. The disease, when it occurs, is met with in debilitated persons, and therefore measures should be taken to improve the general health.

ESCHAROTICS are such powerful chemical substances as when applied to the surface of the body destroy the vitality of a portion of it, this subsequently coming away as a slough or eschar. The most important escharotics are the red-hot iron, the strong mineral acids and alkalies, chloride of zinc, and the strongest acetic acid ; acting in a milder degree they are called caustics. Their chief use is to remove unhealthy growths, or such as by their own malignancy would destroy life, and so to obtain a clean surface after the slough has separated, whereby wholesome growth is promoted. They are also used to completely destroy a part that has been bitten by a mad dog or such animal, to prevent infection from putrid sores and the like.

ESSENCES in the Pharmacopœia are preparations in which the volatile oil extracted from the plant by distillation is dissolved in spirit. The only two essences of this kind are essence of aniseed and essence of peppermint. Essence of almonds is a totally different kind of preparation. The term is frequently employed for a more or less concentrated preparation of the substance whose name is attached to the title.

ETHER, more strictly sulphuric ether, is a liquid obtained from alcohol by abstracting water from the latter. It is a volatile colorless liquid, with a peculiar smell and pungent taste. It is very inflammable, and so volatile as to produce a sensation of cold if applied to the hand. It is most frequently given internally, mixed with spirit. When so taken, or by itself, it is a powerful stimulant, acting more rapidly and passing away more speedily than alcohol. Hence it is useful to dispel wind from the stomach, to relieve asthma, spasms, and pains about the heart. It may be used locally, so as to freeze the part and so give rise to complete loss of sensation. This is sometimes taken advantage of in surgery ; and as it also produces insensibility when inhaled into the lungs, it is occasionally used as an anæsthetic. It is generally considered safer than chloroform, but is less manageable, takes longer time, and is more bulky. Given internally the dose of ether should not exceed half a drachm.

ETHER SPRAY.—Of late years an ingenious method of producing "local anæsthesia" by freezing, has been introduced with a view of rendering painless certain minor and superficial operations in surgery, such as removing small cysts, opening abscesses, extracting teeth, toe and finger nails, etc. The apparatus is precisely similar to that adopted by the chemists for dispersing perfumes, thus one tube dips perpendicularly into a bottle of ether ; another tube is so arranged that a current of air blown through it shall cross the orifice of the first. This creates an upward-suction current in the first tube sufficient to lift the ether, and blows it away in the form of a fine mist or spray. There are various mechanical appliances for providing the air-current, such as working a hollow india-rubber ball or foot bellows, but the apparatus is sold complete by any instrument maker.

EUSTACHIAN TUBE. (See EAR.)

EVAPORATION is the slow production of vapor at the surface of a liquid. It is through evaporation from the earth's surface that wet clothes dry when exposed to the air, and that open vessels containing water become emptied. Aqueous vapor rises in the atmosphere from the evaporation constantly going on from seas, lakes, rivers, and the moist soil. These vapors condense in the upper regions and form clouds, and finally return to the earth as rain, snow, or sleet. Evaporation is much increased by raising the temperature, which acts by increasing the elastic tension of the vapor ; its rate is also affected by the quantity of the same vapor in the surrounding atmosphere ; no evaporation could take place at all in a space already saturated with vapor of the same liquid, while it would take place very rapidly in air free from those vapors. Hence on a damp day evaporation takes place very slowly, while on a fine dry day it occurs readily. It is evident, also, that a breeze, by renewing the air, will increase evaporation, for if the air which surrounds the liquid be not renewed it would soon become saturated and evaporation cease ; the more frequently the air is renewed, the more evaporation goes on. The extent of surface exposed makes an important difference in the rate of evaporation ; the greater the surface the more rapidly this process goes on. It is obvious that a great deal more evaporation goes on in summer than winter. It is possible for 15,000 gallons of water to be evaporated in twenty-four hours from the surface of a lake equal in extent to an acre, on a hot summer's day during a breeze ; in winter, during the same time and from a corresponding area between three and four thousand gallons would pass into vapor.

EVOLUTION is a term variously applied to different changes going on in the body. The enlarged uterus in a case of pregnancy is said to be evolved from its simple elements, and an embryo is also said to pass through different stages of evolution on its way from the cell-elements of which it is at first composed to the complex structures met with in the infant. The term is also used by biologists to signify the development of man at different periods of the world's history.

EXANTHEMATA, a name applied to several febrile and contagious disorders which are accompanied by a rash or eruption on the skin ; the group includes measles, scarlet fever, small-pox, chicken-pox, and erysipelas ; some also include under this head typhus and typhoid fevers, but these are generally spoken of as continued fevers.

EXCISION means the removal by operation of a part of the body ; in surgery the term does not, however, include amputation. The great advancement made in modern surgery has in many instances substituted an intermediate excision or resection in lieu of amputation : thus, the knee joint, elbow joint, hip joint, etc., being the seat of scrofulous disease (white swelling), is *excised* instead of amputating the limb, whereby in many instances a useful member is preserved. The term is also applied to tumors or morbid growths requiring removal, or to any part in which such a growth exists, such as the upper or lower jaws, eyeball, tongue, etc. The remarks made in the article on incision are of course mechanically applicable to excision, the instruments for such proceedings being knives, saws, cutting forceps, scissors, chisels, gouges, elevators, etc. (See INCISION.)

EXCRETIONS.—Whatever is no longer serviceable to the system is an excretion, and is thrown off by one or other of the organs of the body. Secretions are the healthy juices of the body, which enter into its composition, while excretions are the waste and useless parts which pass away either by the bowels, or bladder, or perspiration.

EXERCISE as a remedial agent is too frequently disregarded so far as preventing is concerned. Its degree and kind is too often left to the patient himself. The various kinds of exercise used to be classified as sailing,carriage, horseback, and foot, but practically may be limited to the last three. But these do not include the exercise of all the muscles. To do that gymnastics must be employed. (See GYMNASTICS.) The grand rule in prescribing exercise is this : the patient should never feel actually tired or fatigued, but rest should be grateful after it. Exercise carried to an extent so as to induce much fatigue is worse than useless.

EXFOLIATION OF BONE.—When a superficial layer ,of bone (such as from the shin, for instance) dies and detaches itself, after an injury or disease, and comes away as a scale, the bone is said to exfoliate. It is frequently noticed in the jaws after clumsy tooth-extraction, or in the shins after blows or kicks. A lotion of weak nitric acid is the best application, and when the shell of bone is *thoroughly* loose it should be gently pulled away with forceps.

EXOPHTHALMOS is a name given to the condition in which there is great prominence of the eyes, so that the individual has a marked and peculiar stare.

EXOSTOSIS, a tumor connected with a bone, and composed of true bony substance, which is sometimes very hard and compact, and at other times light and porous. In most instances the unnatural growth is made up entirely of bone, but occasionally is met with composed partly of bone and partly of cartilage or gristle. The former is called a *true* and the latter a false exostosis. These tumors usually occur singly, but sometimes affect simultaneously several bones in different parts of the body. The bones most frequently diseased in this manner are the arm-bone at its upper end, the thigh-bone at its inner surface and close above the knee, the tibia or larger bone of the leg at its inner surface and upper extremity, the collar-bone, and the bones of the skull. Exostoses take the forms of flattened discs, large lumps with broad bases, and oval tumors mounted on a short bony stalk or pedicle. In the first two the structure is generally of ivory hardness ; the growths of this character are seated on the jaws and the bones of the skull ; the oval and stalked varieties are most frequently met with in the bones of the extremities, and their tissue is more open and spongy. The causes of exostosis are very obscure. The growths are sometimes produced through blows or long-continued pressure. Some individuals show a remarkable tendency to the multiple development of exostosis. In patients suffering from advanced venereal disease, and in scrofulous children, hard painful tumors resembling exostoses in form and composition are often met with in different parts of the skeleton. These, how-

ever, differ from true exostoses in the rapidity of their growth, in the pain attending them—very acute in venereal disease, dull and gnawing in scrofula—and in their submission to medicinal treatment. True exostosis grows slowly, and is amenable to no treatment save a surgical operation undertaken for its complete removal. At first it is painless, and often reaches considerable proportions before its existence is discovered. When very large, however, it may be attended with pain in consequence of its pressure on adjacent nerves. Exostoses, when seated on the bones of the trunk or the skull, may cause serious and even fatal consequences through their pressure on important organs.

EXPECTORANTS are medicines or other remedies which promote the expulsion of fluids from the air-passages. This is the strict meaning of the word, though sometimes it is used in a wider sense. The substances included in the group are of a most diverse character, some soothing, some stimulating, some acting directly in altering the kind of the secretion, others in altering its quantity. It is hardly possible, therefore, to give any extended account of the group. (See IPECACUANHA.)

EXPECTORATION is the term applied to the fluid or phlegm which is coughed up from the air-passages ; it varies much in different diseases ; in pneumonia it is viscid, tenacious, sticks to the sides of the vessel, and of a rusty appearance ; in bronchitis it is frothy, abundant, and often marked with black streaks, or it is thicker, and of a greenish-yellow color from the presence of pus ; in catarrh, the phlegm is often coughed up in pellets, which are black or iron-gray or yellowish in color, and generally most troublesome on awaking, as the secretion has been accumulating during sleep ; the color is due to particles of dirt or smoke in the inhaled air. In consumption the expectoration varies from a small quantity of frothy fluid to abundant greenish-yellow purulent phlegm.

EXTRACTS are forms of remedies in which some fluid preparation, infusion, decoction, or tincture has been gradually evaporated until a thick paste is formed. Some substances lose a good deal of their efficacy in the process ; others do not. Remedies so prepared are usually given in the form of pill, the extract serving as the basis of the pill, and having, perhaps, other remedies in the form of powder conjoined with it. A few extracts are kept quite dry, and only moistened as required. This is the case with compound extract of colocynth. Occasionally an extract is rubbed up with some other preparation, to add to its strength. In this way extract of belladonna is not unfrequently added to the liniment or tincture, or itself reduced with glycerine. There are a few liquid extracts which are less dense than the ordinary ones ; these are prescribed in minims instead of grains—sugar of squill, balsam of Peru and tolu, ammonia, ammoniacum, copaiva, tartar emetic, oxide of antimony, creasote, etc. Perhaps the best and most useful of all in most cases is the vapor of water to which other volatile substances may or may not be added. (See MEAT, EXTRACT OF.)

EXTRAVASATION OF URINE.— By extravasation of the urine is meant its unnatural escape from either the kidney, where it is secreted ; the ureter, the duct by which it passes to the bladder, its receptacle ; or from the urethra, the channel by which it is conveyed from the system. The causes of extravasation in each case are either the result of local injury, or the giving way of any one of these structures from special reasons. Usually, however, the term "extravasation of urine" is meant to convey the idea of the giving way of the *urethra*, and the infiltration of the urine into surrounding tissues ; as the result of a blow, a kick, or a fall on the perinæum, thereby rupturing the tube, or from the pre-existence of stricture, and a consequent preternatural distension of the bladder. This case, then, will be described in the first instance. When the urine is retained in the bladder and cannot escape per urethram, from whatever cause, ulcerative absorption, as it is termed, takes place just above the point of obstruction, and its most frequent locality is in the membranous urethra, just behind the bulbous portion. The train of symptoms occurring from the escape of the urine is in general as follows : The sense of fulness of the bladder and inability to make water induce violent attempts at micturition, a sudden yielding takes place, followed by a great sense of relief, much to the patient's pleasure, but to his surprise no water flows from the expected channel, owing to the giving way of the urethra, and the escape of the urine into the tissues. Now, the urine may be extravasated in all directions, upward, when it gets *between* the bladder and pubes or by the side of the prostate, or downward into the scrotum ; it does not advance down the thighs, owing to the deep layer of superficial perineal fascia being bound to the rami of the pubes, and thus limiting its effusion, at all events *at first*, to the scrotal tissues, supposing the giving way to have occurred at a point anterior to the deep perineal fascia. The effused urine soon excites great local irritation and inflammation, and most alarming constitutional symptoms. The scrotum, which is rapidly distended, becomes dark-colored and quaggy to the touch ; the constitutional symptoms are those attendant on asthenic suppuration and gangrene, rapidly becoming more and more typhoid, and ending in fatal collapse if unrelieved. Cases of scrotal extravasation are those in which best hopes are to be held out of the patient's recovery ; although there are cases on record of intrapelvic extravasation, which have got well, after efficient openings have been made, and plenty of nourishment and stimuli administered. But such cases are hopeless if, the urine be extensively insinuated into the pelvic cellular tissue. All relief must be prompt and active — a *free* incision is to be made into the perinæum, through the tissues, into the infiltrated structures, hot fomentations applied, and the patient's strength supported by wine or brandy, etc. It must be borne in mind that the scrotum is not necessarily involved in these cases ; thus, if the extravasation take place behind the bulb, the urine, being temporarily confined by the deep fascia, burrows, and thus the local signs are obscured ; in these cases the glans penis is frequently found to be hard, swelled, and black, indicating infiltration into the corpus spongiosum urethræ, and this is a most alarming symptom, and in such instances free and complete incisions, such as would lay bare the source of extravasation, must be made at once. In cases of extravasation from local injury to the perinæum, as in kicks or blows, etc., scrotal distension is usually very sudden. A full-sized catheter should be passed into the bladder to allow of the escape of urine. In extravasation from *rupture of the kidney*, the different symptoms will be in proportion to the severity of the injury. The symptoms of such an accident are in general collapse, vomiting, pain in the loins and along the course of the ureter, retraction of the testicle, and numbness of the upper part of the thigh. Such urine as would come through a catheter introduced into the bladder will be scanty, high-colored, and contain a large quantity of blood ; this blood is often stringy and wormlike, owing to the form given it by the ureter. If urine escape anteriorly, acute peritonitis is set up almost immediately ; if posteriorly, the symptoms of peritonitis, perhaps, will not be so early marked ; but rigors, high fever, and a general typhoid condition soon make their appearance. Slight cases of ruptured kidney, doubtless, frequently recover ; severe ones, never.

Treatment.—Absolute rest is the first thing, and the free administration of opium, and the urine drawn from the bladder daily until it becomes clear. In cases where the extravasation is posterior, on the possibility of suppuration, careful examination should be made with a view of giving exit to the pus. Extravasation from *rupture of the ureter* has been recorded, but the cases seem somewhat obscure.

EYE.—For an explanation of the structure of the eye, see the article on VISION. The *diseases and injuries* to which the eye is subject may be most conveniently classified thus : 1. Injuries and diseases of the appendages of the eye—*i.e.*, of the eyelids and tear-secreting apparatus. 2. Injuries and diseases of tne eye itself—*i.e.*, of the eye-ball and its contents.

1. Eyelids.—The eyelids are two thin movable folds placed in front of the eye, protecting it from injury. The upper is the larger, and has the most power of action, being provided with a special elevating muscle, the *levator palpebræ*, which is absent in the lower one. The angles of junction of the upper and lower lids are called the *canthi*. At the margin of each eyelid at the inner canthus are seen two small conical elevations, the apices of which are pierced by a small orifice called the *punctum lachrymale ;* it is through this orifice that the tears pass into the nose. The eyelids are composed of the following structures taken in order, as in making a section from the surface : Integument, areolar tissue, fibres of a muscle called orbicularis, tarsal cartilages, fibrous membrane, Meibomian glands, and conjunctiva ; and in the upper lid is the tendon of the muscle before mentioned, the levator palpebræ. The integument is very thin, and continuous at the margin of the lids with the mucous membrane covering the inner surface of the lids. The subcutaneous areolar tissue is very loose and delicate. The tarsal cartilages give the shape and support to the lid. The Meibomian glands, about thirty in number in the upper eyelid, and somewhat fewer in the lower, are situated between the afore-mentioned tarsal cartilages and the conjunctiva, and on everting the lid look like parallel strings of beads ; they are a variety of the sebaceous glands of the skin, and open by ducts upon the free margin of the lids by a number of small orifices ; they secrete a sebaceous matter, which is intended to prevent adhesion of the lids. The eyelashes, or *cilia*, are thick, short, curved hairs, attached to the free margins of the lids ; those of the upper lid curving upward, and those of the lower downward, so that normally there is no interlacement of them.

Diseases of the Eyelids.—Owing to the number of structures which enter into the formation of the eyelids, there is necessarily a great number of diseases to which they are liable ; but in a work like the present, we must content ourselves with describing the most frequent, or at all events such as come most frequently under observation.

Ptosis.—Ptosis signifies a drooping of the upper lid, owing to paralysis of the nerve (*the third*) which supplies the levator palpebræ muscle. The disease is sometimes congenital, but in such cases it is probable that this muscle is absent.

Entropion.—Entropion signifies an inversion of the eyelids, and in its simplest form is sometimes met with in children, who suffer from ophthalmia, owing to spasm of the orbicular muscle. Collodion, painted on the skin of the lower lid, contracts the part into its proper position, while suitable remedies are to be used for quelling the existing ophthalmia. Occasionally old persons are afflicted with a spasmodic contraction of the orbicular muscle ; and in some severe forms, resulting from chronic inflammation of the conjunctiva, the upper tarsal cartilage becomes so contracted upon itself, that the eyelashes are turned inward and sweep the globe, setting up the most intolerable irritation. The cicatrization following burns, acids, caustics, or severe and ill-dressed wounds are the frequent causes of entropion. In such cases, the means of cure lies in operative proceedings, a variety of which have been devised, and which, of course, must be modified to suit the exigencies of the case.

Ectropion.—Ectropion, the reverse of the foregoing condition, signifies an eversion of the lids. As was before mentioned in the case of entropion, a spasmodic form exists, which is seen after purulent ophthalmia in infants, and from the peculiarly unsightly aspect it presents, causes great alarm in those who have the care of the child ; but, as the inflammation subsides, the deformity will cease. In adults a chronic form of ectropion occurs, as a result of thickening of the conjunctiva after purulent ophthalmia, or after burns, exfoliation of bone, etc., and which may be remedied by operation. By far the worst examples we have of ectropion are the result of burns or scalds, or of the ravages of syphilitic ulceration, and for the remedy of which a variety of operative proceedings, forming a branch of surgery termed " plastic," is needful.

Trichiasis.—By trichiasis is meant an irregular growth of the eyelashes, such that in some instances three or four lashes will grow inward against the globe of the eye, setting up a sense of *pricking*, and a constant irritation and weeping of the eye. These lashes should be plucked out from time to time. It must be borne in mind, that considerable care, and no little dexterity is requisite in plucking out these hairs, for if it be broken short off and not completely removed, the broken stump causes more irritation and pain than the hitherto perfect hair. A good well-made pair of forceps, *not too fine* at the points, should be used, and the hair should not be *jerked* or *twisted* out, but gradually withdrawn by a slow steady pull.

Styes.—(See HORDEOLUM.)

Ophthalmia tarsi is an eczematous inflammation of the edge of the lids, associated with a disordered secretion of the Meibomian glands already mentioned, whereby the lids stick together, and become encrusted with the dried secretion during sleep. *Daily* attention to the washing off of the accumulation, night and morning, is of the utmost necessity, as without this, no remedies are of any avail. As very minute sores exist at the roots of the hairs, they should be kept closely cut with scissors ; by this means the formation of crusts is diminished. The edges of the lids should be neatly smeared with the diluted nitrate of mercury ointment, or the red mercurial ointment, or the oxide of zinc ointment, diluted with spermaceti or fresh lard. If a lotion be used, the acetate of lead forms the best, in the proportion of two to four grains to the ounce of distilled water.

Crab lice.—A species of louse (*phthirius*) ruite distinct from that infesting the scalp (*pediculus*), sets up an irritable condition of the eyelids from its presence. This rarely-met with insect gives rise to a condition termed *phthiriasis.* The parasites are readily destroyed by smearing the roots of the lashes thoroughly with the white precipitate ointment.

Epiphora and Stillicidium Lachrymarum both signify an overflow of the tears, but from different causes. In the first case it is owing to an over-secretion of the tears, and in the second it is owing to an obstruction of the little channels situated on the margin of the upper and lower lids (puncta lachrymalia), and which naturally conduct the tears into the lachrymal canal. An ordinary epiphora is usually due to some irritability of the eye, or the presence of some foreign body. Astringent lotions, aperients, tonics, and antacids appear to be the best treatment. In the case where the overflow is dependent on obstruction of the puncta, an operation is necessary.

Obstruction of the Nasal Duct, i.e., of the tube which conveys the tears from the eye to the nose, generally occurs in strumous persons, and it commences with an overflow of tears in one eye, and a dryness of the corresponding side of the nose. This is usually the first stage of inflammation of the lachrymal sac, resulting in abscess, causing *lachrymal fistula,* which requires special surgical interference for its cure.

Injuries.—Ecchymosis, commonly called a *black eye,* is the result of an effusion of blood into the areolar tissue, immediately below the skin, generally caused by a blow on the eye, though not necessarily, as, owing to the continuity of this areolar tissue over the scalp, a blow on the back of the head may lead to the effusion of blood which will gravitate into the lax tissue in the upper lid. The best and readiest way of getting rid of the disfigurement is the application of a poultice formed of the freshly-scraped root of the black bryony mixed with linseed meal or breadcrumbs. The *immediate* application of tincture of arnica is generally of use.

The eyelids are of course subject, as other

parts of the body, to growths of various kinds, which require the assistance of the surgeon, such as cysts, warts, nævi (mother's marks), carcinoma, and epithelial cancer.

Wounds.—In the instance of wounds of the eyelids, the neatest adaptation of the divided surface must of course be obtained, and the greatest care taken to avoid irregularity or puckering of the edges. Very fine needles, armed with fine silver wire, should be used, and the stitches withdrawn directly any inflammation or redness appears around them. Cold water dressings should be lightly applied ; and attention to the bowels and diet are of importance.

Substances in the lids or on the surface of the eye, see FOREIGN BODIES.

2. Diseases of the Eyeball and its Contents.—*Diseases of the Conjunctiva.*—The conjunctiva, the mucous membrane which lines the eyelids and covers the anterior surface of the eyeball, is subject to several severe forms of disease, namely, *conjunctivitis* or *common ophthalmia.* The several forms of ophthalmia and their treatment are described in article on OPHTHALMIA.

Granular Conjunctiva.—That portion of the conjunctiva which lines the lids, and which is reflected on to the globe, is very often the seat of a rough, thickened-looking red papillæ, a consequence of old standing ophthalmia, causing great pain, and disturbing the proper motions of the eye. This has been described as " the complaint of the poor Irish"—perhaps as good a way of describing the malady as any amount of technicality. It is best treated by counter-irritation outside the lids, such as a small blister behind the ear, and by endeavoring to improve the general health by iron, quinine, and, if possible, change of air.

Diseases of the Cornea.—The cornea is the transparent portion of the globe of the eye, through which the rays of light pass to the interior , it occupies about the anterior sixth part of the eyeball, and as seen from the front is somewhat flattened above and below, owing to the overlapping of the sclerotic. In the healthy eye it is perfectly clear and highly polished in appearance, sharply and minutely reflecting any object upon its surface. In *acute corneitis* the originally clear and polished appearance of the cornea becomes hazy, dim, and rough, red, or opaque. The margins adjacent to the sclerotic coat are vascular, and the sclerotic itself at the point of junction is pink, owing to its increased vascularity. There is an abundant secretion of tears, and intolerance of light ; it most commonly affects strumous children, or it may be the result of injury. It is sometimes followed by suppuration between the layers of the corneal tissues, and in some instances ulceration of either the posterior or anterior wall may take place, giving rise in the first instance to a deposition of pus in the anterior chamber (*hypopyon*) shown by a crescentic yellowish fold at its lower part, and in the former case an opening ensues, by means of the perforation, through which the iris protrudes, termed *staphyloma iridis*. Other consequences of corneitis are *opacities of the cornea*, owing to an effusion of fibrin into the corneal substance, or between it and its covering membrane, the conjunctiva, or from a cicatrix after ulceration. In the first case it has received the name of *nebula ;* in the latter, *albugo* or *leucoma.* The treatment of inflammation of the cornea consists in subduing the inflammation by small doses of mercury, given with a tonic, such as quinine and ammonia ; in very acute cases a leech or two to the temple, or a small blister behind the ear, and warm fomentations. All stimulating lotions are hurtful.

Ulcers of Cornea.—The cornea is very frequently the seat of ulceration, which has several varieties, and may result from injury, scrofula, inflammation of the conjunctiva, or conjunctivitis, and from insufficient or non-azotized food. Three conditions are described. Healthy, with a slight opacity from the adhesive effusion necessary to healing ; the inflamed, with a vascular hazy circumference, requiring leeches and counter-irritants ; and a third, clear, transparent, cleanly cut, and indolent, requiring slightly stimulating applications. In very irritable ulcers, where there is great pain and intolerance of light, they may be greatly relieved by being touched with solid nitrate of silver from time to time.

Staphyloma is a condition following perforation or disorganization of the cornea after ulceration, when any of the contents of the eyeball protrude through it toward the surface. (See STAPHYLOMA.)

Conical Cornea is a rare form of disease wherein the cornea is exceedingly convex, in some cases almost approaching to a point, with the apex central.

Diseases of the Sclerotic.—The sclerotic coat of the eye is that which constitutes the apparent body of the eyeball ; a small portion only of it is seen from the front, the central portion of the visible eyeball being that already described, the cornea. It is proportionally, with regard to the eyeball, about five sixths of its entire surface, and contains the whole optic mechanism. It consists of white fibrous tissue, and into it are inserted the muscles controlling the movements of the eyeball ; it is pierced behind by the optic nerve, and covered in front by the conjunctiva already described. It is subject to several forms of disease, and the most frequent is acute *sclerotitis*, or acute inflammation of the sclerotic, a disease difficult to distinguish from corneitis, frequently of rheumatic origin, though not necessarily. It is known by a pinky redness of the white of the eye, generally great intolerance of light, a sharp, stinging pain, general malaise, and severe

supra-orbital pain. With regard to treatment, iodide of potass in small doses, with bark or tincture of colchicum, seem the best constitutional remedies, and perhaps a small blister to the temple or eyebrow. There should be attention to diet ; and beer, port wine, and sugar avoided.

Diseases of the Choroid.—The choroid is the vascular coat of the eyeball, containing pigment ; it extends over the whole of the posterior portion of the eye, and is continuous in front with the iris. It is pierced by the optic nerve. The coat is subject to an acute inflammation (*choroiditis*). It is, however, rarely seen alone, and is very rapid in its course. Its detection is a matter of some difficulty to the most skilful.

Diseases of the Retina.—The retina is the nervous coat of the eye, and lines the choroid. It is the most essential part of the eye, receiving the impression of light, and is very complex in its structure. *Retinitis*, an inflammation of this coat, is very rare, idiopathically, and is caused by exposure to vivid light, the glare of snow, or of burning sands. It is sometimes met in connection with the various forms of ophthalmia.

(The diseases of the *lens* and *iris* are separately considered, under the articles CATARACT and IRITIS.)

Squinting, or strabismus, is a want of parallelism in the position and motion of the two eyeballs. The usual forms are the *convergent* and *divergent.* The convergent is most common in young persons, and is that in which the eye is turned inward. The divergent is more uncommon, and is most frequently met with in elderly persons, the eye being turned outward, generally from partial paralysis of the inner rectus muscle. Various methods have been adopted for mitigating the deformity, and very frequently, if it be of only a few weeks' duration, may be removed by judicious medical treatment ; but if the squint be of long standing and habitual, and if there be inequality of vision, the operation of dividing the internal rectus muscle must be performed.

Short Sight, or myopia, is where the parallel rays of light are brought to a focus before they reach the retina, caused either by the refracting power of the eye being too great, or its antero-posterior axis too long. With regard to treating it, all minute work must be avoided, and carefully adjusted spectacles should be worn.

Long Sight, or presbyopia, is a failure of vision for near objects. "In pure presbyopia the far point is at a normal distance from the eye, parallel rays are united upon the retina, and neither concave nor convex glasses (even after the instillation of atropine) at all improve distant vision. The eye is neither myopic or hypermetropic. There is, in fact, no anomaly of refraction, but only a narrowing of the range of accommodation. The near point is removed too far from the eye, and hence the difficulty of distinguishing small objects." Spectacles should be used, the lenses of which cause the type of a book to appear bright and distinct, but not *larger* than natural, when held ten or twelve inches from the eye.

Hypermetropia.—In this case the parallel rays are brought to a focus behind the retina, and not upon it, and after some time of employment, print becomes dim, the lines run into each other, and the eye feels hot and

dull. It can be remedied by the use of well-chosen *convex* spectacles, which relieve the strain upon the power of accommodation. This affection is frequently associated with convergent strabismus.

Astigmatism is "irregular refraction, in which different meridians of the same eye have different power of refraction." Thus certain lines, for instance, appear clear and well defined, while near ones are indistinct and blurry. (See AMAUROSIS, and for lotions for inflamed eyes or weak eyes, see COLLYRIUM.)

F.

FACE-ACHE is a form of neuralgia, sometimes depending on unsound teeth, and at other times on an anæmic and debilitated state of the system ; it is not uncommon during pregnancy, or during the period of nursing ; it is then associated with general pallor and weakness. A liberal diet and tonic medicines, especially quinine, give the greatest relief. In the case of unsound teeth, the dentist may be consulted ; if the cheek be much swollen, hot fomentations or poultices will relieve, and in some cases leeches may be applied to the gum. (See PAIN and TIC.)

FACIAL PALSY. (See PARALYSIS.)

FÆCAL ACCUMULATION is a not unfrequent consequence of the habitual use of strong purgatives, especially in elderly females. The proper stimulus to the gut is distension ; when it is full at one particular part, it has a tendency to evacuate its contents. Habit has much to do with this. But if the stimulus is unheeded or resisted, then as time goes by the fluid from the fæces is absorbed by the gut and they become hard, and wedged into the bowel apparently; what is technically termed impacted. Many people, the subjects of constipation, only have their bowels opened after they use purgative medicines, and they too often have recourse to the more powerful remedies for this purpose. And it is plain that if the bowel is only called on to respond to such a stimulus as these afford it will not to such a simple one as distension, and so the evil perpetuates itself. Elderly females are particularly prone to this mischief ; their abdomen is capacious from frequent child-bearing, and it is wonderful to what an extent the fæces may accumulate in such subjects. The part of the bowel where these most frequently accumulate are the beginning and end of the great gut, the cæcum and the rectum, though frequently the whole of the large intestine is blocked up in this manner. The principles of dealing with such accumulations are plain. Powerful drastic remedies are to be avoided. Gentler remedies, like castor oil and Glauber's salts, are to be used ; the principle is to soften the

mass of fæces, and if remedies given by the mouth fail to do this, injections must be used. (See CONSTIPATION.)

FÆCAL VOMITING, or the vomiting of substances already converted into ordure, is one of the most disgusting symptoms patients can be subjected to, or their friends be called on to endure. It may arise from various causes, but the essence of them all is arrest of the passage of the refuse food downward, so that it passes upward again when converted into fæces, all downward passage being denied. Hence it is a symptom of various import ; in perhaps the majority of cases it indicates strangulated hernia (which see). But it may also arise from other forms of obstruction, such as twists of the gut, catching in a loop such as may occur when there is much internal motion in the abdomen and any string-like body is adherent to the walls of the abdomen or to the gut itself ; in fact, all the various causes of intestinal obstruction (which see). Even great accumulation of fæces, from the bowels having been long unmoved, may give rise to this symptom. Another cause, though a less frequent one, is a communication between the great gut and the stomach, which does sometimes occur, and then a mingling of the contents of the two takes place. In this way fæcal vomiting occurs from passage of the fæces into the stomach, and undigested food is passed by the rectum. As to *treatment*, that resolves itself into removing the case of the obstruction if possible. Each must be dealt with on its own basis. That form most amenable to treatment is the one due to fæcal accumulation without any distinct cause. (See FÆCAL ACCUMULATION.)

FÆCES, the excrementitious contents of the bowels, on the proper nature of which health very generally depends. The color of this excretion depends on the admixture of bile with the mass of refuse which passes from the stomach into the bowels. When healthy it should be of a light brown color, and moderate consistence : the presence of too much or to little bile is indicated by the dark or light color of the motions, as the fæces are called in common language.

FAINTING. (See SYNCOPE.)

FAINTNESS, though produced by many different causes, may be said to depend in all cases on impaired circulation in the brain, however brought about. In itself it constitutes a peculiar sensation, and people are often accustomed to speak of being faint, when faintness, as we here use the word, is very far from being present. Thus many healthy people use the word to signify the feeling of emptiness which accompanies hunger. This of course is appeased by food, but true faintness is accompanied by nausea, which renders the idea of food intolerable. Such faintness as we now speak of is commonly accompanied by other signs of the condition ; the countenance, including the lips, becomes deadly pale, the muscles relax so that the individual can no longer stand erect and would fall, or does fall, if he does not lie down ; the skin too is relaxed and is covered with a cold perspiration, and there is an uncomfortable beating of the heart indicating imperfect contraction of that organ ; and if the pulse be felt at the wrist it will be found to be either extremely quick and feeble or else imperceptible. All these symptoms may be brought about in various ways. Thus pain, alarm, dread, and a great variety of mental emotions acting on a delicate system may give rise to it. A great number of affections connected with the bowels and other abdominal organs, especially the stomach, give rise to it. Interference with the heart's action, from whatever cause, is, perhaps, the most potent cause of all, and it is in this way these mental emotions act, affecting one special nerve, called the vagus, which in its turn affects the heart. This same nerve is distributed to the stomach, and it is probably through it that any irritation of the stomach causes faintness, not directly, but in a way we term reflex. Thus the irritant, whatever it may be, affects the nerve in the walls of the stomach ; by the nerve this irritation or stimulus is conveyed to the brain, and from the brain a fresh stimulus is sent forth which affects the heart. The stimulus conveyed from the stomach to the brain is called a sensory stimulus or sensation, that from the brain to the heart is called a motor stimulus, as interfering with the heart's motion. For the conversion of a sensory into a motor stimulus some nerve-centre, like the brain, is requisite. Accordingly in this case the stomach, though very near the heart, cannot act on it directly, but only through a nerve-centre. Sometimes, however, when the stomach is very much distended by air, giving rise to flatulence, this distension may interfere with the heart's action and so give rise to a feeling of faintness. One of the most powerful, if not the most powerful, cause of faintness is loss of blood,

from whatever cause. If there is not enough blood in the body to enable it to carry on all its functions, and the brain requiring a good supply to carry on its work, these must be more or less interfered with. All this, however, must be but little apparent while the individual is lying quite flat and at absolute rest, but if he attempts to rise or to sit up, the extra exertion on the muscles of the body and the extra work of the heart in driving the blood to the head may be too much, failure is the consequence and so faintness, which may be deadly. Such an occurrence is, unfortunately, by no means unknown in midwifery practice after childbirth where there has been great loss of blood. Faintness is by no means, however, without certain concurrent advantages. Thus, where an individual is bleeding from wounds, received it matters not how, except the bleeding be artificially arrested he is likely to perish. Should, however, faintness supervene, the lessened force of the circulation, due to interference with the heart, may give the blood time to coagulate and so prevent further hemorrhage, as it is called. This natural coagulation may be simulated in various ways without inducing faintness, but the same process had recourse to by nature is sometimes had recourse to artificially.

Treatment.—For the cure of faintness the first thing is to secure as favorable blood supply to the head as possible, and accordingly the patient should be laid down flat on the ground with nothing under the head. It is better too to place him on the face lest the faintness bring on vomiting, and if the patient was lying on his back and unable to eject the vomited matters from his throat some of it might be drawn by the breath into the windpipe and suffocate the individual. We have known death caused in this way. Accordingly it is best to lay the patient flat on the chest, with the face turned a little to one side. Restoratives may be given, but not till the patient can fairly swallow ; before these, smelling-salts, burnt feathers, or any ordinary preparation of ammonia may be held to the nose, but too violent remedies sometimes leave bad effects, as far as the nose is concerned. We have seen it badly burned, for instance, by the irritants applied to it. If the patient can swallow, it is better to give some stimulant internally, spirit of some kind, or ammonia, especially its aromatic spirit, commonly called spirit of sal volatile. Thirty drops of this on a piece of sugar do well. A small quantity of brandy diluted with warm water may be given if this is not to be had. Spirit of chloroform also, called chloric ether, is another useful remedy. Thirty drops should be given. But it is far more important to let the patient lie quietly at absolute rest, without interference, than to bother him or her with a lot of remedies. Of course there are cases where another rule prevails, where the patient would die if something were not done, but these are not ordinary cases.

FALLING SICKNESS. (See EPILEPSY.)

FALLOPIAN TUBES are hollow canals forming appendages to the womb ; they connect the ovaries with that organ and convey the ovum from the ovaries into the uterus or womb.

FALSE JOINTS.—There are certain bones in man which after fracture rarely become whole again, and the broken pieces of which do not usually unite in the ordinary manner by the deposit around and between them of new bone. The two fragments of a transversely fractured knee-cap, and the detached process of bone at the back of the elbow called the olecranon retain a certain amount of freedom during the lifetime of the individual, and are joined by a thick and flexible structure resembling and indeed closely analogous to normal ligament or sinew. This failure of true osseous union, which is the rule in the knee-cap, the neck of the thigh-bone, the olecranon, and the back part of the heel-bone, occasionally follows the fracture of a long bone where the surgeon usually expects at the end of six weeks or two months to find a hard mass of callous or bony deposit at the seat of injury and restored continuity of the limb. In these cases the ends of the fragments remain movable, and the limb painful and useless. In most instances, after a perseverance in the treatment for another month or more, the usual and expected result takes place, and there is enduring recovery. Occasionally, however, the continuity of the bone remains broken and the fragments glide freely upon each other whenever an attempt is made to use the injured limb. In the former case surgeons say that there is delayed or retarded union, in the latter that there is non-union due to the formation of a false joint. The opposed ends of the two long fragments of broken bone are reduced in thickness, and are connected like the fragments of a broken knee-cap by strong ligamentous bands, or are inclosed in a sac or capsule of similar tissue, the inner surface of which is lined by a smooth and moist membrane resembling the synovial layer found in healthy joints. In this latter case the ends of the fragments of bone are tipped with gristle of cartilage, and glide upon each other when moved by the muscles of the injured limb. Here there is a close analogy to the conformation of a sound and normal articulation, and hence the name of false joint which has been given more especially to this condition. Non-union of broken bones except in those mentioned above is an uncommon event. The bones in which union after fracture most frequently fails or is retarded are the humerus or arm-bone, the thigh-bone, and the bones of the fore-arm. So far as it may be learned

from statistics, the occurrence of non-union does not seem to be influenced by conditions of sex or age. The principal constitutional causes are the presence of diseases such as syphilis, cancer, and scurvy, which cause poorness of blood and general debility, profuse discharges, fevers of a low type, excessive bleeding, and senility. The withholding of an habitual stimulus is often a cause of delayed union : one case has been reported in which a fractured thigh did not unite until the patient was allowed some whiskey, to which spirit he had previously been accustomed, and another in which union was retarded until the patient had returned to his former habit of smoking tobacco. Two cases were observed by Sir Benjamin Brodie in which false joint was formed in patients who had attempted, by placing themselves on a spare diet, to prevent increasing corpulence. The following are some of the local causes to which the failure or delay of union has been attributed : diminished supply of arterial blood in consequence of tight bandaging, wound and division of the nerves of the injured limb, much displacement and overlapping of the ends of the fragments, interposition between the fragments of a small piece of bone and of a piece of tendon or muscle. In the great majority of cases the condition is due to debility and premature removal of splints from the injured limb.

Treatment.—For delayed union of a fractured bone the most effectual treatment is that which consists in improving the general health of the patient by allowing him a full and nutritious diet and in keeping the limb at absolute rest and evenly and firmly compressed by splints and bandages, or by an apparatus of plaster of Paris. In obstinate cases, where some kind of false joint has been formed, a surgical operation is generally necessary, the object to be attained by which is to set up inflammation about the ends of the fragments. Inflammatory processes of bone generally result in the deposit of irregular masses of new osseous tissue. In some cases union may be b ought about by violently rubbing the ends together. In those cases that are less amenable to bloodless proceedings, the introduction of a seton, or of ivory pegs, or the simple puncture of the false joint with a long needle, are often resorted to by surgeons. Occasionally it is thought necessary to have recourse to more severe operations and to cut down upon the rest of the false joint, and to saw off the ends of the fragments, and then, after drilling the bone above and below the breach, to secure the upper to the lower fragment by means of ligatures or metallic pins. False joint sometimes, and especially in children under the age of ten years, obstinately resists every treatment, and finally necessitates amputation of the limb.

FAMINE FEVER. (See RELAPSING FEVER.)

FARCY. (See GLANDERS.)

FARRIS is a name given to a skin disease, usually occurring on the scalp, and sometimes met with in children ; it is due to a vegetable parasite called *Achoiron Schonleinii*. (See EPIPHYTA and PARASITES.)

FAT. (See ADIPOSE.)

FATTY DEGENERATION. (See DEGENERATION.)

FATTY HEART. (See DEGENERATION.)

FAUCES, the back of the mouth and the commencement of the pharynx, extending from the tonsils, and uvula to the root of the tongue and the epiglottis, and sometimes called the gorge. The fauces is often the seat of inflammation, causing sore throat.

FEBRIFUGE.—A medicine to dispel fever, such as quinine, bark, and arsenic.

FEBRILE DISORDERS are complaints in which fever or a rise in temperature forms a prominent symptom. (See FEVERS.)

FEEDING BOTTLE.—There are many kinds of bottles used for the feeding of infants. The best kind is that which, having an elastic tube connected with the inside of the bottle, causes the infant to draw up the last drop of food in the bottle without imbibing air, and thus incurring all the evils of fruitless sucking, which are great. The one point to be observed where a bottle is used by an infant is its scrupulous cleanliness. Much disorder of the bowels is caused in infants by the neglect of this ; for a very small portion of the curd of sour milk which may have been carelessly left in the bottle will taint the whole of the fresh food, and give a fit of illness to the child.

FENNEL or *Sweet Fennel*, the fruit of *Fœniculum dulce*, grows all over the United States. It belongs to a group of plants of which hemlock is the type, but it has none of the properties of that plant. Its properties rather approximate to those of anise, coriander, caraway, and dill, all of which belong to the same group. These are described as being stimulant, aromatic, and carminative. Briefly, they are given mainly for flatulence and for gripes casued by it, especially in children. Fennel water is the preparation commonly made use of, or its oil may be given on sugar like oil of anise, and in the same quantity.

FERN or **MALE FERN** (*Asplenium Filix mas*) is a remedy of very great value in the treatment of intestinal worms. The part of the plant used is what is commonly called the root, but is in reality the underground stem with portions of the stalks of the leaves and of the roots adhering. It grows in this country, and should be collected in the summer. The powder of the root may be employed, but the preparation commonly used is made by steeping the powered rhizome or stem in ether, and then allowing the ether to percolate through it. Partial evaporation of this leaves behind it a thick dark-colored liquid commonly called the oil of male fern, technically it is termed a liquid extract. The ordinary dose of this fluid is thirty drops, which should be taken in any convenient vehicle, such as an aromatic water, the first thing in the morning. Some time thereafter—about an hour or so—a small dose of castor oil should be given, just as much as will gently move the bowels, for the male fern has no effect that way. This will generally bring away the worm quite dead, or if the whole is not removed the greater portion will be, though if the head or smallest portion of the body be not removed there is a great chance of the whole body forming again. This dose may be repeated on more than one occasion if necessary. The worms against which male fern is most useful are of the tape or flat kind. It is useless for small round worms. Male fern is undoubtedly the most useful remedy we possess for the treatment of tapeworm.

FEVER is an abstract term signifying a condition in which there is increased heat of the body accompanied by a quick pulse, furred tongue, headache, and a general feeling of languor. There may be also loss of appetite, thirst, and restlessness. All these symptoms are met with in inflammatory disorders, as pneumonia, pleurisy, and peritonitis, as well as in those cases which are classified under the head of "zymotic disease." Under the different special headings will be given a detailed account of each variety of fever ; it will therefore suffice for us here to give a (1) a classification of the various kinds of fevers, (2) an account of the exciting and predisposing causes, and (3) the general treatment to be adopted.

Classification of Fevers.—For this purpose a simple arrangement may be made. Under the head of each fever will be given its history, mortality, causes, results, and degree of contagion and infection. The following are usually enumerated as fevers :

Continued fever.	Typhus fever. Typhoid, or enteric fever. Relapsing fever.
Exanthemata.	Small-pox, or Variola. Chicken-pox, or Varicella. Scarlet fever, or Scarlatina. Measles, or Morbilli. Erysipelas.

Rheumatic fever.	Febricula.
Puerperal fever.	Yellow fever.
Pyæmia.	Intermittent fever.
Diphtheria.	Remittent fever.
Cerebro-spinal fever.	Milk fever, or weed.

These fevers are spoken of as "zymotic diseases ;" the term "zymosis" means a "ferment," and in these cases the poison from without is supposed to enter the constitution and cause certain changes in the blood where it may develop indefinitely ; this can be but noticed in the case of small-pox, where a minute quantity of the poison entering the blood will multiply enormously, as seen by the numerous pustules over the surface of the body from each of which fresh quantities of poison may be taken. (See ZYMOSIS.) The seven principal zymotic diseases are : Small-pox, measles, scarlet fever, diphtheria, whooping-cough, fever (including those mentioned above as continued fevers), and diarrhœa. Cholera also comes under this division, and other diseases which are less common.

Causes.—There is hardly any subject of more importance than the proper understanding of the causes of fever , every year the returns of mortality show the enormous number of lives lost to the country by epidemic disorders which in many cases might be stamped out, or their ravages at least much lessened by simple sanitary arrangements. Yet it may be hoped that with the general diffusion of knowledge a time may come when the community will take a more active interest in the subject of public health, and that they will bear in mind the old adage that "prevention is better than cure." In treating of the causes of fevers we shall here speak of those only in which a morbid poison is assumed or shown to exist—of those in fact which may attack various individuals at the same time, and be transmitted more or less rapidly from one to another. Of such diseases some are said to be contagious, some infectious, and some partake of both qualities. When a fever is said to be contagious it is meant that another person in close contact with the one attacked is very liable to catch it ; thus measles and scarlet fever are very contagious affections, and often the disease will run through a whole family one after another. When a fever is said to be infectious it is meant that although persons in the house are liable to catch the fever, yet actual contact is not essential ; thus typhoid fever is an example ; but in truth the value of these distinctions is of very little use, as most of these disorders are contagious and infectious, and so no hard and fast line can be drawn. A more practical arrangement might be made, as follows : a person is liable to catch a fever from another person by the poison entering his stomach or lungs, or being taken up by the skin. In all cases we have to assume that particles are given off from the person attacked, and that such particles are conveyed into the system of another individual through the medium of his stomach, lungs, or skin. Now, bearing this in mind, care should be taken to prevent, as far as possible, such poisonous particles from entering. Typhoid or enteric fever is not what is commonly called contagious, yet it may infect a very large section of a community ; the way it spreads is pretty clearly made out ; the stools

of a person so affected may be thrown into a sewer or cesspool. Now supposing that the fluid portion percolates through the soil and drains into a well from which several families draw their daily supply of water, it is quite clear that in this way persons drink in the poison and that therefore it must be absorbed by the stomach or intestines ; it has been shown over and over again that cholera may be communicated in this way. But again, the stools may be thrown into an open drain, such an one, for instance, as may be sometimes seen running the whole length of a country village, and in this case when there is a dry season and the drain is not well flushed, decomposition goes on and may give rise to the disease. Nor does it seem needful in all cases that the poison should have got into a drain or sewer ; it is most probable that decomposition of organic matter itself is a sufficient cause for the disease, and thus sewer gases which leak from a worn-out pipe into a house may give origin to the attack. In these cases we breathe in the poisonous particles, and thus the virus enters by the lungs. In typhus, on the other hand, the poisonous particles are exhaled from the skin and breath of the patient, so that a person in daily contact with him is very liable to catch the fever by inhaling the virus. In yet other cases we may ourselves convey the poisonous elements in our clothes, letters, etc., to a third person, and this is often the case in scarlet fever, and also in puerperal fever. In discussing the treatment of each fever will be mentioned the various means at our disposal of preventing the spread of the disease. Causes are generally divided into predisposing and exciting causes. Under the first class may be enumerated age, sex, occupation, and country. Children are very liable to measles and scarlet fever, and hence it is not often met with in adults. Typhus affects people in middle life, and the majority are attacked after thirty years of age. Typhoid, on the contrary, is met with in the young, and the majority are attacked under thirty years of age. Relapsing fever is met with at all ages pretty equally, but it is very rarely fatal. Small-pox may attack people at any time of life if not guarded by vaccination. Sex has very little, if any, influence, although some are by their occupation more liable to catch some diseases than others. The *exciting* causes are poverty, overcrowding, destitution, bad air, bad food, and bad water. It may be stated as a rule that typhus and relapsing fevers are met with in overcrowded courts and alleys, and originate in a badly-fed family ; but when once the disease has been started in a town or village it may soon spread rapidly among the better class of inhabitants. Typhoid fever has been already mentioned as propagated by bad sewerage which taints the air, or by bad water. The other fevers are not so much affected by these conditions, but they seem to be caught from one to another, and people of all classes are liable to the disease.

Treatment.—This is a subject of the greatest importance ; in this class of diseases more than in any other the greatest benefit may be derived from careful attention to a few ordinary sanitary rules. With regard to *ventilation* a distinction must be made between this and a draught. It is often very injurious for a patient to lie exposed to a draught of cold air, although this is often done under the erroneous idea that this is the way to ventilate. It is very essential that all foul air should be removed from the apartment, but for this purpose the window may be opened three or four times a day while the patient is partially covered over, or if the weather be very warm the window may be open all day. A small fire is of great use in airing a room, as it aids the entrance of fresh air through the crevices of the door and window. The smaller the room the oftener the air requires renewal ; it is therefore wise to make use of a room as ·large as possible, and disconnected from the other rooms in the house where practicable. All useless furniture should be taken away—such as bed-curtains, ornamental hangings, carpets, etc., as these tend to retain noxious emanations, and may cause the spread of the disease afterward to other people. The bed on which the patient lies should not be placed in a direct line between the door and window, as he is exposed to a draught every time any one enters the room. When the room smells close let the window be opened a little way so as to allow the heated, stuffy air to escape. The *light* from the window is sometimes very disagreeable to a patient, as in measles, typhus, etc., hence the blind should be drawn, or the bed be so situated as to prevent any annoyance in this respect. The *temperature* of the room should be about 65° F., or even a little higher when there is any lung complication. When the air is too hot, it makes one feel oppressed and restless ; when too cold, bronchitis is apt to come on. An ordinary thermometer may be kept in the room, and it ought to stand between 65° F. and 70° F. In some cases the air may be too dry ; it is easy to rectify this by keeping a kettle full of boiling hot water on the fire ; the steam then evolved will soon moisten the air.

Cleanliness is also very needful ; for this purpose a mattress is preferable to a feather bed, as the soiled sheets can be more readily taken away, and the patient does not sink so low in the bed. All excreta should be taken away as soon as possible, and after being disinfected by pouring some carbolic acid over them they should be thrown away. The patient's body may be sponged with warm water every morning unless he is very ill, and much comfort is experienced by having the hands, feet, head and neck washed daily. In every case at the commencement of an illness a warm bath is very beneficial, as it opens the pores of the skin and may aid afterward in throwing off the poison ; among the poor, who are often begrimed with dirt, it is most necessary. Another thing of great importance is the subject of *diet*. When a person is attacked with a high fever and has a furred tongue, the appetite for solid food is gone, and therefore all food must be given in a liquid state, as it is then more easily absorbed, and just as much, nay more, nourishment may be given in twenty-four hours in the way of milk and beef-tea than if three or four hearty meals were taken. In all fevers, and until the tongue begins to clean and the temperature begins to go down, milk is the main support we ought to give to the patient. To a child two and even three pints may be given during the day and night, and to an adult three or four pints in the same time. If good fresh milk cannot be obtained, the preserved milk sold in tins is equally efficacious ; one tin will make three pints of good sweetened milk. Cold milk is often more agreeable than hot. Let the patient have something every two or three hours, and in very bad cases, where they can take very little at a time, it may be needful to give it every hour. Tea may be given, when desired, or a rice pudding with plenty of milk. It is bad to drink large quantities of cold liquids ; a little at a time should be given so as to quench the thirst. In summer time lumps of ice may be put in the milk. Beef-tea is a very useful article of food ; to obtain the most nourishment take coarse beef, cut it in small pieces, place them in cold water in a jar, and let them simmer in the oven for a few hours ; then, when cold, remove the fat, and warm up half a pint for the patient to drink. It is a mistake to suppose that because beef-tea is solid when cold therefore it is very strong and nutritious ; the solidification depends on the amount of gelatine present, and this has very slight nutrient properties. When the appetite returns and a desire for solid food is felt, much care should be taken for the first few days ; a small piece of boiled mutton may be given with a little bread or dry toast ; vegetables are not good to take ; jelly, blanc-mange, light puddings made of tapioca, arrowroot, or rice may be given, and an egg for breakfast or tea. As a rule, for mild cases of fever no stimulants at all need be given until the stage of convalescence, when two or three glasses of sherry may be given daily. In some cases, where the disease assumes a very grave form, brandy must be given freely, but no rules can be laid down in individual cases, and such treatment must be left to the medical man. Stimulants should be given with much caution in the early stages ; they frequently tend to congest the stomach and make the patient restless and oppressed, and prevent him absorbing the nutrient food which is so essential for his well-being. If attention be properly paid to all these points much good will arise, and any one who aspires to be a good nurse will make such her careful study ; and let it always be borne in mind that in fever cases good nursing will do more than anything else to expedite recovery. *Quiet* is always advisable ; it is of no use for persons to be admitted into the sick-room, however anxious they may be ; when a patient is delirious the hum of conversation is very annoying, and in most cases they feel quite unable and unwilling to talk much ; a quiet nurse is very valuable. When food is given only a little should be brought at a time, and that in a tempting form ; a large basin of beef-tea or bread and milk is often refused than a small quantity would be taken. The patient's strength should be saved as far as possible by avoidance of exertion or excitement, at least in the severe cases ; they should not be allowed to sit up, and when the sheets, etc., have to be removed the patient should be shifted from one side of the bed to the other without being taken out of the bed. Charcoal may be placed in a pan under the bed, as it has the power of absorbing noxious gases, or saucers containing solutions of carbolic acid may be placed about the room.

FIBRIN is a constituent of healthy blood ; when blood is drawn from the body it separates into a clot and an opalescent straw-colored fluid ; the clot is mainly formed of fibrin, holding in its meshes the red blood-corpuscles, which give it the dark color. It is composed chemically of carbon, hydrogen, oxygen, and nitrogen. It is sometimes found during life in the veins or arteries of the body, and may then give rise to clots in the vessels, forming *emboli* or *thrombi*. Later researches seem to show that fibrin is composed of two bodies, called respectively *fibrinogen* and *fibrino-plastic globulin*, or simply *globulin* ; coagulation of the blood will not occur unless both these bodies are present. (See BLOOD.)

FIBROUS DEGENERATION. (See DEGENERATION.)

FIGS are the ripe fruit, or rather inflorescences of the fig-tree. In medicine they are not much employed. They are supposed to have a slight laxative effect, and are contained in the confection of senna. Split open and heated they have been used from time immemorial as a poultice to boils and such like sores. Now they are rarely used for such purposes.

FILTERS are used for purifying water for drinking purposes. The water of our rivers and ponds contains a varying amount of inorganic and organic matter, and since the presence of the latter is often attended with most injurious effects, it is necessary that careful measures should be taken to rid

the water of these impurities. Rain water and melted snow are very pure, and may be collected in clean vessels and drank with impunity. Spring water often contains lime salts which render it "hard," and it will not then form a lather with soap, and is therefore not useful for domestic purposes ; such water may, however, be taken without injurious effects, although, if the lime salts are very abundant, a form of goitre or large neck may come on. But the case is different when the surface drainage manages to find its way into a well, for then organic matters enter and may give rise to typhoid fever and other diseases. The water supply from wells is by no means adequate to the quantity required daily in large towns, and therefore it is usual to collect water from an adjacent river into large reservoirs, and then pass it through filters for the sake of purification. Various kinds have been proposed. On a large scale water may be received into large reservoirs, and then the heavy substances will settle, while the rest may pass through gravel or sand. Pure animal charcoal is now considered the best filtering material. It should be deprived of all the lime salts by washing it in hydrochloric acid. The particles of the charcoal should be well pressed together, and the water must not pass through too quickly. This substance removes in a great measure dissolved organic and mineral matter, as well as the suspended particles ; it also takes away the color, so as even to make a muddy water quite clear and bright. It is said that the power which charcoal has of removing organic matter is lost after a time, but this power is restored by washing the filter with a little potassium permanganate. Charcoal also appears to exert an oxidizing change on organic matter, and converts it partially into nitrites. Vegetable and peat charcoal are both inferior to animal charcoal. Magnetic carbide of iron, manganic oxide, silica and charcoal may all be used with much benefit. The filters used at the present time for domestic purposes are generally made of animal charcoal or magnetic carbide of iron pressed into blocks, and these can be relied on. As there is a limit to all purifying power, the action of all filters is only temporary. The various substances which have been removed, accumulate and block up the filter, and therefore the filters should be taken to pieces and cleaned every two or three months, or a dilute solution of potassium permanganate may be passed through to get rid of the organic matter, and then a little weak hydrochloric acid to remove the lime salts, then pass through two or three gallons of distilled water, and the filter will be quite fit for use again. When a new charcoal filter is used, the water which first passes through should be rejected, as the substance of which the filter is composed gives off some substances to the water ; a preliminary washing is therefore needed. A pocket filter is useful for soldiers, or for those who are travelling in a country where fresh pure water is not easily obtained. These are usually of the siphon kind, and made of hollow blocks of charcoal, the water filters from without into the central cavity, into which a tube opens and the purified water may be drawn off. In hot countries it is often very rash for people to drink from the stream ; a common plan for purifying the water is to have two barrels of different sizes, one within the other ; the outer one is pierced with holes at the bottom, and the inner one at the top ; the space between is filled with charcoal and sand, through which the water percolates into the inner cask. Water should be boiled first, and then allowed to cool, or tea may be made and the cold tea saved for the next day ; this is a very refreshing drink. When the water supply is very short, advantage should be taken of every rainfall ; salt or brackish water should be distilled, as is done at Iquique and other tropical places on the west coast of South America. In a running stream, men and cattle should be watered at different places, the former above the latter ; all washing should be done lower down the stream, and the excreta must not be allowed to contaminate the drinking water. By adopting ordinary precautions, and making use of charcoal filters, much disease may be prevented ; such filters should be used on board ships, unless there is an efficient apparatus for distilling the water.

FINGERS. — *Supernumerary Fingers.* — This congenital malformation which is called polydactylism is often hereditary and in most instances affects, to an equal degree, both hands. The most common form consists in a small though well-formed and unmistakable digit springing from the root of the fourth or little finger at the inner side of the hand, and attached by a fold of skin. An additional digit is sometimes met with springing from the outer side of the thumb. The presence of a supernumerary finger furnished with a long metacarpal bone is a very rare event. Two cases are on record in which a double hand existed. These irregular fingers are most frequently useless deformities, and cannot be moved at the will of the individual, though sometimes they are furnished with a distinct tendon, and tolerable use may be made of them. The adjacent normal digit is almost always reduced in size. As supernumerary fingers form unpleasant objects, and are generally in the way when the child grows up and begins to use the hands, the surgeon usually advises their removal during the period of infancy, and while the abnormal growths are still small.

Absence or *Defective Development* of one or more fingers is sometimes met with, though much less frequently than the previous condition. It will be obvious that in cases of this

kind nothing in the way of treatment can be done.

Webbed Fingers. — In this condition the fingers, instead of being free and isolated, are bound together. This is a congenital deformity which usually affects both hands symmetrically, and in the same patient the toes almost always are similarly affected. This deformity is due to persistence of a fœtal condition of the digits which is present at the second month of intra-uterine life. A condition somewhat similar is sometimes observed as a result of severe burns or scalds of the hands, but in cases of this kind the affected fingers are surrounded by scar-tissue and not by true and sound skin. In congenital webbing of the fingers the extent of the deformity varies greatly in different cases. The fingers may be bound together along their whole length, or only as far as the first or second joint. The union between the fingers may be very close, or they may be with intervals occupied by true webs, which cause the deformed hand to resemble the extremity of an aquatic animal. The operative treatment of this deformity though not dangerous is attended with much difficulty. Simple division of the web is quite useless, as the fingers always grow together again. This result may be obviated by passing some foreign body into the cleft while the wounds are closing by granulation, or by passing a piece of metal or an elastic band through a small perforation made at the upper and narrowest part of the web, and then after three or four weeks, when the margins of this orifice have cicatrized, along the whole length of the web.

Congenital Contraction. — This deformity is rarely implicates more than one finger. It is due to deficiency of integument, in consequence of which the finger is bent forward toward the palm. When an attempt is made to straighten the finger, a tight ridge of skin starts up along its concavity. It is important to distinguish this congenital deformity from the following similar condition which is produced in adults in consequence of contraction of the fibrous structure in the palm of the hand. The congenital affection may generally be treated successfully by prolonged and continuous extension of the abnormally placed digit.

Acquired Contraction of the Fingers is often met with in adult males as a consequence of increasing rigidity and shrinking of the palmar fascia and other fibrous tissues that intervene between the muscles of the palm and the integument. These changes in the tissues of the palm are supposed to be due either to violent blows or to long-continued pressure ; and this view is supported by the fact that the affection in question is often observed in navigators, gardeners, carpenters, and those whose employment necessitates much compression of the palm. The finger most frequently contracted is the ring finger, and next the little finger. When the ring finger is affected the middle and little fingers become implicated sooner or later, though not to the same extent. The thumb and index finger are never affected. The contraction comes on gradually, the patient at first experiencing some stiffness in the knuckle-joint. The finger is then turned forward and cannot be raised, and after a time the adjoining fingers also become stiff and bent. When the affection is well marked the finger is bent forward at the knuckle-joint, while at the second and third joints it is free and movable. The flexure of the distorted finger is occupied by a prominent and curved fold of skin under which can be felt a tense hard band of the contracted fibrous tissue. This band is tense and rigid, and prevents the finger from being straightened. In a case observed by Dupuytren a weight of 150 lbs. did not bring the bent finger into a straight line with its metacarpal bone. The patient is unable to grasp large bodies, and experiences much pain when he attempts to move the fingers. This deformity cannot be relieved save by a surgical operation. This consists in dividing the skin longitudinally along the summit of the fold in front of the bent finger, in dissecting this back, and then in cutting across the exposed fibrous cord.

Broken Fingers. (See FRACTURES.)

Tumors.—The bones of the fingers are the most frequent seat of cartilaginous tumors. These form hard rounded growths which commence in childhood or youth, and continue to grow slowly for twenty-five or thirty years. Several of these tumors are often collected on one hand, and thus cause much deformity and render the extremity quite useless. For descriptions of other *Affections* and of *Injuries* of the *Fingers*, the reader is referred to articles on NAILS, WHITLOW, FRACTURES, and DISLOCATIONS.

FISH, AS FOOD.—The class of fishes yields a larger number of species used as food by man than either birds or quadrupeds. In some countries the only animal food known is fish. The flesh of fish contains less nitrogenous matter than that of birds and mammals. It usually contains less oil or fat, and a larger quantity of mineral matters. Fish is not so digestible as butchers' meat, and therefore not so nutritious. The muscles of fish contain in larger quantities than other animals a principle called creatin, which also exists in human muscles, and which seems to act favorably on the human system. It is undoubtedly a valuable as well as an agreeable article of diet, and should, where possible, be introduced into all dietaries. Fish, when unfit for human food, acts as a valuable manure. The skins of eels, soles, etc., may be converted into gelatine. Many are used for obtaining the oil which they possess, and this is especially

the case with the menhaden and the cod. The livers of the latter yield the well-known cod-liver oil.

FISH, POISONOUS.—That a certain number of fishes are poisonous is not to be denied, but the exact causes of their giving rise to symptoms of poisoning are by no means clear. Various tropical fishes are poisonous at all times, but in this country shell-fish, as they are called, which at one time are undoubtedly wholesome, at others have given rise to poisonous symptoms. Chief among these are mussels, and it has been assumed by some that those which have given rise to the poisonous symptoms have been fed on or otherwise absorbed poisonous materials, as from the copper on ships' bottoms, but of this there is no proof. The symptoms produced are the same in almost all cases ; they are the symptoms of slight irritant poisoning. Sickness and vomiting, sometimes with purging and marks of prostration, are those commonly observed. These it is not desirable to interfere with, but rather to promote until all the irritant matters have been expelled. After this a slight cordial or carminative draught, with a little iced water, a little brandy and soda water, or such like, to compose the stomach, when the patient will probably go to sleep and awake well. Luke-warm water or mustard and water should be given to aid the vomiting. Occasionally rashes on the skin are observed, but these require no treatment.

FISSURE OF THE ANUS is generally a long and shallow ulcer situated either within or on the verge of the anus. In most cases it is met with on the posterior wall of the anus in the median line of the body, occasionally on one side, very rarely in front. In about nine tenths of the cases it is seated at the posterior verge of the anus. It occurs much more frequently in women than in men. The symptoms caused by this affection are considerable irritation about the anus, and severe scalding pain with a throbbing sensation which comes on immediately after an evacuation, and gradually increases in intensity, and continues for some two or three hours. The suffering in most cases is very acute, and cases have been known in which patients had nearly starved themselves to render the occasions for emptying the rectum less frequent. After the pain has lasted for about three hours it gradually subsides, and the patient remains quite free from pain until the next evacuation. The general health at last becomes affected by the repeated attacks of pain, and the patient becomes weak, indolent, and sallow. There are often dull, heavy pains in the loins and groin. On local examination a small red fissure will generally be found just on the verge of the anus ; this extends upward along the mucous membrane of the rectum and measures from the sixth of an inch to half an inch in length. The ulceration but seldom implicates the whole thickness of the mucous membrane. The edges of the ulcer are generally smooth and level, but in advanced and very bad cases they are thickened and elevated. The outer extremity of the fissure is often covered by a small lump of thickened skin, or by an external pile. The causes of anal fissure are constipation and direct irritation of the mucous membrane of the return by purulent discharges, or by the contact of foreign bodies. It has been stated that the unskilful administration of injections, especially with pointed rough pipes, is often the cause. It remains doubtful whether the violent spasms of the sphincter or constricting muscle of the anus which gives rise to the intense suffering after evacuation of the bowels, be the cause or the effect of the ulcer.

Treatment.—The palliative treatment of this affection consists in the administration of sedative injections, and rest in the recumbent position. The patient should take a small dose of castor oil every morning, and pass the evacuations over hot water. The application of solid nitrate of silver (lunar caustic) or sulphate of copper will often produce a permanent cure. In most cases, however, the surgeon finds it necessary to advise an operation in which the base of the ulcer with more or less of the muscle below is divided by the knife.

FISTULA is a narrow channel or tube leading to a cavity containing matter or dead bone, and lined with a membrane which secretes a puriform fluid. The fundamental cause of fistula is abscess, and the reason of the unhealed tract is an unhealed abscess, where proper outlets to the discharge have not been made, or where some "foreign body" intervenes, such as a piece of dead bone. They usually exist in connection with the rectum, urethra, salivary glands, and bladder. With regard to treatment, all sources of irritation must be removed. If matter forms, it must be let out by what is termed a "counter opening," and the "fistulous tract" stimulated to healthy action by some such an injection as a strong solution of nitrate of silver, or nitric acid. The operation of slitting up fistulæ must be left to the physician.

FITS.—A person may have a fit from several causes ; thus, he may fall down and struggle violently, as in an epileptic attack (see EPILEPSY) ; or he may fall down in a state of intoxication (see DRUNKENNESS) ; or he may become insensible and have an apoplectic stroke (see APOPLEXY). An hysterical woman may have a fit and be for a time unconscious (see HYSTERIA) ; or a fit may occur in the course of Bright's disease, or in some disorders of childhood. A "fit" is a term popularly applied to any condition in which a person suddenly falls down insensible, and has convulsions or not ; but as so many diseases are associated with the so-called "fits,"

the reader must refer to the above-named articles, and also to COMA, CONVULSIONS, and SYNCOPE.

FLANKS, a term corresponding to the lumbar regions, on either side of the abdomen. (See ABDOMEN.)

FLANNEL, though not strictly speaking a remedy, is one of the most valuable means for preserving health we possess. Its great virtue consists in that it prevents the body from being too rapidly cooled after being greatly heated. Flannel is a non conductor of heat, even when saturated with perspiration, and so prevents the heat of the body from being wasted in evaporating the fluid after the body ceases to generate excessive heat. For similar reasons, an individual clothed in flannel will be able to resist exposure to cold and wet better than one clothed in a better conducting material. Such exposure, as we well know, is a frequent cause of rheumatism, and one attack of rheumatism almost invariably predisposes the individual to a second attack. It is, therefore, a good rule for individuals who have once suffered from this malady, never to go about without flannels in future. This is true, even in summer, for it is better to risk a little extra perspiration than to run the risk of a sudden and excessive cooling, which would be apt to bring back the disorder with all its attendant evils and dangers. Those, too, in whom the chest is weak should invariably wear flannels, *especially* when they go abroad in winter.

FLAT-FOOT is said to exist when a person treads on the inner margin of the foot, the toes are turned out and the arch of the foot destroyed, and its cause is a general want of tone in the fibrous structures. In a slight degree it is common in young children, particularly females, in the upper classes, and can be greatly remedied by reducing their standing and walking, the avoidance of fast walking in the company of adults, tonics, attention to digestion, embrocations, and manipulation of the feet, so that inversion and contraction may be prevented. Laced boots, or boots with stiff leather sides, a cork, india-rubber or felt pad under the inner margin of the foot, greatly assists in preserving the arch. It will be noticed that opera and ballet dancers are frequently the subjects of this deformity, owing to the practice, easily taught, of standing upon the toes, so that the calcaneo-cuboid ligaments, or the plantar fascia, naturally designed for the preservation of the arch, become preternaturally stretched and give way. Relief is more easily attained in the child than in the adult.

FLATULENCE, or the undue collection of gas or air in the stomach or bowels, is an exceedingly troublesome symptom. Its accumulation may be brought about in various ways. It may be swallowed, it may be formed from the food, or it may be apparently se-

jreted from the wall of the stomach and bowels. It is a common and exceedingly unpleasant symptom of indigestion (see INDIGESTION), sometimes very hard to get rid of. It is also a very troublesome symptom in other diseases affecting the bowels and abdominal cavity. In a very great number of instances flatulence is due to improper food, or the abuse of certain articles of food, especially tea. Certain substances have a common reputation of being windy, and doubtless in certain instances, not without reason, though in others the accusation is altogether unfounded. The symptoms produced by flatulence are often exceedingly unpleasant. There may be a feeling of faintness, of giddiness, or of choking, accompanied by most troublesome belching. The gases thus expelled are most frequently tasteless and odorless, and if so, are most probably due either to swallowing of air, or to the formation of such simple gases as carbonic acid, or carburetted hydrogen, at the expense of the food. Such forms of flatulence, i.e., flatulence accompanied by tasteless belching, are best treated by dieting, mainly solid food with stale bread, a little dry sherry or weak brandy and water, but no vegetables, tea, beer, or pastry. Flatulence may often be the only symptom of such dyspepsia, and it is often capable of relief by a slight stimulant, as aromatic spirits of ammonia, but spirituous liquors should be avoided. A little acid, or alkali with a bitter, is often of very great service, and nux vomica is an exceedingly valuable remedy in such cases. Occasionally the patient is the subject of horribly nauseous flatulences. He belches up gas of the most horrid odors, disagreeable to himself and every one round him. These gases indicate putrefactive changes in the food, and commonly occur in individuals who have some obstruction preventing the passage of food from the stomach, especially if the obstruction be cancerous in its nature. In cases where there is such obstruction the stomach sometimes expands to an enormous size, and vomiting after food is not unfrequent. In these vomited matters are minute organisms called *Sarcinæ*, and these are supposed to have much to do with the development of the gas, just as the yeast fungus has in the formation of alcohol from starch and sugar with the evolution of carbonic acid.

Treatment.—In all such cases the use of antiseptic remedies to prevent the putrefaction of the food is indicated, and they almost invariably do good. The two most important forms of antiseptic remedies are carbolic acid and sulphurous acid. Carbolic acid may be given in the dose of one or two drops in a wine-glass of water, half an hour after food. Its taste is somewhat disagreeable, but it is exceedingly efficacious. Sulphurous acid may be given in the same way, thirty drops of the

diluted acid in a wine glass of water, or it may be given as sulphite or bisulphate of soda. To the former of these most people will give the preference, as its taste is that of a pure acid ; the taste of the other is far more bitter. Flatulent accumulation in the intestines may be due to any of the foregoing causes, but especially to putrefaction of the food, and apparently in certain cases to secretion of gases from the vessels in the walls of the gut. In children the other variety is not uncommon, especially if they have been allowed to suck empty bottles or breasts, their thumbs, or the like, and they are fruitful sources of gripes. (See COLIC.) Flatus in the intestines often gives rise to very great pain, and the patient urgently demands relief. This can only be obtained by dispersing the wind, as it is called, which is not always an easy task. In most cases it is vain to expect to do so upward, and so we must try to do so downward. Perhaps the best remedy for this purpose, if it can be borne, is turpentine. It tends, however, to upset the stomach, and so it is better given as an injection. If given by the mouth, about a drachm should be given for a dose, if as an injection half an ounce or so, beaten up with an egg in a pint of hot water. At the same time, turpentine applied in the form of stupe to the abdomen is likely to give great relief to the suffering.

FLOATING TUMORS is a term applied to the singular, hard and very movable lump which is sometimes observed in the abdomen, generally on the right side. In most instances the patients are women. No pain is complained of, but only an uncomfortable sensation due to the movements of the lump. Sometimes there is obstinate indigestion. On examination of the floating tumor it will be found to be smooth, very firm, and generally of the size and shape of a healthy human kidney. It is very loosely attached, and can be moved over a considerable extent, both between the ribs and the haunch bone, and from side to side. The nature of this tumor has never yet been clearly made out. From its shape and consistence.it has been supposed to be a kidney, which in consequence of its loose attachment to the spine by fat and membrane, and of elongation of its vessels, has become freely movable. It is probable that in some cases it is an ovarian cyst. It is held by some authorities that the so-called floating tumor is nothing more than a hardened mass of fæces, or some firm body in the intestinal canal, or, in some rare instances, a large concretion lying loose in the abdominal cavity.

FLOODING. (See LABOR and ERGOT.)

FLOWERS OF SULPHUR is a well-known form of that substance obtained by heating the crude substance, converting it into vapor, and afterward condensing the vapor in a cool chamber. This resembles a very fine powder, but minutely examined it is crystalline. (See SULPHUR.)

FLUCTUATION implies the wave-like movement imparted to the hand when there is any accumulation of fluid in a part ; it is often very marked in cases of ascites, and when an abscess is forming.

FLUIDS, ATOMIZED, though recently introduced into practice, have already become one of the standard and most frequently employed means of treating certain diseases at our disposal. They are commonly employed in the treatment of diseases of the nose, mouth, throat, larynx, and windpipe. Occasionally, too, for those of the lungs. They are of course all the more valuable, the more difficult the spot is to reach in any other fashion, and there is hardly any other way in which we can reach the larynx and windpipe. The principle on which the fluids are atomized, as it is called, is tolerably familiar to all, in the shape of a toy for dispersing perfume in a room ; one end of a glass tube is introduced into a bottle containing perfume, the other end being drawn to a very fine point. Another and similar tube is arranged and fastened at right angles to the former, so that its fine point terminates close to and just above the level of the fine point of the tube ending in the perfume. If now, one blows through the tube at right angles to the bottle, the force with which the air is driven from the fine point across the fine extremity of the other, creates a partial vacuum, in which it draws the fluid to the top. Thence it is dispersed in spray in the line of the current of air driven from the mouth. Of course this plan of driving by the mouth would be objectionable in practice, and so two kinds of apparatus have been invented : one, in which steam is driven through a narrow orifice instead of air, the fluid being drawn up from the bottle containing it as before. Another is employed where the air is driven by means of a hand ball made of india-rubber. This, which is attached to the transverse tube by an elastic tube of convenient length, has at either extremity a valve which prevents the air from flowing except in one direction. When compressed by the hand, the air is driven through the tube and atomizes the fluid ; when the hand ceases to contract, the elasticity of the ball causes it to expand again, drawing in the air by its open extremity. The fluids best adapted for use in this way are nitrate of silver in good strong solution, from 3 to 5 or more grains to the ounce of distilled water. This must be kept in a stock bottle, covered over with paper, and only a little poured into the bottle for use as required, otherwise it speedily decomposes and becomes useless. Sulphurous acid of pharmacopœial strength is a most valuable remedy, administered in this way. Tannic acid, 20 grains to the ounce, is

also very valuable. Lime-water of pharmacopœial strength or saturated liquor, or liquor ferri perchloridi of pharmacopœial strength are excellent in their several ways. The diseases best treated by means of the spray producer are, 1st, those of the cavity of the nose ; frequently, for instance, after scarlet fever, there remains a tendency to the formation of purulent matter in the upper part of the nasal cavity. This may go on to destruction of the bones of the nose, and the matter discharged has got a terribly fœtid odor. Smell is often completely and irretrievably lost if the process goes on too long unarrested. For this and all similar disorders of the nose, a good strong spray is the best remedy ; it softens and breaks down the hardened masses, which form troublesome crusts, and after a time brings them away. This done, a bare surface is exposed to the spray, and healing follows. The best fluids for this condition are sulphurous acid and nitrate of silver. If the mouth can be opened widely, as is often better to make use of stronger appliances than the spray, but where the mouth cannot be opened, and its cavity is diseased, it is invaluable. The diseases it is mainly used in are tonsillitis and diphtheria, and syphilitic affections of the throat beyond the fauces. For these sulphurous acid or nitrate of silver is best, but lime-water tends to soften the patches of false membrane in diphtheria. For regions beyond these, there is hardly any means of treatment equal to the spray. Inflammation of the larynx, whether of a common kind or due to tubercular or syphilitic states of the constitution, can hardly be treated in any other way. It is true that the skilled physician, with the help of the laryngoscope can reach them, but the spray ball anybody can use. When there is ulceration, chloride of silver solution is perhaps the best remedy ; if only inflamed, sulphurous acid may be tried ; if œdematous, perchloride of iron will do good. The general symptom, hoarseness and loss of voice, may often be relieved by using hot vapor spray (see VAPORS). In certain diseases of the lungs the same apparatus may be employed, with this additional precaution, that the patient must be made to inhale or inspire deeply, that is, take a long breath at the moment the spray is playing into the throat. But as this would cause violent coughing or spasm of the entrance to the windpipe, were the spray very irritating, only mild applications must be used. The diseases most likely to benefit in this way are croup, as it affects the windpipe, bronchitis, and phthisis, if the lungs are affected. The preparations thus made use of must be carefully adapted for each case. Plain hot water will, however, very rarely do harm.

FLUX implies a flow of fluid ; thus, when the stools are very liquid, in some cases of diarrhœa, the patient is said to have a watery flux ; or a bloody flux when blood flows from any cavity of the body.

FŒTICIDE signifies killing the fœtus while yet in the mother's womb.

FŒTUS is the name given to the child when in the womb.

FOMENTATIONS are warm lotions applied to diseased parts, by means of flannels. The pieces of flannel or blanket should first be cut to the required size, and then soaked thoroughly in water just hot enough to be grateful to the patient ; repeating the process as often as may be required.

FOOD is the term applied to all those materials consumed by man, and which are employed by the body to build up its fabric during growth, and renew the tissues which are lost during the performance of the functions of life. What fuel is to a fire, food is to the body. As fire transforms the fuel into other compounds which it throws off, so the body transforms food into other substances during its vital activity. In the same manner as a fire diminishes as the fuel is diminished, and goes out without fresh fuel, so the human body wastes with insufficient food, and dies from its absence. If a man is weighed without taking food, it will be found that he gradually weighs lighter. The muscles which enable him to sit or stand, the heart which circulates the blood, the brain that enables him to think and become conscious, are all consuming the materials which he has taken as food. Just in proportion to the work a man has to do is the amount of food he consumes. The laborer who digs all day on a railway consumes more food than the tailor or shoemaker. The housemaid who is on foot all day eats more food than her mistress, who sits in the drawing-room all day. Not only do those who work hardest require most food, but just in proportion to the extent of the appetite and the vigor of the digestion will be the ability to perform hard work. As it is with muscle-work, so it is with brain-work. The hard student consumes more food than the idle man who lounges about all day without troubling himself to think. Food is, in fact, taken in order that we may live, and all the great results of our lives depend upon the conversion of the materials taken as food into the forces which make our muscles to contract, and our brains to conceive, think, and will. The ultimate elements of the food we take are precisely similar to the ultimate elements of the human body. (See COMPOSITION OF THE BODY.) The principal elements which enter into the composition of the human body are carbon, hydrogen, nitrogen, and oxygen. These elements are sometimes called organic elements, because they enter into the composition of all the growing tissues of the animal body. It is principally through the chemical relations of these elements that we find the functions of the body carried on. A human body weighing 154 lbs. is found to contain

	lbs.	oz.
Oxygen	111	0
Hydrogen	14	0
Carbon	21	0
Nitrogen	3	10

The other elements which enter into the composition of the body are called inorganic elements. They consist of phosphorus, sulphur, chlorine, fluorine, calcium, sodium, iron, potassium, magnesium, silicon. The compounds containing these elements weigh about 5 lbs. 10 oz. Although in such small quantities, they are none the less important. They are wasted during the processes of life, and unless supplied, disease is produced. For example, one of these compounds is phosphate of lime, and it is well known that unless food containing phosphate of lime is supplied, rickets and softening of the bones, which contain phosphate of lime, comes on. Many of these compounds contain the organic elements ; thus, in the ashes of a human body weighing 154 lbs. there is found 7 lbs. 9 oz., of ashes, which contain mineral compounds. These compounds consist of phosphates, sulphates, carbonates, chlorides and fluorides of lime, potash, soda, magnesia, and iron.

The compounds of the body in which the organic elements exist, and their weight in a body weighing 154 lbs., are as follows :

	lbs.	oz.
Water, containing oxygen and hydrogen	111	0
Gelatine, containing the four elements	16	0
Albumen " " "	4	0
Fibrine " " "	4	4
Fat, containing carbon and hydrogen	12	0

Water is found everywhere in the body ; by its agency all other substances are taken up into the system. All food must contain water, and it is only by being dissolved in the water that the other substances can be used as food. The quantities of water found in 100 lbs. of different kinds of solid food are as follows :

VEGETABLE FOOD.

	lbs.
Potatoes	75
Carrots	80
Parsnips	79
Mangel Wurzel	85
Cabbage	92
Flour	14
Barley meal	14
Oatmeal	13
Indian meal	14
Rye	13
Peas	14
Rice	13
Beans	14
Bread	44
Cocoa	5
Lentils	14
Buckwheat	14

ANIMAL FOOD.

	lbs.
Milk	86
Bacon	30
Veal	62
Beef	50
Lamb	50
Mutton	44

Thus, whether we take water or not in addition to our solid food, we find that large quantities exist there. This will explain why it is some articles of food require more water when eaten than others. Thus it will be seen that while potatoes contain 75 per cent of water, rice and oatmeal contain but 13 per cent. Among animal foods, while milk contains 86 per cent of water, bacon contains but 30 per cent. In the purchase of food, the above facts should be kept in mind. Thus, as far as solid food is concerned, a pound of rice at fivepence would be cheaper than a pint of milk at a penny.

Fat is a very important constituent of the body; it is found diffused around all the tissues. It is the fat that gives roundness and plumpness to the body. When it is deficient persons are said to be "thin" or "lean." When people weigh above their weight (see WEIGHT and HEIGHT), it usually arises from a deposit of fat. In wasting diseases, as consumption and scrofula, the wasting arises from the loss of the normal fat of the body. In such cases it is usual to recommend a fatty diet, and cod-liver or other animal oils (see COD-LIVER OIL) are given.

Albumen and *Fibrine* are two constituents of the body which contain the four organic elements. Albumen differs from fibrine chemically but very slightly. It is, however, soluble in water, and easily separable from it by heat, alcohol, nitric and other mineral acids. It is found dissolved in the blood, where it exists in the proportion of about four per cent. It constitutes the chief compound of nerve-matter, out of which the nerves are formed. It enters into the composition of the eggs of all animals. Its property of coagulating when boiled, forming the "white" of the egg, is well known. Fibrine is found in small quantities in the blood, but is principally distributed over the body, of the muscular tissues of which it constitutes a large proportion.

When fibrine and albumen have performed their duty, and are about to leave the body, they appear to be converted into gelatine. This substance is much more conspicuous in the human body than either albumen or fibrine. It constitutes the cement of the bones, and is the substance out of which the cell walls of all the tissues of the body are formed. It is the waste of these substances that renders food necessary. They do not, however, waste with equal rapidity. Water passes away most rapidly. This is necessarily the case. A man sometimes drinks in ten days as much water as there is contained in his whole body. The water passes away by the lungs, the skin, the kidneys, and the bowels. The fibrine and albumen pass away less quickly than water. Then come gelatine and fat. Last, the mineral matters which are employed in constructing the tissues of the body are removed. Calculating the quantity of material removed daily, it would appear that a period of forty days would suffice for removing the whole of the used material of a human body. Consequently, a man should eat and drink a quantity of food equal to the weight of his own body in forty days. It is food that must supply this perpetual waste of the tissues of the body, and the class of foods which supply the waste of the fibrinous, albuminous, and gelatinous tissues are called "flesh-giving." They all contain also the element nitrogen or azote, hence they are called "nitrogenous or azotized" foods. They do not, however, pass away from the body in the form in which they go in. They are thrown off the body in the form of a substance which is known by the name of urea. This compound appears to be formed in the blood, and is drawn out of it by the kidneys and then passed to the bladder, dissolved in the urine. Urea is composed of the four organic elements, and undoubtedly represents the form in which the nitrogen is got rid of. But there is another function of the body, for the due performance of which food is necessary. The functions of life are attended with the development of heat and the production of force. Although all food which contains carbon is capable of aiding this function, yet there is no doubt that some foods are better adapted for sustaining this function than others. Heat is generated in an animal body by the union of the carbon of the blood with the oxygen of the air. The oxygen is introduced into the blood by the agency of the function of respiration, which consists in the taking into the lungs oxygen gas, and the returning to the air carbonic acid gas. The quantity of carbonic acid thrown out is precisely the measure of the quantity of carbon consumed in the food and the oxygen taken from the air. This is the way in which the carbon taken in with the food is principally got rid of. While the oxygen is uniting with the carbon, an increase of temperature takes place, and the heat of the animal body is thus maintained at a given temperature. This temperature is different in different animals, but in man it is 98° by Fahrenheit's thermometer. This heat is quite independent of external temperature, and whether a man is exposed to the heat of the equator or the cold of the poles his temperature is the same. The great agent by which this is effected is the skin. The skin is copiously supplied with blood-vessels which are distributed over its surface and are influenced by the external temperature, so that when the temperature of the air is great, the water in the blood is converted into vapor, and so delicate is the operation of this structure that the temperature is always kept at the same point, whether the atmospheric heat is great or small. The food possessing this power of maintaining animal heat and force is sometimes called "heat and force-forming." It embraces certain substances not existing in the animal body, known by the names of starch and sugar. It is quite impossible to estimate accurately the quantity of food that any individual ought to take. All the food that is taken is not always digested and made into blood. Many of the materials of food are not converted into chyle (see DIGESTION), and pass off by the bowels. The following table is an attempt to estimate the quantity of food daily taken into the stomach and changed during twenty-four hours :

FOOD ACCOUNT.

TAKEN IN.

GASES.

	oz.
Oxygen	24

LIQUIDS.

Water—	oz.	gr.	
In beverage	68	0	
In food	25	0	
			93

SOLIDS.

Flesh-forming—			
Fibrine	3	0	
Albumen	0	300	
Caseine in cheese	0	137	
			4
Heat-giving—			
Starch	12	0	
Fat and butter	5	0	
Sugar	2	0	
			19
Mineral matters			1
			141

GIVEN OUT.

GASES.

Carbonic acid—	oz.	gr.	oz.	gr.
Carbon	11	0		
Oxygen	24	0		
			35	0

LIQUIDS.

Water—				
By Kidneys	51	0		
Lungs	31	0		
Skin	16	0		
Bowels	5	237		
			103	237

SOLIDS.

Urea	1	200		
Mineral matter	1	0		
			2	200
			141	0

This table must only be regarded as an estimate. One set of food-substances may be substituted for another, and the quantity of food taken may not be digested or taken into the system. The table is drawn up on the supposition that all the food taken in passes into the blood, and is disposed of as indicated by the substances thrown out. The probability is that a large quantity of the matters taken in pass through the bowels without being changed. Persons take food very differently, according to age, height, occupation, climate, and season. Children and young persons take more in proportion to their size than adults, as their food supplies the material of growth as well as waste. The height of individuals determines their weight, and according to the quantity of flesh and the waste is the necessity of supply. Persons employed in sedentary and indolent occupations do not require so much food as those who are more actively employed. Climate also makes a great difference in the demand for heat-giving food. Those who live in cold climates consume more heat-giving food than those who live in warm and tropical climates. The same circumstances also influence the form in which the elements of food are given out from the body. The excretion of carbonic acid is greatly increased in cold weather. Water is also very variously got rid of by the skin, the kidneys, and the lungs, in proportion as the body is exposed to external heat or cold. It is not all food that is taken into the stomach that is digestible. Mixed with the nitrogenous and carbonaceous constituents are other substances. Thus, the cell-walls of all plants are composed of cellulose. This substance, though apparently digested by many of the lower animals, is not digested by man. It therefore passes through the bowels unchanged. Another substance, called gum, is not absorbed in the stomach or bowels, and therefore cannot be regarded as nutritious. Some doubt has also been thrown on the question as to whether the gelatine so largely entering into animal food is taken into the blood and rendered a nutrient agent. At any rate there is one fact known, and that is that gelatine alone is incapable of supporting the life of an animal. Perhaps for the present it is better to regard it as a valuable accessory to our food than as one of the assimilable and necessary articles of food. Besides substances necessary or accessory there are a number of things taken as food which are not necessary, or mixed naturally with nutritious food. These substances are mostly added by choice, or voluntarily sought by man either to gratify his palate or to act upon his nervous system. These substances are called "medicinal" or "auxiliary" foods. Medicinal, because they act like medicines on the system ; auxiliary, because they stimulate the powers of the stomach and aid in the digestion of the food. This class comprises such substances as alcohol, volatile oils, tea, coffee, and tobacco.

In order to get an idea of the various kinds of food and the purposes they supply in the system, some kind of classification must be pursued. The following table is supplied in order to give a general view of foods and their principal action.

CLASSIFICATION OF FOOD.

CLASS I.—ALIMENTARY OR NECESSARY FOOD.

Group 1. Mineral.
 Examples. Water, salt, saline constituents of plants and animals.
Group 2. Carbonaceous, respiratory, heat and force-giving.
 Examples. Starches, sugars, fats, acids.
Group 3. Nitrogenous, nutritious, or flesh-forming; proteoids.
 Examples. Albuminous compounds, fibrine of meat, caseine of milk.

CLASS II.—ACCESSORY FOOD.

Examples. Cellulose, gum, gelatine.

CLASS III.—MEDICINAL OR AUXILIARY FOOD.

Group 1. Stimulants.
 Examples. Alcohol, volatile oils.
Group 2. Neurotics.
 Examples. Tea, coffee, tobacco, opium.

One of the best types of animal food is milk. It is supplied by the mothers of all animals belonging to the group of mammals, and is capable of furnishing all the materials of their growth till they are several months or years old. It must therefore contain all substances necessary for the growth of the body and the maintenance of its various functions. The following is the analysis of cow's milk :

Water	86.0	Mineral group.
Mineral matter	1.0	
Butter	3.5	Heat and force-giving group.
Sugar	4.5	
Caseine	5.0	Flesh-forming group.
	100.0	

From this table it will be seen that with the exception of the accessory group, every class of food is represented in cow's milk (see MILK). The principal difference between the diet of adults and that supplied by nature for the young consists in the fact that the diet of the adult contains less water, and is seasoned with more or less of the group of accessory foods.

We shall now speak of the various groups of foods, as given in the above classification :

1. *Mineral Foods.*—The importance of water in this group is at once evident, and although so large quantities are found in all our solid food, it is necessary to add more for the purpose of dissolving all those constituents which are necessary to the functions of life. Although the group of heat-giving and flesh-forming foods are many of them insoluble in water, they are rendered so during the process of digestion. (See DIGESTION.) Starch is rendered soluble by the action of the saliva of the mouth, by which it is converted into sugar. The proteoids are acted on by the gastric juice, and are thus rendered soluble in water. The fats taken as food are decomposed by the bile and pancreatic juice, and converted into soluble soaps, which are readily dissolved by water and taken into the blood.

Water is taken either cold or hot. It is made into soups, tea, coffee, and chocolate, by the infusion and boiling in it of various substances. Although water is taken cold alone, or with alcohol, as wines and beer, there is a general tendency to take this liquor hot. The savage likes his solid and liquid food hot. It was death to the Roman slave to bring up the water cool or tepid to the table. This craving for hot food and beverages seems to rest on the fact that when cold substances are introduced into the stomach they lower its natural heat, and render it less capable of digesting the food. At the same time the use of cold is refreshing, and in hot, and even cold countries, water is taken in the form of ice. This practice needs, however, to be cautiously indulged in, especially in cases of debility from disease, and in all dietaries of persons exhausted and debilitated from attacks of fever. (See DIET.) In all cases where water is taken pure, the greatest precaution should be taken to render it free from impurities which can generate disease. There is no doubt that diarrhœa, cholera, and typhoid fever spread by the agency of impure water or water contaminated with the poisons that generate these diseases. The water-supply of a house should be well looked to, and when any suspicion exists the water should be boiled and filtered before it is drunk. (See WATER and FILTERS.)

The other substances besides water belonging to the mineral group are common salt, and salts. Common salt is chloride of sodium, and exists in abundance in sea-water. It has the power of preserving vegetable and animal substances from decomposition, and is found in certain quantities in the bodies of all animals. The human body contains about three ounces, which is principally found in the blood. Unless certain quantities are taken daily, diseases characterized by debility are likely to occur. It may be taken in large doses from day to day, and no harm occurs, as that which is not necessary for the use of thh body is got rid of.

The most saline matters found in the human body, and which are excreted by the urine and bowels, are obtained from all forms of food. Animal and vegetable food lose some of these saline matters by cooking; hence the importance of taking uncooked food of some kind or another every day. This should be effected by fruit, or vegetables in the form of salads. (See SALADS.) An instance of the value of fresh vegetables as an article of diet is seen in the treatment of sea scurvy. This disease is brought on by the absence of fruit, vegetables, or fresh meat on board ships. It is prevented by the supply of lemon or lime juice, and vegetables cooked and preserved in tins.

2. *Heat and Force-giving Foods.*—These consist principally of starch, sugar, and fat.

At the same time they may be divided into two groups. The starch and sugar have the following composition :

Carbon.................. 12 parts.
Hydrogen 9............ } Water, 18 parts.
Oxygen 9.............. }

In fact, they contain oxygen and hydrogen in the proportion in which those elements form water, and when taken the carbon is alone oxidized, and forms the heat-giving element. It is different with fat and oleaginous foods. Their composition is as follows : Carbon, 11 parts ; hydrogen, 10 parts ; oxygen, 1 part. Not only the carbon, but a large part of the hydrogen is thus left free to be oxidized by the oxygen taken in during respiration. It is in this way that fats, butter, and oils are much better supporters of combustion in the body than starches or sugars. One pound of fat or oil is equal in combustible agency to two pounds and a half of starch, and two pounds and a quarter of sugar.

Starch is found in nearly all our articles of vegetable food. It exists in very varying quantities, from 80 per cent in rice to 3 or 4 per cent in parsnips and carrots. It is almost pure in arrowroot, sago, and tapioca. It is also contained in a peculiar form in seaweeds and Iceland moss.

Sugar is found in both plants and animals. It is taken as food in the form of cane or crystallized sugar, and grape sugar or glucose. The latter is found in all fruits. (See SUGAR.) It is the only form which undergoes *fermentation*, and is the basis of all fluids containing alcohol. Sugar is found also in animals. The sugar of milk, although differing but little from grape sugar, has, nevertheless, a distinct composition, and, of course, is constantly taken where milk is used as an article of diet. (See MILK.) Sugar is also found in the liver and the blood, and its increase in the system constitutes a disease called diabetes. (See DIABETES.)

Oleaginous foods are those which consist principally of oils, butter, fats, or lards. These foods are not generally eaten alone, but are added to starchy diets. The action of oils on the system is principally to maintain animal heat and force, and, practically, they are most largely eaten by those who do the largest amount of work. They not only act in this way, but they also assist in the digestion and assimilation of other foods. It is on this account that cod-liver oil, pancreatic-emulsion, and butter, cream, and fat have been recommended as articles of diet in cases of consumption, scrofula, and other wasting diseases of the body.

3. *Flesh-forming Foods.*—The substances which lie at the foundation of this group of foods are albumen, fibrine, and caseine. These compounds are found nowhere pure, but are taken in various forms of vegetable and animal food. The most common form in which

the flesh-formers are taken is bread. (See BREAD.) Bread contains fibrine. The flesh of animals, birds, and fishes also contains fibrine. Albumen is found in the white of eggs, and also in the blood of animals. Caseine is found in milk. It is separated with the butter in cheese. (See CHEESE.)

In addition to the flesh-forming principles and fat, animal food contains various other chemical compounds, which are the result of the life of the animal, and act in a beneficent manner on the system. If, for instance, we take the flesh of an animal and squeeze it, we get out a juice called the "juice of meat," and, when evaporated, it is called the "extract of flesh." This compound contains little or no albumen, no fibrine or fat, but it consists of salts and organic substances, resembling in their composition quinine. They are called by such names as *creatine* and *sarcosine*. When this substance is taken with water, in the form of tea, it increases the appetite, and renders digestible the food that is taken with it. It should be remembered that this substance is not nutritious, but that it stimulates the digestive function, and thus renders it possible for food to be digested where none could be before it was taken. (See MEAT, EXTRACT OF.)

4. *Medicinal or Auxiliary Foods.*—These constitute a very large group of substances, which are used for the sake of flavor, and their action on the nervous system and circulation. Their various sources, qualities, and uses will be found under the head of their various names. (See ALCOHOL, BEER, COCOA, COFFEE, CONDIMENTS, SPICES, TEA, TOBACCO, WINE.)

FORCEPS.—Scarcely any instrument has so many different forms or uses as the forceps, suited as they are to almost every operation in surgery. The most common and perhaps most useful, are the ordinary, simple, bowed, or dissecting forceps, which can be most conveniently applied for the removal of foreign bodies, such as thorns, splinters, etc. They should not be too strong in the spring, and should have broad and deeply serrated blades, best with a groove in the centre. Then there are forceps named specially after the operations for which they are used, such as bullet forceps, lithotomy or stone forceps, tooth forceps, polypus forceps, urethral forceps, vulsellum forceps, artery forceps, etc.

FOREIGN BODIES are substances which have been introduced into some structure or cavity in the body, foreign to it in composition. In some cases nature will expel such bodies ; in any attempt to do so, assistance is in almost all cases needed.

In the Windpipe.—Morsels of food more frequently than other substances get into the larynx or trachea, the accident happening when a person is engaged in laughing or talking when the mouth is full of food, the symptoms being sudden spasmodic cough, protrusion of the eyes from the sockets, blood or froth issuing from the mouth and nose ; the patient gasps for breath, turns black in the face, and perhaps falls down insensible. If the morsel of food be light and of small size, it is sometimes expelled during a fit of coughing. Many bodies may find their way into the larynx and trachea, coins, cherry-stones, beans, or in fact anything which may happen to be in the mouth, and their presence sets up precisely similar symptoms. It must be remembered, that although occasionally these substances have been expelled naturally, or with the assistance of an emetic, this remedy should not be recommended, as it has the effect of impelling the substance against the larynx, from the spasmodic action of the stomach and œsophagus, thus endangering the patient's life. Inversion of the body, combined with a shaking or jogging motion, will sometimes cause the foreign body to fall through the larynx, as in the celebrated case of Mr. Brunel. The operation of laryngotomy and tracheotomy are generally needed.

In the Nose.—These are often introduced by children ; such substances as peas, beads, pieces of pencil, etc. They are generally removed readily enough by a small polypus forceps or a scoop, remembering that the direction of the floor and roof of the nasal cavity is nearly parallel with the ground, so that *digging* attempts upward toward the head, are attended with considerable danger. If the body cannot be extracted through the mouth, it should be pushed backward into the pharynx, taking care that it does not pass into the larynx. Very frequently they work out if left to themselves.

In the Ear.—The substance introduced frequently becomes covered and escapes without surgical interference, but in cases where instruments must be used for the purpose, it must be remembered that the ear passage is widest at the external orifice, then narrows somewhat, and again widens toward the membrana tympani. In introducing a small scoop or forceps they should be passed along the *upper* wall, so as to avoid the membrane. The passage should be syringed with warm water, a proceeding which alone frequently removes a foreign substance. A piece of wire bent into a loop and insinuated around the substance is a method sometimes attended with success. Insects and larvæ sometimes lodge in the ear, causing severe inflammation and local suffering, with great constitutional disturbance ; warm oil dropped into the passage until it is filled, or white precipitate, suspended in milk, and injected, will be found sufficient to kill the animals. Oil or water poured into the ear, with the head on one side, will generally bring them to the surface of the fluid, and allow them to be removed.

In the Eye.—The cornea (see CORNEA) should be first examined by everting the lid and telling the sufferer to look up, or down, so that both the upper and lower surface of this part of the globe can be seen. A substance, such as a piece of cinder, or a piece of metal or wood from a turning-lathe, etc., sticking in the cornea, can generally be removed by a silver toothpick or fine forceps, or still better, an eye " spud" or scoop. If the substance be lime or mortar the lids should be everted and the eye well syringed with weak vinegar and water, or oil, or water only. A drop of castor oil or of pure glycerine is a most soothing application in painful cases where the conjunctiva (see CONJUNCTIVA) has been scratched or stripped off the cornea. Very often a piece of dust may be removed by blowing the nose smartly.

In the Rectum.—These consist of the following, viz., those composed of materials which have first passed along the upper part of the alimentary canal, and those introduced into the anus. In the first class we find bones, apple or pear cores, fruit stones, scybala, substances taken as medicines, coins, etc. The presence of these bodies is indicated by pain in passing motions, accompanied by blood, with frequent desire to evacuate, but assuming a something sticking there which will not pass out. These bodies must be removed with care, the bowel must be well lubricated with oil, and a warm water injection used. If this will not dislodge the mass, this must be seized with forceps or broken up with some instrument, and removed piecemeal. A full-sized speculum should be first introduced, so that the bowel be not lacerated or hurt by these attempts at removal.

In the Vagina.—Substances are frequently introduced into the passage, and one of the most common is a pessary ; the strings break and the instrument remains in, setting up a most offensive discharge. Glass bottles are occasionally made use of, and set up ulceration, establishing an unnatural opening into the bladder (vesico-vaginal fistula). These bodies require great care in their removal, which should be attempted with strong forceps, taking care that the margins or extremities of the impacted substance do not injure the mucous membrane. The speculum is frequently required.

In the Throat.—Substances retained in the œsophagus are usually held at the commencement opposite the cricoid cartilage (see LARYNX), or at its lower extremity, just above the diaphragm, as the tube is the narrowest at these points. Various substances have been cited as having been retained in the œsophagus ; thus crust, imperfectly chewed meat, bones, coins, stones, pins, needles, buttons, knives, forks, scissors, spoons, keys, chestnuts, a small apple, fish-hooks, artificial teeth, the handle of a punchbowl, a pencil-case, etc. The symptoms produced by the presence of such a body vary of course with its size ; if it be small it produces considerable irritation, with difficulty in swallowing. It sets up in time inflammation, followed by ulceration of the œsophageal coats, causing most serious consequences and oftentimes terminating fatally. Small, sharp-pointed substances, such as pins, needles, fish-bones, etc., may perforate the walls of the tube and remain firmly impacted, or may perforate the aorta or trachea, and there are instances where such bodies have been swallowed, appearing at the surface of the body after the lapse of years. When a foreign body is impacted in the œsophagus it should be removed as quickly as possible ; in the instance of a small piece of bone it may be propelled onward to the stomach, by making the patient swallow a good mouthful of bread. If the substance is one which on its arrival in the stomach could be easily digested, it may be pushed gently down the canal with an instrument termed a probang. If high up, it may be reached with the finger or long forceps ; gentle pressure with the finger on the side of the neck, opposite to the spot where it is felt, will sometimes dislodge it. Hard, angular masses, such as glass, stones, etc., require removal with forceps. If no other means are at hand the induction of vomiting is occasionally of use. A dilute solution of a mineral acid has been known to be of use in the case of the impaction of a piece of bone in the fauces or œsophagus. It should be taken through a tube to prevent its action on the teeth, and its strength modified to the degree of sensibility of the parts over which it has to pass. In cases where all these means fail the operation of œsophagotomy must be performed.

FORMICATION is the peculiar feeling, like the creeping of ants, which is felt in the onset of some forms of paralysis ; also around the arms, when the patient is suffering from worms, and in the limb when a nerve has been pressed upon and is cramped, as in hanging the arm over the back of a chair.

FOUSEL OIL or *Potation Spirit*, also known as Amylic Alcohol, is contained in greater or less quantity in all forms of crude spirit, from which it requires to be carefully separated by redistillation. Being much less volatile than ordinary alcohol, it comes over last, or may be allowed to accumulate in the last portions of spirit whence all the good spirit has been distilled. It is the substance to which bad spirit mainly owes its noxious qualities. By oxidation it forms valerianic acid, and it is for this purpose only that it is used in medicine.

FOXGLOVE. (See DIGITALIS.)

FRACTURE is a term applied in surgery to the breaking of a bone, which, after incised wound, is probably the most frequent serious accident in civil life. When one or more bones have been broken, while the skin and subjacent soft parts are not torn or wounded, the injury is called a *simple* fracture. When, in addition to the breaking of the bone, there is a large wound through the skin and muscles leading down to the seat of injury, and exposing the fragments, the fracture is called *compound*. When a portion of bone is broken into several small fragments, it is said to be *comminuted*. Fractures, both simple and compound, may be *complicated*—by dislocation at a neighboring joint, by wound or division of a large artery or nerve, or by stripping away from the surface of the fractured bone of a large extent of the periosteum or external membrane. During infancy and childhood fracture is met with in the same proportion in males and females. From the age of ten to that of fifty years, the number of males with fractures greatly exceeds the number of females. From the age of fifty to that of sixty years, the numbers are again about equal, and, finally, above the age of sixty more women than men are the subjects of fracture. The bones most frequently broken are : the clavicle or collar-bone in children ; the bones of the leg and forearm, and the thigh-bone in middle-aged persons ; and the neck of the thigh-bone and the lower extremity of the radius or spoke-bone near the wrist in persons beyond the age of sixty years. The immediate causes of fracture are two : external force applied either directly to the bone at the seat of breakage, or at some more or less remote part, and sudden and powerful contraction of muscles. The latter is a comparatively rare cause ; the bones most frequently broken in this way are the knee-pan, the prominent extremity of the cubit-bone at the back of the elbow, and the heel-bone. The bones of weak and sickly persons can be more readily broken than the bones of those in robust health. In the subjects of rickets, and of a peculiar disease of the bone called mollities ossium, and also in those who have been confined to bed for a long time, fracture may be produced through slight violence. Sometimes infants are born with one or more bones fractured, the injury having been caused by strong contractions of the womb during labor, or by a blow or kick on the abdomen of the pregnant mother.

The *symptoms* of fracture are not very difficult to make out ; a stout muscular man walking along the street makes a false step or slips, and then falls heavily to the ground, with the right leg twisted and bent under him ; in his fall he hears a *sharp crack*, and, on attempting to move, finds that there is great *pain* and loss of *power* in the right limb at a short distance above the ankle. He feels also a peculiar grating sensation—the so-called *crepitus*—at the seat of pain, and finds, on looking at the injured limb, that at this part there is *swelling, distortion*, and *unnatural mobility*. The most decisive symptoms of fracture are mobility of the bone at the injured part, and the peculiar grating noise produced by rubbing together the ends of the fragments; but these may often be absent. In fractures of the ribs, the haunch-bones, the skull, all small and short bones, and the extremities of long bones, it is generally difficult to make out the nature of the injury without submitting the patient to a close and prolonged examination. In fractures of the shafts of long bones, as the thigh-bone, the arm-bone, and the bones of the leg and forearm, all or most of the symptoms can be readily recognized, and the result of the accident learned without delay. In most cases of fracture of the bones of the limbs, beyond the period of infancy, the patient has generally from the first moment of the injury a firm and correct conviction as to its nature. A simple fracture involving the middle portion of one or two long bones of a limb sets speedily, and with good results. A similar injury near the joint end of a bone is not so promising, as the neighboring joint is usually more or less affected, and remains stiff and swollen for a long time after the setting of the fracture. Fractures of bones which are surrounded by thick and powerful muscles are more difficult to cure, without shortening of the limb and deformity, than bones which can be felt close under the skin. When a bone is broken in several places or into several small fragments, the case is much more serious than that of a simple transverse fracture. The greater the obliquity of the fracture the more unfavorable is the case. In debilitated or diseased subjects the setting of a simple fracture takes an unusually long time. Compound fractures are always very serious accidents. When the wound in the skin is large, and the muscles, blood-vessels, and nerves are much lacerated, amputation of the limb will be necessitated. In less severe cases the patient is still liable to the dangers of inflammation, erysipelas, tetanus, and pyæmia. The most severe fractures are those of the skull, chest, and pelvis, on account of the important viscera contained within these cavities, and which may be primarily or consecutively involved. Fractures of the upper extremity are less serious than similar injuries in the lower extremities. The former do not prevent locomotion, or compel the patient to keep to the house until the broken bones are firmly united. Fracture of one or more of the long bones in the arm or leg frequently results after union in puffiness of the skin, wasting of the muscles, and weakness of the whole limb. Stiffness of the joints immediately above and below the broken bone is also a frequent after affection. Those affections retard the convalescence, and sometimes last for several months, but are usually much relieved by

stimulating liniments, shampooing, and the cold water douche.

In the *treatment* of fracture the surgeon has two objects to fulfil; in the first place he removes any displacement that may exist, and returns the fragments to their proper position, both in relation to each other and the parts around. This having been done, he then applies splints or some retentive apparatus to keep these fragments in place, and the whole limb below the fracture in a correct anatomical position until the injured bone is thoroughly set. The process of recovery consists in the effusion between and around the ends of the fragments of a plastic material, which sets as it were into tough gristly tissue, and is finally converted into a mass of true bone, which is called the *callus*. In the long bones of the extremities the fracture is not firmly united until the end of six weeks or two months; in fractures of the ribs, collar-bone, and lower jaw the process of recovery occupies a shorter period. If union be allowed to take place without previous removal of displacement at the seat of fracture, the limb becomes permanently shortened, and perhaps distorted. The removal of the displacement, reduction, or reposition, as it is called by surgeons, is effected simply by keeping the upper fragment fixed, and pulling downward the lower fragment until both are in the same line. Reduction is sometimes unnecessary, and occasionally has to be deferred in consequence of much bruising and inflammation, and also of muscular spasm. In some instances the fragments are so interwedged that the displacement cannot be removed. There are several methods of keeping the fractured bone in position : most surgeons in this country use splints, which are flat slightly hollowed pieces of wood or iron, well padded with tow, cotton-wool, or some other soft material. These are applied in varying number to the surfaces of the fractured limb, and are retained by means of bandages. They have a double object : to keep the broken ends of bone in their proper position, and to restrict the movements more or less of the limb, and to keep the whole of the broken bone together with the joints immediately above and below, at perfect rest. In fracture of the thigh-bone a long splint is usually carried from the arm-pit to the foot along the outer side of the injured limb for the object of preventing shortening.

The management of the patient immediately after the accident, and the precautions to be taken in his removal to home or hospital, are matters of great importance. Much harm is often done by moving the patient without taking any preliminary means to protect the injured limb, especially in fracture of the lower extremities, and the fragments of bone are moved and perhaps forced into the surrounding muscular tissue, causing consider-able pain and subsequent inflammation. In a case of injury to the arm or leg, the first thing to do before allowing the patient to move is to expose the seat of injury. If the pain be great, the clothes covering the limb should be *cut*, and not pulled off as in the usual way, and the sides of the boot divided with a sharp knife. The situation of the fracture will then be indicated by deformity, swelling, and local tenderness. If one or both bones of the forearm have been broken, the limb should be placed in a sling made of a hand-kerchief or neck-wrapper, the ends of which are tied lightly at the back of the neck. In cases of fracture of the arm-bone at some point between the elbow and shoulder, some thick pad—a small pillow or thin cushion will answer the purpose very well—should be placed between the arm and the side of the chest, and the injured limb then fixed to the body by some extempore bandage, the elbow and forearm being supported in a sling. After the patient has been helped into bed, this apparatus should be removed, and the arm laid out on a pillow at an acute angle to the side of the body with the forearm bent. In a case of fracture of the leg or thigh it should be a rule never to transport the pa-tient in a carriage, cab, or any kind of vehicle which will not permit of his lying at full length. If the distance be not very great, the best means of transport will be an ordinary stretcher, or, in the absence of this, a shutter or door, or an extempore stretcher formed by laying boards on two stout poles. A very convenient supporting apparatus for the whole body may be made by fastening with cords the sides of a long piece of canvas or sacking to two long poles. The best temporary ar-rangement of a fractured thigh or leg is to place the limb, when half-bent at the hip and bone, on its outer surface. In this position the muscles are relaxed, and the whole length of the limb is supported. Extempore splints may be made of thin pieces of wood. These should be covered on one surface by thick pads, made of linen, folded into several layers, or of single layers of linen inclosing tow, cotton-wool, feathers, bran, or, if noth-ing else is at hand, and the accident has oc-curred in the country, dry grass. Useful temporary splints may be made of bark, leather, pasteboard, or of wheat-straw or reeds tied tightly into compact bundles. Splints should be fixed over the seat of frac-ture by two or more handkerchiefs, or by a bandage formed by tearing a sheet or table-cloth, great care being taken not to constrict the seat of injury so as to give pain, and to obstruct the upward flow of blood through the veins of the limb. If the skin has been wounded, the blood should be gently wiped away, and a piece of linen dipped in cold water placed over the raw surface. When the

patient has been placed in bed, and the frac-ture again exposed by removing the tempo-rary splints and bandages, cold should be ap-plied to the injured part either through rags dipped in cold water, or through ice placed in a sponge-bag or sheep's bladder. If the injury be seated in the lower extremity, it will be necessary, in order to avoid the pain and inconvenience attending an early shift-ing of the patient from one bed to another, to see that he is placed on a couple of firm horse-hair mattresses, under which should be laid some boards extending transversely from one side of the bedstead to the other. The surface on which the patient has to lie during the treatment should be firm and level, and therefore no feather-bed should be allow-ed. The head-pillow should be removed and replaced by a bolster.

Nasal Bones.—Generally caused by a blow of the fist. As this injury is frequently as-sociated with much bruising and swelling of the soft parts, it is frequently overlooked. The lower fragments are usually displaced backward, and if not returned to their proper position and kept there by plugs of lint or cotton-wool introduced into the nostrils, give rise to great subsequent deformity. A fractured nasal bone unites in seven or eight days, more rapidly perhaps than any other bone in the body. This accident is frequently complicated by a wound in the skin and by bleeding from the nose. During and after treatment the patient may be troubled with ulceration and a discharge of ill-smelling pus, death of bone, lachrymal fistula, impeded res-piration, and impairment of the sense of smell.

Lower Jaw.—Generally caused by a direct blow. The bone is generally broken at some point between the insertion of the middle in-cisor and that of the first bicuspid teeth, the fracture extending through the whole width and thickness of the jaw. Sometimes the jaw is broken on each side of the middle line, so that the piece carrying the incisors, or the incisors and canines, is loose and detached from the rest of the bone, and displaced downward and backward. Sometimes, though not so frequently, the jaw is broken through at its ramus or ascending portion, or at the neck or part immediately below the head, which is forced into the socket in front of the external ear. The symptoms of this fracture are generally well marked. Crepitus can be distinctly felt on moving the fragments on each other ; there is free mobility, and also some distortion ; the pain over the seat of in-jury is in nearly all cases unusually severe ; the gums are frequently wounded, and one or more teeth loosened and perhaps entirely detached. Fracture of the lower jaw gen-erally unites speedily and firmly, although with some distortion along the chin, and ir-regularity of the lower row of teeth. The treatment of this injury is attended with great difficulties, as it is often impossible to keep the bone at rest for any length of time. The patient ought to make up his mind to speak as little as possible, and to live for two or three weeks on fluid nutriment alone.

Collar-Bone.—This, with the exception of the outer bone of the forearm, is more frequently broken than any other bone of the body. An oblique fracture at the junction of the outer and middle thirds of the collar-bone is a very frequent injury in children, and is caused generally by a fall upon the hand when the arm is stretched out. It is sometimes produced in adults by a blow upon the front of the shoulder, as in the recoil of an overloaded gun. In this fracture the shoulder falls downward, forward, and inward. The patient leans toward the injured side, and with the opposite hand sustains the elbow corresponding to the broken collarbone, so as to prevent its dragging downward. The inner fragment of the broken bone is very prominent, and externally to this there is a depression caused by the downward sinking of the outer fragment. The patient feels great pain when he attempts to raise the arm from the body, or to carry the forearm across the front of the chest. He is unable to raise the hand to his head, or to move it forward or backward, without suffering. On drawing back the shoulders so as to bring the fragments of collar-bone into contact, distinct crepitus may usually be felt. A broken collar-bone unites speedily and strongly, but always with some amount of shortening and deformity. It is a very difficult matter to keep this bone at perfect rest, and to restore the outer fragment to its proper position. The most certain method of treatment to insure union without deformity, is for the patient to remain in bed until the fracture has been set, the head being kept as much as possible in one position, and the arms confined to the side of the body. This position, however, to most patients is intolerable. In infants the fracture generally unites well without any treatment, and the deformity and swelling, which at first are very considerable, gradually subside. With children and adults, the usual apparatus consists in a figure-of-8 bandage carried from one shoulder to the other across the back of the chest, a stout wedge-shaped pad in the arm-pit on the injured side, a broad bandage to confine the arm to the side of the chest, and, finally, a sling to support the elbow.

Thigh.—The most frequent seats of fracture in this bone are the upper extremity and neck, the middle of the shaft, and a part about four inches above its lower extremity. The rounded head of the thigh-bone is invested by a loose membranous bag, which is inserted below into the base of a jutting support, which is called the neck of the femur or thigh-bone. Fracture of this neck may occur either within or without this bag or capsule. In the former case it is called intra-capsular fracture, and in the latter extra-capsular fracture. The intra-capsular fracture occurs in old people; is produced by very slight violence—the most frequent cause being a slip on the curb-stone; is attended with very little bruising and not very severe pain; and hardly ever unites by bone, the two fragments being joined together by ligamentous tissue. Extra-capsular fracture on the other hand, is generally the result of great direct violence, as in a heavy fall upon the outer part of the hip, is followed by much bruising and intense pain, and almost always ends in firm bony union. The common symptoms of these two kinds of fracture are, loss of power in the limb, shortening, eversion of the foot. Crepitus is usually absent or very indistinct. The subjects of intra-capsular fracture are in most cases over sixty years of age; extra-capsular fracture may occur at any age beyond thirty. There are several difficulties attending the treatment of fracture within the capsule. The patient is generally old and infirm, and may sink rapidly in consequence of confinement to bed. When this is the case it would be well to allow the patient to get up and make as much use as possible of crutches. When the patient seems to be strong and hearty he should be kept in bed for five or six weeks with the whole limb stretched down between two large sand-bags, or with the hip and knee bent and the thigh and leg supported by a well-cushioned double inclined plane, made of pieces of wood which can be lowered or elevated at will. For extra-capsular fracture the usual treatment is a long splint of wood about four inches in width, extending from the arm-pit to beyond the sole of the foot. The foot having been secured to the lower end of this by means of a bandage, a band, the central part of which is composed of tow or cotton-wool covered by wash-leather, is carried round the inner surface of the thigh at its upper part, and its two ends are then fastened to the upper end of the long splint which touches the arm-splint, the foot of the injured limb having previously been dragged down to the level of the foot on the opposite thigh. The leg and thigh are then bandaged to the splint, and a broad band or sheet is carried round the chest and the upper part of the splint. By pulling at the ends of the band which passes under the upper part of the thigh the whole limb can be extended. When the two limbs are of equal length the injured one is kept stretched by tying together the ends of this band after they have been passed through two holes bored through the long splint near its upper extremity. The shaft of the thigh-bone is most frequently fractured at its middle third. This injury may be caused by the passage of a heavy body across the thigh, by the fall upon the limb of some heavy mass, or by the patient falling from a height. The line of fracture is usually oblique, and there is shortening to the extent of one or one and a half inch. There is much displacement of the bones, and consequent deformity, and crepitus can be distinctly felt. This fracture and also that near the lower end of the bone are usually treated by the long outside splint, applied as in last mentioned fracture.

Knee-Cap.—Of this injury there are two varieties. In one the bone is broken into several fragments, in the other there is a simple transverse line of fracture extending from one lateral edge of the bone to the other. The cause of the first, the *stellate* fracture as it is called, is direct violence, as a blow or fall. The second or *transverse* fracture is usually produced in the following manner : the patient in walking or in going down stairs makes a false step and then attempts by a sudden effort to recover himself ; the knee being half bent and the knee-cap resting by only a part of its posterior surface on the lower end of the thigh-bone, the whole contractile force of the powerful extending muscles in front of the thigh is, in consequence of this effort, brought to bear upon the bone which is thus poised. Transverse fracture then takes place with an audible snap, and the patient falls to the ground. In this injury there is a wide separation of the two fragments, forming a distinct gap, and a depression in front of the joint. The limb cannot be straightened by the patient, and there is generally much pain and swelling of the knee. This fracture, like that of the neck of the femur within the capsule, unites by ligament instead of by true bone. The injured limb always remains weaker than the other, and patients who have been laid up with fracture of one knee-cap often, sooner or later after their recovery, break the bone on the opposite side. This fracture may be treated by keeping the limb stretched on a mattress between two large and firm sand-bags, reaching from the upper parts of the thigh to the sole of the foot. The limb may be kept steady by fastening a long handkerchief around both sand-bags in the middle of the leg. The inflammation and swelling of the knee may be best treated by the local application of ice or of linen rags frequently dipped in cold water or weak lead-lotion. The patient should be kept in bed for at least six weeks.

Leg.—The following are the fractures most frequently met with in this region ; that of both bones at the middle or lower thirds ; that of the shin-bone alone at its upper third ; that of the splint-bone alone at a point about two and a half inches above its lower extremity. In the first mentioned fracture there is generally much displacement and free mobility of the fragments, and crepitus can be easily felt ; it is frequently associated with

much bruising, and large blebs, containing a thin and dark red fluid, are formed on the surface of the skin. In fracture through the upper part of the shin-bone the nature of the injury is not so evident ; there is very little if any displacement, and crepitus is usually very indistinct. Fracture of the shin-bone at its lower third is in most instances marked by a peculiar distortion and outward displacement of the foot ; the lower end of the shin-bone projects very much at the inner surface of the ankle, and the outer edge of the foot is drawn upward and outward, and corresponding to the seat of fracture in the splint-bone there is a well marked superficial depression. When both bones are fractured with much displacement, the limb is generally placed upon an iron splint, and compressed laterally with two well-padded splints of wood which extend from above the knee to the foot. In cases of fracture of one bone only, and when the fragments are not displaced, the best treatment seems to be the application of the starched or plaster of Paris bandage, as the patient may then be allowed to get up and move about on crutches. In fracture near the lower end of the shin-bone, associated with dislocation of the foot, the lower limb should be well flexed both at the hip and knee, and then be placed on its outer surface, either on a bent wooden splint or between sand-bags on a hard mattress. This last injury is generally followed by some deformity and much stiffness in the ankle-joint.

Arm-bone.—This bone may be broken at any point between the head and the lower expanded extremity, but most frequently about its middle. Of fracture near the head there are several varieties, and much surgical skill and experience is required to distinguish these from each other, and to decide whether the case be really one of fracture or of simple dislocation at the shoulder. When there is much bruising and severe pain, and crepitus can be felt distinctly on grasping the upper end of the arm-bone and moving the elbow, the case is one of fracture. The most simple treatment of a fracture in this region is to place between the injured arm and the side of the chest a small pillow or a cushion, arranged so as to form a pyramid, the apex of which is to be applied to the arm-pit ; the elbow and arm are then to be fixed by means of a bandage carried round the chest. When there is much displacement, it will be better to apply a bent leather splint, one limb of which is to be fixed to the side of the chest, and the other to the inner surface of the injured arm, so that the angle occupies the arm-pit. Fracture of the shaft of the bone is generally caused by direct blows. The line of fracture occurs more frequently below than above the middle of the bone ; it is generally very oblique, so that there is much displacement and distortion, rendering the diagnosis of the injury easy.

In this fracture there is generally consider-able bruising, and also some swelling of the whole limb. It almost always results in shortening and a certain amount of deform-ity. Union sometimes fails, and a false joint or ununited fracture is formed. The treat-ment of a broken arm consists in the applica-tion of a long external splint extending from the tip of the shoulder to the elbow, and of two or three smaller splints to the other sur-faces of the arm ; all be'ng well padded and retained in place by bandages. The forearm should then be supported by a sling carried under the wrist and *not under the elbow.* Fracture at the lower and expanded extremity of the arm-bone is a common injury in chil-dren, in consequence of blows or falls on the back of the elbow. Here, as at the upper ex-tremity, there are several varieties of fracture. Sometimes the line of fracture extends into the elbow-joint, and causes much swelling and subsequent stiffness, and impairment of the articular movements. In cases where there is preternatural mobility above the joint, great pain and swelling, and distinct crepitus, a large pad should be placed in the bend of the elbow, and the forearm bent over this, and retained in the same position by means of a bandage. It should always be remembered that fractures in the region of the elbow are very difficult to treat, and often result in permanent deformity and stiffness of the joint, notwithstanding the most skilful surgical treatment.

Forearm.—The prominent upper extrem-ity of the internal or cubit bone of the fore-arm is sometimes broken in adults by a fall on the back of the elbow, the detached frag-ment varying in extent in different cases from a mere shell to the whole of the pro-cess. It is generally widely separated from the rest of the bone, in consequence of being pulled upward by the strong extending mus-cle which runs along the back of the arm. In this injury the movements of the arm are much impaired. In treating it the arm should be kept straight on a padded splint of wood or stout gutta-percha, extending along the front of the limb from the shoulder to the wrist. The fragment is subsequently joined to the rest of the bone, not by bone, as in the union of most other fractures, but by tough flexible tissue resembling ligament. One or other of the bones of the forearm may be broken singly, or both may be broken at the same time. The outer bone or radius is broken much more frequently than the cubit bone. Fracture is of much more frequent oc-currence in the forearm than in the arm above the elbow. In fracture of the shafts of both bones, there is distinct crepitus, and the forearm is much bent. In fracture of the shaft of one bone only, there is less deform-ity but usually much bruising and swelling of the soft parts. Crepitus may in most in-

stances be obtained by holding the upper frag-ment firmly and moving the lower fragment from side to side. The usual treatment for fractures of these bones is the application of two long wooden splints ; one to the poste-rior surface of the forearm, the other in front. The front splint should extend from the bend of the elbow to the ends of the fingers. Both splints should be furnished with pads so made as to be thicker in the middle than at the sides, in order to press between the bones of the forearm and to prevent the broken pieces of bone from falling inward. The lower ex-tremity of the radius is the seat of a very well known fracture, which occurs very frequently in women above the age of forty-five or fifty years, in consequence of a fall upon the palm of the hand. This injury is attended with much pain and considerable deformity about the wrist. The lower fragment of the broken bone forms a marked projection at the back of the limb, and leaves in front just above the line of the wrist-joint a corresponding depres-sion ; the lower pointed extremity of the cubit bone is unnaturally prominent, and the hand is carried backward and outward. In old pa-tients this is a very serious accident, as it often leaves the hand a crooked, unsightly, and permanently disabled member. In the most favorable cases months pass before the patient becomes free from pain and uneasi-ness, and is able to use the hand and fingers. There are several methods of treating this in-jury. The chief point is to keep the hand turned toward the inner side of the forearm. This may be done either by a single curved or pistol-shaped splint applied along the front of the forearm and the palm, or by fixing the hand between a front and a back splint car-ried downward from the forearm. Mr. South recommends ordinary straight splints for the forearm, but the rollers by which the splints are secured in place are not allowed to extend lower than the wrist ; so that when the forearm is suspended in a sling in a state of semi-pronation, the hand shall fall by its own weight to the ulnar side.

Fingers.—The first or long bones of the thumb and fingers extending from the wrist to the web of the hand, are occasionally broken by direct violence, as in a fall or in giving a blow. The bone most frequently broken is that of the thumb. There is usually distinct crepitus, and the end of the lower fragment is often displaced, and projects at the back of the hand. This injury is best treated by causing the hand to grasp a bill-iard-ball, a large circular pad of linen, or an ordinary rolled bandage, and then to fix the fingers over this by means of strapping or a few turns of a bandage. One or more of the bones of the fingers, most frequently the bone nearest the hand, may be broken by direct violence. These fractures are usually associated with much bruising of the soft parts and wound of the integument. When the fracture is simple, that is to say, uncomplicated with wounding, a narrow splint of gutta-percha or thin wood should be applied to the front of the injured digit, and be carried upward over the palm of the hand as far as the wrist. In compound fracture, if there be any chance of saving the finger, the same treatment should be carried out, care being taken not to apply the bandage too tightly. The wound should be covered by wet lint.

Ribs.— One or more of these bones may be broken, either by very great force applied directly or by counter-strokes. The ends of the fragments project inward in the former case, and outward in the latter. These injuries are of frequent occurrence, and are produced very often by the wheel of a cart or some other vehicle passing over the chest, or by crushing in a crowd. Fracture with inward projection of the broken bones is generally a very serious injury, as it may be complicated by wound of the lung or compression of the heart. It is very often followed by pleurisy and inflammation of the lung. The ribs most liable to be broken are the fourth, fifth, sixth, and seventh. The most frequent seat of the fracture is at some point in the anterior third of each rib. The fracture unites in about twenty-five days. The chief symptom of fracture of the ribs is an acute pain over the seat of injury, which is much intensified when the patient coughs or takes a deep breath. Crepitus cannot always be felt. This injury is usually treated by applying broad pieces of plaster to the injured side of the chest, each piece being carried from the spine as far forward as the breast-bone. The plaster should be carried to about four inches beyond the fracture in both the upward and downward direction. Another plan of treatment is to roll firmly a flannel bandage about eight inches in width around both sides of the chest, and to fix it securely by stitching.

FRANKINCENSE is the product of a certain species of pine growing in the Southern States. When it exudes from the tree it is fluid, but speedily solidifies. It has little use in medicine, and its properties correspond with those of ordinary resin. Fine Frankincense, the product of another pine, is not imported into this country. It forms an ingredient in incense, and in fumigating pastiles.

FRECKLES are minute spots or specks of pigment or coloring matter, which are often seen on the skin ; *ephelis* is the technical name given to this condition, while the term *lentigo* is used when the freckles are more permanent than usual. Yellowish-brown, round or irregular spots or patches are thus produced on exposed parts, especially in persons of fair complexion. They are most frequent in those parts which are exposed to the action of the sun's rays, so the face is the part most often affected. The attempt to re-

move freckles by local applications seldom succeeds, but the following may be tried : Take of muriatic acid, one drachm ; rain-water, half a pint ; spirit of lavender, half a teaspoonful ; mix well. Apply it two or three times a day to the freckles with a bit of linen, or a camel's-hair brush.

FRIAR'S BALSAM. (See BENZOIN.)

FROST-BITE varies very much in severity. The simplest form and the most common is the ordinary chilblain ; in a more intense form the affected part becomes cold, livid, and puffy, and feels benumbed. This latter condition, if the cold be no further prolonged, is followed by intense heat and redness, and all the symptoms of acute inflammation, but if no heat or protection be then afforded, passes at once to mortification. This last stage of frost-bite has been very frequently met with in campaigns carried on during the winter. In this country, however, except among the very poor and destitute, mortification from frost-bite is a rare affection. The subjects most frequently affected are old people and those whose circulation is sluggish, badly-nourished individuals, and drinkers. Though met with in a great majority of instances in the winter months, it is not so often produced by frost as by cold and wet together. Continued compression or constriction, associated with cold, is occasionally a cause of local mortification. A case is on record of a boy who, after sleeping during two nights of cold and damp weather with his boots on, suffered from gangrene of the toes of both feet. Gangrene may also be produced by suddenly submitting to heat any part of the extremities that has been exposed for several hours to the influence of cold and has become numb and livid. In the most advanced stage of frost-bite the affected parts are black and dead ; between this portion and the sound skin there is a groove lined by florid tissue resembling that on the surface of a healthy ulcer—the so-called line of demarcation, and beyond this the surface of the skin for a short distance is reddened. In some cases the skin only is mortified, in others all the tissues of an extremity down to the bone. On the formation of the line of demarcation the dead tissues commence to separate, and the subsequent changes are similar to those which take place in ordinary gangrene. The above morbid changes are primarily due to the action of cold which suspends and arrests the flow of blood through the veins. The parts most frequently affected are the toes, the nose and ears, and the fingers ; those structures, in fact, which are most remote from the heart and most exposed to external influences.

Treatment.—The treatment of the mildest and the advanced gangrenous forms of frost-bite should be similar to that of chilblains. A person, when exposed to the risks of frost-bite, should endeavor by active exercise to keep up the circulation of the blood until he obtains some protection against the cold. When a part is livid and cold great care should be taken not to submit it suddenly to heat—to place it in hot water or to place it near a fire. The temperature of the frost-bitten part should be raised gradually, first by friction with snow, if obtainable, then by friction with the hand, and finally by surrounding the part with thick layers of warmed cotton-wool.

FROZEN LIMBS, ETC., are a frequent result of our severe climate and sudden changes of temperature. In treating them, great care must be taken to avoid sudden changes of temperature. Should a person be found quite benumbed with cold, take him first to a barn, or shed, or cool room ; it would probably cost him his life if he were taken direct to a fire or into a very warm room. If the clothes are wet, remove them, and rub the body dry ; then wrap him in blankets and give a little weak spirits and water or tea. After a while remove him to a warm room, but still avoid taking him near a fire. Rubbing the skin is the most important restrative agent ; it should be done either with snow or with the coldest water that can be had. Continue this rubbing for several hours if necessary, till the parts are quite soft, and something like the natural color is restored. Even when this point is reached, rubbing with flannels, continued for some time, will be of great advantage. After this has been done the parts may be anointed with sweet oil or lard, or with lime-water and oil (equal parts), and wrapped up well with flannel. If there should be any sores, dress them the same as burns (see BURNS).

FUMIGATION is often resorted to in order to give still greater completeness to the disinfection required for an infected room or dwelling (see DISINFECTANTS). Dr. Elisha Harris gives the following *Rules for Fumigation :* " Thick woollen stuffs, carpets, etc., to which boiling heat cannot be applied, fumigate with sulphurous acid, thus : First, arrange to vacate the room for twelve hours ; suitably arrange all the movable articles that require disinfection : close every window and aperture, and, upon an iron pipkin, or kettle with legs, burn a few ounces of sulphur ; the quantity required for effectual work will depend upon cubical space of the apartment, allowing 1½ lbs. to every 1000 cubic feet of space, and there should be enough to burn rapidly until want of oxygen in the air shall extinguish the flame. Instantly after kindling it every person must withdraw from the place, and the room must remain closed for the succeeding eight hours, or even for an entire day or night. If any other fumigation is resorted to (as that by chlorine, bromine, or nitrous acid), a sanitary officer or a chemist or physi-

cian should superintend the process. Fumigation should be resorted to in dwelling-houses only under the personal superintendence of a competent medical man, as the disinfecting gases are very poisonous." The term " Fumigation" is also applied to a plan of treatment which consists in bringing the vapors of a medicinal agent into contact with the surface of the skin, either at a certain diseased part or over the whole of the body. The vapors thus applied act locally and at the same time are absorbed by the skin, so that the remedy is diffused throughout the system. Fumigation is seldom carried out save in the treatment of venereal disease by mercury. Here it is frequently used, and is considered by many surgeons to have advantages over the inunction plan or rubbing-in of mercurial ointment, and over the ordinary method of administering blue pills and gray powders by the mouth. A smaller quantity of mercury is required, the action of the medicine may be readily controlled, the stomach is not disordered, and salivation is less frequently produced. The compounds of mercury that are used in this way are the bisulphuret, cinnabar, corrosive sublimate, and, most frequently and most effectually, calomel. The selected powder is placed either on one of the several kinds of lamp that have been specially designed for this purpose, or on a brick or tile heated to redness and deposited in a pan containing boiling water. The lighted lamp or heated brick is placed under a cane-bottomed chair, and the patient, stripped of his clothes, then sits upon the chair and covers himself closely, except over the face and head, with a warm blanket, or a mackintosh, or American cloth coat. In the course of ten minutes light mercurial powder is deposited on the surface of the skin. When all the mercury is volatilized the patient should at once get into bed, taking care not to remove any of the grayish deposit from the skin, as this, during the night, may be partly absorbed. This proceeding is generally repeated every night, or on alternate nights, until the gums become sore. In a general fumigation the amount of calomel usually required is about ten or fifteen grains. This plan of treatment is not well tolerated by every patient, one bath even sometimes causes great prostration and general disturbance. Great care, too, is necessary on the part of the patient to avoid catching cold. Venereal affections of the mouth and throat, and ulcers on certain limited parts of the surface of the body, are often treated by local fumigation, the vapor of calomel or other mercurial agent ascending from the lamp or heated brick, and being directed upon the seat of the disease through tubes specially adapted as to shape and length for this purpose.

FUNGUS HŒMATOIDES.—This is a variety of soft or medullary cancer, in which the tumor is large and of rapid growth, and composed of very soft and pulpy cancerous tissue, mixed with large clots of blood. The tumor often throbs like an aneurism or vas-cular growth, and when ulcerated or broken at its surface bleeds very freely. (See CAN-CER.)

FUSEL OIL. (See FOUSEL OIL.)

G.

GALBANUM is a gum resin that is a mixture of gum and resin of unknown origin. It comes from Western Asia in small agglutinated masses of a greenish-yellow color. It contains a valuable oil something like turpentine and a peculiar resin. Its odor is peculiar, something like that of assafœtida, but less intense, and is not so disagreeable. In its properties, galbanum is supposed to approximate to assafœtida, and is contained in the compound assafœtida pill. Probably it is of little value, but may act by virtue of its oil as a stimulant substance.

GALL. (See BILE.)

GALL-BLADDER, an oval sac or bag, about three inches long, forming an appendage to the bile duct, and situated on the under surface of the liver; it is a receptacle for any surplus bile, and sometimes gall-stones are formed in this cavity.

GALLS, or GALL-NUTS, are small excrescences produced upon the buds of the *Quercus infectoria*, growing in Asia Minor, by means of an insect. This insect deposits its eggs in the young buds of the tree, and around them grows a hard mass, which in course of time becomes the gall-nut. These so-called nuts are more or less globular in shape, and tuberculate on the surface, and are generally about the size of a marble. Galls contain a large amount of tannic acid, and a smaller amount of gallic acid; they owe their properties entirely to these two substances. (See GALLIC ACID; TANNIC ACID.)

GALLIC ACID is prepared by making the powder of gall-nuts into a thick paste with water and keeping it in this state for six weeks at a temperature of 60° or 70°. This paste is then boiled and strained, and gallic acid is allowed to crystallize out of the fluid. After this it requires to be purified. In the pharmacopœia there is a glycerine of gallic acid, which is a useful astringent application in certain forms of sore throat, especially to the tonsils after being inflamed, when they show no tendency to contract to their proper size, and there seems danger of their remaining permanently enlarged. Gallic acid is frequently given internally, mainly for checking bleeding. It is usually combined with sulphuric acid, and may be given in doses of from 5 to 20 grains. It is used this way in bleeding from the lungs and stomach especially. It may be also used in bleeding from the kidney, but with less hope of success. Care must be taken not to order it along with iron, or ink is produced. Soo, too, these reme-dies should not be given closely following one another, or the same change may take place in the bowels, and the patient may be alarmed at the color of the fæces.

GALL-STONES, or solid concretions formed of bile, are usually formed in the gall-bladder, but sometimes, though rarely, also in the bile ducts. During the intervals of digestion in the small intestines the bile secreted accumulates in the gall-bladder where it becomes thickened, both by the fact of the water being absorbed, and by the secretion of mucus from the walls of the gall-bladder. Most gall-stones are mainly made up of a fatty material of crystalline character called cholestrine, mixed with the coloring matter of bile, and may grow to very considerable size. When there is only one gall-stone in the bladder it may grow to the size of a hen's egg, which it somewhat resembles in shape. More frequently a number are formed, and then they have facets or smooth surfaces, corresponding to the points where they have come in contact one with another. They are very light, and when dried, float in water till they have absorbed some of it, and then slowly sink. Gall-stones are always evidence of something wrong in the state of the bile but not of anything in the state of the liver itself. They are more common in women than men, perhaps owing to the modes of life differing considerably, for, of all inducing causes, sedentary occupations and confinement, seem to be the most potent. If the bile have a tendency to form deposits, whatever favors long retention of it in the gall-bladder may lead to the formation of gall-stones. The formation of gall-stones does not seem to be specially associated with any diseases of the substance of the liver, except one, that is cancer; but, inasmuch as that disease in its later stages is frequently associated with obstruction to the flow of bile into the alimentary canal, it is most probably the condition so induced which favors the formation of gall-stones rather than the disease itself. So too, age has some effect in the same way, for gall-stones are rare during the most active period of life, that is, under 30. Their formation is often associated with a tendency to gout, and may possibly be accounted for in the same way, viz., a sluggish life of overeating and drinking. This is hardly the place to speak of the effects of gall-stones in giving rise to diseased conditions of the gall-bladder and its appendages, which are rather the subjects of professional study than of general interest, and so we pass to the modes in which gall-stones may be got rid of. Most frequently when of small size, the gall-stones may be discharged through the natural passages into the intestine, but sometimes they are got rid of by ulceration of the gall-bladder or bile ducts into the intestine lower down in its course. Sometimes, instead of escaping into the gut, the gall-stone may give rise to inflammation or sloughing of the part where it is confined, and so escape into the general cavity of the abdomen. This is followed by inflammation and death, but its occurrence is rare. Usually the inflammation causes adhesion to the wall of some portions of the intestine, and so the two walls giving way. The escape of the gall-stone costs much less pain than does its passage along the natural channels, presently to be described. Gall-stones are at the bottom of a large proportion of ailments of the liver complained of in this country. They may long remain quiet, but when they do begin to give rise to painful symptoms in their passage toward the bowel the pain is frequently horrible. In the gall-bladder, the only symptoms ordinarily produced by gall-stones is a feeling of weight in the right side, or at the lower corner of the corresponding shoulder blade. In the tube which lies between the gall-bladder and the liver they may cause little inconvenience beyond obscure affections of the digestive powers, but in the tract lying between the liver and gall-bladder and the intestine, what is called the common duct, they ordinarily give rise to jaundice. The *symptoms* from the passing of a gall-stone, generally come on quite suddenly, often two or three hours after food, and the pain is described as a kind of spasm. Its situation is, however, peculiar ; it is distinctly on the right side of the abdomen, just below the false ribs, and generally extends through to the back, near the lower angle of the blade-bone, or between that and the spine. The pain is not constant ; if it were, in some cases it would destroy life ; it comes by fits and starts, and, while it lasts, is so severe that the patient writhes in agony, or rolls on the floor, pressing his hands on his side, for pressure frequently relieves the pain. This pain is, moreover, attended with a feeling of constriction in the lower part of the chest, which is frequently interpreted as a difficulty in breathing so that a slight attack may be put down to pleurisy. The fit, as it is called, of gall-stones, produces severe exhaustion, the pulse becomes weak, the face pallid, and the whole body covered with cold sweat. Often the patient questions whether life is worth having on these terms. The pain or irritation in the vicinity of the stomach causes it to contract, and so there is vomiting, which sometimes aggravates, but more frequently relieves the pain. Perhaps the nausea arrests the spasmodic contraction of the bile duct round the

stone to which doubtless the pain is due. Jaundice, as already pointed out, is a common symptom, but not necessarily present in all cases. If the stone be small or angular, it may give rise to some degree of irritation in passing, but may not be large enough to choke up the duct, and so not produce jaundice. Nevertheless, even if these do not pass quickly, the gradual accumulation of bile due to its obstruction may give rise to a considerable degree of jaundice. It is rare for gall-stones to cause death during their passage through the bile ducts, and in the majority of cases, especially if the period of the passage has been short, as soon as the passage is accomplished, the patient is well, though, if the passage has been long delayed, or gall-stone follows gall-stone as sometimes happens, the constitution may be greatly shattered. Once in the intestine, as a rule, all danger is past ; but if the stone be very large it may stick in the intestine, and cause obstruction of the bowels, or if it be very small, it may become fixed in that troublesome spot, the vermiform appendix, and so cause inflammation. Either event is rare. Individuals who have once suffered from gall-stones are unfortunately liable to do so again. This comes in two ways as pointed out ; several gall-stones may exist and only one at a time be passed, or the conditions which gave rise to one may prevail and give rise to others. In all cases it is desirable to secure the stone by carefully examining the fæces, as indications are furnished by it as to the existence of others, or as to a likelihood of the return of the symptoms.

Treatment.—The treatment of gall-stones is a matter of the very greatest importance, for inasmuch as they seldom kill, but give rise to terrible pain, much is left in the hands of the physician. Accordingly, the first thing to be done is to relieve the pain and spasm while the stone is passing, and to attempt to get rid of those still left in the gall-bladder, if any, by dissolving them, and so to prevent new ones forming. For relieving the pain and spasm there is nothing like opium. Formerly it was not easy to give this, as the stomach rejected it, however cunningly concealed, during the vomiting produced by the disease ; now, however, by the hypodermic injection, we can give it as we please. But the sickness is a thing not to be slighted, and so for it we prescribe spirit of chloroform, ice, and the like kinds of remedies. Frequently, however, for this purpose large draughts of hot water and carbonate of soda, may be given partially effervescing with tartaric acid, for the effervescence passes off instantaneously, and does much good. The hot water may be repeated as often as necessary. On the other hand, ice is one of the best remedies we can use. But if hot water inside does good, the hot bath sometimes does

more, especially accompanied by opiate subcutaneous injections. Thus sleep may often
be procured, when such is possible in no other
way. Chloral too, would be well worth trying in good full doses, but meantime, our experience with regard to it in these cases is
almost nil. To get rid of any gall-stones left
in the bladder, various remedies have been
recommended. Chief among these are alkalies and aikaline carbonates, and chloroform
or ether. At one time a mixture of ether and
turpentine was greatly in vogue in France.
On the whole, what applics to this, applies
also to the avoidance of these formations in
future. Chief among preventives are air, exercise, and plain food. Beer should be
avoided, but a fair allowance of light wine
may be taken. The bowels should be moved
daily, if necessary ; in the evening a small
dose of blue pill may be taken from time to
time. If the patient can afford to travel, he
should try a residence at an appropriate
watering place.

GALVANISM is a power like electricity,
named after its discoverer, Galvani, and is
often applied to the body in case of nervous
pains, by means of a small portable machine.
An invention known as "Pulvermacher's
Chains" is a form of machine which is very
portable, and the electric current is continually kept up, passing through the body almost imperceptibly. (See ELECTRICITY.)

GAMBOGE is a kind of green resin, procured from Siam. The juice of the tree is
collected in hollow bamboos ; hence the outside of the pipes or sticks, in which form the
drug is imported, is marked with streaks corresponding to those on the inside of the
bamboo. It is hard and brittle, breaking with
a shiny fracture, bright yellow in color. It
is easily produced, and when from the powder
the resin is extracted, that is found to have
acid properties. Rubbed up with water, the
gum dissolves and suspends the resin, forming an emulsion. It is more used as a pigment
than as a drug. Its only preparation is the compound gamboge pill which is not often used.
Gamboge is a powerful drastic purgative,
giving rise to copious watery motions. Not
much is required to do this, and sometimes
children, with whom it is a favorite plaything,
by merely licking it so as to get it to paint,
as they say, are severely purged as a consequence. It often causes vomiting, and always gaping, and so is seldom given by itself.
Perhaps, if given at all, cream of tartar is the
best adjunct. The two cause copious watery
motions of the bowels. Ginger, cayenne
pepper, or some oil should always be given
along with it to prevent the griping. It is
mainly given as a purgative in dropsies where
the power of causing watery stools is of value.
It is also sometimes used to get rid of worms.
It is the basis of some quack remedies, and
being used injudiciously, as these are apt to

be, has given rise to inflammation of the
bowels, and so to death.

GANGLION in surgery is a tumor connected with the sheath of a tendon, either
arising from a partial sprain, the fibrous and
synovial sheaths being torn, or from the
sheath being attenuated and distended with
the albuminous secretion. It appears as a
fluctuating, translucent swelling, compressible, varying in size and shape from that of a
small pea to a hen's egg, and the swelling,
though tense, distinctly fluctuates. Their
usual situation is at the back of the wrist or
upper aspect of the foot. It will be noticed
in many instances that they appear to be multiple, or that there are several ; this, however,
arises from the fact of the tendons, generally
extensor, passing over them and dividing
them into apparently distinct compartments.
The fluid contained is thick, rancid, and
glairy, like white of egg, or what artists know
as "Macgilp," secreted by means of the
epithelial cells on the surface of the synovial
membrane, and especially by those which are
accumulated on the edges and processes of
the synovial fringes. It consists of water,
mucus, and epithelium, fat, albumen, and
extractive matter, and salts. The *treatment*
consists, if they are very small and have evidently thin walls, in dispersing them into the
surrounding tissues by a smart squeeze of the
thumb, or a sharp blow, friction, and pressure ; these generally cure them. A blister
placed immediately over the tumor often excites a sufficient amount of irritation in the
sac to absorb the contents. In larger cysts, a
fine knife should be introduced flatwise through
the walls of the ganglion, so that by pressure
its contents may be extended into the surrounding structures, and becomes absorbed
after dispersion. A compress and bandage
should next be applied, and if the cyst refills,
the process must be repeated. Sometimes
the tumors may be dispersed by iodine or a
mercurial ointment. Persons who are subject
to these ganglia should wear some firm india-
rubber webbing round the wrist joint, when
about to use the hand much, or if subject to
them on the back of the foot, firm, well fitting
boots, bracing up the instep.

GANGLION in Anatomy is a swelling
consisting of nervous matter. Ganglia are
found in all forms of the nervous system.

GANGRENE is the partial death of a
part of the body—the preliminary step to
mortification, or the absolute death of a part.
(See HOSPITAL GANGRENE.)

GARGLES, or MOUTH WASHES, are
remedies in a liquid form intended for local
application to the mouth and throat. The art
of gargling properly is easily learned, but requires some little skill, and accordingly such
remedies cannot be made use of among children. Among them some adherent form of
remedy should be employed which can be

smeared over the parts. Gargles are intended to fulfil various purposes. Some only to cleanse the parts, some to brace them up, some to allay inflammation, some to heal sores, and so on. Of those intended to cleanse the mouth pure water may take the lead. It should be warm, and in this form too it proves of great service in the acute stage of inflammation of the tonsils, to which many are very liable. Not infrequently it happens that from some cause the covering of the mouth and tongue forms on its surface a mass of decaying material, the odor of which is excessively unpleasant. To remove this something more than water is necessary, and, on the whole, there is nothing better for the purpose than carbolic acid and water. Where there is at the same time inflammation of the throat sulphurous acid and water had better be used. Certain conditions of constitution are accompanied by relaxation of the soft palate and uvula, for which gargles are commonly employed. Chief among the substances used this way are tannin and alum, either separately or in combination. Tannic acid, or catechu lozenges, may be used for similar purposes. If there is ulceration of the mouth, as in children, ether spray must be used or sticky stuff used. Borax and honey is a favorite remedy. If the ulcers are very foul, as sometimes happens in syphilis or after salivation by mercury, the best gargle to use is made by adding hydrochloric acid to chlorate of potass. This may be used freely, well diluted with water. Occasionally a little laudanum or tincture of belladonna may be added to the warm water if the inflammation is very acute, but the practice is not void of risk; and, as a rule, hot water and steam suffice. Atomized fluids, produced by the spray instrument, have to a considerable extent replaced the use of gargles but cannot do so entirely.

GASTRIC CATARRH. (See INDIGESTION.)

GASTRIC FEVER, another term for typhoid fever. (See TYPHOID FEVER.)

GASTRIC JUICE is a thin acid fluid, poured out from the glands of the stomach during digestion. Its acidity arises from the presence of hydrochloric or lactic acids, but in addition to these constituents the gastric juice possesses another called pepsin, to which most of its peculiar action is due. When the food is swallowed and enters the stomach, the movements of that organ, when its walls contract, roll the food about and thoroughly mix it with the gastric juice. If small pieces of meat or hard-boiled egg are placed in acidulated water with which some gastric juice has been mixed, and kept at a temperature of 100° Fahr., it will be found that in a few hours, these substances have been nearly, if not quite, dissolved and reduced to a pulpy state. This is called artificial di-

gestion, and it has been proved that precisely similar changes go on in the stomach of a living animal. There, soluble substances are then rapidly absorbed by the vessels of the stomach while some portions go on to enter the intestines. Advantage has been taken of this fact to aid digestion by giving the patient *pepsin* a short time before a meal; this is usually obtained from the stomach of a pig, and since the active properties of the gastric juice are mainly due to this body, it follows that much benefit may, in some cases, result from its use. (See DIET, DIGESTION, FOOD, and PEPSIN.)

GASTRIC ULCER, or SIMPLE ULCERATION OF THE STOMACH, is an accident of by no means uncommon occurrence, though in its worse aspects it is comparatively speaking rare. It occurs most frequently in youngish persons, especially females of the servant class, though by no means limited to them. Its *symptoms* are mainly these—pain, vomiting, bleeding; in females, absence of menstruation, there is loss of flesh and pallor of countenance, and finally the ulcer may perforate the wall of the stomach, give rise to general inflammation of the cavity of the abdomen, and so cause death. To take each of these in turn. The pain, which is usually the first symptom, is very characteristic. It commonly begins from a few minutes to half an hour after taking food. At first it resembles a feeling of weight, later it resembles burning, later gnawing and sickening, but rarely if ever of the sharp darting character common in cancer. When this pain is severe it often ends in vomiting, after which the pain is relieved. The pain is most frequently situated just below the extremity of the breast bone, and there is often a corresponding pain in the back, between the shoulderblade and spine. The spot where the pain is felt is generally very tender on pressure, but not always so, and the painful spot is rarely of any considerable size. As the ulcer is rarely at either end of the stomach but generally in its middle, and most frequently on its back wall, lying on the face will sometimes relieve the pain very greatly. The pain is, moreover, very greatly affected by different kinds of food. Some articles can hardly be borne—tea and beer, and though, as a rule, brandy and hot water cannot be taken, in some cases it gives relief. Vomiting is a symptom of gastric ulcer of very grave significance, and may itself be the source of very considerable danger. It is very various in amount, sometimes only amounting to a slight regurgitation, as if the food rose in the throat and fell back again, but as a rule when it begins it empties the stomach. Ordinarily it occurs when the pain is at its height, the whole contents are evacuated, and relief to the pain follows. If vomiting comes on speedily after taking food, the food itself comes up little altered, if digestion has advanced somewhat then the matters ejected are sour, and later still they may be mixed with bile. Occasionally the vomiting has no reference to the swallowing of food, it comes on in the intervals of digestion ; if so the substances so ejected may consist mainly of glairy mucus. The danger of this symptom consists partly in that the food being completely expelled none is retained to nourish the body, and so wasting and weakness follow. Besides there is the fatigue engendered by the act of vomiting, which, in a wasted frame, is no slight matter. Moreover, the violent straining of the stomach may favor the occurrence of the two most fatal accessories of gastric ulcer, bleeding and perforation.' Hæmorrhage or bleeding is a very grave symptom, and one which not unusually proves fatal. It is due to the ulcerative process eating through the coats of the stomach until that where the larger vessels are situated is reached. Usually, as the ulcerative process goes on, the vessels which are eaten through become plugged up, and so bleeding is prevented, but it often happens that a vessel of some size on the floor of the ulcer is corroded and gives way before it has time to be plugged, and so the blood escapes into the cavity of the stomach. This is most likely to happen during digestion, especially if the stomach has been distended by a big meal. If the bleeding be small in quantity there may be no sign of it except that the fæces are a little blacker than ordinary, which it is not easy to detect. But should the vessel be of some size and the bleeding free, speedily the stomach becomes distended with blood ; vomiting is set up, and the blood is at once detected. Blood ejected in this way is usually dark, rarely, however, it is bright red. More rarely still the vessel opened bleeds so freely that the patient faints and dies before there is time to vomit. In these cases after death the stomach and bowels are found enormously distended with blood, often clott:d. It does not always follow that because blood is vomited it comes from the stomach. Thus, if there be bleeding from the nostrils behind where they open into the throat, part of the blood may be swallowed, and subsequently causing sickness, is ejected. So, too, in bleeding from the back part of the throat, especially in children, the blood is often swallowed and afterward vomited. In bleeding from the lung, too—what is called hæmoptysis—part of the blood ejected is frequently swallowed and subsequently ejected. All these chances of error must be borne in mind. Usually in gastric ulcer the bowels are constipated, but not always so. The small quantity of food which passes on into the bowels is one great cause of this. There is nothing to void. In ulceration of the bowel itself this is quite different ; in it diarrhœa is the rule, with hardly any exception.

Perforation is at once the symptom and the result most to be dreaded in gastric ulcer. If the process of ulceration go on uninterruptedly by-and-by the coats get very thin ; from some unusual distension or exertion the thinned part gives way and the contents of the organ escape freely into the cavity of the abdomen. The signs of such an occurrence are unmistakable. After suffering more or less severely from the symptoms already described, the patient is suddenly attacked with excruciating pain in the abdomen, spreading rapidly all over it. The patient becomes collapsed, the pulse fails, and in not many hours the patient sinks and dies. Dilatation of the stomach sometimes results from a healed ulcer causing constriction at one part, but this is not very common in simple ulcer.

Treatment.—To begin with, the simplest and the best, ice, especially when there is bleeding, can rarely be dispensed with. The patient may eat it freely, not sucking it only, but champing it and swallowing it in the rough. When there is bleeding, ice should be also kept applied over the pit of the stomach and below the false ribs on the left side. Let it be given and applied freely ; it can do no harm, and is more likely to do good in bleeding than anything else. For the pain, which is often very severe, perhaps opium is the best remedy, and it is best given locally, that is, by the stomach, in this particular case. The best preparation is the extract made into small pills—the smaller the better, half a grain at a time. If these are rejected, morphia in smaller dose, might be tried the same way ; if that fail it must be given subcutaneously. For the pain, when there is no bleeding, bismuth is often one of the best remedies we can give, but it must be in goodly quantity of 20 or 30 grains or more. As regards the vomiting, we have seen it to be a dangerous symptom, and one which consequently demands close attention. It is not easily dealt with. Upon the whole ice is the best remedy—and rest. The stomach will not bear food, and so food must be given in the smallest possible quantity, and of the most unirritating possible quality. Of all foods perhaps the best is milk ; after a time essence (not extract) of meat is to be given, but not at first. In case of vomiting still continuing, all food must be stopped and nutrient injections used. When bleeding occurs the same rules are to be carried out, together with absolute rest. Stimulants are not admissible, except excessively dry iced champagne. Remedies, however, may be given to arrest the bleeding, opium among the number. The most reliable undoubtedly is gallic acid, 20 grains for a dose, along with 10 drops of dilute sulphuric acid, given in the smallest possible quantity of water which will enable the patient to swallow the mass in comfort. By and by, when the patient begins to improve, iron is to be given, but cautiously, beginning with weak non-astringent preparations like the ammonio-citrate or tartrated iron. Infusion of calumba, too, is most useful, especially later on, when a few drops of acid can be given with it to aid digestion. Of course such a mode of treatment confines the bowels, and these must be opened ; first of all by injections later by gentle laxatives, such as compound rhubarb powder or castor oil. Great care in dieting must be observed long after recovery, for relapse is frequent and dangerous.

GASTRODYNIA, with which are sometimes used as synonymous gastralgia and cardialgia, strictly means pain in the stomach, using the word in its strict sense as indicating a particular organ, and not as vulgarly the whole abdomen. In this country, cardialgia is more commonly associated with the idea of acidity, and is used as synonymous with heartburn (which see). As used here, gastrodynia is meant to include all degrees of pain, from the sensations that one possesses in stomach-ache to pain of an almost unendurable character. This pain is not always felt in the same spot, nor is it of the same character. Weight, oppression, and distension are the sensations most frequently complained of ; and this is the form which ordinarily indicates slight forms of disease of the stomach. More severe forms are commonly accompanied by a sensation of burning—not heartburn—and others still by a horrible feeling of gnawing or tearing. Pain in the stomach may come from various causes, such as the presence of irritant matters, disease of its walls, alterations in its own secretions, and perversions of its innervation. Foreign substances of an irritant character may be either indigestible articles of food, regurgitated bile, or corrosive substances swallowed, such as strong acids or alkalies ; these, however, being rare, compared with the former. The diseases affecting the substance of the stomach, and giving rise to pain, are mainly two, simple ulcer and cancer. The pain of the former ordinarily comes on soon after swallowing food, and is relieved when the food is ejected or passes on into the bowel. The pain of cancer, on the other hand, comes on as a rule, either earlier or later, during swallowing, or when the food begins to pass away from the stomach. It is, moreover, somewhat different in character ; sharp and lancinating, instead of dull and heavy. Cancer, too, situated at the end of the stomach next the bowel, obstructs the passage of food into the intestine, and so gives rise to changes in it of a putrefactive kind. These, in their turn, cause flatulence and eructations of a very disagreeable kind. Altered secretions are a very important cause of pain. But most frequently the excessive acidity complained of is not so much due to alteration in the gastric juice it-self, as to alterations in the food. What are technically known as neuroses, that is to say, disorders purely nervous, have much to do with pain in the stomach ; for when there is other disease, this element aggravates it ; and where there is other cause of pain, this abnormal susceptibility increases its effects. This variety of pain is most common in delicate nervous women, and in hypochondriac men, broken down by some debilitating cause. A kind of cramp of the stomach, too, may occur. Pain may be reflected to the stomach from the womb or ovaries in females, or it may be due to pure neuralgia. The meaning of pain in the stomach is far from easy to understand, for pain, exactly the same in character, may have a totally different signification. The great point is to find out which pain is neuralgic, and which due to disease of the organ itself. This is very difficult ; still some clue is afforded by the fact that in pure neuralgia food often relieves the pain, whereas in ulceration this is commonly the reverse. So, too, nervous symptoms, if purely nervous, are seldom limited to the stomach, but manifest themselves elsewhere also. So, too, the effects of pressure and position help, as pressure is generally badly borne where the surface is broken, but does good in neuralgia. Change of posture may often entirely relieve in ulceration ; but, except as increasing pressure, is not likely to influence nervous pain. The reader must not conclude that all pain felt just below the false ribs on the left side, or in the hollow just below the breast bone right in the middle is due to affections of the stomach. Frequently pain in the great gut, which there lies close to the stomach, may be mistaken for pain in the stomach ; but, as a rule, this is generally due to flatulence, and extends to other parts of the abdomen, which pain of the stomach alone cannot do. Rheumatic pain of the muscles covering the stomach might also be confounded with true gastrodynia, but it is rare. So, too, are certain cases of spinal disease, giving rise to pain in the region of the stomach. As to *treatment* of pain in the stomach, that must be entirely guided by the nature of its causation, which may be mainly grouped under three heads : Indigestion, gastric ulcer, and cancer of the stomach (which see). Heat or cold will, however, frequently give relief, and may be applied either externally or internally, or both. Neither of these is likely to do any harm.

GELATINE can hardly, perhaps, be called a remedy, and it is doubtful if it can be called much of a food. In the form of calves' foot jelly, however, it is a favorite article, used among invalids and others. It is used too for the thickening of soups. The gelatine so used is commonly called isinglass, and is the sound of the sturgeon, dried and cut into shreds. It would hardly deserve notice here were it not for the sake of warning the public of the want of nutritious qualities which characterizes it. We do not mean to say that it is absolutely innutritious, but it does not contain the amount of nourishment jellies are commonly supposed to possess, and hence people may be cramming the delicate stomachs of invalids with an almost useless material. It is no uncommon thing for a patient to be crammed with what are supposed to be nutritious articles, which contain little or no nourishment, and leave the patient starving.

GENTIAN, as employed in medicine, is the dried root of the yellow gentian (*Gentiana lutea*) which grows on the slopes of the Alps and Pyrenees. As imported the root is in cylindrical, often twisted pieces, and is very tough. Its odor is sweet, its taste somewhat sweetish, followed by an excessively bitter after-taste. Its active principle is readily given up to water, and the infusion of gentian is a favorite preparation ; it, as well as the gentian mixture, contains, however, other ingredients. The extract is also in use for pills. The tincture too is used, but seldom by itself. Gentian is one of the most valuable simple bitter tonics we possess. It may be given in a variety of complaints ; but, perhaps, is most useful in certain forms of dyspepsia. It is also very valuable in recovery from acute disease, but here many prefer quinine. As it gives rise to no unpleasant symptoms, no irritation of any organ, and almost invariably does good, gentian is one of those remedies best adapted for domestic use. The only precautions to be taken in using it is to see that the bowels are open regularly, and that the tongue is moderately clean, otherwise it will do comparatively little good. The infusion, unfortunately, does not keep well, but this may be overcome by adding to it a small quantity of spirit, such as brandy. In preparing the infusion for home use, half an ounce of the root chopped may be used for a pint of water. After standing for about five-and-twenty hours till then ready for use, two table-spoonsful for a dose. It is better to add a small quantity of spirit and some bitter orange peel when it is set aside to steep.

GIDDINESS. (See VERTIGO.)

GIN. (See DISTILLED SPIRITS.)

GIN-DRINKER'S LIVER, so called because drinking large quantities of gin is one of the causes of chronic disease of that organ. (See CIRRHOSIS.)

GINGER is the root, or rather the underground stem, technically called a rhizome, of a plant growing in both the East and West Indies. Its appearance is tolerably familiar to all—knotted, yellowish-white in color, easily breaking, and possessed of a hot taste and agreeable smell. Its powder is yellowish-white. There are two forms, the white

and the black. The white is scraped, scalded, and dried in the sun ; the black is not scraped, and hence its color. It has in it some volatile oil and some resinous matter, which are probably its active ingredients, but it also contains a lot of starch. Its tincture and syrup are much used, but the powder is, perhaps, more extensively used than both. It is an aromatic stimulant substance, when taken internally producing a feeling of warmth and comfort, and frequently appears to aid digestion. It is accordingly useful as an adjunct to griping purgatives, and to other remedies for indigestion, especially if there is much flatulence. When the flatulence is troublesome the best plan is to carry a piece of ginger in the pocket and when a fit comes on, as it is apt to do under a variety of circumstances, a portion may be broken off and chewed, the saliva being gradually swallowed. Ginger tea, too, will be found excessively useful in the flatulence and indigestion of elderly people.

GLACIAL ACETIC ACID is a form of the acid which is solid at ordinary temperature ; any little elevation will, however, cause it to assume the fluid condition. Its uses are those of ordinary acetic acid, but being somewhat stronger it has slightly marked escharotic powers. It is on this account one of the best applications possible to warts and corns, which have little inherent vitality and are easily destroyed. To that end the top should be shaved off, but not so as to make the part bleed, then the glacial acetic acid may be applied to the spot. The end of a lucifer match is one of the best things for the purpose of applying the acid. It may cause a little smarting at first, but this soon passes away. Layer after layer of the offending growth crumbles away under repeated applications of the acid till the whole is gone.

GLANDERS is a peculiar disease met with in the horse tribe, which may either be spontaneously developed or communicated by contagion from animal to animal. In the horse there are two different forms of glanders. One is characterized by swelling, congestion, and ulceration of the nose, or by a discharge from the nostrils, which at first is thin and watery, and afterward thick and sticky like glue ; and by hardening and enlargement of the glands over the lower jaw. This is *glanders* proper. The other form called *farcy* is characterized by cord-like swellings along the course of the absorbents of the legs, and by hard glandular swellings called farcy-buds, which are observed about the lips, nose, neck, and thighs. In this form as in glanders proper, the animal loses flesh and strength, and generally dies from exhaustion. These two forms are different manifestations of one and the same disease ; the discharge from the nostrils of a glandered horse may reproduce farcy in another horse, and the discharge from a farcy-bud may similarly reproduce glanders. Farcy in its advanced stage is often associated with glanders. This disease is sometimes, though very rarely, met with in man as a result of contagion from an affected horse. It may be communicated from man to man, but is never developed spontaneously in man, as it sometimes is in the horse. It is generally caused by the application of the virus contained in the nasal discharge of the diseased horse to some abraded or raw surface. Cases, however, have been recorded in which glanders was produced by rubbing the face with dirty hands and cloths. In man, the two forms of glanders, viz., glanders proper, and farcy, may be met with, but in most cases these forms are associated. The following are the *symptoms* of the disease : The patient at first suffers from intense febrile disturbance, associated with much perspiration, head-ache and wandering pains in the limbs ; there is often severe shivering, and at times mental disturbance and delirium ; this premonitory stage is then followed by one in which the external and objective symptoms make their appearance ; the glands in the neck, armpits, and groins become swollen ; over the face, neck, and abdomen, there may · be seen a crop of small shot-like papules, resembling very much those met with in small-pox ; the skin covering these breaks down, so as to leave small ulcers ; large and soft superficial abscesses form on the arms and legs, chiefly near joints ; the mucous membrane of the nostril then becomes inflamed and furnishes a sticky, thick discharge of a dirty-yellow or tallowy color, which is sometimes marked by streaks of blood ; the skin of the face and nose becomes swollen and shining ; the inner surfaces of the eyelids also are involved, and their edges are glued together by a thick gum-like discharge ; finally, large patches of inflammation often appear at different parts of the surface ; these increase in size and become livid and gangrenous. The severe external symptoms are associated with diarrhœa, delirium, and coma. Most cases of severe or acute glanders are fatal, death taking place generally between the fifth and fourteenth days. The affection is sometimes chronic, and consists in a constant discharge of viscid and very fœtid pus from the nose, and by swelling and inflammation of the face and eyelids ; these symptoms are attended with much constitutional disturbance, and the patient often dies from exhaustion. In acute farcy, there are hard and painful swellings, extending like thick cords along the limbs ; the glands in the groins and armpits are also inflamed, and there is diffused erysipelatous swelling of one or more limbs. The most characteristic symptoms of glanders are the peculiar eruption on the face and chest, and the thick yel-

lowish and ill-smelling discharge continuously proceeding from the nostrils. In the early stage of the disease, however, and before any external symptoms have been manifested, there is often great difficulty in determining the nature of the case, and in distinguishing an attack of glanders from an attack of rheumatism, small-pox, or typhus-fever.

Treatment.—The treatment should be directed to supporting the strength of the patient by strong broths or beef-tea, milk, and alcoholic stimulants. Great attention ought to be paid to the cleanliness of the patient, and the nostrils should be frequently syringed out with lotions containing creosote, tincture of iodine, chloralum, or chlorate of potash. Poultices of bread-crumbs or linseed meal should be applied over abscesses and inflamed glands, and hot fomentations along the cord-like swelling.

GLANDS are small bodies, mostly of an oval shape, found in the skin, and, in fact, in all parts of the body. They are very liable to enlarge, especially those under the chin, as is often the case in children after an attack of measles or scarlet fever, or when the child is suffering from eczema. They are frequently called " kernels" by the ignorant ; they form a part of the lymphatic system.

GLASS-POCK. (See CHICKEN POX.)

GLAUBER'S SALTS, technically known as sulphate of soda, is a valuable purgative unfortunately gone greatly out of repute. It is contained in sea-water and in most purgative mineral waters. It has been displaced by Epsom salts or sulphate of magnesia, but the change has not been altogether for the better. Perhaps the best thing to do is to combine the two, a couple of drachms of each, which, taken fasting in the morning, will generally be followed shortly after breakfast by a copious loose motion. For ordinary purposes this is what should be aimed at, and anything in excess should be avoided by diminishing the dose. When used in dropsies the case is different.

GLAUCOMA. (See EYE, DISEASES OF.)

GLEET. (See VENEREAL DISEASES.)

GLOBULIN, a substance existing in the serum of the blood, and in some other fluids of the body; it is obtained by passing a stream of carbonic acid gas through serum largely diluted with water ; it aids in causing the coagulation of the blood when that fluid is drawn from the body. (See BLOOD.)

GLOSSITIS. (See TONGUE.)

GLOTTIS, the upper part of the air-passages, also known as the opening into the larynx.

GLUCOSE is found in many plants, and also in the blood of man, being formed to a great extent in the liver. (See GRAPE-SUGAR.)

GLYCERINE is a sweet substance, the basis of fats, being combined in them with the peculiar fatty acid characteristic of each.

Accordingly, when these fats are decomposed by the addition of an alkali, as is done in making soap, the glycerine is set free, and the new combination of fatty acid and alkali constitutes soap. It is also obtained by distilling the fats by means of superheated steam. Thus obtained the glycerine is a sweet liquid, colorless, and syrupy, oily to the touch, yet mixing readily with water. The solution of it in water does not ferment with yeast, and it does not dry up on exposure to heat of a moderate temperature. A high temperature decomposes it and sets free intensely irritating vapors. Its properties are very valuable, it readily dissolves many substances, and not drying up readily it constitutes an excellent basis for applying them to the skin. Heated with starch it forms a " plasma" which can be used as an ointment. It is chiefly as an adjunct to lotions that glycerine is of use. Lotions containing it do not dry up and so the skin is kept soft and moist, and the bad effect of drying in forming scabs is avoided. It is also used in many ways where oil was wont to be used, over which it has the great advantage of being readily removed by washing with water and not being liable to rancid change.

GLYCOGEN, a peculiar substance formed by the liver, and capable of being converted into grave-sugar or glucose. (See Liver.)

GOITRE. (See Derbyshire Neck and Graves s Disease.)

GOLDEN OINTMENT is a bright, yellow ointment in popular use, made of finely-powdered red precipitate and spermaceti ointment. A patent medicine, sold under the name of Golden Ointment, for the eyes, is made with one drachm of the red oxide of mercury to one ounce of spermaceti ointment, and is often very efficacious in chronic inflammation of the eyelids.

GONORRHŒA. (See Venereal Diseases.)

GOOSE-SKIN is a roughness of the skin which occurs when any one is shivering, as in cases of intermittent fever.

GOULARD WATER. (See Lead.)

GOUT is a disease about which much has been spoken and written, whose characters are perfectly well known, and which yet nevertheless retains much of its original mystery. The acute portion of the attack generally locates itself in some joint, and is accompanied by great pain and swelling, general constitutional disturbance, and especially derangement of the digestive organs. It has a very great tendency to recur again and again after intervals at first of apparently perfect health, but afterward of only partial restoration. Most frequently it attacks the ball of the great toe, later also the hands are affected.

Symptoms.—Sometimes, more especially in later attacks, the gout gives some warning

of its approach. Digestion is impaired, the bowels are out of order, there may be some fluttering about the heart, the skin is dry and hot, and the urine becomes very thick soon after it is passed, with a brickdust sediment. Usually, however, in earlier attacks there is no warning. The victim goes to bed well, and is woke up about two or three in the morning with a severe burning pain in the great toe, the ankle, or the thumb. There may also be some shivering, but the pain gradually subsides as morning advances, and the patient may have some sleep. When next observed the toe will be red, excessively painful, and still more tender, and more or less swollen. The patient is exceedingly irritable, and more or less depressed. His tongue is coated with a white fur, his bowels confined, and his urine scanty, high colored, depositing a red brickdust sediment on cooling, or even when passed. This urine is exceedingly irritant, it cannot long be retained in the bladder, and scalds in passing. Each night the patient is worse than during the day ; but the attack does not last long, in about four or five days the patient begins to mend, the swelling abates a little, and scurf is left behind. Presently complete health is regained, and the patient feels better and brighter than he had done for long before the attack. This too frequently entails a return to the mode of life which has ended in the previous attack, and by-and-by, after a longer or shorter interval, the gout returns. This occurs again and again, the interval becoming shorter and shorter, and less and less distinct, until the patient sinks into the condition known as chronic gout. As the disease advances more than one joint is attacked, the small ones having the preference, till almost every joint in the body is seized. Round about the joints a matter, at first fluid, but afterward solid and chalky, is deposited. It consists of urate of soda, and the deposits are called chalk stones or tophi. Small deposits of the same material are frequently also laid down in the ear. These cause much distortion, now and again they suppurate and form very troublesome open sores. But gout is not alone manifested by what might be called gouty inflammation. During the intervals of the attack the patient may suffer from impaired powers of breathing, palpitation, and the like, and he may be subject to chronic bronchitis, which is not easily removed. When the attack comes on it may become what is described retrocedent, that is to say, may leave the limb and attack some internal organ, especially the stomach and heart. Application of cold to the affected limb is very likely to bring this about, and its occurrence very frequently means the death of the patient. So, too, anomalous gout as it is called, may manifest itself in various ways. There is no regular attack, the patient would be better if there were. Instead, there is indigestion with flatulence, heartburn, and constipation, the heart beats painfully and irregularly, there is pain in various parts of the head, the patient is easily fatigued, and is restless and irritable, wandering pains fly about the body, and any little damp in the atmosphere brings them on at once. These symptoms, if they are not relieved or a good honest attack of the malady does not burst out, are apt to end fatally in various ways. Frequently in such patients there will be some form of scaly skin disease, very itchy and troublesome. This coming and going with the other symptoms is a valuable indication of the nature of the malady. Gout is sometimes acquired, but very frequently the tendency to it is hereditary. Luxurious living and little exercise are the two great means of producing gout anew, but in many, with a strongly marked hereditary tendency, no amount of sober-living will avail in keeping off the malady. Nor is it among the rich alone that such forms of the disease are seen. Among the poor, where porters and various forms of beer take the place of port and sherry, gout is perhaps even more frequently seen than among the rich, considering that among them muscular exertion of some kind is the rule and means of life. And there can be no doubt but that those who are the subjects of lead poisoning, especially the workers in white lead, are more liable to the disease than are others.

Treatment.—As to treatment there is a good deal of truth in the saying of the old physician that the best remedies in gout were patience and flannel, but like most such sayings it is only half-truth. In point of fact much may be done for gout both during an acute attack and during an interval. These two naturally separate themselves into two separate headings, of which that relating to the acute attack may be taken first. The remedies may be classified under two headings, specific, and common or ordinary remedies. Of specific medication we have no better example than the use of colchicum in gout. This remedy given in full doses has undoubtedly an extraordinary influence in relieving the gouty paroxysm, but on what principles we cannot tell. It used to be supposed that its use depended on the purging it gives rise to in large doses, but this is not so, for it is best administered in doses which fall far short of those necessary to give rise to either vomiting or purging. Thirty drops of the wine is quite enough for the first dose, and 10 drops every four hours after, until nausea results, which should be stopped. This may be done earlier if the pain abates, and under no circumstances should its use be continued over four-and-twenty hours without intermission. Usually this will stop the attack, but the practice is not without risk, and should not be

lightly undertaken. Besides, when the attack is arrested, it must not be forgotten that the treatment of the malady only begins. It is forgetting this which has brought the treatment by specifics into disrepute. A still more powerful remedy, though a secret one, is Lavelle's Gout Liquid. Its effects are closely allied to those of colchicum, though its constitution is different. As to the lying formula which accompanies each bottle, that is obviously on the face of it false. Most likely its efficiency is due to white or green hellebore. It must be used with great caution and in accordance with the rules laid down for colchicum. Though not without danger its efficiency is undoubted, yet men who have used it for years almost invariably give it up. Its effects are extreme depression of the heart, with a terrible feeling of sinking and prostration, which are far from pleasant. For these symptoms the use both of colchicum and Lavelle's liquid are often objectionable, and recourse must be had to other remedies, which, though less speedy, are more likely to do good in the long run. Chief among these are laxatives, and those are best which best unload the portal system without weakening the patient. For this gentle saline purgatives are best, say a double salted seidlitz powder to begin with, and a couple of drachms of sulphate of soda and sulphate of magnesia, each with a few drops of dilute sulphuric acid, twice or thrice a day. The effervescing form is, however, the best, as agreeing better with the irritable stomach. These remedies must not be allowed to depress the patient, and in some a more comforting draught of senna, rhubarb, aloes, and ginger, is to be preferred. Pullna water is a good remedy, but must not be used too freely. One great object is to get the urine to flow freely, and get rid of the half-metamorphosed material in the system. Alkalies do good in both ways, they tend to increase the flow of urine and they aid metamorphosis. It is best to give the bicarbonate of potass or lithia effervescing with citric or tartaric acid, along with some aromatic spirit of ammonia and tincture of lavender. If the pain is very great it is hardly possible to refuse opiates though they are to be avoided as far as possible, seeing that they tend to aggravate the disease by retarding food and tissue change. If opium must be given then let it be given by the arm, for so, less will be required than by the mouth. But before having recourse to that it is better to try its effects locally. Thus, a warm lotion containing acetate of lead and acetate of morphia will do more good than anything. Extract of belladonna rubbed up with water and glycerine and applied warm will often prove of very great service. In all cases warmth is the great thing, cold having a tendency to drive the gout to some internal organ. Rest must be absolute and the diet plain. It must consist only of milk, arrowroot, and the like. All animal food should be avoided, even at first, save the strength demands it. Toast and water, seltzer water, and the like may be taken freely. After a time fish may be given. If there is need of a stimulant whiskey and water may be given, but for a time all stimulants are better avoided.

Now come the rules as to treatment in the interval. It is by this only that the disease can be cured. Chief among these are exercise in the open air, sufficient to make rest grateful but not to fatigue. Sea-bathing is good, so are early hours and plain food. Indeed this last is imperative, and moderation in quantity as well as in quality must be observed. Claret, chablis, and hock may be allowed, so may whiskey and water. All other liquors are forbidden. The bowels are to be kept open, best by Friedrickshall or Püllna water, and the urine must be kept right as far as possible by alkalies. A visit to a foreign bathing place appropriate to the case is one of the best things to have recourse to. Vichy, Ems, Carlsbad, Wiesbaden, and Aix la Chapelle, are the most appropriate resorts.

GRANULAR DEGENERATION is a term applied to the appearance which some organs assume when undergoing a fatty or fibrous degeneration : applied to the liver, it corresponds with cirrhosis, while in the kidney, a similar change marks a form of chronic Bright's disease. (See CIRRHOSIS and BRIGHT'S DISEASE.)

GRANULATIONS.—On examination of the surface of a healthy ulcer, or of a large wound which has existed for six or seven days, it will be found covered by small and soft nodules of a florid red color. Those nodules are called granulations, and it is by their development and subsequent changes that cicatrization, or scarring of the ulcer or wound, takes place. A wide wound on the surface of the body, when free from excessive inflammatory action or sloughing about its edges, presents on the fourth day a deposit over its surface of a soft, white, and tenacious deposit,through which ruddy granulations project, forming at first isolated red clots, and afterwards a uniform granular surface. The granulations then increase in size, and at the edges of the wounds reach,and in some cases project beyond, the surface of the surrounding skin. The rounded and free extremities of the granulations break down into pus, which is discharged from the surface of the wound, while at the deeper parts the granulating tissue contracts, and is converted into the tough filamentous substance which subsequently forms the scar. Along the edges of the wound, a pink, or chalk-white line is formed, which presents a well-marked border between the moist granulations on the one side, and the healthy skin on the other. This

border increases in width as it encroaches upon the surface of the wound ; it is dry and smooth, and is covered by soft epidermis. By the shrinking of the deeper parts of the granulations, and by the extension of this superficial border of newly-formed epidermis or scarf-skin, the wound is gradually closed, and is at last wholly covered by scar-tissue, which at first is soft and delicate, but finally forms a hard, compact, and contractile scar. Granulations are very vascular, and often bleed profusely on the slightest touch. They vary much in sensibility ; those on a superficial wound or ulcer, generally give very little pain when touched, whilst those lining the cavity of a healing abscess are extremely tender. They possess the power of absorption, and transmit into the system any deleterious substance, such as arsenic, opium, carbolic acid, and mercury, that may be applied too freely to the raw surface. Granulations frequently become diseased, and the wound or ulcer, instead of closing speedily and without trouble, either remains stationary and becomes painful, or increases more or less rapidly in size. The raw surface may become very red and inflamed ; its granulations may be converted into large spongy masses of a bluish color ; or again these bodies may disappear, leaving a smooth and glazed surface. These morbid changes are often due to local irritation, caused by dead bone, foreign material, unsuitable dressings, sloughs, etc., and the local cause is often assisted by some constitutional disease, or a bad state of the system, due to faulty hygienic conditions. These morbid changes of granulations and their treatment, will be found described at length in the article on ULCERS.

GRAPE-SUGAR is a substance in the juice of the grape, and many other fruits ; it possesses the property, when fermented, of decomposing into carbonic acid and alcohol : this is taken advantage of in making alcoholic liquors ; it is made of carbon, hydrogen, and oxygen. When any starchy compound, as bread, potatoes, etc., is mixed with the saliva, this fluid decomposes the starch into grape-sugar ; it is also formed in large quantities in the liver (see LIVER), and it is present in the blood and urine in cases of diabetes. (See DIABETES.)

GRAVEL is the term commonly applied to the small stony concretions formed in the kidney, and which when passed, seem to form a gravelly kind of sediment in the urine. Stones formed in the urinary passages are very various in their composition, but those here referred to, and which are almost exclusively formed in the kidney, are mainly composed of uric acid. Their size varies from those of a grain of sand, to the largest that will pass by the urinary passages. If this size is exceeded, they must remain behind in some part of the track, most probably the bladder, until by their increase of size, and the trouble they give rise to, they either cause the patient's death, or are removed by art. The simple stones passed from the kidney, generally acquire layers of other kinds of material over the original nucleus. Here we have mainly to do with stones formed in the kidney and retained there, or which in passing, give rise to what is called a fit of the gravel. The stone, if it remains in the kidney, is sure to grow, and doing so, gives rise to very troublesome symptoms. There is considerable pain in the back, always increased by jolting, and such accidents are usually followed by bleeding from the kidney, not such bleeding as we see in ordinary kidney diseases, but a well-marked and unmistakable flow of blood, which may coagulate in the passages or in the bladder, and so give rise to very severe suffering. As the stone grows, the symptoms become aggravated, and the health fails. Usually there is great sympathetic disturbance of the stomach, and digestion is imperfect. Little serves to bring on the bleeding, and the bleeding weakens the patient ; moreover, the pain is severe. The stone growing gradually, encroaches on the substance of the kidney, which withers and may finally altogether disappear ; if now, as not infrequently both kidneys are affected, any slight accident happens to disorder the other, life is in great danger. and not infrequently is thus terminated. Sometimes the whole kidney is destroyed and an abscess left, which has been opened and the stone removed, the patient recovering ; sometimes it has ulcerated out. More frequently, however, after the stone has had time to grow to such a size as to be obstructed in its passage through the ureter, it is dislodged, and, carried by the urine, commences its journey toward the bladder. If very small, no symptoms are produced, if very large, it sticks ; between the two are all gradations, from momentary uneasiness to weeks of suffering ending in death. The symptoms generally begin suddenly, sometimes with rigors. There is intense pain in the back and loins, extending down into corresponding groin and testis. Very likely there is sickness and vomiting, partly from the great pain, partly from reflex irritation. And this may go on for days and weeks. Sometimes the urine is suppressed and death by uræmia follows. On the other hand, a few hours' suffering may end in perfect ease, as the stone passes into the bladder, to be followed a few days after, by a short and sharp attack of pain as the stone passes out of the bladder. The symptoms of an immovable calculus in the kidney, are rarely such as can, with certainty, be diagnosed until too late to do anything sure to relieve the symptoms.

Treatment.—We question if the solvent treatment, sometimes spoken of, is ever suc-

cessful. Accordingly, when the existence of a calculus in the kidney has been made out, everything must be done to avoid further increase ; diet must be carefully selected, so as to avoid the formation of uric acid in excess, and to this end too, it ought to be well diluted, and so diluent drinks should be freely used. Alkaline aërated waters too, as a rule, will be specially useful, and liquor potassa, which is ordinarily well borne, may be given in good large doses. As regards the passing of a calculus—that is to say, a fit of the gravel : that is a thing to be carefully studied and judgment shown. The great thing is to relax the passage as much as possible, the pressure of the urine behind will drive the calculus onwards. For this chloroform, or a pipe of strong tobacco may be employed, till the patient is fairly sick. He should also be kept in a warm bath to promote the same object. But these are all expedients which can be employed for a short time only, and the affection is, as a rule, not such as to be so dealt with. We must, therefore, have, as a rule, recourse to opium,provided the urine is not suppressed, and the patient must be put under it, and kept under it till the stone passes. Owing to the irritable state of the stomach, the food given, must be, during this time, small in bulk. The opium had better be given by the arm. (See STONE.)

GRAVES'S DISEASE, also known as Exophthalmic Goître, is a malady ordinarily characterized by these symptoms : Extreme nervousness, protrusion of the eyeballs, a projection and pulsation of the thyroid body in front of the windpipe : the heart pulsates violently, and there is often a bruit at its base coincident with the first sound. Any one of these symptoms may be absent ; the forms most commonly present are the projecting eyeballs, the enlarged thyroid, the palpitation, and the nervousness. Most writers have overlooked the nervous symptoms, and the enlargement of the thyroid body has misled them into connecting it with true goître. With this beyond the enlargement spoken of, it has not the slightest connection, and the remedies which do good in the one are injurious in the other. Thus iodine,which is the best remedy in ordinary goître, is worse than useless in this malady : while digitalis, which is utterly useless in ordinary goître, is most serviceable in Graves's disease. The eyes project sometimes so far that the eyelids cannot close over them, and in a few rare cases the eyeball is destroyed from the pressure. The thyroid is swollen but its swelling is rather that of a pulsatile tumor than the mass of a goître. The vessels in the neck pulsate violently, and the heart beats quickly, violently, and imperfectly. The health is bad, the bowels deranged, and the patient is easily put out by any little excitement. Females are almost invariably the subjects of this disease, and in them it often has come on after a sudden fright. The menstrual functions may or may not be affected at the same time. In dealing with such cases the great object is to improve the health and diminish the local symptoms. Iron and digitalis are the chief remedies, but as the bowels are usually deranged, these require to be sharply looked after. A change of air is good. Usually they terminate favorably.

GREEN HELLEBORE. (See VERA-TRIM VIRIDE.)

GREEN SICKNESS. (See CHLOROSIS.)

GREGORY'S POWDER, or COMPOUND RHUBARB POWDER, consisting of rhubarb, magnesia, and ginger, is one of the best and safest of domestic remedies. Unfortunately it cannot also be called nice, and though wholesome, it is far from agreeable. It is chiefly given to children as a laxative, when the bowels have become out of order from the consumption of forbidden delicacies of an indigestible kind. The dose for them is 5 or 10 grains, and it is usually given mixed up with some preserved fruit, and has not infrequently laid the foundation of a life-long hatred to the preserve so employed. Among adults it may be used for similar purposes, but is not frequently so used ; rather· it is given combined with some other drug in small doses over a long period, for the sake of the beneficial effects of the rhubarb and ginger as stomachics.

GREY POWDER is the popular name for a mixture of three parts of mercury with five parts of chalk. It is most frequently given to children with clay-colored passages, in doses of 2 to 3 grains once or oftener in the twenty-four hours.

GRIFFITH'S MIXTURE or the COM-POUND IRON MIXTURE OF THE PHARMA-COPŒIA, is one of the most valuable means of prescribing iron for delicate stomachs. The iron is contained in it as green oxide, and sugar is added to prevent the changing into the red oxide, but it does so change in a short time. People are apt to suppose, when they see the change, that the medicine has spoiled, and it is true that the red oxide is not perhaps such a good remedy as the green one in certain cases, nevertheless, in most cases it will do no harm. It is the best form of iron for irritable stomachs and for chlorotic females.

GRIPINGS are painful sensations produced by indigestible food in the intestines, and caused by irregular contractions of the bowels and the passage of flatus.

GROCER'S ITCH is a form of eczema which occurs in the hands of those who work with sugar or other sticky substances. (See ECZEMA.)

GUAIACUM WOOD, and the resin obtained from it, grows in the West Indies. The wood is known as lignum vitæ, and is exces-sively hard. The resin is insoluble in water, but soluble in alcohol. It is brownish in color, and produces an irritant taste in the mouth. The preparation commonly employed is the ammoniated tincture, consisting of the resin dissolved in aromatic spirit of ammonia. Taken internally the effects of guaiacum are not very certain. It is described as stimulant and diaphoretic. It seems to do good in certain maladies, especially of a syphilitic taint, which affect the skin, the bones, or their immediate coverings. In the rheumatism of old people, which is relieved by warmth, and which is sometimes therefore called cold rheumatism, guaiacum often does great good. It is supposed to have some stimulant effect on the womb, and hence is given in some disorders of menstruation, but this is doubtful.

GULLET, OBSTRUCTION OF.--This is, as a rule, of two kinds, simple or malignant. Simple obstruction or stricture is most frequently due to the results of some corrosive poison. The poison may not have proved fatal, but may have destroyed the tissues with which it came in contact to some depth, and as a consequence, when the parts were cicatrized, contraction of the cicatrix has taken place, narrowing the calibre of the tube to such an extent that solids cannot pass. The poisons most likely to give rise to such consequences are the strong mineral acids and alkalies. The symptoms of obstruction may not come on for many months after recovery from the immediate effects of the poison. The great thing to be done in such cases is to guard against the contraction which is almost inevitable. To this end an ordinary gullet tube should be passed from time to time. This should be done with the utmost gentleness, and by taking the posterior wall of the gullet as our guide. If any resistance offer, is it to be overcome by firm yet gentle pressure, and any attempt at forcibly driving the instrument past or through it should be avoided. If the resistance cannot be overcome, another and a smaller instrument is to be tried till the obstruction is fairly passed. After this day by day an increase of the size of the instrument should be made, so that stretching the part little by little the calibre of the natural passage may be again attained. Nor should it be forgotten that this is the only end to be attained. For as soon as the distending force is withdrawn the parts will again tend to contract. Indeed, some have worn a tube habitually with very great benefit, and if that is not used a tube must be passed at short and regular intervals, or contraction will speedily begin. There is probably a variety of stricture of the œsophagus or gullet due to syphilis ; this, however, is not quite certain.

True malignant disease of the gullet is not very rare. Its symptoms are at first exactly like those of simple stricture only there is no history of injury to the part. Moreover, it advances more rapidly, and there may be, but not of necessity, pain. The food is swallowed as usual but sticks at the obstruction, and either accumulates for a time or at once regurgitates. The vagus nerves frequently affected, vomiting, cough, and hiccup are common symptoms. The patient wastes rapidly, partly from the character of the disease, partly from want of food. Death rarely results from starvation, as it is wont to do in the other variety of stricture, but the disease spreads to some neighboring part, or bleeding follows, or the like. Very frequently there is produced a communication between the windpipe and gullet. This, too, aggravates the cough. The disease having been fairly diagnosed as malignant, not much, unfortunately, remains to be done. The character of the disease insures the destruction of life in the long run, the only question is a choice of evils. As the disease advances, less and less food can be taken, until the patient is threatened with starvation though surrounded by plenty. This, too, is the case in simple stricture, though in that there is much more chance of the obstruction being overcome. Under these circumstances we must have recourse to nutrient injection, which may be used as long as possible. At the same time the mouth may be washed out with a little water, and kept moist by moistening it from time to time. The question remains, is it right to have recourse to an incision into the stomach with a view to making a permanent opening for cases which cannot otherwise be dealt with. Well, in malignant disease this is merely prolonging torture, supposing the operation to succeed in the first instance; either way the patient must die. Upon the whole, therefore, to make a permanent opening into the stomach through the walls of the abdomen in malignant disease of the gullet is not an operation likely to be followed with success. The only time when there can be a question of performing it is when the stricture is a simple one, but as death has followed in every instance hitherto, the prospect is not an encouraging one. The conditions are totally different from those of opening the stomach to remove something in a healthy individual. Here the opening is to be permanent, and the individual is wasted to the uttermost.

GUM ACACIA. (See ACACIA.)
GUM ARABIC. (See MUCILAGE.)
GUMBOIL is the most common form of alveolar abscess, invariably associated with a decayed tooth, causing inflammation of the periosteum covering the alveolar process and of the bone itself. The abscess causes great pain and discomfort, and frequently considerable constitutional disturbance. In the earliest stage, when the formation of pus is threatening rather than established, the

malady may be cut short by the extraction of the tooth affected, or by the removal of the stopping from some decaying tooth. If the extraction of such a tooth be undesirable, the gum should be freely leeched, the leeches being best applied through a glass tube, or leech glass (see LEECHING), a brisk purgative administered, and hot fomentations applied to the swelling. When pus has formed and it threatens to "point," the walls of the abscess becoming thinned and soft, it should be evacuated by means of a scalpel. This is followed by complete relief, but not unfrequently a pus discharging sinus remains, or the disease may break out again, unless the offending tooth be extracted. Occasionally, in severe cases, the matter will "point" externally on the cheek. The offending tooth must be at once extracted, and a vertical cut be made with a scalpel, between the cheek and the jaw, so as to cut across the pus-containing canal.

GUMS, DISEASES OF.—These structures are occasionally affected with ulceration in consequence of mercurial salivation. In bad cases large and very foetid sloughs are formed, but usually may be observed only redness and superficial excoriation of the gums. The formation of analogous ulcers is often caused by the presence of bad teeth or accumulations of tartar, but then the lips and tongue generally remain sound, and there is no increased secretion of saliva. The best treatment perhaps for small superficial ulcers of the gums is the local application of solid blue-stone or the use of a wash containing alum or borax. In old people the gums frequently become soft and swollen, and separate from the roots of the teeth. The contact of food is painful and causes bleeding. This condition is often associated with disorders of the stomach and liver. The usual treatment consists in washing out the mouth with a lotion containing alum and tincture of kino, and in the application to the affected gums of tannin and glycerine. Attention should be paid at the same time to the digestive organs, and the bowels be freely opened with blue pill or calomel, followed by a black draught. In cases of sea and land scurvy the gums swell and are covered by large spongy outgrowths of a dark red or purple color, which readily bleed when touched. These outgrowths are masses of swollen gum, and generally spring from the small tongues of gum-tissue which project between the necks of adjoining teeth. This morbid condition disappears with the other symptoms of scurvy after the administration of a good diet, comprising fresh meat and vegetables.

Epulis is a firm, painless, and slowly-growing tumor, which appears on the gums, especially over the sockets of the teeth in the upper jaw. The surface of this growth is slightly irregular and lobulated, and resembles in appearance perfectly sound gum. It generally grows forward from the free surface of the gum, and its root is always connected either with a complete and apparently sound tooth or with an imbedded fang. The tumor rarely attains a great size, but in consequence of its situation and of the inconvenience to which it may give rise, it often becomes very troublesome. Sometimes the surface of the growth becomes ulcerated and pours out an offensive discharge. The usual treatment for epulis is early and complete removal. If the tumor be merely shaved off at its insertion into the gum it will almost certainly return. In consequence of the orgin of the tumor from the fang of a tooth and from the inner part of the socket, it is necessary for the surgeon to extract one or both of the displaced teeth, and at the same time to remove with bone-pliers a portion of the corresponding alveolar process of the jaw.

GUNSHOT WOUNDS implies of course, in its first sense, such injury as may arise from cannon shot, splinters of shell, or bullets ; but it must be remembered that injuries inflicted by any explosion, such as the bursting of a boiler or blasting a rock, for instance, possess the same general characters. The immense improvements of late years in artillery and small arms, from the fact of such weapons being rifled, have led to changes in severity, and, indeed, almost in the nature of gunshot wounds, particularly as regards their infliction by small arms. The form of wound is of the lacerated and contused character, followed by sloughing and suppuration. Hemorrhage is seldom very extensive in cases where there is much crushing, as the vessels thereby become twisted and thus closed, although in the case of the puncture that a small bullet would make, some internal vessels may be wounded and bleed internally, while the external wound is very small and no blood flows from it. Bullets frequently lodge. In the instance of a spherical bullet fired from a smooth-bored musket, the aperture of entrance is small, with discolored and inverted margins. The aperture of exit is larger than that of entrance, and its margins are ragged and everted. If the muzzle of the musket were near to the body at the time of discharge the aperture of entrance would be lacerated, generally containing wadding or clothing, and scorched with the explosion. The appearance of injuries from the conical bullet of the modern rifles is different from the foregoing in most cases ; the wound is more like an incision, and if the ball passes through, its apertures of entrance and exit are almost similar. It usually splits any bone in its course, owing to its velocity of rotation. The course taken by bullets, especially round ones, is oftentimes very remarkable, as may be inferred by watching a shot or shell strike the water and rebound indefinitely, and instances

might be multiplied of cases where the apertures of entrance and exit have been exactly in a line, and yet the ball has traversed the entire circuit of the trunk ; or, again, where a ball has struck the forehead and emerged at the occiput, appearing as if it had penetrated the skull, whereas it has made the circuit of the integument only ; and there are cases on record where the ball has returned and emerged at its aperture of entrance, after having completed the circuit. The shock is proportioned to the extent of injury, the importance of the part affected, and the quantity of blood lost, and, as in other injuries, is of two kinds, mental and corporeal. The mental is temporary, the latter is often aggravated by the former and is independent of the mind. Sometimes fatal injuries are effected when there is not the slightest sign of an outward bruise, and bones smashed, muscles and arteries lacerated ; this form of injury was formerly called a " wind contusion," but it is now well known that such injuries must have been effected by the actual contact of the shot. The true extent and danger of wounds inflicted by gunshot, in the case of penetration, can hardly be determined until suppuration has been set up. Sloughs become detached, particularly at the aperture of entrance, as at that point the degree of contusion is greatest, although the aperture of exit is always first healed ; the suppuration of the slough is usually complete in a week or ten days. Many formidable accidents are liable to occur, however, such as inflammatory fever, gangrene, erysipelas, abscesses, hemorrhage, sloughing, phagedæna, nonunion of fracture, necrosis, caries, hectic, tetanus, and pyæmia.

Detection of Bullets, etc., in Wounds. —It is sometimes difficult to determine whether some hard body felt in a wound is a ball or a piece of exposed bone, and for the purpose of making a correct diagnosis a probe, carrying a small piece of unglazed porcelain at its extremity, is one of the best, as the absorbing nature of the porcelain allows of a small stain of the metal being carried on it on withdrawing it from the track (Nélaton's probe). As soon as the injury is inflicted a most careful search should be made for the foreign body before swelling has come on. The best instrument for making an examination is the finger, but if that fails to reach the substance, a long silver probe which readily admits of being bent is required. Bullet forceps, especially made for the purpose, are needed in many cases, but if the ball be near the surface common incision with forceps is sufficient. The external wound must be enlarged in cases where neither the finger nor the forceps can be introduced. Gunshot wounds of the skull are most unsatisfactory and fatal. In the case of simple flesh wounds, if not severe, they will heal under simple dressing and quiet ; if the scalp be severely lacerated, suppuration and necrosis of the outer table of the skull, and perhaps meningitis, may follow. In cases of fracture of the skull without depression, the prospect is necessarily unfavorable, but patients frequently recover, with, perhaps, exfoliation of the external table of the skull. In such a case cold applications should be kept to the head, and the bowels regularly kept open. Perfect rest is enjoined. If severe rigors and head symptoms occur in from a fortnight to a month after the injury, it would point to the probability of the formation of pus. In cases of fracture with depression of bone, and the usual symptoms of compression present (see COMPRESSION), then the surgeon should trephine ; but if there are no symptoms, the operation should be delayed until they present themselves. If balls or fragments of shell lodge in or penetrate the skull, they are almost always fatal. Injuries of the face may be merely superficial, or of considerable importance when the bones are smashed ; care must be taken to relieve any deformity which is likely to arise, by adjusting the parts with sutures, and removing all spicula of bone, and applying a light water-dressing.

Injuries of the Chest.—The several kinds of gunshot injury of the chest may be conveniently classed as follows. for sake of reference. 1, Those in which the thoracic cavity has not been opened ; 2, those in which it has, and a further subdivision is to be made, of injury and non-injury of its contents. In the first class the danger is small, comparatively ; and, in the second, it is serious from hemorrhage and its complications. If the ball has lodged in a penetrating wound the prognosis is unfavorable. The symptoms of wounded lung are—great collapse, blanched, anxious face, difficult breathing, and generally frothy expectoration, frequently emphysema (see EMPHYSEMA), from the fact of a rib having been fractured. The patient should be carefully examined, to find out in the first place if the ball be in the thoracic cavity, or if it has passed out at some counter opening. Splinters of broken ribs must be carefully removed, and some light water-dressing be placed over the wound. No attempt at rousing the patient from his collapsed condition is to be made, as it is favorable to the arrest of internal hemorrhage. He should be placed on his wounded side, so that the escape of pus may be favored and the movement of the ribs quieted. Constitutionally, low diet, perfect rest, and the administration of opium, generally suffice. The unfavorable symptoms which may arise are pneumonia, pleurisy, or empyema.

Injuries of the Abdomen.—These are conveniently divided into non-penetrating and penetrating. Non-penetrating flesh wounds merely require the ordinary treatment of in-

cised wounds. Frequently, however, a non-penetrating wound, such as is derived from spent balls, or portions of shell, or of stones struck up, will cause severe lesion of some abdominal viscus, and cause speedy death, without any appearance of external injury. In the case of penetrating wounds the amount of fatality is very great. If a ball passes through the abdomen without injuring its contents, peritonitis is usually set up. If a large viscus has been wounded, great collapse is the first symptom noticed ; if the intestines have been lacerated or opened, there is severe vomiting, great pain, and passage of blood per anum ; the nature of the discharge from the wound is itself a guide to what viscus or viscera are implicated. With regard to the treatment of these formidable injuries, the first thing to be done is to endeavor to replace any protruding content, avoiding all unnecessary handling of it, and in the case of a wound noticed in any portion of the protruded intestine, its edges should be neatly approximated, by what is known as a *continuous or Glover's* suture, the knots cut off, and the locality of the wound kept as near to the external one as possible in the event of an artificial anus being established. Large and frequent doses of opium are needed, to allay pain and overcome the peristaltic action of the bowels.

Simple flesh wounds of the extremities require the ordinary treatment of incised or lacerated wounds ; if the bones be simply fractured, and there is not much external injury, the limb should be put up on a splint, but severe contusion and lesion of surrounding muscles and deep structures require amputation. The cases of gunshot wounds of the extremities which require removal of the limb are—1, Those in which the limb has been torn off ; 2, Where there is severe laceration of the superficial tissues, with injury to the main artery, vein, and nerve ; 3, Severe compound or comminuted fractures, with destruction of surrounding tissues. The experience of modern army surgeons as to the question of amputation is that when necessary it should be primary.

Gunshot Wounds of Joints.—These are always serious, even though the joint be not opened by an external wound, as in such cases the inflammation set up may, and generally does, terminate in suppuration. Inflammation in the neighborhood of a joint may set up inflammatory action in the joint itself. In the treatment of these injuries it usually results either in amputation or excision. In cases where it seems that the limb can be saved without risking the life of the sufferer, the case may be treated as one of compound fracture, pus being let out by free incision and constant irrigation by cold water.

GYMNASTICS, from a medical point of view, have been far too much neglected in this country, or have fallen into hands little adapted for dealing with the complicated problems often submitted for determination. By gymnastics we mean both physical education and the use of muscular exercises in the cure of disease, and though these are, strictly speaking, totally distinct, it may not be amiss to take them together. Physical education has to a very great extent been left, in this country, to take care of itself. We are no worshippers of the system which would subordinate mind to matter, which would make a well-trained boating-man the most perfect being on earth, but assuredly we do not hold to the other view, that men may grow up misshapen, rickety articles, provided only their mental powers are developed to the uttermost. It is quite true that the perfection of either mental or bodily training is incompatible with the other. Nevertheless, men may fairly strive to improve their muscles as well as their minds, and we strongly hold to the notion that, other things being equal, the strong man bodily will, in the long run, beat the strong man mentally only. In point of fact, strength of body is necessary to strength of mind, and most men of great mental vigor, not necessarily of subtlety and refinement, are also men of bodily vigor. Here it may be as well to say that physical education does not mean what is sometimes described as "hardening" children. You see a miserable little wretch, shivering in the cold of winter, only half dressed, and you are told by his parents they are hardening him. Well, it is true the result may be satisfactory, but it may not; some live and do well, but a good many die in the process. Physical education means taking the material you have got, however unpromising, and making the best of it. To do so you require good food and clothing, air and exercise, and the cleanliness which comes after godliness. True, you may not be able to get all this; if not, you must do your best, but the best in this case will not be the best possible. In physical education the object aimed at is the exercise of all the muscles of the body, none assuming an undue preponderance over others. Most of our ordinary exercises cultivate one set of muscles at the expense of others, and for special training this, no doubt, is what should be aimed at; but in a process of education that is not so, there the endeavor should be to strengthen the weak rather than to foster the strong. The foundation of all physical training is, that a part grows by exercising it. The more it is wanted to do, within due limits, and provided due nourishment be supplied, the more it will be able to do. It grows by exercise; now the part of the exercise which seems to do most good is the motion. Suppose you move your arms backward and forward a score of times: these will do the muscles more good than moving backward

and forward, under greater difficulties, ten times; which leads us to the conclusion that for training purposes, especially among children, apparatus is of little value, save as a means of directing movements. Take the case of dumb-bells. They are intended to strengthen the muscles which protrude and draw back the arms. But if you use heavy dumb-bells, another thing is called into play, viz., the support of their weight, whether close to the chest, or at a distance from the body, that means the use of another set of muscles, which will not be exercised, only strained. So, too, Indian clubs, first-rate things are for opening out the chest, but if you use them too heavy you only drag and strain the muscles, speedily tiring them out instead of exercising them. But the exercises are easier with clubs than without them; with them, too, you can exercise several muscles you could not without them, and so we prefer clubs, but light ones. We had almost said, the lighter the better; at all events not heavy ones to tire the individual during exercise by weight only. It would be useless to enumerate all the various kinds of apparatus used for training. In point of fact, we think, with a pair of light wooden clubs you can do all you want. We want to enforce the point that, what is required is *motion*, and motion, if possible, of every joint and muscle. There is one caution which should not be overlooked; that is, do things by degrees, never attempt violent exercises all at once. The reason is obvious: your muscles may be strong and require little training, but a town life almost inevitably throws out of good training the heart and the lungs, though we do not seem to perceive it until we attempt some unusual exertion. How soon a city man is "pumped," for instance, if he attempts to run any distance. Really for any great exertion, heart and lungs must be trained even more carefully than muscles. And, of the two, more frequently they give way.

Medical gymnastics are totally different things. Their purpose is to train, not the whole body, but some one defective part or organ, to enable it to do its duty aright. Hence medical gymnastics imply a very intimate knowledge of the various organs of the body, and also of those that are most likely to be wrong in a given case, with the means to remedy them. This is really one of the highest departments of medicine, though as yet not fully understood. Let us take an example: but a few years ago it was supposed that squinting was due to one muscle of the eye pulling more toward one side than its antagonist on the opposite side can resist. Then the plan was to cut the muscle which dragged, and let the other get fairer play. Now it is well known that squinting is not due to any one cause, but is symptomatic of many. It may be a sign of paralysis instead of over-exertion, and so requires to be treated in very different ways ; stimulation by a galvanic battery, so as to exercise the muscle at fault, being one of them. That is medical gymnastics applied to one particular muscle. There are scores of other instances we might name, but this will suffice.

GYPSUM. (See PLASTER OF PARIS.)

H.

HÆMOPTYSIS is a term applied to coughing up blood. It comes on in the course of many diseases of the lungs and air-passages. Sometimes it is very small in quantity, and only streaks of blood are found in the phlegm, at other times the flow of blood is so excessive as to cause sudden death. In pneumonia, or inflammation of the lung, blood is always found in the sputum, giving it a rusty or lemon-color look ; in the early stages of consumption blood is frequently coughed up, but never to cause a fatal result ; in the later stages of this disease, when cavities have formed in the lungs, a large vessel may give way and cause hæmorrhage which cannot be stopped. The bursting of an aneurism into the air passages, or rupture of the pulmonary artery through ulceration of the bronchus, may set up fatal hemoptysis. In many cases of heart disease the vessels in the lungs become so engorged from the circulation being obstructed that blood is often found in the sputum during the last few days of life ; nothing can be done in these cases with any permanent benefit ; now and then, bleeding from the arm will, for a time, relieve the over-full vessels. In some forms of bronchitis, hemorrhage occurs, but very rarely, and to no great extent. Warty growths, ulcers, and cancerous disease of the larynx, and trachea or bronchi, may cause the patient to cough up blood ; this symptom generally accompanies cancer of the lung. Under the different diseases here mentioned, the nature of this complaint will be more fully detailed.

Treatment.—The treatment consists in perfect rest in bed, the head and shoulders being generally propped up, as the patient cannot lie down in comfort. Ice broken up and applied in a bladder to the spine or front of the chest is often of service, or small pieces may be slowly dissolved in the mouth. No speaking should be allowed, nor any exertion whatever on the part of the patient ; the room should be warm (60°-65° Fahr.), and the air rather moist, so that any irritation from external cold may be allayed. Inhalation of turpentine vapor is perhaps the most valuable remedy ; a jug may be nearly filled with boiling water and a tablespoonful of turpentine put into it ; the patient should then hold his mouth over the jug and inhale the steam, which will carry with it the turpentine vapor ; better still, to use inhalers which are manufactured for the purpose, and may be had of any druggist. Various astringent medicines have been given internally, as iron, tannin, gallic acid, acetate of lead, etc., but they are more useful in cases of hemorrhage from the stomach and bowels. Opium is of service in quieting the circulation, and in allaying the nervous excitement which is generally associated with hæmoptysis. When the bleeding has ceased, rest should, for some days, be carefully enjoined, and any exposure to cold or other exciting cause should be avoided. There will be anæmia and debility afterward to a greater or less degree, and these must be combated by appropriate tonic medicines.

HÆMORRHAGE is an escape of blood from an artery or a vein, whether as the result of a wound or from some pathological cause, such as ulceration. Arterial hæmorrhage is recognized by the blood escaping in jets, and being of a bright red color ; venous hæmorrhage by an oozing of black blood, although in the instance of some of the large veins being wounded, and the wound opening superficially, the term *oozing* is converted into a *rush*. In the case of a cut or wounded artery, the rush of blood corresponds with the beats of the pulse, and there is a provision of nature to attempt at a closure of such a wound, inasmuch as the divided vessel contracts so as to narrow its orifice, and moreover, contracts into its natural sheath ; again, there is a coagulation of the blood at the point of injury, and the column of blood, inside the vessel, coagulates up to the nearest branch given off from the vessel, and the force of the heart's action is materially diminished by fainting. In the matter of *treatment* of hæmorrhage from an artery, the first indication is obviously to cut off the supply from the heart by applying some method of compression between the wound and the heart (see ACCIDENTS). This may be done either by pressure by the finger, in the course of the vessel (which follows the inner axis of a limb), by a *tourniquet*, or by tying a handkerchief round the limb, with a stone placed in it *over* the artery, and twisting the handkerchief tight with a stick. This arrests *immediate* danger ; the surgeon, however, performs the operation of *ligature*, by cutting down upon the wounded vessel in its track, and placing a hempen or silken ligature upon it above the seat of injury ; an operation, of course, demanding anatomical knowledge and judgment. In the case of bleeding from a superficial small artery, if pressure does not control it, it may be caught up with a pair of forceps and twisted (*torsion*), or its end picked up and tied, or a needle may be placed underneath it, and a loop of silk applied over the ends of the needle, and the inclosed tissue containing the bleeding vessel (*acu-pressure*).   In the case of hæmorrhage from a *vein*, ordinarily it may be restrained by pressure at the spot, either by a bandage, and a graduated compress, that is, a pad made of conical shape, applied with its apex downward, or by unremitting pressure of the finger, or in very severe cases by acu-pressure, or ligature, as in the case of arterial hæmorrhage.

Hæmorrhage from the nose, or *epistaxis*. (See EPISTAXIS.)

Hæmorrhage from the kidneys, is the result of disease, such as calculi, or blows, or the congestion consequent on scarlatina.   From the bladder or prostate gland, by clumsy catheterism, stone, or malignant disease in the instance of the blood coming from the bladder, it continues to flow after the urine is voided ; the microscope will reveal kidney mischief, and local symptoms, such as pain in the back, would assist in the diagnosis. In kidney-hæmorrhage, tinctura ferri perchloridi, or gallic acid, are of chief use, or if there be inflammatory symptoms, cupping, purging, and the administration of acetate of lead.   In bladder-hæmorrhage, a catheter should be passed and tied in, and small doses of turpentine administered.

Hæmorrhage from the urethra sometimes occurs as a result of forcible catheterism, or during chordee ; generally a recumbent position will check it ; if not, pressure, far back in the perinæum, cold, or the injection of tannin or gallic acid are of value.

Hæmorrhage from the rectum is caused either by the bursting of a varicose vein in piles, or from the vascular surface of internal piles, induced by defecation. Should the hæmorrhage result from piles, those piles should be operated on, and astringent applications, such as bark injections, be used.   Internally, bark and sulphuric acid, or copaiba. If the hæmorrhage be very violent, the rectum may be plugged with a cork, having some styptic applied to it, or lint or ice be thrust up the cavity. Occasionally it is necessary to distend the rectum by a speculum, and endeavor to find some bleeding point, to which a ligature or an actual cautery must be applied.   Frequently, hæmorrhage from the rectum is an evacuation, which affords relief in plethoric individuals, to be combated by exercise, temperance, and aperients.

Secondary Hæmorrhage is bleeding which comes on some while after the receipt of an injury or operation. The most simple form is that which occurs after reaction has set in ; thus, after a wound has been dressed, and the small arteries will burst out bleeding. The wound must be opened up again, and bleeding vessels tied, and, if necessary, the surfaces sponged with cold water. Another form is that which occurs from a wound in an artery

(generally of the lower extremity), when a ligature has been placed upon it, above the wound ; although the hæmorrhage is controlled for a while, after the collateral circulation has been established, the blood will find its way back out of the original wound.   The blood is generally of a venous or dark color, and oozes out. If the original lesion, however, be properly treated, this form of secondary hæmorrhage would not have happened, as the vessel should have been secured both above and below the wound.   Again, secondary hæmorrhage may occur from sloughing, or from the imperfect closure of an artery at the point of ligature, at the time when this ligature comes away, which may happen from the roots of the vessel being in a diseased condition, or from some constitutional malady which prevents the proper adhesion of the coats. The extraction of teeth sometimes sets up very severe hæmorrhage, which may arise from the dental artery or from the gum. A small piece of cotton wool, soaked in perchloride of iron, stuffed into the cavity will generally suffice to stop it, but in obstinate cases a very firm plug of lint should be pressed in, with a compress over it, so that by binding the jaws together by a bandage, considerable pressure is exerted upon the bleeding point.

HÆMORRHAGE, CEREBRAL, is caused by the rupture of a vessel in the brain, in consequence of which blood is poured out into the tissue around ; the danger is generally in proportion to the quantity effused, but a small bleeding into the medulla or pons, where most important nervous centres are placed, is nearly always fatal.   The seat of hæmorrhage is, as a rule, in or near the corpus striatum or optic thalamus, and only one side is affected at a time. Very profuse hæmorrhage may escape into the ventricles and kill in a few hours.   The person attacked will then fall down in an apoplectic fit and lie in a state of coma ; or the bleeding may be confined to a very small area, and the patient will recover and find the arm and leg of one side paralyzed. (HEMIPLEGIA.) This condition generally comes in persons of middle life or old age, and is generally associated with diseased vessels ; it occurs in those who have led intemperate lives, and in those who suffer from chronic Bright's disease.

HÆMATEMESIS, or VOMITING OF BLOOD, is a symptom of grave importance. It occurs under various conditions ; sometimes from ulcer of the stomach or from cancer of that organ, or from the mechanical congestion caused by disease of the heart or liver, more rarely from the bursting of an aneurism into the œsophagus or stomach. The blood vomited is generally of a coffee-ground color and appearance from being acted upon by the gastric juice, but if a large quantity is suddenly poured out from a ruptured vessel and vomited at once, it will have a dark clotted character. If due to heart disease, the effused blood is seldom large in amount ; the capillaries and veins of the stomach share in the general congested state of the body, and the coats of some of them become ruptured from over-distension. In such cases there will also be dropsy of the legs and abdominal cavity, and the usual signs of heart disease. Treatment is of little avail, and death speedily occurs, not because of the loss of blood from the stomach, but of the general condition of the body, of which the hæmatemesis only forms a part. In long-standing disease of the kidneys, and more especially in that form which is met with in gouty people, hemorrhage from the stomach is often present ; headache, bleeding from the nose, sickness, a sallow complexion, and the presence of albumen in the urine are generally associated phenomena. The treatment consists in avoiding stimulants, keeping up a free action of the skin and bowels by promoting perspiration and giving aperients, and in this way relieving the kidneys, which are the source of the mischief. Cirrhosis, or contraction of the liver (see CIRRHOSIS), often causes hæmatemesis by mechanically obstructing the flow of blood; all the blood from the stomach passes into the portal vein and then on to the liver, but if the latter organ is so diseased as to prevent its ready transit, the veins of the stomach will become over-distended and rupture occasionally. There will, therefore, be the signs due to a cirrhosed liver before the bleeding occurs ; dyspepsia, loss of appetite, sickness, a jaundiced or sallow skin, and perhaps ascites. The treatment will consist in keeping the bowels open, so as to relieve the liver, giving light and nourishing diet, and making the patient lie down quietly. In these cases, as in those depending on heart disease, the bleeding ought not to be stopped, as it really relieves the dilated vessels, and if it could be stopped in one place it would only break out in another. But if the hæmatemesis depend on an ulcer or cancer of the stomach, it should be put a stop to as soon as possible. (See CANCER and GASTRIC ULCER.) Vomiting is very common in some kinds of hysteria, but there is not often any blood present ; when there is it appears in red streaks, and generally proceeds from the gums, decayed teeth, or back of the throat ; such people do not lose flesh, although they seem to be constantly sick ; they are generally young females, and are suffering from other signs of hysteria. The bleeding is extremely small in quantity, and can do no harm ; the inside of the mouth should be examined carefully, as the bleeding may be done on purpose by scratching the gums, etc., with a pin or with the finger nails ; detecting the imposture is the surest way to cure it. Hæmatemesis from the rup-

ture of an aneurism would at once prove fatal.

HÆMATURIA signifies that blood is present in the urine. It is as a general rule, a grave symptom, as it implies that there is some disease going on in the kidneys. A common cause is some severe blow on the loins, and may be produced by falling backward, or by a direct blow being given over that part. Great pain is at once felt, and is often accompanied by sickness with a sensation of faintness and inability to walk erect. In a few hours the patient finds, on passing water, that there is more or less blood mixed with it, so that it is almost the color of porter. He may need to pass water more frequently, and will most likely have pain extending from the loins down by the groin into the thigh. The blood is present, because, as a result of the fall or blow, some of the vessels of the kidney have become ruptured, just in the same way as the vessels of the skin are ruptured in the case of a severe bruise, only as the kidney is a more important organ, serious results may arise. The patient should be at once put to bed and told to make no exertion whatever for several days, as perfect rest is the best and most useful treatment. To ease the pain, hot flannels, or flannels coming out of hot water and put across the loins, are very beneficial, and if there is much pain and distension of the abdomen, they may be applied in front also. Morphia, or some preparation of opium, may be given under medical advice. The bowels may be kept moderately open, and very plain, simple diet should be given. Nothing should be taken to cause any irritation to the kidney ; as the patient will most likely be thirsty and feel sick, iced-milk and water is very grateful ; beef-tea, arrow-root tea, and barley-water may be also taken. As soon as the urine assumes a lighter color it shows that less blood is being passed, and that an improvement is taking place ; this may occur at the end of a week, and generally in three or four weeks the urine has resumed its usual color, or merely looks a little cloudy when held to the light. So long as any blood is present the urine will give a white, heavy, flocculent precipitate of albumen, when a small quantity is placed in a glass test-tube and boiled over the flame of a spirit-lamp. This is a very simple test, and to insure a good recovery the patient ought to be kept to his bed, or at least to his bedroom, until all traces of albumen are gone. If any exertion is undergone too soon the bleeding is apt to recur, and may, in fact, lay the foundation of serious mischief in the future. For some time the patient will feel weak and will be pale ; to remedy this tonics are of great value ; all preparations containing iron are good, and are useful in checking the hemorrhage ; when the tongue is clean and the invalid feels hungry, solid but light and nutritious food may be given.

Hæmaturia may arise from the presence of a stone in the kidney, and this may be induced by severe exercise, such as riding or driving over rough ground. (See GRAVEL and STONE.)

Hæmaturia is very often met with after scarlet fever, and forms part of the disease known as "dropsy after scarlet fever." It comes on about a fortnight or three weeks after the scarlatinal eruption has appeared, and is often due to a chill being taken while the skin is peeling. The urine suddenly becomes bloody, and rather less than usual is passed; there may be slight aching pains in the loins, but not to any great extent; there is often with this condition a puffiness of the eyes and feet, which is caused by an effusion of serum under the skin of those parts. When the urine stands a short time a considerable amount of dark brown flocculent sediment is observed, and on boiling, as before described, a good deal of albumen will come down. It is of no use applying turpentine embrocations or blisters in these cases, nay, it is actually dangerous, as they may aggravate the congested state of the kidneys from being absorbed into the blood and causing increased irritation. Hot fomentations may be applied and the patient placed in a hot-bath, so as to encourage sweating, and so making the skin do the work of the kidneys. The bowels should be kept gently open. This disease is most common in children, and may often be accompanied with convulsions, but this is more fully discussed under SCARLET FEVER. Hæmaturia may occur at the very onset of scarlet fever, but is not then a sign of much importance; it is found to occur slightly in nearly every severe case of fever, but in such cases rarely calls for treatment. It may come on after taking turpentine or cantharides (the Spanish fly); these are really poisonous agents, and have been occasionally given for that purpose. They cause a stoppage in the amount of urine passed, with great pain and vomiting, and often serious results.

Blood may occasionally appear in the urine from eating indigestible food, from over-exertion, and in some people it is met with in very hot weather, when all the tissues of the body seemed relaxed. In such cases the hæmaturia will only last a few hours, and may be cured by finding out the cause and avoiding it. In the diseases known as scurvy and purpura, blood may appear in the urine in a similar way to the manner in which blood is effused under the skin. More recently the name "intermittent hæmaturia" has been given to a disease where, for a few hours, blood will suddenly appear in the urine and as suddenly disappear; such persons are generally very sallow, and know when an attack

is threatened by feeling a shivering fit come on; in some, getting out of bed and going out into the cold air will bring on an attack at once. Good diet, rest, and improving the general health are the chief things to be done in such cases.

HÆMORRHOIDS are swellings, most commonly known by the name of Piles, which are situated in the region of the anus, and which by their size and their liability to irritation and inflammation, cause much trouble and uneasiness, and sometimes intense pain. These swellings may be formed either by circumscribed thickening of the skin just without the anus, or of enlarged folds of the mucous membrane of the terminal portion of the gut, which folds are often protruded from the anus. In the former case the affection is called external piles, in the latter internal piles. *External piles* consist in a collection just without the margin of the anus of rounded hard tumors covered by thickened skin, and of prominent ridges of skin. These growths at first cause little or no pain, but as they increase in size and number the patients complain of difficulty in passing the motions, of bearing down pains, of a sense of weight about the anus, and of a general feeling of discomfort. After a time one or more of these piles may become irritated and inflamed, and then give rise to very acute pain, with throbbing and a sense of great heat, and to a constant desire to go to stool. These symptoms pass off in the course of three or four days, but the attacks are frequently renewed and the piles gradually enlarge and invade the lower portion of the intestine. This affection orginates in distension of the veins about the anus in consequence of obstruction to the circulation. It is met with generally in those who follow sedentary employments, and those who, in consequence of indulgence in highly-seasoned food and in alcoholic drinks, suffer from congestion of the liver. Much horse exercise, long-continued standing, and constipation, are also causes of external piles. The presence within the anus of large rounded and soft tumors covered by red mucous membrane is attended with more serious symptoms. These *internal piles*, when large, come down through the anus from time to time, generally when the patient is at stool, and become engorged with blood and very painful. Evacuation of the bowels gives rise to a burning or throbbing sensation, and as the piles increase in size becomes more and more difficult. A dull pain across the loins is complained of, and occasionally the urine cannot be passed in consequence of irritation at the neck of the bladder. The most serious symptom is bleeding, which occurs during evacuation of the bowels, when the piles are protruded and compressed by the anus. The blood is red and arterial, and is often passed in considerable quantity. Patients often remain ignorant during a long period of this frequently renewed loss, and finally suffer from extreme debility, become irritable and restless, and present a blanched countenance and a weak and quick pulse. In addition to the discharge of blood there is in most cases a constant flow of thick slimy or purulent fluid. On examination of the region of the anus there will be seen, as the patient bears down, one or more rounded protrusions of a dark red or livid blue color, and varying from the size of a currant to that of a small chestnut. These growths, like external piles, are sometimes inflamed. Then, in addition to intense pain and other severe local symptoms, there is high fever. Inflammation of internal piles sometimes ends in mortification and in expulsion of the mass of abnormal growths from the rectum. The causes of internal are similar to those of external piles. Congestion of the liver causing venous obstruction in the intestines, and direct irritation of the walls of the intestine, are the conditions which most frequently give rise to this affection. The latter condition is often due to an immoderate use of strong purgatives, especially aloes.

Treatment.—The general treatment of piles, both internal and external, consists in removing congestion of the veins of the liver and intestines, in keeping up the strength and health of the patient, and in avoiding or alleviating the results of certain conditions favorable to the development of the disease. The patient should restrict himself to a carefully regulated and temperate diet, and abstain from highly seasoned dishes, pastry, and spirits ; wine and beer ought not to be taken, except in moderation. Walking exercise is to be recommended, and, during the summer months, sitting in the open air. Riding on horseback or in a jolting vehicle is to be regarded as positively injurious. The affected region should be well bathed every morning with cold water and then carefully dried. To external piles may be applied lead lotion or a weak solution of alum. For both external and internal piles the compound gall ointment is a very useful application. When internal piles protrude after every evacuation, they should then be sponged over with cold water or a solution of alum, or be smeared with gall ointment. Great attention should be paid to the state of the bowels, which ought to be kept in daily action by some mild aperient, as rhubarb in the form of a pill to be taken at night, or confection of senna, castor oil, seidlitz or Püllna water, to be taken in the morning before breakfast. In cases of inflammation and great pain in external and internal piles, leeches should be applied to the skin at some distance from the anus, and bran poultices or poppy-head fomentations be placed over the whole of the

affected region. When a patient with external piles complains of almost intolerable pain in one pile which is found to be swollen, tense, and livid, an incision into this with the point of a sharp knife will often let out a small dark-red clot of blood, and give immediate and total relief. By these means the bad effects of both external and internal piles may be much relieved, or, as occasionally takes place, the disease may be permanently cured. When, however, in spite of careful attention to diet and to local ablution, the affection increases in extent and intensity, and becomes to the patient not only a source of annoyance but also a hindrance to the pursuit of his means of support, it will become necessary for him to undergo some surgical operation in order to obtain permanent relief. External piles are generally treated by excision, the tumors, together with the adjacent ridges of thickened skin, being removed with large curved scissors. Internal piles have been treated by various operative methods ; many surgeons apply a ligature round the base or contracted portion of each pile, other surgeons prefer to cut away the pile and then to apply to the raw surface the red-hot iron. Fuming nitric acid is often applied to the surfaces of small internal piles. In cutting operations upon external piles the surgeon, while endeavoring to obtain for the patient effectual relief, is careful not to take away too much of the skin lest contraction of the anus should follow the shrinking of the scar. In these operations, but more especially in those consisting in excision or incision of internal piles, the bleeding is very free, and, if it should recur in the absence of a medical man, dangerous to life.

HAIR, Diseases of.—The hair is subject to alterations in its growth which correspond with those through which the body passes. Thus after fevers or exceedingly acute diseases, the hair, which during the period of the disease has remained stationary in its growth, generally falls off and a new growth begins, which at first very frequently differs in its characters from the hair before illness. Then, too, it usually grows faster for a time. In some individuals the growth of the hair and the character of the hair are very different from those found in others. Usually a good growth of strong hair may be taken as a sign of a vigorous constitution. A thin crop of sandy, that is, imperfectly colored hair, commonly marks one in whom the original force of bodily growth has been deficient. Thinning of the hair may take place from a variety of causes, and very often precedes absolute baldness. (See BALDNESS.) The baldness of old people is generally preceded by alteration in the color of the hair, which becomes gray. It generally falls first from the very top of the head, and thence gradually spreads. Baldness in young people arising from parasitic disease is not preceded by alteration in color. Generally there is some local irritation, and in this spot the hair begins to fall. This is usually on the side of the head, and frequently the hair is broken and stubbly round the bald spot. This form of baldness is commonly called ringworm (which see). Baldness almost invariably follows syphilis during the period of secondary eruption. This baldness is sometimes sudden and very complete, the hair coming out literally by handfuls. Such a loss adds greatly to the mental disquiet of the sufferer, but he may be comforted, for it is almost absolutely certain to grow again. When hair falls through disturbance of the constitution, that must be seen to ; the hair may be left to take care of itself. It is mostly the custom to give arsenic in such cases, and frequently when there is much nervous debility arsenic does good, but not because of its fancied action on the skin, rather because it is a really good, serviceable tonic. Iron, quinine, and strychnine will generally do good. The chance of local measures doing any good will entirely depend on the condition of the hair follicles. If these be totally wasted, local measures will do no good, but if they are in a condition to respond to stimulation, local applications may restore a goodly head of hair. If downy hairs are visible these may usually be made to grow by stimulation ; even if they are entirely absent, good may be done if the scalp look at all natural. If white, shiny, and with little fat below the skin, there is not much hope. If, too, the scalp be swollen or thickened, some local application will be required. The best for this last is tincture of iodine, but it must be used with caution. When there is a chance of getting the hair to grow again, stimulants may be used. If there are downy hairs, let the head be shaved and a blister lightly applied, for of all stimulants to the growth of hair, Spanish flies (cantharides) are the best. The application of a blister may seem a harsh remedy, but experience has shown it to be the best ; if one does not want hair, he need not submit to the blister. These must be used repeatedly if necessary. Where the hair is thinned only, the first thing is to restore the scalp to a healthy condition. The scurf should be got rid of by bathing with tepid water night and morning, and the constant application of glycerine and lime-water in the interval. No fats or oils are to be used at this period, as they are apt to go rancid and so irritate the irritated scalp still more, but glycerine and lime-water, or *fresh* olive oil and lime-water may be used. Cantharides enter into most stimulant applications to the hair, and a very good compound for gently acting upon the hair follicles is to be obtained by adding a few drops of tincture of cantharides to toilet vinegar, and gently

damping the scalp after it has been well washed with the compound.

Most of the advertised pomades and hair restorers are humbugs ; they are got up mainly for enabling hairdressers to impose upon their customers, which some of them are not slow to do. Some of these hair washes, too, contain lead, though said not to do so by the venders and manufacturers ; these are especially to be avoided. In short, the principles above laid down are those which must guide any individual in dealing with the hairy scalp. Each may apply them for himself.

HALLUCINATION. (See INSANITY.)

HANGING.—The cause of death by hanging is suffocation, from the pressure of the rope upon the trachea preventing the admission of air into the lungs, which is common to death by strangulation also, and the explanation that the immediate cause of death by hanging is pressure on the nerves subordinate to the function of respiration, may be rejected on the grounds that such pressure does not prove fatal for many hours. In certain cases of suspension death takes place very suddenly, and this may arise from two causes, fear, producing syncope, or from injury to the spinal cord, by dislocation of the cervical vertebræ, fracture of the odontoid process of the axis, or second cervical vertebra, or rupture of the intervertebral substance. These injuries to the spine are due to the fall of the body from some considerable height, or to a twist given to the body at the time of the fall ; these details are observed in legal executions, with a view of producing death as suddenly as possible. Death from apnœa is next in order of rapidity ; and the least rapid, that produced by apoplexy, induced by the pressure upon the great vessels of the neck.

To restore one found hanging, the body should be immediately cut down, and the knot or loop eased from around the neck. Cold water should be forcibly dashed over the face and chest, artificial respiration (see DROWNING) should be employed. Blood should be taken from the external jugular vein, if there be turgidity of the face, and a galvanic current passed from the nape of the neck to the pit of the stomach, to excite the diaphragm through the course of the phrenic nerves, that is, just in the hollow above the collar-bones. It is obvious, that in the case of dislocation or fracture of the cervical spine, these measures are unavailing, but in cases where the hanging has taken place very recently, or where there has been a very short fall, such measures should be most assiduously applied ; many cases of attempted suicidal hanging have recovered by prompt attention.

HARE-LIP is a congenital fissure of the upper lip, dependent on an arrest of development of the structures forming the upper lip

or its bony support. There are two species
of hare-lip, the *simple* and *complicated*. Of
the simple, there are two or three varieties.
If there is one, it is almost always on one
side of the mesial line, though it has been observed in the mesial line. If two fissures
occur, they are usually lateral, and isolate a
labial segment ; this segment is sometimes
atrophied, sometimes hypertrophied, in either,
attached directly to the tip of the nose or by
a short septal pillar. The varieties of complicated hare-lip are, 1st. A single fissure of
the lip, and a simple fissure of the alveolar
margin of the jaw, the fissure being either
median, or at some one of the lines of junction of the intermaxillary and maxillary segments, existing at the time of development ;
this split in the alveolar ridge corresponds
with that in the lip. 2d. A single fissure of
the lip may be coexistent with a separation
of the opposing edges of the alveolar cleft,
constantly associated with cleft palate. This
variety is most frequently met with on the
left side. 3d. The most frequent variety is
that in which both lateral segments and the
maxillary bone are but imperfectly developed,
while the intermaxillary attains its normal
size and position, a double fissure of the lip
is connected with this condition of the jaw ;
while the intermediate portion is generally
more or less of an oval or rounded shape,
and curling upward toward the nostril, leaves
the incisor teeth uncovered in so unsightly a
manner as to have given rise to this deformity being called "woll's jaw." 4th. With a
double fissure may be entire absence of this
intervening lobe, and of course, absence of
the incisive bone.

Treatment.—Most modern surgeons agree
that operative proceedings for the cure or relief of this condition should be had recourse
to as early after birth as possible ; infants
have been operated upon a few days or hours
after birth with complete success, and although many cases appear, at first sight, to
be impracticable of any relief, so much science and ingenuity has been expended upon
perfecting surgical operations of this nature,
and the physiological and pathological conditions of arrested development have of late attracted so much attention, that no case can
be regarded as incurable, though the various
processes and stages of the operation may
last over some considerable time. Operations on adults are more easy of performance
than on children, for the reason that they
have command over themselves, by the exercise of the will, to keep as still as possible
during the proceeding. Children should always be chloroformed.

HARTSHORN. (See AMMONIA.)

HAY-FEVER, HAY ASTHMA, or HAY
COLD, is a peculiar disease to which some
people are subject in the months of June or
July, or August and September, or when the
hay season is about. It consists in excessive
irritation of the eyes, nose, and the whole of
the air-passages ; producing, in succession,
itching of the eyes and nose, much sneezing
occurring in paroxysms, with a copious flow
from the nostrils ; pricking sensations in the
throat ; cough, tightness of the chest, and
difficulty of breathing, with or without mucous expectoration ; and more or less fever
and constitutional disturbance. This disorder
is confined to a very few people, and only
comes on in them when they are in the way
of hay-fields, etc., for if absent at sea they
will escape an attack. Such people have not
been noticed to be peculiarly susceptible to
catarrh, nor is any satisfactory explanation to
be given of its origin, save that the cause is
contained in the atmosphere and is of vegetable origin. It affects both sexes, and generally occurs in the adult. In some cases the
affection occurs in successive years precisely
at the same period, and has a uniform duration.

Treatment.—The only treatment that
seems effective is such a change as will carry
the patient beyond the reach of the provoking
cause of the disease. A sea-voyage affords
instant and complete relief. There are also
portions of the White Mountains, the Adiron-
dacks, and Colorado that are said to afford
entire exemption ; and Dr. Morrill Wyman
(who has carefully studied the disease) thinks
that one liable to it may secure relief by going
to any point that is 800 feet above the sea.

HEAD. (See SCALP.)

HEADACHE, technically called cephalalgia, may for the most part be looked upon as
a symptom, though sometimes also the most
important part of the disease. Certain pains
in the head, not commonly called headache,
are rheumatic and neuralgic affections of the
scalp, though these are sometimes also spoken
of as rheumatic and neuralgic forms of headache. Pains from syphilitic nodes may usually be distinguished from true headaches by
the great tenderness over localized points ; in
rheumatism of the scalp, too, there is usually
some tenderness, but more d·ffused than in
the preceding. The pain of acute inflammation may readily be distinguished by the other
and more characteristic features of the malady. Pain in the head is a very constant
but by no means invariable symptom of disease of the brain and its membranes. This,
if persistent and accompanied by vomiting,
especially if the pupils be affected, is a grave
sign. Inflammation of the membrane gives
rise to more pain than does inflammation of
the substance of the brain, and it is usually
of a sharper description in the former than in
the latter. Another form of headache is described as congestive or plethoric. It is not
unfrequently occasioned from over brain-
work. It attacks a different part of the head
than do most others, giving rise to a feeling

of tightness across the head, and a fulness and whizzing behind the ears. It also occurs in females of full habit of body, in whom the menstrual function is defective or in abeyance, and is especially troublesome about the period when this stops, when it not unfrequently leads to that form of headache indicative of brain disease by insensible gradations.

Perhaps the most common form of headache is that connected with indigestion. It is sometimes called sick-headache, sometimes bilious headache. Neither is a very good appellation, for there would seem to be also a purely nervous form of sick headache, of even more distressing character than this, and the headache here spoken of is by no means invariably accompanied or followed by biliousness. This form of headache most commonly follows some indiscretion or excess in diet, and is generally worst in the morning ; sometimes, however, it becomes aggravated during the following day. The pain is very severe, of a throbbing or bursting character, and the sickness is intense. Sometimes this sickness ends in vomiting, and then the pain in the head is greatly aggravated for a time, but by and by the sickness having carried off offending matters lying in the stomach, the patient probably falls asleep and awakes refreshed. In point of fact, such headaches seldom last long, and are generally cured by a good long sleep. The best thing to do, in point of fact, is to go to bed, if this is possible, for the work done with such a headache is not very good. Occasionally these headaches become almost constant ; if so, there is permanent derangement of the digestive organs, which must be set right before anything else is done, or before any permanent amendment can be secured.

There is still another form of headache very common. For want of a better name it is called nervous, and is very common in women. In point of fact, just as the preceding might be called the men's headache, so this might be called the women's headache, though it also occurs in weak and delicate males, or those who have been overworked. Not unfrequently this form of headache depends on some distant irritation, as decayed teeth, which must be removed before any good is done. Headaches are sometimes classified by their site, which indeed aids us to determine sometimes their origin. Thus, bilious headaches, if slight, are commonly confined to the forehead, others, especially the nervous forms, are felt more on one side than another. Occipital headaches are perhaps most frequent in the outset of fevers and such like acute diseases. Patients who have resided in marshy districts, and many, who as far as we know have not, are subject to attacks of headache which recur at definite periods. This, however, is not uncommon in pure neuralgia, and so may have no malarious origin. Hysteri-

cal girls are often subject to acute attacks of pain in one particular spot, which has been likened to driving a nail into the head, hence the Latin name clavus has been given to it. Purely neuralgic pains are more distinctly confined to the line of certain nerves than are the others. As headaches are seldom deadly, sufferers do not get always the sympathy this pain deserves, and in many instances they are themselves the authors of their own misfortunes. Broadly speaking, indigestion in some form or other is at the root of a majority of headaches ; if that is present, it should be remedied, for this is usually the first step in getting rid of nervous headaches as well as those due to what is called biliousness.

Treatment.—The treatment resolves itself into two parts, the treatment of an acute attack, and of the general condition which has given rise to it. For a bilious headache, if known to be due to any error in diet or excess in liquor, perhaps the best thing, if the patient can make up his mind to it, is an emetic of zinc and ipecacuanha. Whether this be taken or no, the patient must be kept absolutely quiet, with the eyes shut, or in a darkened room, and pressure on the temples, so as to arrest the beating of the arteries, will often give relief. By and by he will probably fall asleep, however severe the pain may seem, and he will generally awake tolerably well. A cup of strong black coffee, a walk and a fast for six or eight hours will generally suffice to set everything to rights. When the patient is suffering from nervous depression, as often happens, drinking a draught containing a dram of the aromatic spirit of ammonia, with a little gentian, will generally do much good. In the same condition, sniffing a bottle of strong smelling-salts will often relieve the pain. This holds true of the purely nervous headache also. In some cases cold is grateful, in others heat. Cold is best applied by means of eau de Cologne, or some similar spirit, of course still better by ether ; ice or iced water may also be employed. If heat is best borne, let it be applied by heating a quantity of salt in a fire shovel, and binding it in a handkerchief round the temples. If the headaches be purely neuralgic, recurring at different intervals, a full dose of quinine, five grains at least, should be given as well as a little spirit of ammonia. Some sedative should be applied locally, and if necessary, some should be given subcutaneously. For after-treatment in the bilious form, some laxative should be given, blue-pill if there is much tendency to biliousness, ipecacuanha and rhubarb if there is not. Podophyllin is very good for relieving congestion of the liver, and effectually emptying the gall-bladder. A dose of compound rhubarb powder every morning, or just before dinner in smaller quantities, is a very good remedy. Diet too must be regulated ; it must be plain and unstimulating ; beer, port, sherry, and spirits should be avoided, but claret and other light wines may be taken. Exercise should be taken, especially in the open air, and indolent habits should be got rid of. Where the headaches are purely nervous and depend on weakness, a totally different line of treatment must be adopted. The bowels must be regulated, it is true, but the remedies to be used are strengthening, and nux vomica or strychnine is one of the best, especially with nitro-hydrochloric acid. If acids do not suit, liquor potassæ and compound infusion of gentian should be given. Finally, in a considerable number of cases, especially in women, no drug does so much good as sal ammoniac (chloride of ammonium); its mode of action is not quite plain. In these too, if the tongue is foul, rhubarb, ginger, and bicarbonate of potass will do good.

HEALTH. (See HYGIENE.)

HEALTH RESORTS are partly dealt with under CLIMATE and MINERAL WATERS. The places here referred to acquire importance more from their atmospheric qualities than from anything else, and many, we fear, are hardly in a condition to be called health resorts at all, for the unsanitary state of some of them is alike to be dreaded and wondered at. Change of air is in many instances the clew to the benefits derived from health resorts. Thus frequently those living in inland places, remarkably healthy in every way, are often benefited by a visit to the seaside, while those living near the sea are benefited by a visit to the hills. Then, too, for the inhabitants of large cities, the great thing is to get out of town, whether to the country or the seaside.

In this country, aside from the mineral springs, the so-called health resorts are for the most part places which are supposed to benefit those who are suffering from pulmonary consumption ; and their indispensable requirements are warmth, dryness, and equability of temperature. Those most frequented and most often recommended are in FLORIDA. The climate of Florida is one of the finest in the world : though 10 degrees nearer the equator than southern Italy, the temperature is no warmer, and the air far more equable and dry. At Jacksonville the average mean temperature, as fixed by observations extending over several years, is 69.6° ; the highest temperature (in July) is 83.4° ; the lowest (in January) is 52.7°. In the south the temperature scarcely changes the year round, and summer is only distinguished by the copiousness of its showers. The average mean temperature of the State is 73° F., and the difference between summer and winter does not generally exceed 20°, while at Key West it is not more than 11°. The thermometer seldom rises above 90° in summer, and rarely falls

below 30° in winter ; on the average the winters are thirty or forty degrees warmer than in New York, while the summer months of the latter are ten or fifteen degrees hotter than in Florida. Frost is unknown in southern Florida, and is comparatively light even in the northern part of the State. It occurs most frequently between November and March, being most frequent in December and January, and rarely showing itself in October and April as far north as Jacksonville. As a general thing no frost occurs throughout the year below latitude 28° N. Summer being the rainy season in Florida, the winters are usually clear and dry. In addition to the mildness of the climate, it is believed that the immense pine-forests which cover a large part of the State contribute greatly to its healthfulness. The delicious terebinthine odors exhaled by these forests not only purify the atmosphere, but impart to it a healing, soothing, and peculiarly invigorating quality.

Of the principal resorts in Florida those on the St. John's River are *Jacksonville*, *Magnolia*, *Green Cove Springs* (sulphur waters), *Pilatka*, *San Mateo*, *Mellonville*, *Sanford*, and *Enterprise*. On the Atlantic coast are *St. Augustine* (with a singularly equable climate and a mean winter temperature of 58.08°), and *Fernandina* (whose air is bracing but considered too strong for consumptives in advanced stages of the disease). On the Gulf Coast are *Cedar Keys* (with a climate blander than that of Jacksonville, and beneficial to rheumatism as well as consumption), *Tampa*, which is growing in favor, *Manatee*, *Charlotte Harbor*, and *Key West*, where the thermometer seldom rises above 90° in summer and never falls to freezing point, rarely standing as low as 50°. Its mean winter temperature is 69.58°. In the interior, *Gainesville*, *Waldo*, *Lake City*, *Monticello*, and *Tallahassee* (the capital of the State), are much esteemed ; and the entire Indian River country has a delicious climate.

The climate of GEORGIA is less equable than that of Florida. In the lowlands in summer it is hot and unhealthy, and malarious fevers are prevalent ; but in the pinelands farther back the air is salubrious, while in the northern portion of the State the summers are always cool and healthful. The winter climate is delightful, especially in the eastern and southern districts ; the days are bright and sunny, with little variation in the temperature, and the atmosphere is dry and balmy. The mean annual temperature at Savannah, in the south-eastern part of the State, is 66.2°, and of Augusta, in the north-eastern part of the State, is 63.3°. The resorts of most reputation are *Savannah* and *Augusta*, where invalids find the comforts, conveniences, and social attractions of large cities ; *Thomasville*, on the verge of the great belt of pine forest which extends across the southern part of the State ; and *Eastman*, in the central uplands.

Of SOUTH CAROLINA the climate possesses the same characteristic features as that of Georgia, the diurnal variations of temperature being greater than in Florida. Consumptives generally improve faster in the highlands of the western part of the State than in the portions bordering on the Atlantic ; and here, amid vast pine forests, is *Aiken*, the most frequented winter-resort in America. The winter climate of Aiken is wonderfully mild and genial, and the air is remarkably pure and dry. The mean winter temperature is 46.4° ; of the year, 63.1¼°. Rheumatic and gouty patients, as well as consumptives, are benefited by residence here. *Charleston*, on the sea-coast, combines the attractions of a large city with a remarkably bland though somewhat relaxing winter climate. *Summerville*, 22 miles from Charleston, is far enough inland to escape the east winds that frequently prevail on the coast, and is growing in reputation.

The mountain region of NORTH CAROLINA is delightful in spring, summer, and autumn ; and *Asheville* and its vicinity has a winter climate which has been described as "mild, dry, and full of salvation for consumptives." The famous *Warm Springs*, in the valley of the French Broad River, are open to visitors in winter as well as in summer, and have a climate which is considered especially beneficial to consumptives.

All the foregoing resorts are either on or near the Atlantic seaboard. Crossing the continent now to the Pacific coast, we reach the numerous health resorts of CALIFORNIA, which are among the most famous, in the world. The climate of California differs widely from that of the Atlantic slope in the same latitudes, and probably from that of any other country in the world. It is doubtful if any other country has such warm winters and such cool summers ; the mean temperature of the coldest month being only about 10° lower than that of the highest. The coolness of the summer nights is attributed to the extreme clearness of the atmosphere favoring radiation ; and the warmth of winter to the influence of the great Japan current, which performs the same functions in the Pacific as the Gulf Stream does in the Atlantic Ocean. The wind blows for a part of each day from the north and north-west along the coast nearly the whole year. During eight months of the year the prevailing wind in San Francisco is south-west. This wind commences pouring through the Golden Gate toward noon, and increases in violence and chilliness till late at night. Heavy fogs occur during the night in the months of June, July, and August, but are of rare occurrence in winter, when the winds are not so strong. The numerous sheltered valleys near the coast are comparatively free from winds and fogs, and have a delicious and equable climate. In the interior the extremes are much greater, the mercury in the Sacramento Valley often rising in summer to 110° and 112° ; but, owing to the extreme dryness of the atmosphere, this great heat is much less prostrating in its effects than even a considerably lower temperature on the Atlantic slope, and the nights are never so hot as to prevent sleep. In the Sacramento and San Joaquin basin the mean temperature of the winter is about 4° below that of the coast, and of the summer from 20° to 30° above. The greater heat of the summer is supposed to result from the absence of the ocean-breezes and fogs, and the cold of winter from the proximity of the snow-capped Sierra Nevada. Southern California is said to possess a better climate than Italy. South of San Francisco, and in the San Joaquin valley, frost is entirely unknown. Roses bloom throughout the winter, and many trees retain their foliage green the year round. The air, peculiarly *warm* and *dry*, is wonderfully healthful, and highly favorable to consumptives and persons subject to diseases of the throat. Of the most frequented resorts, *Santa Barbara* (mean winter temperature 53.33°), and *San Diego* (mean winter temperature 52.5°), are on the southern coast ; on the coast near San Francisco are *Monterey* (mean winter temperature 51°), *San Rafael*, and *Santa Cruz ;* in the southern interior are *Los Angeles* and *San Bernardino ;* and in the great San Joaquin valley are *Stockton*, *Visalia*, and other popular resorts.

The resorts treated of so far are those which are frequented in winter by such invalids as require above all things a *warm* and equable climate ; but there is another class of invalids—those that are in the incipient stages of consumption, for example—who are more benefited by a cold but clear and dry climate than by one which is warmer and more moist. It is a well-ascertained fact that the *dryness* of the atmosphere is an even more important factor in the prevention and cure of consumption than its *warmth ;* and upon not a few consumptives a climate which is warm without being dry has a peculiarly debilitating and unwholesome effect. For this class it is claimed that MINNESOTA presents greater advantages than any other locality in America. Though its winters are long, and the cold intense and continuous, yet the dryness of the atmosphere and the slight diurnal variations of temperature render the cold far less oppressive than in the eastern Atlantic States, where the moist air and the rasping winds seem to tear weak lungs to pieces. There are few days of a Minnesota winter during which an invalid, if properly clad, cannot go out of doors for several hours at least ; and the bracing air almost always has a decidedly tonic and exhilarating effect. The cold in winter and the heat in summer, though extreme, are subject to but slight variations in short periods, and all the seasons are remarkably free from those sudden changes that are so injurious to persons in delicate health. The places in Minnesota most frequented as health resorts are *St. Paul, Minneapolis,* and *Winona ;* but any of the towns in the central, southern, or eastern portions of the State possess the advantages of climate already mentioned.

Somewhat resembling that of Minnesota in its effects, though very much milder and more equable, is the climate of COLORADO. The winters, especially in the southern portions of the State, are remarkably genial, and consumptives who go there during the earlier stages of the disease are always benefited and frequently cured ; but when hemorrhages have supervened, the extreme thinness and dryness of the atmosphere are likely to aggravate them. Asthmatics obtain benefit in any part of the State, rheumatism and gout are materially alleviated, and hay fever is invariably cured. The places of highest repute are *Manitou Springs, Idaho Springs, Georgetown, Pueblo,* and *Canon City ;* but many visitors camp out in summer.

With the growth in reputation of the resorts on the main land, the WEST INDIES have become less frequented by invalids from the United States ; but their generally delightful winter climate, and the advantages of the sea-voyage, will always preserve for them a certain degree of popularity. *Cuba* has a climate which would prove highly beneficial and agreeable to those suffering from lung and throat diseases from about the first of December to the last of March. The most accessible and attractive points are Havana, Matanzas, Santiago de Cuba, and Puerto Principe. The *Isle of Pines,* which lies in the Caribbean Sea 33 miles off the coast of Cuba, has an exceedingly mild and salubrious climate, and there is no place in the West Indies that can be more confidently recommended to invalids. There are several small towns on it. *Nassau,* on one of the Bahama Islands, is perhaps the most frequented winter resort, and besides its deliciously mild and equable climate has excellent accommodations for invalids. *Jamaica* and *Hayti,* or *Santo Domingo,* are unhealthy, and but little visited by invalids, though perhaps the more elevated portions would prove salubrious. In *Porto Rico* the climate, though very warm, is in general more healthy than that of the other Antilles ; and owing to the diversified surface of the island almost any desired degree of temperature can be secured. The winters are mild, equable, and comparatively free from rain. *St. Thomas* has a warm winter climate, and is recommended for those suffering from either consumption or Bright's disease of the kidneys. *Santa Cruz,* or *St.*

Croix, is considered by many the best resort in the West Indies, especially for those in advanced stages of consumption. *St. Vincent* exactly suits those consumptives who thrive in a warm and moist climate. *Barbadoes* is warm, but the heat is tempered throughout the year by the sea-breezes. *Curaçoa,* near the northern coast of Venezuela, has long been a favorite resort for sufferers from Bright's disease of the kidneys. THE BERMUDAS have a warm, moist climate, which is better for rheumatism or nervous diseases than for consumptives.

The health resorts in the south of France are numerous, but a few words with regard to each will suffice. *Pau* is characterized by calmness and equability. Its climate is soothing and sedative, and so is suited to those who are of an irritable type, but quite unsuitable for cases requiring stimulation. For patients in whom consumption is for the time quiescent no climate could be better ; for those recovering from most acute diseases it is quite the reverse. *Biarritz* is a bathing-place not very far from Pau, and is a good change for patients in the summer. The bathing is not so stimulating as further north, the water being quite warm. *Montpellier* had once a great reputation as a winter station ; that is entirely gone. Its climate is changeable and irritating. Along the south coast of France are a great variety of winter stations, now mainly occupied by English. *Cannes* has a climate intermediate between Pau and Nice, being less sedative than the former, less stimulant than the latter. It is a very good place for a rest if a man breaks down in winter. Cases of indigestion accompanied by nervous and irritable symptoms, do very well here. *Nice* has not the reputation it once had. It is found to be exposed to dangerous cold winds from the east and north-east, which come rushing down gaps in the chain of mountains behind it. The climate is not at all favorable to consumptive patients, but is good in many cases of derangement of the womb, and in children of a strumous habit of body. *Mentone* is one of the most sheltered of the towns along the coast, and the night temperature is mild, so that windows can be kept open. The place is one of the best for patients in the earliest stages of consumption and in chronic bronchitis. *San Remo* approaches somewhat in quality to Mentone.

Malaga is, perhaps, the only spot in Spain which could be called a true health resort. Its climate is excessively mild and equable, and it is neither too moist nor too dry. In winter there is a cold north-west wind sometimes, which is distressing. It is best adapted for those cases where inflammation of the lungs threatens to pass into consumption. It is easily reached by steamship, but the voyage is rather long. Its drainage is said not to be good. With regard to the *cities of Italy*, they are for the most part to be avoided by the confirmed invalid. On the other hand, for a convalescent they are very good, provided one is fit to travel from place to place, but few are adapted for a long stay. Exception may, perhaps, be made to the Bay of Naples, Ischia, and Capri. Patients suffering from kidney diseases often receive benefit there, but the whole peninsula is to be avoided by sufferers from consumption. *Egypt* has recently come into repute for sufferers from consumption, owing to its extremely dry climate. Unfortunately, it is expensive, as patients have to live in boats and take every thing with them if they advance into Upper Egypt. The climate is best adapted for the earliest stage of consumption, chronic bronchitis, clergyman's sore throat, and such like affections. The dust seems to be the most troublesome accompaniment of the journey. *Algiers* is another spot in Africa which has been recommended for a winter resort. It affords an exceedingly interesting winter residence, and is undoubtedly favorable to those suffering from consumption. *Madeira* used to be the favorite place of resort for consumptives. Now it has in great measure been abandoned, the climate being too moist. True, some of the symptoms are relieved, but the disease advances. Certain patients, however, suffering from laryngeal and bronchitic affections, are likely to be benefited. Of other health resorts not much is to be said. *Natal, Australia,* and others have been named, but in their benefits, or the reverse, is to be reckoned a long sea-voyage, which often does good but sometimes kills. The same remark applies to the *Sandwich Islands,* which are popular with American health-seekers, and which enjoy a truly delightful climate all the year round.

HEARING is interesting, medically, chiefly in its absence, which we term deafness. As to the mode in which the ear fulfils its functions, a word or two may be said. Sound consists of vibrations either in the air or some more solid body, and these, for the purposes of hearing, must be transmitted to the sentient nerve, which is called the auditory nerve, as nearly as possible unimpaired. To this end the sound is conducted and reflected from the outer ear inward. It strikes upon the drum of the ear, as it is called, and through it sets in motion a chain of jointed bones which end in another membrane. Immediately beyond this is a collection of fluid which can be set in motion by the membrane, and its undulations affect the nerve, which is spread out something like the keys of a piano. Anything which interferes with the transmission of vibrations to the nerve gives rise to deafness more or less complete. But there is another method whereby the nervous surface may be reached, that is, through the bones of the head, which are capable of conducting the sonorous vibrations. Thus, suppose a man holds a tuning-fork in his teeth while vibrating, the sound will be propagated through the bones of the head to the nerve of hearing. If now a similar fork be tried just outside the ear and no sound be heard, but sound be heard when it is held in the teeth, it is plain that it is the conducting apparatus which is in fault only, whereas, if heard in neither situation, we are fain to confess that the nerve itself is at fault, and the case more hopeless than before. Ear-trumpets and such like inventions are intended to collect a greater number of vibrations, and so render their effect more intense to a dulled ear. (See DEAFNESS and EAR.)

HEART.—The heart is a hollow muscular organ which is the main agent in propelling the blood through the numerous vessels of the body. It is situated in the chest or thorax, resting on the diaphragm, while its upper border is on a level with the junction of the third cartilages with the sternum or breast-bone. Its shape is roughly triangular, the base being directed upward, while the apex points downward, forward, and to the left side ; the apex beat may be felt in the space between the fifth and sixth ribs, and a little within a vertical line drawn through the nipple. The weight varies somewhat, being rather more in man than in woman ; the average weight in man is between 9 and 10 ounces, while in woman it is between 8 and 9 ounces ; in disease its weight may become enormously increased. The heart lies in a sac made up of dense fibrous tissue and lined within by a very smooth membrane called the pericardium ; this membrane is also reflected over the heart itself, so that the movements of the heart are attended with the least possible amount of friction : in the cavity thus formed a little serous fluid is found, just enough to moisten the opposed surfaces. The heart lies between the two lungs, and, for the most part, it is overlapped by their anterior edges, but as the left lung has no middle lobe the heart becomes superficial and appears close to the chest wall. The heart is divided into four cavities, two of which are called the *auricles,* and the two others are termed *ventricles ;* each side of the heart is separated from the other by a muscular partition wall or septum, so that in health the blood on the right side of the heart is quite distinct from the blood on the left side, and since it cannot cross directly from one side to the other, it has to go a further way round by the lungs, and then it arrives into the left cavities. There is an auricle and a ventricle on each side, and the four cavities are therefore named thus : *right auricle, right ventricle, left auricle, left ventricle.* Each auricle is situated behind and rather above its corresponding ventricle ; each has a thin muscular wall, and their cavities are generally smaller than the ventricular cavities. The auricles receive blood from the veins and pass it on into the ventricles ; they form, therefore, a kind of antechamber, and since they contract alternately with the ventricles, they allow blood to flow quietly into them, and are ready to fill the ventricles when these have completed their contraction. The right auricle receives the blood from every part of the body by means of two large veins, known as the superior and inferior venæ cavæ ; the former brings back the blood from the head, neck, and upper extremities, while the latter performs a similar duty for the rest of the body and the lower extremities. This stream is venous and has already passed through the various tissues and organs of the body ; it now requires to be oxygenated and to be exposed to the action of the oxygen of the air ; to do this it must go through the lungs, and the mechanism of the process is as follows : The right auricle sends the venous stream through an opening, which is called the *tricuspid* orifice, into the right ventricle, and the latter sends the current on to the lungs by a large artery called the pulmonary artery ; this vessel is about an inch in diameter and arises at the base of the heart, having a free communication with the right ventricle ; after a course of about an inch and a half it divides into two branches, one going to each lung ; arriving there it breaks up into several main divisions, and these again divide into innumerable small branches which ramify all through the lung substance and spread themselves in a delicate network outside the air-cells or the ultimate extremities of the bronchial tubes. As these arteries diminish in size, so also their coats become thinner, until at length the wall of the vessel appears as a homogeneous, microscopic membrane, which readily allows of the passage of gases to and from the blood. The air cells also have extremely thin walls, so that although the air and blood can never directly mingle, yet the delicate membrane between them allows those changes to go on which convert the venous dark blood into a bright red arterial stream. These minute vessels are called *capillaries*, and they differ from the arteries only in their size and in the simple structure of their walls ; next, these capillaries join together again and form veins, and these uniting one with another form at last four large trunks, which are termed the pulmonary veins ; those enter the left auricle, and this stream passes then into the left ventricle through an opening called the *mitral* orifice ; thence it is sent into the *aorta*, a large artery arising from the left ventricle just as the pulmonary artery does from the right, and then the blood is sent all over the body by means of various large branches, which, after dividing again and again, become finally so small as to form

capillaries ; these again uniting form veins, and at length by trunk after trunk joining to form larger ones, all the blood is brought back to the right auricle, once more to pass through the long circle of the circulation. Arteries carry blood *to* a part and veins carry it *from* a part, while between the two, but continuous with each, are the capillaries ; and the use of the latter is to enable the different tissues to receive nourishment easily and to give up their effete products, which could not be done unless the walls were very thin, and the blood brought into the closest possible connection with the elements of the tissue or organ. Thus everywhere there are intricate meshworks of vessels, some having a close web and some with wider interspaces, according to the requirements of the part. Just as a wide river may be divided into numerous smaller channels by the husbandman so that the land may get more thoroughly watered, so the large arteries break up into little capillaries, that the blood may well water the tissues ; and just as rivulet after rivulet may unite and at length form a broad flowing stream, so these capillaries unite and finally form veins.

This being the course of the circulation, there remains to be considered the means by which the blood is, so long as life lasts, kept in continual motion. Each ventricle will hold four or five cubic inches of blood. Each ventricle is made of strong muscular walls, the right one being about one eighth of an inch thick, while the left one is about half an inch in thickness ; this is because the former has only to send the blood to the lungs, while the latter has to propel the blood through all the remaining parts of the body. The ventricles are lined, as well as the auricles, with a delicate membrane called the *endocardium*, which is also continuous with the lining coat of the vessels which enter into or arise from the heart ; further, there are certain folds or reduplications of this membrane at each orifice which serve as *valves ;* as each ventricle has an entrance and exit, there are therefore four sets of valves, two on each side of the heart. The tricuspid valve is formed of three folds of the endocardium, and is situated between the right auricle and the right ventricle ; it is attached above to a circular fibrous ring round the tricuspid orifice and points toward the ventricle, so that when viewed from the auricular aspect it looks like a funnel ; as the blood flows through it from the auricle to the ventricle, the various segments flap open, while when the flow is from the ventricle into the pulmonary artery, these segments flap to and prevent any backward flow ; they are prevented from being pushed too far back by fibrous cords which are attached to muscular prominences on the inner wall of the heart, called *columnæ carneæ*, on the one side, and to the different curtains of the valve on the other. These cords are often spoken of as the *chordæ tendineæ*. The pulmonary valves are three semicircular folds of the lining membrane which are attached to a fibrous ring at the commencement of the pulmonary artery ; they guard this orifice, and while they readily open to allow the passage of blood from the ventricle into the artery, they close directly afterward, so as to effectually prevent any of it from returning. The mitral valve is formed like the tricuspid, but it has only two curtains instead of three ; it guards the mitral orifice or the opening between the left auricle and the left ventricle. The aortic valves are of similar shape and size to the pulmonary, and perform similar work.

Both sides of the heart, in health, act in perfect unison, and each part has separate duties to perform ; each auricle contracts at the same moment, and each ventricle does the same ; the corresponding valves also open and shut on each side with the greatest precision. Every minute each ventricle contracts some sixty or seventy times, and sometimes a great deal oftener ; after the contraction it rests for a short period while it is being refilled and then contracts again. Any one who listens to the beating of the heart will hear a sound just at the same moment as he feels the heart beat against the chest-wall ; this is called the *first sound* of the heart ; it is followed immediately by a second, shorter and sharper, sound, and this is called the *second sound* of the heart ; then comes a short interval or pause before the first sound is heard again ; each heart-beat, therefore, is divided into three periods, each of which varies slightly in length, although all are very rapidly performed ; it may make it simpler to divide each beat into five equal periods, and give the length of each sound thus :

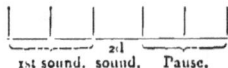

rst sound.   sound.   Pause.

During the first sound both ventricles are contracting, and this is called the *systole* of the ventricles, and the sound is also often called the systolic sound ; the rest of the time their walls are relaxing, and its state is called the diastole ; sometimes the second sound is called a diastolic sound, but this is not quite correct, as the diastole lasts three fifths of a cardiac beat, while the second sound only takes one fifth of the time. Both auricles contract at the end of the pause, and therefore they fill the ventricles immediately before the contraction of the latter ; the rest of the time they are passive and allowing blood to flow quietly in. When the ventricles contract the pulmonary artery and aorta become full of blood, and their coats being elastic, are distended ; directly after the systole they recoil and would send some of the blood back

again into the ventricle if the valves which guard the orifice did not immediately close and prevent it ; the effect of the recoil of the vessels is still farther to propel the blood onward, while at the same time there is great pressure on the valves which thus shut off the blood from the heart while the ventricles are being filled again. This closure of the aortic and pulmonary valves is accompanied by a sharp clicking sound, and it is this which is called the second sound of the heart.

The mitral and tricuspid valves are closed when the other two are open, and are open when the other two are closed ; thus, when the left ventricle contracts, the mitral valve shuts to prevent any of the blood current going back to the left auricle whence it has just come, while the aortic valves fly open so as to allow the blood-stream to enter the artery ; directly afterward the aortic valves close while the mitral valve is open, so that fresh blood may enter the ventricle, in its turn to be propelled onward ; of course similar remarks will apply to the action of the corresponding valves on the right side of the heart. During fœtal life, that is, while the child is in the womb, the circulation is somewhat different ; at that period of existence no respiration takes place, and therefore there is no need for the blood to pass through the lungs ; so the greater part of the venous stream passes directly from the right to the left auricle through an oval opening in the septum called the *foramen ovale ;* there is also a second communication by a small vessel, the *ductus arteriosus,* which joins the pulmonary artery and aorta, but which becomes closed soon after birth ; by this means that part of the blood which, in after life, passes on to the lungs, takes in the fœtus a shorter course, and, avoiding the lungs, is at once carried on to the arterial system of the body. The fœtus derives all its nourishment through the placenta, a complicated structure providing for the free interchange of nutrient elements between the mother and child. The heart is very liable to disease, and important changes may take place both before and after birth. In the former case they generally occur as *malformations.* Sometimes the *foramen ovale* remains open, and persons may live a long time without being much inconvenienced by it ; or the septum between the two ventricles may be deficient and allow of an intercommunication between the venous and arterial streams ; this is a serious defect, and leads to a deficient circulation, coldness, and blueness of the extremities, and shortness of breath. At other times the large arteries arising from the heart may be transposed, or the valves may be deficient or increased in number ; now and then the heart has been developed outside of the body in certain cases of monstrosity. Any defect of development leads to a condition called *cyanosis,* or general blueness or lividity of the skin, especially noticeable in the extremities, where the circulation is more feeble than in other parts. (See CYANOSIS.)

Diseases of the Heart arise from many causes, as follows :

1. *Traumatic causes*, or those caused by external injuries ; falling from a height or from a horse when hunting has ruptured some of the valves and set up heart-disease ; a sword wound, or stab, and a pistol shot, would probably prove fatal at once : sailors are liable to suffer from aneurism of the aorta, or from heart-disease, from lying on their chests while furling up the sails.

2. *Inflammation* may take place in the (a) pericardium or (b) endocardium, or in the (c) muscular wall of the heart. Inflammation of the pericardium or pericarditis causes an alteration in and a roughness of the smooth lining membrane above described ; more or less serum is poured out, and the heart's action is much interfered with. (See PERICARDITIS.) *Endocarditis* is an inflammation of the smooth membrane lining the cavities of the heart and forming the valves ; this disease is far more common on the left side than on the right side of the heart ; it is generally caused by an attack of rheumatic fever, but may occur after scarlet fever, and many other blood poisons, as erysipelas, pyæmia, etc. The change consists in little beads or warts of fibrine which are formed on the valves, and sometimes these form very long and shaggy processes. Their presence, of course, impairs the action of the valves affected, and gives rise to an alteration in the sounds of the heart called bruits, and these are called *systolic, diastolic,* or *præsystolic* bruits, according as they occur during the *systole, diastole,* or *pause.* It has been seen that the proper action of the heart depends upon a perfect rhythm of its movements, and upon a perfect closure of the different valves at the right time ; it is clear, therefore, that if, in consequence of disease a valve will not open wide enough, or will not close accurately enough, grave mischief will ensue ; it is, in fact, like putting a lock-gate across a running stream : all the parts behind the obstruction will get •too much blood, and the parts in front too little ; and the parts that get too much are liable to dropsy, for the serum of the blood exudes from the full vessels and soaks into the different tissues, making them sodden and œdematous. Any change in the valves, whether inflammatory or not, will cause a greater or less mechanical obstruction to the circulating stream, which may end in dropsy and serious impairment of the affected tissues. Each valve has to open and shut, but it may fail in doing one or the other or even in both ; now as there are four valves, they are liable to eight different forms of disease ; four are called *obstructive,* when they will not open

properly, and four are termed *incompetent* or *regurgitant,* when they will not close properly ; but of these eight, three are by far the most common, viz., aortic obstruction, aortic regurgitation, and mitral regurgitation.

3. *Atheroma* of the valves or lining membrane of the aorta may occur ; this comes on in old age, and in those who have lived hard or been fed badly ; it consists of a fatty change which comes on in the tissues in consequence of want of nourishment. (See DEGENERATION.)

4. *Calcification* of the valves is often associated with the last change, and it consists in the deposit of lime-salts from the blood in parts in which living changes have ceased to exist. Both these varieties may be of very small extent and cause no harm ; sometimes they are so marked as to cause great obstruction ; they generally come on in middle life and old age, while inflammatory changes may come on at any period of life, and are often seen in children.

5. *Fibrous thickening* of the valves may occur and cause constriction of the orifice, and in that way obstruct the passage of the blood. The symptoms common to these four changes are shortness of breath, palpitation of the heart, pain in the left side and inability to run fast or hasten up-stairs. These symptoms may go on for a long time, and if the patient is very careful to live quietly, avoid any great exertion, take nourishing food, and avoid bronchitis in the winter, life might be prolonged for years ; such, however, is not the case generally, for very few have the opportunity of living so quietly, yet every precaution should be taken to avoid more exertion than necessary, and so not to increase the work of the heart. The lungs are liable to congestion, and so in the winter, or when the east winds are prevalent, bronchitis is very apt to come on and increase the mischief. Increased shortness of breath is often followed by dropsy of the legs, by a diminution in the quantity of urine passed, which is also darker than usual, and deposits much sand on standing. There may be also pain over the region of the liver, and a slightly jaundiced skin, and at last ascites may come on. (See ASCITES and ŒDEMA.)

The obstruction in the valve of course gives the heart more work to do, and hence hypertrophy of the muscular wall takes place so as to overcome the obstacle ; just as the biceps of a blacksmith's arm becomes hypertrophied or excessively developed by increased use, so in a similar manner will the wall of the heart. Sometimes, however, the patient is badly fed and cannot obtain meat and nourishing food enough to provide for this increased growth, and then the ventricle, unable to withstand the increased pressure, slowly dilates until it attains very large dimensions ; very often hypertrophy and dilatation of the cavity occur at the same time ; at other times the one is in excess of the other. When dilatation occurs, the apex-beat of the heart is much lower than usual, and the heart takes up more space than usual in the chest ; thus the lungs, besides being congested, are also much encroached upon, and so the patient is very short of breath, and unable to lie down ; so at this stage he will be propped up in bed by pillows. *Hypertrophy of the heart* is known by the increased impulse felt by the hand when it is placed over the region of the heart. Much relief may be given to persons suffering in this way by rest in bed, a warm temperature, and light nutrient diet ; in a short time the dropsy will sometimes much subside, and the breathing become easier, and the patient more comfortable. A belladonna plaster placed over the heart (about four square inches in area) and kept on for ten days or a fortnight will give much relief. Tonic medicines, as iron and quinine, with digitalis, are often of much benefit. The great object in treatment is to avoid anything which entails more labor on the overworked ventricle, while at the same time the general health of the body is kept up. There are some other diseases of the heart which, however, are of very rare occurrence, and can only be merely mentioned here. Hydatids have been met with in this organ, and by their rupture have caused sudden death. (See HYDATIDS.) Rupture of the tendinous cords affixed to the mitral and tricuspid valves has been found, and of course it has been attended with a fatal result. A clot of fibrine may be deposited from the blood in the right auricle and right ventricle and cause death by obstructing the circulation ; such a clot is called a *thrombus ;* now and then a portion of a clot is carried through the heart into the pulmonary artery, and it is then called an *embolon.* A more common variety is that known as a fatty heart, where the walls become weak and fragile in consequence of a fatty change taking place in the muscular fibres. (See DEGENERATION.)

HEARTBURN is a sensation of heat or burning in the region of the stomach ; very often accompanied by a desire to belch or bring up some watery fluid which burns the back part of the mouth. It is due apparently to the presence of excess of acid in the stomach, and this excess of acid has two distinct origins. It may be due to fermentative change in certain articles of food, or it may be due to excessive secretion of acid gastric juice. Thus, many substances of a starchy or sugary nature readily ferment and give rise to acetic and lactic acids. With these changes carbonic acid gas is separated, and so there is commonly flatulence at the same time there is acidity. The other, where there is excess of acid secretion, is, as a rule, accompanied by less digestive disturbance than is the other, though the fluid so secreted is deficient in digestive power. The pain produced by excessive secretion comes on sooner after taking food than does the other, and may be induced as readily by a slight stimulus as by a greater one. The pain is generally felt behind the breast-bone, and is worse when the stomach is empty. If at these periods a glass of wine be taken, a fit of heartburn is almost certain to follow, for the gastric juice is induced to flow freely by the stimulus of the alcohol, and it has nothing to act on save the stomach itself. Food, on the other hand, if subject to fermentative changes, is, as a rule, followed by distress. Of course with either kind, nutrition suffers, but more if the alteration is due to fermentation than to excessive secretion. Sometimes the vomiting produced by the acidity is very trying and tends to reduce the patient's strength. As heartburn, however produced, is but a symptom, in treating the condition we must endeavor to determine its cause, to treat it right. Nevertheless, as a symptom, it is very distressing, and not unfrequently urgently demands relief. As it is due to excess of acid, it would be natural to seek relief in the exhibition of an alkali, and for the time being this often succeeds—a dose of bicarbonate of soda being frequently followed by relief. This is only temporary, and other means must be sought. If the acid is due to fermentative changes in the food, a drop or two of creasote or carbolic acid will often do good. If excessive secretion is at the root of the mischief, nux vomica is the most likely remedy. In either condition, large doses of bismuth generally give relief. (See INDIGESTION.)

HEAT and its opposite, COLD, which is only less heat, are both excessively powerful remedies. Much of their power is due to this, that whereas cold causes the vessels to contract, heat causes them to dilate, the one obstructing the flow of blood to and through the vessels of a part, the other favoring it. Cold is applied in a variety of ways, by means of sponging, washing, etc., with cold water, by the use of ice-bags, and again by using artificial means other than water for reducing temperature, as sponging with spirit, or the application of ether spray. Heat is applied mainly by means of hot water as a fomentation, or as a warm bath, but perhaps the most favorable way of applying heat and moisture combined is in the form of a poultice. The warm bath, which is perhaps the most common mode of applying heat after poulticing, is at first a pleasant application to the skin, if not too hot ; but if continued too long, the effects of the heat on the vessels become marked, and there is throbbing in the head and temples, and there is much prostration. The warm bath is sometimes employed in Bright's disease, for the sake of getting rid of some of the effete matter contained in the system, and from which there is risk of poisoning, but its use is of more than doubtful utility, for if it does good in one way, it does harm in another. Some forms of pain are signally relieved by heat. Thus gripes, as they are commonly called, especially if brought on by cold, are better treated by warm drinks and applications externally than any other way. The turpentine stripe is a very good way of applying heat and stimulation at the same time, but a cloth wrung out of water as hot as the hands will bear often suffices. There is another form of pain from which children frequently suffer, a kind of nervous pain or earache. Nothing, as a rule, does this so much good as heat, dry heat from a warming-pan or hot brick, or the like, being best. It is in diseases somewhat similar in certain respects that the hot bath seems to do most good of all. Such are the maladies called a fit of the gravel and a fit of gallstones. Some forms of skin disease benefit greatly by the warm bath. These forms are mainly of the scaly kind, and in the acute stage. The water should be as soft as possible, if nothing be added as a medicament. Sometimes patients have been kept for a very long time in baths for the treatment of these maladies. A similar plan has been recommended for patients the subjects of extensive burns, and also in small-pox. Heat receives a variety of other applications in medicine, which will be incidentally noticed.

HEAT, ANIMAL.—All animals preserve their own natural temperament irrespective of the medium in which they live. The animal heat of various creatures greatly differs. Thus man, birds, reptiles, and fishes have each their special temperature. By the action of the skin human beings are enabled to endure great variations of external temperature, and yet preserve their own animal heat. The proper heat of a human body is 98° ; but excessive external heat is borne and this natural heat remains the same, owing to increased action of the skin and excessive perspiration. Any diminution of the natural heat of the body denotes disease, and is a symptom worth attention.

HEAT-SPOT is a form of eczema, sometimes produced by exposure to the sun on a hot summer's day : bathing with tepid water or lead lotion, and keeping the patient cool, will soon cure it.

HEAT-STROKE and SUN-STROKE, though closely allied, are not exactly the same thing, for there may be heat-stroke without any direct exposure to the rays of the sun. In typical sun-stroke, the individual being exposed to the effects of the sun's rays falls down suddenly, and may almost immediately expire. In heat-stroke, the onset is more gradual, very likely there is a dry skin, prostration, and a tumultuous action of the heart, there may be difficulty of breathing and feel-

ing of restlessness. If now the bodily temperature be tested by the thermometer, it will be found to be far above the normal, perhaps as high as 102° or 104°. This is the typical symptom, and it is this which causes the distress and gives the risk. It must be reduced or there is no safety for the patient.

Treatment.—If the temperature is very high, say 105° or 106° Fahr., sharp means must be used for reducing the bodily heat to a normal level. Ice must be used. Thus it may be used either directly by rubbing the body, which is the most expeditious way, or it may be used to cool the water in which the body is immersed. The head must be looked to specially, and hence a bag of pounded ice ought to be applied to it. If these remedies are not at hand, the douche, represented roughly by pails of water, may be used, or the patient put under a pump and pumped on. No ceremony can be used ; it is a matter of life and death. The water, though of tolerably high temperature, is sure to be far below that of the body, and hence each particle of it can carry away some heat from the overheated surface. Some stimulant may be necessary to keep the heart going. Aromatic spirit is the best, but some iced brandy or iced brandy and water may be given. The patient, too, should be kept absolutely quiet. No exertion of any kind should be permitted, or the heart may stop, and a fatal termination of the disease come about.

These are the most important means to be taken for remedying the effects of heat-stroke, but for preventing them altogether another set are to be observed. In very hot weather it is desirable to alter our habits as well as our dress The dress should be as light and loose as possible, and as we have always a variable climate, flannel should be its material. The head should be covered by some light-colored texture, and ventilated well. Sun-umbrellas of white material should be used if the heat is very great, if not, an ordinary one will do. The heavy meals ought to be taken in the early morning and evening. During the middle of the day exposure to the sun's rays should be avoided. Beer should not be taken, nor spirits. Ice should be used freely with the drinks, and a cold sponge-bath taken once or twice a day. Ventilation commends itself.

HECTIC FEVER, as its name expresses, is an habitual or abiding febrile disorder. It occurs in connection with certain destructive diseases of internal organs, or results from exhausting drains upon the system, either in consequence of a greatly increased amount of normal secretion, as in diabetes, or by profuse and prolonged suppuration, as occurs in death and chronic ulceration of bone. The symptoms are very evident, and the " hectic flush" together with the hot burning hand and the bright sunken eye, are well known among all classes as dreaded indications of pulmonary consumption, and other lingering and fatal diseases. Hectic is observed also very frequently in connection with progressive angular curvature of the spine, associated with a large abscess extending downward to the groin, and in severe and painful diseases of the hip or knee. The most marked *symptoms* of hectic are rapid loss of flesh, great heat of skin, especially of the palms of the hands, occasional chills during the day, and toward the end of the day a distinct fit of shivering, followed by intense fever, and afterward, when the patient is in bed, by profuse and exhausting perspiration. The pulse is quick and very irritable, and is rapidly affected by the patient's movements, and by mental excitement or emotion. The tongue is moist and clean, and always very red. The skin during the day is hot, dry, and rough. During the night it is covered by perspiration. The appetite generally remains good. The bowels are readily affected by any excess in diet, or by the presence in the food of irritating matter, but generally during the early stages of hectic are constipated. The mind remains unaffected. If the primary cause of hectic fever still persists, the patient sinks in consequence of rapidly increasing emaciation and loss of strength. The pulse becomes weaker and the tongue is covered by white patches (aphthæ). The ankle and afterward the legs become swollen and dropsical, while the skin of the arms and body remains thin and shrunken, and is covered by rough branny scales. The cold fits become more and more severe, and the nocturnal perspiration more profuse. When the patient is in this condition, bed-sores may be formed, the discharge from which, associated with diarrhœa, increases the exhaustion of the patient, and hastens a fatal determination. Death takes place slowly and quietly. The following are the diseases that most frequently end in hectic fever : pulmonary consumption, diabetes, Bright's disease of the kidneys, psoas abscess, ulceration of joint-cartilages with death or ulceration of subjacent bone.

Treatment.—In cases where it is impossible to remove the cause of this febrile disorder, the treatment should be directed to supporting the patient's strength by tonics, such as quinine and bark, and by a good diet, with a free allowance of wine and other alcoholic stimulants. The food should consist in broth, soup, beef-tea, and milk, and easily digested flesh, and be given frequently and in small quantities at a time. The bed-linen should be changed often by reason of the profuse perspiration during the night, and in advanced hectic care should be taken to prevent, if possible, the formation of bed-sores. For further information on this latter point, which in this, as in all exhausting affections,

is one of great importance, the reader is referred to the article on BED-SORES.

HELLEBORE (Black) is not now contained in the Pharmacopœia, but has been used in medicine from time immemorial. The parts used are the underground stem and rootlets of the *Helleborus niger* or Christmas rose, dark externally, white internally. Its taste is first sweet, then acrid and bitter. Its powder and tincture have generally been employed in medicine. It was formerly employed for affections of the head, especially for the melancholic condition. It acts as a powerful drastic purgative, and was supposed in these cases to remove the black bile, which our forefathers imagined lay at the root of the disorder. (See VERATRUM.)

HEMIPLEGIA signifies paralysis of the arm and leg on one side of the body ; the loss of sensation is generally very slight, if at all marked, but the loss of motion is most particularly noticeable ; this may be partial or complete. The seat of mischief may occasionally be in the spinal cord, but, as a rule, it is in the brain, and on the opposite side to the paralysis ; thus, if there be hemiplegia of the right arm and leg, the disease will be on the left side of the brain. Any influence which interferes with the due supply of blood to a certain area of the brain will cause hemiplegia ; *white softening*, *cerebral hæmorrhage*, a *clot of fibrine* obstructing the vessels, *disease* of the coats of the vessels from fatty change, and *epileptic* attacks, will all cause this form of paralysis. An ordinary attack of apoplexy in fact, when the patient has recovered from the shock, leaves the individual in this state ; it is, in short, a symptom of the mischief in the brain, and not the disease itself. Hemiplegia may come on suddenly, without any coma or insensibility, as when it is caused by a very small clot ; more generally the two symptoms are present, and when sensibility returns the patient finds he has lost the use of his arm and leg. Sometimes there is stiffness or rigidity of the arm and leg as well as loss of power, and this seems to depend on the nature of the injury to the brain. In most cases the limbs lie useless and flaccid, and, if raised up, drop at once when left unsupported. Improvement may be known by the patient being able to perform simple movements or raise the limb a short distance from the side, but for many weeks or months the strength of that side will be much impaired, and even in favorable cases the patient can hardly expect to fully regain the use of the injured side.

Treatment.—When the individual has recovered from the shock, friction may be used to the extremities, or a galvanic current or rubbing with rough towels after a stream of cold water has been applied ; this should not be done until three or four weeks after the disease has begun, and then only when the patient is in a fit state for it. In every case the treatment and chance of recovery must depend in a great measure upon the nature of the injury. (See APOPLEXY.)

HEMLOCK (*Conium maculatum*) is a plant which grows wild in this country. The fresh leaves and young branches, collected just when the fruit begins to form, and the dried ripe fruit are the parts used. They contain a peculiar substance or alkaloid, called conia. This is volatile, and is easily set free by means of an alkali like caustic potass, with which, when the substance is rubbed, a peculiar and characteristic mousy odor is observed. In many of its chemical properties conia resembles ammonia, but its effects on animals are very different. The best preparation for use is the succus conii, or hemlock juice, got by expressing the juice from the fresh leaves, and adding a little spirit to make it keep. As a remedy, and as a poison, hemlock is of undoubted power, and yet many animals eat it with impunity. It used to be employed in the form of a poultice, to painful sores, especially of the cancerous kind. It is not often used in this way now. Hemlock in large doses seems to paralyze the animals to which it has been given, mainly by acting on their motor nerves. It is chiefly of use this way, in controlling violent muscular movements, as in some forms of chroea, but as yet it has received no very extensive therapeutic application.

HEMORRHAGE. (See HÆMORRHAGE.)

HEMP (Indian) consists of the dried flowering tops of our common hemp-plant grown under the tropical sun of India. The female plant is to be used, and it should be carefully noted that the resin is still present. In India it is used in various forms. Hashish is a form of Indian hemp, but the term seems to be applied to more than one form of the substance. It is the resin, developed by the great heat and powerful sun, which gives its value to the Indian plant. This may be readily dissolved out by alcohol or ether, but the addition of water causes its precipitation. The solution in spirit evaporated to a paste constitutes the extract of hemp. The effects of Indian hemp are very wonderful ; but as the preparations obtainable in this country are not always reliable, they may frequently fail to make themselves observed. The resin of the plant gives rise to a peculiar form of intoxication ; this is always attended with exuberance of spirits, and, if the individual sleeps, is attended by dreams of a pleasing kind. It relieves pain, and in many cases gives rise to sleep, and its after-effects are not unpleasant. There is little languor, and no loss of appetite, neither does it constipate the bowels. Indian hemp is seldom given to allay pain purely ; but in certain cases of painful menstruation it does great good. It

is best given until the patients begin to feel light in the head, after which the pain ordinarily ceases. In cases too where the menstrual flow is excessively profuse, it ordinarily arrests this excess. As a rule, it may be said that if the drug does not produce its peculiar physiological effects, especially lightness in the head, it will not affect the pain. The dose of the tincture should be from 10 to 20 drops, and it is best given in aromatic spirit of ammonia —spirits of sal volatile.

HENBANE, also called hyoscyamus, consists of the leaves of the *Hyoscyamus niger*, gathered when about two thirds of the flowers are expanded. It is supposed to owe its efficacy to an alkaloid called hyoscyamine, about which, however, little is known, if it has even been separated. The preparations mainly employed are the extract and the tincture. As to its action, that seems allied to belladonna and stramonium, but milder. It, like these plants, has the power of dilating the pupil, and it has distinctly a power of soothing irritable conditions of the system, and of preventing the griping action of certain purgatives. Its main use is as a sedative where opium cannot be given for other of its effects, as when the lungs are congested. Hyoscyamus is also used to relieve irritability of the bladder, and for this purpose has been often combined with liquor potassæ ; but the alkali entirely destroys any influence the henbane may possess, probably by decomposing its alkaloid. In large doses its effects are similar to those of the remedies named above —dilatation of the pupil, dryness of the mouth and throat, slight delirium, and partial loss of power. As a substitute for opium its efficacy is doubtful, though frequently prescribed with a view to that substitution. The dose of the tincture is from 10 to 30 drops.

HEPATIC disorders. (See LIVER.)

HEPATITIS, or inflammation of the liver, is a disease which we do not often see in this country. It is exceedingly common in tropical regions, especially in Europeans living there and who are careless in their mode of life. We shall speak here entirely of the form of inflammation which tends to end in softening and the formation of an abscess. Other forms of liver-disease are sometimes spoken of as inflammations, but they have nothing of the nature of that process. Such are the forms of disease which tend to end in contraction of the liver and in gummy deposits in it. The disease would seem to be caused sometimes by free living in unhealthy climates, exposure to marsh or jungle miasms ; and very frequently it is due to dysentery.

The onset of the disease is marked by pain and fulness of the side, with some degree of tenderness, especially on pointed pressure ; and this is all the more marked if the part of the liver affected be near the surface and is easily reached. Then there is fever, the skin is hot, the temperature high ; there is much thirst, and the urine is scanty. The pain in the right side is often severe, but sometimes absent. It is much worse by lying on the left side or by coughing. There may be 'a slight tinge of jaundice, but not much. Sometimes, especially if the part of the liver next the stomach be implicated, there is vomiting and sometimes hiccough. Usually, too, there is a peculiar pain in the right shoulder, especially about the collar-bone. More rarely the left shoulder is affected. If the inflammation go on to the formation of an abscess, as it commonly does, the occurrence of suppuration is commonly marked by shivering, there is increased pain and tenderness very often, especially if the abscess be on the upper part of the liver, a dry cough, and a feeling of weight and dragging in the right side ; the muscles forming the wall of the belly on that side, too, are tense, and kept tight as if to protect the sensitive organs beneath. Sometimes it may be distinctly made out that the liver is enlarged, but often this is not the case, and as time wears on the patient suffers from hectic fever, there is great prostration, and most frequently diarrhœa or dysentery.

Inflammation of the liver may abate, and the patient get well without any abscess being formed ; but generally when an abscess has been produced its contents must be got rid of, or the patient will die. The escape may be natural or artificial, but it is always dangerous. Sometimes they burst into the cavity of the abdomen ; if so death is almost certain, for inflammation is set up, and a fatal termination is not far off. Sometimes they open into the bile-ducts or gall-bladder, and so their contents may escape into and through the gut. Most frequently, as the abscess nears the surface of the liver, an inflammation of its covering is set up, and thus it is glued to the neighboring parts ; if to the wall of the •abdomen, an opening may be made in it, and so the pus escape externally. It may also adhere to the intestines, and an opening be made in them, and the fluid escape that way. The same may occur by way of the stomach ; or if the abscess tries to escape upward it may penetrate the diaphragm, enter the pleura, and finally escape by making its way into the aorta. and so be spat up.

Treatment.—The treatment of hepatitis consists in a considerable measure in letting the patient alone. Mercury, which used to be given freely, is of more than doubtful value ; but the bowels ought to be kept open, but not loose. If confined, and the tongue is brown, and there is much fever, rhubarb and alkalies are likely to do good. If they are too loose, and there is a tendency to dysentery, astringents must be used, especially in combination with ipecacuanha, in full doses. When suppuration has fairly taken place, good nourishing food must be given, probably

also tonics, and we must wait. If there is much restlessness and pain, small doses of chloral or morphia subcutaneously will be best. If the abscess has fairly declared itself as likely to burst on the surface, fluid may be withdrawn by aspiration, or that may be attempted beforehand. If an aspirator be not procurable, it is better to let the abscess burst of itself. After that a stimulant and supporting treatment is solely needed.

HERMAPHRODITE, an individual combining both sexes in the same organism.

HERNIA means any protrusion of the contents of a cavity through its walls ; but in general the term is applied to the protrusion of the abdominal viscera, constituting rupture. The predisposing cause of hernia is a weakness in the walls of the abdomen ; either the natural openings, such as the inguinal or crural rings, are large cr very slightly protected by their coverings, or there may be some congenital deficiency, or the parietes may give way from injury or disease. The exciting cause is compression of the contents of the abdomen by the surrounding muscles, which are very powerful, and are brought into violent action by rowing, lifting weights, pulling, etc. Hernia is divided into the following varieties—reducible, irreducible, and strangulated.

By *Reducible hernia* is meant one returnable into the abdominal cavity, and its symptoms are the existence of a compressible tumor in the abdominal walls, which lessens in size if the patient lies down, or disappears altogether, receives an impulse on coughing, or on any exertion being made, and can be readily returned by pressure. This form of hernia can be treated either palliatively or radically ; the first by means of trusses, and the second by operation for the closure of the aperture through which the rupture passes. A truss consists of a ring of steel, to the extremities of which are attached pads, one of which presses upon the aperture and retains the hernia within the abdominal cavity. There are many forms of trusses, but the measurements to be taken in writing for any form recommended are the same : thus, if an inguinal or femoral truss be required, the circumference of the body at the hips should be stated, midway between the spine of the ilium and the trochanter. Great caution must be taken in the fitting of a truss that it does not excoriate the skin, and that part of the skin upon which the pad presses should be regularly washed and bathed with eau de Cologne or spirit, or dusted with violet powder or fuller's earth. In children, an india-rubber band and pad answers generally.

Umbilical hernia, or *exomphalos,* is most frequent in newly-born children, and presents itself as a protrusion at the navel ; a flat disc of metal, or even a penny-piece, retained against the protrusion with a strap of plaster will retain the hernia. In adults, properly adapted trusses must be employed.

Ventral hernia is a protrusion of bowel through the abdominal walls in the mesial line, or through any parts of the parietes which are not usually the seat of otherwise named herniæ. There are several other forms of hernia which are not likely to be detected except by an experienced physician, which protrude themselves through those natural openings in the pelvic or abdominal walls, which serve to transmit muscles, vessels, and nerves to the limbs, or which may be the result of arrested devlopment or of injury, and which can hardly be mentioned except by name—such as perineal, vaginal, labial, obturator, ischiatic, and diaphragmatic, and for which the reader is referred to special works on surgery.

Strangulated hernia means that a portion of intestine being protruded, there is a total stoppage of its contents, so that they cannot be propelled toward their natural outlet, and, moreover, that the structure of the bowel itself is so constricted that it is itself *strangulated*. The symptoms of this condition are—firstly, those of obstruction of the bowels; secondly, those of inflammation. The individual has flatulency, tightness over the belly, a desire to evacuate the bowels, and an inability to do so. Next, vomiting supervenes, in the first place of the contents of the stomach, of bilious matter, and then of matters smelling strongly of fæces, in consequence of the injesta being detained in the intestinal circuit. In this state of things operative interference is necessary, and that at once, although, until such aid is at hand, some assistance is to be derived from what is termed taxis, from which, even in unprofessional hands, if properly directed, good results may ensue. The patient should be placed in a warm bath, and both the thighs be raised toward the belly and placed close to each other, as a means of relaxing all the muscles and ligaments connected with the abdomen; he should be engaged, if possible, in conversation, so as to relax the respiratory muscles. Next, the visible tumor should be grasped gently with one hand, to empty it as far as possible, and with the other the neck of the tumor should be *kneaded*, with a motion *toward* the abdomen. This operation should be continued for some while, a quarter of an hour or so, if no great pain is produced by so doing, at the end of which time, if the proceeding be successful, a slight gurgling sound will intimate the return of at least a portion of the tissues. In some instances when mere taxis has failed, raising the pelvis and lowering the shoulders have proved effectual; thus, the pelvis being placed on a chair at the end of the bed, and the patient's head and shoulders resting on the bed itself, the legs should be bent upon the body, with a view to relax-

ing the structures which, by their tenseness, assist in strangulating the portion of intestine. Chloroform is a great aid in the reduction of such a hernia, a hot bath (96° 100° Fah.), a large dose of opium, an injection of tobacco (a drachm to a pint of boiling water); cold, in form of ice, or of a freezing mixture placed in a pig's bladder and applied over the swelling. In the event of these milder remedies failing, a surgeon must perform an operation with a view of setting free those structures which are caused by the constriction and delaying the passage of the fæces.

Irreducible hernia means that form of rupture where, from some impediment in the canal through which it passes, it cannot be replaced in the abdominal cavity. In such cases there is no stoppage of the contents of the intestine, and the coats of the gut are not engorged as to their circulation. If an irreducible hernia be neglected, it produces many inconveniences, abdominal pains, vomiting, and general intestinal disturbance, and the contents of the bowel may be obstructed in their natural passage, causing colic and constipation, and, moreover, the chance of the bowel becoming *strangulated* at that point is greatly enhanced. With regard to treatment, it is either palliative or radical, the palliative measures being the application of a *bag truss;* and all violent exertion or excess in diet should be avoided. The *radical* proceedings, which have been before alluded to, are of course only to be attempted by an experienced surgeon. (See TAXIS.)

HERPES is a skin eruption made up of clusters of small vesicles or blisters surrounded by a pink or red areola. It generally occurs on the upper lip in cases of ordinary cold, and it often is found there in those suffering from pneumonia; it is then called *herpes labialis*. It is caused by the irritation of the nasal discharge, and commences with tingling pain and itching and slight redness: in a few hours the clear vesicles appear, and if they are pricked a little serum escapes; they are very harmless, and heal in a few days by drying up, and the scab then falls off, leaving no permanent scar. If rubbed or scratched, increased irritation may be set up, and the patch may become eczematous. A little cold cream applied night and morning, or simple zinc ointment, will suffice for a cure. There is another variety, named *herpes zoster* or *shingles*, the peculiarity of which is that it only affects one half of the body; it is oftenest met with on the right side of the chest, but may occur anywhere; clusters of vesicles with red margins are found along the course of the cutaneous nerves. The eruption commences with pain along the nerves, and in two or three days a copious rash will appear; it may occur at any age, and is perfectly harmless in its nature. It is a common but erroneous idea that when shingles occurs all round the body so as to form a zone or girdle, it is always fatal ; the fallacy lies in the fact that it never does. Bathing with warm water or smearing on a little zinc ointment is all the treatment required, and a cure will take place naturally in a few days. Sometimes obstinate neuralgic pains remain for some time after the eruption has disappeared.

HICCOUGH, commonly called "hiccup," is a very troublesome affection, sometimes of the slighest significance, sometimes of the gravest. It consists of a short, abrupt contraction of the diaphragm, and a sudden jerking, imperfect ejection of the breath. Most frequently it is purely emotional, and brought on no one knows how ; and very often it may be got rid of in the same way by frightening the individual, by exciting the curiosity, and so removing the attention from the hiccup, for that is one of the things most favorable to its continuance. A draught of cold water or sucking a piece of ice will generally get rid of this nervous form of the affection. A variety of it is not uncommon in hysteria, when it may continue for a very long time, apparently resisting every remedy. In such cases the patients either have or do not desire to have any control over themselves, and so if the hiccup does not soon cease it is best to give the patient a narcotic draught of chloral or morphia, or if that be objected to, morphia under the skin of the arm may be given. The application of the galvanic battery to the region of the diaphragm, that is to say, with one pole on the end of the breast-bone and the other on the back where the ribs end, will in most cases put a stop to this hysterical kind. There is a form of very grave origin, however, when gangrene of any part sets in, but especially grangrene or mortification of the gut ; this proves very troublesome, and in most instances it is a fatal symptom.

HICKORY - PICKORY. (See Hiera Picra.)

HIERA PICRA, better known as hickory-pickory, is a popular remedy for constipation. It is not an officinal preparation, but has a large domestic reputation. It consists of a mixture of equal parts of canella bark and aloes. The canella bark is an excellent stomachic tonic, possessing a warm and spicy taste, which is agreeable and comforting. The laxative effects of this remedy are due to the aloes which it contains, and its properties mainly approximate to the properties of that drug. (See Aloes and Canella.)

HIP-JOINT DISEASE.—The large ball and socket, which is formed by the articulation of the head of the thigh-bone with the acetabulum or cup-shaped cavity in the hip-bone, is liable to the following injuries and diseases. Dislocation (which see) and a peculiar form of disease known as Morbus coxæ or coxalgia. It most frequently attacks children between the ages of seven and fourteen, although no age is exempt, and generally prevails in cold moist climates. The first symptom noticeable in a child is the fact of its dragging the affected limb after the sound one, a flattening of the natural fold of the buttocks, and pain referred at first to the knee ; and in standing the patient advances the foot a little, slightly everting the toe, and does not rest his weight upon it. After a while pain comes on in the hip-joint itself, and generally continues chronic for several months. At length the symptoms may disappear, and become far more serious ; thus the affected limb becomes shorter than the sound one, the motion in the joint being impaired or destroyed, and permanent dislocation taking place. Matter now forms in the region of the hip and makes its way to the surface, and then after a tedious illness the patient either becomes hectic and dies, or recovers with a stiff anchylosed joint and a wasted, useless limb.

Treatment.—The treatment in the earliest stage consists in maintaining the limb at perfect rest in the straight posture, and this is best effected by placing sand-bags on each side of the limb, the external one reaching as high as the armpit, and the body and legs kept fixed straight by a stout sheet drawn tightly over them and fastened to the bedstead. Counter-irritation by means of small blisters around the hip, and the internal administration of cod-liver oil and tonics, with residence in sea air, is the best method of treatment to be depended on. In advanced cases, when there is extensive disease of the bone, operative measures, such as removal of dead bone, either partially or by the operation of excision of the head of the femur, must be resorted to. This operation is generally attended with the best results, as shown by the statistics of cases treated in the London and provincial hospitals and private practice.

HOARSENESS, a common term for Aphonia, or loss of voice. (See Aphonia.)

HOB-NAILED LIVER, so called because in cirrhosis the surface of that organ is rough and uneven. (See Cirrhosis.)

HOCK. (See Wines.)

HOME-SICKNESS. (See Mal-de-Pays.)

HOMŒOPATHY is a theory of medicine opposed to that commonly known as Allopathy, and introduced by Dr. Hahnemann, a German physician, about the year 1810. The main principle of the practice of this theory is that "like cures like," and the motto "similia similibus curantur" is adopted by Hahnemann and treated of in his works. Homœopathy professes to cure diseases by the employment of remedies which, if given to a healthy person, would produce symptoms of a disease similar to the one to be treated. The three points on which the fabric of homœopathy may be said to rest are, first, that like cures like ; second, that the curative power of drugs is increased in proportion to their minute subdivision ; and third, as a consequence, that infinitesimal doses of medicine are the proper treatment of all diseases. The reaction against the injudicious and constant use of violent medicines, which was at one time the case in ordinary practice, favored the opposing system of homœopathy, and possibly good has been effected in this way. By degrees most people learn that attention to diet, general habits, exercise, and rest have more to do with the cure of disease than the absolute medicine swallowed, and it is thus doubtless that homœopathy has secured its present amount of popularity ; moreover, the remedies administered by the homœopathist are tasteless, if not pleasant, and find special favor with children and those whose fate has hitherto been to supply the victims of nauseous domestic doses.

HONEY is a substance which enters into the pharmacopœia as well as forming a pleasant article of diet. It is the sweet juice of plants and flowers elaborated by the bee and deposited in waxen cells in the form of liquid sugar. The quality of honey, both in flavor and richness, depends greatly on the character of the country over which the bees roam for food. Honey is used in medicine as an emulsion, expectorant, and laxative, and is often combined with vinegar and squills to make a simple remedy for colds, coughs, and hoarseness.

HOOPING-COUGH. (See WHOOPING-COUGH.)

HOPS are the dried flowers of the female hop plant collected and dried. The flowers consist of scales inclosing a quantity of powder, to which they owe their peculiar effects. This powder may be separated by sifting, and is then called lupulin. Various preparations of hops are in use—the tincture, the extract, and the infusion ; the best is, however, bitter beer. The hops themselves are supposed to be slightly narcotic, and a pillow of hops has been used to give sleep, but there is no evidence whatever to show that any preparation of hops has this particular effect. Hops are besides bitter, stomachic, and tonic, and this wholesome bitter in good bitter ale is often invaluable. In convalescence from acute disease bitter ale often does great good, and in aiding digestion in those with feeble appetite is often of great service to those who would otherwise suffer from indigestion. Much of the beer used, however, owes its bitter to something else than the hop, and in many instances that something is not quite so wholesome.

HORDEOLUM, commonly called a stye, is a small, hard, painful boil developed in the margin of the eyelid. It is of slow growth ;

the suppuration proceeds imperfectly, and as it increases in size it presses on and produces obstruction of some of the ducts of the Meibomian glands. (See EYE.) Suppuration should be promoted by the frequent application of warm fomentations, such as a hot soft sponge, wrung out in boiling water, applied to the eye, and a hot bread-and-milk poultice applied over night. When the pus "points" a very slight puncture may be made, and the warm applications continued. If the margin of the eyelid remains thickened and painful, and the tissues immediately adjacent be indurated, a little citrine ointment should be applied along the margin of the lid. In some instances in which the suppuration is superficial and very scanty, and it enlarges until it opens at a point where the superficial tissue has become thinned, by gently pressing upon its base, a small core or slough is expelled, and after its expulsion the healing is very rapid. Aperients, and afterward tonics and alteratives, are always necessary, as the complaint arises in debilitated conditions. Those affected with scrofulous habits, or who often suffer from chronic ophthalmia, are peculiarly liable to be attacked with stye, and they then occur one or two together or in succession, plainly indicating something wrong in the general health. When stye occurs frequently in relapses in scrofulous children it is readily cured by the administration of quinine. If it occurs in persons of full habit, spare diet and gentle aperients are indicated. If the tumor remains indolent, some stimulating ointment, such as iodine or critine ointment, or nitrate of silver, proves very efficacious in dispersing it.

HOREHOUND, the (*Marrubium vulgare*), a plant belonging to the Labiate family, has long been used in domestic practice. Its uses are ill-defined, but it was supposed to act as a tonic and expectorant, and so was generally used for coughs and colds. It has no place among the remedies of the physician.

HORSE-RADISH (*Cochlearia Armoracia*) is a plant well known for its culinary virtues, though not much can be said for its medicinal properties. It is sharply pungent, and will act as a stimulant to the flow of saliva, probably also to that of gastric juice. Its only officinal preparation is a compound spirit, which is rarely if ever used, though it has been supposed to do good in chronic rheumatism and as a diuretic in dropsies. It is best taken scraped with roast beef.

HOSPITAL GANGRENE includes several gangrenous and ulcerative processes which attack wounds and stumps after amputation, when the patients are collected together in great numbers and are placed under faulty hygienic conditions. Hospital gangrene in all its forms is both contagious and infectious, and seems in some instances to be due to epidemic influences, as it has been known to attack isolated patients under private treatment at periods when it is spreading rapidly from patient to patient in the large wards of hospitals. It is very prevalent among armies during military operations, and when large numbers of wounded soldiers are collected together in buildings unsuitable as to size and internal arrangements for hospital purposes. It attacks small as well as large wounds, and even blisters and leech-bites, but is never met with in perfectly sound individuals. In the most severe form of hospital gangrene a small livid spot or bleb makes its appearance on a stump, or near the margins of a wound, which had previously been closing favorably. This bleb increases rapidly in size, and converts the extremity of the stump or the whole of the wound, with the surrounding healthy skin, into a black and swollen gangrenous mass. The disease spreads rapidly, and is associated with constitutional symptoms of a low typhoid character. At other times a stump swells and becomes hard and very pale, and its surface is marked by large blue veins. This form is also attended with severe general symptoms and much pain. Like the preceding one, it is generally fatal. In the less severe forms the surface of a wound is covered by a thick, yellow, and adherent crust, which increases rapidly both in depth and superficial extent. This thick crust at last separates, and is thrown off at the surface of the deep wound thus formed, again undergoes the same process, until vessels and important structures are laid bare. This disease has been met with chiefly on the Continent, and is there known by the name of diphtheria of wounds. The constitutional symptoms are not so severe as those of the strictly gangrenous forms, and the fever, if it be present, is generally high and of an inflammatory kind.

Treatment.—The general treatment should consist in supporting the strength of the patient by tonics, stimulants, and nourishing diet. In the diphtheritic form, however, the diet should be moderate, so long as there is high fever, and alcoholic drinks should not be given freely. The local treatment is generally directed toward arresting the spread of the gangrene by the application of nitric acid, or the red-hot iron. The affected parts should be frequently cleansed by lotions containing carbolic acid, or tincture of iodine, and after the application of a caustic agent be covered by yeast or charcoal poultices.

HOUSEMAID'S KNEE is a familiar term applied to enlargement of the large *bursa mucosa* situated in front of the knee-cap or patella, and of the tendon immediately below it (*ligamentum patellæ*). This bursa or sac from its exposed position is very liable to become enlarged from kneeling upon it, as in scrubbing, cleaning steps, etc. ; and this con-

dition is frequently noticed in household servants, carpenters, plumbers, carpet-layers, etc. Enlarged bursæ of the patella are frequently attacked by inflammation and suppuration, and usually there is extensive inflammation of the surrounding cellular tissue ; sometimes such large collections of matter are found in the neighborhood of the sac that Sir B. Brodie considered that the suppurating bursa occasionally gave way, and allowed its contents to escape into the cellular membrane. Sometimes troublesome burrowing ulcers remain after these abscesses, which are singularly obstinate, attended with fungous growths, the surrounding skin being dark and unhealthy, with deep burrowings under the integuments of the knee, and a foul, offensive discharge. In severe instances the bone (patella) may become necrosed. (See NECROSIS.)

Treatment.—The treatment consists in the first place of complete rest, and a well-fitting splint must be applied, and all motion of the joint prevented. If a recent enlargement, a stimulating lotion of acetic acid and hydrochlorate of ammonia, or a small blister, will often cause it to subside. If there is considerable thickening, as there always is if the tumor has been of long duration, evacuation of the sac and subsequent counter-irritants will often effect a cure. Some surgeons use a seton (see SETON), composed of a few threads of silk passed through the cyst, and by setting up suppuration and the consequent contraction and granulation the cavity becomes obliterated. When the tumor has become a solid, gristly mass, there is no other treatment than dissecting it completely out. In the cases most commonly brought under observation, rest, leeching, hot fomentations, and purgatives, and failing these a free incision, usually effect a cure.

HOUSES and their construction are of the very greatest importance from a sanitary point of view. Often the health of families is completely lost, very frequently death itself ensues, from defects in household construction. If a man sets to work to build or select for himself a house on sound principles, the first thing he has to satisfy himself about is the site in which it is to be or has been built. We do not, of course, refer to beauty of situation, which will always speak for itself ; but rather with regard to the nature of the soil. There are two kinds of sites—the natural and artificial. Artificial foundations, except they be carefully prepared, are to be strenuously avoided. It is quite true that in a damp soil a good sound artificial foundation is a very great improvement ; but then it must be carefully prepared, not made of materials heaped together at random. As far as site is concerned, the possibility of good drainage ought to be carefully kept in view. A house with damp foundations is an artificial hot-bed for rheumatism, with all its dangers to health and life ; and so a situation below the high-water mark of rivers is to be avoided. A point much studied in selecting a building site, and yet often on wholly erroneous principles, is the nature in the soil. Thus a gravelly soil is commonly supposed to be far superior to a clayey soil, on which to build a house ; and so it is, other things being equal, which is precisely what as a rule they are not. A gravelly soil is good or not according to the nature of the subsoil and the direction of the watershed. If there is a considerable depth of sand or gravel, and a distinct watershed away from it, no better site could be selected ; on the other hand, it is quite possible for a gravelly site to be the very worst site possible. If, as very often happens, the subsoil be gravel or clay, the water which falls on the gravel will sink through it till it reaches the clay, and no further, for the clay is not permeable by water. Having reached the level, it must flow away as it would from a clay surface, only percolating through the soil instead of running above it until it reaches the lowest level in a stream or otherwise.. But if it does not flow away, if there is no watershed, it will accumulate in the soil, just as it might in a reservoir on its surface, and, rising higher and higher, at length reach the foundations of the house, and sap the timbers of its flooring. Such a condition of things is most likely to occur under the following conditions : Suppose, by the agency of the great forces at the disposal of nature, a mighty basin has been hollowed out of the clay and subsequently filled with gravel. This, we know, not unfrequently occurs. Out of this basin there is no escape for the ground water, until it topples over the clay banks of the basin, and so it rises and falls according to the season. Such a gravelly soil would be the very worst site for building purposes it would be possible to select. But if now in this sea of gravel there was a little island of clay, that would be subject to no such variations in the rise and fall of its ground-water. When rain fell it would run off its surface into the gravel beyond ; its own ground-water would be invariable. If, therefore, underneath the gravel there is a watershed which will allow of the free escape of the ground-water, no site could be drier or healthier ; if not, no site could be worse.

The nature of the ground site having been settled, the next thing perhaps, especially in the country, is to consider the direction and nature of the prevalent winds. If possible, the house should have its greatest exposure to the direction whence come the driest winds, and has the best exposure to the sun. Each district and each situation must be considered by itself ; protection from the worst winds and exposure to the most favorable being sought in every case   This, perhaps, is

hardly a proper place to speak of the kind of trees which should be planted round a house with a view to protection ; but such should be carefully selected with a view to shade in summer and protection in winter. It should not be-left to haphazard. The materials of the house itself also merit consideration. Every one may not be able to tell the difference between good brick and bad brick, but there are people who can ; and it is better to pay for such skill than to have one's house constructed of bad material. In districts where stone is used, this too requires to be selected. Some stones are so porous that with a good beating rain on the outside the water pours almost as rapidly inside. Such houses are always damp ; they cannot be kept dry. Of the wood used little need be said, beyond the necessity to provide for future comfort by having nothing but seasoned timber ; otherwise, imperfect fitting doors and window-sashes will try tempers and give rise to drafts. As to plan or elevation, each may suit his own fancy ; but from the health point of view there are certain broad rules to be observed. If the design is fantastical, certain portions of the house must remain unused, the rooms being too small, or if used, they can only be so compatibly with health by a ventilation which can hardly fail of being too free, that is to say, the place must be drafty. Besides, these odds and ends of places are difficult to keep clean. Simplicity of design should as far as possible be aimed at. Again, the rooms should be well balanced. Frequently we see in houses the whole building sacrificed for one or two rooms, and very often these are used almost entirely as show-rooms or reception-rooms.

The living-rooms should correspond to the size of the family, and as a rule the bed-rooms should be larger than the sitting-rooms. We consume a very great portion of our time in bed, not less as a rule than one third, and during that period most people have their windows shut, so that the air can only change by the chimney, and that too is often stuffed, and any chinks left in the walls or door. It is desirable, therefore, that the sleeping-rooms should be so large that the total quantity of air they contain cannot become very greatly fouled, even suppose it is not changed during the period devoted to sleep. There can be no doubt but that breathing the same air over and over again is unhealthy. Houses should always be built, be the plan what it may, so as to admit of through-and-through ventilation. In some parts of the country a horrid plan prevails of building houses back to back, so that the back wall suffices for two streets. Nothing could be more pernicious to health than this, for it is impossible, however desirable, to obtain sufficient ventilation. For this reason, too, houses should not be built with well-squares in the midst of them ;

that is to say, as hollow squares. The stair-cases should be so situated that from it, if not directly, every room of the house may be ventilated. As for the rooms themselves, the best system of ventilation is the natural one, that is to say, by the doors, windows, and grates. All artificial systems of ventilation have hitherto proved failures. The perfection of ventilation is where a room is kept con-stantly sweet and fresh by an insensible change of air. If perceptible, the ventilation is sure, under other circumstances, to become drafty, and drafts are to be carefully avoided. The open fireplace is undoubtedly of great service in ventilation, yet it occasions great waste of fuel, and has the disadvantage of not keeping up an even temperature. The open grate draws upward the cold air which has entered the room, heats it, and causes it to ascend the chimney ; in this way a con-stant current is kept up. Often grates have this great disadvantage, that it is hardly pos-sible with them to keep up an equable tem-perature for the four-and-twenty hours. If a good fire is lit, say at bedtime, it warms the room to begin with ; but as morning ad-vances, and the temperature outside sinks lower and lower, so too the fire sinks, and it goes out just when it is most wanted, that is, in the early morning hours. This is the time which is most trying to those who are sub-jects of chest affections ; it is then when coughs become most troublesome. Such stoves as Arnott's are free from this incon-venience ; but they necessitate ventilation by artificial means. But, besides air, houses should admit plenty of light. The human plant, just like the vegetable one, grows pale and sickly if deprived of air. Hence big win-dows should be procured, if possible ; if the light proves troublesome by its excess, it is easily shut out. Too many creepers should not be trained against the house ; they are picturesque, but they harbor damp.

There is nothing more important in the ordering of a house than the water-closets and drains. In the country, where there is little water-power, and no means of getting rid of sewage, there is only one thing or course compatible with safety—avoid water-closets altogether. Even in towns they are hardly tolerable ; in the country they are intolerable. It has been clearly proved that many diseases are spread by their means, if they do not in-deed arise from them originally ; and these diseases are very fatal : typhoid fever is a good instance. Even in towns the closet should be as far from the living and sleeping portions of the house as possible, and the drains ought not to ventilate through it. Often in the dull days of early winter, when the atmosphere is light and the barometer low, the tense gases rush up through the pipes leading to the closet, and so enter the house. So too in cold weather, when the doors and

windows are close shut in order that the tem-perature may be kept high, often the easiest access of air is from the water-closet. These things are not pleasant to reflect on, but they are facts. The only thing to be done is to let the drains be ventilated by a shaft reaching the top of the house, and having its basis in the drains. The closet is best kept sweet by a teaspoonful of carbolic acid in a gallon of water. In the country earth-closets alone should be allowed in the house, and outside, the same dry-earth system ought to be em-ployed ; closer attention to this rule would save many a life from typhoid fever.

Previous to building a house, it is now very common to have the drinking-water analyzed. This is a very good rule, for bad drinking-water is a sure source of diseases. As a rule a water which contains much nitrate and chlo-ride is to be avoided. They commonly indi-cate contamination somewhere, and this con-tamination may at any time assume a differ-ent character. Chlorides and nitrates are themselves harmless, but their sources may not be so. The storage of the water in the house should be attended to. Lead cisterns used to be the rule ; now galvanized iron ones are coming into use. The lead is dangerous with soft water, if it stands long and is not run off. This risk is avoided by the other. The cistern should always be kept covered, so as to prevent rats, mice, and the like finding their way into it, and being drowned remain in it to flavor the water. Moreover the cis-tern should be readily accessible, so as to be easily cleaned. Finally, the drains should be earthenware pipes : with bricks, rats will get in, make holes, and allow the sewer gases to escape, to the detriment of the health of the inmates. It is well to understand the me-chanism of the traps, so as to know if they are in working order. A very little attention to this slight detail will often save much in-convenience, and guard against detriment to health. (See VENTILATION and WARMING.)

HUMAN BODY, composition of. (See COMPOSITION OF THE HUMAN BODY.)

HUMIDITY.—The air is never free from moisture under ordinary conditions ; for from the surface of the earth, and from rivers, lakes, seas, etc., evaporation is going on con-stantly ; this aqueous vapor, ascending into the higher and cooler regions of the atmos-phere, forms clouds ; and this vapor descends to the earth again as rain, snow, or hail. The amount of evaporation varies much at differ-ent seasons of the year, being much greater in summer than in winter. The hotter the air, the more aqueous vapor will it hold ; this can be shown by bringing a glass of cold water into a hot room, when a dew will be deposited outside the glass ; this is due to the fact that the cold glass cools the air immedi-ately around it, and, being thus cooled, the air can hold less aqueous vapor in suspension. It is to the presence of humidity or moisture in the air that the deposition of dew can take place. Instruments, called hygrometers, have been devised to find out the amount of watery vapor in the air at any temperature and at any time. Hot and damp air generally has a relaxing effect on the constitution, while cold and damp air is unsuited for those who suffer from chest affections, and who are liable every winter to bronchitis and winter cough.

HYDATIDS are cysts formed by the ova of the Tænia Echinococcus or tape-worm of the dog. In the article ENTOZOA it is shown how tape-worms occur in man ; the ova, however, of those tape-worms which infest the human subject will not produce in him the mature worm at once, but it is developed in some other animal, as a Cysticercus ; so in the case of the worm met with in the dog or wolf, the ova passing into man do not develop a worm but a Cysticercus or hydatid. These bodies, minute at first, pass from the alimentary canal into the system and may be carried by the circulation into the nearest organ. The liver is the organ which is most commonly their seat ; but they have been found in the lungs, heart, brain, kidney, pelvis, and bones, etc. When they are deposited in an organ a fibrous cyst (the ectocyst) is formed around, and within this is the endocyst, a clear, gelantinous membrane which lines the former, and itself incloses a large collection of watery fluid holding in solution some common salt and phosphates. These cysts vary in size from a marble to a child's head, and when they attain such large dimensions, surgical interference is called for ; they grow for a long time without the patient being aware of their presence ; at last, perhaps, a firm, almost painless tumor is felt in the abdomen ; it is nearly always globular and elastic, and gives a feeling as if it contained fluid. At times inflammation is set up around it, and then there will be pain, fever, and constitutional disturbance. The natural tendency of these cysts is to grow larger and larger, and they may at length burst ; in this way they have escaped into the heart, pleura, peritoneum, intestinal canal, etc., and nearly always with a rapidly fatal result. No medicines are of any avail in checking their growth or in causing the absorption of the fluid. Various methods have been adopted to empty the cyst by drawing off the fluid, and in a great many cases this is done with excellent results. Now and then inflammation of the cyst takes place and the contents become purulent ; the only chance then for the patient is to have a free opening made and let the matter out. Still more rarely the cyst dies early and never attains a large size ; the contents become of a cheesy consistence, and the cyst may remain in the body for years without giving rise to any symptoms whatever. It has been shown that eating meat is a fruitful source of tape-worms, but the hydatid of the tænia echinococcus seems to be taken in by eating water-cresses or uncooked vegetables, or by drinking water in which the ova of the tape-worm have entered. Fortunately this troublesome disease is of rare occurrence ; when present, it may attack any age and may be met with equally in either sex ; the danger to the patient depends somewhat on the seat of the hydatid ; it is less fatal when it occurs in the liver than when it is developed in the heart or lungs. (ENTOZOA.)

HYDRAGOGUES are remedies of the purgative class, which produce copious watery stools. Some seem to give rise to fluid specially, apart or in excess of their purgative effects ; others seem to do so only incidentally. Elaterium is a remedy of this kind, so are colocynth and most of the purgative salts. Gamboge, too, produces very watery stools. Compound jalap powder is a remedy much used in this way ; so too is compound scammony powder. Hydragogue purgatives are employed mainly to get rid of excessive fluids, as in dropsies, especially of the cardiac kind, or in the earlier and acuter stages of renal dropsies. Some forms of dropsy are not much affected by them.

HYDRATE OF CHLORAL. (See CHLORAL.)

HYDROCEPHALUS is a disease of which the main feature is an accumulation of fluid in the central cavities of the brain. Sometimes the child is born in that condition, and then the dangers of delivery are considerably increased ; more generally the symptoms appear after birth, and become more marked in the second and third years of life. At first, and before the child can walk, nothing particular may be noticed, except that it has a large head. But as it grows older it will be found that the child is not so sharp as others of the same age, that it walks with difficulty, that its teeth are backward in appearing, and that the size of the head is out of all proportion to the rest of the body ; the upper part of the skull enlarges so that the face appears much dwarfed ; the anterior fontanelle remains open ; the eyes are very apt to roll about, and there is inability to look upward ; the skin over the scalp is smooth and tense, and often marked with the superficial veins. The rest of the body is generally badly nourished, and the legs are often bowed if the child has walked too early, and the wrists and ankles are enlarged. The fluid which distends the brain is poured out slowly, and consists of little more than water and common salt ; the ventricles of the brain are very capacious and smooth within, while the brain-substance is stretched out over the fluid, and is, therefore, thinner than usual. The disease is altogether chronic in its course, and gives rise to no pain or any urgent symptoms ; the appetite may be unimpaired, and the general health, although seldom good, does not seem much affected. Such children are more liable than others to catch infantile disorders, such as whooping-cough, convulsions, measles, scarlet fever, etc. When the mischief is but slight the child may grow up to adult life ; but when far advanced death generally takes place before the child has reached five years of age. The *treatment* will consist in giving nourishing food and tonic medicines ; bathing with cold water or sea-water may do good. Bandaging the head has been recommended, and various preparations of mercury have been rubbed in, but very little in this way can be done. This disease is often called *chronic hydrocephalus*, to distinguish it from *acute hydrocephalus*, an affection of quite a different character, and which is described under the head of MENINGITIS.

HYDROCHLORIC ACID, also known as muriatic acid or spirit of salt, is not perhaps so much used as a remedy as is its combination with nitric acid, called nitro-hydrochloric acid. The acid itself is a waste product in the manufacture of common washing soda, and it is cheap enough ; but it is often impure, and requires to be carefully purified to get rid of arsenic and other substances with which it is commingled. The strong acid is not used in medicine. It is a powerful caustic, and produces a white stain in the skin which may be completely destroyed by its application. Several cases of poisoning have occurred through its use, but not so many as might be supposed considering its frequent use in the arts. In the dilute form, given in doses of from 10 to 20 drops well diluted in water, it may be found useful given immediately after food in aiding digestion. It is the natural acid of digestion, being secreted by the stomach for that purpose, and itself has, at the temperature of the body, considerable power in dissolving meat or solid white of egg. It is also of some use as a gargle diluted with water, but it is best given along with chlorate of potass for this purpose, when it has the power of setting free a substance allied to chlorine. This certainly has a beneficial effect in foul ulceration of the throat, and in diphtheria where sloughs are formed and tend to decompose, still further poisoning the system.

HYDROCYANIC ACID. (See PRUSSIC ACID.)

HYDROGEN (SULPHURETTED) is not itself used in medicine, but mineral waters which contain it free and in the form of sulphides of the alkalies are of very great value. The smell of this gas resembles rotten eggs ; and the so-called sulphurous waters do the same. The baths containing sulphur are exceedingly useful in certain diseases of the skin, chronic gout and rheumatism, and chronic lead poisoning. A preparation for destroying the itch-animalcule is made by boiling sulphur and quicklime together. This is very efficacious. Given internally, either as mineral water or as sulphides, this substance is said to benefit scrofulous ulcers very greatly. Sulphur of calcium is commonly used for the purpose, and it is described as exceedingly efficacious. Small doses of the sulphides seem to relax the bowels, as sul phur itself does.

HYDROGEN is one of the gases contained in water, and is notable for its extreme lightness ; as far as we know, however, it possesses no remedial action. A compound of it, called peroxide of hydrogen, has been tried. It possesses the property of freely giving off the oxygen it contains, and so may be useful in certain conditions. Applied to the skin it whitens or bleaches it, and, collected in a somewhat similar way on almond surfaces, has been used to favor the healing of sores.

HYDROPATHY, also known as the Water Cure, is a system of dealing with disease invented by a German named Priessnitz. His doctrine was that plain water outside and inside was all that was necessary to cure disease. Undoubtedly he committed grievous errors, but he was not like a consulting surgeon or physician ; he had his patients under his thumb, and could diet them and manage them as he liked. Undoubtedly, too, he introduced a very potent means of dealing with some disorders which has been too much overlooked by the regular faculty, chiefly on account of its antecedents, and because too much was claimed for it. Now, however, things seem to have reached their proper level, and many practitioners are glad to send their patients to hydropathic establishments provided they can rely on their instructions being carried out. Hydropathic establishments are generally situated in places of great natural beauty, which induces the patients to exercise. The diet at them is usually plain and wholesome, and early hours are insisted on. Such a change in itself often does good, and if to that is added the stimulating effects of the cold-water bath skilfully applied, and the use of appropriate baths for the purpose of relieving that great natural organ for eliminating refuse matter from the system, the skin ; if, moreover, the kidneys are relieved, the liver got to act in a healthy manner, and the bowels regulated, the whole cannot fail to do good. But too much has been claimed for hydropathy ; cases have been sent to the water-cure establishments altogether unsuited for the treatment, and the consequence has been disappointment. The baths mainly used are the shallow bath, in which the individual sits immersed up to the hips, is well laved, and finally has a bucket of cold water thrown over him ; the sitz-bath, where the water plays on the lower portion of the body ; the rain-bath, where every portion of the body is acted upon in the same way ; the shower-bath, where the rain comes only from above ; the douche-bath, where a column of water of varying weight and force is made to play on different parts of the body. Then there is the Roman or Turkish bath where hot air is used to induce sweating. The body is well kneaded, and the perspiration abruptly stopped by a douche or plunge-bath. But one of the chief means for dealing with disease is the wet pack. This is a very valuable means of reducing the temperature and getting the skin to act. The patient must be in good heat. He is stripped naked, laid on a wet sheet, and packed in it like a mummy ; then follow blankets in the same way ; a feather-bed covers over all ; a wet towel is applied to the head, and the patient is left usually to sleep. Presently the cold gives way to heat, the skin is enveloped in one vast poultice, and if there be much irritation of it the relief is magical. A cold douche or plunge or rain-bath ends the process, which should not last much over half an hour.

HYDRO-PERICARDIUM means a passive effusion of serum into the sac of the pericardium, or membrane inclosing the heart ; it occurs in many cases where dropsy of other parts is present, as in diseases of the heart, lungs, and kidneys.

HYDROPHOBIA is the term applied to the conditions which occur in the human being after the inoculation of the saliva of a rabid animal, most frequently of dogs or cats. The term in its derivative sense is not always applicable, as the "dread of water" is not always present either in the patient or in the animal inflicting the injury. Mr. Youatt in his treatise on canine madness thus describes the symptoms of this disease in dogs, from whom the disease is most generally derived : " The disease manifests itself under two forms : the *furious* form, characterized by augmented activity of the sensorial and locomotive systems, a disposition to bite, and a continued peculiar bark. The animal becomes altered in habits and disposition, has an inclination to lick or carry inedible substances, is restless, and snaps in the air, but is still obedient and attached. Soon there is loss of appetite and thirst, the mouth and tongue swollen ; the eyes red, dull, and half closed ; the skin of the forehead wrinkled ; the coat rough and staring ; the gait unsteady and staggering ; there is a periodic disposition to bite, the animal in approaching is often quiet and friendly, and then snaps, latterly there is paralysis in the extremities ; the breathing and deglutition become affected by spasms ; the external surface irritable, and the sensorial functions increased in activity and perverted : convulsions may occur. These symptoms are paroxysmal, they remit and intermit, and are often excited by sight,

hearing, or touch. The *sullen* form is characterized by shyness and depression, in which there is no disposition to bite and no fear of fluids. The dog appears to be unusually quiet, is melancholy, and has depression of spirits ; although he has no fear of water he does not drink, he makes no attempt to bite, and seems haggard and suspicious, avoiding society, and refusing food. The breathing is labored, and the bark is harsh, rough, and altered in tone ; the mouth is open from the dropping of the jaw ; the tongue protrudes, and the saliva is constantly flowing. The breathing soon becomes more difficult and laborious ; there are tremors and vomiting and convulsions." This disease has been noticed in the cat, horse, wolf, and other animals, and is from them communicable to man. A knowledge of the periods at which madness attacks dogs is of great importance, so as to put them beyond the power of causing injury before they become a public danger, and with this object the Council of Hygiene of Bordeaux issued the following instructions : " A short time, sometimes two days, after the madness has seized the dog it creates disturbances in the usual condition of the animal which it is indispensable to know. 1. There is agitation and restlessness ; the dog turns himself continually in his kennel. If he be at liberty, he goes and comes, and seems to be sucking something ; then he remains motionless, as if waiting ; he starts, bites the air, seems as if he would catch a fly, and dashes himself, barking and howling, against the wall. The voice of the master dissipates these hallucinations ; the dog obeys, but slowly, with hesitation, as if with regret. 2. He does not try to bite, he is gentle, even affectionate, and he eats and drinks ; but he gnaws his litter, the ends of the curtains, the padding of the cushions, the coverlid of beds, the carpets, etc. 3. By the movement of his paws about the sides of his open mouth, one might think he was wishing to free his throat of a bone. 4. His voice has undergone such a change that it is impossible not to be struck with it. 5. The dog begins to fight with other dogs ; this is decidedly a characteristic sign, if the dog be generally of a peaceful nature. The numbers 3, 4, and 5 indicated are already very advanced periods of the disease, and the time is at hand when man will be exposed to the dangerous fits of the animal if immediate measures be not taken. These measures are to chain him up as dangerous, or, better still, to destroy him." This notice is suggested to be printed on the back of the notice of the dog tax, on the back of the receipt for this tax, and finally on the back of the permissions for hunting.

When the disease has attacked the human being, we find striking points of resemblance to that already quoted as occurring in the dog, but at the same time several points of difference. At first no symptoms manifest themselves, and it is usually not until some weeks afterward that the effects of the introduction of the poison into the system appear. The first are general, and those of general malaise, fever, nausea, loss of appetite, and restlessness. The peculiar or special symptoms, however, which set in later, comprise an irritation in the locality of the bite, simulating neuralgia. The cicatrix becomes red and swollen, and discharges a thin unhealthy pus. The actions and affections are changed, children, if the objects of the injury, become shy ; adults, depressed, lonely, anxious, and melancholic, and anticipatory of resulting danger. Some, on the contrary, are unusually irritable and ill-tempered. There is a characteristic anxiety, with a sense of weight and pressure in the chest, disturbed sleep, and frightful dreams ; these symptoms, with complications, constitute what may be regarded as the first or primary stage of the disease. The second may be regarded as the imitative stage, having "hydrophobia" as its characteristic. It is ushered in with stiffness of the muscles of the throat, jaws, and tongue, pain in the pit of the stomach, with chills and drowsiness, convulsive spasm of the muscles of deglutition, causing swallowing to be difficult or impossible. There is a great dryness of the mouth with burning thirst, there is spasm of the muscles of the larynx, causing the peculiar hawking or barking noise in the attempts to expel the secretions of the mouth and fauces. There are convulsive paroxysms, and the sight or sound of fluids produces aggravation of them, the mind becomes in a state of fearful agitation, dreadful feeling of despair. Sometimes the mental disturbance may be slight, but generally it is the reverse, bordering on maniacal fury. The third stage or stage of decline, is attended with rapid depression and nervous exhaustion, with incoherency and delirium, and death takes place either from choking, or during a convulsive attack, or from exhaustion. The duration of the disease varies from seventy-four hours to six or seven days, and there are cases on record which have lasted for two or three weeks.

Treatment.—The treatment in the first instance, on the receipt of the bite, must be immediate, and the injured part should be immediately destroyed by some powerful escharotic, which must be used unsparingly over the whole surface and depth of the bite. Thus, nitrate of silver, caustic potash, nitric acid, sulphuric acid, arsenical paste, chloride of zinc, the actual cautery (hot iron), boiling oil, etc., are all of use. In the absence of these means, the bitten spot should be cut out at once ; ligature tightly applied *above* the bite, is only of use till more active treatment can be procured. The probability of cure and recovery should be impressed on the patient with a view to avert despondency. As drugs, stimulants and other antispasmodics, anodynes or narcotics, and tonics, are frequently indicated. Thirst should be alleviated by ice. The course, however, of the disease is usually fatal.

HYDROTHORAX, or, as the name signifies, water in the chest, is met with in cases of disease of the heart and kidney ; it is, in fact, analogous to the dropsy in the legs, which occurs in those affections. Either from an alteration in the quantity of the blood or from a change in its quality, serum is poured out into the pleural cavity, and generally both sides are affected, although not equally so. It is attended by no pain, and its chief result is to cause an increase in the difficulty of breathing, with which such patients are mostly troubled ; this is the case in consequence of the lungs being compressed by the effused fluid, and so there is less room for the air to enter. The *treatment* will consist in the use of purgatives, so as to remove the fluid by the bowels ; by the action of sedatives, if the heart's action be very tumultuous ; and by rest in bed and nourishing food. Any special treatment must be decided upon according to the particular form of mischief in the heart or kidney, and upon the state of the patient.

HYDRURIA means an excessive secretion of limpid, watery urine.

HYGIENE is the science and art of preserving health, by the appropriate nourishment of the body and the proper regulation of its surrounding conditions. It will be seen from this definition that the subject is much too wide to be treated of in a single article, and, in fact, by the special stipulation of Dr. Lankester, the editor, a very large proportion of the present work is devoted to topics connected with Hygiene. (See AIR, BATHS, CLIMATE, CLOTHING, DIET, DIGESTION, DEODORANTS, DISINFECTANTS, FOOD, FUMIGATION, GYMNASTICS, HEALTH RESORTS, HOUSES, LIGHT, SANITARY REGULATIONS, VACCINATION, VENTILATION, WATER-CLOSETS, and WARMING.) In as far as the *prevention* of disease could be dealt with in such a work it has been done ; and any declension from bodily health into disease is treated of under the various diseases.

HYGROMETER is an instrument for observing the dew-point or the amount of moisture in the air, and various kinds have been made for the purpose. The rate of evaporation varies at different seasons, being the greatest in summer and the least in winter. From the surface of water, and during a breeze, it is possible for fifteen thousand gallons of water to be evaporated from an acre in twenty-four hours during the summer months, while in winter time about three thousand five hundred gallons will evaporate from the same area in the same time.

HYOSCYAMUS. (See HENBANE.)

HYPERÆMIA, a technical term for increase in the quantity of blood in a part ; it comes on in every case of mechanical obstruction to the circulation, and precedes inflammation of a tissue. A familiar example of it is seen in the phenomenon of blushing.

HYPERPYREXIA, a term applied when the temperature of the body is very high, as in some cases of rheumatic fever, when 107° or 110° Fahr. may be reached, and a fatal result may be expected ; the only relief at present known is by cooling the patient down by means of a cold bath, or by packing in sheets wrung out of ice-cold water.

HYPERTROPHY, a term applied to an increase of a healthy tissue without any change in the quality of its component parts ; thus a muscle is said to become hypertrophied when it is increased in size by using it, as in the arms of a blacksmith or athlete. (See HEART.)

HYPOCHONDRIA, also known by the old English equivalent of the Vapors, seems to be the correlative in the male sex for what in the female we call hysteria. The conditions have long been well known, though very various causes have been assigned to it, the favorite being for many years the formation and circulation of " black bile," for melancholia means this exactly. Nowadays we assign to it a nervous orgin, and though there may be no actual disease, the condition is one very hard to get rid of. Most frequently there is functional derangement of some part, generally of the stomach, though sometimes there is really alteration in structure. The chief characteristic of hypochondria is a morbid self-consciousness similar in some respects to that of hysteria, but generally taking a different direction. The hypochondriac usually fancies himself the subject of all the ills that flesh is heir to. There is usually a great dread of death, and the patient resents being told there is nothing the matter with him. This is so much the case that it is unwise for a practitioner to deal with such a patient in an off-hand manner ; he cannot be reasoned out of his malady ; for the time he may feel convinced, but presently, when he is left to himself, his fancies resume their old sway. Withal such a man always has something the matter with him—generally a most obstinate indigestion ; and if that be cured the patient is generally in a fair way to be relieved of his mental symptoms. More rarely there is nothing the matter except the mental influences, which become disordered very often in consequence of self-indulgence, until the regularity and controlling power of the head have been utterly lost. Frequently this malady assumes the character of insanity, some member of the body being supposed to be lost, or so altered as to be useless, or worse than useless. Hypochondria seldom occurs in those who lead an active, healthy life in the open air. It is most frequent among those who, living well, take little exercise, and whose lives are what is termed sluggish. Such individuals will often be subject to short attacks of a malady of this kind, which a little laxative medicine and exercise in the open air will soon carry off. It is very frequent, too, among those who, having led an active life, retired to comparatively early rest and quiet, as they think. Such having seldom any internal resources in the way of education and cultivation, have recourse to morbid retrospection ; their own feelings, desires, and aims become their only company, and so each uncomfortable sensation is pondered over until some comparatively slight ailment becomes a thing of the first magnitude. This once established, everything is made to minister to it ; the unfortunate individual takes to looking at his tongue, feeling his pulse, and generally observing his bodily condition a dozen times a day. Those who have long had their minds strained by overwork are liable to a somewhat similar form of disturbance. In them, however, the bodily condition is less the subject of notice—it is the mental ; it is those which suffer most. They become miserable objects for the time being ; they lose their nervous energy, become weak and wretched ; they fear to cross the streets ; they live in constant dread of having done something wrong, or of having wrong attributed to them ; they are the shadows of their former selves. Hard students are frequently so troubled ; among students of medicine this generally takes the shape of fancying themselves the subjects of disease of some greater organ of the body—lung disease and heart disease are the favorites ; but most men who have worked hard in their time will confess to having been at one time or another the supposed subjects of half the diseases in the nosology. We have already hinted at the causes of this malady. These are essentially the continued use of one part of the system, the other being left without due exercise.

Treatment.—In one set of cases mentioned above exercise for the head is wanted ; in another exercise for the body. For the retired man of business something is wanted to keep his mind engaged, and such may often be found in the affairs of the parish or township in which he may be placed. In short, having given up attending with all his might to his own private business, it is well for him and generally well for the public, that he should devote some part of his leisure to that public business which in this country is generally relegated to private enterprise. To the other set of patients mental work is already too severe a burden—they ought to have more relaxation, and this relaxation ought to be devoted to bodily exercise. The selection of the kind of exercise may in great measure be left to each individual. But this is to be borne in mind, that extremely violent exercise for a few moments will not answer the same purpose that moderate exercise for a longer period will. So, too, violent exercise one day and quiet the next will not answer : the great thing is to keep the system equable. Such men as desire to excel in mental work should not attempt to vie with an athlete. The two things are very seldom compatible. These things are, however, rather to be looked upon as means of maintaining health in all these circumstances, or in getting rid of slight attacks of the malady. They will not suffice for more serious ones. When a man is fairly " hypped," as it is called, there is only one satisfactory remedy—total change of scene and pursuits. If the condition has arisen from things bodily or things mental, then the same rule—change of scene and change of occupation. And to this there need be less objection, for a man's work, under such circumstances, is of little value. If he have only a certain routine to get through, which requires no considerable use of the mental faculties, he may succeed ; but if the calling is such as to require a keen use of the judgment, the sooner he gives it up for the time being the better for all concerned. It is wonderful what a difference a few weeks or months may do for a man. Often we have seen a new lease of life gained by a short rest and change of scene. After these the general rules above laid down are to be duly observed ; especially is digestion to be duly looked after, but only by proper food and appropriate exercise, not by medicines, if they can be avoided. There is but one final caution we desire to enforce, and that is this very earnestly. As a rule, hypochondriacs sleep badly—often those, especially, who have too much mental and too little bodily work are troubled with frightful dreams and restless nights. To these we say, avoid opium or other sedative ; if the bowels are not open, try a blue pill and a black draught.

HYPODERMIC INJECTION is a procedure which has been adopted of late years, by which medicines may be inserted under the skin, and absorbed into the blood, without having first to enter the stomach. It has been found in practice to be a very convenient method ; for not only can a smaller quantity of the drug be used, but its action is rendered quicker and more precise, while the stomach is not so much disturbed. Thus there are cases where the patient can take very little food, and opium or morphia swallowed in medicine will bring on sickness and distress ; but if a smaller quantity be inserted under the skin, the stomach will be at rest, the pain relieved and no disagreeable effects follow. The fluid, which is concentrated, so that five or ten drops will suffice, is placed in a small glass syringe so graduated that one can easily see the exact amount to be injected. To the lower end of the syringe is attached a fine and hollow needle, so that the skin can be readily pierced and the fluid introduced. It is not necessary to inject close to the seat of pain ; any part of the skin will do equally well, and a portion which is less sensitive is the best, as the outside of the arm or shoulder. Very slight pain attends the operation, but it should not be adopted without medical advice, as poisonous effects might follow its use.

HYSTERIA is a malady chiefly confined to women, but by no means necessarily so ; though if it does occur in the male it is in the weak imperfect creatures, who approximate to woman mentally and morally if they do not physically. The fact that hysteria may prevail in the male does away with the notion that it is in any way connected with the womb or ovaries. Hysteria is apparently connected with, if not due to, an imperfectly balanced mental and moral system. The controlling faculty is either in abeyance or imperfectly developed, while the susceptibility of suspension is often morbid. It is most common in young women who are unmarried after the ages of puberty up to a very variable period, this period depending in great measure on the time when hope of marriage becomes faint. It is much less frequent in married women who have children, but in married women who have none it is perhaps most common of all. The hysterical tendency manifests itself in very many ways ; sometimes it assumes the form of a regular stereotyped kind of fit. In others it may simulate any disease under the sun, and frequently it appears in the most anomalous shapes it is well possible to conceive. The true hysterical fit or paroxysm commences in various ways ; most frequently the patient is observed for a second or two staring before her with her eyes wide open and then falls to the ground. Here she may lie quiet for a moment as if dead, then suddenly begins all sort of shrieks, screams beating of the breast with clinched fists, tearing the hair or garments, seizing and scratching anything near at hand. Sometimes the limbs seem convulsed and the arms rigid. Presently the patient will be quiet, and suddenly break out into a fit of laughter, beating the ground with her heels ; this again will cease ; she will sob till you think she is heart-broken, and this goes on till she is exhausted, when presently she will come to herself with a very imperfect recollection of all that has taken place. Usually there is presently a profuse discharge of limpid urine, which occasionally indeed is discharged during the attack, but this is only one phase in the numberless forms assumed by the malady. These attacks sometimes closely simulate those of epilepsy, a similar form of malady, and consequently it is of the greater importance to be able to tell which is which, the chances of recovery or the reverse being so very much greater in the one than the other. In epilepsy there is complete insensibility ; during the attack in hysteria there hardly ever is ; the patient generally has a notion of what is going on, and sometimes knows everything perfectly well, as the proposal to saturate her with a pail of water not unfrequently brings the patient round. Moreover the breathing is not interrupted, and the heart-beat is not greatly altered ; the pupil of the eyes always responds to the stimulus of light, and its mode of termination is different. Epilepsy generally ends in deep sleep, and the patient is completely unconscious of everything ; not so in hysteria, as already said. One thing on which we are wont greatly to rely is the state of the tongue. In epilepsy the muscles of the tongue are convulsed, as are most of the others. Accordingly the tongue is thrust forward, while the jaws are ground together, so that it is rare in a case of well-marked epilepsy for the tongue to escape laceration ; it is just as rare or even more so to find it affected in hysteria.

There are two things very common in hysteria : a choking feeling in the throat, and stitches of pain in various parts of the body. The choking sensation seems often due to a ball, and hence is termed the *globus hystericus ;* the sharp pain, especially as it affects the head, often goes by the name of *clavus hystericus.* Certain parts of the body are very liable to this pain ; these are mainly the left side under the nipple, and a corresponding point behind, so there may be pain in the lower portion of the abdomen, in the knee or other joints, in the groin, or the top or side of the head, in fact, in every part of the body. This pain is often of a very serious character, and there may be increased tenderness of the part, but as a rule a slight stratagem will serve to divert the patient's attention, when both pain and tenderness will disappear. It is the opposite condition to this which sometimes enables hysterical women to appear perfectly insensible to all pain or injury to certain parts of the body, and has rendered efficient service to various kinds of impostors. Akin to this is a tendency on the part of some hysterical females to attract to themselves public attention, and so deprive themselves of food in order that they may seem to exist without it. Such usually take to bed, and as they undergo no exertion an exceedingly small quantity of nutriment will suffice to keep them alive. Such a case was once known as that of the Welsh fasting girl, where the unfortunate creature was allowed to die instead of being made to swallow her food. If carefully watched and deprived of the small quantity of food they require, and which they manage to secure unseen, they are bound, as

would be any living thing, to perish. There is but a step from this to the shamming of disease, and there is hardly a disease under the sun which may not and has not been simulated by hysterical women. But there are some which are hardly feigned ; chief among these are cough and shortness of breath or breathlessness ; hiccup, too, is another thing commonly assumed, as is yawning and sobbing, but it is very hard to say where the voluntary and the involuntary impulses begin and end. For it is useless to look upon hysterics as other than a real disease, tormenting alike to the patient and to the patient's friends, and to suppose its assumption is entirely voluntary on the part of the female is a great mistake. True, the symptoms can be got rid of by powerful mental influences, and not unfrequently are so got rid of, but the disease is not cured except the patient be at the same time removed from the mode of life which has led to the loss of controlling power ; the malady is sure to return. Still worse policy is it to yield to such patients one single iota. There is one rule, and one rule only—" kindness and firmness ;" the judicious use of these will overcome the most troublesome cases. There is a form of hysteria very troublesome. The patient complains not of paralysis, but of so much pain that inflammation is suspected, and the pain sometimes remains for years. Occasionally when there has been injury to a joint, the hysterical condition remains long after the injury has been cured. These two classes of cases constitute the great field of diseases to various kinds of quacks, bone setters and the like, those who tell their patients to get up and walk with the desired result.

Treatment.—The treatment of hysteria resolves itself into management during an acute attack and management during an interval. Suppose the patient has a fit and struggles about, the best thing to be done is to remove all tight fastenings about the body, surround her by cool air, and prevent her from hurting herself. No more fuss should be used than is necessary to do this, or the patient may continue her struggles very much longer. If the attack seems likely to continue too long, we must try and stop it by using sharp but not brutal remedies. Cold water plentifully applied to the head and face usually does most good ; strong smelling-salts held to the nose are also beneficial. Above all, the patient should be kept quite quiet, everybody should be removed, save a nurse and the medical attendant, and let her know that her case is understood and that she will gain no sympathy. But the worst of these attacks must be allowed to wear themselves out, for it is in the interval that the physician must aid. The most important thing is to obtain moral control over the patient. To this end it is not necessary to be brusque or hard, but it is necessary to be unyielding. Such patients always have ill health, and this must be seen to. Nervine tonics, like nux vomica and oxide of zinc. usually do good, and as the menstrual function is generally disordered, iron and aloes may be prescribed with advantage. Cold baths and a healthy, quiet mode of life are of the greatest service ; early hours should be the rule, and while the patient should be treated kindly at home, anything like weakness in dealing with her is worse than folly, is criminal. Valerian is commonly given, but about its value opinions do not agree. The food should be good and plain ; anything like fancy in articles of diet should be discouraged ; change of air and scene are almost always beneficial, as assuredly are amusement without excitement, and an occupation in life.

I.

ICE is the name given to water when it is cooled down below 32° Fahr. Its properties and appearance are too well known for description. It is very important as a remedial agent.

ICE-BAGS are made generally of India-rubber, into which pounded ice is placed and applied to the desired spot, or it is equally advantageous to break ice into small pieces and put them into a bladder ; in this way cold can be applied for as long a time as may be wanted without wetting the patient.

ICELAND MOSS is not, strictly speaking, a moss, but a lichen or liverwort, named *Cetraria islandica*. The whole frond, as it is called, or body of the plant, is used. It is collected in large quantities in Iceland, and is used as food by the natives of Iceland and Lapland. The soluble portion of the plant is taken up by boiling water, which thickens on cooling, and deposits a gelatinous-looking substance, which, when dried, forms a semi-transparent mass. The moss contains a bitter principle of acid character, which has been called cetraic acid. The decoction is the preparation used in medicine as far as the substance is used, and that is but little. It is slightly tonic and demulcent, but has no well-marked property.

ICHOR. (See PUS.)

ICTERUS. (See JAUNDICE.)

ICTHYOSIS occurs in two forms. It is met with as a dryness of the skin in both children and adults. It is usually congenital, and occurs in many members of the same family. The skin is dry, harsh, and rough, and it appears as if it were too tight for the body. The epidermis often peels off ; on the neck it is rough and horny, and in the rest of the body the cracks of the epidermis correspond with the lines in the skin. In the other form dry and hard grayish or slate-colored scales appear on different parts of the body, unaccompanied by any redness or heat of skin. Its most frequent seat is on the extremities, and especially on their outer aspect. Alkaline baths will remove the thicker scales, but they are speedily re-formed. Patients who are affected with this kind generally do not enjoy good health. Treatment may relieve, but will not cure this disease ; the part affected may be rubbed with oil, and cod-liver oil and tonic medicines may be taken internally to improve the general health.

IDIOCY might be defined as that form of insanity where the mind from the first is imperfectly developed, and remains permanently in this undeveloped state. This imperfect state of the mind seems due to imperfect development of the brain itself, and this not unfrequently is accompanied by defects in other parts of the body. A distinction which is on the whole useful is commonly drawn between idiocy and imbecility ; an idiot being considered one in whom mental or moral powers can hardly be said to exist ; imbecility the condition where these exist but are defective. Very often the one is confounded with the other. The idiot is distinguished very frequently by peculiarities of countenance, and still more commonly by peculiarity of gait and speech. Having little governing power they are liable to bursts of passion, and in these they may be dangerous, but commonly when excited to passion they give way to tears rather than violence. Notwithstanding the defects of these unfortunate beings, much improvement may be effected in their condition by careful training. The brains of idiots, when examined and weighed, generally speaking present marked deficiencies. Say that the average weight of the brain is 48 oz., the weight of that of an idiot very likely does not amount to more than half that quantity ; in some recorded cases not amounting to more than 15, 13, or even 10 oz. Some of the parts, too, may be wanting. Two very important points need to be borne in mind with regard to idiocy : one is, that some idiots have been known to attain considerable mental powers, especially after injury to the head. The other is still more important. It is a fact finally ascertained that breeding in and in, as it is called, that is, intermarriage in the same family, tends inevitably to lower the intellect, and finally, if the process be not arrested, to produce hopeless idiots.

IDIOPATHIC is a term of no distinct meaning, often used to veil ignorance, and is given as the *cause* of a disease when nothing else is known to give rise to it ; thus one speaks of idiopathic pneumonia and idiopathic

peritonitis, by which one signifies inflammation of the lungs and peritoneum respectively,
in cases where there is no apparent cause for
those disorders.

IDIOSYNCRASY is the professional
term for that condition of mind or body which
is commonly known as antipathy. Some persons are peculiarly affected by certain smells,
sights or noises, and these we call their idiosyncrasies. The smallest possible dose of a
particular drug will, in some cases, produce
the most violent and peculiar effects ; some
articles of ordinary diet likewise ; one man
cannot eat any shell-fish without breaking out
in an eruption all over the skin, another cannot bear the smell of flowers without faintness, and these results are technically called
idiosyncrasies.

ILLUSIONS are sensations without corresponding external objects ; when the eye is
the seat of the sensation it is spoken of as a
spectral illusion, phantom, or hallucination ;
an *illusion* means a mockery, false show, or
counterfeit appearance, and is opposed to
delusion, which is a chimerical thought. An
illusion of the senses, if believed to be a reality, becomes a delusion of the mind. (See
INSANITY.)

IMPETIGO is a skin disease most frequently met with in children ; it occurs also
more especially in those who are badly or
grossly fed, or who have not thoroughly recovered from an illness, as measles or scarlet
fever. Generally found on the face, it begins
as small pustules, slightly raised above the
surface, and surrounded by an angry red
blush ; the pustule is about the size of a pin's
point, and of a yellowish-green color, so that
if pricked a minute quantity of matter will
exude. They are at first separate from each
other, but as the child picks them they run
together and form a bleeding surface covered
with scabs ; this condition is similar to what
is met with in eczema, and has been termed
porrigo. (See PORRIGO.) The angles of the
mouth, chin, and cheeks, are the parts chiefly
affected, and this disease has a great tendency to spread, because the matter exuding
from the pustules is so acrid and irritating
that if any other part is touched with it
another spot rapidly develops. The child
has often at the same time spots about the
arms and legs. The *treatment* consists in
preventing the child from picking or rubbing
the spots, so as to prevent their spread as far
as possible ; the eruption should be washed
twice a day with oatmeal and hot water, and
not with soap ; zinc ointment should be applied night and morning, and the child's health
should be improved by a careful diet and
steel wine twice a day ; the bowels may be
kept open by Gregory's powder, and every
day exercise should be taken in the open
air ; by this method a cure may soon be
effected.

IMPOTENCE, deficient or absent sexual
power, is a subject which in a volume like the
present requires to be handled with delicacy.
Nor indeed would it be touched upon at all
were it not that the plan of systematically
ignoring it has produced disastrous consequences. There are few things which have
given rise to so much mental torture as the
idea of the want of sexual power, and perhaps there are few faculties so seldom absent,
but being peculiarly subject to nervous influence, it may be for the time in abeyance ;
then, too, the idea of impotence tends to per-
petuate the condition. So, too, still more do
efforts to overcome it, until at last the unfortunate individual is sometimes driven to acts
of the rashest self-violence. For such individuals there is nothing like peace of mind
and rest of body. Excitement is most prejudicial. In all probability the condition will
depart of its own accord in due time. Meantime interference is sure to do harm. Rigid
continence and chastity in thought, word, and
deed, should be practised for the time being,
until a more healthy condition of things comes
about. This may seem very hard counsel,
but it is the best ; this refers especially to individuals who have led healthy lives, but whom
overwork or over anxiety have for the time
being incapacitated ; but there is, however,
another class who have had themselves in
part to blame for their condition, real or
fancied. In these, too, the mental condition is the most important part of the malady,
especially the habit of studying each inward
feeling or emotion. Such are often subject
to exhausting discharges which will not cease
until the mind and body both become
healthier. To both of these classes we very
earnestly desire to speak. We counsel them,
if they love their own peace of mind and
future comfort, to have nothing to do with
the class of advertising impostors who prey
on such unfortunates. Let them apply to
the best and most respectable medical practitioner within reach, fearing nothing, for
these things are well known and carefully
studied among such.

INANITION is the condition brought
about by bad feeding, or by giving food
which is deficient in quantity or in quality, or
in both respects ; it is a too frequent cause of
death in infant life ; the child gradually loses
flesh and " wastes to a skeleton," until finally
death takes place by exhaustion.

INCISIONS are a division of the several
tissues of the body, whether made by knives,
scissors, or saws, though the term is generally applicable to such as are made by a sharp
cutting edge. In surgery, rules are laid down
for the various incisions requisite in performing operations ; and it is only by long habit
and a just appreciation of the densities of the
various tissues met with, that freedom and
certainty are acquired. Such incisions must

be made with some definite purpose, with determination and steadiness, so that the operator neither injures his patient nor his assistants. It is obvious that an accurate anatomical knowledge is necessary, and no one should attempt to incise any part of the body without an intimate acquaintance with its structure.

INCOMPATIBLES are remedies which, when mixed together, destroy each other's effects, or materially alter them. It was at one time the universal custom to order several substances in the same prescription, one of which, if care was not taken, might neutralize or destroy the effects of the others. This, of course, had to be guarded against, and so lists of incompatibles used to be given with each remedy. Nowadays we tend more to give remedies singly, and so incompatibles are of less consequence.

INCONTINENCE OF URINE is a troublesome symptom, occurring at different ages from various causes. In children it may arise from bad training, or from some irritation in the penis or bladder ; such children should be made to pass water just before going to bed, and may even be roused in the night for the same purpose ; if punishment be of no avail, search should be made for a cause, and sometimes when proper training will not cure it, some tonic, as iron wine combined with belladonna, will do good ; if it is due to any abnormal condition of the penis, a surgical operation may have to be performed. In adults it may come on from paralysis of the bladder, as in cases of paraplegia, where, from over-distension, the urine dribbles away ; the treatment will consist in drawing off the urine night and morning with a catheter, and in keeping the patient dry and clean so as to prevent the formation of bed-sores. In young hysterical women it sometimes occurs, and the best thing for them is cold bathing, change of air and scene, healthy and useful occupations, daily exercise, and avoidance of hot and overcrowded rooms, late hours, and morbid mental excitement. In old people this symptom may come on from an enlarged prostate, or irritable bladder, and for this condition very little can be done except daily catheterism. In those also who suffer from a fistulous opening into the rectum or vagina, or in those who have a false opening in the urethra, this condition may prove very distressing from the constant flow of urine ; perfect cleanliness must be enjoined, and a piece of sponge may be so adjusted as to catch any fluid that dribbles away ; if a stricture of the urethra is the cause, surgical interference must be resorted to.

INDIA RUBBER, or *caoutchouc*, is the inspissated milky juice of several very different trees of tropical climates. The juice is obtained from incisions in the trunk, and is poured in successive layers into jars, where it hardens. In its pure state caoutchouc is nearly colorless, and the dark blue which most of it has in the crude state of commerce is produced by the smoke to which it is subjected in drying. The elasticity of soft rubber articles is of rather uncertain duration, and manufacturers will seldom guarantee it for more than three months, though it generally lasts much longer. Hard rubber or ebonite is made of pure caoutchouc and sulphur, subjected to high and long-continued heat. It is very strong and elastic, and takes a very high finish ; and optical and surgical instruments are composed in part and sometimes entirely of this substance.

INDIAN HEMP. (See HEMP.)

INDIAN POKE. (See VERATRUM VIRIDE.)

INDIGESTION, or dyspepsia, as it is also commonly called, is mainly due to simple derangement of the powers of digestion, without any eventual change in its organs. This derangement is dependent on weakness ; but the source of the weakness may be local, that is, confined to the stomach ; or general, that is, due to something which affects the whole system. To this group, too, belong the changes which take place in the digestive organs in old age. Another large group of indigestions are connected with inflammatory changes in the stomach. These changes may depend on various causes ; very frequently improper food—improper, that is, in quantity or quality—is at the root of the mischief. Yet another form of indigestion is due to nervous influence ; witness the effects of anxiety, fear, and the like emotions in completely averting not only appetite but digestion.

It is true that the intestinal fluids are also concerned in digestion, and so alterations in the secretions of the liver, pancreas, and other organs, poured into the small intestine, must have a powerful influence in retarding digestion or rendering it imperfect. But though we can sometimes clearly distinguish the particular form of indigestion from that due to imperfect change in the stomach, it is better to consider indigestion first of all as due to the stomach only.

Process of Digestion.—A brief word as to the relative functions of various organs concerned in digestion is necessary to a clear understanding of those defects which constitute dyspepsia. Our food may be taken to consist mainly of three kinds of substances, which are represented to a different degree in almost every article of diet, yet are present in some form or other in most dietaries. These substances are starch or sugar, which can be derived from starch, oil, or fat, and albumen, or white of egg, which may be taken as the type of all kinds of meat. Suppose, then, our dietary is mainly composed of bread and meat, we have in this dietary the starchy element represented in the bread, the fatty in the fat which is contained in all meat, even the poorest, and the albuminous in the lean portion of the same. Bread, too, contains a certain proportion of the albuminous element, being most abundant in brown bread, most scanty in the finest white bread. The object of digestion is to convert these several elements into a material fitted for the nourishment of the body. The first secretion encountered which has any influence on the food as prepared and ready for swallowing is the saliva. This speedily converts the starch of the food, especially if that starch have been cooled, into sugar similar to that found in fruits, and called grape sugar. But the food does not remain long enough in the mouth to undergo this change in its entirety ; it is swallowed, and passes on into the stomach. Now, whereas the healthy saliva as secreted is an alkaline fluid, the fluid secreted by the stomach, and called the gastric juice, is strongly acid. In it the saliva swallowed can no longer act ; but it is not destroyed, its function only remains in abeyance until the fluid again becomes alkaline, when it can again go on. The acid gastric juice is specially intended to alter the albuminous element in the food. It converts it into a substance which can more readily pass through animal membranes than itself can, and all kinds of albuminous food are reduced nearly to the same chemical substance.

When stomach digestion has finished, that part of the food which has not been absorbed by the vessels in the inner surface of the stomach passes on into the small intestine. There it encounters the secretion of the liver called the bile, which promptly puts a stop to all further change such as has been going on in the stomach. Here, too, is poured out the secretion of the pancreas, or sweet-bread, which enters the intestine along with the bile. These are powerfully alkaline, and neutralize the acidity of the gastric juice, so that now the change in the starch can begin again. Another substance, too, remains unacted on, that is the fat contained in the food ; these two substances act upon it, and finally by converting it into a kind of soap, partly by suspending it, the fat becomes ready for absorption, and is taken up by special vessels accordingly. The pancreatic juice has a further influence on albuminous substances. These it alters something in the same way as does gastric juice, and renders them more easily taken up by the absorbing vessels. The refuse of the food is ejected together with certain other waste products. Apparently the most important stage in this process, or, at all events, the one which has been mainly studied, is gastric digestion, and it is its imperfections we are now called on to study.

Symptoms of Indigestion.—The signs that tell us something is wrong with the stomach are partly such as the patient alone is cognizant of, partly such as are appreciable by the skilled practitioner. There are certain special signs which tell us something is wrong with the stomach, such as flatulence, acidity, or heartburn, acid eructations, and perhaps vomiting. The tongue used to be taken as a certain guide to the condition of the stomach and other digestive organs, and so the indications afforded by it were carefully studied. Though of less value than was supposed, still the signs afforded by the tongue are not to be neglected. The "fur," as it is called, is formed from the scaly covering of the lips and cheeks, as well as from the tongue itself. This is agglutinated by the saliva drying, and so gives rise to the furred appearance spoken of. Some people sleep with their mouths open ; these have almost invariably an accumulation of fur on the tongue in the morning. This may mean nothing. Again, any irritation of the mouth may give rise to an unusually copious production of this substance, and so the tongue be furred. This fur often decays and produces putrid gases, which cannot fail to be prejudicial to health. When the fur is due to stomach mischief, that is generally of an intestine or semi-inflammatory character, and thus the appearance of the tongue and mouth affords not only valuable means for making out the nature of the disease, but also helps us to some clew to its treatment. The alterations in hunger and thirst are often of value in enabling us to come to a conclusion as to the existence and nature of stomachic derangement. There may be loss of appetite with dyspepsia ; if so, the malady is generally inflammatory, and rather acute in its nature. But in itself the symptoms are not of great value, being due to an infinite variety of causes, some of which have their seats anywhere save in the stomach. The opposite condition (voracious appetite) is hardly a sign of dyspepsia, though it does occur. More frequently it is associated with such a malady as diabetes or the presence of tapeworms. This is different from the craving or sinking feeling often experienced. Eating food repugnant to ordinary appetites is more a sign of hysteria than dyspepsia. Thirst is most common in irritative states of the stomach, and then mostly manifests itself some hours after a meal. Of all the symptoms derivable from such sources, thirst and loss of appetite are the most valuable.

Flatulence is a very important symptom of indigestion. It is due either to accumulation of gas in the stomach and bowels, or it may be formed there. It gives rise to uncomfortable feelings, and often to colicky pains commonly followed by expulsion of the gases.

Undoubtedly, too, the gas may be absorbed by the blood and given off by the lungs. Even in health there is present a certain amount of gas in the bowels, part of it being swallowed along with the food. That gas which gives rise to flatulent distension is, however, commonly derived from fermentative changes in the food swallowed, especially if these be too long retained in the stomach, as when there is obstruction to its exit. It has been supposed that the bowels can secrete air or gas, and it has been assigned in proof of this that in certain forms of disease the bowels become enormously distended with air, which can apparently only have originated there. There is no proof that the bowels can secrete gas. The distension alluded to is in part at least due to muscular relaxation allowing the gas present at all times to occupy a much larger space than usual.

Acidity is another important symptom of indigestion. It, too, arises from two causes —over-formation and fermentative change. The latter occurs under the same circumstances as does the flatulence produced in like manner. For the fermentation which sets free the gas above referred to produces acids —acetic, butyric, and lactic, when starchy or saccharine substances and milk become altered. These acids are so readily formed that it is not easy to say whether or no they are normal products of the stomach or are the result of fermentative changes. Briefly it may be said that whatever favors the changes which give rise to flatulence from fermentation also give rise to acidity. But another cause of acidity must be admitted in excessive secretion of acid gastric juice. Perhaps more frequently, however, if there be excessive secretion on the part of the stomach, it is of an alkaline mucus, rather than of acid gastric juice. This is especially the case in inflammatory conditions of the stomach.

Heartburn, as it is called, is present as a rule whether the excessive secretion be acid or alkaline, perhaps even more in the latter, though then it may be due to fermentative changes in the food. The sensation is one of burning at the entrance to the stomach, with a desire to bring up something, which commonly ends in a hot burning fluid regurgitating to the back of the throat and sometimes being ejected. If the acid be very great in amount, it may set up vomiting of quantities of acid fluid. Pain is not constant in acidity. If the acidity be due to fermentation, the pain is longer in coming on, and often ends in colicky pains. The pain produced by over-secretion, on the other hand, very commonly occurs when the stomach is empty, and is very readily set a-going by a stimulant, such as a glass of hot spirits and water. Both forms of acidity, if not checked, may give rise to dangerous symptoms. Fermentative change, as interfering more with

digestion in the bowel, is perhaps more dangerous than the other, but both seriously undermine health by the exhaustion to which they may give rise. The vomiting of pregnancy illustrates the danger, for though due to different causes, it is sometimes so severe, and the exhaustion occasioned by it so profound, as to endanger life, or even prove fatal.

Pain in the stomach has already been treated of under the heading of Gastrodynia (which see). It is a symptom of varying import, for some of the severest forms of disease are attended with little or no pain, while some of the more curable varieties are attended by pain of an excruciating character. Pain in the stomach may be due to irritating substances in its interior, derived from without, or perverted secretions derived from within. It may be due to profound alterations in its texture, or to those imperceptible alterations which we ascribe to altered nerve-power. The question how far this pain may depend on altered nerve-power is not easily answered. Usually this is accompanied by some alteration in the secretions. Perhaps, too, there may be a kind of cramp or spasmodic contraction of the stomach, which may give rise to acute pain. Undoubtedly, too, there is a neuralgia of the stomach. In these cases of purely nervous pain digestion may go on well enough during the intervals of ease. Purely neuralgic pain is especially common in females during the earlier period of maturity, when they are also especially liable to other nervous symptoms. Some of the cases which seem to exemplify pain in the stomach are, however, cases of pain of the abdominal walls.

Vomiting is a very complex act, and due to many causes. It is mainly produced by the compression of the stomach against the diaphragm by means of the abdominal muscles ; but the walls of the stomach and various other parts participate in the action. It may be induced by irritating the nerves, either at their centres in the brain, or where they end, in the stomach and neighboring organs. Consequently we have vomiting in head affections as well as in affections of the stomach and neighboring organs. As a means of diagnosis it may be said that vomiting arising from irritation of the stomach is attended with more or less pain, the tongue is furred, and there is a feeling of heaviness and nausea preceding the act. These are rare when the vomiting is due to cerebral symptoms.

Causes of Indigestion.—As to the causes of dyspepsia, the first we shall deal with is unsuitability of food, and the food may be unsuitable both as regards quantity and as regards quality.

The human digestive organs, from the teeth downward, are fitted for a mixed diet, partly animal, partly vegetable. An undue preponderance of either of these, therefore, is likely to lead to injurious consequences if long continued. The excess of saccharine material gives rise to an undue secretion of acid in the stomach, which is not reabsorbed, and disorders digestion lower down in the alimentary canal. Excess of starchy food seems altogether incapable of being digested under ordinary circumstances, and so passes into ferment and undergoes the changes which give rise to flatulence. Again, a certain amount of indigestible material is mingled with all our food. It does good so far by distending the bowels and so inducing in them the movements needful to carry the remains of the food, digested or undigested, out of the body. If, therefore, the food is too nutritious, as in those who live highly, there is not enough of this material in it, and so the bowels become confined. On the other hand, very poor diet, containing little nutriment, is apt to irritate the bowels, and to give rise to indigestion and, perhaps, diarrhœa. Certain, indeed most, articles of food undergo in course of time changes which render them unfit for human use. This takes place under almost any condition, but is more likely to take place under some than others. Thus putrid meat or fish, sour bread, and imperfectly or excessively fermented beer, give rise to irritation of the stomach, or even to worse consequences. One of the most potent causes of dyspepsia is deficient mastication, and this interferes with the digestion both of meat and starchy substances ; of meat, because it is not sufficiently triturated to be acted upon with advantage by the gastric juice ; and of starch, because that is imperfectly saturated with the saliva. Insufficient mastication is a common fault in these days of hurry ; men too often, especially in the morning and mid-day, bolting their food. (This is one grand reason for postponing the principal meal of the day until the hurry is over.) Again, from various causes, people nowadays, lose their teeth earlier than they were wont, and this loss sadly interferes with the due pulping of the food. The remedy for this is a good false set. When from any cause the saliva becomes altered, as it sometimes does, its action on starch may be entirely prevented. This may occur when the secretion of the mouth itself is acid. Such conditions are exceedingly liable, if starch be much used in food, to give rise to flatulent indigestion. The quantity of food is quite as important as its quality. It is notorious that excessive quantities of food, excessive, that is to say, as regards the powers of the stomach, are in a certain class the most prominent cause of dyspepsia. The stomach seems in most cases only to secrete enough gastric juice to digest the food necessary to the wants of the system ; the rest is passed on to ferment or putrefy in the bowel, and so give rise to the tortures of dyspepsia.

When the digestion is good, and this surplus food is digested, obesity results ; or the food so taken into the system is only imperfectly consumed, is not readily extruded, and so the phenomena of gout ensue. Irregularity in taking food is a great drawback to perfect digestion. This is the more so, of course, if the intervals are too small, if the individual is constantly taking "snacks" between meals, just to get rid of the faintness. And this must be remembered, that an interval which is quite large enough for one is altogether too small for another. If a man is working hard in the open air he can digest twice as much food— and many times more of some kinds of food— as on who lives an habitually sedentary life. Deficiency of food, accompanied as this almost invariably is by unsuitable food, is a serious cause of dyspepsia, among the poorer classes. Indeed, among the out-patients of a hospital, three great things may be said to lie at the root of most cases of dyspepsia, and they are sufficiently numerous ; they are tea, hunger, and alcohol. The hunger leads to the use of the others as stays, and the consequence is, most troublesome forms of dyspepsia of the irritative kind. Certain causes of dyspepsia are located in the stomach itself, and these may be mainly referred to such conditions as obstruct the passage of food through it, or to alterations in its secretions. The stomach, like other portions of the alimentary canal, has the power of expelling its contents in due time. But these movements may be impaired, as when the stomach is atrophied and dilated. The movements of the stomach itself may be arrested, as may those of other parts of the alimentary canal ; but this is not likely in simple dyspepsia. So, too, indigestion may result from obstruction at the intestinal end of the stomach ; but that being of a cancerous kind, or of some other new formation, withdraws it from the realm of dyspepsia pure and simple. Adhesions of the stomach to surrounding organs interfering with its movements, and not likely to be discovered save through the signs of indigestion it gives rise to, may be referred to here, but only referred to.

As to altered secretions—these being two in number—deficiency of the active one, that is, the true gastric juice, is often accompanied by an excess of the other, which is mucus, and that is worse than useless. The secretions are undoubtedly influenced by changes in the blood itself, as is well seen in Bright's disease. And alterations in the functions of the other digestive organs, which take effect lower down, will doubtless directly or in directly influence the stomach. The precise mode in which the nervous system influences gastric secretion is not very plain, but that such an influence exists is patent to all, as witnessed in the manifestations of anxiety, sorrow, fear, and even joy. Most probably this takes effect through the sympathetic nerves. There are certain forms of the malady, too, which depend on what has been called reflex irritation. Thus constipation is to many, especially to those not habitually its subjects, one of the surest causes of a temporary indigestion or loss of appetite ; but it is just possible that the same cause which in these has produced the constipation may also produce the indigestion, more distinctly if this reflex character is the indigestion due to the presence of worms. True, in a considerable number of cases, worms rather give rise to increased appetite than to loss of it ; but this is perhaps true rather of long worms of the tænia kind than of the small round worms, which give rise to so much irritation in the parts infested by them.

Atonic Indigestion.—The three main varieties of indigestion are the atonic, the nervous, and the inflammatory. The atonic form is almost invariably chronic, rarely attended by fever or pain, but indicated by a dull sense of weight, uneasiness, and languor, especially after taking food. Very generally, too, there is depression of mind—hypochondria in the male, or hysteria in the female. Weakness of digestion is very often accompanied by weakness of other organs besides the stomach —weakness, too, of a hereditary origin. In aged people this form of indigestion is almost habitual, and is exceedingly easily made worse by any indiscretion of diet, a sub-inflammatory from being thus induced. Such a form of indigestion is, however, still more frequently one's own act, or at all events dependent on one's habits of life, whether these be voluntary or otherwise. The digestion in these cases is excessively slow, and frequently continues from one meal to another. There is also a feeling not amounting to pain, except in hysterical women, but apparently giving rise to imperfect respiration or want of breath, or a feeling as if something had stuck in the throat. There is flatulence, and very likely eructations of acid, or more likely undigested or half-digested food. Frequently the eructations are offensive or acid, and these occur some hours after food. The flatulence is not confined to the stomach, but affects the bowels also, giving rise to troublesome distension. The appetite as a rule is impaired, and certain forms of nutriment, as soups and broths, indeed most kinds of fluid, markedly disagree. The tongue is pale and flabby—marked by the teeth at the edges ; there are marks of general relaxation in the mouth and throat, especially about the uvula, which the patient commonly attempts to relieve by hawking. The bowels are usually constipated, and the gases passing along them give rise to unpleasant noises, though there may be no sensation of them. Frequently

the bowels become distended at one particular spot by accumulations, and this is not relieved by purgatives. The breath in these cases is ordinarily offensive. The evacuations are usually hard and deficient in bile, but in a good many cases this alternates with the opposite condition. In these cases, too, the pulse is slow and weak, but readily raised, so that if the patient be examined immediately after exertion he would seem to have an unnaturally fast pulse ; palpitation, too, is frequently present ; and these two frequently lead the subjects of the complaint to fancy themselves affected with heart disease. There are no marks of fever about the patient ; the skin is soft, flabby, and moist. The extremities, too, are cold, especially after meals. The color of the skin is bad generally, sallow or muddy. For the same reasons the nervous system is affected, and there is languor and lassitude and a sense of weariness in the limbs. Sick headache is frequently present from time to time, and the mental faculties are dulled and incapable of prolonged exertion.

All these general symptoms, as they are called, point to the fact that this form of indigestion is rather a sign of the constitutional condition than that the constitutional signs are dependent on the indigestion alone ; for the stomach itself when examined presents nothing which is unusual. There may be marked transparency and thinness of the walls of the stomach, with degeneration of the tubules which secrete the gastric juice. In slighter forms the particular cells which secrete the juice may alone be affected.

Treatment.—The treatment to be adopted for atonic dyspepsia must have a twofold end : there must be an endeavor to improve what is called the general tone of the system, and special pains must be taken to enable the stomach to do its duty aright. It is especially in cases like the present that due attention to food and drink is necessary, and hence the treatment resolves itself into dietetic and medicinal. As regards diet, we have already pointed out that in a very considerable number of instances the indigestion has been brought on by overtaxing the powers of the stomach by too frequent and too copious meals. It is, therefore, necessary here to beat back, so to speak, to find the least quantity of nutriment which is required by the system, and the greatest which can be digested by the stomach in comfort. Here the conditions of the patient's life must be borne in mind, for there is a mighty difference in the amount of food which will suffice for a listless invalid and one habitually undergoing powerful bodily exertion. Moreover, it is of the first importance to present this food to the stomach in an easily digested form. As already pointed out, soups and broths are rarely tolerated in such conditions, mainly because they dilute the digestive fluids too much. Nevertheless, there is one fluid form of food which can usually be taken, that is milk, and if it cannot be taken fluid it may be coagulated by rennet. When arising from exhaustion this form of dyspepsia necessitates small meals, but these may be frequently repeated, and stimulants may be combined with them. Less care is necessary in the more ordinary forms of the malady, but there is the greater reason that the necessary care should be taken. No salt or preserved meat should be used, and it should be fairly well but not over-cooked. Hence recooked meat is forbidden. Mutton and beef must form the staple diet ; game and fowls may be allowed as a change. Pork and veal are entirely excluded, as are ducks and geese. Fish is also permitted within certain limits ; herrings and salmon are beyond these, and so not allowed ; eels and trout are on the border land. The best for ordinary use are plain boiled turbot, whiting, or haddock. Shell-fish, except oysters, are entirely forbidden. Still more care is necessary with regard to starchy food. All vegetable food should be cooked, none raw, and it should be young, tender, quite fresh, and well boiled. Some authors exclude grease entirely ; for our part we are much more inclined to exclude the starchy foods, as potatoes and over-ripe peas. It is better on the whole to try stale bread, macaroni, and rice, with some green vegetables, than to use potatoes. Such at least is our experience. Light puddings are permissible, not heavy doughy preparations, and all pastry is to be forbidden. Butter may be used with bread, but in no other way, and it should be quite fresh. Fat or oil in any other shape is inadmissible. Fruits must be carefully selected ; but most may be eaten, or rather sucked, provided everything solid—husks, seeds, and woody matter—be spat out again. Nuts of all kind must be rejected.

But briefly, each man must be a rule to himself : whatever he finds disagrees with him should be avoided : very often, though, it is quantity rather than quality which is at the bottom of the mischief.

Three meals a day is perhaps the best rule : breakfast, say at nine, something about one, and dinner at six. Too long intervals are almost as bad as too short ; especially should the sufferer aim at cheerfulness during its digestion and absorption. To aid this there is nothing better than a moderate quantity of good wine. Port rarely suits, and should be avoided ; dry sherry usually does well ; a sweet sherry is most hurtful. Good sound vin ordinaire suits most, if diluted with at least its bulk of water ; while some do best on weak brandy and water. Laying down a rule to be governed by circumstances, we should say all others are better avoided. So, too, is tea after dinner : that almost invari-

ably does harm ; not so black coffee, moderately strong, with a teaspoonful of brandy in it. As a rule that suits, but it must not be swallowed hotter than the temperature of the body, or a little over, so as to feel pleasantly warm.

All the other matters tending to a restoration of health must be observed. Change of air and change of scene, a sea-voyage, etc., will often do great good ; but as the invalid is greatly dependent on food for his cure he should seek to recruit himself where good plain cooking and sound food is to be had. Walking exercise, not the mad scrambles on the Alps of some, but a good steady walk of twenty miles a day with an old-fashioned inn to rest at in the evening, will often do marvels in the way of cure. A cold tub in the morning should be the rule to all who can stand it ; if not, one as cold as possible, but not warmer than tepid should be used. If a settled residence is desired, one of the bracing localities alluded to already should be selected. (See CLIMATE and HEALTH RESORTS.)

To the mass of the public, treatment by diet and regimen is far more important than treatment by drugs, as being far more under their power of control. Nevertheless, even in these most frequent of all maladies, drugs may not be neglected. Chief, in point of fact, among the remedies to be used is iron in some one or other of its forms, nor do some of the vegetable bitters come far behind it. Frequently, however, iron cannot be given from the beginning ; some preparation of the stomach may be necessary to provide for its reception. An alkali and a bitter , as columba, are frequently the best drugs to begin with. Liquor potassium and gentian, too, are useful, but this is mainly when there is no inflammation superadded to the atonic condition.

In simple atonic dyspepsia we can as a rule give iron from the beginning, but rarely such preparations as the sulphate or chloride. Usually we must begin with reduced iron, the carbonate or ammonia citrate, and these may be given along with meals, though best, perhaps, before it. If not well borne they may be given effervescing, but even then they may cause irritation ; if so, a phosphate may be given. Usually we may combine with the iron nux vomica in some shape or other, though perhaps it is sometimes best given by itself. There are certain cases where the tincture of the bark of nux vomica does better than the alkaloid strychnia, though the latter with a mineral acid is a most efficient tonic. Quinine itself does not seem to do particularly well, but some preparations of bark, especially the liquid extract and compound tincture, answer well where there is no irritability. Of the other bitters commonly used besides columba and gentian, cascarilla and chiretta seem of undoubted value. The former, combining slight stimulant properties with its bitter, is very useful in many of these cases. The hop, too, in the form of good bitter beer, is not to be despised. Ipecacuanha, where there is considerable irritability of stomach, is valuable ; but its use, and that of certain other remedies, belongs to another form of dyspepsia altogether. In this form of indigestion, aids to digestion are of prime importance. Chief among these are the mineral acids and pepsin. The normal acid of the gastric juice is hydrochloric, and perhaps on that account we should be led more naturally to prescribe it than any other to aid digestion. Nevertheless, it has seemed to us that a mixture of nitric and hydrochloric acid, which acts much in the same way as does hydrochloric acid, is a better preparation. It should be given in doses of 10 or 15 drops in some bitter preparation, as tincture of orange-peel or infusion of columba, just before, during, or after food ; that is, if the meal contains meat or allied substances, not otherwise.

Pepsin is even a more valuable remedy, though it has been much decried, probably on account of the very inferior preparations abroad. It cannot be made a cheap preparation, unfortunately ; and most attempts in this direction have simply meant impurity or adulteration. It is found useful not only in adults but also in children. For an adult five or six grains will suffice, and it may well be given with hydrochloric acid, as above stated.

Nervous Indigestion.—We have dwelt at this length on atonic dyspepsia both because it is one of the most frequent maladies of the kind encountered, and a knowledge of how to anticipate its effects and to deal with it dietetically is of the utmost importance. The next group is not so easily defined, or, at all events, not so easily appreciated by the public ; we shall call it nervous dyspepsia—dyspepsia or indigestion that is due to interference with the functions of the nerves. In many respects nervous indigestion is allied to atonic indigestion, but there are certain special causes worthy of note. They are much more frequent among women than among men ; but exhaustion or general weakness may reduce a man to a somewhat similar condition. The whole group of symptoms, of which the indigestion is one, are commonly grouped under the heading hysteria. The condition in the male, called hypochondriac, is still more closely associated with indigestion. And the depressing effects of chronic alcoholism, though these give rise to morbid changes of a specific kind, are in part due to nervous influence. The forms of dyspepsia mentioned are, as far as the stomach is concerned, of a nervous origin, especially certain connected with imperfectly known conditions of the ovaries and womb. The pain, which is one of the most promi-

nent symptoms of this form of indigestion, is usually very severe and intermittent. The duration of the attack is variable, from a few minutes to hours. Frequently it terminates in acid emitations or the ejection of an alkaline mucus. Food produces variable effects ; in one class it produces distinct relief, in others it brings the pain on ; most frequently food gives relief, and these are the cases about which there is least doubt. Commonly enough in this class of indigestion, insipid demulcent substances frequently give rise to more pain than do matters of a more irritating kind. This, too, is a very characteristic feature. In many cases digestion goes on readily enough in the intervals of the attack. The most typical form of nervous indigestion is that to which the term gastric neuralgia has been applied. This commonly is accompanied by vomiting. There are many interesting pathological questions connected with the mode in which these nervous dyspepsias originate, but to the public it is far more important to be able to recognize them and know what is best to be done with them. We have briefly alluded to some of their main peculiarities. The pain generally intense, often most so when the stomach is empty, and relieved by food ; the tendency to vomit without much nausea, and from no special change in the food ; especially, the tendency to nervous pains and disturbances elsewhere in the same individual.

Treatment.—As to treatment, nux vomica and iron are the mainstays. In most cases iron relieves the neuralgic pain. The carbonate is the best preparation, and if the bowels are confined, a little aloes and rhubarb may be given at the same time. Next after these comes opium, but that must be used with care. In some of these cases, where there is also constipation, it would almost seem as if opium opened the bowels. Aromatic spirit of ammonia will also be found of great value in many cases. Hydrocyanic acid is also given with benefit, though not, I think, with greatest benefit, in this form of dyspepsia. Of course it should never be given without a physician's prescription and instructions. Where there is much vomiting ice must be employed. (See HYSTERIA, NEURALGIA, and PREGNANCY.)

Inflammatory Indigestion.— The acute form of inflammation, such as seen in other organs, is rare in the stomach ; but it is very subject to such forms of inflammation as occur in mucous membranes, and go by the name of catarrh. This acute inflammation of the substance of the stomach apparently occurs almost only in poisoning by irritant substances. *Gastric catarrh*, as the malady of which we now propose to speak is commonly called, and which may be looked upon as the synonym of inflammatory indigestion, may be either acute or chronic. The acute form of gastric catarrh is most common, perhaps, in young children, as in them few articles of food prove suitable ; and unsuitable food generally brings about an acute attack of indigestion due to this malady. In those, too, whose stomachs have been weakened, from any cause, a slight addition to the inconvenience of digestion may bring about an acute attack of catarrh. Thus atonic dyspepsia, or whatever causes it, starvation, drinking cold water when the system is greatly heated—all may directly or indirectly bring about this condition. Apparently, too, in some epidemics there is a tendency to this malady. Thus, in cholera times, there is a great tendency to looseness of bowels, and often also to inflammatory dyspepsia. Other maladies, too, seem to affect the stomach, as scarlatina, diphtheria, small-pox, etc.

Acute indigestion may assume any degree of severity. Usually it commences with a feeling of fatigue and heaviness, very likely with pains in the back, soon followed by uneasiness in the stomach itself—this sometimes amounting to severe pain. There is also a sense of faintness, with weak fluttering pulse and cold perspiration. Headache affecting the forehead, sometimes with intolerance of light and sound. Nausea and increased flow of saliva follows till the offending substance is rejected, very likely with a quantity of thin acid fluid. Then, after a period of rest, relief follows. Instead of being thus rejected, the offending substance may pass on into the bowels, when follow colicky pains and gripes ; probably diarrhœa comes on, and so the substance is got rid of, though sometimes a purgative is necessary. There is usually in this state a loathing of food and persistent nausea. The tongue is loaded and the breath offensive. There is also much thirst, though few liquids are well borne by the stomach. There is also dizziness, very often palpitation, and if there is much flatus in the stomach the symptoms may almost simulate an apoplectic form seizure. This is most common when the acute attack is superadded to a long-standing one. The form of headache which occurs with this form of the malady is sometimes called sick headache, though there is another form of headache of the same name. (See HEADACHE.)

Very often this form of indigestion occurs some hours after food, frequently after supper, say early in the morning. Besides the symptoms already enumerated, there may be disordered vision, noises in the ears, and throbbing in the temples and eyeballs. There is great depression, sighing, yawning, and shivering. The attack may last a variable period, generally under forty-eight hours, and then passes away in sleep. After waking, the pain is gone, but the patient is weak and nervous and the stomach irritable. Great

care must be taken of digestion for a day or two.

If the irritating matters are not got rid of the condition may last much longer, and prove much more troublesome. Indeed, almost everything depends on the treatment : if properly managed no doubt the case will do well in a short time, but if not it may tend to become chronic. There are, however, still more severe forms of the disorder which may simulate the early stage of typhoid fever. Usually, however, there is a good deal of pain in the stomach, a sensation of burning, and obstinate vomiting, brought on by the smallest quantity of liquid. Mucus mainly is so ejected, sometimes streaked with blood, more frequently mixed with bile, and even after the stomach is emptied the retching goes on. The tongue in these cases is loaded at first, but afterward becomes raw, and sometimes both it and the lips become cracked. There is thirst, not easily appeased, as nothing will rest on the stomach, and the appetite is gone. Shivering is common from time to time and a feeling of cold, though the skin is too hot. The pains in the back and limbs continue, and during sleep there is often delirium. The urine is scanty and high colored. This form of disease rarely lasts long if left to itself or treated properly ; if treated badly it is very likely to give rise to prolonged suffering.

Treatment.—The foundation of the treatment in all these cases is rest. First of all, if there is any substance, as there usually is, to give rise to this troublesome condition, it must be got rid of by emetics. The best remedy in such cases is ipecacuanha, given in 20-grain doses, followed by full draughts of lukewarm water. Sometimes a few grains of sulphate of zinc may be added. Frequently tickling the fauces will be sufficient to set the vomiting a going, and it may then be kept up by warm water. Very often the warm water itself is quite enough. If the substance have passed into the bowels, a purgative must be given. The best is castor-oil, if it can be retained ; if not, a few grains of calomel, followed by a seidlitz powder.

In children, a dose of gray powder with some rhubarb, or calomel and magnesia, is perhaps the best thing to give ; but such must be given with caution, for they are easily weakened, and weakened, are easily carried off by the malady. A small dose of castor-oil will generally suffice to carry off the irritant matters, and careful diet must do the rest.

In the adult, where there is purging, opening medicine should be used with caution, nevertheless its use is sometimes the first step toward recovery. A very small dose of castor-oil, or. perhaps better, of tincture of rhubarb, with a little magnesia, may be given, but nothing more. After that it may be even necessary to use astringents, cf which compound chalk powder is perhaps the best. The irritant matters being lessened, diet must come in ; before, anything of the kind would be useless. The patient must be kept absolutely in bed, and it is better, if the individual is a fairly strong one, to let twelve hours or so pass without any solid nutriment. A teaspoonful of solid beef-tea, frozen if necessary, may be given, but nothing more. In a considerable number of instances this will be all that is required—no other medicine will be needed except attention to diet for a short time. But in more severe cases, where food by the stomach cannot be borne, and food is necessary, nutrient injections may be given. The first food should be milk and lime-water, or milk with a little bicarbonate of soda in it, and soda-water (ordinary soda-water contains no soda), or the beef-tea, or essence of beef, which is better, or, better still, new pulped meat. An exceedingly small quantity of this last will suffice to keep life going for some time. It may be given to infants, too, but sparingly. In them milk is the ordinary diet to be given, but largely diluted with lime-water. We disallow farinaceous food altogether, but some permit and even recommend it. As for spirits, they are to be avoided as far as possible, and all wines but champagne of the driest brand (Bollinger's) are forbidden. If stimulants are required absolutely, the best is either a little soda-water and pale brandy, or the dry champagne alluded to. A mustard poultice, or even a hot simple fomentation over the pit of the stomach, often tends to stop the vomiting ; but cold compress, consisting of a towel wrung out of cold water and applied to the stomach, and covered over with flannel, often does better.

With regard to internal remedies, the first is morphia, and it may be given in small doses either as a pill or subcutaneously. In pill not more than a quarter of a grain should be given, and under the skin not more than one sixth or one fifth. If, however, it be necessary to use emetics or purgatives, it should not be administered until the action has ceased. It is, therefore, of most value in the severer form of the disorder. Hydrocyanic acid is rarely of much good in such attacks—its use is in the chronic irritability, still to be spoken of. Sometimes it does well given in an effervescing draught. Bismuth, too, is perhaps of greater value in the chronic form of the malady ; but even in this it is of great use, provided it be given in full dose. It is best given along with magnesia or carbonate of magnesia, and requires to be given in 20 or 30-grain doses for adults ; for children, 5 to 10. Carbonic acid in the form of an effervescing draught is an excellent remedy. The simplest mode of prescribing it is in soda or seltzer-water, but these must be given with care, on account of the bulk they occupy. For some little time after such an acute attack, the rules laid down for atonic dyspepsia must be obeyed ; but if the malady passes into the chronic form another plan of management must be had recourse to.

Chronic inflammatory indigestion, or chronic gastric catarrh, is one of the most common forms of indigestion. This malady may either originate in some severe irritation insufficient to excite an acute attack, but lasting long enough to excite and keep up a subacute attack, or it may follow on an acute attack, when that has been improperly dealt with. Or yet again, attack after attack, each following on the other at frequent intervals, may finally leave the stomach in that irritable state we call chronic gastric catarrh. Such are very common after cholera.

The atonic dyspepsia, of which we have already spoken is not easily separated from this, the less so that the atonic condition is exceedingly liable to be interrupted by a sub-acute attack of inflammation, or even by an acute attack, which leaves the subacute behind. Diseases which tend to interfere with the venous circulation are very favorable to chronic catarrh. Phthisis, too, is very frequently complicated by it, which accounts for the digestive troubles of the unfortunates who suffer from this disease. So, too, contracted liver and contracted kidneys are both very frequently accompanied by catarrh of the stomach. Habitual excess in eating and drinking, too, very commonly end by producing a chronic inflammatory state ; habitual obstruction to the passage of food from the stomach, so that it is retained and undergoes fermentative changes, which create irritation, is a regular cause of chronic catarrh, this, too, whether the obstruction is due to cancer or to simple stricture. The symptoms of the condition are those of aggravated indigestion. The most notable characteristic is the tendency to exacerbations, and again to fall back into the normal catarrhal state without any very apparent cause. There is at all times a sense of weight and oppression across the chest, a general uneasiness after meals, and a tendency to flatulence, which may be considerable. Food may not cause pain, but it increases the uneasiness. The pain complained of is usually under the left breast, and extends through to the corresponding blade-bone. There is rarely tenderness or pressure. Heartburn and acidity are generally very annoying ; the appetite is very variable, and eating soon brings satiety—even the presence of a meal is often enough to turn the patient against it. Thirst is usually well marked some time after a meal, the tongue is usually furred, and there is a bad taste in the mouth in the morning ; the tongue in most cases is flabby, and the papillæ on its surface raised and reddened. In other cases the tongue is raw red, and inclined to crack, while in yet others it is covered with a thick yellow fur. The bowels, as a rule, are obstinately confined, and there is very often uneasiness of the lower bowel, which is increased by the distension necessary for the passage of the hardened motions. Frequently the hardened masses are covered with a glassy mucus. These motions are pale from deficient bile, and for the same reason the smell is unusually unpleasant. Sometimes the stools are frothy and loose, and this condition may alternate with the former. As might be expected from the habitual constipation, piles are frequent. The skin is dry and harsh, frequently sallow, and often with a tendency to scaly eruptions. The hair is dry, and tends to split and fall off ; the nails are furrowed, brittle, and marked with white lines ; and the teeth frequently decay. There is usually marked loss of flesh and strength, and the circulation, as indicated by cold hands and feet, is imperfect. Nevertheless reaction is easily excited, especially by an alcoholic stimulant, when the palms of the hands and soles of the feet burn, so that the patient presses them against some cool surface. Headache is frequent, but not so regular a symptom as in the acute form, and the judgment is almost invariably not to be depended on. Irritability, timidity, and despondency are generally present. The sleep is disturbed, or there is sleeplessness, and the heart's action is irregular.

Treatment.—The treatment of chronic catarrh of the stomach requires modification to the various customs. To use a rule, the best treatment is the avoidance of the condition ; but if there is an acute attack, tending to become chronic, we must do our best to stop it. Bismuth and magnesia, with or without alkalies, are the appropriate remedies. It is in this condition that nitrate of silver has obtained a great reputation. It does good, but must be used with care. It is best given in pill, half a grain for a dose, along with opium or belladonna. Oxide of zinc is of value in those cases where the nervous symptoms predominate. In some cases, especially where food has been the cause of the irritability of the stomach, blue pill or calomel generally do good ; but in children this, or the gray powder more frequently employed, must be used with caution. When due to obstruction in the portal system, mineral waters are frequently the best things to prescribe, especially if they can be taken on the spot. If not, such water as Pauna, Carlsbad, and Friedrickshall are the best.

For habitual use as purgatives, too, aloes are of great service, as they act specially on the lower bowel, and so tend to relieve the constipation and piles. Small doses of the aqueous extract or of the compound decoction are best given. Castor-oil is also an excellent remedy for relieving oppression and heaviness about the stomach, frequently re-

storing a healthy appetite. Strong purgatives are, however, to be strenuously avoided ; and in many cases a small dose of aloes, iron, nux vomica, and belladonna before dinner will do more to keep the bowels open than anything. If from alcoholic excess, the liver must be seen to, and so mild mercurials and salines had best be given with opium if the irritability of the stomach is great. Even astringents may be given to check the copious mucous secretion from the stomach. The dyspepsia of consumption is most difficult to manage. Prussic acid often does good, more so columba and liquor potassa. In this malady purgatives must be avoided, for unfortunately in its course there is only too much risk from the purgation which arises from the disease. Pepsin and acids are both of great value when the irritability has been removed. The rules for food are much the same as for acute catarrh and atonic dyspepsia.

INDIGO, though an important commercial product, can hardly be called a remedy, even if contained in the pharmacopœia. It is prepared from various species of plants belonging to the leguminous order of plants. The original substance is white ; but from it may be prepared an indigo red and an indigo blue. It is this last substance which is commonly employed. Dissolved in the strongest sulphuric acid, it forms a peculiar compound sometimes called sulphindigotic acid ; this is decolorized by chlorine, and is used as a test of the presence of that substance. Indigo has been used in epilepsy, but its value is not understood. It appears in the urine as a bluish-green compound ; sometimes it is found in abnormal urine.

INDURATION is a term applied to the hardening of tissues around the seat of previous mischief ; thus induration may occur on the site of an old scar or wound, or an old abscess, or in the glands under the chin, etc.

INFANTICIDE is a term used to signify the voluntary murder of an infant, either during the progress of its birth or as soon as it begins to live an independent existence from its mother. In any case where a woman has confessed her condition and made preparations ever so slight for the expected infant, the law looks with a merciful eye upon circumstances which would otherwise lead to a verdict of infanticide. The methods by which the life of a newly-born child may be sacrificed are, in medical jurisprudence, divided into those which consist in omitting the necessary services required by an infant, and by inflicting violence. It should not be forgotten that an infant may die during or immediately after birth, without any criminality on the part of the mother, and in cases of great exhaustion, where disgrace and shame would attend the exposure consequent on calling for help. This is undoubtedly sometimes the case, even where there has been no intention or desire to destroy life. According to the law of Scotland, the earliest period at which the crime of infanticide can be sustained is from the time of quickening. In England the period assigned by law dates from the seventh month, when the child is supposed to be capable of living. Experienced surgeons are able by many signs to judge from the dead body of a child whether it has ever lived or been born dead ; the most certain proof is, however, by *post-mortem* examination. The state of the lungs indicates whether they have ever been inflated with air, and if so it is certain the child has lived ; the next question, therefore, to decide would be, whether its death resulted from natural or from violent causes, and on this decision rests the nature of the verdict in such a case. The objection of a jury to return a verdict of wilful murder in the case of newly-born infants evidently killed is often exhibited. So many extenuating circumstances seem to surround such cases, especially where the mother is discovered, that every possible excuse is urged for appearances which would otherwise be condemnatory ; but in the large number of instances where infants just born are found thrown away in streets, and concealed in all possible places, there can be no doubt that murder has been absolutely committed, and almost certainly by the mother, whose natural instincts are supposed to render such a crime impossible.

INFANTILE CONVULSIONS differ in a good many respects from the convulsions of adult life ; they are exceedingly common in children from the time of birth up to their seventh or eighth year, particularly about the period of their first teething. They are produced in a variety of ways, and' often go off leaving no evil effects behind them. The earliest convulsive phenomena of the kind are ordinarily what are called *inward fits ;* these commonly occur a few days after birth. The baby seems asleep, but its eyes roll so that the whites alone are seen ; it breathes with some difficulty, and the face twitches or is drawn first to one side, then to the other ; this is commonly produced by indigestion, the mother's milk may not agree, or improper food may have been given. A little dill water or a drop of one of the volatile oils, or a little spirit of ammonia will relieve for the time, but this should be promptly followed by a teaspoonful of castor-oil. Almost any irritation affecting the nervous system, local or remote, will in certain children produce fits. The frequency and the fatality of these rapidly diminish as the child grows older. The symptoms of such fits vary. Most frequently the child is suddenly taken, loses consciousness and stops breathing ; sometimes the body is stiffened ; sometimes, and this is rather the rule, it is agitated by smart contractions and relaxations of the muscles.

Usually the hands are clinched with the thumb in the palm, while the face becomes first red, then livid. The contents of the bladder and rectum are voided, and the eyes squint. Presently the limbs begin to relax, and the child gradually recovers, presently falling into a sound sleep. More rarely it passes into a comatose state and so perishes. The action of the muscles is generally more violent on one side than the other, hence one side is dragged to the other. Most frequently there is more than one attack, and sometimes they leave permanent damage of the nervous system behind them.

Treatment.—As regards treatment there is not much to be done during the fit ; everything should be loose about the child, and some water may be sprinkled on the head and face. Should the fit threaten to recur it is best to plunge the child in a warm bath up to the neck and shower cold water on the head. The bowels should be seen to, and a dose of castor-oil given if necessary. Sometimes an emetic does good, and any source of irritation, be it where it may or what it may, should be sought for and removed. Some practitioners recommend the administration of ether or chloroform if the attacks are continuous, with bromide of potassium in the intervals.

INFANTILE PARALYSIS. (See PARALYSIS.)

INFANTS. (See DENTITION, DIET, FEEDING-BOTTLE, CHOLERA INFANTUM, INFANTILE CONVULSIONS, INWARD FITS, LABOR, LACTATION, PARALYSIS, RED GUM, and YELLOW GUM.)

INFECTION is a term used to denote that a disease may be carried from one person to another without either coming in contact. Thus scarlet fever is infectious, as it may be conveyed to a distant part in the clothing, etc., of an affected person. (See FEVERS.)

INFLAMMATION may be roughly defined as an unnatural process, which manifests itself in increased vascularity and sensibility of the part attacked, and which is associated with more or less of constitutional disturbance. This process varies much as to form and degree of intensity, the symptoms being sometimes slight and harmless, as with the transient and superficial redness of a small portion of the skin caused by the bite of an insect, and in other instances, most severe and dangerous, as in acute inflammation of the lungs (pneumonia), or of the thin serous membrane of the abdomen (peritonitis). Very few diseased processes occur in the organism which are not at some part of their course associated with inflammation, and the great majority of diseases are really due to this process. In ordinary cold in the head there is *inflammation* of the mucous membrane of the nose and the adjacent cavities, in bronchitis there is *inflammation* of the mucous membrane lining the air-passages, and in sore throat there is *inflammation* of the mucous membrane of the tonsils and fauces. Inflammation follows all severe injuries, and serves to repair wounds and to unite together portions of fractured bone. When the process runs its course quickly, and is high and severe, we have *acute* inflammation ; when the process is slow and lasts for a long time the inflammation is called *chronic*. Sthenic inflammation is an acute form of rapid progress, which is met with in strong and vigorous persons. Asthenic inflammation is a low and lingering form, the subjects of which are generally debilitated and unhealthy residents in towns. There are many forms of inflammation in which the symptoms of the process are modified by some constitutional disease, as scrofula, gout, syphilis, or by blood-poisoning, as in cases of erysipelas, carbuncle, and the eruptive fevers. In cases of this kind the process is called *specific*. The inflammatory process may terminate in complete restoration of the affected parts, or in effusion of serous fluid, or solid fibrinous material, or of pus. In these latter the inflammation, according to the character of the material effused, is called *adematous, adhesive*, or *suppurative*.

Symptoms.—The chief manifestations of inflammation are the four classical symptoms of redness, swelling, heat, and pain, which were taught by Celsus eighteen centuries ago. *Redness* of an inflamed part is a necessary consequence of increased vascularity. The small vessels are dilated and distended with blood, and in some cases new vessels are formed. This symptom is very manifest in catarrhal inflammation of the conjunctiva (ordinary ophthalmia). Vessels which in the healthy state are so small as to be barely visible to the naked eye become enlarged and filled with bright red blood and form a close vascular network. Another cause of the redness is rupture of some of the over-distended vessels, and effusion of blood into the inflamed tissues. This condition is shown also in some cases of ophthalmia, where one may observe irregularly-shaped patches of a dark-red color. The redness of inflammation varies much in tint and in intensity. In acute inflammation of the surface of the body there is a scarlet blush, which differs from the dusky redness of erysipelas and the coppery tint of syphilitic eruptions on the skin. In inflammation of tissues, which possess no proper blood-vessels, the redness does not extend beyond the adjacent structures : thus, when the cornea is inflamed, it does not itself become red, but is surrounded by a zone of enlarged conjunctival vessels. *Swelling* is due in the early stages of inflammation to the increased quantity of blood contained in the vessels of the affected tissues, and afterward to the effusion of fluid. In superficial

inflammation of the skin, caused, for example, by the application of a blister, a clear yellowish fluid is poured out under the epidermis, forming the well-known blister or bleb. In acute sthenic inflammation of deeper parts of the body and of internal organs, a similar fluid is poured out into the interstices of the tissues. The fluid effusion of inflammation varies in character according to the intensity of the inflammatory process and the health of the patient. In debilitated patients the exudation resembles unhealthy pus, and in chronic inflammation the effusion is fibrinous, and forms a hard and tense swelling. In croup and diphtheria the effusion contains much fibrine or coagulated material, and forms white membranous deposits on the surface of the inflamed mucous membrane. *Pain* varies in intensity, and is influenced more by the structure and relations of the part inflamed than by the severity of the inflammation. It is most acute in inflammation of the structures which do not readily yield to the increased flow of blood and the effusion of serous fluid ; as, for instance, bone and cartilage ; it is also very violent when the inflamed structures are bound by thick fascia. In acute inflammation it is not restricted to the affected parts, but radiates for some distance along the nerves supplying those parts. Pain in inflammation is due to stretching and compression of the nerve fibres of the inflamed tissues. The *temperature* of an inflamed part is increased in consequence of the increased flow of blood, and of the active changes of tissue which take place in connection with the inflammatory process.

No one of the above symptoms is peculiar to inflammation. *Redness* may be due to venous obstruction, and to that process of local determination of blood to a part called hyperæmia. The severe pain of tic-douloureux, and other varieties of neuralgia, is not due to any inflammatory process. Increased heat of the skin of the face and head follows division in an animal of the sympathetic nerve of the neck. Acute inflammation is invariably associated with a train of constitutional symptoms indicating the condition known as feverishness. This condition in cases of iodiopathic inflammation, as pneumonia or facial erysipelas, is termed pyrexia, or symptomatic fever ; and in cases of injury, as compound fracture, surgical or traumatic fever. The intensity of the febrile symptoms varies in proportion to the extent of the inflammation, the importance as to function of the organ attacked, and the previous condition as to health and vigor of the patient. In a typical case, such as one of pneumonia or of compound fracture, the patient first complains of chilliness, or has an attack of shivering. Vomiting sometimes occurs at this stage. The surface of the body then feels very hot, and the mouth and tongue become dry ; the urine is scanty and dark-colored ; the respirations are increased in frequency, and the pulse is raised to 120° or 130°. There is intense thirst, and the patient suffers from headache, has no appetite, and feels very uncomfortable. At times, especially at night, he is "light-headed," and if he has been a drinker becomes violently delirious. These symptoms, as has been just remarked, vary in intensity in different patients. In those who are delicate and sensitive even a slight degree of inflammation will serve to start a train of very alarming symptoms ; while in other patients of less nervous irritability a severe attack of acute inflammation may be followed by but slight constitutional reaction. In cases of inflammation, however, the invariable co-existence of feverishness is shown by these two symptoms, high pulse and high temperature. To this latter symptom considerable attention has been paid by the medical profession for some few years, and the clinical thermometer has now become a valuable means in the diagnosis and prognosis of disease. When kept for about ten minutes on the armpit or mouth of a patient suffering from inflammation, this instrument will indicate an elevation of temperature by four or five degrees. Instead of 98.4° Fahrenheit, the normal point, the temperature of the patient will be 101°, 102°, or 103°, or even higher. As a rule it does not rise very far above 102°. This increased production of heat also occurs in the fevers commonly so called, as typhus and typical, and in the eruptive fevers. The cause of this increase of bodily temperature in cases of inflammation has not yet been satisfactorily determined. According to the views of many influential pathologists the blood as it flows through the inflamed part is warmed, the heat thus acquired being accumulated in the general mass of the blood, and distributed to all parts of the body ; the circulating fluid, the temperature of which has been thus increased, then undergoes active changes and also stimulates every part of the body through which it flows to active textural change. This is produced by over-production and increased excretion of nitrogenous waste-products, processes which are always associated with elevation of temperature. By other pathologists it is supposed that portions of the inflamed tissues are disorganized and destroyed, and that the products of this destruction are taken up by the blood, and produce active changes in that fluid and in the tissues to which it is distributed. It has been proved by experiments that the introduction of dead and putrid organic material into the blood of animals will cause high fever.

In acute inflammation the blood contains more than the normal amount of the coagulable and sizy material called fibrine. When shed into a vessel, as in the operation of bleeding, the blood coagulates slowly, and the red corpuscles run together and form masses which fall to the bottom of the vessel before the clot is formed. The fibrine thus strained, coagulates at the surface of the mass of blood, and forms a yellowish-white clot, which contracts and is depressed at its centre. When these changes have taken place, the blood is said to be "buffed" and "cupped."

Causes.—The causes of inflammation are of two kinds, *exciting* and *predisposing*. Of the former the following are the chief : *Injuries*, as wounds, fractures, the introduction of foreign bodies ; under this head may be included intense heat and cold, the effects of which are shown in cases of burns and scalds on the one hand, and in cases of frost-bite on the other ; catching cold is a frequent cause of inflammation ; the structures most frequently affected by the injurious influence of cold are the conjunctiva, the mucous membrane of the nostrils, and sinuses connected with the nostrils, and that of the throat and air-passages ; another frequent cause of inflammation is the retention within the body of secretions, and of dead and putrefiable material ; a portion of dead bone or of a slough formed after the mortification of a portion of an organ will generally give rise to prolonged inflammation in the surrounding parts ; poverty of the blood, due to chronic disease or to insufficient nourishment, and a poisoned state of that fluid may give rise to inflammation. Professor Billroth of Vienna found that by injecting putrid fluid into the blood of a dog he could set up inflammation of the intestines, of the pleura, or of the pericardium or enveloping membrane of the heart. Inflammation of a part occasionally follows injury or division of the nerve by which that is supplied.

The predisposing causes of inflammation are : advanced age, debility from insufficiency of nourishment, and in consequence of some exhausting disease ; abuse of alcohol, living in damp and badly-ventilated rooms ; great mental and bodily fatigue, chronic diseases of the blood-vessels, certain morbid conditions of the body, as gout, rheumatism, and scrofula. In favorable cases where the affected tissues return at once to their previous healthy appearance and condition, on the subsidence of the general symptoms of inflammation, the process is said to terminate in *resolution*. The serous fluid effused in cases of acute inflammation often coagulates, and is converted into a fibrous and living substance, which in cases of wounds and fractures, serves to bind together the several parts—termination in adhesion. The process sometimes terminates in the formation of pus, which is either discharged from an open raw surface *or* is accumulated in the *midst of living tissues* and forms an abscess.

The process of inflammation as revealed

by microscopical observation has of late years been a subject of increased interest in consequence of the results of investigations made by modern German biologists, and of the prevalence of new views concerning the nature of certain constituents of the blood. The following particulars are gathered from an able and elaborate article on this subject by Dr. Burdon-Sanderson, contained in the fifth volume of Holmes's System of Surgery.

First Stage in the Process of Inflammation.—From microscopical observations of the web of the frog's foot and the mesentery of the frog, it has been found that on the local application of an irritant, the smallest arteries, and subsequently the capillary vessels become detached and slightly increased in width, so that there is more or less contortion of these vessels, at the same time there is acceleration of the blood-current. This dilatation, on the application of most irritants, as dilute sulphuric acid, acetic acid, caustic soda, etc., commences immediately, but where liquor ammoniæ or carbonate of ammonia is used, there is at first contraction of the capillaries, with retarded flow of blood, which after lasting for an hour or two, is followed by dilatation and accelerated flow of blood. Concerning the cause of this phenomenon, there are and have been many views. Some pathologists hold that it is due to a more rapid and violent contraction of the arteries, others that the relaxation of the vessels is dependent upon the fatigue caused by their unnatural irritation ; the most probable view is that it depends upon irritation of the nerves which proceed f om the injured or inflamed part to the brain, which irritation or impression is then reflected through the spinal and sympathetic nervous system, and finally along the nerves which are distributed to the walls of the vessels of this part.

Second Stage.—The blood contains, in addition to its red corpuscles, a much smaller number of colorless corpuscles, which when examined, after removal of the blood from the body, and without any preparation, are found to be round, pale and clouded cells, each containing one or two nuclei of a spheroidal or flattened form. When the blood, however, is submitted to examination in the living state, or under physical and chemical conditions resembling those of life, these colorless corpuscles are found to be small globular masses of contractile protoplasm, which resemble amœloid organisms in their possessing the functions of growth and locomotion, and the power of dissolving and absorbing nutritive substances on which they live. In the second stage of the inflammatory process, the current of blood becomes slower and slower, then oscillates, and finally ceases altogether. If the vessels of the inflamed part be examined in this condition, which is called *stasis*, it will be found that the inner surface of each vessel

is lined by a continuous layer or pavement of white blood corpuscles, some of which pass from the vessel into the surrounding tissue in the following manner : " Here and there on the outer contour of the vessel, minute colorless button-shaped elevations spring, just as if they were produced by budding out of the wall of the vessel itself. The buds increase gradually and slowly in size, until each assumes the form of an hemispherical projection, of width corresponding to that of a white blood-corpuscle. Eventually the hemisphere is converted into a pear-shaped body, the stalk end of which is still attached to the surface of the vein, while the round part projects freely. Gradually the little mass of protoplasm removes itself further and further away, and as it does so, begins to shoot out delicate prongs of transparent protoplasm from its surface, in no wise differing in their aspect from the slender thread by which it is still moored to the vessel. Finally the thread is severed, and the process is complete. The observer has before him an emigrant white blood-corpuscle." These phenomena were first observed and described by Professor Cohnheim of Germany, whose words, as quoted by Dr. Burdon-Sanderson, are here given. Those white-blood corpuscles then accumulate in considerable quantities around the vessels, and together with the serous fluid of the blood, a liquor sanguinis, which is effused at the same time, give rise to the swelling which forms one of the cardinal symptoms of the inflammatory process.

Besides the above changes which have their seat in the blood-vessels, there are others which have their seat in the tissues—textural changes. It has been found by microscopical examination, both of the non-vascular tissues, such as the cornea, cartilage and tendon, and also of the vascular tissues, as connective tissue and muscle, that in consequence of the stimulating properties of the effused liquor sanguinis, with which an inflamed structure is soaked, the permanent cells of the affected tissue which have for their function the maintenance of the unchanging life of these tissues, germinate as it were and become metamorphosed into mobile masses of protoplasm, resembling in all respects the white corpuscles of the blood. These bodies when collected in large quantities form pus, and the inflammatory process then terminates in suppuration. These views, as to the textural origin of these mobile bodies, have not been undisputed. By Professor Cohnheim and his followers it is held that these and the pus cells formed in masses are not the offspring of the permanent tissue cells, but wandering blood-corpuscles which have a tendency to escape from the vessels, as has been stated above, and then to move away from the blood-current in a direction at right angles to the axis of the vessels from which they have escaped.

There seems to be no doubt that pus is formed both by wandering cells and by the cells formed by textural changes, and in acute and rapid suppurative inflammation it is probable that the pus cells are mostly, if not all, wandering blood-corpuscles.

INFLUENZA is a specific and epidemic fever which chiefly attacks the lining membrane of the nose, larynx, and bronchial tubes, lasting from four to eight days, and not preserving the individual from a future attack. This disease has occurred in various countries at different times, and has received a vast number of names; in the seventeenth century it appeared in Italy and first received the name of influenza, because it was attributed to the influence of the stars. It is supposed that this disease has been known centuries before the Christian era; certainly since that time numerous epidemics have been most carefully recorded. The area attacked has also much varied in extent; sometimes only part of a country has been affected, at other times it has spread over a great part of the civilized world. A disease is said to be "endemic" when it is confined to a small area, as a village or town; "epidemic" when it spreads over a country; "pandemic" when it invades a large portion of the earth's surface. Influenza occurs in both an epidemic and pandemic form; it may take weeks for the disease to spread from one country to another; thus in the year 1832 influenza prevailed extensively in Russia, but it took no less than eight months to spread over the whole of Germany. No particular track has been observed which this disease particularly follows; it has travelled in different directions from time to time; some have thought that it usually spreads from west to east. The poison seems to be conveyed by the air, and persons at a distance from land may become attacked. Attacking a community, the disorder generally remains among them from six to eight weeks, but occasionally it has remained longer; the epidemic will completely disappear then for a time, nor is it usual to find an occasional case breaking out in the interval of epidemics. It is common to hear people complain in the winter of having an influenza cold, but this is a misapplication of the word. The onset of the disease is generally very rapid, while the decline is more gradual, and may last several weeks. In 1847 it was calculated that no fewer than a quarter of a million of people suffered; in Paris between one fourth and one half of the population were attacked, and in Geneva about one third. Various physical conditions have been supposed to influence the outbreak of this disease. Volcanic eruptions were once thought to be a cause, but there has been no trustworthy evidence of this; soil seems to have no effect, for the complaint has appeared in every variety of country, in high

lands as well as in low lands, in hill countries as in marshes and plains. Nor does the time of the year seem to have any effect, since it has been prevalent at all seasons. There is also no connection between temperature and influenza; it occurs in high as in low temperatures, nor does any sudden variation of cold or heat seem to produce any effect. Moisture also has an apparent influence, nor is there any evidence at present that any atmospheric condition has any effect on this disorder. The intercourse of human beings does seem to have an influence on the disorder; thus an affected person coming into a village seems to be a centre from which the disease spreads; nevertheless it is very remarkable that thousands may be attacked in the same town in the course of a few hours, while in other contagious disorders the progress is much slower. In this disease, as in other contagious diseases, there seems to be a period of incubation, when the poison seems for a time to be latent in the system, before any of the marked symptoms declare themselves, and although in most cases persons seem to be suddenly struck with influenza, yet there is probably a period of incubation, which may be very short and may last for some days. Most people who have suffered from a contagious disease are not liable to a second attack, but in influenza one attack gives no immunity to another, although persons seldom suffer twice in the same epidemic. Various speculations have been made as to the nature of the exciting causes of this disorder. It cannot arise from contamination of water, as it would then be confined to a particular locality, nor to any kind of food. The rapid way in which it spreads shows that the poison must exist in and be conveyed from place to place by the air, for in this way alone can we account for the rapid transmission of the disorder. But as to the nature of the poison in the air nothing at all is known; were it a gas, it would become diluted by mixing with air and lose its virulence, but this is not the case in influenza, nor does it seem to be made of organic matter, or to be suspended mineral matter. One thing seems clear, that the poison can multiply in the air and reproduce itself. Race and sex seem to be equally attacked by the disease; the young are said to be less liable to it than old people. Overcrowded habitations seem in some epidemics to have increased the mortality, and places which are low, damp, and badly ventilated appear to predispose to it.

Symptoms.—The disease commences with shivering or a feeling of coldness down the spine, with a hot, dry skin, quick pulse, thirst, and severe headache. These symptoms are common to nearly all fevers, and they precede, as a rule, any local affection; sometimes they come on suddenly, sometimes they develop slowly in two or three days. If they come on suddenly, intense frontal headache, with aching pain over the eyes, is generally the first symptom. This feverish state usually lasts four or five days, and then gradually disappears, and its disappearance may be accompanied by profuse perspiration or a troublesome diarrhœa ; in some cases the fever may last several days longer, but then some complication has probably arisen and given rise to inflammation of the lungs or some other organ. The peculiar catarrhal affection usually follows the early symptoms of the fever ; it begins with swelling and dryness of the lining membrane of the nose, and the tissues or cavities of the forehead, causing great frontal headache and frequent sneezing ; the mucous membrane of the eyes or the conjunctiva are generally affected in a less degree, and a thin acrid discharge takes place ; now and then bleeding occurs from the nose ; this condition then occurs all the way down the air-passages, even down to the smallest branches of the bronchial tubes. This affection may occur in the whole tract of the membrane at once, or, beginning in the nose, it may spread downward into the lungs. The inside of the mouth and the tongue and pharynx may also become implicated, but in a less degree. The discharge from this inflamed surface is at first thin and acrid, and at times bloody ; it then becomes thicker, tenacious, and purulent. The patient sneezes, has a troublesome and violent cough, pains in the side, and a difficulty in smelling. There is great distress in breathing, and the pallor of the face and lividity of the lips show how great is the obstruction to the circulation in the lungs, for the blood becomes in such cases imperfectly aerated, and, owing to the accumulation of carbonic acid, flows through the vessels with difficulty. In most cases the catarrh is at its height by the third or fourth day, and generally declines from the fifth to the seventh, but in severe cases it may last longer. Coincident with the fever and catarrh, and perhaps in próportion to the severity of the former, is a peculiar state of the nervous system. There is great depression and loss of spirits, with aching pains in the muscles and neuralgia pains in various parts of the body or extremities. The mind is often affected, and the patient may become stupid or delirious. These nervous symptoms often last some little time after the fever and catarrh have subsided, and thus leave the patient very weak and retard his convalescence. The temperature of the body appears to be raised in most cases, but no exact observations on this point have yet been made. Sweating of the skin often occurs during the defervescence of the fever, or at the time when a descent of the temperature takes place, but rarely in the early stages. Crops of minute transparent vesicles, or little blisters containing fluid are often seen on the skin. Menin-

gitis, or inflammation of the membranes of the brain, and otitis, or inflammation of the ear, may come on now and tnen. Great delirium, as well as intense headache, is a dangerous symptom. Neuralgic pains are met with in many parts of the body, and there is also a remarkable prostration of the muscular strength. The cough comes on in paroxysms, and may be so severe as to bring on a rupture, or even abortion in pregnant women. There is but slight expectoration at first, and then the phlegm which is expectorated is stringy and often bloody, then it becomes more consistent, opaque, and purulent. Bronchitis, or inflammation of the bronchial tubes, pneumonia, or inflammation of the lungs, and pleurisy, or inflammation of the serous covering of the lungs, are present in some cases, and add to the danger ; however, the frequency of their occurrence varies much in different epidemics ; their presence may be detected by a careful examination of the chest, and by the increase in the distress of breathing. Vomiting and nausea often come on at the commencement of an attack ; diarrhœa, as a rule, occurs later on in the disease, when the fever begins to abate. In some cases the skin assumes a yellow tint, and bilious vomiting comes on. As in most febrile affections, the urine is at first high-colored and scanty, and afterward it often deposits a pink or reddish sand, made up of lithates. Occasionally there is complete or partial suppression of the urine. Now and then swelling of the glands under the chin or in the neck has been observed. Convalescence is often retarded by rheumatic-like pains in various parts of the body, and by prolonged debility or unusual nervous depression.

Mortality.—The mortality varies much in different epidemics. Yet at no time is it very high, for in 1837, which was looked upon as a severe epidemic, not more than two per cent died. Age seems to have an influence on the death-rate, and it is more fatal among the old than among the young. It is higher, too, in those who have suffered from heart disease, bronchitis, or emphysema, and especially in those who have weak and fatty hearts.

Treatment.—No means are yet known by which influenza may be prevented. Overcrowding, and low, damp, or badly ventilated houses seem to favor the disease. When the disease has declared itself, the patient must be kept in bed, and the room should be cool and well ventilated, although draughts are to be avoided ; when a person suffers from catarrh or a common cold, he will prefer to be in a warm room, but those suffering from influenza like a lower temperature. As the appetite is bad, not much solid food can be taken, and, in fact, for the first three or four days it had better be avoided. Plenty of cold drinks may be given, and if there is much fever they may be iced ; those are the more grateful which are made slightly acid ; barley-water and lemon-juice, raspberry vinegar, oranges, and cream of tartar water may be given with benefit. Stimulants, except in the case of old and feeble people, need not be given ; claret or hock, in combination with seltzer-water, may be given if necessary. But if the heart's action fail, and there be great distress in breathing and duskiness of the face, brandy and ammonia must be given, and pretty freely. Beef-tea is not of much use, but the patient may take plenty of milk. Too much food should not be given, as it will only tend to derange the stomach ; too much fluid should not be given at one time, as it may cause painful distension of the stomach and flatulence ; nor should any acid drink be taken just before or after the administration of milk, as it will only cause clotting of the latter and perhaps subsequent vomiting. The patient may be allowed solid food as soon as the severity of the fever has passed away ; some well-boiled mutton may be given at first, or any easily digestible food ; in these cases, however, the diet should be similar to that which is recommended in the article on FEVERS. The air of the room should be kept moist by keeping a kettle of boiling water on the fire, or by putting boiling water in shallow vessels about the room ; in this way the steam passing into the air in the room keeps it moist and eases the cough. Drugs are not of very much use, and in slight cases nothing is required, but some cooling saline or effervescent medicine may be given to check the febrile symptoms and to allay thirst. In the early stages a purgative is often given with benefit, but this must not be carried too far ; a dose of calomel, followed by a saline purgative, will generally suffice to open the bowels enough and relieve the patient, but the persistent use of purgatives is injurious. In the case of children a dose of gray powder may be substituted for the calomel, or a little castor-oil may be given for the same purpose. In no case should repeated doses of any mercurial preparation be given so as to produce salivation. Some have recommended emetics, but no marked benefit has resulted from this practice, and it seems, therefore, undesirable to give them ; for this purpose tartar-emetic has been advocated by some, but it is too lowering for the patient ; nevertheless, if any bronchitis or pneumonia come on in the course of the illness, ipecacuanha, combined with ether and ammonia, will prove of service. Opium should be administered with the greatest caution ; in all cases where there is a tendency to congestion of the lungs this drug seems to intensify the condition, so that its use may seriously increase the danger of the patient ; but if there be not much congestion and the cough be very violent, then small doses of Dover's powder may be given or preparations

containing belladonna. If there is great tightness across the chest or a severe stitch in the side, hot linseed meal poultices will be of great use in alleviating the pain ; flannels wrung out of hot water may be applied with the same object, or, if the pain be very severe, the application of three or four leeches will be found very useful. Later on, when the expectoration is more profuse, ammonia, chloric ether, and senega may be ordered with advantage. Quinine seems to be very useful as soon as the acute stage is passed ; it tends to promote convalescence, improve the appetite, and lessen the severity of the neuralgic pains. Blisters do no good, and only increase the patient's sufferings. Inhalation of steam may be tried, either by using an ordinary inhaler, or by breathing in the steam from a jug of boiling water. If the diarrhœa be moderate it should not be stopped, as it is beneficial to the patient, but if excessive it should be checked. Should suppression of urine come on, a hot hip bath should be used, and flannels wrung out of hot water ought to be applied to the loins ; dry cupping is also useful, and the patient should drink plenty of linseed-tea or barley-water. An ice-bag to the head will often relieve the intense frontal headache, and this may be applied by putting pounded ice into a bladder or india-rubber bag and kept on for two or three hours. Sometimes a single piece of linen dipped in vinegar and water may be applied to the forehead with relief, or a wet towel may be laid on the pillow, so that whenever the patient turns his head it will come in contact with a cool surface. During convalescence iron and quinine should be given to promote an appetite and to act as a tonic to the system. A generous and nutritious diet must be given, including milk, meat, and some beer or wine. For some time after an attack the patient should avoid exposure to cold, and wear flannel next the skin.

INFUSIONS are preparations of remedies which yield their properties to hot water. The substance is, as a rule, reduced to coarse powder or roughly comminuted, placed in a pipkin, and covered with a measured quantity of distilled water. In most of the preparations boiling water is used, and the time of infusion varies from ten minutes to four hours, according to the time requisite to separate the active ingredients. In some few cases water at a lower temperature is used ; with chiretta and cusparia, water at 121° F. is used ; while with yet others, as columba and quassia, absolutely cold water is used. In any case the great point with an infusion is that no heat shall be applied after the first. Infusions are mainly used as vehicles.

INFUSORIA are minute microscopic organisms developed in water in which animal or vegetable matter has been dissolved by steeping ; they are readily produced in a few days by making an infusion of common hay.

INGROWING NAILS. (See NAILS.)

INHALATION is a method of introducing remedies into the system which has only recently come into vogue. Of course every substance is not capable of being so used, it being essential that the material should be capable of being volatilized, and at the same time of inflicting no injury on the lung substance. Some of the substances given in this way are nominally in a gaseous state, and only reduced to the fluid condition for the sake of convenience in stowage. Such is nitrous oxide, the favorite dental anæsthetic. Others, again, are substances which very readily pass from the fluid to vaporous condition ; such are chloroform and ether. These substances are used to deaden pain or overcome sensibility generally, but what are technically termed inhalations are substances which, nominally solid or fluid, can be made to evaporate, and so act as local medications to the lungs and air-tubes. Conium, creasote, and hydrocyanic acid are given this way. Of course it is essential that these and all similar preparations which are intended to act on textures so delicate as those of the lung, should only be applied to its membranes at a suitable temperature. To enable the substances to be volatilized, and these, too, at a suitable temperature, proper vessels have been made, called inhalers. We generally tell the patient to put some of the material to be inhaled in a jug with hot water ; to place it close to the mouth, and cover head and all with a towel. The plan answers well. The smoking of stramonium, fumigation by sulphurous acid, and inhaling the fumes of nitrate of potash paper, are all modifications of this process.

INJECTIONS.—There are two kinds of injections, one of which is treated of in the article on ENEMA, and the other in the article on HYPODERMIC INJECTIONS.

INOCULATION really means the introduction of any poisonous matter, particularly if that be of an animal origin, beneath the skin. It has come, however, to mean in great measure the introduction of the small-pox virus into a healthy system. The term vaccination is on the other hand limited to the introduction of cow-pox into the system in the same way. The plan of inoculating with small-pox has been known in the East from a very remote antiquity, and was introduced into England from Turkey in 1717 by Lady Mary Wortley Montagu. The great advantage of the plan is that a time may be selected for undergoing the disease, and the malady so induced is not nearly of the gravity small-pox commonly assumes. Besides, a mild case may be selected whence the matter to be inoculated is taken, and it would seem that this has something to do with the virulence or comparative mildness of the subsequent attack. Nowadays, when vaccination is in this country enforced by law, the practice of inoculation has fallen into desuetude, and has in point of fact been prohibited by law. Nevertheless, occasions may arise when its practice would be sound wisdom. Should the plan be had recourse to, every individual of the family must be inoculated, provided they have not suffered from small-pox, otherwise the disease would spread in the ordinary way, and run in the non-inoculated individuals its ordinary course. The plan of inoculating is as simple as that of vaccination ; any instrument sharp enough to scratch the skin is to be selected, dipped into the small-pox matter, and the skin of the upper arm scratched until the blood begins to appear. This should be done in one or two places slightly apart. Very probably no other eruptive spots will appear.

INSANITY, being a word of negation, is not easy to define. Doubtless we may speak of it correctly enough as any condition which is not that of sound mind, but as this soundness of mind cannot be judged absolutely but only relatively, in giving such a definition we are simply tossing the ball from one hand to the other.

Each case, in point of fact, must be considered by itself and as a whole. Broadly it may be said that the tests of lunacy, which are commonly used in the same sense exactly as insanity, are, in each case, incapacity to manage property or danger to the public in criminal law ; however, it is broadly laid down that the test of sanity is the knowledge of right from wrong—a test, as has been well said, which, applied to our lunatic asylums, would set at liberty three fourths of their inmates. It would be useless here to enter into metaphysical speculations as to the connection between mind and brain, or the alterations in brain substance which are most commonly associated with the insane condition : suffice it to say that the current belief is, that in the great majority of cases of insanity there is a change in the brain substance just as there are changes in the lung in diseases of that organ, or of the heart when that part of our body is affected. For just as the function of the lung is respiration, and that of the heart circulation, so is the function of the brain the manifestation of mind. As we find in other parts of the body, however, when the self-balancing power is lost or in abeyance, there may be disorder of its functions without any marked, or, at all events, protracted signs of local change, so we may have in the case of the brain temporary insanity without any permanent disease of its structure such as gives rise to the more permanent form of the malady.

Causes.—The causes of insanity are generally assumed to be of two kinds, as is usual in medicine—predisposing and exciting, but the so-called predisposing causes mean merely a state in which the individual is more likely to become insane than if the same set of circumstances were operating on him in any other state. Accordingly, the term predisposing cause may be looked on as synonymous with *tendency*, and the origin of these tendencies has here to be discussed. By far the most potent of these *tendencies* is derived from hereditary transmission, or, as would sometimes seem, transmission from collateral branches of the family. This last is simply a modification of the former, for though the immediate parent may not have been insane and his brother or sister may, yet the inheritance is common to the stock of the grandparent or great-grandparent, and so descended in the ordinary way after all. It is of the greatest possible importance to fairly understand and to face the tendency of insanity to become hereditary, for an individual with such an inheritance, if duly guarded, may pass through life fairly able to fulfil its duties, whereas if the fact of this inheritance be ignored and the individual left to face the world like men of stronger mental equilibrium, it is more than likely that at some crisis the equilibrium will be upset more or less permanently. The most difficult question arises, however, when marriage comes into play. Too often these things are kept profoundly secret, or even intentionally hidden away, especially where property is concerned ; the result in many cases is unfortunate, the more so that the consequences of the deception frequently fall on the guiltless. We may, however, lay it down as a rule that if one has once been insane—be the individual male or female, though the rule is more binding on the latter than the former—marriage should determinedly be put out of the question. Much more difficult is it to decide in the case of those who belong to an insane family, but who have not themselves shown any signs whatever of the malady. There is the twofold question to solve—depending on the tendency these conditions have to die out, and the tendency they have to be transmitted from generation to generation. There is always a certain r sk, and this must be fairly faced, but the risk is less the further removed the insanity is from the individual concerned. Thus, an insane uncle or aunt would be a matter of much less moment than an insane father or mother, and an insane father or mother portends less risk than does an insane brother or sister. This heredity, however, tends to obliterate itself in course of time in two ways. Intermarriage with a healthy stock gradually diminishes the tendency to insanity in the survivors, and there is besides not only a natural but an artificial tendency to put an end to the heredity from the increasing numbers thereof. Thus, a certain number of those tainted will probably be incapable of propagating the race, and a certain number more being locked up will have no opportunity, and so between the two the insane members of the family tend to die out, while the stronger having intermarried with a more healthy stock, in course of time become like other people. Age has something to do with the liability to insanity. It is greatest between 25 and 40, least in the first ten years of life. Then, too, the nature of the insanity varies with the age at which the individual is attacked. In the earlier years of life there is much more violence connected with insanity than there is with the later ; if this rule is reversed, the likelihood of recovery is very greatly lessened. Sex, as already hinted, has a good deal to do with the liability to insanity, though not in the way one might have conceived. When men have grown up they are exposed much more to conditions likely to disturb the mental equilibrium than are females ; but, on the other hand, if there is a tendency to insanity in married females, especially the time of child-bearing, etc., are likely to have full effect. It is of course plain that men living as we do in a highly complex state of society, with a highly complex brain and mind, should be more liable to lose their mental equilibrium than in a more primitive state of society. But, on the other hand, it is perhaps interesting to know that the poorest counties of England contribute the largest proportion of insane to the district asylums.

Next come the causes called *exciting*, which are special to the individual and not to any class or group, though practically it is found that the same causes do operate in a very considerable number of instances. Chief among these are the moral causes of insanity, which may operate suddenly in the way of mental shock, or they may act over a number of years. Doubtless the former are the more potent in destroying mental equilibrium, especially in a mind which is badly balanced by heredity or by means of the individual's own habits and training. These last, however, belong rather to the group of physical exciting causes. The most important, according to all accounts, is the inordinate use of alcohol.

Symptoms.—The symptoms of insanity differ greatly from the signs of any other disease. They consist in great measure of the sayings and doings of the insane individual, either acquired by the observer from direct inspection or by hearsay. Very frequently indeed you get the cue from the attendant, or those who have been in contact with the insane individual ; for it is wonderful how many of them contrive to conceal their insanity until their delusion be touched upon, and then it is just as difficult for them to conceal it. There are certain words used in connection with the mistaken beliefs of the insane, which, though in ordinary parlance used synonymously, yet, strictly speaking, have got totally dfferent significations. These are *delusion, illusion,* and *hallucination.*

A *delusion* is a false belief relating to something which has a real existence, but to which the insane individual supplies attributes totally false. Thus he entertains a belief that some one, probably the least likely to do so, desires to swindle him ; that he himself is a prince possessed of boundless wealth, etc., etc.

An *illusion* is a false interpretation of the senses. There is something to be and something to have, but the patient gives them a totally false significance. A few rags are gorgeous robes ; pebbles, pearls of great price ; a few words spoken in an ordinary tone, a command to an army, etc., etc.

An *hallucination* is, on the other hand, a mistake on the part of the senses. The eye or the ear itself seems to be at fault ; the patient hears and sees things where there is nothing to see or hear. The word delusion is that commonly used, so as to cover both the other terms, but should be limited to the mistaken imaginings of one whose brain is disordered. These delusions are sometimes of a gloomy description. The patient is depressed or nervous, and proceeds to account for this feeling in the way most congenial to his fancies. A rich man may imagine himself a beggar ; a good and worthy man damned to all eternity. Moreover, everybody knows it and treats him accordingly, or he has some special tormentor who will never leave him alone. Mere delusions have reference solely and entirely to himself, and he has only the power of directing attention to one small part of his condition at a time ; were he able to take a more extended view he would be able to see that the delusions are delusions, and not, as they seem to him, actualities. Some patients entertain delusions of a totally different character ; these delusions are exalted delusions. They fancy themselves rich, powerful, and they are happy ; and yet the bodily condition of some of these poor patients is most miserable. They very frequently indicate a form of brain disease which advances through what to the bystander are exceedingly painful stages to certain death. Such delusions are most frequent in the condition known as general paralysis of the insane. Again, there are patients whose delusions take a different turn ; they live in fear and dread, but under which they are not passive ; they are prepared to fight, do anything for their life and liberty. Such are among the most dangerous class of lunatics.

But it is not only by means of ideas, it is also by means of *acts* founded or not on these that we judge of a man's sanity. These acts have reference either to himself, and then they are often associated with delusions of exaltation, or to the public, and then they are often founded on delusions of dread ; but in neither instance is the rule invariable. Among the most notable acts of the insane are indecent exposure, which very often occurs in the early stage of general paralysis, and stripping off of clothes, which has a most variable signification. Very frequently the removal of clothes is had recourse to out of revenge for not being allowed to do as the patient pleases. In other cases the patient cannot bear the feeling of clothes on the surface, and so tears them off him to get rid of them forever. In either case it is a troublesome and an expensive symptom. In many cases it is hopeless to cope with it. Give the patient the strongest materials, fastened on ever so carefully, by and by they will be torn off and torn to pieces ; blankets are torn in the same way. Very frequently in a female ward two patients, on being contradicted or opposed, will promptly set to work, and in a few minutes will stand naked as they were born, in the middle of a heap of rags. For the former class, that is to say, those who destroy clothing not knowing what they do, nothing well can be done ; but for the others some sort of punishment has to be devised, for they know perfectly well what they are about ; and if these fail, constant watching, which generally puts a stop to the nuisance. Fantastic ornaments and dress are not such frequent marks of insanity as writers who are ignorant of the subject would have the public believe ; yet some little peculiarity in dress or appearance is frequent enough. Perhaps the reason why there is not more oddity is that there is in asylums such scanty material for it. If, however, an individual has been ill for some time before being admitted to an asylum, and has been allowed to have his own way, it will generally be found that either in his dress or his apartments is something not quite compatible with the sane condition.

Suicidal acts or acts of self-mutilation are frequently committed by the insane. In the form of insanity called suicidal melancholia, where from the depression of mind life has become unbearable, it is frequently hardly possible to prevent the patient from destroying his life. He will watch his opportunity for years, and the first opportunity is sure to be taken advantage of. Very frequently in these patients the homicidal is closely associated with the suicidal impulse. The two things, however, ought perhaps to be made more clearly distinct ; for there is homicide and suicide founded on delusion, and undoubtedly there is, too, homicide and suicide founded on mere motiveless impulse, when the individual previously has exhibited no signs of delusion or of insanity, other than general weakness of mind. The subject of homicide, however, brings us to the consideration of those acts which are directed toward others rather than to the patient himself. Not unfrequently homicide or suicide is the result of overpowering terror. This perhaps is the most frequent form assumed in the insanity of drunkards. The patients in dread of their lives attempt to escape, and are killed in the attempt, or in their desperation and dread of attack turn upon the attendants and kill them. A goodly number of the murders committed by the insane, are from delusions. A man thinks his wife and children are going to starve, and so thinks it better to kill them at once ; or he fancies he has got a command from on high to sacrifice them, and does so. Yet again it may be done from sheer wantonness, as by an imbecile. In all of these cases there is, as a rule, no difficulty in making out the insanity ; it is not concealed, and may otherwise be only too apparent ; but there is yet another group of cases, which are of a much more doubtful category. It is well known that the great majority of confirmed epileptics sooner or later become totally insane. These constitute the very worst class of insane patients. Utterly untruthful, not a word can be depended on. Nor is an attendant's life safe with them. Before the onset of the epileptic fits, if they have them at intervals, they generally go through a stage of excitement, in which they are exceedingly dangerous. Now the stage is sufficiently well marked long before the mind of the patient is so far gone as to require to be sent to an asylum, and during these periods they are at any time liable to commit murder, and so it may be said of them just after such an attack. It is, however, with regard to paroxysmal insanity that there is most discrepancy between the opinions of alienists and the public at large. In the latter the idea is not pleasant that a man may go on all his life quietly and decently ; yet suddenly an uncontrollable impulse comes on him to murder some individual, after which he returns to his normal state. Yet most physicians, who have studied the subject, are agreed that this is so ; and it has now apparently been admitted by the bench. Perhaps, however, we ought to make a distinction between momentary insanity, giving rise to homicide, and the homicidal impulse with no insanity, momentary or otherwise. We fear the distinction is too fine to be generally appreciated by those liable to such attacks.

Homicidal mania, on account of its great importance, is not unfrequently elevated into a special form of insanity. So too are certain others, one which, however, we generally hear of when affecting some of the higher classes of society. These are kleptomania, entomania, and pyromania.

By *kleptomania* is meant an uncontrollable tendency to steal things when there is no object whatever to be gained by it. Most frequently it seems to manifest itself in ladies of a tolerably good position ; but sometimes also it seems to affect gentlemen in easy circumstances. Among them the stealing is taken as a mark of insanity. Stealing by insane persons, which is common enough, is quite another thing. Frequently when thefts are committed by men, especially from the ages of 25 to 40 or 50, it is one of the early signs of the general paralysis already alluded to. For with such patients stupid thefts, like indecent exposure, are early frequent signs of the onset of the disease. There are, too, very considerable numbers of our great body, who really, did they belong to a different class of society, would be included in the ranks of these kleptomaniacs. They are lads originally imbecile or nearly so, who have had a thorough bad training, and no governing power whatever. These had far better be sent to an asylum at once than be allowed to prey on society, in the wasteful manner which is habitual to them when not locked up. With regard to most cases of so-called kleptomania in ladies, the less said the better.

Entomania can hardly be called a special form of insanity ; inasmuch as the patient almost invariably labors under other signs of brain weakness. Nevertheless, in some insane patients, the lucid instincts are the most prominent of these symptoms. Such instincts are exceedingly common in many cases of insanity, especially in the early stage of general paralysis. It is, however, in women that the form of the malady is most marked, especially in young women, and it is this tendency on their part which renders the refractory female wards of a large infirmary the most horrible scene it is ever possible to conceive.

So of *pyromania*, undoubtedly a good many patients have an inclination to set things on fire, but such a tendency is hardly to be elevated to the rank of a special form of insanity. In many insane people the impulse to destroy everything they can lay their hands on is very great, and a very convenient way of so getting rid of things is to set them on fire. With such insane individuals, there is of course no regard for what the fire may do : it may destroy numberless lives, but they only desire to destroy property ; sometimes to escape from the asylum in the confusion, or it may be for the purpose of self-destruction. But in all of these patients there are other signs of insanity than a tendency to set a light to everything they can. Many cases of incendiarism, however, are the work of imbeciles who sometimes have no better motive than that of seeing a good blaze. Often in the country, in the case of hay and straw ricks, this is associated with the idea of some evil having been wrought by the farmer on the incendiary, so that the motive of revenge, out of all proportion to the magnitude of the supposed injury, enters into the mind of the imbecile incendiary. Roughly, and in such a way as will well suit our purpose, we may divide most cases of insanity into two divisions : those in which there is apparent exaltation, and those in which there is depression, and these two we shall describe as *mania* and *melancholia ;* but both of these tend in the long run to end in a condition characterized by absolutely no mind—what we term *dementia*, though there is a condition not inappropriately termed acute dementia. (See DEMENTIA.)

Most forms of insanity are preceded by a period during which the patient is not quite himself ; he is odd in his ways ; there is confusion of intellect ; bad sleep at night, and the patient is easily excited. The advance of the malady depends a good deal on the amount of sleep taken. If he sleeps well, it is just possible he may recover ; but if not, he gets more restless, more irrational, and more easily excited. His delusions, at first mere momentary fancies, become fixed and insuperable, and drive the patient to acts of insanity. Then most likely the medical practitioner is called in, and the patient is probably moved to some place of refuge. At this time the patient will probably complain of headache, very likely with slow pulse and confined bowels ; if a woman, the menstrual function is generally impaired, or there may be pregnancy, recent parturition, or suckling. All these may be removed and yet the patient does not get well ; we cannot restore the mental balance.

Treatment.—As a rule, the first thing is to remove the patient from home, and surround him with new attendants who will take him duly in charge in every way, when he will be removed from the causes of aggravation be they what they way. With this change of scene and pursuit there should be a change of diet to a nutritive one, if it has not been so before, and then everything must be done to secure good digestion, and a due nutrition. The bowels must be properly looked after, and sleep must be got. Opium is not good, in such patients it often excites rather than soothes, and increases the headache. Chloral is better, and had better be given in good full doses, 30 grains or so. It does not confine the bowels. If there is a tendency to epileptic fits, bromide of potassium had better be given them too, in full doses of 20 or 30 grains, three times a day. Tonics, especially strychnine, in careful doses, given so that the patient can never command enough to do himself harm, should be administered.

All these things require very careful superintendence, and as it is quite possible that the patient may get worse instead of better, when constant action may require to be taken, it is always better to select a place for the change of air and scene where there is a good prac-

titioner on whom you can rely, not only for medicine, but also for what in such cases is more valuable, namely, advice how to act. Most probably, if the case assumes a confused character, it will also assume the phase either of melancholia or of mania. If *melancholia*, then the utmost depression overpowers the unfortunate individual. Everything that happens round him seems to be connected with his evil fortune. Very likely he thinks he has committed sins too black for him to hope for forgiveness. No argument will get him out of these notions ; it is useless at this stage to attempt it. The appearance of many such patients is very striking. Woe-begone in the extreme, he may stand for hours in one spot, never moving, or he may be restless and trying to wander away, so to speak, from his evil fate. Usually such patients suffer a good deal in health ; they become thinner from want of food and sleep. The pulse is slow and weak, and the general condition of the patient indicates imperfect nourishment and bodily change. Suicide is greatly to be apprehended in a good many cases, perhaps the majority ; and this tendency may be so suddenly developed as to defy anything save the greatest caution from the commencement. For this reason, skilled attendance is of the first necessity, but it does not greatly matter whether that be carried out at home or in an asylum. Of course, in an asylum, the means of superintendence are much more readily available than in a private house, and the expense is correspondingly small. Besides, take it all in all, a patient is better in a small asylum, or in a large one, if classified, than in most private homes. Food and sleep are the two great remedies for this state, with absolute mental quiet. For sleep, chloral is best ; but if this does not suit, morphia may be given under the skin. The diet should be carefully selected, so that nourishment in abundance may be given. Sometimes these patients refuse their food, and when that is so they must be made to take it, either by the stomach-pump or through the nose. The bowels must be moved and kept open. First had better be administered a turpentine injection, after which a dose from time to time of castor-oil, or a small quantity of aqueous extract of aloes, daily at dinner-time, will suffice to keep them open. The *moral treatment* of such patients needs to begun as early as possible. The great thing is to draw their attention from themselves, and that must be done carefully and judiciously. Once they are brought to take an interest in anything outside themselves, they will generally do well ; this is the first step toward recovery. At this stage, any sudden event which necessitates or ought to necessitate exertion will frequently suffice to complete a cure ; but if there has been no improvement it may do harm. Perhaps the majority of melancholic patients improve, and a great proportion get quite well ; but sometimes it passes into a kind of melancholy called melancholia altonita, which rarely does well, or the patient may even become frenzied as in mania. Certain of the peculiarities above alluded to as characteristic of melancholy are much better marked in the form of the malady known as acute melancholy. The patient becomes actually frenzied from fright. Such patients have very strong suicidal tendencies, and require the most careful watching. These, too, are the patients who most commonly refuse food, and who require to be fed forcibly. They also refuse to lie in bed at night, and especially to be covered by bedclothes. This, too, must be forcibly combated. Patients, the subjects of this form of disease, generally end badly. They are sure to be badly nourished, and a very little superaddition to their troubles in the way of acute disease finishes them. The lungs are especially liable to be fatally affected by low forms of inflammation.

Mania, accompanied by delirium, is perhaps that form of the malady which is taken as the type of madness by uneducated people. The patient may be suddenly seized with this form of the malady, and may as suddenly become free from it. It is most frequently caused by violent passions, disappointed love, violent grief, and the like, especially if the patient be weak-minded or hysterical. The importance of such an attack must depend very greatly on the soil in which the bud is cast ; if there be hereditary taint, the attack may be a final and complete one, whereas under more favorable circumstances it may speedily pass away. As a rule, too, the more marked the symptoms of onset, and the longer they have shown themselves before the actual malady bursts forth, the more severe is it likely to be. It is not always desirable to hurry these patients to an asylum, for, as said, they may recover perfectly in a day or two ; but frequently it becomes absolutely necessary to do so.

Acutely delirious patients generally behave much in the same way. They sing and shout, and will not rest a minute. Commonly they are utterly incoherent, jumbling their words together, or they repeat one word or phrase like a parrot *ad infinitum*. They show less delusions than do many other insane patients —their condition is indicated more by gesture and behavior than word. Sometimes they are full of glee, laughing and shouting ; at other times they are angry and outrageous, but not nearly so dangerous as some who are quieter in their demeanor. As in most similar conditions, the great object here is to get sleep and rest, for which chloral is the best medicament. Opium generally does harm. Some prefer digitalis to all other remedies, or give its active principle digitalin under the

skin. The wet pack is a means of treatment greatly commended by some. Its mode of application will elsewhere be given. (See PACKING.) As is well known to those who have experienced this mode of treatment, it produces a soothing calm which can hardly be described, and at the same time it implies as absolute rest in being swathed as a mummy. It acts, moreover, as a strait-waistcoat, while acting medicinally ; but it is not to be used overmuch, for too long exposure to its influence greatly weakens the patient— a result we do not desire to obtain. It is chiefly with regard to these cases, or to the occasional outbursts of chronic lunatics, that the question of restraint or non-restraint arises. To some it may seem superfluous to speak of there being nowadays a question between the two. Nevertheless, no asylum can be carried on without some system of restraint. It is itself a system of restraint, and the only question is how best to restrain the patient ; if that can best be done by living force, let it be applied in the form of the male or female attendant's hand. If such is likely to do more harm than good, or even if it cannot be used with such advantage as can some other form of restraint, let the other, even if it be the strait-waistcoat, be applied. During the very acute attack there will be sometimes an entire absence of sleep for days and nights. Women can stand this much better than men ; but both men and women require to be well sustained by food during the sleepless period. Rest, food, and sleep are the great remedies, and the means of procuring the last have once more to be examined. Once more chloral stands at the head of the list ; once more opium has only to be mentioned to be forbidden. Indeed, before chloral came in, drug treatment had been a good deal abandoned, and treatment by baths relied upon, so general was the distrust against opium. No doubt the baths did good by soothing, but they also weakened the patient. The bath, to do any good, must be hot, and a stream of cold water or an ice-bag should at the same time be applied to the head. The best temperature for the bath is about 92° or 93° F., and the patient must be kept in it for a considerable time—half an hour or so. Shower-baths are *not* to be given. If the bowels be confined, a good dose of calomel may be given ; but this had better not be repeated. Frequently, during the acutely delirious period, when many of these purgative drugs have been administered, no action follows until the patient begins to come round, and then the action may be so violent as to occasion danger to life, and under any circumstances will occasion an important loss of strength. For similar and other reasons bleeding and blistering are to be avoided.

There is still another form of *mania*, which may be acute, and yet there is *no delirium*.

This insanity may consist of delusions, but more frequently manifests itself in actions usually of a violent and dirty description. Frequently they have their wits about them in an almost surprising fashion, quite baffling the medical man who endeavors to examine them, so as to sign a certificate for their admission to an asylum. It may be as plain as a pike-staff that they are mad, and yet it may hardly be possible to obtain enough to place on the schedule to justify one in signing such a cer-tificate. Yet, as soon as the practitioner is gone, they are dirty and abusive as ever, shameless in their conduct, tear up clothes and sheets, break windows, chairs, and the like—in short, act like the veriest demons. Their incoherence might sometimes be mis-taken for delirium ; but it is totally different. The health of these patients is fairly good ; they eat well and sleep well apparently when they like. At all events they will have good rest one night, and the next day they will dis-turb the whole ward throughout the whole night. Sometimes they may go on like this for long periods together, and, as they are exceedingly troublesome, care must be taken to get them quieted. This was the class, and they constitute a goodly proportion of our asylum folks, who used to be dealt with by bodily punishments. The plan did not suc-ceed. The plan now adopted, which as a rule but not invariably answers, is to give to those who are quiet and well-behaved, and who do any work, some trifling reward—extra beer for dinner, tobacco or snuff, which are always greatly relished ; the privilege of excursions and the like. Much they do is like the acting of ill-trained children, and they must be taught, if that be possible, the habit of self-restraint. Work in the field or garden is the best means of keeping such maniacs out of mischief. Formerly tartar-emetic was given freely to such patients, and un-doubtedly it kept them quiet for the time being ; but it had no tendency to cure the malady as work in the open air has. Very often these patients are allowed to run on without care or attention until too late, pro-vided they are not especially troublesome, for the malady tends to a chronic course if not speedily cured, and the only hope of cure rests in seclusion.

There is a variety of insanity to which the name *monomania* has been given, and of it kleptomania, entomania, pyromania are com-monly adduced as examples. But it is rare, if indeed such a thing ever happens, to find a man mad on one point and not on others also. At all events, this almost invariably happens—one permanent feature of their madness may for a time be most marked, but by and by, as time passes, the madness is seen in other features of the patient's charac-ter, and he probably ends by becoming a chronic maniac of the class just described, or

a melancholic, but without the characters of either division being very strongly marked—that is to say, the patient being chronically insane, is neither very excited nor greatly depressed, when the health is ordinarily good. Such patients may, in most respects, behave themselves seemly enough, and yet may on some particular point be as mad as possible.

Such are the main divisions of insanity ; but there is one other so peculiar that we are fain to give a brief sketch of its history. The malady is commonly called general paralysis of the insane. We have briefly alluded to insane impulses, to what is called moral in-sanity, where there are no delusions—only acts which may or may not be insane, and these need not further detain us ; but to this a few remarks may be well devoted. This malady is one of the most fatal known, for no case has been known to recover, and though to the individual concerned there is little pain connected with the disease, yet the condition it produces in him is such as to create ex-treme disgust among his friends. The malady is commonly described as constituting three stages, of which the first is such as may give rise to little anxiety on the score of insanity, though the individual is often greatly altered from his former self. The second period is one of acute mania, with exalted delusions, and the third one of complete dementia, with complete prostration of mind and body. In the first stage, or stage of inert, a general paralytic is usually a prey to exalted notions of his own importance and power. If he has money he scatters it broadcast, fancying his supply of it is unlimited. He asserts himself as some great dignitary, not unfrequently God himself ; but if this position is denied, he will not take the trouble to argue the question—he will let the objector go in what seems to him his besotted ignorance. As a rule, too, sexual ideas take hold of him ; he exposes himself in any situation, or assaults women in the most unlikely neighborhoods. He is restless, and above all forgetful. He takes an interest in nothing, or if he do it is laid aside in a moment, all about it being forgot-ten. The man mentally seems falling into the dotage of old age, and yet he is in the prime of life, most likely between thirty and fifty. At this period, too, in some cases, though not in others, there may be observed a tremulousness about the upper lip and a slowness of speech which are very character-istic. Both these signs, are, however, much more marked in the second stage, when the patient becomes fairly the subject of delusions, These delusions, as already pointed out, are all of an exalted character. He can do won-ders in every way. But he has no fixed de-lusion ; that varies in degree, if not in kind, day by day ; but it ever goes on increasing, and that too rapidly. All his surroundings, though of the most trumpery kind, are interpreted as being of the grandest character ; his power is immense, and his bodily strength, though like that of a a child, he thinks incomparable. As to physical signs, as already pointed out, they commonly begin with slowness of speech, or rather a kind of interval between each syllable, with a kind of stutter or drawl something like the utterance of an intoxicated man. There is, too, that tremulous motion of the upper lip which is so peculiar ; but in some there is in its stead a kind of stiffness and swelling instead of the tremulous condition. The tongue, too, trembles when thrust out, and it is thrust out with a jerk, as if the patient had not full command of it.

As the malady advances the delusions of these unfortunately get worse. At the same time they are liable to break out in fits of violence of a most dangerous character. They are altogether unreasoning, and they are generally men in the prime of life, and so they are not easy to manage when they break out in fits of violence. By and by they become subject to fits of a peculiar kind, not the true epileptic convulsion, but most frequently epileptic in form, and not seldom of the kind resembling the slighter attacks of epilepsy called *petit mal*. The walk alters, it becomes vacillating about the hips, and the legs are not moved as usual, but are rather thrown forward with a kind of jerk. The hand-writing, too, becomes imperfect, both as to mechanism and material. Words or letters are omitted or inserted wrongly, the same word is repeated over and over again. The whole is nonsense. The food is eaten voraciously, sometimes bolted, but in other cases, especially as the malady advances, there is difficulty in swallowing from paralysis of the fauces. Such patients are very destructive and very dirty, but they tear up their bedding without knowing what they are doing, and they dirty themselves very frequently for ornament. From such a condition the patient may for a time partly recover, and in all cases there is great variability of the symptoms—well today and worse to-morrow—but there is only one end, that is dementia and death.

By and by the patient gets worse ; he can hardly walk or shuffle round the room ; he loses power over the bladder and rectum if not constantly attended to. His face has lost all expression, and yet it seems fat and puffy. He can hardly hold anything in his hand, and if he is confined to his bed, sores form which are hard to heal. Grinding of the teeth is very often a marked symptom. His appetite is still good, but he has lost the power of swallowing comfortably, so he crams his mouth and throat, so then there is risk of suffocation if this is not seen to. At this time all such patients require to be fed. In point of fact everything must be done for them. But even in this state they may sur-vive a good long time if care be taken of them, and if they are protected from cold, to which they are very sensitive. Much, therefore, depends on the care taken how long the disease may last, but the average duration from the onset to the end does not exceed as a rule two years, while it may be much less.

The causes of the malady are hard to determine. As already pointed out, it generally occurs in the prime of life, and most frequently in males. In a certain number of cases it can be traced to overwork of the brain, but as the malady is more common among the laborers than the rich, this will not account for nearly all. Another cause assigned is sexual excess. This, of course, is not very easy to make out, but irregular lives have been noted in a considerable number. As to treatment, that is useless ; we must just do our best to keep the patient quiet, clean, and orderly. We must try to feed him well, and as soon as any difficulty in swallowing appears no food must be given in the solid state—the pulpy condition is best. If they are confined to bed for a day or two their backs must be carefully watched, and, if necessary, washed with some weak spirit with a little corrosive sublimate in it. Stimulants are usually necessary, especially in the later stages—in the maniacal stage they must be given with caution. In these cases during a maniacal paroxysm digitalis or digitalin often does good, but only for subduing the paroxysm ; nothing does good permanently. Meanwhile, as far as we know, the malady invariably ends in death.

Feigned insanity has already been alluded to, but briefly. Often as insanity is assumed, the fiction rarely succeeds. The would-be lunatic, as a rule, overdoes his part : most likely he has never seen a lunatic, and his only conception of one is a raving maniac. As the penalty of insanity is something considerable, it is rarely feigned save by criminals before or after conviction ; before conviction to escape the consequences of his crime, after it to escape from the heavy labor of penal servitude. The means of detecting feigned insanity are not too numerous : each case must be dealt with on its own merits ; and there are some men known to be sane who have for years succeeded in keeping up an ostensible insanity.

INSECTS (POISONOUS).—It may be useful to point out the different classes of insects which are venomous, the countries where they are found, and the treatment to be adopted in the event of injury.

(a) Of the *Invertebrata*, the scorpion is perhaps the most formidable. Its sting is the claw with which it is armed at the end of its caudiform abdomen ; this claw is perforated and connected at the base with poison-glands. The symptoms produced by its attack very much resemble aggravated forms of wasp-stings. It is found in the hotter regions of the globe, and a small species in southern Europe. The best remedy is the external application of ammonia, as well as its administration internally.

(3) Centipedes (*Scolopendrida*). The poison of these creatures is conveyed by some curved fangs connected with the mandibles, which are perforated and probably communicate with poisoned glands.

(4) Spiders (*Araneida*). Of these there are a few species deserving of special notice. The tarantula (southern Italy) has long enjoyed a reputation for the extraordinary effects said to be produced by its bite. Direct experiment, however, has shown that nothing beyond slight local irritation is produced ; in fact, most of the tales connected with spiders' bites are fabrications. The bites of insects are comparatively innocuous ; but it is otherwise with their stings. Stinging insects belong chiefly, if not exclusively, to the order *Hymenoptera*, in which the sting, in the sterile females, represents the modified ovipositor. The instrument consists essentially of two exceedingly fine sharp darts inclosed in a tubular sheath, at the base of which is placed a special venom gland or sac, whose contents are injected into the wound made by the serrated or barbed darts.

Treatment.—Ammonia in the form of sal volatile is the best application for allaying the smarting and inflammation produced by the stings. If a person has been stung sufficiently to cause faintness, cordials and opiates must be administered without delay. The point of injury should be examined minutely, and the sting, which is frequently left in the wound, removed with a fine forceps. In Sydney, a poultice of ipecacuanha is considered to be a specific for almost any kind of venomous bite. It sometimes happens that a wasp or bee may be swallowed in fruit or drink ; the danger then is very urgent, from the rapidity with which the fauces swell up the moment the sting enters : leeches should be applied externally, and hot salt-and-water gargle used frequently. The operation of laryngotomy, however, is usually the only available remedy.

INSOLATION. (See HEATSTROKE.)

INSOMNIA. (See SLEEPLESSNESS.)

INSPISSATION is the process of thickening any liquid solution by evaporating part of the water over a fire. Its object is to render an infusion or a solution stronger. It is best performed in a water-bath, to prevent burning.

INTERMITTENT FEVER, or AGUE, is a specific fever occurring in paroxysms and characterized by a cold, a hot, and a sweating stage, followed by a period of complete absence from fever. The exciting cause of this malady seems to consist in certain invisible effluvia, or emanations from the soil of

marshy districts, to which the term *malaria* has been applied. The malaria is, therefore, a distinct and specific poison, producing certain specific effects on the human constitution. It is important, therefore, to know in what countries and in what parts of a country this malarious influence is liable to exert its pernicious qualities. Of the nature of the poison we know nothing ; whether gaseous or aeriform, it undoubtedly exists in the atmosphere of particular districts, but nothing as yet has been made out of its physical or chemical qualities. (See MALARIA.) Climate seems to exert a marked effect upon this malady ; in the Arctic circle it is not known to exist, nor is it found in the colder seasons of more temperate climes. Seldom met with above the 56th degree of latitude, it occurs in its most pernicious and injurious forms as the torrid regions are approached. Observers are not yet agreed as to the exact nature of the soil which breeds the mischief ; some have thought that decaying vegetable matter was a cause, but in vegetable markets, where rotting leaves are abundant, ague is not found ; and, again, others have shown that this fever is produced where there is no vegetable matter to decay. Heat and moisture together are not sufficient, for the disorder does not prevail among sailors at sea in whatever climate they may be ; the air and water of affected districts have been examined, but at present with no result. It would seem that those places are the most dangerous which have been flooded and then become dry ; so the edges of swamps or the banks of drying or half-dried rivers become dangerous, according to the season and the amount of water in them. It has often happened that those living in the lower parts of a district are very liable to the fever, while those in the same locality, but perched upon a hill, escape altogether. Ague and aguish fevers are, therefore, in temperate climates, more common in the autumn than at any other period of the year, and follow after the heat of summer. Persons who live in marshy districts become acclimatized to it, while strangers are readily affected. Although the natives in such places are not so liable as new-comers to catch ague, yet they seem to be chronically affected by the unhealthy atmosphere ; they are not a strong race, being short in stature, of a sallow complexion, a melancholy and short-lived people. It is a very remarkable fact that the negro seems to enjoy a marshy swamp : places which to the white man are most pernicious seem to have no effect on his darker brother, while conversely the black can find no enjoyment on the hilly lands where the white man flees for health and safety. Persons are much more liable to catch the disease at night than during the day, so much so that many places which are safe during the day are pernicious at night-time. Even now travellers at Rome are warned against crossing the Pontine marshes after sunset. Again, the malarious poison, whatever it may be, likes to keep near the ground, so that those who live in the upper rooms of a house are safer than those who sleep on the ground-floor, and this may be the reason why lying down in the open air in such places is so dangerous. The malaria movable by the wind ; sometimes it hangs over a district like a thick fog or milky vapor near the surface of the ground. The malaria seems to lose its effect by passing over a sheet of water, so that while troops on shore are having ague, those in the ships a few hundred yards off may be perfectly free. The marsh poison seems to cling to the foliage of thick and lofty trees in those districts, so that it is very dangerous to go under them, and still more to sleep there ; and this fact has been made use of, for, in Guiana, where large trees abound, the settlers live fearlessly and unhurt close to the most pestiferous marshes, and to leeward of them, provided that a screen or belt of trees be interposed. Some curious notions were entertained in earlier times about the salutary effect of an ague. Dr. James Sims, a London physician, when suffering from an illness, which afterward proved fatal, felt convinced that he should recover if he could catch an ague ; for this purpose he went down into a marshy district, but came back, complaining that there was no ague to catch, and that the country had been spoiled by draining. "The superstitious Louis XI. prayed to the Lady of Selles that she would confer upon him a quartan ague"— a request with which it does not appear she complied. There is an old English proverb, "an ague in the spring, is physic for a king ;" in spite of this, James I. died of it.

Symptoms.—The person who has been exposed to malaria will generally suffer for a few days from premonitory symptoms. There is nausea and loss of appetite, with a slight feeling of chilliness ; often, too, there are muscular pains in the back and lower extremities ; these symptoms may last only for a few hours or may be prolonged for several days. Then comes on the regular attack, beginning with a cold stage, in which the patient lies shivering in the bed with chattering teeth, gathering the clothes round about him to keep himself warm ; yet there is only a *feeling* of cold, for in reality the patient is hotter than usual, and the temperature in this stage will rise from 98.5° Fahr. to 101° or 102° Fahr. This is followed by a hot stage, in which all shivering has ceased, while the temperature goes on rising, so that in three or four hours from the commencement of the fit it may have risen to 105° Fahr. or 106° Fahr. ; thus in that short time the whole mass of the human body has been heated up seven or eight degrees in many cases. When this hot stage is on there are flushes of heat

about the face and neck ; the coldness ceases ; the skin, which before was shrivelled and pale or even livid, now returns to its natural color, and the face assumes an ordinary appearance ; then a reaction takes place and the face becomes red, the skin hot and dry ; there is a violent headache and throbbing of the temples ; the pulse is full and quick and strong, while the breathing is oppressed ; the patient feels very miserable and restless. But presently the skin becomes softer and breaks out into a gentle perspiration on the head and face ; this quickly increases, so that soon the whole body is bathed in sweat, and great relief is experienced ; the thirst ceases, the tongue is clean again, the pains go away, and the heat and discomfort pass away, so that in a few hours he will feel as well again as ever until the next recurring fit. During the sweating stage the temperature is falling, and at last reaches the normal line of 98.4° Fahr. If any one draw on a chart the state of the temperature he will find an ascending and descending curve, rising at its highest point seven or even eight degrees above the ordinary line of health (98.4° Fahr.). During the first part of the ascent the patient shivers and has a rough skin—goose-skin as it is called—during the second part of the ascent he is burning hot, but directly sweating begins the thermometer shows a fall in the mercury, and this continues until the third stage is over, when the patient generally falls asleep and wakes feeling much better. But the cold stage may occur and never be followed by the hot, and again there may be a hot stage not preceded by a cold one ; those who suffer in this way often speak of it as *dumb* ague, and they generally have suffered from the ordinary ague previously. This set of symptoms is exceedingly well marked, and no one can mistake the disease ; yet there is great misery endured by the patient during an attack, although it is only temporary, and in this country very curable. A most characteristic feature of ague is its *recurrence*, and hence it is called an *intermittent* fever ; the intervals between the attacks are marked by a total absence of fever ; when these intervals are imperfect and the patient remains ill between the paroxysms, the disease is called *remittent ;* this form is met with, as has been said, in tropical climes, while the former is most common here. There are three principal types of ague. When the paroxysm comes on every day, it is called *quotidian ;* when it comes on at the same time every other day, appearing and remaining absent day by day alternately, it is called *tertian ;* and when two whole days intervene between the paroxysms it is called *quartan ;* the two latter terms ought to be "secundan" and "tertian" respectively, but old observers gave the names and they have been kept in the language. As a rule, the quotidian ague comes on in the morning, the tertian at noon, and the quartan in the afternoon, although there are many exceptions, and it often happens that instead of coming on at the same hour the fits may be postponed or accelerated ; that is, they may begin a little later or a little earlier. These types differ from each other somewhat, not only in their respective intervals, but in the duration of the paroxysm and in the period of the day at which the paroxysms commence. In the quotidian form the fit lasts ten or twelve hours, while in the tertian its duration is from six to eight hours usually, and in the quartan it may be only four or six hours. Of these various forms the tertian is the most common in this country, although the others are far from infrequent. Besides these forms there seem to be others in which two types are mingled ; thus a double tertian or a double quartan may be met with, and a patient may have two attacks in the same day.

Treatment.—Ague is a very curable disease, and in this country at least very simple measures may be taken. During the cold stage the patient will naturally prefer to go to bed, and wraps himself up as warm as possible, and he may be allowed to drink any simple fluid that he likes, such as tea, barley-water, weak wine and water, etc., and no more need be done than that. There is no occasion to be over-fussy during a paroxysm, as the patient will come out of it all right. Of all the remedies in the Pharmacopœia none are more valuable than quinine and arsenic in the treatment of ague. Expensive as quinine is, it is yet preferable to arsenic, because not only is it more efficacious, but it is not poisonous, and can be left about with safety. Given in large doses it is apt to produce giddiness, singing in the ears, and deafness, but these effects will pass away of themselves. Four or five grains of the sulphate of quinine taken every four or six hours during the interval of an ague fit will generally cure the patient. Very often the cure is immediate and the patient has no more attacks ; more often he has one or two very slight ones and then becomes convalescent. In hot countries larger doses may be required, and for travellers in aguish districts nothing is more invaluable than a plentiful stock of quinine. Bark used to be given in such cases, but when the good effects of the bark were found to be due to the small quantities of the alkaloid in it, then physicians preferred giving the latter, as it was better borne on the stomach and is much pleasanter to take, besides occupying less compass. If a relapse take place, quinine must again be taken. Sometimes quinine is disliked, and then arsenic may be tried, but it is very poisonous, and must only be given under medical advice ; the injurious effects that should be looked for are a soreness of the throat, vomiting, diarrhœa, pain in the abdomen, smarting and redness of the eyes and

nose. Numbers of other remedies have been tried, but no remedy is so powerful, effectual, and simple as quinine. The diet should be liberal and nourishing ; the patient, being much weakened by the attacks, will not bear any depressing influences ; strong beef-tea, milk, and some wine may be given daily, and meat or fish, etc., if the patient care for it. After an attack, a nutritious diet, abundance of exercise, pure air, and pure water are very important ; the individual, if possible, should leave the neighborhood or live at a higher elevation. After repeated attacks, the liver and spleen are liable to become enlarged, and the latter condition is known as ague-cake. An ointment of the red iodide of mercury has been strongly recommended for this affection : a portion the size of a hazel-nut, may be rubbed into the left side and exposed to the heat of the fire ; its application should not be too often repeated, as it is apt to make the skin sore.

Brow-ague is another disease allied to an intermittent fever, and produced by malarious influences also : it consists chiefly in an intense pain confined to one side of the head, lasting for several hours and gradually passing off ; it comes on in paroxysms with varying intervals, and is chiefly met with in sensitive and nervous people, and more especially if they have had any mental trouble or wrong previously. From affecting half the head only, brow-ague has been technically called *Hemicrania ; Migraine* is also another term for the same affection. It generally occurs in the adult, and is more common in women than in men. This form of headache is very different to that which comes on after a debauch or from eating too good a dinner, and which is generally accompanied by disturbance of the functions of the stomach or liver. The true sick headache is a purely nervous affection and quite independent of excessive eating or drinking ; it occurs in very temperate people and is often hereditary in families. Any cause which produces a strong impression on the nervous system of those who are predisposed to it will bring on an attack. Exposure to heat and fatigue, breathing the hot and impure air of a theatre or concert-room, working late at night by gas-light, working with a microscope, and any special mental worry or excitement will cause this painful affection. Those who suffer in this way wake in the morning feeling more dead than alive, unable to swallow any food, and perhaps actually sick ; the head throbs, and any movement or conversation is avoided as the pain is increased by doing so ; the patient begs to be left alone and to be kept quite quiet, as the only means of obtaining sleep. The sufferer looks ill and pale, has contracted pupils, and a dark line under the eyes. The head feels hot, and the application of cold is most refreshing. The appetite is gone, the mouth feels clammy, and there is a feeling of nausea. Hot tea and coffee seem to allay the nervous system and give relief, and a little wine or ammonia may be given. The only relief during an attack is to be found in a wet bandage round the head, profound quiet, and a darkened room. Medicines have been over and over again tried and found useless, nor is it to be cured by remedies which act as purgatives or on the liver and stomach. Aconite and belladonna have been employed locally but with no good result : sometimes relief has been obtained by giving bromide of potassium. During the intervals the general health must be kept up by quinine and tonics, but most good will be done by the sufferer carefully avoiding any of those causes which are found by his or her own experience to bring on an attack. Very recently guarana has been recommended for the cure of th's affection. This drug has long been known, but it has not yet come into general use. It consists of the seeds of *Paullinia sorbilis*, a tree growing in Brazil, and belonging to the natural order *Sapindaceæ*. The seeds are ground into powder, and contain an alkaloid which is said to be identical with that found in tea and coffee. The seeds, roasted, bruised, and pressed into cylindrical masses, form the guarana paste, which, when finely pounded, is then known by the name of Paullinia powder. It is light brown in color, has an odor faintly resembling roasted coffee, and a bitter astringent taste. It contains tannic acid, and a principle called guaranin, which has much the same effect on the nervous system as tea or coffee. This drug has been much used in France for sick headache and various forms of neuralgia, and is given in doses of ten to fifteen grains once or oftener in the day.

INTERNAL BLEEDING. (See CANCER, GASTRIC ULCER, and HÆMORRHAGE.)

INTERTRIGO, or ERYTHEMA INTERTRIGO, is the name given to the local condition known as a chafe or fret, and which consists in redness and excoriation of a part of the skin. This condition is caused by friction and prolonged contact of two adjacent surfaces of skin, by the friction of portions of dress, or by the contact of irritating discharges. It is met with generally in corpulent persons and infants, and in those who perspire freely. Its development is favored by the accumulation of sweat, and occurs in persons who pay very little attention to cleanliness. The most frequent seats of intertrigo are the inner surfaces of the thighs, the inner portions of the buttocks, the navel, the armpits, the back of the neck, and about the genitals. It may be often met with in the folds of the skin of fat infants, and especially in the flexures of joints. It occurs especially in warm weather. It is often produced in persons who take active exercise, as in making walking tours during the sum-

mer or autumn, and thus causes much annoyance and distress. The friction of the skin induced by any of the above-mentioned causes produces a raw and inflamed surface, from which there is a thin discharge. The perspiration also from the adjoining portions of skin is increased and becomes very fœtid. There is much itching of the affected parts, which is increased by movement and friction.

Treatment.—Intertrigo may in most persons who are liable to be troubled by the affection be prevented by frequent ablution, followed by careful drying, and by occasionally bathing the most likely situations with diluted spirits of wine or weak lead lotion. When a raw surface has been formed the patient should keep it at rest as far as possible, and prevent further friction of applied surfaces of skin. The excoriated portion of skin should be kept very clean and be frequently dusted with absorbent powder, such as starch, lycopodium, or oxide of zinc. Fuller's earth and a weak solution of lunar caustic or of alum are useful applications. In some cases, where the patient is out of health and has taken much exercise in spite of the development of very painful intertrigo, the raw surface becomes much inflamed and is covered by a thick ill-smelling discharge of pus mixed with blood. In some cases of this kind perfect rest of the patient becomes necessary, and the affected parts should be brushed over with a strong solution of lunar caustic, and then be treated by zinc or chalk ointment, attention at the same time being paid to the general health.

INTESTINAL WORMS are parasites which infest the intestines. (See ENTOZOA and PARASITES.)

INTESTINES form a long, hollow channel from the stomach to the anus, and allow the passage of food along them. Different names have been given to different portions ; the first part nearest to the stomach is called the *duodenum*, and is about twelve inches long ; in this portion the liver and the pancreas empty their secretions and mix with the food which has just left the stomach. The next portion is called the *jejunum*, and is about two feet long, and it is abundantly supplied with vessels ; this leads on to the *ileum*, which is several feet in length, and forms the greater portion of what are called the bowels ; these three parts make up what is known as the small intestines. The next portion is of larger calibre, but made of similar materials, and is called the large bowel or the large intestine ; it commences in the right iliac region of the abdomen as the *cæcum*, and then, making a large curve round the abdomen, it descends by the left side of the abdomen into the *pelvis* and ends at the *anus*. In its course it is called, respectively, the ascending, transverse, and descending colon, while the lower twelve inches are called the rectum. Although having different names, the intestines form one continuous channel of great length, but they are kept in position by a membrane called the mesentery, so that the different portions all lie coiled together and can move freely upon one another. The coats of the intestines are three in number : 1. A serous smooth external coat, called the peritoneum, which allows the bowels to glide over each other smoothly and without pain. 2. A muscular coat, by which contractile movements go on in waves, so that the food is propelled gently from one end to the other. 3. Most internal is the mucous coat, a membrane lined with epithelium and richly supplied with vessels, so that all the soluble parts of the food can be absorbed as it passes along ; this coat is really much longer than the intestines, as it is arranged in folds, like the tucks of a dress, so that the surface available for absorption of the food is thus vastly increased. In this coat are numerous glands, which empty their contents into the canal. The movements of the bowels caused by the contraction of the muscular coat are called *peristaltic*, and when abnormally increased in force and frequency they cause griping pains and diarrhœa. At the lower end of the cæcum is a curious little tail or appendage, known as the *appendix vermiformis ;* its function is not known, but serious effects have ensued from foreign bodies, as pins, cherry-stones, etc., becoming lodged there on their way down the bowels. The intestines are very liable to disease, and these affections are treated of in the articles on CHOLERA, CONSTIPATION, DIARRHŒA, ENTERITIS, FLATULENCE, HÆMORRHOIDS, HERNIA, INTUSSUSCEPTION, OBSTRUCTIONS, TYPHRITIS, TYPHOID FEVER, and VOMITING.

INTOXICATION. (See DRUNKENNESS.)

INTUSSUSCEPTION is the name given to a condition in which a portion of the bowel is protruded, something like the finger of a glove half turned inside out. It is very frequently met with in children, and it seems wholly independent of any symptoms of disorder of the bowels during the patient's lifetime ; sometimes only a single intussusception is found, but oftener there are several ; and as many as ten or twelve have been found in the same subject. They are generally confined to the small intestine, and are most common in the ileum. The great frequency of this occurrence, the absence of any symptoms during life, or of any indication of disease about the intestines after death, lead to the opinion that this invagination may often occur in the process of dying. Although this form of intussusception is so common, yet there are some rare cases in which a portion of the large intestine may become invaginated, and cause a serious, if not fatal, obstruction. This accident takes place generally in infants under one year of age, and often less than six months old.

Sudden and violent vomiting, followed by loud cries and signs of general uneasiness and pain, recurring at uncertain intervals, accompanied by violent straining and effort to empty the bowels, are the earliest *symptoms* of the mischief. At first some fæces may be voided by these forced efforts, then mucus is discharged tinged with blood, or even pure blood in considerable quantities. If an injection be given the fluid generally returns at once, as its passage upward is obstructed, and the obstacle may be felt sometimes when the finger is passed up the rectum. Vomiting commonly comes on, and is renewed whenever any food or medicine is administered. The pain comes on in paroxysms, alternating with intervals of quiet ; the child is often thirsty, and will take the breast or bottle readily, although the sickness is persistent. It is seldom that anything can be made out from the external examination of the abdomen. As the obstruction continues there is exhaustion of the infant's strength ; its pulse grows more and more feeble, the face becomes anxious and sunken, and it falls in the intervals of pain into a quiet, sleepy condition. In many cases convulsions come on a few hours before death, which takes place within a week, and often in two or three days. In some rare cases the symptoms abate, the pain and the sickness may cease, the bowels act of themselves, and a speedy recovery of the little patient occurs ; but this is a state of things which can be rarely hoped for.

Treatment.—The treatment must consist in stopping the use of any purgative or active remedies to remove the constipation, and in adopting soothing measures ; as soon as the symptoms of intussusception occur, no aperient medicines must be given by the mouth, as they will increase the action of the bowels and make the obstruction worse than ever. Hot fomentations should be applied to the abdomen, and a little opium may be given to allay the pain ; but this drug must be given with the greatest caution, as even the administration of two or three drops may prove dangerous to an infant under twelve months of age, and then the remedy will be as fatal as the disease. If the symptoms do not disappear within twelve or eighteen hours, injections must be given by the rectum so as to unfold the invaginated intestine, just as the introduction of a finger will push back the above-mentioned glove. Injections of warm water may be used, and even inflation with air has in some cases been effectual ; this plan is useful in cases where the large intestine is involved in the mischief, but not where the small intestine is implicated ; however, this last occurrence hardly ever occurs so as to cause any symptoms, and nearly always it is the large bowel which is the seat of the obstruction. Surgical interference has in some cases been successfully resorted to, and the operation of gastrotomy has been performed, where the abdomen is opened by an incision in front, and then search being made for the obstruction, the invaginated bowel is restored to its proper condition ; this is, of course, a very dangerous operation, but it may be required when there is no other chance to save life. For an account of intestinal obstructions in the adult, see OBSTRUCTIONS.

INUNCTION.—The remote and constitutional effects of certain remedial agents may be produced by rubbing into some parts of the surface of the body ointments containing these agents. Thus belladonna ointment rubbed into the skin of the forehead or temple will produce enlargement of the pupil and relieve the severe pain attending certain forms of inflammation of the eye, and friction with sulphur ointment will in some cases relieve the articular pains of rheumatism and rheumatic gout. The agent which has been most frequently administered in this way is mercury, and mercurial inunction is still with many surgeons a favorite way of treating syphilitic affections. The patient is ordered to rub in half a drachm or a drachm of strong mercurial ointment every night, or on alternate nights, until it is thought necessary to discontinue the use of this powerful agent. The most favorable parts to which the ointment can be applied are the armpits, the inner surfaces of the thighs and arms, the hams and calves, and the front of the abdomen. The ointment ought not to be rubbed into the same part on two nights in succession. The patient should keep up friction for twenty minutes, and then put on a thick flannel shirt, or flannel drawers and retire at once to bed. No attempt should be made to wipe away any of the ointment which remains on the surface of the skin after the friction. By this proceeding all the general effects of mercury may be produced, and after a period varying in different individuals, the stomach becomes irritated and the gums sore. The presence of the latter result of the inunction treatment is an indication that the mercury has commenced to act upon the system and that it is necessary to discontinue the frictions or to perform them less frequently. This mode of administering mercury is as effectual as any other, and for many patients more convenient, but it has the disadvantage of being dirty and causing much trouble.

INVERSION OF THE WOMB is said to have taken place when that organ is either completely or partially turned inside out ; it is of rare occurrence, but may come on after a labor, if there is an adherent after-birth and too forcible attempts are made to remove it. The only remedy is to push it back again into its proper position.

INVOLUTION is a term, opposed to evolution, implying the return of an organ or tissue to its earlier state ; thus there is said to be involution of the womb after a confinement, when that organ becomes small again, after having expelled the child. The womb, as soon as the fœtus begins to grow, develops an immense quantity of muscular fibres, and the various vessels and nerves also increase in size ; but after pregnancy is over, these muscular fibres become fatty, waste away, and are absorbed ; the vessels also become much smaller, and in the course of a month or six weeks the womb returns to its natural size. In some cases, however, this process does not take place properly, and although the womb contracts a great deal, it may still remain much larger than it ought to be, so that the patient has a feeling of weight in the lower part of the abdomen, bearing down pains in the back, and inability for any exertion. Menorrhagia or hæmorrhage from the womb is also a very frequent symptom, and thus there is great debility. When this takes place, *sub-involution* is said to have occurred. It is liable to be met with in women of delicate health, in those who get up too soon after a confinement, in those who do not suckle their children and in those who are not properly attended to at the time of the birth of the child. Rest on a couch in a horizontal position should be enjoined, and especially when there is any bleeding ; no exertion should be taken, and even a carriage drive, unless upon a very even road, will cause distress. Tonics must be given to support the general strength, and the diet must be light and nourishing. An abdominal belt will often give great comfort, and enable the patient to take a short daily walk. Such cases are very troublesome to cure, and weeks may elapse before any marked improvement may take place ; but, although for a long time an invalid, recovery generally occurs sooner or later.

INWARD FITS are often caused in an infant by indigestion or by flatulence, and are relieved when the intestinal disorder is relieved, and a great quantity of wind has been passed. The child thus affected lies as though asleep, winks its imperfectly closed eyes, and gently moves or twitches the muscles of its face ; this convulsive twitching, due to irritation in the course of the alimentary canal, is spoken of by poets as the "angel's whisper," which makes the babe to smile ; it is more correctly ascribed by nurses to the "wind."

IODINE is prepared from the ashes of seaweed as collected on the west of Ireland and Scotland. It occurs in the form of scales which have a metallic lustre, but have not the weight of a metal. They readily pass, when heated, into a beautiful violet vapor. The solution of iodine strikes a blue color with starch, which is very characteristic. It may be dissolved up either by alcohol or in a solution of iodide of potassium. The most important preparations are the tincture, ointment, and liquor, or strong tincture. When applied externally iodine acts as an irritant, or if its vapor be kept in, even as a vesicant ; at the same time, if well rubbed in, it is absorbed and affects the system. It is applied to the skin for a variety of purposes ; but its application is painful and sometimes can hardly be borne. The liniment or liquor is applied to the chest as a counter-irritant in chronic pleurisy, and it is painted on the collar-bones for the chronic pleurisy of consumption, giving rise to pain and cough. But as a rule other counter-irritants are preferable. It very often does good applied in the same way to the joints when enlarged and tender from chronic inflammation. As an ointment iodine is a capital application to chilblains, if applied before the skin is broken. So too in certain neuralgic or rheumatic pains of the chest the ointment often does great good, especially if the part be tender on pressure, and yet due to the kind of malady above alluded to. If the skin be sore or tender on pressure, belladonna suits better. Iodine, in some form or other, is the great application for swollen and indurated glands, and no remedy applied locally certainly does so much good. In the accumulation of fluid in the scrotum, commonly called hydrocele, it is usual to inject tincture of iodine after the fluid has been withdrawn ; some say plain spirit suits quite as well. On similar principles, iodine has been injected into the cavities of joints, the subjects of chronic inflammation. So, too, it has been injected into the cavity of the chest after fluid has been withdrawn, and even into cysts of the ovaries. In all of these cases the solution must be rather weak or mischief may follow. In chronic pulmonary consumption an inhalation of iodine of the strength given in the pharmacopœia night and morning will give relief. Similar inhalations have been commended in diphtheria and chronic bronchitis. Chronic inflammation of the nostrils and upper part of the air-passages often yields readily to a little iodine inhalation. Iodine being readily volatile, easily passes into the state of vapor with the heat of moderately hot water, and the steam which escapes at the same time with the iodine seems to mitigate the violence of its action. Iodine is not nowadays given internally. As a rule, one of its compounds, most likely the *iodide of potassium*, is prescribed. This salt gives rise to all the constitutional effects of iodine, but is not nearly so irritating. Nevertheless it does irritate, and cannot in many cases be given very long without indicating some irritability of stomach.

Both iodine and iodide of potassium taken in excess give rise to certain peculiar phenomena called *iodism*. These are the pain in the forehead, smarting of the eyes, etc., already referred to. The disease in which iodide of potassium is most largely and with most benefit given is syphilis ; but it is useless until near the third or tertiary stage ; in it the drug works marvels, especially when the bones are diseased and nodes have formed on them. Here the local application of iodine to the swelling and iodide in full doses internally will often give sleep and rest when they have long been absent. Other thickenings of the covering of the bone, not syphilitic, also yield to the same remedy. Some forms of rheumatism do so rapidly, others not at all. Those do best which are worst at night, especially if there be any syphilitic taint in the system. Iodide of potassium is the best remedy for chronic lead poisoning after the bowels have been freely opened, and provided they are kept open. Certain forms of gout, too, yield readily to its influence, though others are not affected by it. Iodide of potassium given internally and iodide externally are the great remedies for goitre of the endemic kind.

There are many other remedies for which iodide of potassium has been given, but these are the chief for which it is nowadays prescribed. It requires to be given in very different doses. In tertiary syphilis there is no use playing with the malady, and so iodide of potassium must be given in doses of 15 or 20 grains, or even more, three or four times a day. In other kinds of disease this would do great harm. In rheumatic gout, as it is called, similar large doses require to be given. The ordinary dose for other maladies is 5 grains three times a day. It is best given with a vegetable bitter; some like giving it in milk.

IPECACUANHA, or Ipecac, is the dried root of a plant, *Cephælis Ipecacuanha*, growing chiefly in the Brazils. It belongs to the same valuable group of plants as yields us Cinchona bark. The root itself looks as if marked with rings, with a hard woody axis surrounded by a dark brown woody substance. Its powder is pale brown, and is the preparation chiefly used. Its compound powder, also called Dover's powder, contains opium, ipecacuanha, and sulphate of potash. This is a powerful sudorific. Ipecacuan wine, which consists of wine in which a quantity of the root has been soaked, is also a good deal used, especially for children. Given in large doses, 20 or 30 grains, ipecacuan acts as an emetic, causing some sickness, but not so much as tartar-emetic. In smaller doses it promotes the secretions of the alimentary canal and respiratory organs, and also upon the skin. It is therefore laxative—though slightly—expectorant, and diaphoretic. Dover's powder has the last quality in the highest degree. The drug has a peculiar odor, which to some is unpleasant, and causes sneezing, cough, and watering from the nose and eyes. Ipecacuan is often of great value in allaying irritability of the stomach, as in the vomiting of pregnancy and imperfect menstruation. It also seems to do good in many forms of constipation. It is, however, mainly used, so far as the alimentary canal is concerned, in dysentery, especially that of tropical climates. In some cases it acts like a charm, in others it does no good. Large doses of 40 or 60 grains require to be given in this form of malady, and they may be given and repeated without any nausea being produced. In some cases of dysenteric diarrhœa in children the wine given in small doses frequently repeated often cures the malady very speedily, especially if the stools be slimy. In asthma—hay or common asthma—ipecauan wine often gives great relief. Here it must be given in good large doses, not sufficient to give rise to vomiting, but enough to give rise to a certain degree of depression. It must be used to cut short the paroxysm, and must therefore be given at the onset of the attack. In whooping-cough ipecacuan is a great remedy. It may be given to lessen the severity of the paroxysms as well as their frequency. Ipecacuan acts best as an emetic given in small doses frequently repeated, and with plenty of water.

IRRIGATION is a term applied by surgeons to a mode of local treatment in which the temperature of an injured or inflamed part is kept reduced by the continual dropping on its surface of cold water or some cooling lotion. It is practised for the purposes of preventing or relieving inflammation, and of cleansing wounds and ulcers. It is especially useful in cases of severe sprains and injuries to joints, and for suppurating wounds in connection with fracture. In the latter class of cases the discharge is continually being washed away, and the wound and its dressings are kept clean and free from smell. The most ready way of applying irrigation is the following : A wide-mouthed bottle filled with iced water is suspended over the injured part, a long skein of cotton or an ordinary lamp-wick, having been dipped in water, is so placed that one end rests in the wide-mouthed bottle while the other hangs over and almost in contact with the surface which is to be irrigated. A kind of siphon is thus formed, from the outer end of which there is a continual dropping of fluid. In cases where the injured surface is extensive, two or more skeins may be used.

IRIS is that portion of the eye which, to outside view, gives to the eye its peculiar color ; anatomically, it is a structure, partly vascular, partly muscular, and loaded with pigment, which separates the anterior from the posterior chamber ; it may be considered as a prolongation of the choroid coat ; the pigmentary element is wanting in Albinos. The black portion in the centre of the iris is an opening, in man, circular, called the pupil, which is capable of contraction or dilatation, according as light is required for the

illumination of the retina, acting under the stimulus of reflex nervous action, and bearing pretty much the same relation to the optical structure of the eye that the diaphragm in the stage of a microscope does to that instrument. The muscular fibres are of the involuntary or unstriped variety, and are arranged in directions both circular and radiating.

IRITIS, or inflammation of the iris, may be divided into acute and chronic, and its constitutional modifications into syphilitic, gouty, rheumatic, and scrofulous. The general *symptoms* of iritis are that, in the first place, the fibrous texture of the iris loses its color, becomes confused, the pupil loses its movements and becomes contracted and irregular ; next, lymph is effused, in some forms in small nodules, in others as a film over the pupil. The sclerotic redness is marked in the form shown by the vessels passing in straight lines, running from the circumference toward the centre of the eye, and possessing the characteristic pink vascularity immediately round it. There is intolerance of light, and frequently deep-seated pain or aching about the brow and orbit, with great dimness of vision. In a typical syphilitic case we notice that it is distinguished by effusion of lymph on both surfaces of the iris, in reddish or brown *nodules*, causing the pupil to become irregular ; pain most severe at night ; generally associated with other secondary syphilitic affections. In the rheumatic variety, there is less tendency to the deposit of lymph on the iris, and what deposit there is is not nodulated, and there is a haziness of the cornea absent in the syphilitic variety. The pupil is contracted, and more or less irregular, in consequence of the effusion of the lymph taking place between the edge of the pupil and the capsule of the lens. The surface of the eyeball is often very much inflamed and injected, so that the well-defined vascular ring, seen in syphilitic iritis, is not so evident ; another great characteristic is its tendency to return. Scrofulous iritis signifies either idiopathic iritis occurring in a scrofulous habit, generally combined with corneitis, or else a deposit of cachectic lymph in the iris, which leads to scrofulous suppuration of the eyeball or atrophy. Traumatic iritis signifies an inflammation of the iris set up by a penetrating wound of the eye.

Treatment.—With regard to the treatment of iritis, we must endeavor to subdue inflammation, and thus prevent the effusion of lymph ; endeavor to promote the absorption of what is effused, to preserve the form of the pupil and allay pain. The bowels should be well cleared out, the diet unstimulating, and blisters, which are frequently of great use when the most acute stage is over. To arrest the effusion of lymph, mercury seems of the greatest value, given in the form of calomel, 1 to 2 grains with $\frac{1}{4}$ or $\frac{1}{2}$ grain of opium every

6 or 8 hours. Better still is the administration of 2 or 3 grains of the hydrarg. c. creta with henbane four times in the twenty-four hours. If the patient be very debilitated or scrofulous, iron (the potassio-tartrate), quinine, or cod-liver oil is preferable. The pupil may be kept dilated by dropping a solution of sulphate of atropine upon the eye in the strength of 1 grain to 1 oz. of distilled water. Extract of belladonna smeared round the orbit is of great use also in relieving pain. In gouty iritis colchicum should be administered in small doses. Pure air or sea air, good animal food, and warm clothing, with general attention to diet and the general mode of life, are most important adjuncts to treatment.

IRON is one of the most valuable remedies we possess. It is used in a great variety of forms, commencing with the metal itself. Of metallic iron two forms are used in medicine —namely, iron wire, soft, easily flexible, and non-resilient, and reduced iron. Iron wire is used for the preparation of the aromatic iron mixture, a very valuable preparation, and the wine of iron not the less so. The vinum ferri, or iron wine, is made by macerating iron wire in sherry. Reduced iron is the metal prepared by passing hydrogen gas over the peroxide of iron in a red-hot state. The hydrogen abstracts the oxygen, and leaves metallic iron in the form of a powder. This powder is steel gray, inclining to black in color, and is strongly attracted by a magnet; by pressure it may be made to show metallic streaks in a mortar. Iron enters into the composition of the living body in considerable quantity. Especially is it present in the red blood corpuscles, into whose coloring matter it enters. When, from whatever cause, this proportion of iron is deficient, or the coloring matter itself is not present in sufficient quantity, ill-health, accompanied by pallor, weakness, shortness of breath, and various other signs of imperfect nutrition, is seen to follow. And though it is by no means clear that the giving of iron internally alters this directly, it certainly does indirectly, favoring nutrition and the formation of more healthy blood, bringing back strength, color, and mental and bodily vigor. This iron acts as a tonic apparently directly on the blood, but indirectly on other tissues, especially the nervous system, when deranged. Most preparations of iron are astringent, some more so than others, hence they do good by restraining discharges when these have grown chronic, and the parts whence they are derived flabby. When, on the other hand, certain normal discharges, as the menstrual, are in abeyance, no remedy is so useful, provided the stoppage arises from weakness, for bringing them back as iron. This, however, they do not directly, but indirectly, by improving the quality and character of the blood. Almost all preparations of iron in passing

through the bowels, become blackened, so that the motions of one taking iron may seem unnatural, where they are perfectly natural. For the same reason the tongue and teeth are colored black by them.

Metallic Iron being tasteless, is probably one of the pleasantest forms in which to take iron. It has no astringent effect in the mouth, but being dissolved in the stomach it is rapidly absorbed, and acts splendidly as a blood restorer. It sometimes, however, gives rise to unpleasant eructations. It is best taken in doses of from 2 to 5 grains during a meal, or in the intervals it may be given between a sandwich. It is tasteless, but unfortunately rather expensive. It is a very valuable preparation when the stomach is irritable and there is anæmia. The *wine of iron* is a favorite prescription for children. It contains a combination of tartar and tartrate of iron.

Carbonate of Iron is another very valuable compound; unfortunately, though easily prepared, it will not keep, but passes rapidly into the condition of rust. The most important preparation of this salt is the compound iron mixture, also known as Griffith's Mixture. This, though it contains certain ingredients which might be dispensed with, is perhaps the most available and cheapest form in which we can prescribe iron to an irritable stomach. (See GRIFFITH'S MIXTURE.)

Iodide of Iron, which is made by combining iodine and iron directly, is a greenish preparation. It, too, keeps badly. Its properties partake partly of those of iodine, partly of those of iron; but those of the former are predominant, so that it is commonly given rather when iodine is to be given without the irritant effects of the simple substance or the weakening influence of the potass. Hence it is commonly prescribed generally as syrup of the iodide of iron, in scrofulous diseases, in some forms of rheumatism, in syphilis happening in broken-down subjects, and perhaps best of all in children of a scrofulous tendency, who are threatened with brain disease. The dose of the syrup is for adults 20 drops to a drachm; for children from 2 to 15 drops.

Sulphate of Iron is made use of in two forms as ordinarily sold—dried and granulated; the first of these being most commonly exhibited. It is green in tinge, and crystalline, but the crystals are imperfect. These also tend to break down and assume another hue from the formation of a persulphate. This salt is a powerful astringent, besides having the ordinary action of iron salts. It is, therefore, useful in chronic discharges and relaxed habits of body. Three or four grains may be given for a dose. It is usually given as a pill.

Arseniate of Iron is a combination of arsenic acid and iron; it is not often given, but when so it is with a view to combine the effects of arsenic and iron as a tonic in skin diseases.

Phosphate of Iron is of greater importance ;

it is a slate-blue preparation, and is chiefly used as syrup. This is a really valuable preparation, especially ,for children, who readily take it. It is not astringent, does not bind the bowels, and it may be given usefully in certain maladies where the other preparations give rise to too much irritation.

Magnetic Oxide of Iron is an oxide intermediate between the green and the red. It is a brownish-black powder, and has not much taste. It has been used instead of reduced iron, but is not considered so good. It is not much used. Its dose is from 3 to 5 grains.

Peroxide of Iron is used in two forms—one moist, the other dry. The moist peroxide is used chiefly in use in case of strict emergency, which rarely occurs—viz., poisoning with arsenic, for which it is a kind of antidote. The dried oxide is more irritating, and is often given when not intended when the compound iron and other mixtures are long kept.

Perchloride of Iron is the preparation most frequently used as tincture or liquor. It is a powerful astringent preparation, somewhat inclined to irritate ; but if astringency is desired, and the irritant qualities not objected to, no preparation of iron better fulfils its object. It is used in poisons, bleedings, and discharges, and applied locally for a similar purpose. Internally it is given in water with good effect in erysipelas, pneumonia, and other inflammations of a low type.

Ammonio-Citrate of Iron is an exceedingly mild and very valuable preparation. It exists in beautiful red scales, and as it possesses little astringency, is often one of the best remedies of an iron kind along with a vegetable tonic, when a patient is recovering from acute illness, especially if the stomach has been troubled.

Tartrated Iron is in many respects similar to the ammonio-citrate. It may be given with effervescing alkaline preparations. The dose of these is from 5 to 20 grains.

Citrate of Iron and Quinine contains both quinine and iron in a palatable and digestible form. It is unfortunately rather expensive, and cannot be given with alkalies. It is one of the favorite modes of prescribing iron to delicate patients. In some cases of neuralgia this preparation is invaluable, though the carbonate is commonly prescribed for that malady. Other nervous maladies may be benefited in the same way.

IRRITANTS are substances which, being applied externally or internally, give rise to marks of inflammation of a greater or less degree of activity. They include all substances called rubefacient or reddening, epispastic or blistering, and pustulant or producing pustules. The chief are mustard, turpentine, cajuput, corrosive sublimate, iodine, croton-oil, etc., which act both externally and internally, besides a great variety which only act internally. The blistering substances are mainly cantharides, or Spanish flies, in some form or other, or glacial acetic acid. The pustulants are croton-oil, tartar-emetic, and nitrate of silver. They are used for various purposes, mostly for the relief of internal inflammations, though sometimes of pain of a different character. They are supposed to have what is called a derivative action, their effect being to neutralize the inflammation within ; more probably, however, their influence is much more closely connected with the nervous supply of the part, especially with the nerves called the vaso-motor.

IRRITATION is a term employed in medicine to denote a variety of ill-defined conditions and actions. Health consists in a due balance of all our functions whereby they are carried on almost imperceptibly ; but for the performance of each a certain stimulus, varying in different cases, is necessary to the action of the parts. When this balance is lost, and the stimulus becomes excessive, we call the action irritation. Suppose we apply an irritant substance, say an acid, to the surface of the body, the consequence is an excessive irritation of the part which is out of all proportion with that of any other in the system. The condition is called irritation if it stop short of that which we commonly call inflammation. (See COUNTER IRRITATION.)

ISCHURIA, a technical expression for suppression of urine.

ISINGLASS. (See GELATINE.)

ISSUES are artificially-produced wounds which are kept raw and open, so that there may be a constant flow of pus from the surface. It is employed in surgical practice either as a counter-irritant in certain local affections, as caries of one or more bones of the spine, joint diseases, and inflammation of the eyes, or to keep up a constant drain from the system in certain constitutional derangements. It is often thought necessary, whenever an old ulcer upon the leg has been dried up, to substitute for it a smaller wound, from which a constant, though less abundant, discharge of pus may flow. An issue may be made either by transfixing a pinched-up fold of skin with a knife and cutting through this, by blistering the surface of the skin, or by making a slough by the application of strong caustics or the red-hot iron. Whenever the surgeon has a choice of situation, he avoids regions where the skin is thin and stretched over prominent surfaces and angles of bone. and selects such parts as the outer surface of the arm below the shoulder, the calf, and the inner surface of the thigh immediately above the knee, as here there is much muscle and a thick layer of cellular tissue between the muscle and the skin. The wound made with a knife is dressed for the first three or four days by a pad of dry lint, which is lightly pressed upon its surface by means of sticking-plaster. At the end of this period a raw granulating wound is established which resembles the issue-wound formed by the detachment of the eschar after the application of a caustic or the red-hot iron. The issue is then kept open by keeping some foreign body in constant contact with its surface, in order to irritate the granulations, and to cause them to dissolve into pus instead of forming scar-tissue. The bodies used for this purpose are either peas or small solid glass beads. The former cause irritation in consequence of their swelling ; but whenever the wound can be kept open without difficulty, the glass beads are to be preferred on account of cleanliness and the comparative ease and freedom from pain with which they may be worn. One or more of these beads, according to the size of the issue, are placed upon the raw surface, and then strapped lightly down by sticking-plaster. When there is free discharge, they should be removed and cleansed every day, and the edges of the issue should be frequently bathed with some weak lead lotion, or spirit and water. Over the plaster should be placed a thick pad of ordinary cotton-wool, or of the chloralum-wool. When the beads or peas are removed and changed, the surface of the wound should be syringed with a weak solution of carbolic acid. Sometimes, notwithstanding the presence of a foreign body, the issue-wound heals : it then becomes necessary to prevent this by applying some stronger irritant to its surface in the form either of blistering fluid or of caustic potash, or by merely smearing the foreign body with a salve containing iodine or some other stimulating agent. (See SETON.)

ITCH is a most troublesome skin disease, caused by the presence of the *Acarus Scabiei* or itch insect. (See ECTOZOA.) These little creatures burrow their way into the skin and the female deposits the eggs ; at night time especially they crawl very actively along the skin and cause intolerable itching ; the patient to relieve the distress is sure to scratch the part, and so pustules are formed and numerous scratch-marks. Close by the pustules may be seen an oblique line in the cutis, which is the mark of the burrow. This disease it very catching, and is a frequent accompaniment of dirt ; the little animals readily pass from one body to another, so that children sleeping in the same bed or using the same clothes will readily transmit it to each other. It is more common in children than in adults, as the insect prefers a tender and delicate skin ; and while in old people the rash is generally confined to the arms and between the fingers, yet in children it may be all over the body ; as a rule the head and face are rarely, if ever, attacked.

Treatment.—A cure may be readily effected if care be taken to rub sulphur ointment every night into the skin of the affected part : this must be done so as to make the part glow ; next morning wash the patient with coarse soap and hot water, and rub the soap well in with a flesh brush. At some places sulphur baths are given with advantage, but this can only be done in hospitals or other large institutions. The clothing which the patient has worn next the skin may be kept on during the treatment, and then it should be burned to stop the spread of the disorder. Although devoid of danger, this disease is so liable to extend to other people that when one person is affected he should be kept separate from others until he is well. When this disorder appears in a school, isolation should be at once practised, although those similarly affected may be kept together and placed under similar treatment. It is entirely a local disorder, and no internal remedies need be given, although it is a popular impression that the itching arises from some impurity in the blood, so loath are people to admit the faults arising from uncleanliness. The treatment should be repeated every night for a week in order to insure a cure.

J.

JACTITATION, a term applied to the unconscious movements of a patient when in the delirium of a fever.

JALAP is the dried root of a plant growing in Mexico, mainly near the city of Xalapa, whence the name. The roots are somewhat egg-shaped and pointed, untinged, about the size of an orange. They are brown externally, and yellowish-gray internally ; sometimes they are sliced. There is also in use a resin procured from the jalap-root called jalap resin. This is dark brown in color, and very bitter. It is produced by means of rectified spirit, in which it is freely soluble, but is not all soluble in water. This is the jalap of the shops. Jalap itself, as powder, has got a sweetish yet nauseous odor and taste. Jalap resin from the true plant contains a substance called convolvulin, which is strongly purgative. Jalapine is found chiefly in a false variety of the root, but is also found in the true one. The preparations of jalap mainly used are its powder and compound powder. The latter consists of jalap, cream of tartar, and ginger, and is a most valuable remedy in many forms of dropsies, when it is desired to pump the water out of the system. Jalap itself is a brisk purgative, producing watery motions. It is not so irritant as scammony, and seems to act more on the small intestine than on the large one. It has a tendency to gripe, and hence is seldom given alone ; usually some substance like ginger is given along with it to prevent the pain. Frequently it is combined with calomel, the two constituting the favorite purgative powder of many old practitioners. It is frequently given to children to get rid of worms. The use of jalap as an habitual purgative is to be deprecated, as its use frequently gives rise to subsequent constipation. The dose of jalap powder is from 5 to 20 grains ; of the compound powder about half a drachm.

JAMES'S POWDER is a secret preparation long in vogue. It has been a good deal employed in fevers, but it is most serviceable in incipient colds. It acts as a sudorific, and is suitable when Dover's powder, on account of the opium it contains, is not admissible. The preparation is generally understood to be oxide of antimony with phosphate of lime.

JAUNDICE can hardly be looked upon as a separate disease, though as a symptom of disease it is so grave and important as to be ranked in nosologies as a distinct malady. The one essential of jaundice is a yellow color of the skin, due most frequently to an absorption of the coloring matter of the bile and its circulation along with the blood. By some it is also supposed that there are cases of jaundice due to non-separation of the coloring matter of the bile from the blood and its accumulation there, but this is not generally taken as fact. Whatever, therefore, obstructs or prevents the flow of bile into the intestine will give rise sooner or later to jaundice. Chief among these are narrowing of the bile-ducts from whatever cause. Thus inflammation of the lining membrane of the ducts may do so. Still more likely is pressure to do so, and this indeed is the most common cause of jaundice. In this way fæcal accumulation in the colon may give rise to jaundice, tumors about the orifice of the bile and pancreatic ducts, abdominal aneurisms and cancer of the glands in the great fissure of the liver. Again, we may and often do have obstruction from gall-stones, plugs of thickened mucus, hydatids, etc., blocking up the ducts. These same ducts may be blocked up by ulceration from gall-stones and subsequent contractions, or in a variety of other ways. Within the liver any pressure on the main tubes containing bile may give rise to jaundice, so too may inflammatory conditions extending to them. In certain degenerations of the liver jaundice is found, but by no means in all. It occurs in acute yellow atrophy, and in cancer if the mass happens to press on a duct, but not otherwise. Yet again there are certain maladies where the liver need not be specially affected in which jaundice prevails ; such is the case in relapsing fever, still more in yellow fever and ague ; while in not a few cases jaundice is entirely due to emotional causes, as fright or excessive anxiety, or the sufferings from

wounded pride. The *symptoms* come on gradually or suddenly. If gradually, then there is progressive loss of appetite headache, and depression ; there is also some nausea and a sense of weight in the stomach. If it comes on suddenly, the patient may make the discovery in the morning that he is yellow. This color is most marked in the whites of the eyes. At the same time the urine becomes of a rhubarb tint and stains the linen, while the fæces are whitish or clay-colored. The skin itches, and there is a bitter taste in the mouth. Digestion is interfered with, and sometimes every object seen seems of a yellowish hue. When the malady lasts long the brain power is weakened, and there may be stupor or delirium, while the nutrition of the patient suffers and he becomes thin and weak. Sometimes there is a tendency to bleed from various parts, and most frequently there is some bleeding. All this may speedily pass away, or become more and more aggravated, till the patient becomes almost black. At the same time there may be excruciating pain, particularly if a gall-stone be the cause of the jaundice ; or pain may be entirely absent.

Treatment.—The treatment to be adopted for the jaundice will depend entirely upon the cause of the obstruction to the flow of bile. But suppose we take a common case of obstruction from catarrh of the bile-ducts, or obstruction from some emotional cause. This last form of the malady will pass away spontaneously, but both may be aided by medicine. Of all remedies adapted to the complaint rhubarb and soda or potash seems best. Then, as there is ordinarily some stomach-ic derangement, a little ginger added is an improvement, and some spirit of chloroform aids to make the whole sit easily. Sometimes sulphate of magnesia, with sulphate of soda, does good ; but the treatment must vary with each individual case. The food should be light and nutritious, and stimulants should be avoided. If any are required, claret and water or very weak brandy and water is best.

JAW, BROKEN. (See FRACTURES.)

JOINTS (DISEASES OF).—The most common affection to which a joint is subject is inflammation of the synovial or the thin delicate lining membrane. This membrane contains and secretes the synovia or joint oil which lubricates the joint. Now, if, from any cause, such as blows, strains, or from local injury, or from exposure to cold, rheumatism, gout, etc., a severe aching pain in the affected joint comes on, and great swelling very soon after the pain, attended with redness of surface and constitutional fever, a condition exists called *Synovitis*. The knee is the most frequently affected. The shape of the joint is altered, owing to the effusion into the synovial cavity, which consequently bulges at those portions of the joint which are least protected by the natural coverings, ligament, tendon, or muscle. In the case of the knee, the affection can be distinguished from the inflammation of the bursa over the knee-cap, from the fact that in synovitis the knee-cap can be distinctly felt floating as it were upon the fluctuating swellings, which are situated on either side of the joint ; whereas in the latter case the swelling is in front of the knee-cap and of the ligament tying it down to the shin-bone (tibia).

Treatment.—Perfect rest is indispensable, and the joint must be confined either by splints or by a piece of gutta-percha or stiff leather or card-board, made pliable in boiling water, and moulded over the joint. It should be lined with some soft leather, and capable of being laced up or let out as the condition of the joint requires it. Leeches may be applied to the joint ; ice-evaporating lotions, hot fomentations, or a linseed-meal or bran poultice (sprinkled over with laudanum), are the best local applications. A dose of calomel, and afterward a saline purge, with an opiate at night, form the constitutional treatment. In rheumatic cases, and when the urinal sediment is red, ammonia and potash, and afterward iodide of potash may be given. If the disease has been very acute, the joint may become permanently stiffened. Chronic rheumatic inflammation of a joint (arthritis) is generally met with in old persons ; it is characterized by racking, gnawing, rheumatic pains in the joint affected, aggravated by changes of weather. The joint is stiff and swollen, very painful if touched, the muscles become wasted, and on any attempt made to use the joint either actively or passively, a sort of cracking or creaking sound is audible.

The joints most frequently affected are the shoulder, hip, and articulations of the hand and the spine. The treatment is to give iodide potass, or ammonia, or guiacum, with generous diet, anodyne embrocations, and vapor or Turkish baths, but rarely much relief is obtained.

JUNIPER (*Juniperus communis*) is a common enough plant. It has berries, which, being distilled when green, yield a colorless or pale green oil, having in a high degree the odor and warm taste of the fruit. The berries themselves are about the size of black currants, of a dark purple color, with a bloom on the surface. Their interior is filled with a brownish-yellow plup. Their odor somewhat resembles that of turpentine, but is more agreeable. The oil usually contains a little resin from its own change by oxidation, and, mixed with rectified spirit, constitutes Spirit of Juniper. This is largely used in medicine on account of its action on the kidneys, which it stimulates and causes to pass through them a larger quantity of urine than natural. This property is valuable in certain forms of dropsy, and in these the remedy is used. It is sometimes given alone, but most frequently it is combined with other remedies, especially broom-tops and cream of tartar. A capital combination for country use is thereby formed. The same ingredient is found in Hollands, and in less quantity in ordinary gin. For this reason Hollands may be substituted for the ordinary Spirit of Juniper, and, added to broom-tops with cream of tartar, will be found a most efficient diuretic and stimulant in slow cases of heart disease.

K.

KAMELA or WURRUS is a drug comparatively recently introduced into European practice as a remedy against worms. It is an orange-red powder which adheres to the capsules of the *Rottlera tinctoria*, an Euphorbiaceous tree growing in India. The powder hardly mixes with water, but is almost entirely soluble in alcohol. It has long been employed in India as a remedy for tapeworm, but has not been much used in this country. It usually purges severely, and this may in certain cases be an objection. It is best given in doses of 30 grains, or a drachm, in some thick substance or in spirit.

KELOID is the name given to a disease of the skin in which there is hardening or thickening of that tissue, so that the part very much resembles that seen after a burn. It occurs on the back and upper extremities chiefly, and seems to be an incurable disorder. The word "keloid" or "kelis" derived from a Greek word signifying a crab's claw. There seems to be two kinds or varieties of this disease. The one appears as hard, shining tubercles or small nodules of a dusky or deep-red color, and generally attended with itching, pricking, shooting, or dragging pain in the part. These tubercular elevations gradually increase in size until they are as large as a horse-bean or even an almond, and about one tenth or one sixth of an inch above the general level of the skin. They are hard, firm, and elastic, but after a while they become broader and more irregular. Some delicate whitish, glistening lines appear on the surface, and from each there is a claw-like process from a quarter of an inch to an inch in length, which appear to cause a puckering of the skin. Growth may go on for months and even years, but they only cause local inconvenience and do not impair the general health. The other form of keloid does not begin with tubercular elevations, but as white, roundish patches of skin, very slightly raised and surrounded by a zone of redness. At first there is no pain nor uneasiness; afterward there is itching and pain with a feeling of tightness in the part; at length the part becomes hide-bound, and the skin is hard and rigid, so that the movement of the part is impaired. The fingers are very liable to be affected in this way. After a time the skin shrinks, becomes red or yellowish, and may go on to ulceration. If the affected part be extirpated it often returns, and no treatment seems to be of any avail.

KERNELS. (See GLANDS.)

KIDNEYS (THE) are two in number, and lie in the back part of the abdominal cavity, one on each side of the spine. Each is about the same shape, but nearly twice the size of an ordinary sheep's kidney; each is supplied with a vessel (the renal artery) which brings the blood from the aorta to the kidney, and with another vessel (the renal vein) which brings the blood to the inferior cava after it has passed through the kidney. Soon after the artery enters the organ, it breaks up into a great many small branches, and these again divide, so as at length numerous fine tubes are formed which have extremely delicate walls and enable the blood to come into the closest contact with the kidney tubes; there are two sets of these capillaries, the one being arranged in an intricate net-work around the tubes, the other being arranged in clusters and surrounded by the dilated commencement of a tube; thus one set is outside and the other inside the kidney tubes. In whatever way these different sets are formed, they finally join and form the renal vein. The rest of the kidney is made up of tubes, or hollow canals, lined with epithelium, commencing at first in dilated extremities and inclosing fine blood-vessels; then they pursue a tortuous course, surrounded by capillaries, till, joining each other by degrees, the tubes open into a funnel-shaped opening called the pelvis of the kidney; this in turn ends in a narrow tube which conveys the urine to the bladder. The kidneys are, in fact, so arranged as to form filters, which abstract from the blood water and various constituents which make up what is known as the urine. The kidney is the only organ in the body which takes away materials from the blood without giving anything in return, and hence the blood in the renal vein is probably the purest to be found in the body. It is essential for the proper performance of the renal function that a certain amount of healthy blood should pass through this organ in twenty-four hours, and so any alteration in the quantity or quality of that fluid will cause some derangement of the kidneys. In cases of heart disease, or of empyema, the circulating stream is obstructed and the venous system becomes too full; the renal vein shares in this fulness, and hence the kidneys become over-distended with blood in their vessels, but the blood is more or less stagnant and not

removed as it ought to be, so that very little urine is passed, and that excreted is dark in color and deposits some sand on cooling; often it contains also a trace of albumen, because the serum of the blood escapes from the tense vessels into the tubes and so into the urine, and wherever serum is present albumen will be found, for it is one of its constituents. Now and then blood will be seen also in the urine. This state of things generally comes on toward the end of the disease; the patient suffers also from dropsy of the legs or abdominal cavity, or from jaundice, and shortness of breath and palpitation of the heart, or from a combination of these symptoms. An alteration in the quality of the blood will also induce kidney disease; this is the case in many fevers, and more especially after scarlet fever, when acute Bright's disease is by no means an uncommon complication. Those who are scrofulous or rickety, and those who suffer from consumption, syphilis, or gout, and those who have led intemperate lives, are liable to get disorganization and destruction of their kidney substance, and serious mischief may ensue, leading sooner or later to a fatal termination. For an account of these changes and the symptoms which accompany each variety, the reader is referred to the article on BRIGHT'S DISEASE. The kidney is also liable to various other diseases, which may here be briefly mentioned. A blow across the loins, a stab, or a gun-shot wound may cause rupture of the organ; great pain over the seat of injury, with sickness and faintness followed by the appearance of blood in the urine, are the chief symptoms. (See HÆMATURIA.) Cancer of the kidney may occur; this is a rare form, and is generally associated with cancer of other organs; a large tumor may be developed on one or the other side of the abdomen, blood may appear in the urine, or albumen may be present more or less persistently; the pain, emaciation, loss of flesh and strength, are to be found in these as in all other cases of cancer, and finally lead to a fatal result. Death, however, takes place less rapidly than when the stomach or the liver are the seat of this malignant disease. Various poisons, other than the fever-poisons mentioned above, may cause disease of the kidney; in this way phosphorus, arsenic, turpentine, and cantharis or Spanish fly have been known to cause blood in the urine, and even suppression of urine. The inhalation of arseniuretted or sulphuretted hydrogen has caused similar results; these gases are evolved in the process of separating certain metals from their ores; in their pure form they are rarely met with, except in the chemical laboratory. A stone, or calculus, may exist in the pelvis of the kidney without bringing about any untoward result; at other times it may block up the channel so as to prevent the flow of urine, and then the kidney will become distended into a large cyst and be rendered quite incapable of performing its functions ; or, again, the stone may pass down the ureter, or the canal which conveys the urine from the kidney to the bladder, and cause great distress ; intense pain in the loins and down into the groin and thigh, faintness, and vomiting are the chief symptoms ; a hot bath to alleviate the pain, and the administration of cholroform to diminish the spasm of the ureter, are the best means for relief ; tea, water, or any diluent drinks may also be given so as to wash the stone down into the bladder. (See OBSTRUCTIONS, RENAL).

Cancer of the bladder, or the presence of a tumor or stone in that cavity, may cause also disease of the kidney, by pressing on the ureter and distending that tube, so that from this cause the kidney may become cystic or converted into numerous dilated cavities ; inflammation, too, of the bladder may exist in these cases (CYSTITIS) and add to the mischief, for this process is very apt to extend up the ureter to the kidney itself and cause the formation of pus and a total destruction of the kidney ; this change is often a cause of death in those persons who have long suffered from a stone in the bladder. Abscesses may form in the kidney in cases of pyæmia, but no symptoms of any marked importance attend this change ; more frequently an abscess may form around the kidney and burst either in the loin or in the intestinal canal. If the state of things is known during life, and the abscess point in the loins, an opening may be made by the surgeon and the matter be let out ; but these cases are very obscure and very difficult to make out accurately. Lastly, malformation of the kidney may occur, but give rise to no harm in consequence ; they may be joined together in front of the spine by their lower extremities and so have the form of a horseshoe, or, more rarely, both are developed on the same side of the spine ; these peculiarities, however, do not seem to interfere in any way with the healthy performance of the renal function.

KING'S-EVIL. (See SCROFULA)

KINK-COUGH. (See WHOOPING-COUGH.)

KINO is the juice of a tree belonging to the Leguminous group, the *Pterocarpus marsupium*. The trunk is incised, and as the juice flows it hardens in the sun, forming brownish or reddish black tears. It is generally seen, however, in broken pieces, more or less angular, translucent, and ruby-red at the edges, shining and brittle. It has no odor, but its taste is powerfully astringent and turns the saliva blood-red. Kino contains a kind of tannic acid, and another astringent principle called catechin, together with red gum. Its most important preparation is the compound kino powder, which contains kino, cinnamon, and opium—one grain of opium in twenty of

the powder. Kino is a very powerful astringent, and may be given for the tannin it contains. It is not so soluble as catechu. It is often, and perhaps chiefly, used in diarrhœa, for stopping which the compound powder is a very excellent preparation. It may also be chewed for relaxed sore throat. These are the chief uses of this substance. It mainly differs from tannin in being less soluble.

KIRSCHWASSER is a spirit distilled from cherries in Germany, and resembling brandy From the quantity of prussic acid it contains, extracted from the kernels of the cherry, it is dangerous to take any amount of it inadvertently, but when mixed with water it forms an agreeable stomachic, and is a good substitute for a better stimulant.

KLEPTOMANIA is a form of madness in which stealing is a prominent and singular feature. (See INSANITY.)

KNEE-JOINT is the largest articulation in the body, and is composed of three bones : the thigh-bone (femur), shin bone (tibia), and knee-cap (patella). These bones are held together by a great number of strong ligaments, and the movements of the joint are controlled by numerous muscles. Like all movable joints, the *articular* surfaces are covered over with cartilage, and a large and complex synovial membrane is insinuated between the structures forming the joint. A remarkable feature about the articulation, it has in common with one or two more in the body, viz., the inter-articular fibro-cartilages, or as they are here called semilunar ; their office is to defend the joint from severe and sudden concussions, and their mechanism is so adjusted, they are always between the ends of the bones when, and at the point at which, the greatest pressure is experienced. From the complex nature of this joint, its size, and exposed situation, it is obvious that it must come in for a large share of injury, and it is peculiarly subject to disease. The natural movements of which this joint is capable are flexion, extension, and partial rotation outward and inward.

Diseases of the Knee-joint. — The several affections to which the knee-joint is subject, are—1. Fracture. 2. Dislocation. 3. Synovitis. 4. Bursitis. 5. Scrofulous disease. 6. Rheumatic affections. 7. Loose cartilages. 8. Malignant diseases. 9. Hysterical affections. 10. Deformities.

Fractures connected with the knee-joint are treated in detail in the article FRACTURE.

Dislocations of the Knee are fully treated in the article on DISLOCATIONS.

Synovitis of the Knee-joint is treated of in the article on JOINTS, DISEASES OF.

Bursitis, or inflammation of the bursæ in connection with the joint, viz., the bursa patellæ, the bursa on the tubercle, of the tibia, and those between the condyles of the femur and the gastrocnemian muscle. have been already treated of in the articles BURSA, HOUSEMAID'S KNEE.

Scrofulous Diseases of the Knee-joint, White Swelling.—This condition of the joint, or of the structures forming the joint, always occurs in those of scrofulous constitution. It is probably more common in children than in adults, and it commences with slight lameness, swelling of the joint, and, from the pain or stiffness of the articulation, the muscles are not brought into play, and so waste or atrophy. The general train of symptóms is much as follows : In the first place there is either history or evidences of scrofula ; occasional pains are noticed in the joint, becoming gradually worse, especially at night ; swelling is rarely noticeable at first, and the peculiar form subsequently taken by this swelling—a sort of *globular* enlargement—is owing to the infiltration of the structure surrounding the joint, rather than of effusion into it. If the disease proceeds unchecked, some disorganization of the joint ensues, and from having been kept so long bent, or becoming bent by the hamstring muscles, at last, in many instances, dislocation of the tibia backward takes place. The morbid conditions occurring in the knee affected with scrofula are identical with those of other joints, and will be more fully discussed under the article SCROFULA. The *treatment* in the early stages is, locally, to procure rest, and to endeavor to prevent deformity, leeching, fomentation, and poultices, if there be much pain. Counter-irritation—blisters, iodine, or issues, Scott's dressings, are all of value. Constitutionally, cod-liver oil, iodide of iron, quinine, good food, and sea air. In severe cases, operative interference is necessary, such as excision of the joint, or even amputation. The results following excision of the knee-joint are very satisfactory, provided a proper case for the operation be selected. There are many instances of a perfectly useful limb being retained, and frequently a shapely one, and one on which an individual can follow his ordinary occupation as well as formerly.

Loose Cartilages.—The knee, in common with other joints, is sometimes the seat of these bodies. They are usually of an irregularly oval form, but vary ·in structure and density ; they vary also in size from a pea to a plum-stone. Their surface is generally smooth, they seldom occur singly, usually two or three are found in the joint. They are attached by means of a delicate pedicle to the capsule of the joint. The symptoms of their presence are excruciating pain from their suddenly getting between the ends of the bones, when the limb is rendered rigid, and motion arrested suddenly, and these symptoms will continue until the substance has been manipulated back again from its position. If they

do not cause very much inconvenience, palliative treatment is useful ; thus, an elastic bandage or a tightly fitting knee-cap should be applied. and the patient kept in a recumbent position. Should this fail, an operation for their removal must be had recourse to, and it must be borne in mind that no operation is expedient except in troublesome cases ; and considerable precaution must be taken, as, even in the most skilful hands, it has been fully shown that it is not altogether free from risk, by setting up serious inflammation in the joint.

Malignant Disease.—Occasionally the knee-joint is the seat of malignant growths, particularly of cancer, and any soft tumor springing from the lower end of the femur, or head of the tibia, is to be viewed with anxiety. (Non-malignant growths, such as exostoses, fibrous or enchondromatous tumors, are sometimes met with.)

Hysterical Neuralgia of the Knee-joint depends upon some morbid condition of the uterus, stomach, or rectum. The term hysterical affection is taken rather in a general sense, and it is often the custom to refer anomalous nervous affections of joints to hysteria. The *treatment* in these cases may be both constitutional and local. If there be emaciation and debility, iron, bark, and cod-liver oil are of service. In plethoric persons purging and low diet are indicated. Locally, hypodermic injection, aconite and belladonna liniments should be rubbed into the part affected. Galvanism in a continuous current passed through the joint, and the application along the spine of ice-bags, are remedies which should be tried. Division of the main nervous trunk is not advisable, as, although the relief is instantaneous, the attack speedily returns as severe as ever.

Knock-Knee.—This affection consists in an inward projection of the lower extremities of the thigh-bones, and a more or less considerable outward divergence of the legs and feet. Great deformity is thus produced, and the patient experiences much difficulty in walking. The knees constantly strike against each other, and the foot is turned outward, so that the inner edge is applied to the ground. Knock-knee is caused by weakness and yielding of the ligaments and sinews about the inner aspect of the joint, and the affection is increased by walking and standing. It is met with in workmen who carry heavy loads, or in those accustomed to wheel heavy barrows. It may also occur in youths who grow very fast, just before the age of puberty, but it affects most commonly weak unhealthy children of the poor classes, who live in towns, and is due in those subjects to general weakness and poverty of the blood, engendered by bad quality or insufficient supply of food. It then often shows itself while the child is still in arms, but is made much more apparent and increases rapidly in extent after the child begins to walk. Knock-knee and rickets are often associated together. When undue use is made of one leg, and too much weight thrown upon it in consequence of disease or injury in the other limb, óne-sided knock-knee may be produced. This deformity when it occurs in rapidly growing and overworked young people may be remedied by rest and cold douches. If discovered early in infants the best *treatment* is fresh air and good and suitable nourishment. The patient should not be allowed to move about on the floor until the unnatural prominences at the inner surfaces of the knees have disappeared. In severe cases, in older children, it will be necessary to apply to the outer surface of each limb either a padded wooden splint long enough to extend from the hip to the foot, to which the limb is to be bandaged ; or irons, furnished with a ratchet screw, and fixed by means of buckles. The treatment demands much care and patience, and, to be effectual, must, in severe cases, be continued for eighteen months or two years. The deformed limbs should be well rubbed and bathed with cold water every morning, and the child's general health should be kept up by good living, and, if possible, by a prolonged sojourn in the country or by the sea-side.

KOUSSO. (See Cusso.)

L.

LABOR is the common term for a confinement or delivery. It usually takes place at the end of the ninth month, or at the expiration of 280 days from the time of conception. If the birth of the child takes place before six months it is called an abortion, or miscarriage, and when between six and nine months it is known as premature labor. A labor, as a rule, is a perfectly safe and natural process, and attended with very little danger to either mother or child, if properly conducted. It commences with pain in the lower part of the abdomen, gradually settling down in the back, and known as bearing-down pains ; this is accompanied by contractile movements of the enlarged womb, by which the child is gradually expelled and brought into the world. The duration of a labor varies from six to twelve hours, in most cases being longest in those who are having a child for the first time ; the pains may begin much earlier and may be of a grinding character, but in general these are not attended by any expulsive effort ; they are caused by an error of diet sometimes and are removed by giving a purgative, followed by an opiate draught. In at least 99 cases out of every 100 the head of the child comes down first and is the part which emerges into the world the soonest ; the rest of the body soon follows, and the main object of care is to see that the womb well contracts as soon as the child is expelled. In from ten to twenty minutes, but sometimes longer, after the child is born, the placenta or after-birth comes away, and then, seeing that the womb is still well contracted, a wide binder may be placed around the abdomen, and the mother should be allowed to rest quietly for a time, after removing the soiled linen around her. The child, when born, is generally for the first few moments rather livid in the face, but soon begins to cry out lustily ; after wiping the mouth and nose, it should be wrapped in warm and soft flannel until the nurse is ready to wash it in warm water and dress it. If the child do not breathe at first, it may be gently slapped on the back or held out for a minute in the open air, or hot and cold water may be alternately dashed over it, and efforts made to keep up artificial respiration. After-pains are usually the worst in those women who have had several children, and are very troublesome the first twenty-four hours ; an opiate is the best remedy. The mother's diet must be light and nourishing ; usually, the first day, gruel or tea is preferred, but afterward a small chop or a piece of fish may be taken. The child should be put to the breast as soon as possible, but it is seldom able to suckle much the first day or two. The mother must be kept in bed for nine or ten days, and then may be allowed to get up, still keeping the horizontal position for a few days longer.

The details of management in a midwifery case are obviously out of place in a book of this nature. Sometimes instead of the head of the child presenting, the feet may come down first, but this is not of much matter, although it may prolong the labor. If, however, the arm or shoulder come down first it is a sign that the child is lying in a wrong position, and there is then said to be a cross-birth ; in such cases skilled interference is at once to be sought. There are some cases of such great deformity of the pelvis that premature labor has to be induced, but this must be only done after consultation, as, if done with a criminal intent, the operator is subject to severe punishment. In other cases operative interference is required to save the life of the mother or the child, or both, but for its proper performance great skill and experience is required, and can only be attained by a proper education and practical acquaintance with the subject. For a disease which may follow a difficult or premature labor, see Pelvic Cellulitis.

LABURNUM, a beautiful ornamental shrub, known to the botanist as *Cytisus Laburnum.* It yields seeds of an acrid and poisonous nature which by possibility may be eaten by children, and produce vomiting, cramps, purging, and all the symptoms of an irritant poison. The remedies in such a case are to give an emetic of mustard and warm water, or ipecacuanha, or white vitriol, and afterward to support the patient with ammonia and brandy.

LACTATION is the name given to the period of suckling a child after a confinement. After the birth of a child the breasts of the mother, which during pregnancy increase in size, secrete a large quantity of milk for the sustenance of the offspring during infant life. The process is a perfectly natural one, and it is by far the best means of rearing a child ; for the first day or two, especially after a first confinement, the milk flows in very small quantities, but after that time, when the secretion is well established, the production of milk goes on uninterruptedly in many cases. It is the duty of every woman, if in good health, to give the child the breast, yet there are many, and those chiefly in high life, who from indolence or apathy, or some other cause, prefer their children to be brought up by a wet-nurse or by the bottle. Many, again, are incapable of suckling their children from ill-health, or by the cessation of the secretion of milk in the breast, and then, of course, other means must be taken to bring up the child. The mortality of infants not brought up by suckling is vastly higher than in those who are kept to the breast for the first nine or ten months of life. If the child die soon after birth there may be some trouble in checking the distension of the breasts, which may often be very painful. The breasts may be drawn by a syringe, or by another child being put to them ; a popular method is to apply a soda-water bottle previously warmed to the nipple ; on cooling, the air within the bottle is more rarefied than that outside, and the difference of the atmospheric pressure will cause a flow of milk. The secretion may in such cases be generally stopped in a few days by applying a large belladonna plaster to each breast, and giving a saline purgative. Inflammation of the breast may take place during lactation, and end in an abscess ; the breast will be found hot, enlarged and painful, and perhaps a swelling, tender to the touch, may be noticed at one part ; when it is clear that matter has formed, an incision must be made with a lancet to let it out ; the breast must be slung in a towel from the shoulder, so as to relieve the patient of its weight, and hot linseed-meal poultices should be often applied. Unless the opening be a free one the matter is apt to accumulate again and require a second or even third incision. When an abscess forms, the health of the woman suffers ; she loses her appetite, becomes faint and weak, and loses color and strength. Tonic medicines must be given her and a

nourishing diet, and rest and quiet enjoined ; it is not always necessary to wean the child, as it can feed from the opposite breast. A child should be weaned at the end of nine months as a rule, but this is very seldom observed, and many poor women go on suckling for a much longer period, until, perhaps, the child is eighteen months or even two years old. This is bad for the mother as well as the child ; to the former, because it is a great drain upon her strength, and to the latter, because when the teeth have appeared a more solid food may be given. Women who have a family fast, who wean their children at a late period, and who, perhaps, live badly all the time, are very liable to suffer much in their health ; they lose their appetite and strength, and become low and nervous ; they are very liable to headache, pain in the back and left side, and very often they suffer from leucorrhœa. The proper treatment for such cases is to wean the child and improve the mother's health by tonics, good diet, and rest ; stout is often recommended to mothers as being more nourishing than beer, but at such times all stimulants should be taken in moderate quantities, and reliance should be placed on a more liberal diet. Over-suckling is a very frequent cause of ill-health in a woman, and lays the foundation of future illness by bringing on a state of debility.

LACTEALS are very minute vessels or absorbents which arise in small conical projections of the mucous or lining membrane of the intestines ; joining together, they finally form larger branches, which pass up by the mesentery into the mesenteric glands, and then on to the *receptaculum chyli*, a small chamber lying in front of the spine in the abdominal cavity. The function of the lacteals is to absorb various soluble portions of the digested food or chyme as it passes along the intestinal canal, and chiefly the fatty portions of the food ; the fluid, thus absorbed, is milky in appearance, is called *chyle*, and this, passing through the mesenteric glands, undergoes these various changes ; finally, it goes through the receptaculum chyli, and then on by the thoracic duct as lymph, to join in the blood current at the root of the neck. The lacteals are to the intestines what the lymphatics are to the rest of the body. They are often diseased in children, and many cases of *marasmus*, or wasting away, are due to affections of the small glands and lacteals of the intestines ; the mesenteric glands are also frequently associated in the change, and become swollen and enlarged, and add to the general mischief. Such children waste because the food cannot be properly absorbed, and the blood loses in quality and quantity because it does not receive its due supply of lymph. Diarrhœa, too, is a common symptom, and this, with the emaciation and general weakness, often brings about a fatal result. The diet must be carefully looked to ; all solid or thickened food should be avoided, while milk and beef-tea or chicken-broth may be be given. If there is any sickness, lime-water may be added to the milk ; but any error of diet may again bring on diarrhœa. Cod-liver oil is not well borne by the stomach in such cases, and therefore it may be rubbed into the skin night and morning. A little steel wine may be given daily, but no other medicine ; and no purgative should be administered.

LARD is hog's fat deprived of its membanes and purified by heat. It is used in making ointments, and is often better for application to a blister or sore place than the more skilfully prepared and expensive ointments. A piece of ordinary lard put into boiling water and allowed to cool and settle, and then taken out free from all impurities and kept in a stone jar, is a very useful and pleasant application.

LARDACEOUS DEGENERATION is another term for waxy degeneration ; it may affect the liver, kidney, spleen, and intestines. (See DEGENERATION.)

LARYNGEAL PHTHISIS is a form of consumption in which the patient suffers from hoarseness and loss of voice ; it is very common in the later stages of phthisis. (See CONSUMPTION.)

LARYNGISMUS STRIDULUS, also known as spurious croup or child-crowing, is a spasmodic form of disease commonly afflicting children during the period of their first teething. A considerable number of children die of the malady, being mostly under one year old. This disease might be and often is mistaken for croup ; but there is no fever, almost the only symptom being the interruption of the breathing. The first attack may often come on in the night, the child having been put to bed apparently well. There may only be one or two prolonged crowing inspirations, and the patient fall asleep again. In other cases the child may have been been irritable and restless for a day or two, when suddenly it is seized with difficulty of breathing, and kicks and struggles, unable to draw a breath. Presently, however, it is enabled to draw in a breath with a long crowing or shrill whistling sound, the chink or opening into the windpipe being much narrower from spasm than usual. This may end the attack, but it may return shortly in a few hours or sometimes in a day or two. In other cases the attack resembles epilepsy more than croup, the face being swollen and flushed, the veins starting out with convulsive movements of hands and feet. The child may even seem dead for a few moments, until by and by there is a gasp, and then a thin sounding breath, the patient gradually recovering. This is not always the case, for not unfrequently the little patient does die in one of these paroxysms. The cause of this spasm undoubt

edly lies in some irritation acting through the nerve which supplies motor power to the larynx. It has been supposed to be due to enlarged glands pressing on and irritating this nerve. In reality it may originate in a variety of ways, being most of them reflex, but all unite in acting through this nerve. It may originate in teething, indigestion, constipation, or the reverse condition, catharsis, which, acting through the brain and spinal cord in subjects whose nervous system is easily put out of gear, is thus manifested as spurious croup.

Treatment.—Fortunately much may be done by way of treatment, especially if that is done at once : and there is no disease in which a general knowledge of the principles of treatment are more important to the public. The best thing is promptly to put the child in hot water up to its chin, and pour cold water on its head. Some propose the same remedies as when a child seems still-born, such as slapping the nates or even artificial respiration ; but the conditions are totally different. When the child is born so, the only difficulty is to induce its muscles of respiration to act ; there is no obstruction to the entry of air. Here it is quite the reverse ; the muscles of respiration are ready to act, nay, are acting ; but the air cannot enter, the way being barred by the closure of the entrance to the air passages. Before any good can be done, these must be relaxed, and accordingly such treatment as we have recommended, or a few drops of chloroform swallowed, will do what is necessary. It is useless to think of the inhalation of chloroform when the air itself cannot make its way into the lung. Afterward the child's bowels should be carefully looked to, and the teeth and mouth examined. Change of air is of the utmost benefit. Belladonna in very small doses sometimes is useful, but the great thing is careful nursing. The child should have nourishing food, but it must not be overstuffed. Overstuffing and improper food are a very important cause of the malady. The best food for a young child is its mother's milk, but sometimes they will not suck, and then the best substitute must be procured—a good wet-nurse, if possible ; if not, asses' milk, or the milk of a healthy cow diluted with water.

LARYNGITIS, or inflammation of the upper part of the windpipe, is a troublesome and even dangerous affection ; it may be produced by exposure to cold, and also by inhaling any irritant gases, as ammonia, chlorine, or hydrochloric acid. There is more or less of a croupy noise during inspiration, the breathing is short and hurried, the patient can only speak in a hoarse whisper, and will point to his throat as the seat of distress. The symptoms are much the same as if a foreign body were accidentally to get into the air-passages, and there is a feeling of impending suffocation. When occurring in children it is commonly called croup ; but every case of croupy breathing does not show that laryngitis is present, as it may be only spasmodic. This disease is liable to come on in those who are subject to phthisis or bronchitis. Steam must be at once inhaled ; hot sponges must be applied to the throat, and a warm, moist atmosphere must be kept up. In some cases it may be needful to perform the operation of laryngotomy or tracheotomy ; but in all cases it is imperative to have medical aid at once. (See CROUP, LARYNGISMUS STRIDULUS.)

LARYNGOSCOPE is an instrument used by physicians to explore the larynx and upper part of the windpipe. In its simplest form it consists of a small reflecting mirror, which is mounted on a long and slender stem so that it can be passed to the back of the throat. Upon this are thrown rays of strong artificial light, reflected from another mirror, which is carried on the forehead or over the right eye of the surgeon. In the latter case the second mirror is perforated at its centre by a small hole, so that the rays reflected from the mirror at the back of the throat may pass through to the physician's eye, placed at the centre of the reflection. The physician sits in front of the patient, and on introducing the mirror into the mouth pulls the tongue well forward. The light is placed behind and to one side of the patient. The laryngoscope is used for investigating the nature and extent of diseases affecting the larynx, and also as an aid in applying remedial agents directly to the lining membrane of this organ, and in removing by operation warty growths, polypi, and other tumors.

LARYNGOTOMY should be performed only in cases where *great* urgency demands on opening of the windpipe, and when the proper apparatus for tracheotomy is at hand. The operation is thus performed : The patient is seated in a chair, with the head well thrown back and kept steady ; the finger of the operator is passed over the front of the neck, and the crico-thyroid depression felt for ; then a vertical incision, about an inch in length, is made in the mesial line over this spot, and the crico-thyroid membrane is divided sufficiently to allow of the introduction of a tube. The operation can be performed readily enough with a penknife, and if no tube be at hand, a quill or a thin piece of wood, turned on its axis, will admit sufficient air to the lungs. Care must be taken not to mistake the thyro-hyoid space for the crico-thyroid. Laryngotomy may be performed in cases of lodgment of some foreign substance in the larynx. The only casualty in the performance of the operation is the division of the crico-thyroid artery, a small branch which runs across the membrane, and might, if large, cause considerable trouble. In a

young subject the chief difficulty of the operation is the recognition of the parts, after the superficial wound has been made, from the smallness of their size, and this difficulty may be greatly increased if the skin and cellular tissue be inflamed and infiltrated with serum.

When the operation has been performed, it is of the greatest importance that the proper after-treatment be adopted. An experienced nurse or attendant, or at all events some one who can be trusted, should be left with the patient. He should be placed in a room where the temperature is warm and equable, as the introduction of cold air into the trachea is very liable to set up inflammation of the air-passages. The warmth should be a *damp* warmth, best so rendered by steam. The bed should be surrounded with curtains or blankets, and a jet of steam admitted from the end of a tube connected with a tea-kettle. It must be borne in mind that blood and mucus very readily collect in the tube and obstruct it and if no one be at hand to remove it the patient runs imminent risk of suffocation. The nurse or attendant should remove it carefully, as soon as any difficulty is noticed in the breathing, with a fine feather or camel's hair brush. This collection of mucus is especially liable to occur during the first few hours after the operation ; and should the patient drop off to sleep, and the above precautions not taken, he may be suffocated. If the patient desires to cough or to speak, he should be told to *draw* in a full breath, and then close the orifice of the tube with his finger, when expectoration can be performed, and the voice may be heard. Care must be taken that the tube be securely fastened round the neck with tapes, but not too tightly. The margin of the incision made in the neck very frequently becomes seriously inflamed ; a small poultice should be placed over the affected part. The attendant must be cautious in the administration of the patient's food, to give little at a time, and to take care that, in the increased difficulty experienced in swallowing, no morsels pass into the trachea.

LARYNX (THE) is a complicated structure, surmounting the windpipe, serving the double purpose of being an air-passage and of containing that mechanism which produces sound during expiration. The *note* is formed by the approximation or the divergence of two margins of membrane, which, in a state of quiescence, resemble in mutual relation the letter V. These membranous margins are called the *true* vocal cords, in contradistinction to two somewhat similar folds placed some way above them, called *false* vocal cords. These true vocal cords are acted upon by a series of muscles, which place them in the proper position to make either sharp or grave tones. In order that these muscles may act upon the vocal cords they are themselves attached to the several parts of the framework of the larynx, which are formed of cartilage. The prominent cartilage in the neck (Adam's apple) is the triangular front edge of the *thyroid*, or shield-like cartilage ; this cartilage expands behind, and at the inferior part of it are two little hinges, one on each side, which turn upon another cartilage placed partly within the thyroid and partly below it. This is very much like a signet ring, with the signet part of it backward, and receives its name from this fact ; it is called the cricoid cartilage. Two muscles, one on each side (crico-thyroid), pull down the thyroid upon the cricoid cartilage, at the same time slightly advancing it. Within the triangular voice box, formed at the sides by the thyroid and below by the cricoid cartilages, and resting upon the upper part of the signet of the cricoid, are two little triangular cartilages called arytenoid, from their supposed resemblance to an ancient pitcher. To these two arytenoid cartilages are attached the legs of the V, and the point passes across the larynx to the inner surface of the Adam's apple seen in front. Now, as there are muscles which drag the thyroid down upon the cricoid, so there are muscles which draw the arytenoid cartilages apart, and also approximate them. The former, which are attached to the sides of the signet of the cricoid and to the arytenoid, are called the *crico-arytænoidei-postici*, and the latter, which are also attached to the same cartilages, but the fibres of which act in a contrary direction, are called *crico-arytænoidei-laterals*. The arytenoid cartilages have special muscles of their own, called *arytænoidei*, which, according to some, draw their cartilages together, thus closing the opening of the vocal cords (*rima-glottidis*) ; according to others, rotate them upon a pivot and so open the chink. Immediately underneath and external to these true cords are two muscles, one on each side, attached to the thyroid and arytenoid cartilages, which approximates the vocal cords by shortening the distance between the arytenoid and thyroid cartilages (*thyro-arytenoid*). These vocal cords are not only employed in the production of voice for the object of speech or song, but can be completely closed, so that the air may be, as it were, imprisoned in the lungs during the phenomenon of *effort*. They, moreover, enter into vibration, in coughing, hiccough, sobbing, and laughing. The sound is always produced by the vibration of the lower or true vocal cords, whether the air enters the larynx from above downward, as in hiccough, or whether it passes from below, as in other acts. The larynx grows after birth, as do other organs, both in girls and boys. But at the time of puberty, the larynx all of a sudden develops rapidly, more especially so in boys. In the male sex, in fact, the glottis is twice the size,

both in length and breadth, and the Adam's apple (*pomum adami*) becomes conspicuous at this period. It is at this age that the voice "cracks," that is to say, that it becomes deeper, correspondingly with the modifications of the glottis. But as the muscles of the larynx are not as yet accustomed to this disproportion of the vocal organ, they contract irregularly, with inability, so to speak, and producing those singularly inharmonious sounds which are peculiar to this period.

Foreign Bodies, such as morsels of food, sometimes get into the *rima-glottidis*, and by sticking there may cause death speedily unless search be instantly made with the finger in the pharynx to dislodge it, or the operation of tracheotomy be performed. (See FOREIGN BODIES.)

The diseases of the larynx are laryngitis, croup, diphtheria, œdema of the glottis, chronic inflammation and ulceration ; tumors, warty excrescences, and epithelial growths, polypi of a fibro-cellular, fibro-plastic, or epithelial nature, have been met with in the larynx, epiglottis, and trachea. For the detection and treatment of these diseases the laryngoscope is necessary. The glottis is sometimes scalded, from the effects of swallowing boiling water, such as in the case of a child, in the nurse's absence, putting its mouth to the spout of a tea-kettle. Leeches, ice to the throat, and opiates, or the administration of chloroform, are of use, unless the symptoms are so urgent that tracheotomy be necessary.

LATERAL CURVATURE is an affection of the spine in which there is a curvature either to the right or to the left, so that one shoulder is lower than the other, and more commonly it is the left shoulder which is the lowest. It is often due to girls carrying children or any heavy weight at too early an age. Unlike angular curvature, which is a grave symptom of disease of the spine, it is a habit which may may be overcome by drilling and gymnastics. As the curvature occurs in both sexes during the period of youth and childhood, means must be taken to remedy the defect, as it is of no use trying to alter it when the person is grown up. The carrying of any heavy weight on one arm should be avoided, while daily drill should be enforced at school, so as to make the individual erect. For boys dumb-bells may be used, or the elementary drill of a soldier ; if able and strong enough, he should go through the various gymnastic exercises so common now at all good schools, whereby he will not only gain in muscular strength, but he will expand his chest and improve in health. For girls the dumb-bells should be lighter, or they may use elastic bands, or even do gymnastics on a small scale. The great point in all these exercises, is to begin at first very gradually ;

never tire the muscles ; then every day a slight progress may be made, until, in the course of a few months, a marked improvement will be found. Too often, under the present system, such exercises are continued so as to tire the child, and even to give pain ; but this is a great mistake, and such exercise ought to be made pleasurable instead of being looked upon as a punishment. In this way girls would develop into much stronger women, and there would be less need of stays and other articles of dress which are required by women, not only to improve what they term their figure, but to prop up a too feeble spine. (See GYMNASTICS.)

LATERITIOUS URINE is urine in which there is a sandy deposit of lithates on cooling. (See URINE.)

LAUDANUM. (See OPIUM.)

LAUGHING GAS. (See NITROUS OXIDE.)

LAUREL - CHERRY. (See CHERRY-LAUREL.)

LAVENDER, the *Lavandula spica*, is a plant from which is distilled an oil which gives fragrance to the plant. This is either colorless or, if it has been long kept, pale yellow, and has a hot aromatic taste. This oil dissolved in water is much used as a perfume. In medicine two preparations are used, the spirit, which consists of the oil dissolved in spirit, and the compound tincture. This last, which is almost a liqueur, contains lavender, rosemary, cinnamon, and nutmeg, the whole colored with sandal-wood. It is greatly used as a carminative and stimulant in hysteria and such nervous affections. It is also employed in flatulence and colic. The oil may be given, a drop or two on sugar, for a dose. The spirit from half a drachm to a drachm, and the compound tincture in about the same quantity. The well-known *lavender water* is made thus : To a pint of proof spirits of wine, add one ounce of essential oil of lavender, and two drachms of the essence of ambergris. Put all in a quart bottle, and shake it up well daily.

LAXATIVES are remedies which gently open the bowels, so that they are inclined to be loose, but no more. There is thus a distinction drawn between such and purgatives, which purge, and cathartics, which are supposed to act still more strongly. As it is highly desirable in all cases that the least power should be employed, laxatives should be employed when it is necessary to open the bowels artificially, if this will suffice. In many instances, however, they will not, and something stronger will be necessary ; but powerful opening medicines are apt to be followed by the very condition they have been used to get rid oft, and so the latter end is something worse than the first. Sometimes a change of diet will act as a laxative. Thus, if the food has been too concentrated, that is

to say, if there has not been a fair amount of indigestible matter in it, the bowels are apt to become confined. In this way the use of brown bread instead of white bread will often suffice to regulate the bowels and procure a daily motion. Figs and prunes are inclined to be laxative, especially the latter. Manna, tamarinds, and cassia, more so. But the most convenient for use are flowers of sulphur, castor-oil, and magnesia, or its carbonate.

LEAD in the metallic form is not used in medicine, but as acted upon by water it not unfrequently gives rise to slow lead poisoning.

Oxide of lead, or litharge, consisting of heavy orange-red scales, is never given internally : it is only used for the preparation of the plaster, which is so commonly used for fastening up wounds, etc. Technically this is known as lead plaster, but much more commonly is called mere sticking-plaster, or *diachylon plaster*. It is prepared by boiling together oxide of lead and olive-oil ; these, after boiling some hours with constant stirring, form a thick tenacious paste, which is applied to calico, and so the plaster is formed. The lead unites with the fatty acids off the olive-oil, forming a kind of soap. It is due to the presence of the lead that this plaster blackens over putrid wounds. Some prefer a plaster made of less irritating materials, and certainly, were it not for its tendency to harden, isinglass plaster is infinitely superior to the litharge plaster.

Iodide of lead exists as a bright yellow powder, or in fine scales, which is soluble in boiling water, forming in it a colorless solution, which in cooling allows the iodide to fall as crystals. It alters and loses its brilliant color by exposure to light. Two preparations are in use—a plaster seldom used, an ointment much more frequently employed. It acts when applied externally as a very mild stimulant, and is used as an application to scrofulous joints. It gives to these a yellow stain, which may be objectionable in an exposed part of the body. It is seldom used internally.

Acetate of lead, also known as sugar of lead, is prepared by dissolving oxide of lead or litharge in vinegar or weak acetic acid, and afterward evaporating. It is generally seen in white spongy masses, composed of interlaced needle-shaped crystals. It has a sweetish vinegary smell and a sweet metallic taste. It is readily soluble in water, and when exposed to the air tends to give off water, and fall down in the form of powder. The solution of sugar of lead in distilled water is clear, or almost so. Its main preparations are lead and opium pill, a very valuable preparation, consisting of acetate of lead, opium, and confection of roses ; one grain of opium in eight of the pill mass. In small doses acetate of lead acts as a sedative and astringent, diminishing especially mucous discharges, drainage from relaxed vessels and surfaces, and the like. It produces constipation and thirst. This, as well as other preparations of lead, interfere with the normal condition of the blood, diminishing the number of red corpuscles, and so giving rise to anæmia. It also paralyzes the muscular coat of the intestines. Lead poisoning, however brought about, tends to favor the production of gout, and gout in such cases is rather intractable. Acetate of lead is used as a remedy in internal hæmorrhages, and is one of the best we possess. It is also used in consumption, to check diarrhœa and perspiration. It is a capital remedy, especially as lead and opium pill, for diarrhœa accompanied by pain and a tendency to dysentery. It is also largely used externally as a sedative and astringent. Solution of the subacetate of lead is made by adding litharge to the ordinary solution of acetate of lead and boiling. It is a clear colorless liquid which tends to become tinted on standing by the formation of carbonate on its surface. This solution, under the name of " Goulard water," has long been known and valued. It is chiefly used as an external application, and is so used more than the acetate. A combination of it with acetate of morphia is a singularly soothing preparation for inflamed spots, if the surface is not broken. Its use with a broken surface might be dangerous. There is also an ointment which is used in the same way.

Carbonate of lead is mainly used in the arts ; not much in medicine. There is, however, an ointment of it which is applied to whole surfaces as an astringent and sedative. Sometimes also it is used as a powder along with starch. It is the most poisonous form of lead salt.

Nitrate of lead is only employed in the manufacture of iodide of lead ; it is not used medicinally.

With regard to the general uses of lead salts, we may say, first of all, that lead applied to a raw surface forms a kind of precipitate on its surface which protects it from the air for the time being. Hence, when surfaces are raw or weeping, a lead lotion removes the burning and itching, and stops the discharge. It matters not where such a surface is, except perhaps the eye, for it has been found that prolonged applications of lead lotions to inflamed eyes, especially if the clear part or conjunctiva is affected, tend to form a deposit of the metal and to produce a permanent opacity. In summer diarrhœa acetate of lead acts as a sure and certain astringent, especially if a few drops of laudanum, or, what is better, a fraction of a grain of morphia is added.

If the acetate can be looked upon as an irritant poison at all, it must be considered as peculiar, inasmuch as it produces constipation rather than diarrhœa. Frequently acetate of lead may be given for weeks or months without producing any signs of lead poisoning. This is especially the case where lead is used to avert the wasting diarrhœa of consumption.

LEAD COLIC. (See LEAD POISONING.)

LEAD POISONING may be brought about in a variety of ways. Painters, and other workers in white lead are its most frequent victims. It has been produced by sleeping in a newly-painted room ; from taking snuff which has been wrapped in lead ; and it used to prevail extensively in Devonshire, its cause being the action of the apple-juice on the lead used in forming the cider presses.

Lead Colic, which is the primary symptom of lead poisoning, is a variety of colic characterized by intense twisting pains about the navel. Frequently, too, there is retraction of the walls of the abdomen and pain in the back. At the same time there is obstinate constipation, and if the gums be examined there will commonly be found a blue line extending along the gum at its junction with the teeth. These are the prominent symptoms of the first stage of lead poisoning. If, however, the malady is not arrested, but goes on, by and by the nutrition suffers. First to so suffer are the extensor muscles of the fore-arm ; those, that is, which lift the back of the hand, so that if an attempt is made to raise the hand that way, the *wrist drops*. The muscles themselves waste, and though at first they respond to the stimulus of electricity, later on they do not. By and by the muscles of the upper arms also fail, so that the muscles which raise the whole arm, and even those which are attached to the shoulder-blade. may waste and become useless ; the remedy for such a state of things is, first of all, to get rid of the poison, and, secondly, to restore the paralyzed parts to their several functions.

Treatment.—In the stage of colic with or without palsy, the bowels must be well moved, and for that purpose nothing is so good as Epsom salts, with some dilute sulphuric acid given freely until the bowels are opened well. Half an ounce of the Epsom or Glauber salts should be given for a dose, and repeated in a couple of hours. Jalap is sometimes given, but it is a mistake. Castor-oil may do good, but the best means is common Epsom salts. A warm bath frequently gives great relief, until the bowels have acted, and very likely aids in moving them. If not speedily moved, an injection of soap and water will help. After the bowels are moved freely, a quarter grain of the extract of belladonna may be given to relieve the pain, but the salts must still be continued, though in smaller and less frequent doses. The application of electricity to the bowels often aids in opening the bowels, and otherwise gives relief, but it is not to be trusted to solely. Neither indeed is Epsom salts, for as soon as the bowels are fairly open, iodide of potassium must be given in good full doses (10 grains), hoping thereby to remove the lead still remaining in the system. This must be continued for some time. If there is paralysis as well as colic, electricity must be freely applied to the weakened muscles, so as to exercise them, and aid in recovering their contractility. This should be applied at least once a day, and undoubtedly is of great benefit. Sometimes the patients make use of sulphur baths, but this is hardly needed, and is of questionable benefit.

LEAVENED BREAD. (See BREAD.)

LEECHES (*Hirudo officinalis* and *medicinalis*) are species of the class Annelida, or worm-like animals. They are elongated, tapering to either extremity, and either extremity has a muscular disc or sucker. This is larger in the hinder extremity. The mouth on the anterior extremity, is tri-radiate, and contains three jaws, each furnished with two rows of teeth. These in cutting into the skin leave a permanent triangular mark, which is characteristic of their having been used. The intestinal canal is straight, but has a number of chambers on either side, in which blood may be stored, and used up at leisure. Hence these animals, if fully fed, do not require another meal for a long period. (See LEECHING.)

LEECHING is the most useful and most convenient method of local blood-letting, the blood being drawn from the capillaries or small vessels by the incision and subsequent suction of the leech. (See LEECHES.) The part to which leeches are to be applied should be well washed with warm water, and, if hairy, shaved. If the leeches will not stick, the skin should be smeared with milk, sugar and water, or some saliva, or should be pricked at two or more points with a sharp needle in order that a few drops of blood may be shed. If the part to be leeched is on the body or one of the limbs and no delicate structures or natural cavities are close at hand, each leech may be taken by its hind part between the thumb and finger, and its head or thinnest part applied to the surface of the skin. When several leeches are to be applied over a small extent, they should be covered with an inverted tumbler or cupping-glass until they are fixed. When two or three leeches are used they may be covered by an inverted chip-box. When the parts to be leeched are situated near to delicate mucous membrane, as on the face near the lips, nose, and eyelids, care must be taken to prevent the leech from wandering by placing it in a proper leech-glass or in a cylinder formed by rolling up some pasteboard or thick paper. Leeches before being applied should be well dried in a clean cloth. When fully distended the leeches usually drop off, but should they remain longer than is necessary the bodies may be sprinkled with a little snuff or common salt. A good leech will take about two teaspoonfuls of blood. After the leeches have droped off the bleeding may be kept up for some time afterward by applying linseed-poultices and hot moist flannels. In some instances, however this after-bleeding is too prolonged and too excessive, and with young children and weak and delicate persons becomes a source of danger. In a case of persistent hemorrhage from one or more leech-bites the following plans may be successively carried out : to cover the wounds with small pledgets cf dry lint, taking care to apply the rough surface of the lint to the bleeding part, and then to keep these pledgets in position by pressure with the fingers ; to apply ice to the bleeding surface ; to press firmly into the bleeding orifices small pieces of lint or cotton-wool dipped into the tincture of perchloride of iron ; to touch the wounds with a red-hot knitting-needle ; to transfix the base of each wound with a sharp sewing-needle, and to surround the skin beneath this by some stout silk-thread wrapped tightly round in four or six turns. The needle should not be removed for twenty-four hours.

LEG, BROKEN. (See FRACTURES.)

LEMON is the fruit of the *Citrus limonum* or lemon-tree, growing in the more sheltered parts of Southern Europe. Its bark, its juice, and the oil extracted from its fresh peel are all employed in medicine. The rind contains a valuable oil, which gives the well-known fragrance to the fruit. The juice contains a considerable proportion of citric acid, which has by some, but erroneously, been supposed to be the principle on which its value depends. It also contains a considerable quantity of the salts of potass. The preparations properly so called are the syrup and tincture. The peel is fragrant and stomachic, while the juice is cooling and possessed of most valuable antiscorbutic powers. The lemon-juice has very frequently lime-juice, the product of the *Citrus limetta*, substituted for it. Lemon-juice may be given effervescing along with bicarbonate of potass, and constitutes a very valuable and very refreshing drink for patients ill and parched with thirst. Lemonade, too, made from the lemon sliced into cold or hot water and sugar, is exceedingly refreshing.

LENS.—The lens is a transparent doubly-convex, crystalline body, placed immediately behind the iris and in front of the vitreous humor, and is separated from both of them by a transparent capsule. The use of the lens is to enable one to distinguish the form or outline of objects, and act on the rays of light by concentrating them, or bringing them to a focus after they have passed through it, exactly at the surface of the retina, which may be regarded as a kind of sensitive screen upon which they fall. (See EYE.)

LENTIGO is the name given to a disease of the skin in which the freckles are more permanent than usual. (See FRECKLES.)

LEPOID is commonly seen on the face, nose and forehead of elderly persons, usually males of a delicate, florid complexion, with tendency to congestion of the capillary vessels, and having light eyes and hair. It generally makes its appearance as a small speck about as large as a mustard seed, and of a dirty grayish color, soon becoming covered with a rough brownish scale resembling the bark of a tree. The first scale or crust falls off and is succeeded by another, and so the disease may go on for years. At length ulceration sets in, and a red glossy surface is left, secreting a thin pus. The disease is attended with itching, but not with pain. It is best not to interfere with the growth; the crust may be softened by covering it with a mixture of one part of castor-oil and two parts of collodion.

LEPRA is a dry skin disease occurring in circular red patches, and chiefly on the elbows and knees. (See PSORIASIS.)

LEPROSY is a malady of very great interest in many ways. Fortunately for us, however, not because it is common in this country—though apparently some cases have occurred. Most of the cases seen in this country have originated in patients who have at one time lived abroad. The technical name of the malady is *Elephantiasis Græcorum*, and it is of two kinds—one where the surface is marked with tubercles, and the other in which the surface is smooth, but in which there are ordinarily a number of spots in which there is no feeling. Leprosy does not begin at any particular age—sometimes children are its subjects, sometimes old people. Of the two forms, the tuberculated and the anæsthetic, the tuberculated seems to kill the sooner, for it is stated to last but from nine to ten years; the anæsthetic between eighteen and nineteen. In both varieties, but especially in the non-tuberculated, the morbid action seems to be sometimes stationary for years. Many lepers die from other diseases, as chronic diarrhœa, dysentery, diseases of the lungs, like bronchitis and pneumonia, or kidney disease. When the leprosy destroys life it does so by attacking the opening of the windpipe, or by deep ulcerations, laying bare some important vessel, and causing hemorrhage, or yet again by convulsions or coma.

Leprosy is generally considered hereditary, but sometimes it overleaps a generation as in other hereditary maladies. What the maladies are which prevail in the intermediate generation are not noted. As usual, the influence of the mother seems greater than that of the father in giving a hereditary taint;

and it seems more inclined to spread among those of the same family, than from parent to offspring. It does not seem, notwithstanding the loathsome sores it produces, to be able to spread by direct contagion. It prevails most among the lower classes of society, being greatly fostered by dirt, insufficient food, badly ventilated and damp dwellings, and especially by malarial districts. Generally the sufferers are most wretched in every way. The disease sometimes approaches very gradually, sometimes more rapidly. There appear, with or without fever, one or more pinkish or purplish-red spots, which may be isolated or in patches. These may go, and nothing be seen for a time, but again there is a feverish attack, and the red spots again appear more extensively than before. This may happen several times; but at length the spots begin to harden and to become prominent, so that they appear as hard semi-transparent tubercles. The skin at the same time becomes brawny and coarse-looking, while swelling of the tissues beneath makes it pit on pressure. Patches of brown appear here and there on the skin, and by and by some of its tissues waste, so that white patches appear as if a wound had formerly existed on the spot. The face, too, alters; the cheeks, lips, and ears become swollen and bloated, and there is a copious watery discharge from the nostrils; the eyes look watery, and little nodules form on the edges of their lids, which are turned outward. The hair changes in color to a dirty white, and often falls in considerable quantity. The membrane lining the mouth and nostrils swells and looks flabby, pimples often forming on it. The glands in various parts of the body swell, especially in the groins, and a peculiar greasy sweat comes from all parts of the body. As the tubercles enlarge and spread over the body, the mind becomes torpid, and the extremities are swollen and useless. This is the period of complications, which often carry off the sufferer.

The *non-tuberculated* variety commences with a few small patches on the hand or face. These are shining, wrinkled, and paler than the surrounding skin. On these not even a red-hot iron can be felt. Ulcers often form, and heal after a time; but the affection continues to spread. The mental faculties are dulled, the surface is cold, and the appetite voracious. By and by frightful ulcers form, without any pain, sloughs form and fall off, exposing the interiors of joints till bone after bone drops off, leaving behind only the stumps of the arms and legs. These by degrees become useless, the patient being only able to crawl. The temperature falls, and the whole surface exhales a loathsome smell, which is more troublesome to the spectator than the miserable patient.

Treatment.—Cases of recovery from leprosy are not unknown ; but they are rare. However, the first thing to be done to insure anything like a successful treatment is to remove the patient from an unhealthy to a healthy locality ; if a European abroad, to send him home. The diet, too, must be improved ; high-seasoned or long-preserved meat, especially salted provisions and fish, are to be carefully avoided, and a plain nutritious diet, containing a due supply both of fresh meat and fresh vegetables, must be insisted on. Personal cleanliness is of the first moment, and so baths must be used regularly and frequently. As for internal remedies, these have been used of almost every kind —none seem very decidedly to do good. Some have given aperients, some alteratives ; arsenic has been largely used, and is now employed by the Arabs, especially in the form of yellow sulphuret. Cupping-glasses all along the line of the spinal cord have been used and recommended, but their use is more than doubtful.

The latest treatment which has been reported on as moderately successful, or even more than moderately so, is one invented by Dr. Beauperthuy, and by him applied first in the West Indies, and after that in Guiana. One imporant part of it, if not the most important, is the application of the stimulant cashew-nut oil to the tuberculated parts on the lower extremities. At the same time the diet was attended to, and made more than usually nutritious. The results are reported as good. Unfortunately Beauperthuy died before he had time to fully test the value of his supposed discovery.

LETTUCE, the *Lactuca sativa*, is well known as a spring and summer salad, and is very wholesome and good in diet. It is remarkable, however, for yielding, when fully ripe, a quantity of sticky, milky juice, which, on exposure to the air, becomes of the consistency of cobbler's wax and possesses a narcotic principle resembling opium in its effects, and known as lactucarium or lettuce opium. Dr. Duncan, of Edinburgh, first discovered this drug, and for some time it was constantly used in practice as being less exciting than opium. It is now, however, seldom employed, being superseded by morphine.

LEUCORRHŒA, commonly known as "the whites," is a disorder frequently met with in women. It is met with either as a thin, watery discharge, and is then merely an increased flow of the ordinary secretion of the vagina, or as a thick, yellow discharge which generally comes from the womb. This state is accompanied by debility, pain in the back and loins, pain and difficulty in passing water, and anæmia, or pallor of the skin. It often occurs during pregnancy, more often when the mother is weakened from having had a large family, and from over-suckling. Any disease

which has a debilitating effect on the system may cause this complaint, so that it is not uncommon after a fever or a protracted illness. There are also local as well as constitutional causes ; many diseases of the womb, growths in the vagina, stone in, or disease of the bladder, and many affections of the rectum, will cause this symptom.

Treatment.—The treatment must consist in keeping up the general health, and in removing any irritating cause ; the object in the first case may be maintained by giving tonics, of which iron and quinine are the most valuable ; moderate exercise, fresh air, and a generous wholesome diet are also required. The removal of any irritating cause will in each case depend on the nature of the mischief, and for each a special course may have to be adopted. Astringent lotions should be injected two or three times a day with a syringe ; lotions containing tannin or alum or sugar of lead are the best. The parts affected may also be freely bathed with cold water night and morning.

LICE. (See ECTOZOA.)

LICHEN is a skin eruption consisting of a number of small pimples or papules arranged together in clusters, and occurring anywhere on the surface of the body ; the frequent rubbing in of zinc ointment is the best thing to do. but the rash is often very troublesome to heal. The patient should not scratch the heads off, or he may produce an irritable sore which may become eczematous ; when the rash is due to syphilis, the treatment will consist in taking iodide of potassium in conjunction with some tonic infusion, as gentian or sarsaparilla. The prickly heat of tropical countries is really an aggravated form of lichen. This disease is not contagious ; it is more common in women than in men, and in those of a nervous and excitable temperament. It is more common in spring and summer than in autumn or winter.

LIEBIG'S EXTRACT. (See MEAT, EXTRACT OF.)

LIFE, CHANGE OF. (See MENSTRUATION.)

LIGATURE is a cord or thread employed in tying a blood-vessel or tumor, consisting either of milk or strong hempen twine, cut gut, horse-hair, or other substance. In applying a ligature to a bleeding artery, its orifice is first laid hold of, and pulled out from the surrounding tissues, either with a pair of forceps or a tenaculum, so that a loop formed by the ligature may embrace it and it alone. The length of the ligature for ease of application should be about a couple of feet. The knot used for an artery is what is known as a reef knot, and, in bringing the bight tight, care must be taken to run the cord over the tips of the forefingers, so that all pressure is made at the bottom of the wound, and that the vessel be not "tugged" at, to the imminent danger both of tearing it through and of pulling the ligature away from the vessel. One end of the ligature, in the case of tying an artery, should be cut off, and the other left hanging out of the wound, so that when the sloughing of the arterial coat has taken place, it may be readily withdrawn. The time that ligatures require to separate varies from a week to three weeks or a month.

LIGHT is as important to health almost as air. Its want is noticeable in those who work in dark workshops and underground kitchens. These have complexions as devoid of color as a piece of blanched celery or asparagus, and for exactly the same reason, both have been deprived of light. In the vegetable kingdom light is absolutely necessary to convert the white shoots into green leaves and branches. So, too, in the human being the circulating blood requires an exposure to light to give it its true vivifying qualities. Such individuals, therefore, who have not a due exposure to light are what is technically called anæmic, and though in some the health seems tolerably good, in others there is a tendency to passive dropsies, owing to what is called thinness of blood, and a want of breath when called upon to undergo exertion. The exact nature of the constitutional change is not known, but probably is connected with some change in the intimate structure of the red corpuscles or the chemistry of their coloring matter, hemoglobin.

LIGHTNING STROKE, or a flash of lightning, will generally strike the most prominent object near it, and if this chance to be a conducting body, it is carried off to the earth and may do no harm ; in other cases it may strike the chimney of a house and do serious harm to the walls and inmates, and especially if there is a conductor, the supports of which are made of metal and inserted between the stones or bricks of which the wall is made. *Lightning* is the dazzling light emitted by the electric spark when it shoots from clouds charged with electricity ; sometimes the flash is zigzag, and moves with great velocity and sharp outline ; sometimes the flashes, instead of being linear, fill the whole horizon without having any distinct shape. There is also the so-called heat lightning, which illumines the summer nights, without the presence of any clouds above the horizon, and without producing any noise. The lightning discharge is the electric discharge which strikes between a thunder-cloud and the ground. The latter, by the induction from the electricity of the cloud, becomes charged with contrary electricity, and when the tendency of the two electricities to combine exceeds the resistance of the air, the spark passes, which is often spoken of as a thunderbolt having fallen. The discharge generally falls on the nearest and best con-

ducting objects, and, in fact, trees, elevated buildings, and metals are more particularly struck by the discharge. Hence it is imprudent to stand under trees in stormy weather, especially if they are good conductors, such as oaks and elms. The lightning discharge kills men and animals, inflames combustible matters, melts metals, and breaks bad conductors in pieces. After the passage of lightning, a very singular odor is often produced. This odor is attributed to the formation of ozone, a peculiar modification of oxygen, first discovered by Schönbein in 1840. The return shock is a violent and sometimes fatal shock, which men and animals experience even at a distance from the place where the lightning discharge has passed. It is caused by the inductive action which the thunder-cloud exerts on bodies placed within the sphere of its activity. These bodies are then, like the ground, charged with the opposite electricity to that of the cloud ; but when the latter is discharged by the recombination of its electricity with that of the ground, the induction ceases, and the bodies reverting rapidly from the electrical to the neutral state, the concussion in question is produced. A lightning conductor consists of a rod and a conductor ; the rod is a pointed bar of iron, fixed vertically to the roof of the building to be protected ; it is from six to ten feet high, and its basal section is about two or three inches in diameter ; the conductor is a bar of iron or copper which descends from the bottom of the rod to the ground, which it penetrates to some distance. Strands of iron or copper wire may be used instead of a rod. A conductor, to be efficient, must satisfy the following conditions : 1st, the rod ought to be so large as not to be melted if the discharge passes ; 2d, it ought to end in a point, to give readier issue to the electricity disengaged from the ground ; 3d, the conductor must be continued from the point to the ground, and the connection between the rod and the ground must be as intimate as possible ; 4th, if the building which is provided with a lightning conductor contains metallic surfaces of any extent, these ought to be connected with the conductor, or else lateral discharges may take place between the conductor and the edifice, and the danger may be increased. Death by lightning is instantaneous, and leaves nothing to be done ; there is generally a mark as of being burned, and articles like a watch or coins may be partially destroyed.

LIME, the oxide of the metal calcium, is used for a variety of purposes. Quicklime is prepared by burning chalk or limestone, and so driving off the carbonic acid, quicklime is left behind, retaining something of the shape of the original blocks. If, however, water be added, the whole mass heats and breaks down into a fine powder, and this, which is slaked lime, dissolved in water to saturation constitutes *lime-water*. If sugar be added to the water the lime is taken up much more freely, and a stronger solution--saccharated solution of lime—is produced. Lime given as lime-water acts as a powerful antacid both on the alimentary canal and after absorption. It is also astringent and tends to diminish secretions, and so is very useful in many forms of diarrhœa. Lime-water is also used with advantage to check abundant discharge in the skin disease called eczema. In some forms of vomiting, especially in children, lime-water is a most valuable remedy. It is, perhaps, of most use in chronic vomiting. It is best given mixed with milk. It prevents cow's milk from coagulating, and so obviates fertile risks of stomach-ache and diarrhœa. If the bowels be constipated bicarbonate of soda should be used instead.

Carbonate of lime as precipitated from a solution of a lime salt, or as prepared thus as well-washed chalk, is a useful remedy. Two preparations are available : chalk mixture and aromatic powder, or aromatic powder and opium. Chalk is an antacid and astringent. It is mainly used in diarrhœa, seldom alone, but generally with other remedies astringent and aromatic. If given too long it is apt to cause concretions in the bowels which may be troublesome. Chalk is also often used as a dusting powder, when sores form, as in the creases of fat children. Sometimes these are better dealt with by using greasy preparations, but in a goodly number chalk does well.

Chloride of calcium is a remedy which is very variously estimated by different observers. It is introduced into the pharmacopœia mainly as a drying agent, and as concerned in the manufacture of chloroform and ether. Some, however, esteem it highly because it is supposed to exercise a special influence on the glandular system in scrofula.

Chloride of lime. (See CHLORINE.)

Phosphate of lime, or bone earth, is a very important salt in the animal economy, constituting the main basis of our bones. It is obtained from bones by a slight process of purification. This salt is necessary not only to the growth of bone, but also to the growth of other tissues. It is most useful in the anæmia of young and rapidly-growing individuals, and women weakened by frequent child-bearing. It is also a very good thing to give to women while suckling, if they have been previously weakened in the same way. In rickets it is a most valuable remedy, giving hardness to the softened bones. The time best adapted for its use is just after the acute stage of the malady when the pain and tenderness of the bones has ceased. Not too much should be given for a dose, as given in excess it hinders digestion. Most of it passes into the intestines, where, if much be given, it is apt to form concretions. In various

forms of chronic diarrhœa, especially in children, this remedy is of value. It may be given along with iron in the form of syrup. Most of the phosphates necessary for the welfare of the body are taken in the food.

LINIMENTS, from the Latin, *lino*, I smear, literally mean those remedies which are smeared on the skin and left there. Nowadays we commonly include in the term those also which have to be rubbed in—really embrocations, and the common idea of a liniment is something to be rubbed in. Liniments are made use of for all sorts of purposes, and many are included in tne pharmacopœia.

LINSEED, as commonly used, consists of the pounded seeds of the common flax plant. These seeds contain a valuable oil obtained by expression, the linseed-oil, which is also officinal. Not infrequently the substance which remains after this oil has been expressed is ground down and the powder made use of as linseed for poultices. It is not, however, so good as the pounded seed, being deprived of its oil. Linseed is mainly employed for poultices, and it furnishes one of the best materials for these (see POULTICES), and in this form is commonly applied to open and suppurating sores. Internally linseed is given as *linseed tea* or infusion of linseed. This is an old-fashioned and useful remedy, being employed when there is irritation about the bladder or urethra. It has also been employed with less benefit in diarrhœa and dysentery.

LEPOMA. (See NOSE, DISEASES OF.)

LIPS, WOUNDS OF.—The most common cause of wounds of these structures is a blow or fall. The lip is driven backward against the teeth, and laceration of its soft tissues produced. These injuries may vary, from a slight wound on the free margin or posterior mucous surface of the lip to a large cleft, involving its whole thickness, so as to expose the teeth and corresponding portion of the jaw. The lips are sometimes bitten by sudden closing of the lower jaw, and when any of the incisor teeth are sharp or broken at their free ends the wounds may be deep and serious. Wounds of the lips generally bleed profusely, as their structures are traversed by large blood-vessels, and the internal lining of red mucous membrane is very vascular. As there is usually much gaping, even in small wounds of the lip, it is almost always necessary to apply sutures. When the whole thickness of the lip has been involved, the separated parts should be brought together by sutures twisted round long needles, as is done by most surgeons in operating for the relief of hare-lip. Care must be taken to bring the corresponding portions of the wounded lip into contact, and to preserve the line of red margin. For the simple stitching together of the margins of superficial labial wounds, thin silk or thin silver wire are the best materials to use.

Cracked Lip.—The most common form of this troublesome affection is a superficial crack at about the middle of the red portion of the lower lip. This is raw and painful, bleeds readily when touched, and is generally associated with a sensation of dryness in the whole lip. It is met with in weak and unhealthy individuals during the winter months. The upper lip may be affected, but cracks are much more common on the lower one. Deeper and larger cracks are occasionally observed on the lips of scrofulous children. The best and simplest treatment consists in applying blue-stone or sulphate of copper to the base of the ulceration, and then to keep the sore and surrounding portion of lip moist with glycerine, and to protect these parts from cold and external irritation by means of cotton-wool.

Ulcers.—Non-cancerous ulceration of the inner surface of the lips is generally due to an extension of some similar disease affecting all parts of the lining mucous membrane of the cheeks and gums and the surface of the tongue. Ulcers on the lips are in most cases superficial grayish patches surrounded by a zone of inflamed and swollen mucous membrane. The following are the most frequent causes of oral and labial ulceration : dentition ; rough and broken teeth ; action of mercury ; disordered stomach ; venereal disease. The white spots called aphthæ, which are so frequently observed on the tongues of unhealthy and badly fed infants, may also attack the lips. In the management of cases of ulceration of the mouth and lips, the medical man directs his attention, in the first place, to the constitutional origin of the disease. The best local treatment in most cases consists in the application of blue-stone to the surfaces of the small ulcers, and in the patient's sucking frequently during the day small pieces of alum. Borax and honey is a well known and very useful application. (See APHTHÆ ; CANCRUM ORIS ; CHAPS ; HARELIP ; SALIVATION.)

LIQUEURS. (See DISTILLED SPIRITS.)

LIQUOR SANGUINIS is the opaque straw-colored fluid or plasma in which the blood-corpuscles float when in the living body. (See BLOOD.)

LITHARGE. (See LEAD.)

LITHATES or URATES, form the red or pink deposit which settles from the urine on cooling ; it is often found in cases of dyspepsia, or when too little water is passed, or when the urine is very acid. (See URINE.)

LITHIA is an alkali closely allied in its properties to potass. It has recently been brought into practice as a remedy in gout, it being supposed to favor the passage of uric acid from the system more than potass. Its preparations are the carbonate and the citrate.

It is somewhat expensive, and is not much better than potass.

LITHOLYSIS means an attempt at solution of the stone within the bladder, which may be endeavored in two ways—by medicines given by the mouth, and by injections into the bladder of the former class of remedies. The alkalies are the most useful, especially the carbonates of soda and potash, given in copiously diluted doses. Of the natural waters, Vichy appears to be the best. The oxalate of lime calculus, however, resists its influence. The uric formations, however, are benefited in two ways by their administration ; alkalies thus given tend to correct the *diathesis* whereby the calculus has arisen, and at the same time they have an undoubtedly sedative and corrective effect on the urinary organs. These remedies should be continuously given, and in small doses copiously diluted. In the case of the phosphates, they seem to have the effect of gradually disintegrating the stone by solution of the animal matter whereby its particles cohere. Solvent injections into the bladder have been in use since 1792, both acids and alkalies ; acid injections appear to be not without their efficacy, especially in phosphatic stones. Carbonate of lithia has been proposed for uric concretions, and the salts of lead in phosphatic.

LITHOTOMY, the operation of the removal of a calculus, or stone, from the bladder, has created the greatest interest from the earliest ages ; but its history, though one of the most interesting in the annals of surgery, would be manifestly out of place in a work of this character. The instruments required for the operation of lithotomy in our own time are remarkably few and simple—a knife, a grooved staff, forceps or a scoop. The patient having been previously prepared, and placed under the influence of chloroform, the method at present in use—usually the lateral operation—is by a skilful operator a matter of a few seconds. After the stone has been extracted, a morphia suppository may be advantageously administered, and the patient should lie on his back with the shoulders elevated, a napkin applied to the perinæum to soak up the urine, and the bed protected with mackintosh drawsheets. Pain is to be allayed by opium, the bowels kept open with castor-oil, the wound perfectly clean, nourishing diet, and the wound made by the surgeon, generally speaking, heals entirely in four or five weeks, the urine commencing to flow by the urethra in about a week. The cases in which lithotomy is preferable to lithotrity are in case of children, or when the urethra is strictured, irritability of the bladder, great enlargement of the prostate, and great size of stone to be removed.

LITHOTRITY is the proceeding whereby a stone in the bladder is crushed or broken by means of some instrument introduced through the urethra, so that its *debris* may either be extracted by the said instrument, or may pass with the urine in the act of micturition. As may be surmised, lithotrity is a.an operation requiring, in the first place, great perfection in diagnosis, so as to determine when this operation is to be preferred to that of lithotomy or cutting ; secondly, it requires immense practice, and a peculiar attention to detail. The operator must have originally great nicety of touch, so as to discriminate the tissue between the jaws, and to this gift must be added great experience. Each operation has its advantages. The advantages of lithotrity (which, of course, are determined by the surgeon), or rather the cases in which lithotrity should be resorted to instead of lithotomy, are, when the patient is an adult with a full-sized urethra, when the prostate is not enlarged, when the stone is small and friable, when there is good general health, and the bladder healthy and free from irritability, and cases where the stone is single, not large, or very hard. It must be borne in mind that the greatest benefits are likely to be derived from this operation at the earliest possible period after the descent of the stone from the kidney into the bladder.

LITMUS or LACMUS is a peculiar blue coloring matter extracted from a variety of lichens. It is prepared extensively in Holland. As it is readily affected by acids and alkalies—turned red by acids and blue by alkalies—it is used as a test of acidity or alkalinity. It is used for no other purpose. *Litmus paper* is made by spreading the substance on a piece of ordinary paper, and afterward cutting it up into slips.

LIVER (THE) is the largest gland in the body, and weighs generally between fifty and sixty ounces,being greater on an average in the male than in the female. It is seated on the right side of the abdomen, just below the diaphragm, in the right hypochondriac region. (See ABDOMEN.) It stretches across to the left also, crossing the epigastric region, and reaching as far as the spleen in the left hypochondriac region. It commences as high as the fifth rib in the line of the nipple on the right side, while its lower border comes down as low as the lower margin of the rib. Its length, however, varies a good deal even in health, and in women who wear tight stays it may come down an inch or two lower. In front of the liver, in the erect position, is the abdominal wall ; behind is the right kidney ; while above there is the arch of the diaphragm. The organ is covered all over with a thin,smooth, serous membrane,called the peritoneum, by which it is held in its place. The liver can move up and down slightly with each movement of respiration, for during inspiration, when the diaphragm descends, the liver is pushed down also, returning to its old position when the diaphragm ascends during expiration. It is of firm consistence and of a dark red color, smooth and convex in front ; the hinder surface is flattened and irregular ; the upper border is thick and rounded, while the lower edge is thin, and can be felt in thin people. The liver is divided into two main lobes or divisions, called the right and left lobes ; the division is marked by a deep notch in the lower border ; looked at from behind,the lobed arrangement is very evident, but from a front aspect the two lobes seem, and are, continuous. The right lobe is by far the larger of the two, while the smaller, left lobe, occupies chiefly the epigastric region ; its lower border, too, does not come down so low as that of the larger lobe. This gland contains various vessels, and also a secreting structure formed of myriads of cells. The portal vein brings the blood to the liver, and then breaking up into numbers of small branches the blood-stream is carried through very fine tubes known as the hepatic capillaries ; here the current is brought into the closest proximity to the liver cells, and they take from the blood the elements necessary to form bile and glycogen. These hepatic capillaries next join together again and finally form the hepatic vein, which carries the blood into the inferior vena cava, and so on to the right side of the heart. The portal vein collects all the venous blood just after it has passed through the stomach, spleen, and large and small intestines, so that this blood, highly charged with nutrient material from absorption of the elements of food in the alimentary canal, is at once carried on to the liver. The hepatic artery is a branch from the aorta and sends several branches to the liver, and probably is the nutrient vessel of that gland. The hepatic duct is a tube which conveys the excreted products of the liver away into the alimentary canal, and enters that channel in the duodenum. It has been said above that hepatic capillaries form a very close network of thin-walled vessels, and among the meshes lie the liver-cells ; these communicate readily with the commencing branches of the hepatic duct, which, uniting together, finally form branches of considerable size ; along the ducts the bile flows in health and runs down into the intestine. Thus there are two currents, each in an opposite direction ; the blood is brought from the intestines to the liver ; the bile, then formed, flows from the liver down to the intestines. In addition the liver is supplied with lymphatics and nerves, which play an important part with regard to the functions of the organ ; branches of the vagus or pneumogastric are the chief nerves which supply the liver. Opening out from the main hepatic duct is a dilated reservoir, called the gallbladder ; it forms a large oval sac, and varies in size according to the amount of its contents.

The liver removes from the blood certain substances, and also supplies that fluid with new compounds ; it is thus a source of loss as well as a source of gain. The blood loses, because bile is being constantly formed, and passing down the hepatic duct, and experiment shows that there is a difference in the blood which enters the liver and in that which leaves it, in consequence of this loss. The total quantity of bile secreted in twenty-four hours varies, according to different authors, the probable average being between two and three pounds. It is a greenish-yellow fluid, slightly alkaline, and of an extremely bitter taste. It consists chiefly of water, holding in solution from 10 to 17 per cent of solid substance. The solid matter consists chiefly of *bilin*, a resinous substance composed of carbon, hydrogen, oxygen, nitrogen, and sulphur, in combination with soda. Chemists have separated this *bilin* or biliary matter, into two acids, the *taurocholic* and the *glycocholic*, each of which exists combined with soda. In addition to this constituent, the bile contains also a substance called cholesterine, which is very soluble in alcohol, and crystallizes out in thin quadrangular plates ; it forms a great part of what are known as gall-stones. Of these bodies, the water, saline matter, and cholesterine have been found in the blood, and probably the liver-cells simply abstract them from the stream as it flows along, and certainly the blood in the hepatic vein is poorer in water than that in the portal vein. Bilin has not yet been discovered in the blood, and thus this substance must be formed in the liver itself. But the liver is also a source of gain to the blood. If the blood in the hepatic vein be examined it will be found to contain a large quantity of glucose, a kind of sugar formed in the liver, while the blood in the portal vein, or hepatic artery, contains a very much smaller quantity, and sometimes none at all. Experiment has shown that an amyloid substance called *glycogen* is formed in the liver ; this substance is made up of carbon, hydrogen, and oxygen, and much resembles starch, dextrine, and gum in chemical composition. Further, this glycogen, like starch, can be acted upon by ferments, and so is converted into hepaose or liver-sugar. This ferment exists, under ordinary circumstances in the liver. So that it would appear that the liver forms glycogen from the blood with which is is supplied, and also a ferment which, at the ordinary temperature of the body, will convert the slightly soluble glycogen into very soluble sugar, and this is carried away by the hepatic vein on to the vena cava, and so through the right side of the heart into the lungs.

The bile, as has been said, enters the duodenum, or that portion of the small intestine next to the stomach, and here it mixes with the chyme, or partially digested food ; the

bile neutralizes any free acid, and, perhaps, aids in digesting any fatty matter ; it also increases the peristaltic action of the bowels, and thus acts as a purgative. If from any cause the flow of bile down the hepatic duct is obstructed, the bowels often become constipated and the motions are of a pale clay color. At the same time the liver becomes rather larger than usual, because the ducts are full of the retained bile, and the patient feels pain over that region, which is worse on pressure. Since the bile cannot flow in its usual course, some of it is absorbed by the blood, and thus some kinds of jaundice are produced. Then the urine becomes much darker in color and has a dark olive-green hue, the conjunctivæ also are tinged yellow, and in a short time the whole skin assumes a yellowish tint. Exposure to cold, indigestible food, and sedentary employment may set up an inflamed condition of the bile-ducts and so cause temporary obstruction, but this form is very amenable to treatment ; any great emotional disturbance or fright also causes jaundice. Sometimes the jaundice is more permanent, as when a growth or tumor presses on the duct, and such cases may prove fatal. Or a gall-stone may become lodged in the canal and cause intense pain in the right hypochondriac region, with sickness and faintness ; these symptoms are relieved when the stone has passed into the intestine. (See JAUNDICE and OBSTRUCTIONS.) Excessive drinking and the immoderate use of ardent spirits cause congestion of the liver and finally produce profound changes in that organ, as in cases of cirrhosis and fatty liver ; inflammation of the liver, ending in abscesses, is uncommon in this country, but it is frequent in tropical climates ; abscess of the liver in this country is generally associated with pyæmia, or ulceration of the bowels.

The liver, like the kidney and other organs, is liable to various acute and chronic diseases. Among the *acute* changes may be classed catarrh, or inflammation of the bile-ducts, acute atrophy of the liver, congestion and inflammation of the liver, and the presence of gall stones in the hepatic duct. *Catarrh* of the bile-ducts has been briefly mentioned above ; there is, in addition to the jaundice, a loss of appetite, a coated tongue, slight sickness, and a feeling of retching ; the motions are pale, the urine dark, the skin and eyes become yellow, and there may be, in some cases, a troublesome itching of the skin. Pain is not a very troublesome symptom, and it is generally felt in the right shoulder-blade and along the lower edge of the liver, being often worse on pressure. The best *treatment* is to open the bowels freely by means of purgative medicines ; a dose of calomel at bedtime with a rhubarb draught twice a day will generally suffice. The diet must be very light, and capable of being easily digested ;

all rich food should be avoided, while milk, broth, beef-tea, toast, and biscuits, or a light pudding may be taken ; no stimulants need be given, as they would only tend to increase the congestion of the liver. Effervescing solutions may be given with benefit, as they allay thirst and sickness ; those containing soda salts are the best, and those also which have an aperient action ; for this reason effervescing Carlsbad waters often prove beneficial. In three or four days a mixture containing extract of dandelion, hydrochloric acid, and gentian, may be given three times a day, and the bowels must be kept open daily ; active exercise should be taken daily, if the patient can bear it, and for some time care must be taken to avoid indigestible food. This disease is not a dangerous one, and with early and proper treatment is easily cured.

Acute atrophy of the liver is a very formidable disease, and fortunately is of rare occurrence. The patient becomes hot and feverish, vomits often, and the skin assumes a deep yellow tint ; the liver rapidly shrinks inside so as sometimes to lose half its weight. Headache comes on quickly, followed by delirium and insensibility ; the patient lies in a prostrate condition, and there is picking of the bed-clothes and low muttering delirium ; bleeding may take place from the nose or mouth, and small hemorrhagic spots may be seen in the skin. Death generally occurs in four or five days, and treatment is not of much avail.

A ' sluggish" or congested liver is generally associated with catarrh of the bileducts, and arises often from want of exercise and eating or drinking too much ; but congestion may go on to inflammation in tropical countries and end in the formation of an abscess. This may be known by the pain over the region of the liver, the swelling of the abdominal wall on that spot and the frequent shiverings ; the patient loses flesh, strength, and appetite, and his skin becomes of a sallow tint ; such people generally come back to this country invalided, and if they get over the illness they seldom recover their former state of health.

A *gall-stone* in the hepatic duct will cause great pain over the liver, chiefly referred to one spot, much sickness and distress, and a feeling of faintness ; a hot bath and the administration of chloroform will ease the pain, while purgative medicines may be taken, and all means used to get the stone to pass onward to the bowel. Jaundice will come on from the obstruction to the flow of bile, but this will disappear when the stone has escaped. Sometimes the stone will remain in the canal for weeks and become imbedded there, but generally there is a passage left by the side for the escape of bile. In some cases of hysteria very analagous symptoms

are met with, but in such cases there is no jaundice ; the disorder occurs in nervous young women, and there is a dark areola under the eyes.

Among *chronic* changes may be enumerated cancer, cirrhosis, fatty and waxy degeneration, passive congestion, syphilitic deposits, and the presence of hydatid cysts.

Cancer of the liver is a most fatal and serious disorder, carrying the patient off within a year or a year and a half from the first appearance of any symptoms. There is at first loss of appetite and pain over the abdomen ; the latter begins to swell as the cancer increases in size, and becomes extremely tender ; rapid emaciation goes on, but the temperature is generally no higher than usual and there is no attendant fever. The loss of flesh, the hollow temples, the great prostration, the pain and swelling or enlargement of the liver are the chief symptoms, and these gradually become worse and finally cause a lingering and painful death. Jaundice is not often present, nor does the patient suffer from shivering. Cancer of the liver may occur in both sexes and be met with at any period of life ; more frequently, perhaps, between thirty and fifty years of age. This terrible disease is not often confined to the liver, but may attack the stomach or parts adjacent, and so add to the distress ; sometimes the peritoneum is also implicated, and then there is more or less dropsy ; œdema or dropsy of the feet is also of common occurrence. The *treatment* must be directed to the relief of the patient, as no cure can be looked for. The pain may be alleviated by the administration of opium or morphia, and this may be given internally as a draught, or a small quantity may be injected under the skin with a syringe. Chloral is of much use in easing the pain. The diet must be light and nourishing, and must be varied from day to day to please the fancy of the patient, whose appetite will be but small and capricious.

Cirrhosis of the liver comes on more generally in middle life ; at first it may be mistaken for cancer, as there is loss of flesh and appetite and pain in the abdomen, but the symptoms come on more gradually, the liver does not increase in size, but rather shrinks, and dropsy of the abdomen soon comes on ; jaundice also is very common, and the distended abdomen becomes marbled over with blue veins as the stream of blood through them is impeded. (See CIRRHOSIS.)

Fatty degeneration of the liver is common in many disorders, and there is hardly any affection of the liver in which more fat than usual is not found ; to a slight extent it occurs in health, and especially after a meal. The liver may be very fatty and give rise to no symptoms, as in cases of consumption, and in those who drink a great deal of beer. Fatty livers may attain a great age and become double the ordinary weight ; they occur often in scrofulous people. Not only is the liver larger than usual, but the abdomen may swell from the presence of dropsy, which is very common in these cases ; often, also, there is a similar affection of the kidneys, so that albumen is present in the urine, and there is less of that fluid passed than usual ; frequently, also, there is dropsy of the legs. The patients do not lose much flesh, at any rate at first, nor is the appetite much impaired ; there is no pain, or if present, it is very slight, seldom any jaundice or shivering ; the symptoms come on very gradually, and the liver is generally much diseased before any notice is taken of the mischief ; the disease is often very chronic and will last for years, unless there be much mischief in other organs ; dropsy is a bad symptom, and, when general, will frequently point to disease in the kidneys. Constant rest in bed is not required, unless the patient be too weak to go about ; attention must be given to the diet, and any indigestible food should be avoided. Stimulants need not be given, but a pint of beer a day, or a glass or two of sherry or claret, will do no harm. If dropsy be present purgatives must be given, so as to remove the fluid, and the general health must be kept up by tonic medicines, as iron and quinine.

Waxy degeneration of the liver is a less frequent disease ; it rarely, if ever, occurs alone, and is generally associated with similar disease in the kidneys, spleen, and intestines. It occurs in persons who have long suffered from diseased joints and chronic abscesses, in the scrofulous, and in those who have suffered from syphilis or ague, and some other wasting disorders. As in the case of the fatty liver, there is seldom pain or jaundice, or loss of flesh ; in each the appetite is good, or but slightly impaired, and in each the mischief may go on for a long time and cause no symptoms. In waxy change, however, dropsy seldom occurs, diarrhœa is often present, the spleen on the left side enlarges, and the patient passes a large quantity of pale, limpid water, in which is contained a good deal of albumen. The liver, also, attains larger dimensions than in the case of a fatty change, and its lower border comes lower down, and can usually be easily felt. The *treatment* will consist in improving the general health by liberal diet, and by the administration of tonics ; and attention must be directed to any other disease on which the waxy change may depend.

Passive congestion of the liver often occurs in heart disease, and some disorders of the lungs, and depends upon the fact that since the course of the circulation is disturbed at those points, the veins become too full all over the body ; now the hepatic vein shares in this fulness, and so the liver is stuffed with blood and the stream flows through sluggishly. From a similar cause the veins in the leg and kidney are full, and so there results dropsy of the lower extremities and a scanty flow of urine, which will contain a variable amount of albumen. There will be pain over the liver, but not of marked intensity, and frequently there is some yellowness of the skin from the presence of jaundice : after a time dropsy of the abdominal cavity will come on, and then a fatal result often follows. Since this state of liver depends upon the disease of the heart or lungs, the *treatment* must be directed to allaying any tumultuous or irregular action of the heart and removing any dropsy by purgatives or small punctures in the leg ; then by diminishing the quantity of fluid in the circulation relief may be temporarily given.

Syphilis will produce various changes in the liver, and cause a hardening of that organ and thickening of the capsule ; sometimes rounded masses, something resembling cancer, are met with in that organ ; the health in such cases must be improved by a visit to the sea-side, if possible, or a sea-voyage, by liberal diet and regularity of living ; preparations containing iron and quinine are valuable, and may be given in conjunction with iodide of potassium. In such cases the patient is generally of sallow complexion, feels low and nervous, and is in a feeble state of health ; to improve, therefore, the general condition of the constitution is the chief indication.

Hydatid cysts occur more commonly in the liver than in any other organ, although they are by no means very often met with. They may occur in the liver either as small, round and firm tumors, formed of a fibrous capsule, with putty-like contents ; these are hydatid cysts which have undergone spontaneous cure, and can do no more harm ; or as cysts with a tough, fibrous capsule, inclosing much fluid, and a greater or less number of smaller cysts floating about. The fluid is limpid, clear, of low specific gravity, and is, in fact, chiefly made of water holding common salt in solution. These cysts may attain a great size, from a walnut to a child's head ; they form a rounded, abdominal tumor, firm yet elastic, and giving a peculiar thrill when tapped. They are seldom attended with pain, unless there is inflammation outside setting up adhesions ; the general health is seldom affected, so that the nature of the disease is chiefly recognized by the presence of a tumor in the liver, and the absence of any constitutional symptoms. The *treatment* will consist in having resort to surgical aid, whereby the contents may be evacuated and the cyst allowed to shrink ; in most cases this is very successful treatment, and, with certain precautions, it is not difficult to perform. If allowed to grow, such cysts may cause death by bursting into the abdominal cavity, or into some neighboring organ. (See HYDATIDS.) *For an excellent liver pill*, see COLOCYNTH.

LOBELIA INFLATA is a plant growing in North America, where it has long been in use among the native tribes. The whole herb is employed in medicine. It has a peculiar odor and a burning taste, not observed until after the medicine has been chewed for a time. Two preparations are in use, a tincture and an ethereal tincture. In small doses it is expectorant and diaphoretic ; in larger, emetic and cathartic. In still larger doses it causes death. This has not unfrequently followed its use by a medical sect appropriately called Coffinites. It closely resembles tobacco in its action. It has been chiefly used for asthma and other diseases of the respiratory passages accompanied by spasm. It is sometimes smoked. If it is to do any good, lobelia must be given in large doses, and very carefully watched. Sickness and vomiting are often so produced, put pass away. It is also useful in whooping-cough.

LOCK-JAW. (See TETANUS.)

LOCOMOTOR ATAXY. (See PROGRESSIVE LOCOMOTOR ATAXY.)

LOINS. (See LUMBAR ABSCESS.)

LONG-SIGHTEDNESS. (See EYE and VISION.)

LOTIONS are medical preparations used as outward applications for bruises, burns, or hurts of any kind, for allaying local inflammation, or for stimulating indolent sores or ulcers. They are of various kinds, such as astringent, sedative, stimulating, evaporating, or refrigerant, according to the effect they are intended to produce. *Arnica lotion* (made by mixing one part arnica with five to eight parts of water) is an example.

LOUSE. (See ECTOZOA.)

LOW FEVER. (See TYPHOID FEVER.)

LUCIFER-MATCH-MAKER'S DISEASE. — Lucifer-match-makers are frequently affected with necrosis of the jaw-bones, especially the lower, owing to the action of the fumes of the phosphorus used in their trade. Thus " Phosphorus disease" was not known to have any existence until the extensive use of the modern style of match so much prevailed, and there cannot be the slightest doubt that it is due to the introduction of phosphorus in some form, and that is " applied to the periosteum, or what is equivalent, some raw surface in immediate connection with the nutrition of the bone, and that its application must be prolonged, and be under peculiar circumstances of temperature, and probably of oxidization." If the pulp of a carious tooth is exposed to the influence of the poison, the resulting necrosis is that of the jaw-bone. It is a matter of speculation in what manner the phosphorus-oxide may be absorbed, but the fact of phosphorus itself entering so largely as

it does into the formation of the skeleton is a suggestive circumstance ; and perhaps if it be accumulated by the periosteum, it may generate upon the bone's surface a condition of chemical superphosphate inconsistent with osteal vitality. Efficient sanitary measures should be adopted to prevent the disease,and it has been suggested that " there should be a periodic and rigid scrutiny of the mouths of all those employed. Those whose teeth are bad should be excluded from the rooms where the obnoxious fumes are being developed (the *dipping* and *drying* rooms). All carious teeth should be extracted or stopped, and a simple and effective respirator, having its centre composed of a porous diaphragm, such as a sponge or some woven fabric, linen or cotton, which should be daily dipped in a solution of one of the fixed alkalies or of their carbonates, should be worn over the mouth by those employed, or the respirator devised by Mr. Graham for persons exposed to carbonic acid vapor,would be probably as efficacious. It consists of a mixture of fresh-slaked lime and sulphate of soda, through a cushion of which it is easy to breathe. The acid vapor might be neutralized or rendered innocuous by keeping the atmosphere of the apartment ammonuretted." (S. J. A. Salter.) The *symptoms* of this disease do not differ in any essential particular from ordinary necrosis not produced by phosphorus. They usually commence with a feeling of toothache, and the pain is referred to a decayed tooth, by which channel the poison enters. The disease takes a slow course at first, the gums become red and sore, and there is general pain and extreme tenderness. The mucous membrane of the cheek becomes involved. The teeth become loose, appear elongated, and cause intense pain when brought against those of the opposing jaw. After a great deal of suffering, matter-forms, and points either internally about the fangs of the teeth, or externally on the outside of the jaw. The matter is peculiarly fetid, and a probe introduced into the opening made either naturally or artificially, is long, tortuous, and burrowing, leading to portions of dead bone, or sequestra. It is worthy of remark here, that in necrosis of the *lower* jaw-bone, whether from phosphorus or any other cause, we have a wonderful exhibition of the *vis medicatrix naturæ*, in fact that there is an immense amount of repair, which does not exist in the event of a similar affection of the upper jaw. In milder cases, if the disease progresses favorably, the dead bone loosens and becomes detached, and the teeth fall out, and in very severe cases, and when the extent of the disease is very great, the patient may have intense constitutional disturbance, the local condition being peculiarly distressing from the secretion of fetid matter and loss of tissue ; œdema of the face and neck may supervene, probably accompanied by erysipelas, and terminating in an agonizing and long delayed death.

Treatment.—With regard to the treatment of this form of necrosis, the first indication is obviously in the early stage, or where it is anticipated to remove the individual from his work, pure air, cleanliness, attention to bowels and secretions, and all bad teeth taken out. If the disease has made any progress, and the extreme pain, swelling and thickening of the soft parts, show themselves, active measures for the relief of the periosteum and bone should be resorted to, leeching and general antiphlogistic treatment, free vertical incisions into the tender, soft places in the gums, carrying it clear down to the bone, so as to afford relief to the loose *overloaded* periosteum. In advanced cases the treatment is that generally adopted in necrosis from other causes. (See NECROSIS.) With regard to the peculiar region affected, which of course disables the powers of mastication, and so consequently interferes in great measure with digestion, and the great length in duration of such cases, suitable food, such as mashed meat, cod-liver oil, etc., are to be abundantly given, and iron.

LUMBAGO is a form of chronic rheumatism specially affecting the lower part of the back and loins. The pain is sometimes muscular, but sometimes also seems situate in the broad and strong ligament situate in that region. Chronic rheumatism is rarely a malady of youth; it is a totally different complaint from acute rheumatism, and mainly affects old people who have been exposed much to cold and wet. The pain sometimes called lumbago which may affect young people who stoop much at their work, or who have to raise heavy weights, is merely the pain of tired muscles, and demands the same remedy, rest. True lumbago is quite different; there is no feverishness with it, as in acute rheumatism, and it is not relieved by rest, as tired muscles are. The individual moves stiffly, as if he were tired, but night and day the pains continue. Sometimes the malady gets better from the application of cold, much more frequently it is improved by heat, so that a roll of flannel means to such positive comfort. There is not much difficulty in the diagnosis, nothing, in fact, can well be confounded with it, but the making of the diagnosis is no great comfort, for the malady is often a most intractable one. Broadly it may be said that internal remedies are of little use. Hot or tepid baths, applied even locally, do good, especially if salt water is used; local applications are, in point of fact, the best remedies in true lumbago, and as a rule they are best applied hot. Turpentine, ammonia and oil, blisters, iodine paint lightly used, belladonna, and chloroform with opium, may all be tried. In a considerable number of cases, but these

are neuralgic, the subcutaneous injection of a small dose of morphia will act as a charm. Sulphur is by many praised as a local remedy, wrapped up in flannel, which should be habitually worn. All exposure to damp and cold should be avoided, and the diet should be carefully regulated.

LUMBAR ABSCESS.—In the region of the loins acute abscesses are not met with so frequently as slowly-growing and almost painless purulent collections, which in the course of time acquire large proportions. The former resemble acute abscesses in other parts of the body in being due either to injury or to acute inflammatory action. The chronic lumbar abscess generally has its origin in disease of the vertebræ of the back and loins, or in suppuration in the loose areolar and fatty tissue about the kidney. The former, however, is the frequent cause, and the presence of a slowly-growing and fluctuating tumor in the right or left lumbar region, paleness and debility, and a peculiar sickening pain on tapping the sharp posterior spines along the lumbar portion of the spinal column. are almost sure indications of vertebral caries. When angular curvature is present, together with the above symptoms. there can be no doubt as to the cause of the lumbar abscess. Occasionally, though rarely, a chronic abscess forms in the loins of patients whose spine and kidneys are both quite free from disease. It seems to be due, then, as most spontaneous chronic diseases are, to general debility, and a slow inflammatory action in the areolar tissue of the region affected. A lumbar abscess generally terminates, after it has been growing for some time, and has attained a certain size, in pointing and subsequent outward discharge of the contained purulent fluids. In some few instances the pus contained in the abscess becomes converted into a shrunken semi-solid or cheesy mass and the external mass subsides. Occasionally the pus contained within the lumbar abscess makes its way into the thorax and lungs, and is discharged through the air-passages.

Treatment.—The treatment of lumbar abscess differs very slightly from that usually carried out in cases of chronic abscess in other parts of the body; and whenever there is a suspicion of disease of the spine, it is thought advisable by most physicians not to open the swelling until there is advanced pointing, and the integument over the most prominent part of the abscess has become very red and thin.

LUNACY. (See INSANITY.)

LUNAR CAUSTIC, or nitrate of silver, is much used by surgeons for cauterizing purposes. By melting with it a certain proportion of chloride of silver the "stick" is now rendered flexible instead of brittle. (See CAUSTICS.)

LUNGS are the organs by which the process of respiration is carried on, and where those changes occur by which the carbonic acid is removed from the blood while oxygen is supplied to that fluid. The lungs are two in number, the right and the left ; they are seated in the closed cavity of the thorax or chest and occupy most of the space ; the right lung is subdivided into three lobes, while the left has only two ; each is surrounded by a smooth thin serous membrane called the pleura, which is reflected at certain points from the surface of the lungs and lines the chest wall ; this surface is kept constantly moist by the secretion of a small quantity of fluid, so that the lungs can glide upon the thoracic wall with the greatest ease and the least amount of friction. Each lung consists of a bronchus, which allows of the passage of air to and fro, of an artery, which brings the venous blood from the right side of the heart, of capillaries, which surround the air-cells, and of veins which carry the purer blood on to the left side of the heart. (See AIR PAS-SAGES.) The trachea divides into two branches, called bronchi, and one bronchus goes to each lung ; as soon as it enters that organ, it divides into four or five main branches, and then again into very numerous sub-divisions, too fine to be seen by the naked eye : finally, these very small branches end in dilated extremities with extremely thin walls, called the air-cells or vesicles of the lung. At first the bronchus has pretty thick walls, which consist of an internal mucous coat lined by epithelium, of a middle coat, made partly of muscular fibres and partly of cartilaginous plates, and lastly, of an outer fibrous coat ; these various coverings become thinner by degrees, until at last, when the ultimate ramifications are reached, nothing is seen but a nearly homogeneous membrane of extreme thinness and lined by epithelium. The artery which supplies each lung is a branch of the pulmonary artery, which is a vessel of great size and arises from the right ventricle of the heart ; in this way, all the blood which has passed through the various vessels of the body is carried to the lungs ; the artery, like the bronchus, breaks up into a vast number of branches which at length end in a fine network of capillaries surrounding the air-cells ; so that, although there is no direct contact in health between the air and the blood, yet the two are by this means brought as nearly together as possible, and all the necessary changes can take place through the moist and thin-walled air-cell. The wall of the artery in the first part of its course consists of an inner epithelium coat, of a middle coat chiefly made of elastic fibres and partly containing involuntary muscular fibres, and thirdly of an outer coat of ordinary white fibrous tissue ; in its smaller branches there is less fibrous tissue and the

muscular coat is relatively the thickest, while in the smallest branches of all, these various structures disappear, and only a thin homogeneous membrane is left with a few oval nuclei in its walls. The veins are formed by the union of the capillaries, and these uniting to form still larger branches end by forming four large trunks which carry the blood to the left auricle of the heart. In addition to these various important structures, the lungs are supplied by various nerves and lymphatics, while the aorta gives off numerous small branches, called the bronchial arteries, which supply the lung-tissue with nourishment.

It will thus be seen that the lungs are spongy, elastic bodies, and they are capable of much distension, as may be seen by inflating a lung after death. During life the chest-wall is constantly moving up and down with each inspiration and expiration, and corresponding movements at the same time take place in the lungs, which closely follow the chest-wall, so that there is always a varying amount of air in the lungs. Inspiration and expiration follow each other in health with the greatest regularity, and the two actions make up what is known as respiration ; each movement is repeated fifteen to eighteen times in a minute on the average when the individual is sitting quietly, but they occur much faster during a period of active exercise, as in running or rowing, etc. The structure of the lungs is such as to admit of a very large amount of blood being exposed to the air, and the movements of the chest in respiration are to enable fresh currents of air to be constantly brought while the impure air is also removed. The expired air differs from the inspired air in these particulars : 1. Whatever may be the temperature of the external air, that expired is nearly as hot as the blood, or varies between 90° Fahr. and 100° Fahr. 2. The expired air is quite or nearly saturated with aqueous vapor, however dry the outer air may be. 3. Ordinary air consists of 79 parts of nitrogen and 21 parts of oxygen, with a trace of carbonic acid gas, in every 100 parts ; expired air contains more than 4½ parts of carbonic acid gas, between 15 and 16 parts of oxygen, and about 80 parts of nitrogen ; so that while the quantity of the latter gas is not materially altered, there is, on the other hand, a great loss of oxygen and a great gain of carbonic acid gas. But carbonic acid is very prejudicial to health, and hence the need of a movement of the chest-wall to expel it from the lungs ; in ordinary expiration, the normal elasticity is enough for the purpose, and very little muscular force is used. From three hundred and fifty to four hundred cubic feet of air are passed through the lungs of an adult man, taking no active exercise, in the course of twenty-four hours, and this amount must in that time become deprived of five per cent of oxygen and be charged with five per cent of carbonic acid. Thus it has been calculated that " if a man be shut up in a close room, having the form of a cube, seven feet in the side, every particle of air in that room will have passed through his lungs in twenty-four hours, and a fourth of the oxygen it contains will be replaced by carbonic acid." But carbonic acid is a compound of carbon and oxygen in the proportion of 32 parts of the latter to 12 parts of the former, and hence the quantity of carbon eliminated every twenty-four hours and calculated from the amount of carbonic acid given off is equal to a piece of charcoal weighing eight ounces. The amount of water given off varies very much in the twenty-four hours ; about half a pint is the average quantity, but it may be much more or less. The lungs during life can never be emptied of air, however forced an expiration we make ; the amount of air which cannot be got rid of is called residual air, and varies from 75 to 100 cubic inches in amount. After an ordinary, but not forced, expiration, about as much more remains, and this is called supplementary air. In ordinary breathing, from 20 to 30 cubic inches of air pass in and out of the lungs, and this is called tidal air. Thus about 230 cubic inches of air are contained in the lungs after an ordinary inspiration, but this may be increased by another 100 cubic inches, if a very deep inspiration is made ; this extra-supply is called complemental air. Since the lungs can contain 230 cubic inches of air and the tidal air amounts to only 20 or 30 cubic inches, it follows that only ½ or ⅓ of the air in the lungs is renewed with each inspiration, so that the remaining air acts as a buffer between the incoming fresh air and the blood in the capillaries ; it plays, as Professor Huxley has shown, the part of a middleman between the two parties, the blood and the fresh tidal air, who desire to exchange their commodities, carbonic acid for oxygen and oxygen for carbonic acid.

Experiments have been made by means of an instrument called a spirometer, with reference to the power of persons taking air into the lungs. The person first inspires to the full extent, and then breathes into the instrument as much air as he can, and it seems that the height of the individual has much to do with the result. On an average a person of 5 feet breathes 174 cubic inches ; one of 5 feet 1 inch will breathe 182 cubic inches, and for every inch of height up to 6 feet will breathe about 8 cubic inches additional. Weight seems to have much less influence than height, and tends to diminish the respiratory power when beyond a certain limit. In males of the same height the respiratory range increases from 15 to 35 years of age ; but from 35 to 65 it decreases nearly 1½ cubic inches per year. The activity of the respiratory process is far greater in children than in old age, and this activity is also modified by other circumstances. Cold greatly increases the quantity of air which is breathed, the quantity of oxygen absorbed, and of carbonic acid expelled ; exercise and the taking of food have a very similar effect. There is more carbonic acid excreted during the day than during the night ; during the day also much more oxygen is given out than is absorbed, while at night-time much more oxygen is absorbed than is excreted as carbonic acid during the same period. Air may become unfit to breathe therefore in two ways, viz., by the deprivation of oxygen and the accumulation of carbonic acid ; both will give fatal results, but when acting together, death is of course much hastened. Asphyxia will take place when the proportion of carbonic acid in tidal air reaches 10 per cent, provided that the oxygen is diminished in like proportion ; life could be carried on with 10 or even 15 per cent of carbonic acid, so long as the supply of oxygen is simultaneously increased ; hence it will appear that carbonic acid is not of itself so poisonous, but that its fatal effects are due in a great measure to its taking up the room that ought to be occupied by oxygen. Thus it is most essential for health that every human being should have fresh air and plenty of air ; every man ought to have at least 800 cubic feet, a cubic space of rather more than 9 feet to the side, and this air should be constantly renewed from the external atmosphere. Lassitude, uneasiness, and headache come on when the due amount of oxygen is by any means diminished, and there is in time a great loss of vital energy.

Lastly must be considered the changes which take place in the blood in its passage through the lungs. The blood in the pulmonary artery is venous, as has already been stated, and is of a dark purplish color ; the blood in the pulmonary veins, on the contrary, is of a bright scarlet color and arterial in character. Now it is known by experiment that when venous blood is mixed with oxygen it becomes brighter in color and resembles arterial blood, and when the latter is mixed with carbonic acid it becomes darker in color and resembles venous blood. Now in the lungs the interchange can only take place in the thin-walled pulmonary capillaries, and here the carbonic acid is removed from the blood and fresh oxygen is supplied ; this oxygen in its turn combines with the carbon from the tissues to pass away again as carbonic acid. It must not be supposed that all the carbonic acid is removed from the blood during its passage through the lungs ; on the contrary, arterial blood always contains a certain amount of carbonic acid and all venous blood contains a little oxygen. The cause of the change of color during the process of respiration is not yet well made out ;

the blood contains myriads of rounded bodies, called corpuscles ; these are rendered somewhat flatter by oxygen, while they are distended by carbonic acid ; in this way, by reflecting more or less light according to the convexity of the surface, the changes may be due ; on the other hand, it has been shown that solutions of blood-crystals free from blood-corpuscles change in color from scarlet to purple according as they gain or lose oxygen. In this way, those changes are constantly going on in the blood by which effete materials are carried away in part by the lungs, and other organs of the body, while fresh oxygen is constantly being absorbed and carried by that fluid to nourish every tissue and organ in the individual. (See BLOOD, CONSUMPTION, PLEURISY, and PNEUMONIA.)

LUNGS (WOUNDS OF.)—Wounds of the lung are, of course, common enough in military practice, though somewhat rare in civil. When arising from external wound they may be either incised, punctured, lacerated, or gun-shot, or they may be produced by fractured rib, which generally causes a lacerated wound. The simplest form of a wound of the lung is a punctured one, such as a stab ; next in severity are the lacerated, the lung substance having great power of retractibility ; and the most dangerous are the incised, on account of the hemorrhage and escape of air. When a lung is wounded three conditions are observed : (1) Hemorrhage ; (2) escape of air from divided vesicles and tubes ; (3) collapse of lung. The symptoms indicative of wounded lung are : (1) The escape of blood through the external wound of a pale red and frothy character ; (2) the issue of blood mixed with air and mucus from the mouth during the efforts of coughing ; this is always to be regarded as a dangerous symptom, as the blood accumulating in the tubes produces a choking sensation which may suffocate the patient ; (3) a deeply fixed pain in the chest, and a good deal of irritation of the larynx, producing a constant desire to cough ; (4) dyspnœa and difficulty of respiration. The constitutional symptoms are, at first, collapse, though, unless this extend over any considerable time, it need not excite alarm ; inflammation, pleurisy, and pneumonia are next to be apprehended. In all cases, the prognosis of wounds of the lung is unfavorable. Wounds from projectiles (gun-shot) are exceedingly dangerous. They are fatal from hemorrhage, causing exhaustion and suffocation ; from pleurisy, irritation, fever, or from accumulations of blood, pus, or serum in the pleural cavities. In the case of gun-shot wounds, the usual train of symptoms above enumerated as characteristic of this injury must not be always constantly expected to exist, and it is by no means easy to decide whether the lung is wounded in perforating wounds of the chest-walls. Serious bleeding

rarely occurs from any vessels external to the cavity of the chest. Although hæmoptysis indicates *injury* to the lung, it does not prove penetration. Dyspnœa is a frequent accompaniment of penetrating wounds. It was formerly thought that escape of air by the wound necessarily indicated laceration of the lung tissue, but it must be remembered that external air may pass into the external wound during contraction of the lung, and be expelled during inspiration, but, as was above stated, if air with frothy blood and mucus be expelled, there is no doubt that the lung itself is perforated.

Treatment.—All hemorrhage from superficial vessels in the chest-walls should be first arrested ; these vessels are cutaneous, muscular, or most frequently intercostal, which lie *under* the *lower* margins of the ribs, and are very liable to be lacerated by some splinter of bone from an adjacent rib. All foreign bodies, dirt, pieces of clothing, or wadding, should be carefully removed, and the external wound closed as quickly as possible with some light dressing. The natural motions of the chest should be restrained by broad strips of adhesive plaster, or by broad bandages passing round it. The patient should be left where he is found, or very carefully carried to some more convenient place. In the first stage, that of collapse, no constitutional treatment need be attempted, but hæmoptysis, dyspnœa, or chest complications must be carefully watched for, and total abstinence from food or stimulating drink must be enjoined for the first few days. Ice to the chest, or iced-water to drink, are useful in checking the hemorrhage, and in severe hæmoptysis, venesection, with a view of producing artificial collapse. Dr. M'Leod remarked that, during the Crimean war, he noticed that those cases of gun-shot wound of the lung did best in which early, active, and repeated bleedings were had recourse to. Dilute sulphuric acid or acetate of lead is frequently useful in checking violent hæmoptysis

Complications.—There are frequent complications associated with wounds of the lung which demand particular attention. These are : (1) *Emphysema*, or an escape of air into the subcutaneous cellular tissue ; (2) *Pneumothorax*, when air has escaped into the cavity of the pleura (usually associated with the foregoing) ; (3) *Hæmothorax*, or hemorrhage into the pleural cavity ; (4) Foreign bodies, such as bullets, buttons, clothing etc. *Emphysema* comes from a broken rib penetrating the pleura and entering the lungs, thus allowing air to escape ; or from penetrating wounds of the bronchi or lungs, when, on inspiration, the air received into the lung escapes from its wounded part into the chest, and on expiration is forced out through the external wound, thereby getting into the cellular tissue. The symptoms of emphysema are a swelling of the integument, beginning at the seat of injury, and increasing in all directions. There is no change in the color of the skin; the swelling crepitates or crackles under pressure, and there is no pain. The prognosis is generally favorable, except in cases of extensive wounds. With regard to treatment : If it be found that moderate pressure with the hands on the seat of injury affords relief, a roller-bandage should be applied ; and in cases where the air has diffused itself over a large surface, punctures are useful. Constitutionally for the relief of dyspnœa in such cases, antimony and ipecacuanha appear to be the most useful remedies given in full and repeated doses. Bleeding may be resorted to if there is lung congestion and oppression of the circulation. In *Pneumo thorax* the symptoms would be distressing dyspnœa ; on percussing the chest a tympanitic resonance, and a ringing metallic resonance on auscultating the chest, supposing the lung is not too much compressed. The treatment consists either in enlarging the external wound so that the air may escape, or in puncturing the cavity with a small trocar or canula (as in paracentesis). In *Hæmothorax* the symptoms present depend in a great measure on the quantity of blood poured out in a definite period. If large quantities are effused suddenly, as in a wound of a large vessel, death speedily follows from loss of blood and pressure on the lung. The treatment consists in closing the external wound, and thus allow the effused blood to coagulate if possible, so as to form a plug to prevent further bleeding. Paracentesis is sometimes necessary. Some surgeons recommend the external wound to be kept open, so that the blood may escape, while others prefer enlarging the external wound to let the blood escape speedily,-and in several recorded cases this has been successful. If the difficulty of breathing be very urgent, the trocar and canula may be used, and the wound dressed with carbolic acid dressing.

LUPUS is a name given to several forms of obstinate inflammatory and ulcerative affections of the nose, cheeks, and lips, which give rise to much disfigurement, and often to destruction of soft parts, and deformity. Some of these diseases are known by the name *Noli me tangere*.

The following are some of the chief varieties of lupus : In the affection called *erythematous lupus*, which is the least troublesome, the skin of the nose or face presents numerous deep-red or livid patches slightly elevated above the general surface, and smooth and shining. These increase in size and run together, forming large purple patches, which, if not treated, become covered by thick crusts of scarf-skin. This form of lupus is attended with but little pain or itching, and does not result in ulceration or loss of substance. It

is, however, very obstinate, and often resists for a long time all kinds of treatment.

In another form of the disease, called *lupus non-exedens*, numerous small reddish-yellow and waxy nodules are set upon a dark-red base of thickened skin. These nodules increase in number and size, and become capped by small horny-like scabs. No open sores are formed, but the disease leaves behind a very distinct scar, which is tense and depressed below the level of the sound skin. This disease has a tendency to heal at the centre of the patch whilst fresh nodules are formed about the circumference.

The most severe form is that called *Lupus exedens*, or *Noli me tangere*. This generally commences at the tip or edges of the nose, and often attacks simultaneously the skin and the internal mucous membrane. Red or brownish-red nodules are first formed, which increase in number and run together, and then crack down into a jagged ulcer, which is covered by a thick adherent crust, under which pus collects. This ulcer, after a time, commences to heal at one part of its circumference ; but at the same time fresh nodules and ulcers are formed, and the disease, if not arrested by treatment, spreads slowly and insidiously, until a considerable part, or even the whole of the nose, with its bones and cartilages, has been destroyed. The subsequent disfigurement is made worse by the presence of large pale-red scars, traversed by tough bands of a white color, which are very contractile, and cause by their shrinking considerable displacement of parts of the face. In Lupus exedens there is generally a tendency to an early relapse.

Lupus is seldom met with in patients over thirty years of age, and occurs more frequently in the country than in large towns. It is more common in females than in males. Lupus in all forms is generally associated with scrofula, and occasionally with advanced or tertiary syphilis. The patients in the majority of cases have fair, delicate skin, and light eyes and hair. Like other local affections dependent upon a scrofulous or syphilitic taint, it is met with chiefly among the poor.

Treatment.—In the treatment it is necessary first of all to attend to the general health, to support the strength by tonics, good diet, and wine or malt liquor ; to keep the digestive organs free from irritation and disease, and to improve, if it be possible, the hygienic circumstances of the patient. In this disease, especially the form of *Noli me tangere*, the remedies used in cases of scrofula are especially useful ; of these the best are cod-liver oil and the citrate of iron and quinine. In some cases it is necessary for the patient to take arsenic or mercury ; but those agents ought not to be administered except by medical advice. In *Lupus erythematosus* and *Lupus* *non-exedens* the safest local applications are sulphur ointment, tar ointment, and tincture of iodine, which may be applied by means of a camel's-hair brush. Obstinate and severe cases of these two forms of lupus, and the slowly spreading ulceration of Lupus exedens, are treated by the application of caustic potash, chloride of zinc, Vienna paste, lunar caustic, and the actual and galvanic cauteries, powerful and very painful remedies, which necessitate in many instances the administration of chloroform, and, in all, delicate surgical manipulation and careful control.

LYMPH is an alkaline fluid which fills the absorbents or lymphatics ; it differs from the blood in containing no red corpuscles, and in having a very small proportion of solid constituents ; lymph may be looked upon as blood diluted with water and deprived of the colored corpuscles.

LYMPHATICS are vessels distributed throughout the body, generally closely accompanying blood-vessels, but also pursuing a solitary course. They are intended to retain that portion of the nutrient fluid which has poured out from the smaller blood-vessels, and which does not return by the veins. This fluid is colorless, and is called lymph. These vessels are of great importance in the spread of some maladies. Thus it is fairly established that cancer spreads from one organ of the body to another mainly by these means, and it is very likely that tubercle does so likewise. The lymphatic system is especially liable to invasion by inflammation when that seizes upon any part to which they are richly supplied, and especially if the inflammation is of a bad kind. Thus a fresh wound of the hand, into which putrid animal matter has entered, speedily gives rise to an appearance resembling a number of irregular red cords running up the arm. These seem all to run to the armpit, and there enlarged lymphatic glands soon are felt, as hard painful knots are soon to be felt. Most likely these will suppurate, and may constitute a new form of disease. If the lower extremities are affected, the glands in the gums are affected in like manner.

LYMPHOMA or **LYMPHADENOMA,** is a name given to a disease in which there is great enlargement of most or all of the lymphatic glands of the body. The growth takes place gradually and without much,if any,pain; the patient becomes pale and weak. Children and young people are most commonly affected ; the glands do not soften and form an abscess, as they do in scrofulous cases, but are firm and retain their rounded outline ; they are most easily seen at the root of the neck on either side, and sometimes large prominent tumors are in this way formed.

M.

MACARONI is a well-known Italian food made of the best wheaten flour, and formed into long thin pipe-shaped lengths about the size of a quill. Macaroni is a highly nutritious and digestible article of diet, and if properly cooked may be eaten by an invalid with a delicate digestion with advantage. It is pleasant either well boiled till it becomes quite soft, and served with gravy from roasted meat, or plainly boiled in milk and eaten with salt or sugar.

MACE is a well-known spice, the product of a tree growing in the Molucca Islands, known to botanists as *Myristica officinalis.* The fruit of the tree is of the size and form of a peach, and when ripe the fleshy part bursts in halves exposing the kernel, which is the nutmeg, surrounded by an arillus or scarlet net-work sort of fibre, which is the mace. It is a valuable and powerful spice, and realizes a high price. It contains a large quantity of aromatic oil, the taste of which is pungent and sharp. It is used in medicine as a stimulant, and is imported into this country for that purpose in considerable quantities.

MADEIRA. (See WINES.)

MADNESS. (See INSANITY.)

MAGNESIA, or *oxide of magnesia,* occurs in two forms, one more bulky than the other. It is obtained by burning the carbonate, and then appears as a white powder with hardly any taste, almost insoluble in water, and slightly alkaline in reaction. When introduced into the stomach magnesia acts as an an acid, and its antacid properties are considerable. If not all neutralized, what remains passes on into the stomach, where, if given incautiously, it is apt to accumulate. In the small intestine it acts as a gentle laxative, in very large doses having considerable power. It also passes into the blood and tends to render the urine clear if previously turbid from urates. It is given as an antacid in heartburn, and is still more useful for the acidity of the intestines which gripes, and so is useful along with rhubarb in the early stages of diarrhœa. It is also useful when there is acidity with a tendency to constipation, as in gouty subjects. In these patients it does remarkably well. On account of alkalinity, too, magnesia is often given as an antidote to poisoning by mineral or vegetable acids, and for this it is well suited. It neutralizes the acid and protects the stomach from injury. Many metals are also precipitated by it and rendered nearly insoluble. Magnesia has also been given as a remedy for vomiting when that has seemed to depend on excess of acid, but other remedies are more powerful.

Carbonate of magnesia also exists in two forms

—heavy and light. It is prepared from the sulphate of magnesia by precipitating by carbonate of soda. The powder so thrown down is a white almost tasteless substance, insoluble in water, and nearly neutral in reaction. A solution of the bicarbonate of magnesia has long been in use under the title of fluid magnesia ; it is an admirable preparation. This may effervesce slightly ; when opened the liquid is clear and is not bitter. Carbonate of magnesia acts in much the same way as magnesia itself, only, when introduced into the stomach, and it meets with an acid, it gives off its carbonic acid, which may be unpleasant. Sometimes, however, the carbonic acid gas so set free is pleasant to the stomach. The great disadvantage these remedies labor under is their bulk, so much requires to be taken ; but the objection does not apply to the fluid magnesia.

Sulphate of magnesia, or Epsom salts, is a well known remedy. (See EPSOM SALTS.)

Citrate of magnesia, the effervescent preparation popularly used as a laxative, contains, besides the magnesia, citrate or tartarate of sodium or potassium. It is pleasant to take and mild in its operation.

MAL-DE-PAYS is a condition of mind which assumes the form of a disease in people, who, having been born in mountainous countries are removed from the scenes of their early childhood. It is especially frequent among the Swiss, and is sometimes so uncontrollable as to prevent them ever settling in distant regions.

MALARIA is a term used for those badly defined agencies which give rise to fevers of the remittent and intermittent type. There are commonly supposed to be certain effluvia or miasms given out from marshy grounds, especially from salt marshes, which entering the system give rise to the well known phenomena of ague. What, however, these effluvia are no one knows ; the air has been examined and nothing found, and as the malady is equally rife in the dry regions of Central India and the Sahara, it is plain that they do not depend on marshes. True they are most powerful in tropical regions, especially near the mouths of great rivers, or among mangrove swamps where decaying vegetable matter and heat most abound, but the same phenomena are frequent far north, in temperate regions especially, in certain seasons of the year. Some have been constrained from the phenomena, and from the situations in which they are produced, to infer that these extremes of heat and cold are the main causes or constituents of malaria ; that is to say, great heat during the day and cold during the night are the concomitants of severe attacks of aguish maladies. Certain it is that the individuals who are attacked are more readily attacked during the night than during the day. Thus people can work in the Roman Campagna during the day, but cannot remain in it all night without suffering. The poison lies low, too, for a man standing upright may escape it while one lying down will not. Another peculiarity of it is that it does not prevail at any distance from the shore, and even blowing over a wide river will interfere with the malarial influence, be it what it may. A row of trees, too, has been found to give protection when planted between marshy districts and inhabited places. All these facts seem to point to the conclusion that the poison after all is something material that is closely connected with the ground, and water of any extent interferes with its spread. (See INTERMITTENT FEVER and REMITTENT FEVER.)

MALFORMATION.—During the development of the fœtus in the womb it sometimes happens that some parts are not properly formed, and there is an arrest of growth or else an union of parts which ought to be separate. In this way many of those monstrosities are formed which excite the wonder of the ignorant. The Siamese twins were two individuals who were united together by a band of skin, and this change took place in the womb at an early period of fœtal life ; sometimes the union is more complete and may extend along the whole length of the spine, as in the case of the " Two-headed Nightingale," twin sisters who were exhibited some time ago ; of course this peculiar malformation can only exist when there are twins, and the union takes place along the middle line of the body, either in front or behind. There are, however, other cases of malformation which affect only one child, and these are always congenital, that is to say, are met with at birth and produced at some period of fœtal life. Malformations may result in various ways.

A. Those resulting from incomplete development or growth of parts. 1. Of the body generally. The head may be absent or rudimentary, and the fœtus is then said to be *acephalic ;* it is either born dead or lives a very short time ; the arms and legs may be defective, or the hands are joined to the shoulder-blade, and the feet to the thigh-bone, so that the arms and legs are absent ; the fingers may be too many or too few in number. 2. Of the nervous system. The brain and spinal cord may be absent, or exist only as rudimentary formations. 3. Of the organs of special sense. The eyes may be absent or imperfect, or the eyelids may remain united ; the ear may suffer in a similar way, and deafness is the result ; sometimes the nose is absent or deformed and resembles a proboscis. 4. Of the vascular system. The heart may be absent, or the cavities of the heart may be deficient in number ; sometimes two or more valves are joined together, or they are too numerous ; at other times the orifices between the different cavities of the heart are closed or unduly large, or the vessels which carry the blood from the heart into the system are wrongly placed. Such cases generally die early and the infant often suffers from cyanosis or blueness of the skin in consequence of the impaired circulation. (See Cyanosis.) 5. Of the respiratory system. The lungs may be absent or only one may be present, or the lobes may be deficient in number, or the air passages may be absent or imperfect. 6. Of the digestive system. The intestines may be deficient in various regions or impervious, or the liver may be unduly small. 7. Of the urinary system. The kidneys may be (one or both) absent or united together so as to form a horseshoe shape. 8. The organs of generation. These may be absent or malformed so as to cause a doubt in some cases as to the sex, thus giving rise to hermaphroditism.

B. Malformation resulting from the incomplete union of lateral halves of parts which should become conjoined. The ordinary cases of hare-lip and cleft palate are deformities of this kind ; sometimes there is a fissure of the abdominal walls so that the bladder is visible, or there may be a fissure of the urethra, producing the conditions known as *epispadias* and *hypospadias*. On the posterior surface of the body, there may be a fissure of the skull, or of the spinal cord, causing spina bifida.

C. Malformation resulting from joining together of the lateral halves of parts which should remain distinct. Examples of this class occur when the fingers or toes are joined together so as to give a web-like appearance to the extremities, or the lower extremities may be joined together.

D. Malformation resulting from duplication of parts in an infant. Examples of this variety are seen when the child has extra fingers or toes. In addition to these varieties, there are other occasional malformations, as transposition of the internal organs and herniæ of the intestines, brain, heart, and lungs. Some of these malformations cause the death of the fœtus in the womb, others are of such a nature as to prevent it coming to maturity, while some are so slight as not to give rise to any symptoms during life. It is, in fact, a matter of surprise, when the complex structure of the organism is taken into account, how seldom any deformities occur, but when malformations do happen, they are due to some defect of devlopment in early life, and such monstrosities are capable of explanation on ordinary scientific grounds, although to the ignorant and uneducated they form food for awe and superstition.

MALIGNANT DISEASES are those which are very rapid in their course, which always end fatally, and for which all human aid seems powerless. In an epidemic of

typhus or scarlet fever it may happen that some cases will be attacked much more severely than others, and be knocked down or prostrated from the virulence of the poison ; thus, a person may die of malignant scarlet fever in twenty-four or forty-eight hours. Either the individual has received an unusual quantity of the poison of the fever, or he may be in a bad state of health at the time, which has rendered him more susceptible to its influences. Such cases seem hopeless from the very commencement, but fortunately they are rare ; the usual symptoms of a disease are intensely exaggerated ; there is great prostration, low muttering delirium, and sometimes bleeding from the nose and gums, or petechial spots over the body. Scarlet fever is the most likely one to become malignant, but it may occur in measles, typhus, and typhoid fevers ; it generally is noticed at the commencement of an epidemic. The term "malignant" is also applied to those tumors of which cancer is the best-known example, where there is very little hope when the disease is internal, or has once made much inroad upon the constitution ; such cases go on for some months, and finally die of exhaustion and emaciation. (See Tumors.)

MALLOW is a plant known to botanists as *Malva sylvestris*, belonging to the natural order *Malvaceæ*. The whole plant, but especially the root, yields, when boiled, a plentiful tasteless mucilage, which is useful in some cases of internal irritation. Decoctions of the leaves are employed in dysentery, and they are used in fomentations, poultices, etc.

MALNUTRITION is said to take place when the body is badly nourished, and supplied with impure air and food. The unhealthy state of the children in our large towns is largely due to this cause, and may be seen in the stunted and rickety condition which they present. The evils thus taking place in early life influence the future development of the individual and are more likely to render him susceptible to some forms of disease than those who are more healthy. For the prevention of such mischief the diet should be regulated according to the rules laid down in the article on Diet, and our sanitary officials should see that there is a plentiful supply of good water and air. In this respect it is very important that all adulteration of milk, bread, and other necessaries of life should be severely punished ; that the water-supply should be abundant and wholesome, and that in every large town open spaces or parks should exist for the children to play about in.

MALT LIQUOR. (See Beer.)

MANDRAKE is a powerfully narcotic plant belonging to the genus *Atropa*, known also as *Mandragora*. It was at one time thought to have a sort of supernatural efficacy, and was gathered with great solemnity,

with incantations. From the forked appearance of its roots, and its fancied resemblance
to a man, it was in superstitious times supposed to have an influence on the health of a
person against whom it was used, so that, as
the root withered away, the life of the doomed
victim would gradually wane also.

MANGANESE can hardly be said to be
used in medicine. The black oxide is largely
employed for the production of oxygen gas,
and the sulphate has been given internally.
In very large doses it gives rise to purgation,
and in smaller does it has been supposed to
act in a fashion somewhat similar to iron.
It has been given in anæmia, therefore, but it
is not a standard remedy.

MANIA is that variety of insanity characterized by delusions of exaltation, with or
without delirium. Sometimes such patients
become very greatly excited, very destructive
and dangerous, but not so much so as do certain others. Mania has also been used as a
generic word to imply all forms of insanity.
(See INSANITY.)

MANNA is the hardened exudation from
the incised bark of various species of *Frax-
inus*. These trees are cultivated for the purpose of producing this substance in Calabria
and Sicily. Manna of the best description
forms pieces not unlike stalactites, about six
inches long and one or two broad, hollowed
out and discolored on one side where attached to the tree. This is called flake
manna, and is porous and friable. It also
occurs in smaller masses, or in broken and
colored fragments. This substance, when
pure, has a sweetish odor and taste, but is,
withal, bitter. Manna is soluble in water and
alcohol, and consists almost entirely of a kind
of sugar called *mannite*, which differs from
ordinary sugar in not fermenting with yeast.
Manna itself is a very mild laxative, generally
given to children or added to other purgatives,
as senna, to sweeten them ; sometimes it
gripes. The dose is from a drachm to half
an ounce.

MARASMUS is a technical name given
to the wasting disorders of children. It occurs as a symptom in cases of bad feeding,
malnutrition diarrhœa, constitutional taint,
and in diseased conditions of the intestines
and mesenteric glands. The term marasmus
corresponds to the word emaciation or wasting. (See LACTEALS and TABES MESEN-
TERICA.)

MARJORAM, the *Organum vulgare* of
botanists is an indigenous plant, yielding a
volatile oil, and possessed of properties very
similar to rosemary. It is not officinal, but
is sometimes employed. Formerly it was
contained in the Pharmacopœia, but has been
expunged.

MARRIAGE.—The marriage-rate of the
country is to a great extent an index of the
prosperity of the nation. When the neces-

saries of life are cheap, and when there is a
great demand for labor and when high wages
are given, the number of marriages in a year
is much greater than when the opposite conditions exist. Generalizing from this fact,
the historian, Buckle, laid down the proposi-
tion that the marriage-rate in any country is
in exact proportion to the price of corn. The
statistics gathered for a series of years by the
English Registrar-General seem to prove
that married people live somewhat longer than
the single.

MARSH-MALLOW, the root of *Althea
officinalis*, is no longer officinal. The outer
covering is usually removed, so that the substance looks yellowish externally, and white
internally. The syrup of marsh-mallow had
at one time a great reputation for allaying
coughs, etc., and even now in France, under
the title *Guimauve*, it is greatly used. Specific properties it has none.

MASTICATION is the process by which
the food when taken into the mouth is chewed
into small pieces by the teeth and thoroughly
mixed with the saliva. Pieces of meat, etc.,
are thus finely divided, and are so rendered
more easy to be acted upon by the gastric
juice when they are swallowed and enter the
stomach ; farther, since starchy foods, as
rice, potatoes, bread, etc., when mixed with
the saliva, become converted into sugar, it follows that if mastication is properly performed,
all the insoluble starch will thus be changed
into the soluble form of sugar and made
ready for absorption. It will thus be seen
that it is a very important thing to eat a meal
slowly, and not swallow the food hastily, as
some are apt to do ; the latter fault is very apt
to cause indigestion, and be a source of much
distress if the habit be persevered in. (See
DIGESTION and DIET.)

MASTICH is a resinous exudation flowing from a plant of the turpentine family, a
native of the countries bordering the Eastern
Mediterranean. The best mastich consists
of small masses called tears, which are light
yellow and friable, but becoming soft and
ductile on chewing. The surface of the
masses is often covered with a whitish dust,
produced by rubbing one against the other.
Larger masses are formed by the agglutina-
tion of several tears. It has an agreeable
odor. Mastich is not much used save to give
a pleasant odor to the breath when chewed
and to stop teeth.

MATCH MAKER'S DISEASE. (See
LUCIFER-MATCH MAKER'S DISEASE.)

MATICO is the leaf of a kind of pepper
plant growing in Peru. The leaves are oblong and pointed, marked on the upper surface, downy and reticulated beneath. Their
colors are green, their taste aromatic, warm,
and slightly astringent, the odor is pleasant.
Matico contains some tannic acid, and a
peculiar substance called aranthic acid. The

only officinal preparation is an infusion ; but a tincture is also in use. The leaf, in substance or in powder, applied to small bleeding surfaces, as leech-bites and the like, acts as a powerful styptic. Given internally, it is said to act as an astringent in the urinary ways and on the rectum, but this is by no means clear.

MATTER in a medical sense is synonymous with pus, and means the fluid humor which is contained in an abscess or sore tumor.

MEAD. (See DISTILLED SPIRITS.)

MEASLES is a contagious febrile disorder, and forms one of the group of the exanthemata. It is nearly always more or less prevalent in this country ; but at times it spreads with great rapidity, and carries off a large number of victims. As a rule, children and young people are attacked, but the exemption of adults and older people probably depends on the fact that they have had the complaint in early life, and so are not subject again to the influence of the poison ; yet in a few rare cases persons have suffered twice from this disorder. This disease is more fatal in the autumn and early winter than in the spring and summer. Measles varies much in malignity ; in some years, although many are attacked, the mortality is moderate ; while at other times the disease is fatal in a much larger ratio. Measles is essentially a contagious disorder, and often attacks all the members of a family one after another. In measles, as in other contagious disorders, there is a period of incubation ; and by incubation is meant the time which elapses between the exposure to the contagion and the first appearance of symptoms. This period varies in different fevers ; in the case of measles it seems to be ten to fourteen days. Such a question as this is usually very difficult to answer, as there are many sources of fallacies when the disease is prevalent in a town, and a person may be exposed to contagion unawares.

Symptoms.—Before the appearance of the rash there are some precursory symptoms ; the patient feels languid and hot, and there is shivering, followed by a rise of temperature, a quick pulse, thirst, loss of appetite, and sickness. Such, in fact, are the usual symptoms which precede most febrile attacks. But in addition to the above signs, there is super added an inflamed condition of the mucous or lining membrane of the air-passages ; and this state is so marked as as to be very characteristic of this disorder. The eyes become red and watery, and give the appearance of a patient having cried ; the membrane which lines the nose, throat, larynx, and trachea is red and swollen, and pours forth a watery secretion ; thus the affected person seems to have a severe cold, with running from the eyes and nose ; hence there is generally much

sneezing, with a slightly sore throat and a dry, harsh cough. In addition there may be diarrhœa, with pain in the stomach, and a good deal of vomiting ; but these cease when the eruption appears. Convulsions occasionally occur in children, and the younger the patient the more liable is it to have a fit.

After these uncomfortable symptoms have lasted about three or four days, the rash appears. Although the rash may come out as late as eight or even ten days from the first appearance of the symptoms, the fourth day is by far the most usual for it to come out. The rash begins in very small papules or minute pimples which rapidly multiply, and then run together into patches which have a tendency to a horseshoe or crescentic shape, while the portions of skin between are of a natural color. Commencing on the face and neck, it spreads to the arms, then the trunk of the body, and gradually reaches the lower extremities ; this process takes two or three days. The same order is observed when the rash fades ; it is generally out fully for three days on the face, so that the whole duration is at least six or seven days, and it disappears on the upper part of the body, while a few faded spots may still be seen on the legs. At first the eruption has a dark pink or mulberry color, but toward the end it becomes browner. The face is generally bloated and swollen, and if the finger be passed gently over the surface of the skin, the rash may be felt to be slightly elevated. When the eruption has disappeared, the part of the skin affected becomes covered with a dry scurf, and seems covered with a branny powder ; the cuticle does not come off in large flakes, as in scarlet fever. The fever does not diminish on the appearance of the rash, as in cases of small-pox, and there is no proportion between the abundance of the rash and the danger to the patient ; indeed in some cases, where the rash is late in appearing and not very plentiful, the individual may be in much danger.

This disease is known, then, by the catarrhal affection, or appearance of a cold with which t is ushered in, and by the peculiarity of the rash. Scarlet fever mostly begins with a sore throat, and the rash comes out earlier ; and in small-pox there is more vomiting and much pain across the loins ; these are the two disorders with which measles is most liable to be confounded, but if any doubt exist, the case will be cleared up when the rash is seen. Of late years cases have been seen where the eruption has appeared without the fever and catarrh, and to this variety the name of *rubeola sine catarrho* has been given ; but it confers no protection from measles, and often the latter disease appears in its regular form a few days afterward. In some cases a measle-like eruption comes out in the early stages of small-pox, and may give rise to mistakes ; but in two or three days the regular rash of variola will appear. A rose-colored rash sometimes appears in children while they are teething, but it is more diffusely spread over the body, and there are no signs of running at the eyes and nose, and the fever is but slight.

The temperature in measles rarely rises to more than 103' F., and in mild cases may not be more than 101 F. or 102° F. ; this is a guide to the severity of the disease, for the higher the temperature the more danger there is to the patient. To find out the temperature, a delicate thermometer should be placed in the arm pit, and the arm held closely to the side, so that the skin perfectly surrounds the bulb of the instrument ; it should be kept in this position for at least five minutes, and then the point to which the mercury has risen in the graduated tube can be read off. The patient should be in bed, and have his arms covered up half an hour before the observation is made, or otherwise the skin of the arm-pit will be unduly chilled. Since the ordinary temperature of the body is 98° F. or 98 5° F., it follows that any degree of heat observed higher than that indicates a state of fever.

The pulse at the wrist beats quicker than usual, but is usually not so high as in scarlet fever ; from 120 to 140 beats in a minute is a common occurrence. The tongue is generally furred, and has a moist, white appearance ; in very bad cases it may become dry and brown ; it begins to clean at the tip and edges from the fourth to the eighth day of the disease, and the rest of the tongue is clean in twelve or fifteen days, unless some other disease arises. Small superficial ulcerations may in some cases be seen on the lining membrane of the mouth and gums. The throat may be red and swollen, and the act of swallowing rendered in consequence rather painful : but this symptom is of very slight importance. Still more rarely, there may be some difficulty of hearing. In some cases the glands behind the jaw, or down the neck, and even those in the groin, may become large, swollen, and painful : but this varies in different epidemics, is rarely severe, and is far more common in scarlet fever.

Vomiting is frequent enough at the outset of the attack, but is very seldom met with afterward if care be taken with the diet. Diarrhœa, now and then, is very troublesome ; in a moderate amount it is not injurious, but sometimes it is very exhausting, and blood may be found in the stools ; it is worse in those cases where the children are weak and sickly previously. The urine is generally scanty, and in fact in most febrile disorders less urine is passed than usual. On standing and allowing to cool, a light, yellow, sandy sediment is deposited ; this deposit need never cause any alarm, as it is merely due to certain salts in

the urine being less soluble in the cold, while when the water is first passed at the ordinary heat of the body they are soluble.

More characteristic symptoms are the cough and expectoration. The cough is at first dry and hacking, very frequent and annoying ; in a few days it comes on in occasional paroxysms or fits, which give much distress, and even cause retching and vomiting : it generally disappears when the rash fades, but in some cases it may remain for some time. At first there is not much expectoration, and it is clear and viscid ; in a few days it is more abundant and frothy, or even of a greenish-yellow color.

Complications.—These are very liable to occur, but they vary in their nature, in their severity, and in various epidemics. The rash generally disappears when the finger is pressed upon it, but in some malignant cases the spots turn a dark purple and will not disappear ; hence the name of *black measles* has been given to this variety, and this form is usually fatal. Or the rash may suddenly disappear instead of gradually fading away, and this may indicate some internal mischief ; this also is a serious symptom. Convulsions at the commencement are usually without danger, but if they come on at the end of the disease they may lead to a fatal issue. Inflammation of the larynx or upper part of the air passages may give rise to harsh, croupy breathing, and sometimes the inflammation becomes chronic and very obstinate, being always liable to return whenever the patient takes cold. Inflammation on the lungs is very common in measles, and unless great care be taken may prove very dangerous ; the breathing is hurried and the temperature and pulse rise ; the patient may lie in great distress, and in children there is often dilatation of the nostrils at each inspiration. Children under ten years of age rarely expectorate, and so the bronchial tubes often become choked with phlegm. Wheezing sounds may be heard all over the chest if the ear is placed there, and may be felt when the hand is placed over the back or front of the chest. Bronchitis often proves fatal to very young children, and in all cases adds to the danger. Whooping-cough is very frequently an accompaniment of measles ; indeed, an epidemic of each is generally prevalent at the same time. Any children who have just suffered from measles, and are living in a house in which or near which whooping-cough is prevalent, are almost sure to take it. Consumption sometimes follows measles, and especially in those who are liable to be subjects of it. Inflammation of the ear now and then occurs ; the child then cries a great deal and puts its hand to the side of its head ; at first nothing may be seen, but afterward a discharge flows, and this will then give great relief.

Measles, as a rule, is a mild disease, and

the great majority of cases recover; if attacking children in previously good health, the result is nearly always favorable. In those who have bad health, and chiefly in those who are liable to diseases of the chest, the danger is greater. Cold and damp weather increases the mortality by favoring the development of affections of the lungs.

Treatment.—Since there is no drug which can cure the disease, attention must be directed to those means by which we can relieve the patient and avoid any complications. In the first place he must be kept in bed, as in this way a more equable temperature can be kept up, draughts can be avoided, and so any liability to inflammation of the lungs may be lessened. The room should be airy and well ventilated, but great care should be taken that the patient is not exposed to any draught All offensive excreta or dirty linen should be removed and disinfected, and the way to do this has been already fully discussed under the article on Fevers. A temperature of 60° F. or 65° F. may be kept up by having a small fire in the room, but the heat should never be oppressive. It is as well to keep down the blinds and to allow the patient to lie with his back to the light, as the eyes are generally inflamed, and a strong light causes much discomfort. Any feeling of dryness or tingling of the skin may be relieved by bathing the part with tepid water, but for this purpose do not expose the patient all at once, but bathe and then dry one part at a time. In all cases it is advisable to give the patient a hot bath at the very onset of the disease, then dry the surface of the body and put him to bed directly; no chill need be feared, and it may help to bring out the rash; if the children are dirty it is all the more needful, not only for the sake of seeing the rash clearly, but to aid the skin in performing its proper functions. All sources of annoyance or irritation and all noises should be avoided, and the patient should be kept quite quiet so as to try and induce sleep. Food of the simplest nature should be given, and this point is important, as a favorable result much depends upon it. In all cases of fever, care should be taken to give food which is at the same time nourishing and easily absorbed. At first no solid food will be cared for, and the thirsty patient may drink milk, or milk and water, or tea, chicken-broth, beef-tea, or toast and water. The quantity given should be moderate, and the child should only drink enough to quench its thirst. A pint and a half of milk, with half a pint of beef-tea, free from fat, will suffice for a child of three to five years old, and if requiring more drink, some simple fluid as barley-water or tea may be given. The quantity, however, will vary with the age of the patient. As too much food should not be given at once, the meals must be given every two or three hours, and something should

always be ready in the night or early morning when the child awakes, as it is often much required then. Acid drinks are very grateful and agreeable; lemonade with a little sugar, or raspberry vinegar, may be given in moderation, but should not be taken at the same time as milk is taken, as they apt to make the latter curdle in the stomach and be vomited. Stimulants are rarely needed in children, and should only be given under medical advice. When the fever subsides, a small piece of chicken or mutton may be taken; toast or bread and butter, with a fresh egg may also be given, and as the tongue cleans and the appetite returns, the patient may return to his ordinary diet. If there is any inflammation of the lungs, a hot linseed-meal poultice may be applied to the back and front of the chest, and in all cases the chest may be kept covered with cotton wool and all exposure to cold avoided. But in all cases of any complications arising, medical advice should be taken, as it is impossible to lay down any rules for the treatment of what may occur in any individual.

Although children generally recover rapidly, yet there are times when much debility ensues and the general health becomes impaired, although the fever has quite left. Such children as are in bad health are liable to lumps or glandular swellings of the neck and under the jaws, or they may remain weak for a long time or be subject to some skin eruptions. Steel wine and similar tonics may here be given with great advantage. If the child be growing fast, too much exercise should be avoided, and plenty of rest allowed. Should the weather be warm and genial, outdoor exercise is beneficial, but the child should wear flannel next the skin and be protected from cold and wet. Regular hours, plain nutritious food, and a very moderate amount of stimulant should be given. A cold bath may be given every morning, but it should be discontinued if, after being well dried, the child should not feel a healthy glow; salt water is the best for this purpose, but if not procurable, a handful of sea salt may be put into the bath. Nor should the child remain long in the water, as depression and chilliness may ensue; generally two or three minutes will suffice. A visit to the sea-side is to be recommended if possible, and if there is any tendency to enlarged glands or discharge from the ears, a moderately bracing place should be chosen. If a child is timid and afraid to bathe in the sea, sea-water may be procured for its morning bath; no child should be forced into the water, but should be coaxed and encouraged to bathe, and a sea-bath should be given to delicate children about two or three hours after breakfast, and not before that meal. Baths, to do good, should act as tonics; if they depress and make the patient feel worse they are doing harm.

Morbilli and *Rubeola* are techincal names which have been given to measles. The disease is contagious when the rash is out, but other children are probably safe to mix with affected patients a week after the rash has disappeared. Isolation is the only way to stop the spread of this affection ; if by leg-islative measures a quarantine could be established round an infected house or district, the disease might become· stamped out, or its spread vastly diminished. We cannot do much in the way of *curing* measles, but a great deal might be done to prevent its coming among the people and pursuing its ravages.

MEAT, EXTRACT OF.—The substance commonly sold as Extract of Meat is also known as Juice of Meat and Liebig's Extract. The name of the late Baron Liebig, the great chemist, is especially connected with this compound, as he undoubtedly was one of the first to call attention to it as a valuable article of diet. In his " Familiar Letters on Chemistry" he devotes a letter to vegetable and animal food, and gives an account of their various chemical components. He shows that all animal flesh contains, besides fibrine, albumen, gelatine, and fat, certain other constituents which may be separated from the rest by a simple process of infusion, straining, and evaporation. The substance thus obtained is extract of flesh. This compound was not unknown to chemists before Liebig drew special attention to it, but they regarded it only as a remedy for disease and exhaustion, and recommended it as a resource for extremities of nature, and especially for the sick and wounded soldier on the field of battle, with sinking and exhausting powers. That which at one time was considered to be a last resource is now an article of daily consumption in our hospitals and households, and is almost as commonly used as tea, or any other beverage. A frequent inquiry is, in what consists the efficacy or advantage of Liebig's extract ? and the popular idea is, that being a concentrated extract of pounds of flesh, it cannot fail to be extremely nutritious. But it is not so, and it will be surprising to those who believe in this doctrine to hear that the extract of meat contains little or nothing of what may be said to be at all nutritious. The substances which go to form nourishment for the body are fibrine, albumen, and fat ; but these are not present in the extract of meat. One hundred parts of beef contain the following constituents :

1. Fibrine................................. 4
2. Albumen.... 4
3. Gelatine............................... 7
4. Fat.................................... 30
5. Mineral matter......................... 5
6. Water................................. 50
                                          ___
                                          100

Let us contrast with this the composition of a hundred parts of Liebig's extract of meat :

1. Creatine, Creatinine, Inosic Acid, Osmazome.. 51
2. Gelatine................................ 8
3. Albumen................................ 3
4. Mineral matters........................ 21
5. Water.................................. 17
                                          ___
                                          100

The difference will be at once seen. The water has diminished by half, the albumen is less, and there is four times the quantity of mineral matter, and a set of substances is introduced which occupy half the bulk of the compound, which are not noticed in the composition of beef itself at all. If, then, the extract of meat differs from beef, and all other nutritious articles of diet, it is not in containing nutritious matters, but in the fact that chemical compounds and mineral matters just mentioned are found in large quantities. It is to these, therefore, that we must ascribe the marvellous powers which the extract of flesh exerts on the human system. The chemical action of these products on the human body are imperfectly understood ; but it is certain that when albumen or fibrine is partaken of alone it will not digest or support life ; but when in combination with these mineral matters found in the juice of meat, and of course present in every pound of meat, they are digested and appropriated to the nutrition of the body. It therefore follows that Liebig's extract of meat, if partaken of alone, would in no way support life ; but, if in combination with bread or eggs or any ordinary food, it enables the stomach to assimilate all the nourishment contained in these articles and provides sustenance for the failing powers at a much less cost to the digestion than if it had unassisted to extract what nourishment they contain.

The juice of flesh or extract of meat, it will be seen, contains no new product after its manufacture, but simply those constituents in a concentrated form, which are ordinarily present in the flesh of animals. The great advantage it confers is, that it is already fit for use. A teaspoonful of the extract in a pint of hot water is a stock for any soup, and admits of any variety of flavoring. For the dyspeptic, whose stomach cannot bear tea or coffee, it is an excellent beverage and assists materially in the digestion of any solid food that may be taken with it. Hence it is a most valuable adjunct to the invalid's table, or to the ménage of the ordinary cook,' in whose hands it may be made to form the basis of many rich and well-flavored soups. A mutton chop eaten alone, or even with tea or coffee, will frequently prove most indigestible, and the dyspeptic who seeks for nourishment will give it up as impossible. Let him, however, try with it, instead of tea or coffee, a cupful of Liebig's extract, with salt and pepper, and he will find his chop nutritious and pleasant and usually require no other addition besides a little bread to his excellent meal. A portion of this preparation partaken of at proper times will often render recourse to alcoholic stimulants unnecessary and supply the needful refreshment to the system.

MECONIUM is a word used for the inspissated juice of the poppy, a substance resembling treacle. It is also, and more frequently, applied by medical men to the first dark slimy discharge from the bowels of a newly-born infant, which generally passes within an hour or two of birth, but is not entirely expelled till the babe has partaken of the mother's first milk, which is in itself of a cleansing and slightly purgative nature, and is adapted to remove all these impurities from the infant's body.

MEDULLA OBLONGATA is the name given to an important and central part of the brain which is situated at the lower and posterior part of the skull, just where the spinal cord joins the brain ; it is the centre from which emanate most of the principal cranial nerves ; an injury here causes sudden death. (See BRAIN.)

MEDULLARY CANCER is one variety of cancer. (See CANCER.)

MEGRIMS means a variety of headache to which women are often liable, especially if they have been subjected to weakening influences, such as prolonged suckling or profuse menstrual flow. It is also common in badly nourished women from whatever cause. Occasionally it will come and go almost like an aguish complaint ; more commonly, however, it persists for a time. The best means of getting rid of such headaches are stimulants, especially ammonia, for the time being, but they invariably return. Good food and tonics are necessary, therefore, for the relief of such headaches, and without these and the removal of the cause of the weakness little good need be expected. Perhaps the remedies best adapted are bark and ammonia, followed by iron. But if these headaches have lasted a long time, something more is necessary. In such cases we have found the chloride of ammonium (sal ammoniac), combined with the perchloride of iron, both in full doses, the best remedy attainable. The sal ammoniac must be given in doses of 20 or 30 grains, and the iron in doses of 30 drops. At the same time the bowels must be attended to. (See HEADACHE.)

MELÆNA is the name given to hemorrhage from the bowels. It may occur under various conditions sometimes from bleeding into the stomach, when some of the blood naturally finds its way down the intestines. (See HÆMATEMESIS.) Cirrhosis of the liver occasionally causes it, because the circulation

through that organ is so obstructed, and the bleeding in such cases is a source of relief to the distended vessels ; more frequently piles are produced in the course of this disease of the liver. (See PILES.) Melæna is not uncommon in the course of typhoid fever, and when very profuse may cause a fatal result ; yet when the quantity is but small the danger to the patient does not seem to be greater in consequence. Turpentine in small doses is valuable at such times, but it should be given with great care ; iced milk, but no solid food, may be given. (See TYPHOID FEVER.) In some forms of Bright's or kidney disease, melæna may occur in the early stage, but it seldom is profuse, and does not call for special treatment. (See KIDNEY.) Ulceration of the intestines is often accompanied by melæna, and may occur in cases of phthisis or in scrofulous disease. Rest in bed, cold drinks, light and nourishing food should be given, and also some astringent medicine, as iron, tannin, or gallic acid, sugar of lead, and turpentine ; but the quantity to be given and the choice of the remedy must depend on the special peculiartiies of the case. When the melæna proceeds from the upper par' of the bowel, it passes away mixed with the excreta, and has a brown or coffee-ground color ; when from the rectum or the lower part of the bowel, it has the usual appearance of clotted blood ; in these latter cases injections containing iron or tannin, are very useful, as then the astringent fluid can be applied directly to the part. When admissible, opium is often of service in keeping the patient quiet and in allaying nervous excitement, and also in preventing any undue movement of the bowel itself. In cases of melæna it is very rarely advisable to give purgatives, and they should always be taken with much caution, as their presence may give rise to irritation and increase the flow of blood.

MELANCHOLIA is that variety of insanity which is characterized by delusions of depression. Sometimes the patient becomes excited over these, but not very frequently. There are several varieties of melancholia, but the above definition applies to almost all of them. (See HYPOCHONDRIA and INSANITY.)

MELANOSIS is a disease characterized by the deposition of black or dark-brown coloring matter in various textures and organs. Almost any form of tumor may become melanotic, but one variety of cancer has a special tendency to do so. True melanosis has its site most frequently in the skin and tissue just below it ; but it is also frequently present in and beneath mucous membranes. In certain parts of the body pigment of this kind is normally present ; thus there is always some in the skin even of a European, much more in those of the darker-colored races.

So, too, it is present in certain parts of the eyeball, the lungs, and other internal organs. In general melanosis we have usually a number of masses scattered through different parts of the body. These are of very various sizes, from a pin's head to that of a walnut, and may invade even the tissues of the heart and the bones. It is commonly supposed that these melanotic masses are malignant in their nature—cancerous, that is to say—and tend to shorten life ; but this is by no means certain as yet. There are certain forms of false melanosis of importance. Thus, there is the lung of those who inhale large quantities of carbonaceous matter. These individuals often suffer from a form of consumption characterized by certain peculiarities. After death their lungs are found quite black ; the bronchial glands, too, are blackened, though that is nothing unusual. Most lungs of those who live in cities tend to acquire, in course of time, more or less pigmentation. Blood may be blackened. Thus the action of the gastric juice on the blood in the vessels of the stomach after death may give the stomach an appearance of mortification. Blood, too, which has been extravasated into the gut becomes quite black before it passes from the bowel, if it is allowed to remain in it long enough. As to *treatment* little can be said. In true melanosis, if the mass can be reached it should be removed ; but if truly melanotic it is most likely multiple, and represented elsewhere.

MENINGITIS signifies inflammation of the membranes of the brain. Under article BRAIN it has been shown that this organ has three coverings, in the following order, from without inward : 1, A dense, fibrous structure the *dura mater ;* 2 A thin delicate membrane, the *arachnoid ;* and 3, a tissue full of vessels, the *pia mater.* The last two membranes are those usually affected in the process of inflammation. There are several varieties of inflammation of these coats, depending chiefly on the cause. 1. *Traumatic meningitis,* or meningitis dependent on a blow or injury to the head, which may be also accompanied with fracture of the skull. In the course of two or three days after the accident, severe pain will be felt over the seat of injury, and the patient will be feverish and thirsty ; constant sickness and a moist white tongue are also very prominent symptoms. The pain increases in intensity, the head is hot, and the face often flushed, and the patient restless and disliking the light ; generally, convulsions come on, or the limbs are affected with convulsive starts ; delirium at night is generally present, and the patient may lie on his back in a prostrate condition, moaning at times, and picking at the bedclothes. By degrees he becomes drowsy and stupid, with sometimes a flushed face and suffused eyes, taking little, if any, notice of

what is going on around him. As the drowsy state deepens, the pain is not felt, and gradually he passes into a state of coma or deep insensibiliy, from which he cannot be roused ; the pupils, at first small, generally become at this stage larger than usual, and squint may be also present ; now the face is pale, and the pulse frequent and often irregular. Very few cases recover when the disease has advanced so far, and, in fact, the greatest danger always exists whenever these serious symptoms come on after an accident or injury to the head ; nevertheless, death may not take place for weeks in some cases, the patient lying meanwhile in an unconscious state. The *treatment* will consist in perfect rest after the injury, however slight it may seem to be at first ; he should lie in bed or on a sofa in a cool and rather darkened room, and avoid all kinds of excitement. No stimulants should be given, nor are any required, as they only tend to flush and excite the patient ; the boweis should be kept open, and the forehead may be cooled by means of vinegar and water, or an evaporating lotion, in which small pieces of ice are melting, or an ice bag may be applied to the head. Medical advice should always be sought for early, as very little hope can exist when the disease has made much progress.

2. *Tubercular meningitis* is another variety which is very fatal to children, and equally so, but not so commonly met with, in adults. The same tissues or covering are affected as in the preceding variety, but the nature of the inflammatory process differs somewhat in this kind. *Acute hydrocephalus* is another name for this disease, but it should be avoided, as persons are apt to confound this disease with chronic hydrocephalus, whereas there is no connection between the two. Some have also styled this *brain fever.* Children are usually attacked between two and five years of age. The symptoms begin by their feeling listless and disinclined to play about ; they are fretful, and wish to lie in their nurse's or mother's lap ; generally they complain of pain in the head. In a day or two these symptoms are more marked and the pain increases, accompanied by very constant vomiting and generally convulsions, although in some cases this latter symptom is wanting. The tongue is white and moist, the abdomen generally concave instead of convex, and the thumb of each hand is turned inward. The child dislikes the light, and the pupils are smaller than usual. At times squinting is met with ; the face may be pale or occasionally flushed, and if the finger be lightly drawn across the forehead, a red blush or wheal will at once appear. The bowels are usually confined, and the pulse quick, and the heat of the body greater than usual. Gradually and in the course of a few days the child becomes semi-unconscious, lies in its bed, taking little,

if any, notice of what is going on around, and the symptoms become still more marked. The stupor generally goes on increasing, the pupils dilate, the pulse becomes irregular, often the child utters a low moaning cry, and by degrees it passes into a state of complete insensibility until death puts an end to its sufferings. No case recovers after the disease has once clearly developed itself, but many cases may be mistaken for it ; thus, a child when teething may present at first many symptoms similar to those met with in the early stages of this disease, and much alarm may at first be created in consequence. To accurately distinguish these cases requires a good deal of experience and knowledge. The children most liable to be affected with tubercular meningitis are those of a nervous and excitable temperament, and such as are percocious for their age ; those also who are suffering from diseases of joints and enlarged glands are subject to it. Death usually takes place in the course of three or four weeks from the onset of the malady. The *treatment* will consist in following the same plan as has been mentioned above for traumatic meningitis. A pleasant way of applying cold to the head is to pound ice in small pieces, and place them in a bladder ; this can then be suspended from the head of the bed, and placed on the child's head for a couple of hours at a time ; in this way the pillow and bed-clothes are prevented from getting wet. Very little nourishment can be taken in these cases, and usually a little iced-milk is most grateful to the patient.

3. *Meningitis* may come on as a result of exposure to cold or to great heat, or in the course of many febrile disorders, as pyæmia, septicæmia, etc., or from the presence of tumors in the brain, or from disease of the bones of the skull as a result of scarlet fever and syphilis. The symptoms are such as have been mentioned above, and show a serious disturbance of the functions of the brain. Nearly all such cases commence with pain in the head, sickness, fever, intolerance of light, and are followed by convulsions, which lead on to stupor, coma, and insensibility. In nearly all these instances, too, a fatal result may be expected in consequence of the serious injuries which ensue to such important structures. They all depend probably in the first place on some altered and poisoned condition of the blood, and the *treatment* to be adopted, in addition to what has been above recommended, will consist in the various remedies appropriate to the special cause which gives rise to the disease.

Another form of meningitis also affects the spinal membranes as well as those of the brain ; it occurs in an epidemic form, and is very fatal. (See CEREBRO-SPINAL FEVER.)

MENORRHAGIA is a profuse discharge of the catamenia at the menstrual period,

attended with more or less debility, pallor, and discomfort. It generally occurs in those who are out of health, and who have been weakened by having had a large family, by a difficult labor, or in whom there is some disease of the womb, or ovaries. The flow may be more profuse than usual, but only last the usual number of days, or it may be more or less persistent for several days or weeks, but not very great in amount. After delivery, the womb does not always return to its usual proportions, and this will often be a source of menorrhagia, as the vessels are then congested and readily bleed. A tumor in the womb, or a polypus growing from it, or cancer of that organ, or displacement of it, will all cause this malady. Those who have suffered from lactation, or who have had miscarriages, are also liable to this affection. The patient will also have pain in the lower part of the back, and perhaps down the thighs, pain or difficulty in passing water or in evacuating the bowels, and a general inability to walk or undergo any exertion. The amount of pallor will be great in proportion to the loss of blood, the health of the patient is but indifferent, there is very little appetite, no fever or thirst, but a general feeling of languor and prostration. It is sometimes difficult to distinguish between this affection and an abortion, but in the latter case there will be the fact of pregnancy, and the expulsion of undeveloped portions of the fœtus.

Treatment.—The treatment must consist in rest in the horizontal position on a couch, and this may have to be enforced ·for several days or even weeks. Astringent medicines, especially containing iron or tannin, gallic acid or ergot, may be given with much benefit. Cold applications and the injection of cold water are very useful, and must be given when the hemorrhage proceeds from a tumor in or cancer of the womb. The latter disease generally comes on in women over forty-five years of age, and generally produces a fatal result in a year and a half or two years. Emaciation, great pain in the back and abdomen, a sallow, cachectic appearance, a fœtid discharge and occasional menorrhagia are the main symptoms of this disorder. In all cases of menorrhagia rest must form a chief part of the treatment, the diet must be light and nourishing, constipation must be avoided ; a gentle drive may be taken in fine weather by those who are able to do so ; if not, a short walk should be taken daily, so as to obtain a little fresh air, but if it bring on a fresh discharge of blood this must be discarded for a time. The malady is very apt to recur, and the more so after every succeeding confinement. Patients so affected should not get up too soon after a labor, as an erect posture tends to cause congestion of the womb, and to induce a fulness of the vessels which predisposes to menorrhagia.

MENSTRUATION is a function performed by women between the age of puberty and middle life, this forms the child-bearing period, and usually lasts about thirty years. Various names have been given to this function : it is spoken of as the menses, the period, the catamenia, etc. When not performed at all, the patient is said to have *amenorrhœa*, when the function is performed with difficulty or pain it is called *dysmenorrhœa*, and when the discharge is very profuse, the individual is suffering from *menorrhagia*. Each of these is treated of in its proper place. The appearance of menstruation is generally accompanied by more or less pain in the back, headache, and lassitude ; often, also, the patient loses color and has a dark ring round the eyes. From thirteen to fifteen years of age is the average time when menstruation commences, but it may come on a year or two earlier in some cases, or it may be much delayed in others. The periods are frequently irregular at first, and some months may elapse before the function is carried on with regularity. When well established, an interval of about four weeks elapses between each period, but sometimes only three weeks intervene. The blood that flows differs from ordinary blood in being acid instead of alkaline, and in not clotting unless poured out in large quantities. This function is always suspended during pregnancy, and is, in fact, the chief symptom from which a woman dates the expected time of her confinement. Many causes will tend to cause irregularity in the performance of this function during the child-bearing period of life—exposure to cold or wet, mental emotion or worry, acute diseases, consumption, cancer, and many other diseases may either cause menstruation to stop altogether, or be diminished in quantity, or to occur at irregular intervals. The period of life when menstruation ends is known as the climacteric period or *change of life*, and the cessation of the function is often accompanied by more or less distress ; the patient becomes nervous and is easily worried, suffers from lowness of spirits, pain in the back and between the shoulders, pain also frequently in the left side, and headache ; the temper may be irritable and the appetite capricious. These symptoms arise in a great measure from a disturbance in the nervous system, giving rise to various neuralgic pains. Such symptoms, however, though often troublesome, are not attended by danger, and subside when the function of menstruation has quite ceased. The flow generally ceases gradually, and becomes more and more scanty ; sometimes it is for a short time much increased in quantity. Tonic medicines, such as iron, and quinine, may be given with advantage, and if there is much nervous derangement, assafœtida or valerian may be given in addition. (See EMMENAGOGUES.)

MENTAGRA, or TYNEA SYCOSIS, is a disease of the beard, moustache, whiskers, and inner part of the nostrils, in which a little fungus or vegetable parasite finds its way into the root of the hair. Its presence sets up inflammation of the hair follicle, and a little matter forms around the hair ; the part around becomes hardened, and brownish thick scabs form among the hairs. It may be mistaken for acne or impetigo, but the presence of the parasite under the microscope will clear up the doubt. The scabs should be removed by moistening them with oil and then applying a hot poultice, or the part may be washed with hot oatmeal and water. Each hair should be pulled out and some substance applied which will destroy the life of the parasite, as acetic acid, perchloride or pernitrate of mercury. The general health must at the same time be kept up by a light and nourishing diet, daily exercise, and the administration of tonic medicines.

MERCURY, in Latin *Hydrargyrum*, is a remedy about which many doubts have been raised. Of its power none can doubt, but of the appropriate cases in which to allow its power to be exerted there is still much hesitation and doubt.

Metallic mercury is mainly obtained from its red sulphide, cinnabar, which, being distilled, yields mercury. This requires re-distillation, however, and washing with dilute hydrochloric acid. When pure, metallic mercury is a brilliant white, metallic-looking liquid, becoming solid at 40° below zero F., and volatilizing at a heat below redness. Rolled on paper, pure mercury forms globules which leave no stain ; if amalgamated with other metals it generally does. Liquid mercury is seldom used in medicine ; it has been given in obstinate constipation, with the idea that its weight would force a passage, but in vain. There are, however, several preparations of mercury where the metal is only in a finely divided state, and these are very efficacious. Thus there is *mercury with chalk* or *gray powder*, consisting of metallic mercury rubbed with chalk till the globules disappear. *Mercurial pill* or *blue pill* is prepared in the same way, by rubbing metallic mercury with confection of roses till globules can no longer be seen. *Mercurial ointment* or blue ointment is prepared in like fashion, by rubbing mercury with lard and suet. These are all admirable preparations. Mercurial plaster, mercurial liniment, and mercurial suppositories all have metallic mercury as their basis. The vapor of mercury acts powerfully, as used to be seen in its effects on the makers of looking-glasses. By rubbing metallic mercury into the skin till the exceedingly small globules make their way through its pores, the full effect of the metal may be produced. Given in repeated small doses, mercury first of all increases the various secretions—the saliva, the bile, the intestinal juices, etc. The increase of the saliva is well marked, and salivation is one of the best established actions of mercury. The saliva increases, the gums become sore and tender, till they can hardly close on a morsel of solid food. Round the bases of the teeth they seem swollen, red, and spongy, and the whole yield an exceedingly disagreeable fœtor ; the inside of the lips may suffer also. This is, as a rule, the first indication of the full effects of mercury. On the liver mercury seems to have the power of increasing the flow of bile for the time being, or, at all events, of emptying the gall bladder. The metal ordinarily acts as a purgative, producing copious high-colored soft motions, the increased fluid coming partly from the intestinal canal. Frequently, too, the kidneys act better, getting rid of a larger quantity of urine than usual.

Mercury, in producing its effects, always finds its way into the blood, and may also be detected in various secretions. It influences the nutrition of the blood for the worse, especially if long continued or given in weak subjects. Given in over-doses or too long, mercury produces serious mischief. The body wastes, and a kind of fever may be induced. This is sometimes marked by skin eruptions. There are also tremors or shaking, beginning in the hands and arms, whose movements lack precision ; gradually they extend to the whole body. They cannot be controlled, and being excited last some time ; even the respiration may become spasmodic. Salivation may be absent with tremors. Sulphur baths and iodide of potassium are the remedies for the condition. In other cases there is terrible salivation, the tongue swollen so that nothing can be taken into the mouth ; the teeth may fall out and the jawbones die. Occasionally pints of saliva flow per day. In these cases there may be excessive purging. Salivation is most likely to follow the swallowing or inunction of mercury. Mercurial tremors more frequently occur after inhalation. Children are rarely salivated. Mercury used to be given in all cases where acute inflammation existed, with a view to arrest its effects. Its preparations were constantly given to salivation with this view ; now they are seldom so used, except in inflammation of the eye, and in inflammation of some serous membranes, as the peritoneum, which lines the abdomen. They are also used with a view to the removal of the deposits caused by inflammation, as in effusions into the pleura, in pericarditis, etc., but assuredly in some of these its effects are not only useless but injurious. In acute rheumatism mercury is still given by some practitioners, but they are not numerous. The general opinion is that few remedies, if any, affect the cause of the disease, which must be watched for com-

plications. In these, mercury may be of use ; in the ordinary disease, never. In certain forms of dropsy mercury may be of great use, especially if from antecedent inflammation, but in the majority it is worse than useless. In syphilis mercury used to be given invariably ; now, not in all, but in a certain number of cases. On the whole, a case of syphilis is better for mercury, provided the health of the individual will bear the course, but not otherwise. It is best, too, perhaps to give it by vapor, as a mercurial vapor-bath, or by injections, as that will not interfere so much with digestion. Its use should in no case be continued too long, or harm will follow. It is a common practice, and on the whole a satisfactory one, to give a blue pill and a slightly purgative draught for bilious headache, or when there are the usual symptoms of biliousness, furred tongue, foul breath, etc. Gray powder is much given in the affections of children ; in many it requires caution, especially in rickets. The same preparation is extremely valuable in certain forms of diarrhœa, where the motions are green and slimy, frequent and offensive, as occurring in children. In other cases, corrosive sublimate, especially in adults, does better. The gray powder is usually given in doses of from a grain to five grains. The blue ointment is rubbed into the skin at the groins and arm-pits to produce the effects of a mercurial preparation.

Calomel, also known as the sub-chloride of mercury, is one of the best known preparations of mercury. It is prepared from the sulphate of mercury, and is a heavy white powder, insoluble in water, ether, or spirit, but volatized by heat : it is quite tasteless. Added to lime-water, calomel yields the well-known *black-wash* (*lotio nigra*), which consists, however, of the suboxide of mercury mainly. The *compound calomel* or *Plummer's pill* is another well-known preparation ; it contains calomel, sulphurated antimony, guiacum resin, and castor-oil. Calomel ointment consists of lead and calomel. Internally, calomel does not irritate, but generally gives rise to nausea, and purges if given in sufficient dose. It acts apparently on the liver and intestines, and is largely used in the treatment of certain of these maladies. In children it produces green stools. The compound pill is mainly used as an "alterative" in chronic skin disease, especially of a syphilitic origin. Black-wash is used for sores of a syphilitic character. As a purgative calomel is given in doses of from 2 to 10 grains. Its chief advantage is its tastelessness. If given to affect the system, only small doses, $\frac{1}{4}$ or $\frac{1}{2}$ a grain, are given every four hours. Calomel may also be given as a fumigation. Calomel ointment is often used for the relief of the itching of some forms of skin disease, and is blown into the eye in certain forms of ophthalmia.

Corrosive sublimate is the perchloride of mercury. (See CORROSIVE SUBLIMATE.)

White precipitate of mercury, or ammoniated mercury, is obtained by precipitating corrosive sublimate by ammonia. It contains ammonia itself, and is a white amorphous powder, capable of sublimation. It is never used internally, and its only preparation is an ointment which is chiefly used for destroying vermin ; for this it is well fitted. It is also used for unpleasant-smelling discharges from the nostrils.

Green iodide of mercury is obtained by causing iodine and mercury to combine directly. It does not keep very well. It acts similarly to calomel, but does not purge nearly so much ; it is therefore chiefly used for the constitutional effects of mercury, and with many is the favorite preparation for syphilis. It is also used as an ointment for skin eruptions.

The *red iodide of mercury* is of more importance. Internally it is usually combined with some other substance, iodide of potassium and corrosive sublimate being mixed to form it. It is largely given in the advanced stages of syphilis, with excess of iodide of potassium. It should not be given in any substance containing an alkaloid. Goitre and enlarged spleen have been successfully treated by its ointment, especially in India.

Red oxide of mercury, also known as red precipitate is usually seen as red shining crystals, entirely volatilizing by heat. Its ointment is only used externally, as an irritant to the eyelids in ophthalmia, to destroy vermin, and the like. It is also applied to indolent ulcers of a specific kind.

The *acid nitrate of mercury*, or its solution, is made by dissolving mercury in nitric acid. This is a colorless and highly acid solution, of which the only preparation is an ointment which used to be called citrine ointment. The solution itself is a powerful caustic, and has been applied to arrest the disease called lupus. It is not given internally. This ointment, too, is irritant or stimulant, and is used in some eye diseases, especially inflammation of the lids, and in chronic scaly eruptions about the hands, especially from syphilis.

The *sulphuret of mercury* (artificial cinnabar), better known as vermilion, is not now officinal. It may be employed for local fumigation, and also as an inhalation off a hot brick, or from a lamp specially contrived, for syphilitic sore throat. It is useless in the earlier forms of sore throat, but may do good when there is ulceration ; dilute solution of corrosive sublimate applied as spray is, however, better.

Sulphate of mercury is only used as the basis of these preparations.

MESENTERY (THE) is a double fold of peritoneum which retains the small intestines in their place in the abdominal cavity ;

it is fan-shaped in form, and attached to the front of the spine at its narrow end. Around its longer margin the bowels are arranged, so that perfect freedom of movement upon each other is allowed, while yet each portion keeps in its proper place. Between these two folds run some vessels which take blood to and from the intestines ; these are called the mesenteric vessels, and consist of arteries and veins ; they are also accompanied by various nerves. There are, besides, a great many glands in the mesentery, called the *mesenteric glands*, and these are often liable to disease. Through these glands passes an alkaline, opalescent fluid called the *chyle :* this chyle is collected in the intestinal walls by a vast number of small vessels called *lacteals*, which are very analogous to the lymphatic vessels in other parts of the body ; these lacteals join together and form larger branches, until, having passed through the mesenteric glands, they convey the chyle, altered by that process, to the *receptaculum chyli*, a dilated tube lying in front of the spine, and serving as a kind of reservoir for that fluid, which, afterward passing up the thoracic duct, enters the blood at the left side of the root of the neck. The mesenteric glands become much enlarged in typhoid fever and in some cases of consumption : very frequently also they become diseased in children, and this may occur very early in infant life ; when this occurs, the nutrient material of the food which ought to be absorbed in the intestine is diminished in quantity, and so the chyle, being altered perhaps in quality as well as in amount, and obstructed by the disease in the glands, is unable to pass on into the blood as usual, and thus emaciation and death may ensue. Mesenteric disease is most common in scrofulous children ; such infants have large stomachs, a doughy skin, perhaps the enlarged glands may be felt through the abdominal walls, as a hard, firm mass in front of the spine ; sickness and emaciation attend the disease, and frequently diarrhœa and constipation alternate ; at times dropsy comes on and adds to the suffering. In such cases great care must be taken in giving the child a light but nourishing diet, consisting chiefly of milk, with an occasional egg and some beef-tea or good mutton broth ; solid food should be avoided, as it may irritate too much the delicate mucous membrane of the stomach and intestines. Preparations containing iron, as iron wine or the syrup of the phosphate of that metal, are very valuable, and may be given in small doses twice a day. Lime-water should be mixed with the milk in the proportion of one part of the former to three or four parts of the latter, if there is much sickness or diarrhœa present. Great benefit will be derived from rubbing cod-liver oil into the skin night and morning ; it often happens that a child dislikes or cannot take cod-liver oil by the mouth, and this plan of rubbing it into the skin is very efficacious, and prevents any sickness or distress arising from taking it by the mouth ; its action is increased in value. if about one tenth part of strong solution of ammonia is added to the oil ; the skin is thereby stimulated and absorption takes place with greater ease. Children suffering from this disease often derive benefit from going to the sea-side, and although they are seldom in a condition to bathe in the sea, yet a daily bath in sea-water in the nursery is of much service. Should dropsy come on in these cases, an operation for removing the fluid from the abdomen may be resorted to. The procedure is simple enough, and consists merely in introducing a small hollow tube through the skin, and allowing the fluid to run through it. The after-treatment will consist in a soft flannel bandage being passed round the abdomen, in rest in bed for a while, and a diet similar to what has been mentioned above. Disease of the mesenteric glands attended by wasting is called *tabes mesenterica*. (See TABES MESENTERICA.)

MESMERISM is a term usually applied to the phenomena of animal magnetism, after the name of its first propounder, Anton Mesmer, a German physician, born at Baden in 1739. Perhaps the time has not yet come when the combined physiological, pathological, and psychological phenomena of mesmerism can be rationally explained. At any rate it involves a series of facts in relation to the human system which historically have a high interest. Many of these are new since the time of Mesmer, but he first gave a systematic character to the phenomena and sought to refer them to scientific principles. Since the death of Mesmer animal magnetism has had directed toward it a great amount of attention, and has been investigated by physiologists of eminence, and used as a curative agent by some medical men. A great impulse was given to the theory of mesmerism by a series of letters to the *Athenæum* from Miss Martineau, who attributed her cure from a long-standing ailment to the influence of animal magnetism. A correspondence on the subject took place at that time in the pages of the *Athenæum*, in which Miss Martineau's conclusions were shown to be mistaken to the satisfaction of the majority of readers. Writers on animal magnetism distinguish many stages. The following classification is by Kluge, a German writer, on the subject :

First Degree.—Called waking, when the intellect and senses retain their ordinary powers and susceptibility.

Second Degree.—Half sleep, or imperfect crisis. Most of the senses retain their activity, that of vision only being impaired, the eye withdrawing itself from the power of the will.

Third Degree.—The magnetic, or mesmeric sleep. The organs of the senses refuse to perform their respective functions, and the patient is in an unconscious state.

Fourth Degree.—The perfect crisis, or simple somnambulism. In this stage the patient is said to "wake within himself," and his consciousness returns. He is in a state which can neither be called sleeping nor waking. but which appears to be something between the two.

Fifth Degree.—Lucidity or lucid vision. This is called in France and mostly in this country clairvoyance ; in Germany, *hellsehen*. In this state the patient is said to obtain a clear knowledge of his own internal, mental, and bodily state ; is enabled to calculate with accuracy the phenomena of disease which will naturally and inevitably occur, and to determine what are their most appropriate and effectual remedies. He is also said to possess the same faculty of internal inspection with regard to other people who have been placed in mesmeric connection (*en rapport*) with him.

Sixth Degree.—Universal lucidity. In this state the lucid vision becomes greatly increased, and extends to objects whether near or at a distance.

Such is the system as recognized by mesmerists, and many volumes have been written on each phase and condition. Many who practise mesmerism are themselves sceptical with regard to the real existence of the last two degrees, although such cases are recorded. Many theories have been propounded in order to embrace the facts of animal magnetism, and numerous aspects given to the question by inquiries both in confirmation and refutation of the supposed facts elicited by inquiry. The whole series of phenomena known as electro-biology, table-turning, spirit-rapping, and odylic force are based primarily on this condition of mesmeric sleep or influence. There can be no doubt that the condition of mesmeric sleep does exist, and that some persons are much more susceptible to this condition than others. When under this influence they are readily open to obey the will of another, and become as it were the slaves of suggestion and the victims of the operators. They exercise their volition unconsciously, and attribute it to the existence of a mysterious force. The known fact of the great increase of force that takes place in normal conditions of the system when the whole attention is concentrated on one idea, serves to explain the feats of strength performed by persons in the sleep-waking state. Although so many of the phenomena of mesmerism admit now of a rational explanation, it is still practised as a mystery, and large numbers of persons give credence to its marvels. Exhibitions are constantly made before the public, professing to be tests of the power of mesmerism or electro-biology, which are, in fact, but the feats of clever conjurors, and present no remarkable phenomena at all. Such attempts do but injure the reputation of those who may be earnestly inquiring into natural and curious conditions of the nervous system with a view to discover truth.

METACARPUS is a name given to the bones which lie between the wrist or carpus and the fingers or phalanges.

METASTASIS means change, transposition. This is a medical term used by physicians to express that change which sometimes takes place in the seat of a disease, as when, in gout or rheumatism, the heat and pain suddenly leave the foot and take up their abode in the hand or fingers, or go from an external to an internal organ ; such a condition of a disease is called a metastasis, and is always to be apprehended, as a disease migrating from an external to an internal part may be more or less dangerous.

METATARSUS is the name given to the bones which lie between the tarsus or ankle and the toes ; it corresponds to the metacarpus of the hand.

MEZEREON BARK is the bark of the *Daphne mezereon*, a shrub well known. Two plants, however, yield the bark of commerce. This bark, is thin, flat, or curled, tough, brown outside and white within. It is not easily broken. When boiled, an acrid vapor is given off. There is an ethereal extract of the bark, which, however, is seldom given internally in this country. It is a powerful local irritant, and even blisters. Internally, it causes vomiting and purging. It has been used in chronic rheumatism, syphilitic pains and skin diseases. It is contained in the compound decoction of sarsaparilla.

MIASM. (See MALARIA.)

MIDRIFF is another term for the diaphragm, or muscle which is attached to the sternum or breast-bone just above the stomach, entirely dividing that portion of the trunk into two cavities. The upper, the thorax or chest, and the lower, the abdomen or belly. It is this muscle which is liable to a spasmodic affection known as hiccup, or *hiccough*, occasioned by some slight derangement of the stomach, and usually very transient.

MIGRAINE, or BROW-AGUE, is a painful disorder generally seated on one side of the forehead, and causing, while it lasts, great distress to the patient. (See INTERMITTENT FEVER.)

MILIARIA are minute vesicles or little blisters, which at first are transparent, but soon become opaque and purulent in appearance. They are often seen on the trunk and extremities in cases of rheumatic fever. They differ from sudamina in being pointed, in their opacity soon after they appear, and in the narrow red halo around ; however, some look on the two as identical. They usually come on in summer time, and are connected with profuse perspiration.

MILK is the liquid formed in the breasts of all the mammalian tribe of animals, the object of which is the support of their young ti l the time comes that they can take other food. From the earliest time man has used the milk of the domesticated mammalia for the purpose of supplying himself with food. The milk of all forms of mammalia is more or less alike, and contains substances necessary for the nutrition of the whole body. Milk is, in fact, the type of all food (see FOOD). Although man in various countries has recourse to the milk of the horse, the ass, and the goat, the milk which is most frequently used as man's food is that of the cow.. For this purpose the cow is extensively fed and pastured in the various countries of Europe. The following is an analysis of the milk of the cow in 100 parts :

Water	86.0
Casein	5.0
Butter	3 5
Sugar of milk	4.5
Mineral matter	1.0
	——
	100

or

Water	86.0
Flesh and force producers	5.0
Heat and force producers	8.0
Mineral matter	1.0
	——
	100

Not only does milk contain food in an easily digestible form, capable of becoming the food of infants, but its easy digestibility does not interfere with its being used as the food of strong men. The value of milk as an article of diet may be stated in an abstract manner, in the ascertained fact that one pound of cow's milk, when digested and oxidized, is capable of producing a force which would raise 390 tons one foot high. This force, if it could be exactly realized, would enable a man to raise 70 tons one foot, or to perform an amount of work with his brain and muscles equal to the act of raising 70 tons a foot high. This is done through the agency of the oxygen of the air acting upon the carbon and hydrogen contained in the heat and force producing constituents of the milk. The flesh-forming constituents in a pound of milk, the casein, if all digested and appropriated, is capable of making eight tenths of an ounce of dry muscle or flesh. Although cow's milk contains the same general constituents as human and other milks, there is a considerable difference in the quantity of these constituents. The following table presents the different quantities of the substances contained in 100 parts of woman's, cow's, and ass's milk :

MILK MILK 313

	Cow's Milk.	Human Milk.	Ass's Milk.
Water	86	89½	90
Casein, or flesh and force producers	5	3	2
Butter } heat and force producers	3½	3	1½
Sugar }	4½	4	6
Mineral matter	1	½	½
	100	100	100

It will be seen from this table that cow's milk contains less water and more casein, butter, sugar, and mineral matter than mother's or ass's milk. Hence, when cow's milk is used for the feeding of young children, it is usual to add a certain quantity of water ; one tablespoonful of water to two tablespoonfuls of milk is usually recommended. This is, however, not needed when the milk presents less than 8 per cent of cream by the lactometer. When, however, a third of water is added to the milk, the sugar of the cow's milk is reduced below the quantity in human milk, hence it is desirable to add a little sugar, say half a drachm, or half a teaspoonful, to three tablespoonfuls of the watered milk. A better substitute for mother's milk than cow's milk is undoubtedly ass's milk. It should be remembered that ass's milk is altogether a feebler milk than mother's milk. It contains more water, less casein and less butter. It contains, however, more sugar, and this may make up for the deficiency in butter. The milk of the goat is very like that of the cow, and is extensively employed in the mountainous districts of Switzerland, where the goat is more easily grazed than the cow. In Sweden and Denmark the milk of the sheep is used as an article of diet ; in Lapland the people use the milk of the reindeer, and in Tartary mare's milk is employed.

Cow's milk varies in its quantity and composition, according to various circumstances, so that no standard can be given by which genuine milk may be ascertained. Thus, milk is known to vary according to the age of the cow, the age of the calf, and other circumstances in the life of the cow. The time of the day at which the cow is milked makes some difference. It is found to be richer in solids in the morning. The kind of feeding also produces a difference, beet-root and carrots, for instance, are known to increase the sugar. There are different varieties of cows which are known to give milk of different quality. Thus, Alderney cows give more butter, and long-horns give more casein. Milk as sold in the large towns of this country is frequently adulterated. The most common and frequent, because the easiest, form of adulteration, is that of the addition of water. Although when in large quantities water may be easily dectected, yet within the limitations of the natural varieties of milk it is difficult to detect the addition. The easiest way of detecting adulteration by water is to take the specific gravity of the suspected milk. Instruments are sold in the shops by which the specific gravity of the milk may be easily ascertained. The quantity of cream afforded by milk after standing, is a good rough test of the presence or absence of added water. The percentage of cream, which may be ascertained by the use of a long glass divided into one hundred parts, varies from 5 per cent to 40 per cent, the larger percentages having been known to be given by Alderney cows. The average quantity of cream found to be given by cows at Aylesbury is 13 per cent. The milk may be as low as 5 or 6 per cent of cream and yet not be adulterated, but if this low percentage is attended with low specific gravity, then the milk is undoubtedly adulterated. Starch is sometimes added to milk to give consistency to the water which has been added. This may easily be detected by the microscope or the addition of iodine. Salt is added to keep up the specific gravity, and may easily be detected by nitrate of silver, throwing down chloride of silver. The brains of animals have been added to thicken with, but this fraud is easily detected by the microscope. It may be said, however, that the adulterations of milk otherwise than with water are very infrequent.

Milk, after being allowed to stand for some time, is very liable to decompose and become acid. In this condition it is quite unfit for the food of young infants. Much of the diarrhœa that prevails in the summer among children in large towns seems to be due to this condition of the milk. Boiling the milk before allowing it to stand will to a certain extent prevent this tendency. To prevent this, as well as to render adulteration with water impossible, milk has been evaporated, and sold under the name of " *condensed milk.*" This article is now manufactured on a large scale, both in this country and in Europe. Condensed milk is easily converted to the condition of ordinary milk by the addition of cold or hot water. The only difference is, that the condensed (such as is sold in cans) contains proportionately more sugar than ordinary cow's milk. The addition of sugar is rendered necessary in order to prevent decomposition. This really proves a recommendation of condensed milk for infant's food, as the addition of the sugar brings the milk in point of sweetness up to the condition of mother's milk. Condensed milk, we think, may be confidently recommended, not only where new milk cannot be had, but in all cases where the milk sold is suspected of adulteration. It can also be converted into milk for use at any moment, and conse.

quently is free from the suspicion of any injurious decomposition. New milk has also been the means recently of conveying typhoid fever by means of the water used in adulteration, or for cleansing the cans. This evil is entirely prevented by the exposure to heat in the preparation of condensed milk.

MILK FEVER, known also as Ephemera, often comes on two or three days after a confinement, but generally passes off in a few days, leaving no evil effects. The symptoms are languor, heat of skin, furred tongue, restlessness, pain in the stomach and breasts, and loss of appetite. The secretion of milk does not take place regularly, and the breasts may have to be drawn. Any febrile symptoms coming on soon after a labor, are apt to cause alarm, as in many cases danger may be apprehended. Milk fever comes on very soon, while puerperal fever, a very serious disorder, does not generally appear for a week or ten days after delivery. The *treatment* of milk fever consists in giving cooling saline medicines, a diet of milk, gruel, and broth, etc., and keeping the bowels open. The child may be kept to the breast, and these must be drawn if the milk does not flow freely.

MILK LEG. (See PHLEGMASIA DOLENS.)

MINERAL WATERS are such as contain an unusual amount of mineral substances in solution, from which they derive important healing properties not possessed by ordinary water. All water, except that which has been distilled, or which falls from the clouds in wide open spaces far removed from towns, contains a certain amount of mineral matter in the shape of salts of various kinds, to which they owe in great measure the pleasant taste which characterizes good drinking water. (See WATER.) Such a water becomes a mineral water when these saline ingredients are present in excess. Wells yielding medicinal waters derive, as do most other springs, their water orginally from rain-fall. This water permeates the soil, carrying with it a greater or less quantity of the salts which it encounters in its passage, until it reaches the surface of the soil at some lower level, or is artificially raised to the surface of the soil, there to be made use of. According, therefore, to the qualities of the strata though which the water percolates will be its qualities when the surface is reached. Various salts of sodium, especially the chloride, the sulphate, and carbonate, salts of lime, iron and magnesia, with various other less widely distributed ingredients, are found. Very often these are held in solution by the help of carbonic acid gas,' which gives the water a sparkling quality, or they contain sulphur in the shape of foul-smelling sulphuretted hydrogen. Different kinds of water have different uses. For the most part they are used either externally or internally, very often in both fashions.

Those usually applied externally are as a rule above the temperature of the surrounding atmosphere; what are commonly called hot springs. They increase the circulation through the parts, and favor the removal of any effete material which it may be needful for the general health to remove. Given internally, their action varies with their constitution, some acting as tonics, others as eliminatives, as the case may be. One very powerful agency in effecting cures by means of mineral waters is the regimen laid down, and the total change of air and scene, as well as of habits of life, necessitated by removal to the spot where they are to be obtained. The existence of these subsidiary influences accounts for the failure which commonly follows any attempt to secure the benefit of a particular water at home, and to procure the full benefit from them they must be taken on the spot.

Mineral waters are of use only in chronic disorders, and certain forms of these seem to be much more benefited than others. Skin complaints, scrofulous disorders of various kinds, stiff joints, gouty and rheumatic affections, and neuralgic pains of certain descriptions, diseases of the liver and kidneys, disorders of the bowels, and certain abnormal conditions of the womb, are those most likely to receive benefit from a "course," as it is called, of mineral waters. It is important for the invalid very clearly to understand that immediate relief does not follow the change. Very likely he feels worse in the first instance, for he has been dislocated as to his old habits and has not become accustomed to the new; but by and by he will reap the benefit of the change, and this will follow him even when he has returned to his wonted way of life. Broadly speaking, no invalid should go to a watering-place without consulting a medical man of skill; and having selected the spot most suitable, it is generally desirable to place one's self under the care of a local practitioner who is acquainted with the specific property of the waters and how they are taken with most advantage. Very often the rules laid down by these local men seem frivolous, and we are not prepared to say that they never are so, but following them in a good many instances means reaping the full benefit of the waters, abandoning them no advantage at all. It is a notion, unfortunately, but a mistaken one, that the greater the quantity of water drank the greater the advantage reaped; no notion could be more erroneous. In all cases the patient should begin with a moderate quantity of the water, say two or three glasses in the morning before breakfast, and one or two in the evening before supper. The patient, if strength will permit, should rise early, walk to the springs or pump-room, swallow slowly a tumbler of the water in a luke-warm state, neither too hot nor too cold, walk for a quarter of an hour, return for another glass, renew the walk gently ; a third glass should follow, if permitted, and a gentle saunter home to breakfast. If there is bathing, that is generally done in the forenoon, about two hours after breakfast. Dinner should be early and light ; an excursion may be made in the afternoon ; in the evening, drink as before a tumbler or two of water, and to bed before ten, a light supper having been partaken of some hour or so before. A little attention to diet is necessary, and, as a rule, during the period that the patient is drinking the waters his stimulants should be restricted to some light wine or well-fermented bitter beer. Above all things, regularity and persistence are to be cultivated at such water-cures ; regularity in rising, in eating and sleeping ; persistence in the object of the cure. Taking twice as much of the water one day will not make up for total neglect on the next, and so indiscretions in the way of diet or stimulants may at particular periods of the course undo the work of weeks. The grand rule is *festina lente*, for too great a hurry to get well may undo the whole good acquired or acquirable.

Mineral waters are commonly divided or grouped according to their constituents. Chief among these are the saline, the chalybeate or iron, the sulphurous, and gaseous. Some are faintly acidulous, others are alkaline. The saline waters are the most numerous. Some contain mainly purgative salts, like sulphate of soda or sulphate of magnesia ; such are Leamington, Cheltenham, Seidlitz, Püllna Carslbad, etc. Others, again, contain more common salt and so act less on the bowels ; such are Wiesbaden, Baden-Baden, Homburg, and Kissengen. The chalybeate waters are sometimes also slightly laxative—a valuable combination. Most, however, retain the iron in solution by means of carbonic acid. These are especially useful in cases of debility, where the patient seems bloodless and weak. The sulphurous waters are a tolerably numerous class. They are largely patronized for a variety of chronic disorders, skin eruptions, liver and womb diseases, gouty and rheumatic ailments. In many of these the efficacy is increased by the warmth of the waters. Gaseous springs can hardly be recognized as a distinct group, inasmuch as both saline and chalybeate waters frequently obtain their palatable character owing to the carbonic acid gas they contain. The thermal waters of Vichy are salines of this class. The waters of Kreuznach are peculiar in containing both iodine and bromine in considerable quantity. Hence they possess considerable efficacy in dealing with scrofulous disorders, but, being weakening, require to be prescribed with caution. We shall now proceed to give a short account of the most important springs

in this country and in Europe, with their most important properties.

The most famous group of mineral waters in America are the SARATOGA SPRINGS, in New York State, comprising 28 springs in all ; some chalybeate, others impregnated with iodine, sulphur, and magnesia, and all powerfully charged with carbonic acid gas. *High Rock Spring* is the oldest of the group, and its medicinal properties were known to the Iroquois Indians at the period of Jacques Cartier's visit to the St. Lawrence in 1535. Toward the end of the eighteenth century its curative value began to be appreciated by the whites ; the first hotels were erected in 1815 ; and since then the fame of the springs has spread so widely that, in addition to the hosts of visitors, immense quantities of the waters are bottled and sent to all parts of the United States and Europe. The High Rock water contains 69.5 grains of mineral constituents and 51 cubic inches of carbonic acid to each pint, and is strongly cathatric and tonic. Perhaps the most popular of the Saratoga waters is that of the *Congress Spring*, which contains 75 grains of mineral constituents and 49 cubic inches of carbonic acid gas to the pint, and is carthartic and alterative. The *Columbian Spring* is strongly impregnated with iron, and acts as a diuretic and tonic. The *Washington* is less ferruginous than the Columbian, but is an excellent tonic and pleasant to take. The *Hathorn* is a powerful cathartic, and acts also as a tonic and diuretic. The *Hamilton* is mildly cathartic and alterative, and the *Pavilion* is one of the best of the cathartic waters, the *Star* being also excellent. The *Putnam* is chalybeate, and is used for bathing. The *United States* is tonic and alterative, and agreeable to drink ; and the *Empire* closely resembles the Congress. The *Saratoga*, the *Excelsior*, and the *Eureka* are mildly cathartic. The *Red Spring* contains an unusual proportion of iron ; and the *Eureka White Sulphur*, strongly impregnated with sulphuretted hydrogen, is chiefly used for bathing. Of the "spouting springs," the most famous one is the *Geyser*, which is so highly charged with carbonic acid gas that it foams like soda-water when drawn from a faucet, and exhilarates like champagne ; the *Glacier Spring*, which spouts high above the pipe and is strongly cathartic ; and the *Seltzer Spring*, which closely resembles the celebrated seltzer of Germany, and is a sparkling and invigorating drink.

Other celebrated springs in New York State are the *Ballston Spa*, near Saratoga ; the *Chittenango White Sulphur*, iron and sulphur ; the *Clifton*, sulphurous in character and remedial in bilious and cutaneous disorders ; the *Richfield*, a numerous group, considered especially efficacious in diseases of the skin ; the *Sharon* group of four springs, comprising a chalybeate, a magnesia, a white sulphur, and a blue sulphur; the *Vallonia*, whose waters are impregnated with sulphur, iron, and magnesia, and are beneficial in cutaneous diseases; and the *Avon*, saline-sulphurous, and good for rheumatism, indigestion, and cutaneous diseases.

After the Saratoga Springs, the most famous group in America are the VIRGINIA MINERAL SPRINGS, which are also the most numerous in the world. The region through which they are scattered covers the entire northwestern part of the State, and comprises nearly a hundred springs of every quality and variety. The most famous of the Virginia Springs are the *Greenbriar White Sulphur*, which have been resorted to for more than a hundred years. The water is alterative and stimulant and is beneficial for dyspepsia, affections of the liver, nervous diseases, cutaneous diseases, rheumatism, and gout. The *Blue Sulphur* is near the White Sulphur, and is said to be beneficial in chronic hepatitis, jaundice, chronic irritation of the kidneys and bladder, and diseases of the skin. The *Old Sweet Springs* are situated in Monroe County, and are said to be the oldest watering-place in Virginia. The water derives a peculiar briskness from the carbonic acid which predominates in it, and is recommended for dyspepsia, diarrhœa, dysentery, and general disorder of the system. The *Red Sweet*, near by, is chalybeate and tonic. The *Salt Sulphur* is also in Monroe County, and the water is prescribed for chronic affections of the brain, for chronic diseases of the bowels, kidneys, and bladder, and for neuralgia and the various nervous diseases. The *Red Sulphur*, near the Salt Sulphur, is strongly charged with sulphuretted hydrogen, and is remedial in cases of scrofula, jaundice, dyspepsia, and chronic dysentery and diarrhœa. The *Rockbridge Alum Springs* are situated in Rockbridge County, and are highly esteemed for chronic dyspepsia, diarrhœa, scrofula, gastric irritation, and diseases of the skin. The *Rockbridge Baths* are impregnated with iron and as a tonic bath are useful. *Jordan's Alum Springs* possess similar properties.

The thermal baths of Bath County, Va., are a remarkable group, unrivalled by any yet discovered either in Europe or America. The *Warm Springs* were discovered by the Indians, and have long been a popular resort. The water is used chiefly for bathing, and so used is beneficial to gout, rheumatism, swellings of the joints and glands, paralysis, chronic cutaneous diseases, and calculus disorders. The *Hot Springs* are said to be the hottest baths in the world, the temperature reaching 110° Fahr. Their most marked effect is in cases of rheumatism and torpid liver. The *Healing Springs* closely resemble the famous Schlagenbad and Ems waters of Germany, and are excellent for skin

diseases, scrofula, chronic thrush, neuralgia, rheumatism, ulcers of long standing, and chronic dyspepsia. The *Bath Alum* waters are tonic and astringent, and are recommended for scrofula, dyspepsia, eruptive affections, chronic diarrhœa, nervous debility, and various uterine diseases.

One of the most famous and popular of the Virginia watering-places are the *Berkeley Springs*, situated in Morgan County, West Virginia. The waters are not remarkable for their curative properties, being but slightly impregnated with mineral ingredients, but the bathing is highly invigorating. The water of the *Capon Springs* is more valuable, and benefits gravel, diseases of the intestinal canal, various forms of dyspepsia, chronic diarrhœa, and affections of the nervous system. The *Rawley Springs* are a compound chalybeate, alterative and tonic in their effects. The water of *Coyner's Springs* is recommended for imperfect or painful digestion, indolent liver, chronic disease of the bladder or kidneys, and an enfeebled condition of the nervous system. The *Blue Ridge Springs*, in Botetourt County, have a special reputation for the cure of dyspepsia. The *Alleghany* water is cathartic, diuretic, and tonic, and is considered remedial in costiveness, dyspepsia, scrofula,* jaundice, and incipient consumption. The *Montgomery White Sulphur* are a strong sulphur, acting as a mild cathartic. The *Yellow Sulphur* possesses valuable tonic properties.

The foregoing are the most prominent and popular of the Virginia springs, but there are others which deserve mention, as the *Bedford Alum*, the *Sharon Alum*, the *Pulaski Alum* the *Grayson White Sulphur*, in Carrol County, the *Fauquier White Sulphur*, *Eggleston's*, a powerful sulphur, *Jordan's*, in Frederick County, the *Shannondale*, in Jefferson County, the *Holston*, in Scott County, and the *Huguenot Springs*, in Pinkerton County.

The *Warm Springs*, in western NORTH CAROLINA, have a wide reputation. They are sulphurous, with traces of lime and magnesia, and in the form of baths are excellent for dyspepsia, liver-complaint, diseases of the kidneys, rheumatism, and chronic cutaneous diseases.

There are no mineral springs in SOUTH CAROLINA or GEORGIA which have more than a local reputation, but some, perhaps, deserve to be better known. In South Carolina are *Glenn's Springs*, a strong sulphur, the *Limestone Springs*, a chalybeate, *Chick's Springs*, a mild sulphur, and *Williamston Springs*, a valuable tonic. In Georgia there are several which are better known. The *Warm Springs*, in Meriwether County, are an excellent sulphur. The *Chalybeate Springs*, in Talbot County, have fine tonic properties. The *Indian Springs*, in Butts County, are highly esteemed sulphurous waters. The

Madison, in Madison County, is a fine chalybeate. The *Red Sulphur* are in Walker County, and the *Powder*, sulphur and magnesia, in Cobb County. The *Catoosa Springs*, in Catoosa County, *Rowland's*, in Bartow County, and *Gordon's* in Murray County, are chalybeate.

In PENNSYLVANIA are several valuable springs. *Bedford Springs* have long been popular. The waters are a saline-chalybeate, and are beneficial in dyspepsia, diabetes, incipient consumption, and skin diseases. *Cresson Springs* are a saline and an alum, situated nearly on the summit of the Alleghany Mountains. The *Gettysburg* water resembles Vichy, and is considered remedial in gout, rheumatism, dyspepsia, and affections of the kidneys. The *Mt. Holly Springs*, near Carlisle, are mildly sulphurous in character and tonic in effect ; and the *Perry Warm Springs*, also near Carlisle, are aperient and diuretic when taken internally, and as a bath help diseases of the skin. The *Doubling Gap Springs* comprise two springs, a sulphur and a chalybeate. The *Minnequa Springs* are impregnated with iron and hence tonic in effect. The *York Sulphur Srings* were once famous, but have declined in popularity.

Some of the most celebrated mineral spings in the country are in VERMONT. The *Sheldon* and *Missisqnoi Springs* are alkaline in character and are very efficacious in cutaneous diseases. The *Highgate Springs* are also alkaline, and are thought to be beneficial in dyspepsia and cancer, as well as in eruptive diseases. The *Alburgh* water is prescribed for cutaneous diseases. The *Clarendon Springs* are alkaline, highly charged with carbonic acid gas, and much resorted to. The *Middletown Springs* are impregnated with iron, and are highly esteemed.

The *Stafford Spring*, in CONNECTICUT, is one of the best chalybeate springs in the United States. The *Sand Springs* and the *Berkshire Soda*, in MASSACHUSETTS, have some reputation.

The ARKANSAS HOT SPRINGS are among the most famous of the kind in the world, and a great sanitarium has grown up around them. There are 66 springs in all, varying in temperature from 93° to 160° F., and the waters are used chiefly as a bath. Taken both internally and as a bath, they have worked many wonderful cures of rheumatism, rheumatic gout, stiffness of the joints, mercurial diseases (arising from the effects of mercury in the system), malarial fevers, scrofula, and diseases of the skin.

The *Hot Sulphur Springs*, in Middle Park, COLORADO, are rapidly growing in reputation, and the *Manitou Springs* are perhaps the most famous in the West. The waters contain sulphur, soda, and iron, and are recommended for their tonic effects in all diseases of which general debility is a feature. The

Idaho Springs (hot and cold) have fine tonic properties and are frequented both in summer and in winter. The *Boiling Springs*, near the foot of Pike's Peak, are 6350 feet above the sea.

In CALIFORNIA are several springs enjoying considerable local repute. Among the best of these are the *Hot Sulphur*, near Santa Barbara, which are said to cure rheumatism and various diseases of the skin. The *Paso-Robles Hot Springs* are a famous bath, and are considered remedial in gout, rheumatism, and chronic diseases of the skin. The *Congress Springs*, near Santa Clara, contain carbonate and sulphate of soda, chloride of sodium, lime, iron, silicate of ammonia, and magnesia, and are regarded as a specific for rheumatism. *Harbin's*, the *White Sulphur*, and the *Napa Soda Springs*, in the northern part of the State, are valuable. The *Geyser Springs* are rather a curiosity than a sanitarium, but a properly-directed course of the waters is said to afford an almost certain cure for gout, rheumatism, and skin diseases.

Well-known springs in Canada are the *St. Catharine's*, which are used chiefly in the form of warm baths, and are remedial in rheumatism, neuralgia, and gout ; the *St. Leon*, on the Riviere du Loup ; and the *Caledonia*, the water of which is largely exported under the name of " Plantagenet Water," and is recommended for rheumatism and skin diseases.

Of the English mineral springs the oldest are those of BATH. The Bath waters are thermal, their temperature being always over 100° F., sometimes as high as 120° F. They contain sulphates of lime and soda, chloride of sodium and magnesium, some carbonate of lime, silica and iron, all held in solution by carbonic acid ; other gases are contained in the waters, chiefly oxygen and nitrogen. The waters are sparkling in appearance, owing to the presence of these gases, and they are generally drank in quantities of half a pint morning and afternoon. They usually raise the temperature and quicken the circulation, increasing certain of the secretions, especially those of the kidneys. The waters are also largely used for bathing purposes, all kinds of baths being provided. Bath is chiefly frequented in late autumn, winter, and early spring, for, situated as it is, in a hollow, it is warm, though, as a rule, damp, owing to the steam of the hot waters permeating the soil. The main diseases for which the Bath waters are adapted are rheumatic and gouty affections of a chronic character, neuralgic affections, especially lumbago, rigid joints, and some forms of paralysis. Certain skin diseases, too, are benefited by them.

CHELTENHAM is perhaps more of a health resort than a watering-place. The waters are cold and all saline, except one, which is chalybeate. The chief spring is the *Mont-*

pelier Spa, whose waters contain chlorides and sulphates, with a little iron. These are used both internally and externally, but chiefly internally. They are supposed to be especially valuable in torpidity of the liver and bowels, and in gouty disorders. Patients commonly resort to the waters at the time when the regular inhabitants are absent, that is to say, in the summer months.

LEAMINGTON, in many respects, resembles Cheltenham, but is more beautifully situated. The composition of these waters resembles that of the Cheltenham springs. They contain chlorides of sodium, calcium, and magnesium, with sulphate of soda. They contain also carbonic acid and nitrogen and oxygen. On the whole they are more powerful than are those of Cheltenham, and so better adapted for those who suffer from torpid liver and bowels, in the first instance at least.

TUNBRIDGE WELLS, situated in one of the most beautiful districts in England, is largely visited, not so much on account of its waters, which are almost neglected, but because the air is mild yet bracing, the walks are fine, and the place one well fitted for a pleasant sojourn. The waters are chalybeate, but only feebly so, and require to be taken for a good long time ; nevertheless, drank regularly, and combined with exercise taken in the open air, they may be relied upon as being most efficacious in cases of anæmia, such as occur in young females of sedentary habits.

One of the most important of the English watering-places, certainly the one which is most visited for its waters, is HARROGATE. The soil is sandy, the air pure and bracing. The waters are all cold, but are usually warmed before being drank. Springs of the most various kinds are found here, some strong sulphurous, some mild sulphurous, with alkali combined, some saline chalybeate, and some purely chalybeate of a most unusual kind. The strong sulphurous waters are the typical waters of Harrogate, for which it has received its reputation. They are taken internally in doses of about a pint, in divided doses, every morning before breakfast, and are also used as baths. These are mainly used to stimulate the liver and bowels ; used as baths, the skin ; they also favor the secretion of urine, and are especially useful in certain forms of skin disease, and gouty and rheumatic affections. The mild sulphur springs contain less sulphuretted hydrogen and chlorides of sodium and magnesium, but they have carbonate of magnesia in addition. These are antacid as well as alterative. The saline chalybeate contains carbonate of iron, so that these waters are tonic as well as alterative. Again, some are purely chalybeate, one especially, of a very rare kind, containing a protochloride of iron.

BUXTON is totally different from any of the preceding. The springs are situated among the Derbyshire hills, and are exceedingly bracing. The climate is only adapted for summer and autumn ; at other times it is often cold and badly adapted for invalids. The salts contained in the waters are small in quantity, and are mainly salts of sodium, magnesium, and calcium, with a trace of iron. They contain much carbonic acid and nitrogen. The waters are chiefly used for bathing, and douche-baths are perhaps the favorite form of applying them. The waters are useful for stiff joints, especially when these are due to gout or rheumatism, to old sprains or muscular contractions.

There are a few other places in England where waters are drank, and also a few in Scotland ; most, however, have been abandoned for their more fashionable continental rivals, and indeed some of those here described are not in much better case. It would be impossible to give in this slight sketch any full account of foreign watering-places, but equally so would it be impossible, while pretending to deal with the subject of mineral waters, to conclude this sketch without reference to these.

SPA in Belgium, whose title has become generic, being applied to almost all watering-places, is situated in a valley of the Ardennes. The waters are of a temperature of 50° F., and contain much carbonic acid. This holds in solution salts of soda, magnesia, lime, and iron, so that these partake of the qualities of alkaline and ferruginous waters. The dose given is considerable, as much as three pints a day in divided doses, but beginning with a couple of glasses. They are valuable as chalybeates. The season is from May to September.

In the Pyrenees are a multitude of springs which we cannot describe individually. Chief among these are the Bagnères de Bigorres, Barège, Bagnères de Luchon, Cauterets, Eaux Bonnes, and Eaux Chaudes. The *Bagnère de Bigorres* are saline, sulphurous, and ferruginous. The *Barège* waters are sulphurous, and are of three kinds—hot, temperate, and tepid. A peculiar pellicle floats on their surface, which is supposed to be peculiarly beneficial in chronic rheumatism. It is called glairine, zoogene, or barégine. These waters are highly esteemed. *St. Sauveur*, four miles off, has waters similar, but less active. *Cauterets* is more sheltered than Barège, has many sulphuretted springs, the warmest having a temperature of 122° F. ; barégine is also present in these waters. These are mainly used for skin diseases of an obstinate kind, scaly and pimply, in chronic rheumatic and gouty affections, stiff joints, etc. Some maladies of the womb are also greatly benefited by them, so, too, is scrofula and threatened phthisis. *Bagnères de Suchon* and *Eaux Chaudes* are also sulphurous waters. The latter contain or deposit a substance called *sulfuraire*. *Eaux Bonnes* are mildly sulphurous waters, of which the supply is scanty. They are supposed to be specially efficacious in threatened consumption. They are situated 2400 feet above the level of the sea, and the air is exceedingly fresh and pure.

VICHY affords perhaps one of the most important mineral waters known to us. It is situated in central France, in a wide open valley. The air is temperate, and the season lasts from May till September. The springs are nine in number, they are all warm, alkaline, and gaseous. They contain mainly carbonate of soda and carbonic acid. They also contain, however, some potash and ammonia, and barègine. The springs mostly employed are the Grande Grille, the Celestins, and L'Hôpital. They are used for diseases of the lungs, especially catarrh, for irritability of the digestive organs, gravel, catarrh of the bladder, diabetes, chronic gout and rheumatism, etc. The Grande Grille is supposed to be most useful for liver complaints ; the Celestins is mainly given for urinary disorders ; the Hôpital spring for gastric catarrh. Their taste is someth ng like soda-water. They are largely exported, and are given in doses of half a pint to two pints.

AIX-LA-CHAPELLE lies to the westward of Cologne, between the Rhine and the Maas. Its waters are partly warm sulphurous, partly cold chalybeate. The waters are extremely disagreeable, though less so than some of the Pyrenean sources ; there is abundance of the rotten egg flavor, but the barègine is wanting. They are not much given internally, but are chiefly used in baths, douches, shampooing, and kneading, and in these cases are of great use in curing old standing sprains, stiff joints, contracted muscles, and the like. Other maladies, like skin diseases, may also be benefited by them. The season is from June to September.

KREUZNACH is a spa of singular value. The waters are bitter and contain chlorides of sodium, calcium, and magnesium, bromides, iodides, and some iron. The waters are chiefly used wth a view to procure the absorption of tumors. It is drunk at first in small quantity, sometimes mixed with hot milk. The baths are taken tepid, and quantities of the substance which remains behind when the salts have been crystallized from the waters are added according to circumstances. The waters of Kreuznach have obtained their chief reputation in maladies of the womb, especially in chronic inflammation, with hypertrophy and induration. Scrofulous ulcers and glands, too, are frequently relieved.

NEUNAHR is situated in the valley of the Ahr, not far from Cologne. It contains much carbonic acid, with carbonates of lime and magnesia, some sulphate and chloride of sodium, with a little iron, alumina, and silica. The waters are useful in rheumatism and a

tendency to the formation of gravel, in maladies of the throat and lungs connected with these.

EMS lies in the valley of the Lahn, not far from Wiesbaden. It has long been a noted place of resort, and the beauty of its situation, is in itself almost a sufficient attraction. The waters are warm or hot—86° to 133° F., and are saline, alkaline, and gaseous. The waters contain chloride of sodium, carbonate of soda, and magnesia, with smaller quantities of lime, iron, manganese, potass, and lithia. Hence they are alterative, mildly diuretic, and laxative. They are recommended in catarrhal affections of the lungs and air passages, and in dyspepsia when there is a tendency to consumption ; also in the form of skin disease known as eczema. For gouty subjects they are also valuable, but less so than Vichy.

WIESBADEN lies on the southern slope of the Taunus mountains, and is greatly frequented. The season extends from June to September, but the climate is good much later. The water contains chloride of sodium in large quantity, with potass, lime, iron, magnesia, some arsenic, and bromine. The carbonic acid is in very large bulk. The taste has been compared to weak chicken broth slightly salted. The waters have to be cooled before being taken, and then three or four glasses produce a slight diuretic and laxative effect. These waters do good in gout or rheumatism, with congestion of the liver and piles, and also in some skin diseases.

HOMBURG, too, has derived its attractions from other sources than its waters ; nevertheless these are worthy of mention. This place lies not far from Frankfort, and its air is invigorating and bracing, but variable. The waters are cold, and contain chlorides along with carbonic acid. The flavor is fairly agreeable, though saltish and somewhat bitter.

BADEN-BADEN is another of the German watering-places where a gambling table and the French demi-monde offered the chief attractions. It is situated in the Black Forest, in a delightful valley. Its mineral waters are weak, but they are said to contain a good deal of lithia. This place, now that gambling is suppressed, has a doubtful future.

KISSINGEN, in Bavaria, stands in a totally different category. It is one of the favorite bathing-places of Germany, and is situated about thirty miles from Würzburg. Its waters are cold and gaseous. The chief salts are chloride of potassium, sodium, lithium, and magnesium, carbonate of lime and sulphate of magnesia, with iodine, bromine, and iron. The waters are useful in habitual constipation, with congestion of the liver, in dyspepsia with flatulence, and in tubercular disease. Gout and gravel are also benefited. Baths are also used.

GASTEIN, in Austrian Tyrol, not far from Salzburg, is another favorite place of resort for the sake of its waters. It is situated 3200 feet above the level of the sea, and the air is extremely bracing. The season is limited to July and August. The springs are thermal, but weak; sulphate of soda is the chief ingredient. The waters are used for baths after cooling, and some derive great benefit from them; the prematurely old, the hypochondriac, and paralytic are said to be most so. The waters of *Toplitz*, in Bohemia, resemble those of Gastein, but the town lies much lower, in a situation of great beauty, the climate being exceedingly agreeable.

FRIEDRICHSHALL and PÜLLNA both supply a water which is highly laxative. The waters are bright and clear, with a slight tinge of yellow, and are largely exported, previous to which, however, they are somewhat concentrated. They contain sulphates of soda, lime, and magnesia, with chlorides, carbonates, and bromides.

CARLSBAD is situated in Bohemia, some distance from Prague. The season extends from May to September. It is 1200 feet above the sea. There are several springs; the principal rises some feet in the air, and gives off clouds of vapor; its temperature is 165°. It contains sulphates of soda and potash, chloride of sodium and carbonate of lime, with some iron, alumina, and silica. The waters are mainly given for abdominal complaints, as in diseases of the liver, engorgement, and the like, dyspepsia, hypochondriasis, constipation, diabetes, gout, and rheumatism; also in jaundice from gallstones.

Still higher up is MARIENBAD, in the same valley as Carlsbad. The waters are mixed, saline and chalybeate, with some carbonic acid; by standing they become turbid. The waters are valuable for chronic diseases of the digestive organs, combining laxative and tonic influence. The water, made into a paste with peat soil, is used as a mud bath or poultice, which is useful in healing chronic ulcers and dispersing glandular swellings. Gas baths, consisting of carbonic acid with a little sulphuretted hydrogen, are also used to remove pains from the muscular and nervous systems.

In the same district is FRANZENSBAD, whose waters are cold; they are acidulous, and contain mainly alkaline salts. Here, too, besides being used in the ordinary way, the mud and gas baths are high in favor. Various chronic skin diseases, indolent ulcers, gouty deposits, etc., may be thus removed. The water taken internally improves digestion and the nervous system.

AIX-LES-BAINS, in Savoy, is a watering-place well worthy of the attention of the invalid. Its greatest fault is its remoteness. The springs are warm, one containing sul-

phuretted hydrogen, the other none. The waters are chiefly used externally as douches. The climate is very mild, and admits of a stay from April to October. Chronic rheumatism and stiff joints are the forms of disease most benefited by the treatment.

ZONECHE stands high in the valley of the Rhone. The waters are hot, and contain mostly sulphate of lime and other sulphates. Scrofulous enlargements, eczema, and gout and rheumatism are chiefly benefited by the baths here.

PFEFFERS is also high above the sea, in the Grisons. Its waters are conducted in wooden tubes down the heights to Rogatz, in the valley of the Rhine. Sometimes, however, the waters can only be obtained at Pfeffers. Baths are chiefly used, but the waters are also drunk. They are useful in hysteria and nervous excitability.

TARASP, in the Grisons, has springs something like those of Marienbad, cold and gaseous. The springs are 4300 feet above the sea; they are said to be useful in early phthisis, and when the abdominal organs are out of order. Doubtless, the elevated regions and the pure mountain air have much to do with the benefit. This is still more markedly the case with ST. MORITZ, in the Upper Engadine, which lies 5863 feet over the sea level. The waters, which are situated on a still higher level than the village, are chalybeate, with free carbonic acid. They are used both internally and externally. The place seems of most value in the early stages of consumption.

SCHINZNACH and its neighborhood contains many springs, those of Schinznach resembling those of Zoneche; those of *Wildegg*, close by, being more like those of Kreuznach.

MINERAL WATERS, ARTIFICIAL, are imitations of mineral spring waters, made by dissolving the salts which constitute a basis of the natural mineral waters in distilled water impregnated with gases, especially carbonic acid gas. They are prepared according to analyses which represent the natural mineral waters when in their best condition; and consequently, if properly made, have some advantages over natural waters. The supply of the latter exported from the European springs is inadequate for the demand, and they are apt to be diluted; and most natural waters lose materially by bottling. The springs, too, are subject to many changes, and frequently vary in the quantity or the relative proportion of their mineral ingredients. Artificial waters, on the contrary, are always the same in composition, in consequence of the technical perfection of their manufacture, and they produce the same general effect as the natural waters. They are also more highly charged with carbonic acid gas than the latter, which insures their keeping in any climate, and renders them more pleasant to the taste.

MISCARRIAGE. (See ADORTION.)

MIXTURES are perhaps the most favorite forms of remedies—one or two substances intended to aid each other's action being combined and given in some pleasant vehicle. It is not desirable that too many objects should be aimed at in any mixture, so that its composition should be as simple as possible.

MOLES, called also liver stains, mother's marks, pilous and pigmentary nævi, are congenital marks of a light or dark brown or black color, situated on the surface of the body. They are formed by circumscribed thickening of the scarf skin with excessive deposit of organic coloring-matter, and are covered by numerous thick, stiff hairs. They vary much in shape, size, and situation. Most frequently one or two small marks of a rounded form are met with, either on the face or on the back of the neck, but in some cases a mole covers several inches of surface, and is very irregular in form. The usual seats of moles are the face, the back of the neck, and the back. Moles, in consequence of friction, often become sore, and sometimes ulcerate. It is believed, too, by many surgeons that they are often the starting-points of cancerous growths. For these, as well as for cosmetic reasons, it is advisable to have a small and isolated mole cut out. If the surgeon's knife be carried in the direction of the folds of the skin, the scar will cause very little if any disfigurement. For large moles very little can be done ; removal of the hairs by tweezers is soon followed by renewed growth, and the use of depilatories, or hair-destroying applications, does much more harm than good, in consequence of their irritant and caustic action on the skin.

MOLLUSCUM is a disease of the skin, characterized by round elevations of the skin, varying in size from a hemp-seed to a hazelnut, and marked on the summit by a dark point and a depression in the centre. The color of the skin over them is sometimes translucent, or of a pinkish color. Some of the growths have no black mark and no depression. These little tumors may increase slowly in size without undergoing any change, or they may ulcerate and discharge their contents. There seem to be two kinds : (1) *Molluscum fibrosum,* which consists in an increased formation of the fibrous tissues round the hair follicles ; and (2) *Molluscum contagiosum,* which is due to an increase in the sebaceous follicles, so that the contents of each tumor have a cheesy appearance. The usual seats of molluscum are the back or front of the trunk, the neck, face, and scrotum. Its presence is not attended with any constitutional disturbance ; it may exist at any age, but is most frequent in children. There are generally several of these small

tumors present at the same time. The *treatment* is purely local ; the tumor should be laid open, the contents squeezed out, and the inside touched with caustic ; if attached by a thin stem to the skin, the growth may be snipped off with a pair of scissors, and the cut end touched with caustic.

MONKSHOOD. (See ACONITE.)

MONOMANIA. (See INSANITY.)

MORPHINE. (See OPIUM.)

MORTIFICATION means the death of a part of the living body. There are several varieties of this process, styled by surgeons gangrene, sphacelus, sloughing, mummification. *Gangrene* is that stage in which the part is hot, swollen, and livid, but not yet quite dead. The term *sphacelus* expresses that condition in which the part is cold and black, and utterly deprived of life. *Mummification* is dry gangrene, a condition in which a portion of an extremity is dry and shrunken. By *sloughing* is meant a limited death of skin and soft structures at the surface of the body. Death of bone is called *necrosis,* and destruction and breaking down of brain tissue is called " ramollissement" or softening. The following are the most common exciting causes of mortification :

1. *Mechanical or Chemical Action.*—Severe injuries of the extremities, and especially compound fractures, associated with much crushing and contusion, occasionally result in death of the injured parts, the blood supply of which has been cut off. The application of actual flame or of hot fluids to the surface of the body may also cause destruction of the skin and subjacent soft parts. The contact of sulphuric, nitric, and other strong acids, and of the caustic alkalies, as ammonia and potash, will also produce mortification. Under this head may be included the action of living and putrid animal fluids, as the poison of snakes, and the fluid causing the bad result of certain dissection wounds.

2. *Stagnation of the blood in a limb due to an obstruction to the circulation through the veins.* —If a limb be tightly constricted at any part by a bandage or handkerchief, and this constriction be kept up for eight or ten hours, the extremity of the limb will swell, become cold, senseless, and livid, and at last mortify. These changes sometimes take place after a limb has been tightly bandaged for fracture of one or more bones. A common instance of mortification produced by compression of some part of the surface of the body and obstruction to the circulation of the blood through the veins, is the *bed-sore,* the formation of which, however, is much favored by other conditions, as the exhausted state of the patient, and the prolonged contact of the skin with urine, sweat, and other irritating fluids.

3. *Arrest of the Supply of Arterial Blood.*— When a large artery is torn through by a frac-

tured bone or ruptured by external violence, mortification of the parts supplied by this vessel will often result, though not always, as a sufficient amount of blood may still be supplied by smaller collateral vessels. When, in addition to division of a large artery, there is much effusion of blood and contusion of soft structures, mortification will certainly take place. Mortification occasionally results from the blocking of an artery with a clot of fibrine detached from the lining membrane of the heart, and carried along with the current of blood. In feeble old persons, spontaneous mortification sometimes attacks the toes or fingers ; this affection, which is called senile gangrene, is generally due to a combination of two or more of the following causes : weakness of the heart's action, ossification and contraction of the arteries, sluggish circulation, diminished nerve force, exposure to cold and wet.

4. *Injury or destructive disease of nerves* sometimes causes mortification. In a case under the care of Sir Benjamin Brodie, mortification of the skin of one foot was observed within a few hours after injury to the spinal cord. Opacity and ulceration of the cornea or transparent membrane of the eye, occasionally follows division or compression of the fifth cerebral nerve.

5. In France and other countries on the continent where rye is an article of food, mortification has been observed to follow the use of this grain when diseased, in consequence of the growth of a fungus in the ovary. The grain when in this state is called ergot, spurred rye, cock-spur rye, and. by botanists, *Secale cornutum*. It then contains a poisonous active principle called ergotine, which gives rise, when black bread made of the diseased rye has been eaten, to severe cramps, itching of the skin at some parts of the body, numbness and loss of sensibility at other parts, deafness, and dry mortification of the extremities of the limbs.

6. For mortification resulting from the effects of cold and moisture, (see FROST-BITE.)

The predisposing causes of mortification are general debility due to senile decay, to insufficiency or bad quality of food, and to exhausting and severe diseases—to disease of the heart, and diabetes. To these may be added the local predisposing causes, such as inflammation and congestion of a part of the body.

Moist mortification generally occurs when the circulation of blood in a part ceases suddenly ; the dead tissues are then mixed with the stagnant blood and serum. Dry mortification, or mummification, results from slow death of a part, due to deficient or obstructed arterial supply. When moist mortification attacks a limb, or a superficial part of the body, the skin becomes distended and livid.

Blisters containing a dark fluid then form on the surface, and the epidermis becomes moist and can be readily detached. Large black and purple patches then appear, and these increase in size and run into each other. The affected part is cold and insensible ; it is much swollen, and crinkles under the finger in consequence of the presence of gases. The soft parts under the skin are black, putrid, and rotten, and soaked in a thin, ill-smelling fluid. After a bad compound fracture or gun-shot wound of a limb, the mortification often spreads very rapidly toward the trunk, and its course is not arrested before it carries off the patient. In less severe cases, and when but a small portion of an extremity mortifies, in consequence of a deficient supply of blood, the edges of the gangrenous patch become sharply defined, and are separated from the surrounding and living integument by a bright red groove, moistened with purulent discharge. This groove is the so-called line of demarcation, and when present always indicates an arrest of the mortification. It increases in width and depth, and ruddy granulations are formed, from which there is a healthy discharge. The deeper portions of the mortified part then become detached from the tissues beneath, which are also covered by granulations ; before the dead portions are detached, the blood-vessels are closed by coagulation of the blood at their extremities. After the whole of the mortified tissue has been thrown off, the resulting wound closes rapidly by granulation and scarring. The tissues, which are less readily detached, and which generally remain when dead for a long time after the removal of other soft parts of the gangrenous patch, are tendon and bone. By the formation and subsequent extension of this line of demarcation, a foot or hand, or even the greater part of a limb, may undergo spontaneous amputation. This process of separation is generally attended by a profuse and exhausting discharge of pus. The intensity of the constitutional disturbance varies according to the extent and the cause of the mortification. In cases of rapidly-spreading death of a limb, in consequence of injury, the patient falls into a very serious condition, and presents all the symptoms of low typhoid fever. If the mortification be limited to a small part only of a limb, there will generally be heat of skin, thirst, headache, nausea, a rapid pulse, and other symptoms of high inflammatory fever.

Treatment.—The general treatment of mortification should consist in husbanding the patient's strength by giving very nutritious and easily digestible food, with port wine and brandy and water as drinks ; opium is administered in most cases for the purpose of relieving pain and allaying nervous irritability. When the symptoms are those of a low typhoid condition of the system, bark, ammonia, and chloric ether are indicated. The local treatment, in cases where the mortification is limited to one or more fingers or toes, or to a small part of the surface of the body or of a limb, ought to be such as would favor the speedy separation and throwing off of the dead tissues, and prevent as much as possible putrefaction. Loose shreds of skin should be at once removed, but care must be taken not to pull away with force any dead tissue that may still adhere at its deeper parts. By rough proceedings of this kind troublesome bleeding may be caused. The mortified part may be covered or surrounded by strips of lint dipped in a weak solution of carbolic acid or of permanganate of potash, and then enclosed in a thick layer of prepared oakum or of cotton wool dusted with carbolic acid powder. This dressing ought to be frequently renewed during the day, and the gangrenous tissues and surrounding line of demarcation be well syringed with a solution of carbolic acid, permanganate of potash, or chloralum. In some cases of mortification, where the pain is very great and the discharge from the seat of disease very profuse and fœtid, cold charcoal poultices may do good. For further information concerning the varieties and special modifications of gangrene, the reader is referred to the articles on BED-SORES, CANCRUM ORIS, FROST-BITE, HOSPITAL GANGRENE, and PHAGEDÆNA.

MOTHER'S MARKS. (See NÆVUS.)

MOTES IN THE EYE. (See MUSCÆ VOLITANTES.)

MOUTH, DISEASES OF.—Under this title it is proposed to deal with some of the chief affections of the structures contained within that cavity which is bounded in front by the lips; behind by the soft palate and fauces; above by the hard palate; below by the tongue; and on the sides by the cheeks and parts of the upper and lower jaw bones.

Contraction of the Opening of the Mouth.— This is met with very rarely as a congenital deformity, but often results from the effects of injury or disease. In the latter case the opening may be much reduced in size, the lips distended, and the interior surfaces of these structures and of the cheeks closely united to the mucous membrane covering the jaws. This condition not only produces a most unpleasant deformity, but often, in consequence of the contraction of the opening of the mouth and of the restriction of the movements of the jaw, leads to serious disturbances of digestion and nutrition. The most frequent causes of this unfortunate condition are cancrum oris, lupus or epithelioma, ulceration of the cheeks and gums from mercury, deep burns of the face, and sloughing wounds. For affections of this kind there is no treatment short of a well-planned and carefully-performed plastic operation.

Spasmodic Contraction of the Masseter, or large muscle which closes the jaws, is occasionally met with in connection with painful cutting of a wisdom-tooth. In consequence of the ulceration of the gum and the irritation about the crown of the appearing tooth, the masseter muscle falls into a state of painful and persistent spasm. The jaws are kept closed and cannot be separated except by using a wedge. A similar spasmodic condition is sometimes associated with caries of a molar tooth. The only treatment for this affection, and one which is always effectual, is removal of the offending tooth.

Tumors.—Cysts are frequently met with on the floor of the mouth. They may be congenital or the resut of obstructed and distended ducts. These growths are described under the head of RANULA. Solid tumors are not often observed on the floor of the mouth. A fatty growth occasionally makes its appearance under or within the tongue. Large calcareous masses, which are supposed to be salivary calculi, seated within the ducts, sometimes form under the mucous membrane at the floor of the mouth. The most formidable morbid growths to be met with in this cavity are those which spring from the upper or lower jaw, in most cases from the former. These generally grow rapidly, involve important organs, and necessitate sooner or later some capital surgical proceeding.

For other affections of the mouth see articles on APHTHÆ, GUMS, HARE-LIP, QUINSY, RANULA, SALIVATION.

MOUTH-WASHES. (See GARGLES.)

MOXA is a method of applying actual fire to the surface of the body for the purpose of producing counter-irritation, blistering, or cauterization, or to form an issue. It is an old and common plan of treatment in the East, but is seldom applied in this country. The affections for the treatment of which it is generally employed are diseases of the spine and joints, and muscular rheumatic pains. A moxa is composed of some material which will burn readily, as cotton wool, lint, German tinder, dry rotten wood made into a paste with spirits of wine, blotting paper, etc. The most convenient plan is to take a cylinder of cotton-wool, previously dipped in a solution of nitrate of potash, and then well dried, and to place this in a small pill-box without the lid and the bottom of which has been removed. This box should then be held over the selected part of the surface of the body by means of a loop of thick wire, and the wool ignited. If it be intended to produce only redness of the skin or slight blistering, the moxa should be held at some distance from the skin and be left to itself to burn out, but when a slough is to be formed and a deep wound to act as an issue, the ignited wool should be applied close to the surface and combustion be kept up by blowing upon the flame through a long tube, or by working a small pair of hand-bellows. The surrounding skin must be protected by layers of lint dipped in cold water. When the moxa has been consumed, the patch of burned skin should be covered by a fold of moistened lint, which after twenty-four or thirty hours is to be replaced by a bread poultice.

MUCILAGE is a solution of gum in water, used as a medicine sometimes. (See ACACIA.) The mucilage of gum-arabic, so commonly used for household purposes, is made by dissolving gum-arabic (the best is nearly colorless) in either hot or cold water till the whole is of the consistency of cream. It must be kept from contact with the air.

MUCOUS MEMBRANE is the interior lining of the human body, which, commencing at the lips, nostrils, eyelids, and ears, and after lining the several organs from which it starts, unites at the back of the mouth or in the pharynx. One portion descending through the wind-pipe and bronchial tubes, finally terminates in the air cells of the lungs ; while another proceeds down the œsophagus, or gullet, lines the whole length of the alimentary canal, and finally ends in the outlet of the bowels. This membrane is extremely delicate, and when irritated or inflamed gives rise to many troublesome ailments.

MUCUS is a thin glairy fluid secreted by the mucous membrane of the body, and always present in a certain quantity, which is greatly increased by inflammation, and altered in character if it be deposited from unhealthy tissue.

MULBERRY CALCULUS is a name given to a stone in the bladder, which has a rugged surface and is a deep purple color like a mulberry ; such stones give great pain and distress from their rough outline ; they are formed principally, if not wholly, of oxalate of lime. (See STONE.)

MUMMIFICATION. (See MORTIFICATION.)

MUMPS, also known as PAROTITIS or CYNANCHE PAROTIDEA, is an inflammatory affection of the salivary glands, especially of that one lodged between the jaw and the ear, called the parotid. It seems to be contagious or infectious, and often spreads through a family or district. It begins with some degree of fever, and soreness and swelling about the angle of the jaw. This swelling gradually extends toward the ear and toward the chin, so that the whole side of the face is swollen. The swelling interferes with the movements of the lower jaw, so that the mouth can hardly be opened, and even the slightest attempt at opening gives rise to acute pain. From accumulation of the saliva, etc., in the mouth itself, it decomposes and, as a consequence, the fœtor of the breath is very troublesome. The appetite at the same time is usually good, and much pain is experienced in endeavoring to satisfy it. The

disease does not last long, in about four days it reaches its height and then gradually declines, rarely going into suppuration. Occasionally it is said that, during the abatement of the disease other organs may be affected, especially the testes in male subjects ; but this is by no means necessarily so. It is true that these organs are so affected sometimes, but the attack is rather simultaneous or following close upon the mumps than any true metastasis, that is, passage of inflammation from one spot to another, leaving one organ and attacking another. Such a change of site is rare, and this can hardly be said to be an example of it.

Treatment.—The remedies to be used for mumps are simple. The bowels should be well opened, best by some saline medicine, say a dose of Rochelle salts, in the form of a Seidlitz powder. The patient had better be kept in-doors and kept warm, with a piece of warm flannel round the throat. The diet should be milk as nearly as possible. Warm opiate fomentations or dry heat may be used to relieve the pain of the inflamed parts.

MURIATIC ACID. (See HYDROCHLORIC ACID.)

MUSCÆ VOLITANTES, or MOTES, are the small dark bodies and beaded strings which appear to some individuals floating across the field of vision, and give rise to much trouble and often to unnecessary alarm. Sometimes but one or two small bodies, like specks of dust, are observed rising and falling, and then when the person's attention is directed to some external object these suddenly disappear. At other times, by a sudden movement of the eye, the field of vision is crowded both by specks and by beaded strings, which glide about for a time and then slowly sink. These bodies are analogous to the highly-refractive globules which are observed both singly and arranged in strings when one looks through a microscope, the field of which is brightly illuminated. Muscæ are generally observed the first time at about the age of twenty-five, or between this and the age of thirty or thirty-five. At first they are few in number but soon increase. They are usually observed by myopic or short-sighted persons, but do not indicate, as is too often supposed, any deep-seated and progressive disease of the eye. Although these floating bodies increase in number as the patient gets older, and give much trouble, they are generally associated with a normal continuance of good, and in many cases even acute sight. Muscæ, however, must not be confounded with the mistiness of vision and the floating opacities which are among the symptoms of advanced disease of the membranes and humors of the eye, or of hemorrhage into the interior of the eyeball. The floating muscæ or motes, are loose portions of the delicate filamentous tissue of the transparent and apparently structureless viscid material which fills up the interior of the eye, and which is called the vitreous humor. Muscæ volitantes cannot be completely removed by treatment. If at any time they should become unusually numerous and give rise to much irritability, the application of a solution of atropine dropped into the eye will give temporary relief.

MUSCLE forms a very important tissue of the body. There are two kinds of muscle —1. That which is found in the muscles of the trunk and arms and legs, and by which we have the power of moving about ; this kind is made up of bundles of fibres and by the contraction of these fibres, under the influence of the will, movement is performed. The biceps, for instance, is a muscle of this class ; it forms the well-known prominence in front of the upper arm ; it is attached above to the shoulder-joint, and below to the humerus or bone of the upper arm ; now, when it contracts, the two ends tend to approach each other and so the arm becomes flexed or bent. In the body, therefore, there is an immense number of muscles, by means of which, when they act either singly or in combination, every movement can be performed. All these muscles are supplied by nerves which are in direct communication with the brain or spinal cord, and by which the will can act upon any part of the system. Since these muscles are under the influence of volition, or the will, they are called *voluntary* muscles. These muscles may become *hypertrophied*, or increase in bulk from active use, as is well seen in the limbs of an athlete, and in the arms of a blacksmith ; or they may waste from disease, or become *atrophied*, as in cases of paralysis, or in cases of long-standing disease, as cancer or consumption. Local wasting of the muscles of the fore-arm may come on in some cases of lead poisoning, and a general and gradual wasting of these muscles is an important part of the disease known as progressive muscular atrophy. Spasms, or irregular muscular contractions, which occur in the course of tetanus and some other nervous disorders, are due to an altered condition in the nerves which supply them. 2. The other class of muscles is known as the group of *involuntary* muscles, because although they are supplied by nerves and have the power of contracting, they are not under the influence of the will ; they also are formed of bundles of fibres, but of different structure and arranged in a different way. The heart, the womb, and the muscular coats of the stomach and intestines are made up of this involuntary muscle ; in the heart, the blood is propelled through the body by the muscular contraction, and it may increase in frequency from disturbed nervous influences, as is seen by the palpitation which comes on from fright, etc. This tissue may become hypertrophied

in some cases of heart disease, and cause that organ to increase vastly in size. In the womb, hypertrophy of the muscular coat takes place in pregnancy, and it is by the contraction of this tissue that the infant is finally expelled from the womb ; the muscular fibres afterward become fatty and waste away. In the intestines it is by means of the gradual wave-like contraction of the muscular coat that the food is propelled along the intestinal canal ; if the coat is torpid and will not contract, constipation ensues ; while if its action is irregular and excited, diarrhœa and griping pains may come on.

MUSHROOMS are one of many species of Agaricus, of which few are usually eaten ; and although much has been written and said to prove that many more may be safely used as food, we are inclined to think that such experiments are better left to those who are botanically acquainted with the peculiar structure of each species, for of late many serious accidents have occurred to unskilful judges, who have eaten poisonous mushrooms instead of those which are really edible. The common mushroom, *Agaricus campestris*, is readily known by its fragrant odor, which is its chief characteristic, and the absence of which is very suspicious. When in a very young state it resembles little snow-white balls which are called buttons ; afterward it acquires a stalk, separates its cap, and becomes shortly conical with liver-colored gills, and a white thick fleshy cap, marked wth a few particles of gray. At a more advanced age the cap is concave, the color gray, and the gills black ; in this state it is called a flap. As a ruie, the colored varieties of mushroom are unfit for food, and such as have a milky juice should be avoided. In case of *poisoning* by spurious mushrooms, take a liberal dose of any emetic that may be at hand, and after the stomach has settled take 2 or 3 tablespoonfuls of castor-oil.

MUSK is the peculiar secretion of the musk deer, a native of Thibet and Central Asia. It occurs in irregular reddish-black grains, bedded together and soft to the touch. The odor is powerful, diffusive, and persistent. This substance is described as stimulant and antispasmodic. No physicians care to prescribe it. Most of the musk commonly encountered is spurious. There is a root known as *Sumbul* or *Musk Root* which has a similar odor. This comes from Siberia— otherwise its origin is unknown. It seems to act something like valerian, and has been used in Russia for low fevers. Its use has not been attended by much success in this country.

MUSTARD is the seed of two kinds of *Sinapis*, one of which yields the black, the other the white mustard. These yield an oil, which is also officinal. The black seeds are smaller than the white. The two together, reduced to powder, constitute the officinal substance. The composition of these seeds is peculiar. The black give off a volatile oil, which it does not contain ready made, but which is formed by the union of two substances contained in mustard if water be present. This volatile oil is light yellow and very pungent, while mustard does not yield the oil, but contains a non-volatile crystallizable compound exceedingly irritant.

There are two preparations of mustard: first, the well-known poultice, which should be made with luke-warm water, and no spirit or vinegar should be added. In the officinal preparation, linseed-meal is added. One of the best plans is to spread a little mustard paste over a moderately hot linseed-meal poultice, with a thin piece of cambric over that. Second, the compound mustard liniment, a new preparation. Mustard seeds and flour act as a powerful stimulant. In good large doses it causes speedy vomiting, useful in narcotic poisoning; in smaller doses, as a mild stimulant, it aids digestion. Externally,

mustard acts as a powerful stimulant, useful in local pains of various kinds, especially slight inflammations. It is frequently used in foot-baths. Various plans are employed for applying mustard to the skin; the best are those hinted at above, but by far the most convenient is in the shape of Rigollot's mustard leaves.

MYALGIA. (See RHEUMATISM.)

MYOPIA, a condition in which the vision is altered. (See EYE and VISION.)

MYRRH is a green resin, exuding from the *Balsamodendron Myrrha,* a tree of the turpentine group, growing in the East. It occurs in irregular reddish fragments, the surface often covered with powder. It has a peculiar odor and taste. Myrrh is contained in a variety of pills, for the sake of its warming and stimulant properties. It is frequently given along with iron and aloes especially. Occasionally also as an expectorant. The tincture of myrrh may be used for that purpose, as well as for an application to spongy gums, aphthous mouth, etc.

N.

NÆVUS, called also erectile tumor, or commonly, mother's mark, is, anatomically, a tumor composed of dilated blood-vessels, and those small ones; or generally it is a diseased formation, in which the vascular tissue bears the most prominent part. The simplest form, a congenital one, is an affection of no danger, rather a deformity than a disease. It is very superficial and hardly projects above the level of the skin consisting of a patch of dilated capillaries. Although usually congenital, it may develop later on; frequently the patches disappear of themselves, but more generally increase in size, and vary in shape. The most common localities are the true skin of the face, head, neck, back, and buttocks. Another form of nævus is found in the subjacent tissues, consisting of dilated veins, causing an elastic, livid tumor. The true erectile tumor is one composed of capillaries and arterial branches largely dilated, with strong, thick, vascular walls. There is free communication between these vascular tubes, which are attended with large tortuous veins. The growth is usually congenital, and its most common situations are beneath the integuments of the face, head, neck, back, and buttocks, orbit, bones or viscera. Pathologically considered, there is no aneurism in the proper sense of the term in this growth, as there is no degeneration of coats, but dilatation with hypertrophy, and increased function as well. A bruit is heard in it, sometimes associated with a vibratory thrill, and though it pulsates synchronously with the action of the heart, this pulsation is less distinct and has less expansion than true

aneurism (See ANEURISM). It is worthy of note, that in the female adult these tumors may be the seat of vicarious menstruation, the tumor becoming dense and full at the return of each period, and the blood slowly distills from some fissure or sore on its surface. The treatment of erectile tumor may be conducted on three principles: (1) Removal; (2) Diminution of arterial supply; (3) Effecting change of structure.

1. *By Removal.*—In cases where the skin is involved, the removal by the application of ligature is the most successful. The tumor is transfixed by a stout needle carrying a hemp or silk ligature, which is left in by withdrawing the needle; the loop is then cut, and both sides of the mass included in each portion of the thread, which is tied up as tight as it will bear; by this means the nævus is strangulated or killed, and brought away. In complex nævi several such loops must be used. In some instances transfixion with hare-lip pins, and ligatures twisted round them, suffice to destroy the tumor. In large nævi, where the integument is uninvolved, it may be dissected from off the mass and reflected, so that the nævus may be strangulated subcutaneously, and the flaps being replaced, there will be an avoidance of the puckered scar frequently left after ligaturing these tumors. A method of ligature devised by Professor Wood, whereby skin is saved and scar prevented, is thus performed: A slightly curved needle on a handle, with an eye near the point, is armed with a fine smooth hempen thread; "it is first passed under the skin round half the circumference

of the morbid tissue, entering and emerging through the skin at the opposite pole of the tumor. The short end of the thread is left in the puncture and the needle withdrawn, carrying the long end. Next it is passed under the base of the tumor across its diameter, entering and emerging at the punctures first made. The loop at the eye of the needle is then caught and held while the needle is withdrawn, carrying the free end of the thread. Lastly, the needle is passed round the remaining half of the circumference, under the skin, through the same two punctures, and the ligature thread detached. There are now a loop and two free ends emerging from the farther puncture, the thread being entirely sunk into the puncture nearest the operator. The ends are then made each to pass through the loop, and tied very tightly in a loop knot, so as to leave the power of tightening it as the parts enclosed shrink under the ulcerative process. As the thread is tightened the loop recedes into the puncture, but is held there by the ends passing through it, and the pressure exercised throughout is everywhere equalized. By this method the suppuration and slough formed by the nævus escapes by the punctures along the thread. The small spots of cicatrix remain as the only evidence of the operation." Extirpation by the knife is only advisable when the tumor is small, and can be lifted up from the parts beneath i:,and its shape and size accurately determined. Two elliptical incisions should be made so as to include the whole of the diseased growth and a little of the sound tissue surrounding it.

2. *By diminution of the arterial supply.*—If the tumor is so situated that it is inaccessible by the knife or ligature, the main arterial trunk may be ligatured ; as, for example, if within the orbit, by ligature of the common carotid, or, in some instances, the feeding vessels can be obliterated with a twisted suture.

3. *By effecting change of structure.*—Consolidation of the contents, converting the texture into a compact mass, or by converting it into abscess. The means adopted to this end are : Pressure, potassa fusa applied lightly, so as to induce ulceration, thus imitating the process of spontaneous cure. A red-hot needle or galvanic cautery introduced frequently and freely through the mass. The injection of some coagulating fluid through a fine syringe. The seton. If very small and superficial, the repeated application of nitric acid will destroy it, taking care in applying the acid that the parts immediately surrounding the nævus are not implicated. Vaccination is sometimes of use, in children, in the site of the tumor. By obstructing both the arterial and venous supply, the contents of the tumor may be caused to suppurate, and after the evacuation of the pus the nævus will,

in some instances, entirely disappear. Amputation has been necessary in some instances, as for example, where the greater part of the foot or hand, or a finger or toe, have been involved. There is likely to be considerable hemorrhage in such a case.

NAILS, DISEASES OF.—Acute inflammation of the soft structure into which the root of the nail is implanted frequently occurs in the toes, in most instances the great toe, after much walking in tight or short boots. The whole toe becomes red and swollen, and even slight pressure upon the free edge of the nail causes acute pain. The crescentic margin of skin at the root of the nail is retracted, and in the course of three or four days there is from this part and from under the nail a scanty discharge of dark-colored and ill-smelling pus. The whole nail then becomes thickened, and is finally loosened from its attachment to the top of the toe, and is thrown off, leaving a new short and delicate nail which grows slowly and is for several months overhung by the structures forming the swollen extremity of the toe. The usual treatment for this affection is absolute rest, and the frequently-repeated application of lint dipped in cold water or lead lotion.

Psoriasis of the nail is not often met with. It attacks the nails both of the fingers and toes, and is in most instances observed in patients who have had the venereal disease. The affected nail becomes brittle, thickened, and very rough on its surface. It is also very convex on its upper surface, and has been said to resemble in miniature the outside of the concave shell of an oyster. The free edge is rough and broken, and the skin at its root and along its sides is swollen and reddened. The nail is finally thrown off, and leaves a raw and inflamed surface upon which a small and malformed nail finally grows. In severe cases the inflammation extends and involves the whole of the finger. In acute and severe cases, warm fomentations or poultices may be applied with advantage. The chronic form is usually treated by the internal administration of mercury, arsenic, and bark, and by the local application of glycerine or tar ointment. The roughness of the surface of the nail may be removed or reduced by friction with a small file or with sand-paper. The affection in most cases is very obstinate.

Onychia maligna is an unhealthy ulceration of the bed of the nail affecting one of the fingers or toes of a scrofulous child. The end of the digit is much swollen, and of a deep red color. The nail is expanded at its free edge and is in-curved laterally ; it also loses its color and is lifted from its bed by a layer of a dark thin and very fœtid discharge. The fold of skin at the root of the nail is retracted, and at last the nail itself is thrown off, leaving a deep irregular ulcer with jagged edges. This ulcer increases in extent, and the surrounding skin becomes redder and more swollen. The progress of the ulceration is attended with severe pain. This disease is often excited by a slight injury to the digit. The simplest local treatment for this very obstinate disease is the application every second or third day of a strong solution of lunar caustic or blue-stone. Arsenic also is a useful agent, but ought not to be employed except with great caution. The nail should be removed as soon as it is loose. The general treatment should consist in placing the patient in good air, giving plenty of nourishing food, and in the internal administration of chlorate of potash with small doses of bichloride of mercury or gray powder.

In-growing toe-nail, or growth of nail into the flesh, occurs in almost all cases on the outer side of the great toe. It is a very troublesome and painful affection, due to over-paring of the nail, or to compression of the foot and toes caused by walking in tight boots. The soft and delicate integument at the outer edge of the nail, in consequence either of compression or of irritation, swells and becomes inflamed. The swelling does not subside but is kept up and increased by the contact of the sharp edge of the nail, which, as this structure grows, is imbedded into the overlapping fold of skin, and at last causes ulceration and a discharge of pus. From the raw surface thus formed there often springs a prominent mass of proud flesh or exuberant granulation, which is so extremely tender that the patient cannot wear a boot. Exercise, or even an attempt to walk, will often produce redness and swelling of the whole toe. When the end of the toe is much inflamed, and the edge of the nail is overhung by a large fungous mass of proud flesh, the foot should be kept at rest and uncovered by shoe or stocking, and the inflamed parts be bathed with some cooling lotion, as a solution of muriate of ammonia in water and spirits of wine or lead lotion. In slight cases, and when there is little or no inflammation present, the nail, after having been softened in warm water, should be scraped very thin with a sharp pen-knife, and under its outer edge should be then pressed in a small quantity of scraped lint, so as to form a soft and yielding pad between the irritated skin and the ingrowing nail. The scraped lint should first be oiled and then introduced little by little with the end of a small probe. This treatment should be repeated every second or third day, until the edge of the nail no longer presses against the side of the toe. In severe cases where the ulcer is large and very tender, and the margin of the nail is deeply imbedded, removal of the outer half of the nail is the most effectual mode of treatment. This is a very painful operation, and should always be preceded by the administration of chloroform, or by what acts quite as effectually, the local application of ether-spray.

NARCOTICS are remedies which procure sleep. There are usually other remedies associated with these which may procure sleep in another way, by relieving pain. These are technically known as *Anodynes,* and include such substances as belladonna, stramonium, and aconite, which have directly no influence in procuring sleep. The only real narcotics, however, are opium and its chief alkaloid, morphia, and chloral. The latter substance has only been recently introduced, and had better be given in the form of the syrup of the hydrate of chloral. Bromide of potassium and Indian hemp are commonly included in the list, but have no direct influence in giving sleep. Sleeplessness may, however, arise from various causes, and so a remedy which is useless at one time may be of value at another. In the main, however, if we desire to procure sleep, we must have recourse either to opium or to chloral. Opium seems to have this inconvenience : it is apt to be followed by headache, and it constipates the bowels ; it also disorders the digestive organs generally ; and so it is better to avoid its use in a good many cases. Chloral was introduced with the notion that, being absorbed, it would in the blood become converted into chloroform, and so produce a kind of anæsthesia. Experience, however, shows that it acts rather as a simple narcotic, than as an anæsthetic, and its effects seem to be less felt afterward than those of opium.

NASAL HEMORRHAGE. (See EPISTAXIS.)

NAUSEA is a common symptom of dyspepsia and disorders of the stomach. When occurring in women in the early morning during the child-bearing period, it is one of the signs of pregnancy. In diseases of the liver and kidney, as well as in those of the stomach, nausea may be a prominent and disagreeable symptom. It may be produced also by nervous and emotional influences. (See INDIGESTION.)

NAVEL. (See UMBILICUS.)

NECK (STIFF) is nothing more than a cramp or rheumatic affection of the muscles of the neck, and is caused by sitting in a draught ; but its persistence and the delicacy of the part affected make it necessary to treat it with care. Relief may generally be obtained by warm fomentations and the warm bath ; warmth should also be applied by means of hot flannels wrapped round the neck. This will usually effect a cure in a few hours at furthest, and in the mean time it is best to keep quiet and especially to avoid any sudden starts or wrenches to the neck. Any attempt to place the neck in its proper position by manual force is attended with danger.

NECROSIS means the death of bone, analogous to gangrene of the soft parts, by which the shaft of a long bone (generally) dies from injury or inflammation, and is enclosed in a case of new bone. Exfoliation is a form of necrosis, but it is the death of a thin superficial layer not encased in new bone. The dead portions of bone are called sequestra. Necrosis may be divided into simple and complicated ; the former when it is unaccompanied by any disease, the latter when associated with fracture or caries. It is further divided into idiopathic, when arising without any assignable cause ; traumatic, when the result of injury. Necrosis is also acute and chronic ; the first act, so to speak, the local death, being acute, and the subsequent process of the throwing off of the old and formation of the new, being chronic. Necrosis varies very much in extent, its simplest form being, as above-mentioned, exfoliation, which is an external form ; the internal form is where the inner portions of bone die, and the external retains its vitality. General necrosis is where both internal and external portions of the bone perish simultaneously. Necrosis generally stops at the articulating extremities of bones, and thus the cancellous structure is less liable to necrosis than the compact, being, however, at the same time, more liable to caries (see CARIES). The bones most liable to suffer are those most exposed ; viz., the tibia, femur, humerus, cranium, lower jaw, clavicle, ulna, etc. Acute necrosis is more prevalent in the young than in the old. The peculiar form of necrosis affecting the jaws has often, as its specific cause, the phosphoric fumes of lucifer-match-making ; this disease has been discussed in the article on Lucifer-Match-Makers' Disease. The process of necrosis can be related as follows : In the first place, *Inflammation*, in the second *Death*, in the third *Separation*, in the fourth, *Reparation*.

(1) *Inflammation.*—This inflammation may be the result of a wound, bruise, or fracture ; or it may be apparently of spontaneous origin. The periosteum is removed, and inflammation of the bone supervenes, and death of the bone is probable, although it does occasionally recover ; but if the internal periosteum be removed or perish, the death of the bone is certain. Exposed bone often retains its vitality, though apparently dead ; in these cases, it is of a brownish color, is dull on being struck, somewhat slippery to the feel, and if pulled about with instruments, bleeds. If all vitality be gone, it is white, dry, resonant on being struck with a probe, and is perfectly bloodless.

(2) *Death.*—Death of bone is often very rapid, sometimes occurring in a few hours ; while at others it takes a considerable time. The appearance of the necrosed portion varies in accordance with the period during which the bone is perishing. If very rapid, it hardly appears different from healthy bone.

(3) *Separation.*—The separation of the dead portion from the living is slow, and its process is somewhat similar to the detachment of sloughs in the soft structures. There is great activity in the structures immediately surrounding this dead portion. The vascularity is increased. It is painful to the touch, and bleeds, the blood being florid and arterial in color. A line of separation forms at the junction of the dead with the living, and the periosteum at this point is thickened. The living osseous substance along the line of junction, by becoming transformed into a soft granulation material, at length becomes a continuous trench ; this goes on deepening, and the above-mentioned substance is firmly connected with the living bone. The formation of this trench is accompanied by suppuration, and the pus formed in the neighborhood of the dead parts makes its way to the nearest surface, and in so doing interrupts the formation of the new bone which is being formed, and leaves sinuses, or, as they are called, *cloacæ ;* these cloacæ correspond with sinuses in the soft parts, the pus passing out through them. A probe passed down any of these cloacæ detects the peculiar feeling of the dead portion or sequestrum at the bottom of the openings. The separation of dead bone, and the formation of the new bone, are processes which advance together ; consequently the sequestrum is often entirely surrounded by a shell of newly-formed bone. The sequestrum is, after separation, to be regarded as a foreign body, of no use to the bone, and unconnected with it, and must consequently be treated as such, and its removal assisted. The sequestrum is always smaller than the recess in which it lies, not from its absorption, but from the transformation of more or less of the living bone into soft texture, whereby separation is effected.

(4) *Reparation.*—The process of reparation advances both superficially and deep, the former carried on by the periosteum, a membrane invested with special ossific power, and the latter consists of osseous production from the living bone beneath the loosening sequestrum. If the necrosis be internal, and a part of the cancellous structure only has become dead, as soon as the sequestrum is extruded, reparation rapidly follows. The pus escapes through the laminated portion of the bone by cloacæ, which have been formed by ulcerative action, and the process is a very tedious one, if the sequestrum be small, and the original inflammatory action limited, the resulting suppuration being slow and slight. When the necrosis is general, the processes of separation and reparation advance in the same way. There are instances where long bones have been almost entirely reproduced. Great care must be taken throughout the

whole period of treatment to keep the periosteum as entire as possible, and not to remove
the dead shaft too soon. Short bones, if
wholly necrosed, are never reproduced, and
reproduction is rare in the flat ones.

Symptoms of Necrosis.—In the first place,
there will be signs of some local injury, with
suppuration of the soft textures round the
affected bone. In old-standing cases there is
great thickening of bone and superficial tissues, with sinuses leading to the dead bone.
The dead bone, however, can only be felt by
probing or by seeing it. If it can be seen,
it is either white and dry or black, or yellowish and bare and hard to the touch.

Treatment.—In the early stages of necrosis, active measures are not admissible.
When abscesses point, they should be opened,
all sequestra should be assisted to escape, and
the health supported by nourishing diet and
stimulants. If the new shell of bone is not
able at first to support the weight of the limb
or muscular tension, splints must be employed. No attempt is to be made to detach
the sequestrum, unless nature has done her
part of the work by entirely freeing it from
the living tissue. During this process the
surgeon's duty is to mitigate the symptoms,
to prevent the extension of the evil, and to
favor the advancement of repair. From
time to time the rate of progress of the separation is to be gently tried by careful probing, and the sequestrum having been determined to be loose, steps for its removal must
be taken. An incision is to be made over it,
and if the natural openings or cloacæ, etc., be
sufficient without further interference, the
mass may be withdrawn by properly-devised
forceps. If, however, such openings are not
large enough for its evulsion the new bone
must be cut away, with gouges, chisels, saws,
trephines, etc., taking care that as little new
bone as possible is removed, so that it may
be brought to the surface. All unnecessary
violence must be avoided, from the risk of
doing damage to surrounding parts. When
the sequestrum has been removed, the wound
must be moderately stuffed with lint, to arrest
bleeding, and to insure granulation from the
bottom. Antiphlogistic treatment must be
maintained to keep under any accession of
inflammation, and perfect rest insisted on, as
it must be borne in mind that even in the most
favorable cases, a considerable time must
elapse before the necessary consolidation has
been accomplished. Amputation is rarely
demanded ; but in young subjects, when violent inflammation is followed by hectic, or in
very chronic cases, when separation has become far advanced, but does not complete itself, and when the system has been long
battling with the exhaustion of irritation and
discharge, and there is evidently no chance
of a continuance of the struggle, the cause
must be removed. Again, amputation is

necessary when in the case of the extensive
death of a bone throughout its whole thickness, the expected reproduction fails.

NEPHRALGIA is a technical term for
pain in the kidney. (See KIDNEYS.)

NEPHRITIS signifies inflammation of the
kidney ; the disease may be either acute or
chronic ; but for the various kinds the reader
is referred to the article on BRIGHT'S DIS-
EASE.

NERVES play a most important part in the
phenomena of disease, but the exact nature of
the influence so exerted is far from being fully
understood. The nervous system consists of
certain central parts called ganglia or nerve
centres, and certain cords connected with
these, which we commonly call nerves. These
centres are distinguished by being made up
of small masses, more or less irregular in
shape, and called cells. These cells commonly send off one or more projecting portions of their substance, like tails. These,
which are excessively minute, serve to connect one cell with another, so as to bring all
into accord, or, after being carried over by a
kind of sheath, go along with many others of
a like kind to form one of the strings or cords
already alluded to, and called nerves. Nerve
centres, then, are mainly composed of nerve
cells. Nerves or nervous cords are made up
of bundles of finer cords directly communicating with these central nerve masses. The
functions or duties of these two are totally
different ; the nerve cords can only conduct
impressions, whether they originate without
the body or within it. The centres, on the
other hand, take note of these impressions
and convey a knowledge of them to our
understanding. Moreover, should these
impressions conveyed from without be of
such a nature as to demand active exertion
on the part of the body, the appropriate command, so to speak, comes from the central
organ of the nervous system, and passes to
those muscles which have to execute the
order. Now, the set of nerve fibres which
fulfil the one function will not fulfil the other,
and so we have two sets of nerve fibres, those
which convey impressions to the brain, called
sensory, and those which convey impulses to
motion from it ; these are called *motor*. Sometimes we find nerves entirely made up of one
kind of fibres, motor or sensory, as the case
may be, but most frequently nervous cords
are made up of both kinds. If we take a
simple illustration of the two kinds of nerves
and their respective functions, we shall better
understand them. Suppose by chance we
touch a piece of hot iron, the flame of a
candle, or anything of the kind, this conveys
to the part of the body touching an impression ; this impression is conveyed by the
nerves to the brain, and is converted into a
sensation of pain, but promptly, as the result
of experiencing this sensation, a stimulus is

conveyed through the motor nerves to the part in contact with the hot object, which causes it to be promptly withdrawn. So speedily, however, all these various acts follow the one on the other that they seem to us simultaneous. This is not the place to enter into proofs of this position, we merely make the assertion, knowing that it can be easily proved.

There are several nervous cords, however, which have a function quite different from these. They provide all the various movements of life, especially respiration, circulation, and digestion, and these have a most important bearing on the subject of disease, especially those presiding over circulation, which are sometimes grouped under one heading and called the *vaso-motor* nerves. One of the most important bearings nerves have on disease, however, is that exercised through sensory nerves ; these, whenever over-stimulated, it matters not by what means, give rise to a peculiar sensation called *pain*, and this pain, which is a purely nervous adjunct to most diseases, sometimes rises to the dignity of disease itself. But the pain thus experienced and referred to some particular part, is not really felt there ; it is felt in the brain, but the origin of the impression is ordinarily, but not always, referred to the diseased or injured spot. But this painful impression in its turn affects other parts ; the nerves which rule the action of the heart are affected, and so that peculiar sickening feeling and palpitation or tremulous action of the heart is produced, which is commonly the result of great bodily pain. The sickness is referred to the stomach, the palpitation to the heart, but it is the same nerve which produces them both, and the same stimulus which affects the nerve. In recent times this mode of arguing has had wider application. The set of nerves already alluded to as vaso-motor run along the blood-vessels, and with them penetrate into every part. The blood-vessels possess a distinct muscular coat which is ruled by these nerves, insomuch that when the nerves are stimulated, these muscular fibres contract and so diminish the calibre of the vessels, and consequently the supply of blood sent to a part. On the other hand, when the power of these nervous cords is relaxed, the muscular coat of the vessels dilates, and so their calibre is increased, thereby, of course, increasing the quantity of blood sent to a part. Some seek to account for the phenomena of inflammation in this way, but in the meantime our knowledge is too scanty to admit of any wide generalization ; nevertheless, such facts show how much nervous influence is connected with disease, and that it is a fact never to be overlooked.

No part or organ of the body can be isolated or considered apart from its neighbors ; all are mutually dependent, so that when one is ill the others are sure not to be well. This is especially true of the nervous system. Sound health requires that it should be in a carefully-balanced condition, and any organ out of order is apt to give rise to this imperfectly-balanced state of the nervous system. Say, for instance, that an individual suffers habitually from indigestion, so that his body is ill nourished, his brain of course will be so also. In him comparatively slight causes will bring on nervous manifestations, owing to this condition of his system ; he may be impatient of slight sounds or garish colors, in point of fact slight stimuli give rise to inadequate results. This is the condition which in ordinary parlance is called nervousness, or the individual is said to have " the nerves." Rest, quiet, and good nourishment are the appropriate remedies.

NETTLE-RASH, or URTICARIA, is a troublesome, stinging, skin eruption, which gives rise to a sensation resembling that felt after being stung by a nettle. It appears on the skin in red and white wheals, slightly raised above the surface, and producing tingling and itching. It may come on very suddenly and disappear in a few hours, and then again temporarily appear ; eating muscles or periwinkles, or some kinds of oysters, will sometimes produce this disorder, and the whole of the face and the body may in a very short time become swollen and marked with the eruption. An emetic is the best remedy in such cases, so that the irritant matter in the stomach may be at once expelled ; bathing the skin with warm water will allay the tingling. Some people are very susceptible to this rash, and in some it comes out after taking beer or any stimulant ; some, again, have it after eating mutton, or pickles, or any acid substance. The eruption is disagreeable but harmless ; the treatment must consist in removing the cause and in avoiding any article of diet which will produce it : when it appears, bathing with warm water or lead lotion will give relief, while a purgative should be administered so as to remove any irritating cause in the intestinal canal.

NEURALGIA is a term given to pains sometimes following the tracks of nerves, sometimes lying apart from them, which cannot be referred to any distinct morbid change going on in the part. Nevertheless, though there be nothing in the spot to account for the pain, this may be of the severest possible character, and we are fain, for want of a better causation, to refer it to the nerves of the part. Now, nerves, when cut, are not more painful than other parts. There is absolutely no pain experienced in slicing the brain ; in flammation in tendons, ligaments, periosteum, and the like, where there are few or no nerves, is intensely painful. Sometimes, however, pressure on a nerve will give rise to pain in distant parts of a most intolerable character. Such is seen in aneurisms of internal vessels pressing upon nerves, and giving rise to frightful neuralgias. But in ordinary neuralgia we have nothing of the kind, we have only most intolerable pain and no apparent cause for it. In point of fact, wherever we see pain long continued and aggravated, and at the same time we can make out no definite cause for the pain, we term it neuralgia. Neuralgia, strictly speaking, however, ought to be restricted to pain in the course of nerves, and some nerves are affected much more frequently than others. Thus, neuralgia of the nerves which give the face sensation is by no means infrequent : it is most commonly called tic-douloureux. Sometimes the head on one side may be affected, or yet again into the sciatic nerve be its subject, giving rise to the condition called sciatica. But internal organs may be affected in like manner, so that we may have neuralgic pains of the heart, stomach, or intestines, still more frequently of the womb, the bladder, and the rectum ; pain not due, that is, to any inflammatory or other local change, but due merely, as far as we can say, to irritation of the nervous filaments supplied to the parts. It is hard, however, to determine whether such and such a pain is neuralgic, where internal organs are concerned, so here we shall limit ourselves to an account of the simple and better-known varieties of neuralgia.

Facial neuralgia, or *tic douloureux*, is perhaps the most common of all, and as the nerve attacked is made up of three branches, any one of these may be affected. One of these branches goes to the eye, and a part of it passes out from the orbit and turns up over the forehead. This is often the seat of pain, and when so, the neuralgia generally affects one side of the forehead, extending upward toward the hair. The next branch of this nerve comes below the eye and extends over the cheek and on to the side of the nose. This, too, may be affected, and very often is so, especially when the teeth on the corresponding side are decayed. The third branch of the nerve extends along the lower jaw, and is not so often the seat of pure neuralgic pain as the others. For instance, the branch on the forehead may be affected without any definite cause being ascertainable, that is to say, purely neuralgic, and very possibly dependent on some change in the nerve centre, rather than in the branch itself. But in the other branches we are much more likely to find a cause in some decayed teeth or some condition of the jaw which gives rise to irritation, and though we include such maladies under the heading neuralgia, they are rather instances of pain produced in one spot appreciated by the sensory centres in another. Neuralgic attacks commonly affect one side only, and they are often attended by or end

in attacks of nausea and violent vomiting. The pain varies in severity from a slight twinge now and again to unbearable agony. Not unfrequently this pain becomes periodic in its onset, appearing at certain hours of the day and departing at others—thus simulating the effects of malaria. Whether such periodic neuralgias are really produced by malaria is not plain, undoubtedly cold and damp do exercise a powerful influence over them. Certain forms of neuralgia appear to be associated with anæmia, and a goodly number with imperfect digestion, so that frequently bad teeth and the consequent imperfect mastication of the food setting up dyspepsia, go hand in hand in producing face-ache. Undoubtedly, bad teeth are one of the most prominent causes of face-ache, neuralgic and otherwise, insomuch that when a patient comes complaining of pain in the face, especially the cheek, the first thing we do is to examine the teeth, and if any be decayed we send him or her to the dentist before trying any local remedy. But even with this precaution we sometimes miss our mark, for the source of the malady may be the teeth, and yet that source be not apparent, for one of the common causes of this form of face-ache is an outgrowth from the tooth itself, what is technically called an exostosis. This is often difficult to detect, though pressure on the affected portion of the gum gives rise to pain, and even when detected, the patient, having had no toothache, and having nothing the matter with the tooth, may refuse to have it removed. To those not conversant with the influence exercised by the teeth, particularly the molars or grinders, on the neighboring parts when diseased, it would be incredible the amount of suffering which may be produced by an apparently slight cause. In some instances so obstinate is the pain produced by irritation of the jaw, that all the teeth may have to be removed for the relief of the malady, and even this may not suffice, for we have known violent face-ache produced by a badly fitting set of false teeth. Exposure to a draught of cold air falling on the side of the face is not an uncommon cause of neuralgia. This, perhaps, most frequently occurs in crowded halls or assemblies, when a window or door has been partially opened for ventilation, or driving home in a carriage, with the windows open, or the like. These causes are all the more likely to set the pain going if any such permanent irritation to the gum as bad teeth exist, or if the patient is liable to attacks of neuralgia of the face. The pain frequently prevents all attempts to sleep, but once the patient has fallen asleep, he may rest soundly, for the pain no longer torments him. Neuralgia of the head presents nothing special. Fatigue or debility is its most frequent cause. (See HEADACHE.)

Sciatica is a form of neuralgia which in a

good many respects differs from those already noticed. (See SCIATICA.)

Treatment.—One thing is to be noticed with regard to all neuralgic attacks, they are most frequently brought on by fatigue, mental or bodily, or if not produced by these causes, are most readily produced by any other cause in subjects who have been exposed to these, or are suffering from the weakness produced by them. This is a most important indication in treatment ; indeed it may be said to be its basis. Rest and nourishment—nourishment and rest are the foundations of our means of dealing with neuralgia. Nourishment may be taken with tolerable ease, but in certain cases where there is sickness and vomiting, there may be some difficulty in improving the general condition. This form of neuralgia is often associated with bad teeth, as already pointed out, and these must promptly be removed. Frequently, after this is done, nothing except tonics are required. Stimulants, too, are as a rule necessary, but must be carefully selected to meet the wants of each individual case. Rest is all-important, but rest is not always attainable without something being done for the patient. Where the pain is very severe, the best thing that can be done after having had the bowels well cleaned out (this in all cases is an indispensable preliminary), is to give the patient the fifth of a grain of morphia acetate under the skin of the arm. This will relieve the pain, and procure rest so much needed. Many, indeed, look upon this as being the treatment in all instances, but its true value we have just pointed out—it procures rest, and so allows time for self-recuperation.

For opening the bowels in the first instance, saline purgatives are perhaps best, but they may not act sufficiently powerfully, and so calomel or even croton oil is required, best of all use the repeated doses of a laxative mineral water, such as Hunyadi Janos water. Aloes, especially as watery extract, is a good preparation, particularly in sciatica, where there is a suspicion that the cause of the malady is habitually over-distended bowels. In most cases iron and cod-liver oil are essential. Iron is best given in some mild form, as carbonate, or as reduced iron, or as peroxide. Cod-liver oil is best given just after food, iron along with food. In reality they are both forms of food rather than of medicine. If the malady is periodic, as it often is, whether this depend on malaria or not, quinine had better be given, at first in full doses (5 grains or so), after in smaller quantities, say 5 grains of the citrate of iron and quinine three times a day. If there is indigestion, that too, must be seen to, rhubarb and soda or potass being usually the best remedies.

In rheumatic cases, especially in sciatica dependent on this cause, iodide of potassium is the remedy. Full doses of 10 grains or more must be given. Bicarbonate of potass is useless. In neuralgic headaches, sal ammoniac often does the greatest good. That too, must be given in large doses, 30 grains or so, and it does not always succeed. If the pain be very intense, chloroform may be given, but as we can generally procure rest by the administration of opium subcutaneously, that is to be preferred. Chloral is a totally different thing—doubtless it will be of the greatest use, but its exact value remains to be determined. (See NARCOTICS.)

In sciatica, local remedies have, as a rule, more power than general ones. This is hardly true of real neuralgia, but even in that one local remedy seems to surpass all others ; moreover it seems to be best adapted for those very cases which we cannot treat otherwise ; we allude to the continuous current of electricity. This form of electricity is to be carefully distinguished from those commonly in use, which is termed the interrupted current, and which not only is of no use, but may positively do harm. The continuous current is that which does most good in true frontal neuralgia. Other local means have been tried—blisters, red-hot irons, ointments containing aconite. belladonna, veratria, and other powerful drugs—all have been tried,and too often tried in vain. As a rule, if the neuralgia depend on a local cause, it will disappear with the removal of that cause, and will not disappear until it is removed. Thus, it is utterly useless to apply soothing remedies ointments, lotions, or what not, for a neuralgia dependent on a bad tooth or diseased jaw ; once these are seen to, the neuralgia will go. (See SCIATICA.)

NEURITIS (OPTIC) is a condition of the eye not yet fully understood, and only comparatively recently discovered. It affects the back part of the eye, and is only discoverable by means of the instrument which allows of the posterior part of the eye being examined during life. If this instrument is so held as to throw a beam of light into the eye while the observer's eye is situated at the proper point of observation, there will generally be seen a kind of reddish glare, indicating the reflection of the retina, or fine nervous net at the back of the eye, with its blood vessels. At one particular point, however, if the instrument is in proper focus, a white patch becomes clearly defined, and in its centre are to be observed blood vessels, to and from which branches of other vessels are seen to ramify in every direction. This is the optic disk, or papilla, the point where the optic nerve enters, and inflammation of this spot is what is called optic neuritis. The marks of this form of disease are commonly laid down as being an irregular, hazy, or woolly appearance of the margin of the disk instead of its clear, sharply-defined edge, as seen in health ;

the surface, too, seems swollen and the whole seems to merge into the surrounding parts. The most peculiar point connected with optic neuritis is, that it may exist in the most marked degree without any interference with vision, so much so, that most frequently this condition is first discovered by ophthalmoscopic examination. The great value of optic neuritis seems, indeed, to be as a sign of intra cranial disease—disease, that is to say, under the skull, which imght not otherwise be detected. Tumors of the brain are commonly so indicated, but many points have to be made clear with regard to it. Indeed, its mode of causation is by no means clear, some supposing that the influence is propagated along the nerve of vision, that is, the optic nerve and its covering ; others rather by means of the blood vessels which pass from the interior of the skull to the interior of the eye.

NEUROSIS is a word employed by modern physicians to indicate a malady which depends on some perverted nervous influence rather than on merely local change. There may be local change, but this would probably depend on the perverted nerve force rather than the perverted nerve force should depend on the local change. A goodly number of cases of palpitation of the heart are of this character ; that is to say, dependent rather on perverted nerve influence than on disease of the heart itself.

NICOTINE. (See TOBACCO.)

NIGHTSHADE, DEADLY. (See BELLADONNA.)

NIPPLES, AFFECTIONS OF.—*Excoriated or Sore Nipples.*—This troublesome and painful affection, in almost all instances, is met with during suckling, usually of the first child, and is due to irritation and ulceration of the delicate skin of the nipple. The mother first notices one or more deep cracks, which are extremely tender and bleed when touched. The skin around these cracks, fretted by the suction of the child's mouth and constantly bathed by milk and discharge, becomes inflamed and raw. In debilitated and unhealthy women the inflammatory mischief extends deeply into the tissues of the affected nipple and also to the skin covering the breast. When the nipple has become so tender that the mother is compelled to suckle the child almost entirely on the opposite and healthy side, the breast corresponding to the excoriated nipple is engorged with milk and finally attacked with acute inflammation, which results in the development of a large mammary abscess. The usual *cause* of sore nipples is the disregard of cleanliness and neglect on the part of the mother to keep the surfaces of the nipples dry. The tender cuticle covering these structures is readily excoriated by the contact of fluids, and by the friction of moist and dirty linen. In most cases the affection is produced by the suckling of

the first child, and never occurs again ; but some mothers, in spite of their careful preparation of the nipples before lactation, and constant attention to those parts after the birth of the child, are troubled with painful cracks and all the inconveniences to which these give rise during every period of lactation. Some mothers suffer from sore nipples in consequence of their allowing the child to retain the nipple in its mouth for too long a time. It is very probable that a diseased condition of the child's mouth may also be a cause of this irritation. In order to prevent sore nipples, the delicate skin should be hardened by the frequent application of diluted spirits of wine, weak lead lotion, or a lotion containing one drachm of alum to half a pint of water. A strong infusion of green tea, with the addition of about one-fourth of brandy or gin, will also be found a good astringent wash. The use of the lotion should be commenced early in the seventh month, and the nipples should be bathed night and morning. A bottle with the mouth and neck just large enough to admit the nipple should be half filled with the astringent lotion and then be applied over the nipple and inverted, so that the lotion may fall down upon this organ, and bathe it at every part of its surface. During suckling, the nipples should be kept as dry as possible, and when not used be covered by cotton wool dusted with lycopodium or starch. When ulcers and cracks heal, and there is no inflammation of the surrounding skin, these should be lightly touched with a pointed crystal of blue-stone, but when the nipple is hot and tender, the application of caustic must be deferred until the irritation has been allayed by warm fomentations. When a considerable extent of skin both of the nipple and on the breast is red, raw and moist, the best local application will be a solution of five grains of lunar caustic in one ounce of water. This should be brushed over the affected parts every morning. After the application of any active or caustic agent to the nipple, the mother must take care to keep the child from the breast until all traces of the local remedy have disappeared.

Retraction of the nipple when associated with a hard and painful swelling of the breast, in a woman above the age of thirty-five years, is diagnostic of cancer. This condition is occasionally met with in young women whose breasts in all other respects are perfectly sound and healthy. This, when the patient becomes a mother, gives rise to much trouble, as the flow of milk is obstructed, and the secretion accumulating in the gland sets up acute inflammation, which may terminate in the formation of a large abscess. Simple and uncomplicated retraction of the nipple in young females is caused in most instances by tight lacing and compression of the breast, which flattens the nipple and prevents its full development. In cases of this kind an attempt should be made at the commencement of the period of lactation to produce protrusion of the nipple by means of a breast-pump, or by the suction of a strong infant of five or six months, if this can be made to take a strange breast. The infant just born should be handed over to a wet-nurse until the state of the mother's nipples has been improved.

NITRE, or NITRATE OF POTASS, is a remedy of considerable value. It is procured, by washing the soil or beds of vegetable matter specially prepared for the purpose, and is mainly used for the manufacture of gunpowder. In medicine it is chiefly used as a cooling remedy, for, dissolved in water, it has a cooling saline taste, which is sometimes very grateful. It is also given with a view to acting as a diuretic, but its efficacy here is doubtful. *Sweet Spirits of Nitre* is treated of under NITROUS ÉTHER.

NITRIC ACID is one of the mineral acids used in medicine, but not, perhaps, so much as the others. It does not, for instance, seem so stringent as sulphuric acid, nor does it aid digestion so well as hydrochloric acid. It is procured from nitrate of potass by distillation with sulphuric acid, and should be quite colorless. Usually, however, it contains nitrous acid, which gives it a green or yellow tint. The mixture of this acid with hydrochloric acid, called aqua regia, or nitro-hydrochloric acid, is much used. Nitric acid is used externally as a caustic more than the other mineral acids. It is applied by means of a piece of stick to the sore, which it completely destroys. The sores so treated are usually of an unhealthy description, and this destructive agent is applied for the purpose of procuring a fresh and healthy surface with a prospect of healing. A similar plan is adopted for getting rid of piles. In these cases the surrounding skin must be protected ; if the acid touches it, the skin is stained yellow, which is characteristic of the acid. It has been injected into the bladder in a very dilute state, for the purpose of neutralizing the evil effects of alkaline urine. Internally, the acid is mainly given as a refrigerant, and to remedy phosphatic urine. Sometimes it seems to do much good in cases where mercury cannot be given, in syphilis, and also in some liver diseases, especially those of tropical climates. Dilute nitro-hydrochloric acid is perhaps preferable as an internal remedy. It acts better as a tonic, and as a remedy in dyspepsia arising from chronic catarrh. When the urine is free of phosphates, this acid may be given with advantage, certainly with more good than can be obtained from either acid singly. It is largely used in liver mischief, especially in chronic inflammation, as well as in the cachexia of syphilis. The ordinary dose of either acid in its dilute state, as sold by the chemists, is 10 or 15 drops in water, if freely diluted.

NITROUS ETHER is only used in the form of spirit, commonly called *Sweet Spirit of Nitre*. It is made by a somewhat complicated process, and the product is not uniform in quality. The basis of the product is alcohol ; this is treated and decomposed by sulphuric acid, so that ether is formed. At the same time nitrous acid is set free by decomposing nitric acid by means of copper ; the two unite together and form nitrous ether. This liquid is clear, transparent, sometimes with a slightly yellow tint, and a fruity odor ; usually, too, it is acid. The therapeutic properties of nitrous ether are not very clear. All kinds of things have been said of it, but as the preparation is of most uncertain strength, and sometimes contains no nitrate of ethyl, its supposed active principle, it is somewhat hard to tell what these really are. It has been used mainly as a stimulant diaphoretic and a diuretic. It has been the custom to order it in slight febrile cases, to open the skin, and it has been given in dropsies. Its well-defined properties are limited to its refrigerant action. Hence, mixed with water, it is an exceedingly pleasant refrigerant to the lips of one parched with fever. It is a good deal used for this purpose, and may very well be combined with nitre itself. The dose should be about a drachm or two drachms, freely diluted with water.

NITROUS OXIDE—PROTOXIDE OF NITROGEN, also called LAUGHING GAS—is a transparent, colorless gas with faint, sweetish smell and taste. It is easily made by submitting crystals of nitrate of ammonia to heat, when the protoxide of nitrogen will pass over. Sir Humphrey Davy first discovered that this gas was respirable, and that it produced intoxicant effects upon the human system. One of its peculiarities is, when given in small quantities, to produce uncontrollable laughter ; hence the name laughing gas. It was found subsequently that not only would this agent produce excitement, but anæsthesia, in the same way as ether and chloroform. After the discovery of the use of the last agents, nitrous oxide was almost discarded, but it is again getting into use, as it has been found by experience that it is less likely to produce fatal effects than either ether or chloroform. It is now principally used in the operations of dentistry. (See ANÆSTHETICS, ETHER, CHLOROFORM.)

NODES are a term used to denote certain tumors in connection with bone and periosteum. It may be either scrofulous or syphilitic In the scrofulous node there is scrofulous matter confined between the carious bone and its periosteum, and is due to an affection of the bone. The *true* node, however, is the syphilitic, and is caused by the effusion of lymph between the bone and periosteum, and is due to inflammation of its deeper layers.

The inflammation in nodes is not always limited to the periosteum, and the deeper structures are thus implicated. The subperiosteal effusion either ossifies, or softens, giving rise to caries. The most common place for these tumors is along the shin-bone, or the radius and ulna, and on the clavicle and cranium. They frequently appear so hard as to seem osseous, but in reality it is the semisolid effusion beneath the dense thickened periosteum which gives rise to the feeling. Nodes sometimes soften, and pus forms, the indications of which are the fact of the skin becoming shining, dense, and thinned. The formation of subperiosteal abscess usually ends in exfoliation of the bone. (See Ex-FOLIATION.) The first indication of the appearance of nodes in a person affected with syphilis is tenderness of the affected bone, and severe pain and nightly exacerbations. Soon roundish or oblong swellings are noticed on the bones, usually commencing with the skin ; they are tender, and convey a sense of obscure fluctuation. If by treatment the disease be arrested, syphilitic exostosis is the result ; if not, a quantity of glairy serum is effused between the periosteum and the bone, forming a very painful tumor. Extensive exfoliation may ensue, causing intense suffering to the patient, if situated in the skull frequently terminating fatally.

Treatment.—The treatment consists in the administration of a regular course of mercury or of the iodide of potassium. The nightly pain is best relieved by leeching or blisters, and the application of strong iodine paint. If very tense, fluctuating, and painful, subcutaneous incision, made by passing a narrow knife under the skin, and across the tumor, gives great relief. But it generally will be found that such remedies as iodide of potass, sarsaparilla, and blisters will be sufficient to produce absorption and allay pain.

NOLI ME TANGERE. (See Lupus.)

NOMENCLATURE treats of the names and classification of diseases. Unto every disease a name is given, and as our knowledge increases year by year names are added to signify either new complaints or fresh groups of symptoms. It is obvious that some diseases are more closely allied than others ; thus all those disorders which are accompanied by a high temperature are called fevers ; these, again, are divided into those which are catching and those which are not. Some diseases are caused by the presence of parasites, others by accident or design. To arrange diseases according to any precise plan is, however, extremely difficult. If we knew accurately the causes of every disease, some scientific arrangement might be carried out, but our knowledge on this point is as yet very imperfect. Then, again, there are many "causes of death" registered which are not diseases at all, but symptoms ; convulsions and diarrhœa, debility and wasting, are examples of this kind ; the true cause of the death is really the cause of these symptoms, if in all cases it could be ascertained. The English Registrar-General has adopted a nomenclature which has been long in use, and is very well adapted for its purpose ; he divides all diseases into five great classes, and these in their turn are divided into orders, while under each order are placed the diseases as known by their general name.

CLASSES.	ORDERS.	
I. Zymotic diseases..	1. Miasmatic diseases	Small-pox, measles, scarlet fever, diphtheria, quinsy, croup, whooping-cough, continued fever (comprising typhus, typhoid and simple continued fever), erysipelas, puerperal fever, carbuncle, influenza, dysentery, diarrhœa, cholera, ague, remittent fever, rheumatism, and other zymotic diseases.
	2. Enthetic diseases	Syphilis, stricture of urethra, hydrophobia, glanders.
	3. Dietic diseases	Privation, want of breast-milk, purpura, scurvy, alcoholism.
	4. Parasitic diseases	Thrush, worms, parasites, etc.
II. Constitutional diseases	1. Diathetic diseases	Gout, dropsy, cancer, cancrum oris, mortification.
	2. Tubercular diseases	Scrofula, consumption, hydrocephalus.
III. Local diseases	1. Diseases of nervous system	Apoplexy, paralysis, chorea, epilepsy, convulsions, brain diseases, etc.
	2. " organs of circulation	Pericarditis, aneurism, heart disease, etc.
	3. " " respiration	Bronchitis, pleurisy, pneumonia, asthma, etc.
	4. " " digestion	Gastritis, peritonitis, ascites, hernia, intussusception of intestines, etc., jaundice, diseases of stomach, liver, and spleen.
	5. " urinary organs	Nephritis, Bright's disease, cystitis, kidney disease, etc.
	6. " organs of generation	Ovarian and uterine diseases, etc.
	7. " " locomotion	Synovitis, arthritis, and diseases of the joints, etc.
	8. " integumentary system	Phlegmon, ulcer, diseases of skin, etc.

CLASSES.	ORDERS.	
IV. Developmental diseases........	1. Diseases of children..............	Premature birth, cyanosis, malformation, spina bifida, etc.
	2. " adults.................	Childbirth.
	3. " old people............	Old age.
	4. " nutrition.............	Atrophy and debility.
V. Violent deaths....	1. Accident or negligence............	Fractures, contusions, gun-shot wounds, cuts, stabs, burns and scalds, poison, drowning and suffocation, etc., by accident or negligence.
	2. Homicide........................	Murder and manslaughter (homicide).
	3. Suicide.........................	Suicide by any method.
	4. Execution.......................	Hanging (execution).

NOSE (DISEASES OF.) (1) *Of the external nose. Fracture of the nasal bones.* (See FRACTURES.)

Incisions.—Accurate adjustment of the cut surfaces must be obtained at once, as owing to the extreme vascularity of the integuments of the nose, union takes place very rapidly, and unless the adaptation be careful, deformity may be the result. In cases where there has been removal of the nose, either partially or entirely, the portion, after washing and neatly stitching on, has completely united, leaving scarcely any scar. The same remark as to the neat approximation of the edges of wounds is of equal importance in any part of the face.

Hypertrophy (Lipoma).—The integument of the nose occasionally, as the result of acne rosacea of long standing, becomes irregularly enlarged, and "fleshy excrescences" appear. They give rise to great disfigurement. They are of a dusky purple color, cold and greasy to the touch, usually occurring in individuals over middle age, and most frequently in males. They consist anatomically of hypertrophied skin and connective tissue, with dilatation of the small veins and enlarged sebaceous follicles. They are not dangerous or painful, only inconvenient and unsightly, and of slow growth. Removal is the only remedy. The mass must be entirely shaved off, an operation requiring some dexterity ; and the growth does not recur, except in rare instances. The term lipoma, frequently given to this skin disease, is pathologically a wrong one, as it contains, as a rule, little or no fatty tissue.

Lupus.—The various forms of lupus are discussed in the article LUPUS.

Deficiency from disease or accident.—Cases where the whole or part of the external nose has been destroyed by disease, such as scrofulous ulceration, lupus, syphilis, etc., can be in great measure remedied by what are termed rhinoplastic operations, a department of plastic surgery which consists in the transplantation of integument from an adjoining part of the face. Such operation requires considerable ingenuity of plan and skilfulness of performance, but it must be borne in mind that if the entire bone or cartilaginous framework of nose be wanting, it is almost useless to attempt anything of the sort, as the points of support are gone. The most common method of supplying the deficiency is by what is termed the *Indian method.* A piece of integument, the shape of the nose laid out flat, is drawn upon the forehead, and the edges of the mutilated nose pared, the flap is dissected off the forehead and brought down and attached to the pared edges just mentioned, by sutures. Flaps may be dissected from the cheeks, or from any convenient spot where the integument is healthy. Occasionally such a flap has been taken from the arm.

Diseases of the Internal Nose.—In order that the internal nose or nasal fossæ may be completely inspected, with a view to find out accurately their condition, a proceeding termed rhinoscopy is had recourse to. Rhinoscopy is anterior or posterior. *Anterior rhinoscopy* is an inspection through the nostrils. The nostrils must be dilated, and a good strong light brought to bear upon them. A bivalve nasal speculum, made on purpose, is used. This consists of an instrument having two slightly curved blades, these blades are introduced into the nostril, and when separated by handles they dilate the parts, and, being highly polished, they throw light into the cavity.

Foreign substances in the nose. (See FOREIGN BODIES.)

Posterior Rhinoscopy.—In rhinoscopy, or inspection of the nasal cavities from behind, a somewhat more complicated mechanism is necessary. The instruments required are those similar to what are used in examination of the larynx (see LARYNGOSCOPE), and consist of a mirror, perforated, and worn round the head, hand-mirrors, somewhat longer and a little more bent than those used in laryngoscopy, and a blunt flat hook with which the uvula and soft palate are supported and drawn forward. In using the instruments the patient is seated with his back to the light, and the head well thrown back, the mouth opened to its fullest extent. The operator, seated opposite to him, by means of the spatula, held in the left hand, raises and draws forward the uvula and soft palate, and directs the light reflected from the perforated mirror, bound round his head, down the pharynx. When the light is made to shine upon it, some portion of the walls or contents of the nasopharyngeal or nasal cavities may be distinctly imaged on the speculum. If a good view be obtained, the two superior meatuses can be seen and their contents, and the mucous membrane of the three turbinated bones, a considerable portion of the septumnarium, some portions of the posterior surface of the velum pendulum-palati, the lateral wall of the naso-pharyngeal cavity, and the orifices of the Eustachian tubes. In order to see all these structures, however, at all satisfactorily, great patience and self-control on the part of the patient are essential.

Diseases of the nasal cavities. Nasal Calculi or Rhinolithes.—These concretions are generally found in the inferior meatus. They consist of phosphate and carbonate of lime and magnesia, chloride of sodium, and mucus or some animal matter. These bodies can be removed with forceps, a proceeding of some difficulty very often, and the nasal cavities must be afterward thoroughly cleaned by syringing or douching.

Epistaxis or *Nose-bleed* has been discussed in the separate article on that subject. (See EPISTAXIS.)

Coryza is an excessive discharge of mucus depending upon catarrh, struma, syphilis, or the presence of a polypus. Best treated with mild astringent washes, or the insufflation of powders, such as tannic acid, or some astringent, warm, dry atmosphere, good living, and cod-liver oil or iron. A severe form of coryza, due to syphilis. is occasionally met with ; it commences as an ordinary cold in the head, with increased secretion of the mucus, which, on exposure to cold or to alcoholic excess, becomes thicker, more profuse, and greenish in color. There is great uneasiness and tenderness in the nostrils, with continuous desire to blow the nose. It is frequently attended with headache, alteration of the tone of voice, and impairment of smell. With regard to *treatment,* it is necessary to administer mercury, both internally and by the exhalation of its vapor, and mild astringent and detergent lotions.

Ozæna, or rhinorrhœa, is a purulent or sanious discharge, giving rise to most offensive fœtor, rendering the sufferer unbearable both to himself and to those around him. It may be either (1) Catarrhal ; (2) Strumous ; (3) Syphilitic. The catarrhal is met with in patients of delicate constitution, and after a long and troublesome cold the discharge will become very fœtid, generally worst in the morning. The discharge is accompanied with headache, relaxed throat, cough, and great depression of spirits and deafness. The *treatment* consists in sending the patients to a dry, bracing atmosphere, the inhalation of steam, carrying with it vapor of creosote. Constitutionally, bark and mineral acid tonics.

Scrofulous or strumous ozæna, generally begins during childhood, and depends upon a strumous taint, causing ulceration of the nasal mucous membrane. The discharge is very offensive, and if not treated may implicate the bones and cartilages of the nose, setting up destructive ulceration, whereby the most dreadful deformity is occasioned. The *treatment* consists in the administration of bark, iodide of potass, cod-liver oil, phosphate of iron, etc. ; of the washing out of the nasal cavities with a large syringe containing a little chloride of zinc in solution, or a little dilute citrine ointment to be applied to the ulcers with a camel's-hair brush. With regard to the syringing out of the nasal cavities, it may be borne in mind that when one side of the cavity is entirely filled through one nostril with fluid by hydrostatic pressure, while the patient is breathing through the mouth, the soft palate completely closes the chordæ, and does not permit any fluid to pass into the pharynx, while the fluid easily passes into the other cavity, mostly round and over the posterior edge of the septum narium, and escapes from the other open nostril, after having touched every part of the first half of the cavity of the nose, and a great part, certainly the lower and median canals, of the second half. The syringe should have a long slender nozzle, with a bulbous extremity, perforated by a rose of small holes. The best apparatus, however, for the purpose is the nasal douche.

Syphilitic ozæna has precisely the same general character as the preceding, and must, of course, be attacked on the general principles indicated for the treatment of constitutional syphilis.

Polypi.—The most common are the simple mucous or benign. The growth is soft, of a rather tough consistence, and yellowish-gray in color, bleeding slightly when touched, usually growing from the inferior spongy bone, never from the septum. It may occur at any period of life, though most frequently in middle age. It obstructs respiration, causes a stifling sensation in the head, seriously affects the senses of smell and taste, and often occasions deafness. The treatment consists in its removal by forceps. The root of the growth is to be seized with them, and, by a movement of twisting and pulling, the mass is withdrawn from the nostril. The next variety of polypus is the medullary, bleeding or malignant, occurring usually at the middle or later periods of life ; the growth causes great pain, and it occasionally grows and increases with great rapidity. Owing to the malignancy of this growth it is impossible to remove it thoroughly, although attempts may be made from time to time to clear away the mass from the nostril. The third form to be mentioned is the fibrous variety, distinguishable by its great firmness. It is distinctly fibrous in composition, and occurs most frequently in young male adults. It seems to adhere to the bone, but not to depend upon any diseased condition of it. The operation of removing a mass of this nature is always a very serious and formidable proceeding. These growths occasionally grow backward down the pharynx instead of forward, adding, of course, greatly to the difficulty of any operative procedure.

NOSE-BLEED. (See Epistaxis.)

NOSOLOGY, the systematic arrangement of diseases. (See Nomenclature.)

NOSTALGIA means home-sickness. (See Mal-de-Pays.)

NUMBNESS, a peculiar sensation felt at the end of a nerve, and caused by some altered condition of the nerve, either at its origin or in its course. It may be caused by an injury or by pressure ; it may be also a sign of brain disease ; it is also caused when a drug, like aconite, is rubbed into the skin.

NURSING of the right kind is rapidly coming to be recognized as even more important in the treatment of many diseases than all the drugs that can be administered. Only doctors know how many valuable lives are saved by good nursing, and on the other hand how many are thrown away by the want of it ; but fortunately its importance is beginning to be understood and appreciated by the public at large. In those cases which are lingering, or where bodily strength is required for proper attendance on the patient, it is better to employ a professional *sick-nurse*. Such a nurse knows, or ought to know, how to perform many important duties of which those less accustomed to sickness are ignorant ; and she will be quick to detect changes in the patient's symptoms which others might overlook, and which it may be important for the doctor to be informed of. In cases of ordinary sickness, or where professional nurses are not obtainable, the members of the family usually do the nursing ; and here many mistakes are made for want of a little homely and practical knowledge, such as is contained in an article entitled " A Few Hints on Domestic Nursing, by the Mother of a Family," which appeared lately in one of the English magazines. This article is written in very familiar language, but its suggestions have been indorsed by competent authorities as eminently valuable and practical. The article is as follows :

" There are many little useful hints in nursing the sick to be gained only by personal familiarity with illness ; and as my boys have obstinately persisted in having almost every form of infantile infectious disorder, I shall jot down a few of the points which I found to be most useful to me during the long weary time we were kept in the nursery.

" When my little boys Percy and Louis were suffering from scarlet fever, I had every article of furniture save Percy's little bed, Louis's cot, two chairs, and a boxful of toys —which were afterward burned—carefully removed, all curtains and carpets rigorously excluded ; while I had two print dresses,

which I wore alternately during the dreary time. I used to be so sorry for the little patients ; for of course all visitors were strictly prohibited, and children naturally like a change both of people and places. It is always advisable, if it is possible to have a choice of rooms for illness, to choose a large airy apartment with a south aspect ; for there is nothing like sunshine for keeping one cheerful, as well as acting most beneficially upon the health of the patient ; besides it is invaluable as a disinfectant, worth bushels of chloride of lime. In cases of infectious disease, people cannot be too careful in communicating with the outer world. Many and long-continued were the efforts I made to prevent the spread of scarlet fever ; and truly thankful I am that I never heard of any one catching it from us. I placed an old saucer nearly full of cold water, in which I poured a little carbolic acid, in each room, on the hall table, stair-head, window, in fact, on every available spot in the house. At first the strong gaseous odor was highly offensive, but that soon wore off, and in a very short time its presence was almost unnoticed.

" My boys used to hate the sight of their ' bokkies,' as they called their medicine ; so I placed a little round table, which I covered with a clean napkin, outside the room door ; and thereon I put the bottles, spoons, liniments, etc. which were needed, and found it such an improvement on the old plan of keeping them promiscuously on the chimney-piece that I have adhered to it ever since. Every utensil as soon as used should be carefully removed and well rinsed out—cups, spoons, glasses, all should be at once cleansed, and not suffered to lie about in disorder. It is rather more trouble ; but surely the little extra labor will not be grudged when the comfort of the patient is increased.

" In these enlightened days it is almost an insult to write about the value of fresh air, yet there are some people who carefully keep their rooms shut up ; and what a fatal mistake it is thus to exclude one of God's best gifts to man ! If the patient be kept warm and free from draughts, plenty of fresh air may be admitted without the slightest danger. In most modern houses the upper window-sash lets down, and may be kept open a few inches. If there is the slightest draught, it may be prevented at a very trifling cost, by having a light wooden frame from six to eight inches in width made to fit the upper part of the window, and a single thickness of flannel tacked on each side of it.

" I find it a capital plan to fold a sheet in two, lay it across the bed, above the under sheet, with the upper edge just touching the pillows, and the ends tightly tucked in under the mattresses. It does not wrinkle or crumple up, as single sheets will do ; while crumbs can be readily brushed off, and it can much more easily be changed than a large one. It is best to fold the upper end of the quilt *under* the blankets before turning down the top sheet, as it helps to keep them in place ; and as there is nothing more fidgeting to a healthy person than to have the chin grazed by blankets, the annoyance must be doubly great to one lying on a bed of sickness.

" The greatest care should be taken to keep the beds clean ; so the linen ought to be changed twice, and the blankets once a week ; those that have been removed hung in the open air for a few hours, then thoroughly dried in a warm room, and put away to replace those in use which must be similarly treated. There is nothing easier to an experienced nurse, or more difficult to an inexperienced one, than to change the bed-linen when a patient is in bed. I once noticed a capital plan in an American paper, which I have followed in scores of cases, and never found to fail. I shall copy it here *in extenso* for the benefit of those who may be placed where such a scrap of advice may be useful : ' In the first place, everything required must be at hand before beginning ; then move the patient as far as possible to one side of the bed, and remove all but one pillow. Untuck the lower and cross sheets and push them toward the middle of the bed. Have a sheet ready folded or rolled the long way, and lay it on the mattress, unfolding it enough to tuck it in at the side. Have the cross sheet prepared the same way ; lay it over the under one, and tuck it in, keeping the unused portion of both still rolled. Move the patient over to the side thus prepared. The soiled sheets can then be drawn away, the clean ones completely unrolled, and tucked in on the other side. The coverings need not be removed while this is being done ; they can be pulled out from the foot of the bedstead, and kept wrapped round the patient. To change the upper sheet, take off the counterpane, and lay the clean sheet *over* the blankets, securing the upper edge to the bed with a couple of pins. Standing at the foot, draw out the blankets and soiled linen ; replace the former, and put on the counterpane ; lastly, change the pillow-cases.'

" I found it most refreshing to my little patients to sponge the entire body with vinegar and warm water, and was very careful not to let them catch cold while doing so, just sponging over a small portion at a time ; while the bed and the patients were equally protected by a large blanket, which I carefully pinned round their shoulders. It is a great mistake to have large quantities of fruit, biscuits, etc. lying about a sick-room. A very few grapes, an orange peeled and divided, and two or three milk or water biscuits are quite enough to have displayed at one time. The same may be said of food. I have often been pained, when visiting some of my sick pensioners, to see their friends, with well-

meant but mistaken kindness, bring large basinfuls of horrible compounds, which they dignify with the name of gruel or sago or tapioca, as the case may be. The mere sight of the food seemed to set them against it. Whereas if a little care had been bestowed upon its preparation, and a small cupful provided instead of the large quantity I name, they probably would have partaken of it with pleasure.

" Another error, committed with the best of intentions, is to keep asking the patients what they would like ; if they could take this thing or the other. The sickened, wearied expression I have often seen flit over the faces of people who are recovering from a lingering illness, when their officious relatives come teasing them as to their requirements ! During the lingering illness of a dear relative, I verily believe we made her often eat, just by providing dainty morsels of food, displaying them temptingly arranged, and taking them to her bedside quite unexpectedly ; when if she had been asked, *could* she eat anything, I feel confident the answer would have have inevitably been, ' No ; thanks. I don't feel at all inclined to eat.'

" A very simple and expeditious way of cooking a little bit of chicken or fish is to butter a paper thickly, place the food to be cooked within the paper, and place it on the gridiron over a clear fire. A very short time suffices to cook it thoroughly ; and I have often found that to be eaten when all other modes of invalid cookery have been tried in vain.

" I always find Percy and Louis take refuge in milk when they are ailing, and truly thankful am I that such is the case. Once when Percy had a very severe attack of bronchitis, I felt in despair, for all the tempting food I could contrive failed to make him eat ; for several days—eleven if I be not mistaken —he lived almost entirely on milk ; and when I mentioned to our medical attendant my fear that the child would die of starvation, he quite laughed at the idea, and said, ' As long as he can take the milk, the child will do very well.'

" In conclusion, I would earnestly impress upon my readers the great importance of having every article in the shape of body or bed linen thoroughly well aired. The slightest trace of damp may undo the careful work of days or weeks, may even cause all our nursing and attention to prove in vain."

NUTMEG is the seed of the *Myristica Moschata*, and is better known as a spice than as a medicine. It contains a concrete oil solid at ordinary temperatures, and a volatile oil, to which it mostly owes its property. Nutmeg itself is mainly used as a flavoring ingredient in various important preparations. The solid oil is used in some plasters, and the volatile oil, dissolved in spirits, acts as a gentle stimulant and carminative. It is also contained in aromatic spirit of ammonia and aloes pill. It is said in very large doses to produce drowsiness or even stupor. The dose of the spirits is about a drachm given in sugar. (See MACE.)

NUTMEGGY LIVER is produced in some cases of heart disease when that organ becomes gradually congested and full of blood ; slight jaundice may come on, and there is generally a dull pain over the liver. It is also the result of drinking alcohol. (See LIVER.)

NUX VOMICA is the seed of the koochla-tree, growing in the East Indies. It and the fruit of another plant called the St. Ignatius bean, owe their properties to the strychnine which they contain. The fruit of the tree is a round berry like an orange, filled with these peculiar seeds in its pulp. The seeds of the nux vomica are hollow on one side and raised on the other, as if pinched by the thumb. They are very tough and hard, so that they are not easily crushed. They are covered with hairs of a velvety character. Two alkaloids are contained in these, viz., brucia and strychnia, combined with an acid, igasuric acid ; the properties of the drug, however, depend almost entirely on the strychnia. The preparations of nux vomica are an extract and a tincture, and there is an officinal solution of strychnine. Of these the doses are as follows : of the extract, a quarter of a grain to 2 grains ; of the tincture, 10 to 20 minims ; of the solution of strychnine, 10 to 20 drops. The alkaloid itself may be given, but never in doses which exceed the one twenty-fourth of a grain ; even these may give rise to trouble-some symptoms. Given internally strychnine produces spasms, mainly by acting on the spinal cord. It does not seem to influence the brain to any extent, for in cases of fatal poisoning through its means, the mind is usually clear to the last. The mode in which it acts would seem to be the induction of an over-sensitive condition of the spinal marrow, so that impressions which under ordinary circumstances would produce little or no effect, give rise to violent convulsions. The first symptoms of an overdose are twitching of the muscles, often of the lower extremities, and a kind of choking sensation about the throat. These gradually spread and increase in intensity till the whole body seems rigid from violent muscular contraction. These in many ways resemble those of tetanus. The body is often so contracted by the powerful muscles of the back as to resemble an arch, and to rest only on its head and its heels. The muscles of respiration are likewise convulsed, so that no respiratory movement is possible, and the face becomes first red, then livid, and almost black. Such attacks are not continuous, but last for a minute or more ; during the interval the patient is quite sensible, but the slightest motion of those round about him, sometimes even a breath of air, will suffice to bring the attack on again worse than ever. The great thing, therefore, is to avoid disturbing the patient if the attacks be slight, but if they be severe, or seem to increase in severity, prompt remedies must be employed or death will speedily ensue. If in point of fact the fatal issue can be postponed for three or four hours after the beginning of the attack, there may be good hopes of ultimate success. The cause of death may either be a too prolonged interference with respiration or exhaustion caused by repeated convulsive attacks.

Used remedially nux vomica or strychnine acts as an incentive to digestive action, gives rise, that is, to the sensation of hunger, and probably aids digestion. It is one of the best of its class, especially in chronic catarrh of the stomach, when the tongue is loaded and the bowels irregular. Two or three up to ten drops of the tincture of nux vomica given in a little water will often effect great relief in the way of temporary symptoms. Flatulence from the same or allied causes it also relieves more than any one drug. The dose is as before, and should be taken just before meals. Some varieties of headache it also remedies, especially those connected with gastric disturbance and foul tongue. Extract of nux vomica is a favorite remedy in constipation, especially that which is habitual. It is, however, seldom given alone, but is combined with other substances, like aloes, rhubarb, and steel. All cases, however, do not answer to the stimulus. The action of nux vomica or its alkaloid is, however, still more marked in diseases connected with the nervous system, especially those of a functional character. What is commonly called nervous exhaustion, from whatever cause arising, is better treated by this than any other drug whatever. In business men who have been exposed to much mental worry, and partly broken down, its use is attended with singular benefit. In females affected with low spirits and hysteria this is often a most valuable remedy, more so than any other ; but on the whole it seems better adapted for the coarser maladies of the male sex.

In poisoning by either nux vomica or strychnia, the remedies must be prompt removal of the poison where possible, either by the stomach-pump or vomiting, but when the paroxysms have set in, an attempt to use either generally brings on convulsions. Then it is best to trust to choloroform inhalation.

O.

OAK BARK, the bark of the stems and small branches of our common oak tree is mostly used as an external remedy. It should be collected in spring. Its smell when moistened is somewhat peculiar, but its taste is almost purely astringent, owing to the tannic and gallic acids which it contains. The quantity of these in different species of bark vary a good deal, especially with age, season, and the part of the tree from whence the bark is taken. Its only officinal preparation is a decoction which is only used externally. It may, however, be given whenever tannic acid is indicated. It is best suited for astringent lotions and injections, and may be combined with various other remedies, provided these are not incompatible with tannin.

OBESITY. (See BANTINGISM, and WEIGHT AND HEIGHT.)

OBSTRUCTIONS may occur in many of the organs of the body and set up a train of symptoms of a very serious nature, but in each case it will depend much upon the nature and seat of the obstruction. The most important obstructions are those which take place in the intestinal canal, in the liver, or in the kidney, or in the course of the circulation.

Intestinal Obstructions.—The causes of this mischief are very various, and most of them are very difficult to make out, and also to relieve ; hence the treatment is uncertain, and often unsuccessful. The causes of intestinal obstruction may be divided into two classes : 1. Those which come on suddenly, pursue an acute course, and which will prove fatal if relief be not quickly afforded. 2. Those which come on gradually and pursue a chronic course, and produce symptoms which may subside more readily under the aid of medical or surgical measures. Under the first division may be mentioned :

a. Congenital stricture or malformation.

b. Foreign bodies impacted in the intestines.

c. Loops formed as a result of inflammation which may entangle portions of the bowels.

d. Invagination or intussusception of the intestines.

Under the second division may be classed :

a. Constipation, habitual or accidental.

b. Inflamed intestine, the result of injury.

c. Chronic inflammation of the peritoneum.

d. Tumors pressing on the bowels.

e. Simple stricture of the bowel, the result of ulceration generally.

f. Cancer of the bowel.

Congenital malformation of the intestinal canal is generally confined to the rectum or the lowest portion of the large intestines ;

this deformity, as its name implies, occurs before birth, and the only thing that can be done, if the bowel is closed up, is to make an artificial opening for the passage of the excreta. (See ANUS, ARTIFICIAL.) Sometimes the malformation is higher up in very rare cases. Few symptoms come on within the first twenty-four hours, but after that constant vomiting comes on and continues until relief is obtained, or until death ensues. When the deformity is in the upper part of the intestines nothing can be done.

Foreign bodies may become impacted in the intestines, either by being accidentally swallowed or introduced up the rectum, or as a result of external injury, but this subject has been dwelt upon in the article on FOREIGN BODIES.

Twisting of the intestines, so as to form loops, which prevent the passage of the fæces, is not a very common occurrence. The symptoms are very urgent from the first ; great pain is suddenly experienced in a small circumscribed spot of the abdomen, and obstinate constipation begins from this time ; the part becomes much distended and painful on pressure ; vomiting is generally present, and often constant ; the pulse is small and the countenance is expressive of pain and exhaustion. Medical aid in such cases must be at once sent for.

Invagination, or intussusception of the bowel, is often the result of worms or of some other irritant cause acting on the bowels. (See INTUSSUSCEPTION.)

The second division of obstructions offers a more hopeful chance for treatment.

Habitual constipation may go on for so long a time that a hard mass of fæcal matter forms in the intestines, and cannot be dislodged by the natural efforts ; the mass generally forms in the large bowel, and may be sometimes felt as a tumor ; it may occur in those who take large doses of opium, or any drug which has a constipating effect. Injections of warm water must be given until the mass is softened, or portions may be removed by a scoop from the rectum. (See CONSTIPATION.)

Inflammation of the intestines, as a result of injury, may cause portions of the bowel to adhere together and set up obstruction ; if high up, little, if anything, can be done ; if low down, an artificial opening may be made in the loin so as to give relief ; in all these cases pain in the abdomen, distension, and vomiting are the most marked symptoms.

Obstruction may result from adhesion of two portions of the intestines in cases where there is a tubercular deposit on the peritoneum, or where two ulcers have set up inflammation around them and caused adhesions; sometimes an abscess may then be formed; in such cases there may be no constipation, but often diarrhœa, and especially at first; it is very difficult to find out the nature of such cases during life.

The most common cause of obstruction is *stricture of the bowels ;* it follows generally as a result of ulceration ; rarely met with in the upper part of the intestines, its common seat is in the rectum, or within a foot of the end of the canal. The lower the stricture is the more distended becomes the abdomen with flatulence ; there is great pain, vomiting, anxious countenance and constipation ; often a copious vomit relieves the pain for a time ; surgical aid must be sought early, and if the strictures cannot be overcome, an artificial opening may be made in the loin so as to give great relief, and perhaps save life. The operation is called colotomy, and it is of course only useful in those cases where the mischief is below the seat of the operation. The patient will afterward pass the excreta through the new opening, and must wear an apparatus for the purpose.

Tumors pressing on the bowel may cause obstruction, as hydatids, ovarian tumors, etc., but these cases will vary so with each individual state that no description of them would be useful. Cancer may affect the bowel and chiefly the rectum, and cause obstinate constipation and obstruction ; the emaciation, pain, loss of flesh, and gradual onset of the symptoms will help to reveal the nature of the case, although it may be mistaken for simple stricture ; usually the disease may be made out on examining the rectum. Colotomy will often give great relief, although it cannot save life.

Hepatic Obstructions.—The only obstruction in the liver that need be mentioned here is due to the presence of a gall-stone in the duct or tube which conveys the bile from the liver to the intestinal canal. The symptoms of a gall-stone in the duct would be great pain in the right hypochondriac region and over the liver, vomiting, anxious expression, thirst, loss of appetite, and a yellow tinge of the skin ; when the stone has passed into the intestinal canal great relief is at once experienced, and the urgent symptoms pass away. If the jaundice is intermittent and persists for some time, it may be due to several smaller stones passing at different times. The *treatment* of such obstructions during the attack must consist in putting the patient in a hot hip-bath, applying hot poultices or hot fomentations to the seat of pain, and in giving some purgative medicine ; should the pain be very severe, chloroform may be administered with caution, and opium may be given so as to try and procure sleep. After the severe symptoms have passed away the patient should lead a regular life, be careful of

his diet, take plenty of exercise, avoid intoxicating liquors, and see that the bowels are kept regularly open. Tumors pressing on the bile duct, as in some cases of cancer, may produce obstruction, and so will any inflamed condition of the bile duct itself ; in the first instance, nothing can be done, but the last cause will generally be removed by treatment. The administration of purgatives, as rhubarb draughts, and an occasional dose of calomel, the regulation of the diet and active exercise will generally bring about a cure, although such cases may persist for a long time. The main symptoms are a dull, aching pain over the liver, constipated bowels, pale fæces, a jaundiced skin, dark-colored urine, loss of appetite, a furred tongue, and dyspepsia ; under treatment these symptoms gradually subside. (See LIVER.)

Renal Obstructions.—A stone or calculus in the kidney is one of the most troublesome obstructions that can occur in that organ, for if situated in the pelvis of that organ it prevents the flow of urine to a greater or less degree into the bladder ; great pain in the back in one loin or other so severe as to double the patient up, and pain passing downward to the groin, nausea, vomiting, and often blood in the urine are the main symptoms. They may occur at intervals, and generally come on after exertion, as riding on horseback or in a jolting vehicle. Often small pieces of stone become detached and pass away in the urine, but their passage is accompanied with very severe pain at the time. A hot hip-bath, or hot poultices and fomentations must be used to ease the pain, and, if needful, opium may be given or chloroform may be cautiously inhaled. If the stone pass down into the bladder, it may then be recognized and removed by the operation of lithotomy or lithotrity ; very often it remains in the pelvis of the kidney, and then causes dilatation and subsequent destruction of that organ. Tumors growing in the abdomen and hydated cysts may cause an obstruction to the flow of urine, so also will cystitis or inflammation of the bladder, a stone or tumor in the bladder, and a stricture of the urethra. In all these cases the flow of urine is more or less impeded and the parts behind the obstruction become distended, and finally cause a serious disease in the parts behind.

Obstructions of Circulation.—Obstructions may take place in the course of the circulation ; thus a vein may be plugged with fibrine, and if this occur in a large vessel the parts below will become swollen and œdematous ; in this way gangrene of an extremity may be caused ; in other cases an artery may be blocked up, as occurs in some cases of hemiplegia or paralysis, because the supply of blood is then cut off from the part ; if the block exist in the brain, it may cause serious consequences, but if a small vessel be blocked in other parts of the body very little harm is done. A plug which is formed at the spot of obstruction is called a thrombus ; a plug which is carried from a distant point to the seat of obstruction is called an embolon. (See EMBOLISM and APOPLEXY.) In very rare cases a plug forms in the right side of the heart, and may cause death in a short time ; very often, in the process of dying slowly, clots form there, but give rise to no symptoms to call for remark. Any foreign body met with in any part of the body may be looked upon as an obstruction ; thus a marble or coin in the air-passages, a piece of meat or false teeth lodged in the œsophagus, a bean or pea in the nostrils, are all instances of obstruction. (See FOREIGN BODIES.)

In addition to the above obstructions there are some which are of a much more minute character, by which very small vessels get blocked up, and as a consequence of which very serious mischief may be set up in the organs thus affected. It now and then happens that when the heart has become affected after an attack of rheumatic fever, vegetations or growths of fibrine which are then found in the valves of the heart become washed off by the stream of blood, and block up vessels in different parts of the body ; in such cases there is generally pallor of the skin, a fluctuating temperature, as shown by the thermometer, and much prostration ; such cases generally prove fatal. In all cases of disease of the heart and in many affections of the lungs, as emphysema, etc., the general course of the circulation is impeded, and often dropsy may ensue in consequence. (See DROPSY and HEART.) A diseased condition of liver, as in cirrhosis, will also cause an obstruction to the circulation, and since the blood cannot flow freely through the portal vein, ascites or dropsy of the abdominal cavity will ensue. Also see GULLET.

ŒDEMA is the swelling caused by effusion of serous or inflammatory fluid into the loose areolar tissue lying under the skin or mucous membrane. A well-known example of œdema is the diffused and soft swelling which occurs over the feet and ankles, either as a result of general debility or in connection with dropsy due to disease of the heart and kidneys. Pressure upon the veins of a limb, and consequent obstruction to the flow of blood toward the heart, constitute a frequent cause of œdema. In the last stage of cancer of the breast, the arm often becomes enormously swollen in consequence of the pressure of the enlarged and cancerous glands in the armpit upon the veins which return the blood from the upper extremity to the heart. A similar result is sometimes produced at the extremity of a limb, in consequence of tight bandaging after fracture. In inflammation a modified serous fluid is generally poured out, which causes swelling and

œdema of the affected part. True œdema, caused by effusion of fluid, always forms an inelastic swelling, which retains for some time any marks made on its surface by compressing it with the finger. It may be thus distinguished from the hard, solid effusion produced by chronic inflammation.

The following are some of the chief forms of œdema :

Dropsical œdema, as may be met with in the swollen limbs of patients suffering from Bright's disease of the kidneys, from disease of the heart, and from exhaustion. The legs sometimes become much swollen, and the distended skin smooth, glistening, and sometimes red and inflamed.

Inflammatory œdema is caused by the effusion of a fluid containing fibrine, which coagulates spontaneously on exposure to the air ; fluid of this kind is formed whenever the surface of the body is inflamed by the application of a blister ; but here the effusion is quite superficial. When a similar fluid is poured out into the loose areolar tissue under the skin in connection with irritative or inflammatory processes, the swelling is called inflammatory œdema. The extent of the swelling depends upon the amount and the character of the areolar tissue found in the region inflamed. In the loose and abundant subcutaneous tissues of the eyelids a considerable quantity of fluid may be readily effused. Hence the rapid and extreme swelling of those structures in cases of erysipelas and inflammation of the face. The tissue under the conjunctiva is also very loose, and swells up rapidly, in some severe cases of ophthalmia, constituting the condition known by surgeons as *chemosis*. The most dangerous form, probably, of local inflammatory œdema is the effusion of fluid into the loose tissue at the upper part of the larynx, which is often produced in children who have inadvertently swallowed some very hot fluid. The narrow orifice leading to the larynx and windpipe is speedily closed by the swollen tissues, and the patient, if not relieved by surgical treatment, soon dies from suffocation.

ŒSOPHAGUS is a muscular tube which connects the pharynx above with the stomach below, so as to allow of the passage of food from the mouth into the intestinal canal. It is lined by a smooth epithelial membrane, and is capable of expansion according to the size of the food swallowed. It is commonly known as the gullet. It is sometimes the seat of cancer and stricture, and it may become seriously injured in cases where children swallow boiling water by mistake. (See GULLET.)

OILED SILK. (See SILK, OILED.)

OLFACTORY NERVES are special nerves emerging from the brain, one on each side, which, spreading out over the interior of the nose, enable man to have the sensation of smell. The nerve filament may be stimulated, as in cases of taking smelling-salts, and, if seriously injured, the sense may be lost. (See BRAIN.)

OLIBANUM is a kind of gum resin obtained from a plant called the *Boswellia serrata*. It is not now officinal. It occurs in small masses called tears, of an oblong shape, and having a peculiar odor. The resin in it enables it to burn with a peculiar odor, and the gum to form an emulsion with water. It is a stimulant like myrrh, but is mainly used as incense in Roman Catholic places of worship.

OLIVE OIL is perhaps better known as an article of food than of medicine, yet it is valuable as both. This oil, which is obtained from olives grown in Southern Europe, by pressure, and commonly called salad-oil, is of a pale straw color with a tinge of green in it. It tends, at a low temperature, to become solid, apparently by the crystallization of its bases, oleine and palmitine. When brought into contact with an alkali these bases are decomposed, the acids, oleine, and palmitine uniting with that to form a soap, the glycerine, which is the normal base, being set free. (See SOAP and GLYCERINE.) The oil itself is used in making several liniments, plasters, and ointments, and is sometimes given internally. Internally, in large doses, whether given by the mouth or as an enema, it tends to open the bowels, and to act as a laxative. For this reason the plentiful use of salad with olive oil will not infrequently tend to open the bowels regularly. It is also used externally for lubricating the surface. When cod-liver oil cannot be taken, olive oil often can, and if so, is sure to do some good. It is not so easily digested as cod-liver oil ; nevertheless, its pleasant flavor and taste render it superior to the former in a certain number of instances. It may be used with advantage by inunction in some wasting diseases of children, the smell produced being not nearly so unpleasant as that of cod-liver oil used the same way.

OINTMENTS are forms of remedies in which the active substance is wrought up with lard or some similar fatty substance, which, being smeared on the skin or raw surface, keeps the part moist and prevents evaporation. Formerly a distinct kind of ointments, called cerates, was employed ; in these a considerable quantity of wax was mixed up with the other substances, so that their substance was harder and firmer than those of ordinary ointments. The name is now done away with, but the substance remains, for a good many of the ordinary ointments contain wax, and are essentially cerates. Ointments have been long in favor as applications to wounded surfaces, and doubtless in many cases they do good ; but in certain instances,

especially when the discharges tend to decompose, they do harm : the fats break up and the fatty acids are set free, and so the application becomes a curse rather than a blessing. The same untoward results follow the use of a single application of ointment too long. (See GOLDEN OINTMENT.)

ONYCHIA. (See NAILS, DISEASES OF.)

OPHTHALMIA is a term applied to inflammation of the conjunctiva or thin mucous membrane, which covers the front of the eyeball and lines the inner surfaces of the lids. In some forms, however, of ophthalmia, there is inflammation also of the cornea, and of the anterior part of the strong fibrous coat of the eye which is called the sclerotic. Ophthalmia is a very frequent affection, presents very many forms, and originates from one or more of a great number of local and constitutional causes. Its simplest form consists in slight and temporary redness and itching of the surface of the eye, due to the presence of a particle of dust or to the prolonged exposure of the eye to strong light. At the other extreme of a long list of ophthalmic affections is placed the acute purulent or Egyptian ophthalmia, in which there is intense inflammation of the conjunctiva, attended with profuse suppuration and constitutional irritation, and, in many cases, terminating in rupture of the eyeball and total loss of vision. The following are some of the principal and most common *causes* of ophthalmia : the presence between the lids and the surface of the eyeball of foreign bodies, such as particles of dust, sand, and other matters. Particles of steel and iron, when impelled with much force, adhere to or are imbedded in the tissue of the cornea or conjunctiva, and so long as they remain keep up inflammation. An inverted eyelash, by irritating the conjunctiva on the front of the eyeball, often causes ophthalmia. Exposure of the eyes to a strong draught and the prolonged action of a heated atmosphere are common causes, and also much and long-continued exercise of the eyes on minute objects, especially if this be carried on under artificial light, and in close, badly-ventilated rooms. To these conditions may be attributed the frequent occurrence of ophthalmia among watchmakers, working jewellers, compositors, needlewomen, reporters, and clerks. The eyes are usually much irritated by very bright artificial light, whether direct or reflected, and by the reflection of strong sunshine from very extensive light-colored surfaces, as the sea, a long stretch of sand, or snow. There are certain constitutional diseases which render their subjects liable to attacks of ophthalmia ; of these, the principal are gout, rheumatism, scrofula, and inherited syphilis. Individuals who, in consequence of high living and of indulgence in alcoholic drinks, suffer from dyspepsia and congestion of the liver and other digestive organs, are much predisposed to inflammation over one or both eyes. Ophthalmia is very common among the very poor, and in bodies of men who are crowded together in foul and close rooms, and who are badly fed. Under these circumstances the ophthalmia is caused directly by the presence, on the inner surfaces of the eyelids, of firm and rounded swellings called granulations.

Simple or Common Ophthalmia is produced by slight injury or by exposure to a draught. The symptoms are, redness of the conjunctiva, " watering" of the eye, and a feeling of smarting and stiffness. These, in most cases, soon pass away after the application of a cooling lotion, care having been taken to protect the eye both from light and the action of cold.

Catarrhal Ophthalmia is so named because it is caused by exposure to those external and climatic influences which give rise to the symptoms of the affection known as catarrh, or common cold, but which here attack the mucous membrane of the eye and lids exclusively, or to a greater extent than that of the nose, fauces, and air passages. This form of ophthalmia is met with in patients attacked by measles, and occurs in some cases of scarlet fever and of erysipelas. The symptoms resemble those of simple ophthalmia much aggravated. The eyelids feel stiff, and the patient complains of a feeling as if " sand or dust had got into the eye." There is a bright scarlet redness of the conjunctiva, disposed not regularly over the whole surface, but in irregularly formed patches. There is a discharge from the eye, which at first is clear and thin, but afterward of a yellow color, and thick and viscid. During sleep this discharge collects at the edges of the lids and dries there, gluing together the eyelashes. The lids become red and swollen. The general health gradually becomes disordered, and the patient complains of headache, fever, dryness of the mouth and throat, and loss of appetite. In ordinary cases the affection generally lasts for about ten days or two weeks, but when the inflammation has been allowed to proceed without treatment, it often passes into an obstinate and dangerous purulent ophthalmia. In most cases both eyes are affected. In old people this form of ophthalmia often becomes chronic, and is then very rebellious to treatment. In ordinary cases of catarrhal ophthalmia, where there is not very much local irritation, frequent bathing of the eyes with cold water and the application of alum lotion (one grain to one ounce of water), or of one or two drops of a solution of lunar caustic (one grain to two ounces of distilled water), will generally be found effectual. The application of the lotion or drops should be made thrice daily. When, however, the patient complains of severe pain, and the eyelids are red and inflamed, light

poppy fomentations should be applied, and afterward, if these give no relief, a leech to each temple. The edges of the lids should be anointed every night at bedtime with glycerine or olive-oil. The patient should keep to a light diet, and the bowels be kept freely relieved, if necessary, by the administration of calomel and black draughts. The eyes should be protected by a dark green shade.

Purulent Ophthalmia sometimes attacks new-born infants, and under these circumstances is regarded as a distinct affection, which has been styled *ophthalmia neonatorum*. The purulent ophthalmia of adults, or the Egyptian ophthalmia, as it is called, in consequence of its prevalence in the French army after the campaign of 1801, sometimes attacks individuals who have been collected together in numbers under faulty hygienic conditions, and breaks out occasionally in large schools of young children. The symptoms of purulent ophthalmia at first resemble those of the catarrhal form, but they rapidly increase in severity, and in the course of twenty-four or thirty hours the eyelids become of a deep-red color, and swollen to such an extent that they cannot be opened. The patient is much alarmed by these symptoms, and, as he cannot obtain a glimpse of any object, or even tell whether it be day or night, believes that he is blind. Now, between the swollen lids there is a constant discharge of thick purulent fluid, which, if applied even in minute quantity to a healthy conjunctiva, soon sets up purulent inflammation. The conjunctiva is reddened and much swollen, so that it forms large rolls which cover over a greater part of the surface of the cornea. The patient complains of acute pain, which shoots from the eye to the corresponding cheek, forehead, and temple. There is considerable constitutional disturbance and the patient is generally very nervous and fearful of permanent blindness. The affection, if unchecked by treatment, causes ulceration with perforation of the cornea, and, in some cases, sloughing of the whole of this transparent membrane. In the latter case there will, of course, be complete loss of vision, with slight ulceration and even perforation ; the sight, though not destroyed, will in most cases be seriously impaired. Occasionally the purulent ophthalmia extends with great rapidity from the conjunctiva to the other coats, and even to the interior of the eyeball.

The purulent ophthalmia of infants, generally occurs on the third or fourth day after birth. In many instances, and especially among the poor, it is not noticed for the first day or two, and until irreparable mischief has been produced. In the first stage the lids are slightly swollen, and are stuck together by some dried mucus. There is intolerance of light, and the infant's brow is generally much contracted. At a more advanced stage the lids become red and puffy, and are separated from each other by the protrusion of rolls of inflamed and swollen conjunctiva. From the surface of this membrane there is a profuse and continuous discharge of thick yellowish fluid, which is sometimes stained with blood. The effects of this affection, when severe and if it be allowed to take its course, are sloughing of the cornea and ulceration of this transparent membrane, and subsequent opacity. Purulent ophthalmia is more amenable to treatment in new-born children than in adults, and in the former class of patients, unless the cornea has been already involved, speedily and completely subsides without any bad results, after the application of suitable remedies.

The adult subjects of purulent ophthalmia are usually pallid and weak, and should not be treated on any lowering system ; the strength ought to be kept up by good but easily digestible food ; beer, wine, and in very bad cases brandy may be given in moderate quantities. The most useful medicinal agents are quinine and opium. The local treatment consists in incising the masses of swollen conjunctiva, and in applying some strong astringent, as lunar caustic in strong solution, or in the solid stick. The eyes are then to be frequently syringed with a solution of alum. There is probably no other local affection in which early professional assistance is more necessary than purulent ophthalmia, whether in the adult or young infant. But in all cases of this kind, much responsibility is thrown upon the nurse or attendant. The eyes have frequently to be bathed, the face must be kept clean, and above all, great care must be taken to wipe away at once the purulent discharge, as the contact of this with the conjunctiva of a healthy eye will almost certainly set up fresh inflammation and suppuration. The affected eye should be covered by a layer of cotton-wool fixed by a bandage. This covering should be frequently renewed, and when removed should at once be burned.

In the purulent ophthalmia of infants the local treatment need not be so severe. The frequent application of a solution of alum (15 grains to one ounce of water) will in most cases prove an efficacious means of arresting the course of the disease.

Granular Ophthalmia generally presents the following appearances : the edges of the eyelids are red and swollen, the upper lid droops over the front of the eye, and the lower lid is slightly everted ; the conjunctiva is reddened and on exposure to bright light there is a free discharge of tears, and the lids are closed spasmodically ; the cornea is pitted on its surface and more or less hazy, and near its circumference is invaded by a well-marked zone of dilated blood-vessels. On everting the upper lid it will be found that

the conjunctiva lining its inner surface is very red and vascular, and studded with numerous soft and ruddy projections resembling the granulations observed on all healthy ulcers. In consequence of this resemblance the soft growths which are enlarged follicles and papillæ of the conjunctiva are called granulations, and the inflammation to which they give rise by friction over the surface of the eyeball is called granular ophthalmia. These granulations are different in form, size, and consistence in different cases. In some cases the inner surface of the lid is studded by minute and pale gray granules, which have been likened to soaked sago grains. The precise nature and origin of this affection has not yet been made out. It is of frequent occurrence among sailors and soldiers, and in large parochial schools. It is very common among the peasantry in some parts of Ireland. The subjects of prolonged granular ophthalmia are usually pale, weak, and out of health. It is a very chronic and obstinate disease, and often causes dense opacity of the cornea and incurable blindness.

The *treatment* of this affection consists in supporting the strength of the patient, and in attempting to rub down and destroy the granulations by astringents and caustics. The applications most frequently used for this latter purpose are blue-stone, lunar caustic, acetate of lead, liquor potassæ, tannin, and quinine. These are all very powerful agents and necessitate great care in their application.

Scrofulous or Strumous Ophthalmia differs from the preceding forms of ophthalmia in being an inflammatory affection of the cornea and not of the conjunctiva. It is met with generally in ill-nourished and unhealthy children and young women. It is often associated with pustular affections of the scalp, and with eczematous scabs and excoriations about the nose and ears. Although called scrofulous ophthalmia, this affection is not met with exclusively in individuals in whom there is any morbid disposition of a scrofulous or tuberculous character. The most morbid symptom of this kind of ophthalmia is great intolerance of light (photophobia). The patient generally lies with the face downward and the eyes covered by the hands or arms, and when he is raised and brought to the light, the eyelids are closed spasmodically, and the whole face is much contorted. There is a profuse flow of tears, which irritate and redden the lower lid and the cheek. When with much difficulty the eyelids have been separated, the observer will find at first sight but very little to account for the acute pain and intolerance of light. The conjunctiva is generally clear and free from swelling and redness. On examining the cornea closely, it will be found somewhat clouded and studded by a few small superficial pits or small ulcers, and at the margin may generally be seen one or more whitish specks surrounded by distended blood-vessels.   In advanced cases there is deeper and more extensive ulceration, with dense clouding of the cornea.

The essential point in the *treatment* of this affection is the improvement of the general condition. If possible, the patient should be sent to the sea-side and be allowed to take exercise in the open air.   The diet should be nutritious and easily digestible, and a small quantity of wine may be allowed. Preparations of steel and cod-liver oil are especially beneficial in cases of this kind.   The intolerance of light and the pain in the eyes may be much relieved by applying small blisters one after the other to the temple and forehead.

OPHTHALMOSCOPE is an apparatus used for exploring the interior of the eyeball and the posterior portions of the retina and choroid. It was invented by Professor Helmholtz in 1851, and has since proved itself an invaluable agent in the diagnosis of affections of the organ of sight arising from local morbid changes, and also from constitutional disorders. Surgeons skilled in its use are now able to detect inflammatory and hæmorrhagic changes in the membranes and humors of the eye, and occasionally discover intraocular indications of disease of the brain and of Bright's disease of the kidneys, of the existence of which serious disorders there had previously been no suspicions on the part either of patient or medical attendant. The simplest form of ophthalmoscope is a round concave mirror perforated at its centre by a small orifice, through which the surgeon can look directly upon the fundus of the eyeball illuminated by the rays of a bright light thrown upon the mirror and reflected to the patient's retina. These rays are reflected back from the bottom of the patient's eye, and converge at the surface of the reflecting mirror, to the back part of which the eye of the observer is applied. In addition to the mirror a small convex lens of short focus is often placed before the eye of the patient, in order that the observer may obtain a clearer and magnified view of the retina. The examination is made in a darkened room, and a gas-burner or oil-lamp is so placed that it is on one side of, slightly behind, and on a level with the head of the patient, who sits facing the surgeon. Before the examination, atropine is usually applied to the surface of the eye, in order to dilate the pupil. The use of the ophthalmoscope seldom causes any pain or uneasiness. The temporary disturbance of vision, which sometimes follows the examination, is usually due to the action of atropine.

OPISTHOTONOS is a technical term used to designate those convulsions in cases of tetanus or hysteria, etc., in which the patient

is arched backward, so that the head nearly touches the heels. (See TETANUS.)

OPIUM is perhaps the most important drug in our pharmacopœia. Various kinds are in use, but all are obtained in the same way. The white opium poppy is allowed to mature its capsules only for a very short period, only indeed for a few days after the flower leaves have fallen. Then incisions are made in its texture, so deep as to reach the sap, but not so deep as to reach the interior of the capsule. The sap exudes as a milky juice which speedily hardens and becomes brown, forming little masses. These are carefully gathered or scraped off and wrought up into balls or cakes, and usually covered over with some leaf. The preparations are many and various, comprehending a confection, a plaster, an enema, an extract, a liquid extract, a liniment, a pill (commonly called compound soap pill), a lead and opium pill ; aromatic chalk powdered with opium, compound ipecacuanha powder, compound kino powder, compound powder of opium, tincture of opium or laudanum, compound tincture of camphor, also known as paregoric elixir ; an ammoniated tincture of opium, opium lozenges, ointment of galls and opium, and wine of opium. Of course the doses of these vary according to the effect it is desired to produce, but supposing it is intended to give rise to an effect comparable to that produced by a grain of opium, that is, an ordinary full dose, they would be as follows : of confection of opium 5 to 15 grains, of the extract about a grain, of the liquid extract 25 drops, of laudanum 25 or 30 drops ; of compound tincture of camphor about the same, and of ammoniated tincture of opium rather less, of opium wine rather more than 30 drops may be given ; of chalk and opium powder 30 to 40 grains, of compound ipecacuanha powder 10 grains, of compound kino powder 15 grains, of compound soap pill 4 or 5 grains, of compound powder of opium 3 grains, of lead and opium pill 4 grains, of opium lozenges 1 to 4. Opium contains a great variety of substances of a crystalline character, and possessed of distinct properties. Its chief acid is one called meconic acid, its chief base is morphia. But besides morphia it contains codeia, papaverina, thebaia, or paramorphia, narcotine, narceia, meconine, or opianyl,opianine, and porphyroxine, with perhaps a variety of others. On the whole, not much is known with certainty of the action of any of these bases except morphia. Two salts of morphia are used, the hydrochlorate and the acetate. Of the former we have a solution, suppositories, lozenges, and another form of lozenge combined with ipecacuanha. Of the acetate merely the liquor or solution is officinal.

In opium, to exercise its free influence, it is necessary that it should be absorbed into the blood, but it does not greatly matter by what way it is introduced, whether by the stomach, the bowel, by a raw surface, or, as is now extensively practised, by subcutaneous injection. If in any of these ways an ordinary dose of opium, or its alkaloid, morphia, is introduced, there is first of all a stage which might, though incorrectly, be called one of excitement. The mind becomes quiet under its soothing influence, the pulse quickens, the mouth becomes somewhat dry, but the moisture of the skin increases. By and by the pulse slackens, the breathing is long and full, and the patient sleeps. When he awakes there is generally thirst, some nausea, and very often headache ; the tongue is furred and the bowels confined. Should a large dose have been given the effects are more marked, the preliminary stage is hardly noticed, sleep of a heavy kind speedily comes on, and the breathing is often stertorous, while the pulse is slow. This condition may be induced by very different quantities of the poison in different individuals. Children are unusually susceptible to its action, in somuch that there is danger in giving them the very weakest preparation of opium in the smallest quantity. On the other hand, certain individuals can hardly be affected by its use except in large quantity. Such a condition of system is especially brought about by prolonged use of the drug.

After a *poisonous dose* of opium, the stage of excitement is hardly noticeable, and narcotism comes on almost at once. There is a craving for sleep which can hardly be overcome, and sleep if permitted soon passes into complete insensibility ; the surface at first pale and covered with sweat, becomes cold and livid, the breathing, exceedingly slow and stertorous, gradually becomes more and more shallow till it ceases. The pulse from being full and firm becomes smaller and smaller, slower and slower until it ceases to be felt. The muscles of the whole body are relaxed, there is complete loss of sensibility, the patient can no longer be roused. The rattle begins in his throat, and gradually death ensues. One of the most marked peculiarities of the action of opium is its influence on the pupil, which it contracts powerfully, so that when the patient is fully under its influence the pupil may seem no larger than a pin's point. This is an important diagnostic as to the cause of insensibility in poisoning by opium, and the insensibility produced in other ways. As to the remedies to be employed in poisoning by opium, these are chiefly meant to prevent sleep, as the system, after having once fallen under its influence to the full extent, is not easily roused. Shaking, flicking the soles of the feet, etc., are commonly resorted to. But these are only to be had recourse to after the stomach has been emptied of the poison. Perhaps the best

thing here is the stomach-pump, as it admits of the stomach being washed out, but if that is not at hand a stimulant emetic—mustard is the best—should be given, after which black coffee should be freely administered from time to time till the patient gets well.

There are few diseases in which complications demanding the use of opium may not arise. In fact its uses are legion. Thus in fevers, though we cannot hope to cut short the malady, we may obviate certain of its most distressing symptoms by means of opium : want of sleep especially in typhus often gives rise to delirium of a low muttering sort, with picking of the bed-clothes and wandering. Here opium judiciously given may save the patient, who in such cases is in very great danger. Graves used to give opium combined with tartar emetic in these cases, apparently with the best results. He gave three or four drops of laudanum, with a little tartar emetic every two hours till the patient was quieted. In any malady accompanied by this form of delirium, where the strength is at the lowest ebb, the tongue brown and dry, the pulse hardly perceptible and too quick to be counted, opium given in this way along with a certain quantity of brandy, to be administered as carefully as the opium, safely may be obtained almost when past hope. But the opium must be given in small doses, frequently repeated, and the brandy in teaspoonfuls. In acute mania opium, with or without tartar emetic, is of great service. Sometimes it is best given under the skin.

But the great use of opium is to relieve pain. For this purpose it is now mostly given hypodermically, that is, under the skin. For thereby the digestion is less disturbed, and the patient is free to take food ; moreover, a smaller quantity suffices. However, this must be borne in mind, that the quantity required to procure ease rapidly increases, so that what would suffice at one time will not at another, some time thereafter ; and the same holds good of its internal administration. It is best, therefore, to alternate its use with that of other sedatives, especially chloral, so that the system has time to recover from the use of the one before it is necessary to return to it. In this way opium is of the greatest service in gall-stones, the passage of urinary calculi, cancers, painful ulcers, etc. A single injection may suffice to cure sciatica and other forms of neuralgia, if applied on the spot, but as the same result used to follow acupuncture in certain instances, we cannot be quite certain of the efficacy of the morphia. The same means may be employed to cure pleurodynia, that is, pain in the side, if the pain be deep seated. The same form of the remedy may be of use in the vomiting of pregnancy, or to assist persistent hiccup. When the pain is in the stomach itself and the vomiting arises from disease of that organ, of course it is better to give the opium by the mouth, provided it be not rejected, as it too often is. If so, either a very small piece of morphia and sugar may be given, or it may be administered subcutaneously. In certain forms of heartburn, too, it may be employed with advantage, and may be combined with tonics. On the bowels it acts much as it does on the stomach, arresting their secretion and motion. Hence constipation is one of the most certain consequences of giving opium, even in small doses. This property becomes of great value in disease, when it is desirable to restrain inordinate action of the bowels. To effect this no substance is so useful as opium, especially when it is desirable to allay irritation as well as to arrest action. In diarrhœa, therefore, both acute and chronic, opium is of great value, especially after the irritant substance which has given rise to the diarrhœa has been removed ; previous to that its use is inadvisable. Hence, too, a prescription of use in many forms of diarrhœa with griping—10 drops of laudanum in half an ounce of castor-oil. This combination insures the ejection of the irritant matter, and the immediate action of the opium to follow it. In some forms of colic allied to the diarrhœas we have already spoken of, the castor-oil and lau-danum is the best remedy. In peritonitis, where the motion of the intestines is provocative of harm, opium is the best remedy ; so, too, in injury to the intestines, especially rupture, from whatever cause. When the bowel is affected, especially in its lower portion, it is often the practice to administer opium, i.e., laudanum, by injections. When so administered the injection ought to be of the smallest possible bulk, not exceeding an ounce, and ought to be of the temperature of the body, i.e., about 100° F. This is very effectual in some forms of diarrhœa, especially in those dangerous forms which carry off children rapidly. In diarrhœa from tubercle or typhoid a similar law prevails. Opium may be exhibited in this manner with great success when it is desired to relieve pain in the neighborhood of the rectum, especially in the bladder and womb. Commonly suppositories are used in such cases instead of injections. Mixed with gall ointment it is one of the best remedies we possess for ulceration of the rectum and piles. Fissure of the anus, one of the most excruciating of maladies, too, may be relieved, if not cured, by a similar application. Given internally, or by the skin, opium or morphia are of the very greatest service to patients the subjects of delirium tremens. . Frequently it is advisable to add tartar emetic or aconite to it : but if, on the other hand, the patient has long been without food it is necessary to feed him carefully, and even to administer stimulants. In these cases ammonia is invaluable. In whooping-cough opium is often of signal service, if

swallowed slowly, as by sucking a lozenge, it relieves the irritability of parts, and when introduced into the system seems to relieve the irritability which gives rise to the whoop. As, however, whooping-cough ordinarily occurs in young people, and these bear opium badly, care must be taken in its administration. A small dose of opium, especially in the form of Dover's powder, will frequently check a cold if it is as yet in the shivering stage. It should be taken at bed-time, 5 or 10 grains for a dose, and care taken to secure a good perspiration afterward. When morphia is given hypodermically, the acetate is commonly used, as nearly as possible in a neutral state, and some prefer giving a little atropine with it. The solution should be so regulated that one or two drops suffice ; not more than five should ever be given, and the quantity ought not to exceed the fifth part of a grain.

OPODELDOC is the name commonly given to the soap liniment of the pharmacopœia. It consists of hard soap, camphor, oil of rosemary, spirit, and water. Its chief use lies in enabling us to rub a part with ease, obviating unpleasant friction, and at the same time acting as a slight stimulant to the parts. Its chief value is in sprains after they have ceased to be acute, and when rubbing is of value, tending to remove stiffness and swelling, and so rendering the joint supple again It is also a most useful basis for other liniments which it is desirable to rub into a part, when these contain no oil or soap, such as is necessary when much rubbing is intended.

OPOPONAX is not now contained in the pharmacopœia ; but was so in that of London up to 1836. Its properties are similar to other fœtid gum resins, perhaps most closely approaching galbanum.

OPTIC NERVES are two in number, and one is supplied to each eye, enabling man to have the sensation of sight. (See EYE and VISION.)

ORANGES are the fruit of several species of *Citrus* belonging to the family *Aurantiaceæ*. To the same order belong the lemon, the lime, and the shaddock. These fruits are all distinguished by containing citric acid. The orange juice contains in addition sugar. Hence their use as fruit for eating. In all cases where citric acid is indicated, oranges may be used. As a refreshing article of food in the sick-room, there is no fruit superior. The peel of the fruits of the *Aurantiaceæ* contains in little receptacles a volatile oil, which is a pleasant stimulant and flavorer. This oil is often separated and sold under the name of neroli oil, oil of lemons, etc. (See CITRIC ACID, LEMONS.)

ORTHOPNŒA means that condition of respiration which compels the individual to sit upright. It is one of great discomfort, and often is of dire significance. Like most other symptoms, it may depend on a variety of causes, some of them having apparently nothing to do with respiration. In many cases of disease of the heart, the patient, for a very long period before death, is quite unable to lie down. The only sleep that can be procured is got while the patient is propped up by pillows. In dropsy, too, though not dependent on heart mischief, the patient is often compelled to sit up continually, any other position interfering so sadly with breathing as to necessitate instant change, and altogether precluding sleep, except in that posture. In point of fact, whenever there is difficulty in obtaining breath, the patient instinctively starts up, for in the upright position he is able to call into play many powerful muscles, not ordinarily employed in respiration. Moreover, the weight or pressure of the contents of the abdomen against the lower boundary of the chest is removed, and the powerful muscle of respiration called the diaphragm, or midriff, may be called into play with more advantage.

In various maladies affecting the respiratory organs this condition is noticeable. Thus it may be seen when, from whatever cause, the air is prevented from entering the chest freely, as when in any malady affecting the air passages. Perhaps spasmodic asthma furnishes as good as example of extreme orthopnœa as does any disease, for in it the patient may be compelled to lay hold of something over his head, so as to fix his arms, besides assuming the upright position. When, too, the pleuræ are filled with fluid so as to interfere with the movements of the chest, if the condition be symmetrical, that is to say, affecting both sides of the chest, we may have orthopnœa very markedly.

It is, however, in heart disease that we commonly see the condition called orthopnœa in its extreme form, to an extent most distressing to the patient, and even to onlookers. In these unfortunates, owing to causes we cannot here explain, the circulation of the blood is sadly interfered with. The blood current, especially in the veins, is dammed back and obstructed, so that these vessels become overloaded and over-distended. As a consequence, the fluid portion of the blood passes through their coats into the tissues beyond, and accumulates there. This is dropsy. Most frequently these transudations begin in the feet, and gradually creep upward ; the ankles are affected, then the legs and thighs, and then the abdomen, too, is filled. By this time, too, the circulation of the blood through the lungs is sadly interfered with from the same cause. The lungs are congested, and the blood cannot pass freely from the right side of the heart to the left. Now to give a sensation of comfortable easy breathing, when the process is a pleasure rather than otherwise, it is quite as necessary that there should

be a due flow of purified blood from the lung as of pure air into it. Now in the condition of which we speak the flow of blood from the lung is sadly interfered with. Moreover, the accumulation of fluid in the abdomen prevents the use of the diaphragm as a muscle of respiration, and so the movements of the chest-wall must accomplish all. But as matters advance apace, the fluid from the distended vessels begins to accumulate in the pleuræ, which in its turn sadly interferes with the indrawing of air into the lungs. Thus there is the condition of the circulation already alluded to as a cause of difficult breathing, and a condition of the respiration arising from the former, and intensifying its evil effects added on to it, the consequence being orthopnœa of the worst kind, and in too many instances only to be terminated with the end of the patient's life.

It will be seen that the explanation here given is mainly a mechanical one. The remedies, too, are mainly mechanical. In most cases the orthopnœa depends on interference with the circulation as indicated by mechanical congestion. In the olden time, men used to remedy that by the lancet ; now we seldom use that instrument. Hot-air baths and hydrogogue purgatives take its place.

OSSIFICATION is a term applied to any of the parts of the body in which calcareous or other matter is deposited in the tissue, so as to produce hardness and a bony-like aspect and character. (See DEGENERATION.)

OTORRHŒA signifies a discharge from the ear ; it is often seen in children, and chiefly in those who are scrofulous. The ear should be syringed with warm water four or five times a day, and then filled with cotton-wool and sweet-oil. (See EAR.)

OTITIS is a technical name for inflammation of the ear. (See EAR.)

OVARIAN DROPSY is the name given to that disease in which a large cyst or cavity, filled with fluid, grows from the ovary and fills the abdomen ; these cysts may grow as large as an adult head, or even larger ; the walls are tough and fibrous, and contain generally fluid of a dark color. (See OVARIES.)

OVARIAN IRRITATION is sometimes produced when the ovaries are congested at the time of menstruation ; pain, nausea, and faintness are often the chief symptoms. A hot hip-bath, or the application of two or three leeches over the seat of pain, and then rest in the horizontal position will generally give relief.

OVARIES are two in number and are situated one on each side of the uterus or womb, with which they are at certain times connected by means of the Fallopian tubes. Each ovary is about the size and shape of an almond, and contains within itself numerous round cellular bodies, called ova, which are of much importance for the development of the ovary. This substance is liable to congestion as each menstrual period comes round, and it is often the seat of much pain and suffering in cases of dysmenorrhœa. The ovary is liable to inflammation, and then adhesions may be set up with surrounding structures, and sterility ; great pain over one or other side of the lower part of the abdomen, a feeling of languor and nausea, and pain in the back, are among the chief symptoms. The ovaries are liable to cystic disease, and in some cases enormous tumors are formed in the abdomen, and the case is commonly called one of ovarian dropsy. The tumor is generally of slow growth, commences on one side of the abdomen, gradually filling it up and making it tense and convex. The cyst may be tapped, so as to allow the fluid to escape, or in some cases it may be removed altogether by the operation known as ovariotomy.

OVARIOTOMY is the operation by which a surgeon removes an ovarian cyst or tumor from the abdominal cavity ; formerly it was thought a very formidable operation, but of late years it has been frequently performed with success.

OVUM is the small cellular body which exists in the ovary in great numbers, and which when they have passed into the womb and become impregnated, are developed into the future embryo.

OXALIC ACID is an organic acid found present in many plants. It gives the acidity to sorrel and rhubarb, hence these plants are used as articles of diet. Oxalic acid, however, is a poison, and is often mistaken for Epsom salts, or used by the suicide. The best remedy for poisoning by oxalic acid is carbonate of lime—common chalk. The lime forms an insoluble compound with the oxalic acid, and renders it innocuous. (See POISONING.

OX-GALL, or **BILE,** is not very often used in medicine. It is purified after being taken from the gall-bladder of the ox by adding to it spirit ; this throws down the mucus, which is afterward separated by decantation. In color it is yellowish-green, with a peculiar odor. Its taste is at first sweet, afterward intensely bitter. It is soluble in water and spirit. Bile contains many things, but the bile acids seem to be the most important sub-

stances. These are reabsorbed under ordinary circumstances, and undergo further changes. Bile prepared thus is supposed to act as a laxative, and also to aid in preserving the contents of the alimentary canal from putrefactive change. Thus it is of use where the entrance of the bile of the liver into the alimentary canal is prevented from whatever cause, and in constipation supposed to depend on insufficient bile flow. Meanwhile its value is mainly speculative. (See LIVER.)

OXYGEN is one of the elements, is a gas and a supporter of combustion. It exists in the atmosphere in the proportion of twenty-one parts to seventy-nine of nitrogen. In the atmosphere it becomes the means of all kinds of combustion, and the principal agent in putrefaction. It is the sole means of supplying the oxygen that is required for the oxidation of the tissues of animals, and the maintaining of animal heat. (See AIR, HEAT (ANIMAL), OZONE, RESPIRATION.)

OXYMEL consists of a mixture of honey and acetic acid. This preparation is but little used, even as a vehicle, though it is rather an agreeable one. The only oxymel of importance is that of squills, which is largely used, especially among children, for the purpose of. procuring the effects of squill. (See SQUILL.)

OXYURIDES is a word used for threadworms or ascarides. (See ASCARIDES.)

OZONE is a peculiar substance discovered by Schönbein in 1858. It may be prepared by passing a succession of electric sparks through atmospheric air or dry oxygen, when a peculiar odor will be perceived. Ozone is much denser than oxygen itself ; it may be destroyed by a heat of 550° Fahr. ; it is insoluble in water and in solutions of acids or alkalies ; when present in the air it acts as an irritant to the air-passages. This substance possesses considerable bleaching properties, acts as a powerful oxidizing agent, and corrodes organic matters. Its presence may be detected by moistening a slip of paper with starch and iodide of potassium ; the ozone, if present, will liberate the iodine from the iodide of potassium, and the free iodine will color the starch blue. Its influence on man is not yet understood. (See AIR.)

OZŒNA. (See NOSE, DISEASES OF.)

P.

PACKING, as it is technically termed, is of two kinds, *wet* and *dry*, but the latter is so uncomfortable that it is seldom had recourse to. Wet packing has become almost entirely an instrument in the hands of hydropathic practitioners, but most certainly it is worthy of a wider appreciation. One reason for its want of popularity is really a want of knowledge of how and when to apply it. A great

number of slighter maladies, such as incipient colds, etc., may be cured by it ; it marvellously removes fatigue, and withal may as easily be given in a private house as in the best appointed hydropathic establishment. Perhaps the simplest form of wet pack is the local one for sore throat, which our great-grandmothers were wont to employ. A stocking fresh removed from the foot, and so somewhat damp from perspiration, was applied to the throat, the damp part or sole next the skin, and the whole then wrapped round the throat and kept on all night. In the morning this was removed, the parts washed with cold water, and very probably the pain was gone. This was modified somewhat by employing a towel or piece of linen wrung out of cold water, wrapped round the throat and covered over with flannel. This, too, is very successful, and well worthy of a trial. The wet pack is the same in principle, but applied to the whole body instead of to a part of it only. As a preliminary the patient should take a smart walk or some similar exertion, not enough to tire, but sufficient to put the surface in a nice warm glow. When he returns his bed should be found prepared, by removal or folding down of all the bed-clothes, including the feather-bed, if any. On the mattress should be spread a piece of waterproof sheeting, if desired, but this is not absolutely necessary. Over this or in its place may be spread a thick blanket, and when the patient is ready, this, in its turn, is covered with a sheet loosely wrung out of cold or nearly cold water, according to the season of the year. On this the patient is stretched quite naked, and then the sheet is tucked up tight all round about him, so·that he lies swathed in the sheet like an Egyptian mummy ; over this blankets are tucked in, and the whole may be covered up by the feather-bed, if there is one, if not, a due supply of blankets must be used. The head is carefully wetted, a wet towel placed over the forehead if desired, and the patient left to himself for half an hour. Though cold at first the bodily heat soon begins to exert itself, and the whole mass becomes heated, so that the wet sheet acts like a kind of gigantic poultice applied to the whole surface of the body. Should, however, the reaction not take place of its own accord, it will be necessary to insure its appearance by the use of hot-water bottles. The result is a copious but imperceptible transpiration from the skin, which tends to open it more than anything else. At the end of half an hour the patient is to be stripped and well bathed with cold or nearly cold water, and the process is at an end. During the period he is left he most frequently sleeps or dozes, so much tranquillity does it give. Dry packing is simply sweating induced by a heap of bed-clothes.

PAIN is one of the most common symptoms in disease, but it may arise from a great many conditions, and may therefore require a different mode of treatment. No greater relief can be afforded than to adopt some means by which a patient can be made easier and free from this disagreeable symptom, and whereas in many cases it may be impossible to cure the disease or avert the fatal end, yet it is often in our power to modify the severe symptoms and give a vast amount of relief. All pain is felt in the nerves, whatever may be the cause which gives rise to the sensation. The causes of pain may be divided into two great classes : 1. Those depending upon too much blood in the part, and where there is an increased tension in the vessels. 2. Those depending upon an impoverished state of the blood and an altered condition in the nutrition of the nervous centres ; such pains are more commonly known as neuralgic pains, although, strictly speaking, all pain must be neuralgic. Under the first head may be included the pain caused by inflammation of any part, and more especially of the serous membranes, as in pleurisy, peritonitis, and pericarditis, also in the joints, as in gout and rheumatic fever ; these cases are attended with more or less fever, and the pain is caused by the over-distended vessels in the inflamed part interfering with the nerves distributed there ; of this kind also is the pain met with in an abscess, and it is well known how much more painful is an abscess under the tendon of the finger or in the gum than in some more lax tissue, and this is due to the tension of such a part, for where the skin is loose the swelling does not hurt much, but where the abscess is bound in by firm walls the pain is much greater. This accounts, too, for the relief sometimes experienced after the face has become swollen after a toothache, for the fluid has escaped then from the vessels, and the tension is diminished. Pain, when due to this cause, can be relieved in several ways, but all the methods adopted have in common the object of relieving the distended vessels ; a few leeches over the affected part will give relief by withdrawing some of the blood ; hot fomentations, made by wringing out flannels in hot water, turpentine stripes, hot linseed-meal poultices, and cotton-wool, are all most useful means of locally allaying the pain ; sometimes continuous cold, applied by placing pounded ice in a bladder, will relieve, at other times a hot bath will do good. Other measures may be adopted, as the hypodermic injection of morphia, the local application of belladonna, or aconite, and the internal administration of opium. The second class includes tic-douloureux, sciatica, hysterical pains, and what is commonly known as neuralgic pains. They are generally associated with pallor and debility. Nothing is more common than to meet with such cases as the following : A woman who has borne several children and suckled them for some time finds herself losing strength and flesh ; her appetite is bad, and she generally has been unable to get sufficient nourishment, perhaps only having meat once a week, while all the time her strength should have been well supported while nursing her baby. In time, besides feeling weaker, she is nervous and low-spirited, has pain across the forehead and over the top of the head, dimness of vision, occasionally giddiness, pain in the left breast and left side, pain in the back and either across the loins or between the shoulders. Now and then there is pain in the limbs ; she is pale, and may suffer from leucorrhœa, is unfit for much exertion, and although still feeling ill, is obliged to attend to her children and household work. For such cases relief can only be obtained by supplying them with nourishing and wholesome food, by rest in a horizontal posture, by a short walk daily on a fine day, so as not to tire themselves, and by moderating the quantity of stimulants taken daily. It is also most needful to give tonics, as iron and quinine, so as to improve the general health. Iron is not often borne well at first, and then the mineral acids with some bitter infusion may be given. Change of air and scene is very valuable, but in such cases few can afford it.

Pain of this kind is very common in pregnancy, and then one side of the face is generally affected ; there is no swelling nor redness, and leeching the gum or extracting a tooth is a perfectly useless proceeding in such cases ; quinine and some chloric ether is the best remedy. The pain in brow ague or migraine (see INTERMITTENT FEVER) is of a similar nature, and may be also caused by malarious influences. Removal from the damp locality, and the internal administration of large doses of quinine are the most likely measures to give relief. In these people, as in the previous cases, there is always pallor and anæmia, and the mischief is not in the nerves but in the nerve-centres—as the brain and spinal cord—which are not properly nourished. In many fevers and in cases of syphilis, where the blood becomes gravely altered in quality, neuralgic pains are very common ; in the former, the fever must be treated ; in the latter, iodide of potassium will do much good. Lastly, there are certain muscular pains which come on because the muscle is tired ; of such a nature are the pains felt by one after a long day's ride without being used to it, the pains caused by a troublesome cough in the intercostal muscles on each side from the violent exertion, the aching pains caused by laughing immoderately, and the pains brought on by any unusual exertion, and generally known as stiffness. The treatment must be rest for the affected part ; when the cough is distressing, means must be taken to relieve this, and a warm and wide flannel bandage should be fastened round the waist. *Pain in the side* is a frequent symptom and is treated of in the article on PLEURODYNIA.

PAINTS produce diseases arising from working with lead which are described in the article on LEAD-POISONING. It may suffice to state here the danger arising from children using toys painted green, or from the habit children have of sucking paint-brushes and

the cakes of ordinary color-boxes. Most green colors contain arsenic, and the poison may cause a sore throat, running at the eyes, purging, sickness, and pains in the abdomen. Similar results are brought about by having a green paper in a room, and serious results may ensue. (See POISONING.)

PAINTER'S COLIC is commonly met with in those who work with lead and its preparations, and more especially with those who deal in white lead. (See LEAD POISONING.)

PALATE (THE) may be considered under the separate portions of the hard and the soft palate. *The hard palate* is that portion of the roof of the mouth immediately posterior to the gums and teeth ; it is supported by the bony arch of the palate and upper jawbones, and is covered with a tough, dense mucous membrane, inseparably united to the periosteum of the above-mentioned ·bones. There is a median· ridge, which marks the position of the congenital division of the parts, and there are numerous transverse ridges on either side of it. *The soft palate* is a soft movable substance, attached above and in front to the hard palate, while behind and below it terminates in a thin, free crescentic edge, from the centre of which the uvula hangs, thus dividing the edge into two semi-lunes. This velum is situated somewhat obliquely, its fixed edge being superior and anterior to the bone, the surface looking downward and forward toward the mouth and tongue, the opposite surface looking upward and backward. The mucous membrane contains a good many glands, and is covered with ciliated epithelium on its upper surface, and squamous in its inferior. In the act of deglutition the velum and uvula are raised so as to touch the back part of the pharynx, and thus prevent the food from ascending into the upper and nasal part of the cavity, from which it might regurgitate into the nares. The soft palate is each side attached to the tongue and pharynx by muscles named palato-glossus and palato-pharyngeus, forming the anterior and posterior *pillars of the fauces*, and between these pillars lie the tonsils, vascular glands which secrete a viscid mucus, expressed at the moment of deglutition, and which lubricates the food on its downward passage to the œsophagus.

Affections of the Palate.—By far the most frequent affection of the palate, which is met with either in the hard or soft, or both combined, is *cleft palate*. This is a congenital fissure, arising from an arrest of development of the natural vault of the palate, the nature of which is alluded to in the article on Hare-lip. (See HARE-LIP.) If the hard palate be extensively deficient congenitally, it is very difficult to remedy it surgically, at least by operative proceedings. The dentist, by applying a metal or vulcanite plate (termed an obturator), to take the place of a natural pal ate, may do great good ; such a deformity, however, usually is coexistent with fissure of the lip, and in such cases the lip affection should be first attended to. When the chasm is very wide, probably an operation will not do much, but first-rate advice should be taken at the earliest stage of the child's existence. A mere fissure or crack often closes spontaneously, uniting during adolescence. It is in cases of fissure of the soft palate that the surgeon has it in his power to render such valuable service. It has been already mentioned that the normal soft palate is arched and vaulted, and subject to varying degrees of tension during deglutition, etc. The muscles which raise the soft palate are the levatores palati ; those tending to stretch it and make it tense, the tensores or circumflexi palati ; while others, the palato-glossi and palato-pharyngei, likewise put great tension upon it downward and laterally. Now it is evident that the actions of these muscles must tend to keep apart the pre-existing fissure, and this fact having been determined, the surgeon has a plan of action before him. The operation for the relief of this affection is termed staphylorhaphy generally, and it consists of three stages : the first is the setting free of muscular tension ; the second the preparation of the edges of the existing fissure thus set free ; and, thirdly, the putting in of such sutures or stitches as are necessary for the purpose of securing contact between such prepared edges. The first stage requires a long time generally for its preparation, such, for instance, as the accustoming of the patient to keep the mouth open for a long time at a stretch, the rendering the palate less sensible to the tickling, niggling proceeding to be hereafter practised, the determination not to swallow saliva, if the patient be of a sensible age (young children should have chloroform). Then the muscles, levator and circumflexus palati, and sometimes the palato-glossus and palato-pharyngeus, are divided by a peculiarly formed knife or scissors ; afterward the edges to be approximated are pared of their mucous membrane, so as to admit of their union by adhesion ; and, lastly, the sutures to fix them are introduced, and this is by far the most difficult part of the procedure, requiring as much steadiness on the part of the patient as skill on the part of the surgeon. The patient should be thoroughly acquainted with the nature of the operation, as far as he can be, and must be convinced of its utility and chance of success. The operation, however, should be proposed, and, if possible, undertaken at infancy, with mechanical assistance and chloroform, in order to obviate any defective articulation, which will have been acquired if the operation be put off till puberty or after.

Dropping of the Palate. (See GARGLES and UVULA.)

Ulceration and Exfoliation.—The mucous membrane is liable to ulceration, usually as a result of syphilis. This form of ulceration, however, invariably affects the bony palate as well, causing perforation, exfoliation of bone, and adhesion of the soft palate by cicatrix, occluding the buccal and nasal portions of the pharynx. Such a state of things is associated, of course, with tertiary syphilis and constitutionally requires iodide of potass, tonics, etc., and locally nitrate of silver, nitric acid, and chlorinated lotions. If the hard palate exfoliates (see EXFOLIATION), the separation must be patiently waited for, and not hurried by rough attempts at pulling the piece of dead bone away. As the whole thickness of the palate perishes an aperture will exist between the nasal and buccal cavities, and if this cannot be closed by spontaneous cicatrization an obturator must be adapted ; a mere fissure or sinus can be closed generally by the repeated application of a heated wire or cautery.

Tumors of the soft palate.—These may be : (1) fibro-cellular ; (2) cysts ; (3) warts. (1) The fibro-cellular are usually pendulous in character, painless, usually attached to the free border or upper surface of the soft palate. They are painless, and being inconvenient they must be removed by scissors and forceps. (2) The cysts are generally obstructed muciparous ducts ; they are to be treated by free incision, and the subsequent application of nitrate of silver or nitric acid. Sebaceous cysts occasionally occur, appearing of a yellowish-white color through the mucous membrane ; free incision and a drop of nitric acid on a probe destroys them. Abscess occurs sometimes, and should be immediately opened.

PALMA CHRISTI is the name by which the castor-oil plant is known in this country. It is sometimes cultivated in gardens on account of its beautiful leaves. Its botanical name is *Ricinus communis*. It is a native of India, but is widely distributed over the warmer regions of the globe, and throughout the Mediterranean region. In our climate the stems of the Palma Christi do not attain a height of more than from three to five feet ; in India they grow from eight to ten feet, while in Spain, Crete, and Sicily the plant is said to become a small tree. The stem is pointed, of a purplish red color, and covered with a glaucous bloom like that of a plum. The leaves are large stalked palmate ; deeply divided into seven lance-shaped segments, and at the junction of the blade with the stalk of the leaf is a small saucer-like gland. The flowers are in spikes. There are several varieties of this plant, differing chiefly in the size of their seeds. It is stated that the best oil for medicinal purposes is derived from the small seeds ; that procured from the large seeds is coarser, and in India is only used for lamps and veterinary purposes. The oil is extracted by boiling the seeds, and by press-ure under a hydraulic press ; the latter pro-cess without boiling the seeds yields the most esteemed oil. After expression the oil is pu-rified by being allowed to stand, by decanta-tion, and by filtration. In India the oil, after having been obtained by pressure, is mixed with a certain proportion of water and boiled till the water has evaporated. In France the oil is obtained by macerating the bruised seeds in alcohol, but the process is expensive and the product inferior. (See CASTOR OIL.)

PALPITATION is the name given to the beating of the heart, when that ceases to be insensible and becomes obvious to the feeling of the individual. The two things which seem to have most influence in producing this al-teration are increased violence of the heart's action, and perhaps even mere irregularity of action. Under ordinary circumstances the motion of the heart is so even and regular that one can only detect its beating by placing the hand over the spot where its apex strikes against the ribs, but in certain cases of heart disease the beating may be so violent as to shake the bed in which the patient lies. Pal-pitation, though very often a sign of heart disease, is by no means invariably so ; per-haps, out of all the cases of palpitation one sees, the majority are in individuals not the subjects of heart disease, for as already pointed out, anything which interferes with the regularity of the heart's action produces the painful or unpleasant feeling of palpitation, and that may readily be done in many ways without the substance of the heart being affected.

Disorder of the motion of the heart, which may be taken as synonymous with palpita-tion, is commonly due to some alteration in the functions of the heart necessitating more violent effort on its part, or to some other cause interfering with its movements, such as disordered nerve influence. Like other muscular structures, the heart is directly under the control of the nervous system, but its nerve-supply is more than ordinarily compli-cated. The nerve-supply is drawn from the brain, as is the case with other organs, but it reaches the heart in two ways : one by a nerve called the vagus or pneumogastric, which passes downward from the brain through the neck to the chest to end finally in the ab-domen ; the other by way of the spinal cord, which at different places gives off branches which ultimately reach the heart and control its motions. The former of these, i.e., the vagus, is mainly engaged in controlling or regulating the heart's action ; the latter nerves are rather devoted to stimulating its substance to act. Any increase or diminution of the action of the vagus is likely to give rise to alteration in the motion of the heart itself, to produce quicker or slower motion or irregu-lar motion, in short, the phenomena we call palpitation. Now this nerve has a very wide series of connections, supplying many organs besides the heart itself, and any affection im-plicating these is likely to derange the nerve influence not only as affecting the diseased organ, but also as influencing the heart. Hence it is that palpitation is very fre-quently brought about by affections of the stomach, the vagus supplying both. Palpi-tation or some other form of irregularity in the heart's action, say irregularity or even intermittence of the pulse, may be brought about by nerve action in a totally different way. Thus, as is well known, anxiety, fear, and various other mental emotions produce beating of the heart, i.e., palpitation, where the stimulus arises in the brain, and is con-ducted to the organ where it is manifested, i.e., the heart, by means of nervous influ-ence. In a goodly number of cases, however, the palpitation is due to change in the heart itself. This change very likely is in the first instance valvular, that is to say, connected with the floodgates of the heart. The altera-tion in the valve interferes with the heart's action, chiefly in that the heart never is able to empty itself properly, or if it does the cavities are promptly filled again, so that the chambers become habitually over-distended. At the same time in many cases the substance of the heart increases in thickness, and its beat in force, so that the ordinary work of the heart is, so to speak, accomplished with greater violence than usual, this violence being manifested as palpitation.

Treatment.—It is quite plain that as the cause of palpitation varies, so must its treat-ment. If it depends on disease of the sub-stance of the heart, then treatment must be directed to remedy that, and digitalis is most commonly the best remedy. If from other mischief, as indigestion or the like, that must be seen to, but in the majority of cases the palpitation yields to a stimulant, as aromatic spirit of ammonia.

PALSY is the common name for paralysis. (See PARALYSIS.)

PANCREAS is a gland lying in the ab-dominal cavity in front of the spine and be-hind and below the stomach. It consists of a main tube from which branch off multitudes of small tubes, each of which has a blind ex-tremity or dilatation ; these tubes are lined with epithelium, and around the tubes are to be found vessels and nerves which supply the gland with nourishment and regulate the amount of its secretion. The secretion from the pancreas is called the *pancreatic juice ;* in conjunction generally with the bile-duct it opens into the duodenum or first portion of the small intestines. Except at the time of digestion the functions of this gland are not called into action, but when the food has passed from the stomach and become *chyme,* the secretion from the pancreas mixes with it and converts it into what is called *chyle*. The pancreatic juice seems to have the property of subdividing the fatty particles of the food into very minute particles, so as to make an emulsion and cause an easier absorption of the fatty matter by the vessels which lie in the walls of the intestines. (See DIGESTION.)

PANDEMIC is a term applied when a disease has spread all over a large continent at the same time : thus an attack of cholera or influenza may affect all Europe in any given year, and then the disease is said to be pandemic.

PAPULES, or pimples, occur on the skin in some diseases of that tissue ; small-pox generally begins with a papule, and then goes on to become pustular. Lichen is also a papular disease, and so is the small pimple caused by a flea-bite ; strophulus or red-gum and prurigo are also papular diseases.

PARACENTESIS means an operation for removing fluid effusion from the interior of the body. The common and expressive word for this operation is *tapping*. The region in which it is most frequently performed is the abdomen. Dropsical fluid, or the fluid effused in connection with ovarian disease, often accumulates to such an extent as to interfere seriously with the respiratory movements, and to threaten death by congestion of the lungs and suffocation. Paracentesis by withdrawing the fluid gives great relief, and in some cases strength, but in comparatively few assists a radical cure of the dropsy. The spot at which the surgeon generally taps is in the middle line of the anterior wall of the abdomen, and about three inches below the navel. The instrument used is a thick, sharp-pointed trocar, which slips through a tube called a cannula. After the abdominal wall has been punctured the trocar is withdrawn, and the dropsical fluid is discharged in a full stream through the cannula. The chest is frequently tapped for the relief of the lung-mischief caused by accumulation of dropsical or inflammatory serous effusions and of pus. The operation is usually performed with a small trocar and cannula, and the surgeon selects either the space between the fifth and sixth ribs at the side of the chest, or the space between the eighth and ninth ribs in a line with the lower angle of the blade-bone. In cases where pus has made its way outward from the chest, an incision is usually made at the place where it *points*. Tapping in some few cases been performed for the relief of dropsy of the *pericardium*, which is the loose fibrous bag inclosing the heart and the roots of the large blood-vessels. This operation is an extremely dangerous one, and is not resorted to save in the presence of critical cardiac symptoms. The head is occasionally tapped for the relief or radical cure of hydrocephalus.

The surgeon uses a very fine trocar, and punctures the head either at the anterior fontanelle or at some other open place away from the middle line of the body. But a small quantity of fluid is withdrawn at each puncture, and the child's skull is then compressed by an elastic bandage. This operation ought only to be performed in almost hopeless cases of the disease, as it is by no means a safe proceeding, and is often followed by convulsions and other serious symptoms of nervous irritation.

PARADISE, GRAINS OF.—A name given to the larger cardamom seeds—a beautiful aromatic carminative ; but the lesser seeds and of a smaller variety are supposed to contain more aromatic qualities than the grains of paradise, hence they are generally preferred.

PARALYSIS, with which the word Palsy is often used synonymously, signifies a loss of motion in any part of the body ; but as the nerves supplying most parts of the body are of a mixed character—that is, motor and sensory—the idea generally conveyed implies also a loss of sensation. Paralysis may, however, be *motor* or *sensory*, or both. Moreover, it may be *complete*, when there is a total loss of power and sensation, or *partial*, when these are partly, not wholly, lost. Sometimes the word partial is used to imply that only certain parts of the body are affected, but for this purpose the term *local* is perhaps preferable. General paralysis implies that the whole body is affected, but the term general paralysis of the insane expresses one particular form of malady, which is accompanied by insane delusions. Occasionally the term *Acinesia* is used to signify paralysis of motion, *Anæsthesia* being employed to indicate loss of sensation ; but most frequently the idea of paralysis is limited to loss of motion, anæsthesia being the corresponding term made use of with regard to loss of sensation.

The two most common forms of paralysis are *Hemiplegia* and *Paraplegia*. Hemiplegia is that form of paralysis which affects one lateral half of the body without the other side being affected. Hence hemiplegia is right or left. Paraplegia, on the other hand, means paralysis of the lower half of the body, but there is no right or left paraplegia—it must affect both sides, if not quite equally, at all events to some extent.

There are certain other peculiar titles given to varieties of paralysis, such as *Amaurosis*, which used to be bestowed on any form of blindness supposed to depend on disease and paralysis of the optic nerve. Loss of hearing was called *Cophosis*, and loss of smell *Anosmia*. Moreover, certain forms of paralysis have the distinguishing character of proceeding from bad to worse. These forms are described as *progressive*, but this is a title of little value. There are besides these many other forms of paralysis, the chief of which we shall briefly record.

General paralysis, as seen in ordinary practice, means practically double hemiplegia. Both sides are affected, especially the extremities ; but, of course, respiration and circulation go on, otherwise death would ensue. In general paralysis the patient is motionless, and very often unconscious ; but the heart and the lungs having a nerve supply not affected by what may render the limbs motionless, go on, the diaphragm becoming the sole organ of respiration. This cannot continue long—either the patient recovers or dies. If he recovers gradually, it is seen that one side has been affected more than the other, so that the one usually gets well before the other, and the case resolves itself into one of hemiplegia. For general paralysis of the insane, see INSANITY.

Hemiplegia is the most common form of paralysis. It is fully treated in the article on HEMIPLEGIA. For full particulars relating to it, see HEMIPLEGIA.

Paraplegia, or paralysis of the lower half of the body, is generally due to disease or injury to the spinal cord. (See PARAPLEGIA.)

From what has been hinted rather than said above, paralysis may depend on disease of the nervous substance itself, or pressure on it interfering with the due fulfilment of its functions. But in a great number of cases, perhaps the majority, the paralysis depends rather on the latter than the former cause. Say a man is advanced in years, with weak arteries, and from some cause or other too much pressure is applied to them. They give way, blood is poured out, a clot is formed, and stops the bleeding. As a consequence of this accident, he has what is sometimes called a " paralytic stroke ;" but this loss of power of sensation and consciousness is due merely to the pressure of the clot, not to any disease of the nerve substance. Subsequently the clot may soften, and surrounding portions of the brain substance soften with it ; but the original cause of the paralysis was pressure only. There are, however, other causes of paralysis—the nerve substance itself may decay or soften, as it is called ; and if it does so, there is little hope of its recovery. Or yet again the nervous tissue may gradually waste away. Both of these forms of diseases are such as give rise to progressive symptoms ; but this is the grand rule in studying nervous maladies—that the kind of lesion, except as giving the symptoms a progressive or retrogressive character, is of but little importance in producing symptoms. That depends almost entirely on the site of the mischief. Thus injuries to or disease of certain parts of the brain give rise to symptoms of one kind, and those of others to symptoms of a totally different kind. We have described certain forms of paralysis,

which most commonly originate from injury to or pressure on nerve substance ; we may now briefly allude to certain of a different kind, such as locomotor ataxy and wasting palsy.

Locomotor ataxy, or, as the malady used to be called, *tabes dorsalis*, is a form of disease apparently depending on wasting of the posterior portions of the spinal cord or of the nerve-roots arising thence. (See PROGRESSIVE LOCOMOTOR ATAXY.)

Wasting Palsy, also known as *Progressive Muscular Atrophy*, is a malady in some respects similar to the former, in some totally different. It is fully treated of in the article on PROGRESSIVE MUSCULAR ATROPHY.

In *Hysterical Paralysis* there is of course no disease to be detected, and on probing the story told by the patient it will usually be found that there are discrepancies in it altogether irreconcilable with the malady being of an ordinary kind. On inquiry, too, it will be found that the patient has at some former time suffered from hysteria. Sometimes the supposed paralysis has been brought on by some definite cause, as fright or over-excitement—causes quite inadequate to the production of paralysis, but quite sufficient to evoke some characteristic symptoms from an hysterical woman. The forms of paralysis assumed are of all kinds, but perhaps paraplegia is the favorite. There is no difficulty about it ; the patient has only to refuse to move the legs, and there it is. Hemiplegia is different : there are little points about the eyes and mouth which are not easily mastered, the tongue too is a difficulty which these patients cannot get over ; nevertheless we have seen instances of patients the subjects of hysterical paralysis who had managed to impose not only on friends—though that is not saying much—but also on medical attendants. These are the class of patients who make the fortunes of quackish impostors if they fall into their hands, and so wise men will beware of allowing them to do so. Frequently the relations are fully confirmed in their belief as to the reality of the malady, and any brusquerie on the part of an attendant is likely to secure his dismissal—not, however, if they are wise. Far better is it, having made out the malady to be what it is, quietly to take the patients in hand and subject them to the remedies most likely to benefit their hysterical condition. These of course are good nourishment, steel, cod-liver oil, galvanism especially, quinine and nux vomica. The bowels have to be seen to carefully, cold or tepid baths given regularly, applying friction to the limbs or parts affected. But moral control is the great thing : let the patients fairly know that you are aware of any attempted imposture, let them feel you know it, and afford them every opportunity of giving it up without creating anything like scandal, and success is tolerably certain. Nevertheless there are some of these patients who do not get better, who take to their bed and keep there, making a little court round about them : any encouragement to this kind of thing is most pernicious, and ought to be instantly put a stop to. Make such patients invalids by all means, but furnish them with no enticements to continue such. These patients are rarely if ever wholly well—that is not to be forgotten ; they are really ill, and must be treated kindly not harshly. On inquiry it will generally be found that the womb is wrong or the menstrual function is disordered. Usually, too, there are pains in the back and loins. There is indigestion, and in all probability constipation. All these must be remedied before the patient is well, only it is mainly—but not entirely—to these and not to the paralyzed parts that attention is to be directed.

Rheumatic Paralysis is a form of the disease not very well understood. It affects the muscles of the extremities for the most part, especially those of the lower extremities or the muscles which raise the arm. It is by no means clear, however, that this paralysis is due to nerve change, many facts pointing rather to the conclusion that the change lies in the muscle itself or the nerve sheath. There is usually pain along with the inability to move the parts, and this pain is increased on pressure. It may come on suddenly after exposure to cold, or it may creep on more gradually. These cases are best treated by hot mineral waters, but if dealt with at home the hot douche and friction with stimulating liniments do most good. Iodide of potassium in 5-grain doses and cod-liver oil should be given internally.

A third form of paralysis of uncertain nature is that which is called *Diphtheritic Paralysis*, from its following the disease diphtheria. This paralysis, which mainly affects the throat or the muscles of swallowing, may come on during the disease, perhaps just after the throat is cleared and convalescence sets in, or it may appear later when the patient is getting about. If, however, it comes on so late as this, it merely affects the extremities, so that for a time the patient can hardly walk. The main point is, that if the symptoms can for the time be overcome, the patient is almost sure ultimately to get well, though for a time the health may be delicate. When, however, paralysis comes on in the last stage of the malady, it is a somewhat serious matter, because it is just then that nourishment is most valuable, and the paralysis sadly interferes with the power of taking it. The soft palate is paralyzed, and so are the muscles of the upper part of the gullet, but the tongue and cheeks are not. Hence, as the mouthful of food and still more of liquid is passed backward, the soft palate does not close the posterior orifice of the nostrils, while the muscles of the gullet refuse to pass it on ; it is therefore compelled to regurgitate through the nostrils. This of course is excessively unpleasant and not a little alarming. The only remedy is to pass the food so far downward as to be beyond the influence of the paralyzed muscles. This may be done by means of a tube, but it is not pleasant to have to pass a tube down a passage of raw flesh, as is the condition of the throat just after the diphtheritic sloughs have peeled off. Accordingly should the paralysis come on so early, it may be desirable for a time to give nutrient injections. Care must be taken in feeding those the subjects of this form of paralysis, as the sensibility as well as the motor powers of the parts may be lost, so that there is risk of cramming food into the gullet so as to press on the windpipe fatally, rather than into the stomach. As for remedies, the chief are time and good nourishment—food and port wine—with strychnine and the use of galvanism.

Infantile Paralysis, also called the essential paralysis of children, no immediate cause of it being ascertainable as far as the brain and spinal cord is concerned, is a disease peculiar to childhood, and though not fatal to life, is often of a most inveterate description. This paralysis often seizes upon one limb, less frequently one arm, or a single group of muscles. Sometimes it gives rise to hemiplegia, sometimes, more frequently, to paraplegia. This form of paralysis is not infrequently the source of very great deformity, for it begins early in life, and the parts not only cease to grow in due proportion, but also waste. If this condition becomes permanent, the limb is utterly dwarfed compared with the other ; or if a group of muscles are affected, their antagonists are so much more powerful as to drag the limb over to one side. This form of paralysis may come on suddenly ; the child will be noticed to drag the limb ; if it is lifted, it will fall as if dead, and though sensation remains, yet it seems useless. Very often this condition comes on after one of the eruptive fevers, when the dregs of the fever are described as lodging in that l'mb, or, yet again, it may appear during teething, when the nervous system generally is in an irritable state. Such paralyses have been mistaken for hip-joint disease, but the absence of pain is a sufficient guide to discrimination. Sometimes the unfortunate child is supposed to be playing at make-believe, and punished. Fortunately, this is not very often the case, at all events with parents interested in the welfare of their offspring. When paralysis occurs at the period of the first dentition, and the child recovers, there is a risk that it may occur again with the second. If so, the second attack is more likely to be permanent. Infantile paralysis passes away in the majority of cases, but in a good many it does not, and as there is no very good test for those cases which are likely to get well ; anything like a forecast must be received, not exactly with doubt, but with a knowledge that it may prove wrong.

As to *treatment*, one remedy here is of undoubted efficacy, and has done more good than, perhaps, all the others put together ; that is electricity, in the form of continuous currents. If this be not available, the health must be attended to, the parts daily bathed, and kept as near their normal state by friction as possible. The limb should be wrapped in flannel, to keep it warm. The food must be good ; tepid salt bathing is most useful : finally, strychnine or nux vomica in small doses is likely to give rise to good results. Cod-liver oil and steel must not be neglected ; but with all, recovery may fail to take place, and permanent deformity result.

Facial Paralysis in some respects resembles infantile paralysis, inasmuch as what is called facial palsy, in contradistinction to that paralysis of the face which depends on disease or injury to the brain, seems to depend rather on local than central change. The nerve affected is the motor nerve of the whole face, and its power over the muscles being destroyed, that side of the face is a blank, while the muscles on the opposite side being unopposed, that side is more or less contorted by dragging. The angle of the mouth is accordingly dragged over to the sound side, and on the side which is paralyzed the patient is unable to purse up the eyelids as in grinning. In the same action, too, the mouth is drawn up on one side, and not on the other, giving the individual a very peculiar appearance. The mouth droops at the paralyzed side, and the patient has difficulty in pronouncing lateral consonants. Double-facial palsy is rare, and most commonly the palsy on one side is due to exposure to cold air, especially draughts. Sometimes, like neuralgia, it may be due to stumps or bad teeth, so the mouth should be carefully examined. Inflammation of the cavity of the ear, or of the bones behind the ear, may lead to a less tractable form of the malady. This simple variety of facial palsy does not last long, and a little attention to the diet and bowels, a few doses of strychnine, with the application of the galvanic current, will most probably soon restore the functions of the part.

Labio-glosso-laryngeal Paralysis is a very peculiar form of paralysis, well illustrating certain facts already alluded to, illustrative of the phenomena of paralysis generally. As its name implies, this form of paralysis affects specially the lips, tongue, larynx, and gullet, but often it may affect many other organs in the chest, the heart especially. In it the voice is lost, the power of swallowing is lost, the lips can no longer retain the saliva, which

constantly dribbles from the mouth. In other respects the patient may be said to be well. Ultimately, however, in most cases the malady carries him off, sometimes from one cause, sometimes from another. It is hard to say what is the most prominent symptom in the disease. To the patient himself it is loss of the power to swallow, for not even the saliva, which is abundantly secreted, can be got over ; it dribbles from his mouth ; any attempt to give him food occasions great discomfort, and even risk of choking. If it comes on suddenly, then all these things appear at once and there may be risk of starvation ; in that case the patient must be fed by the stomach-pump passed downward far enough to reach the œsophagus, but not the stomach. It is useless to speak here of the remedies for such a malady ; this must depend on the causes.

Scrivener's Palsy is an exceedingly curious form of nerve affection, which, fortunately, is not very common. It is also called *writer's cramp*, apparently from the fact that it most commonly attacks those who have long used the pen ; but it is by no means confined to these, and may attack any handicraft worker almost. The mischief seems to lie in a want of co-ordinating power in the muscles, which have long been accustomed to fulfil one definite function. Each one seems to act independently, and so it becomes quite impossible to call them into simultaneous or concerted play. At first there is merely unsteadiness or stiffness after a long day's work, which speedily passes away, but by and by this makes mischief, for the writer scrawls at some point perhaps where he wishes to be most particular, the pen darting away out of his hand. As the disorder advances the patient gets worse ; as soon as the pen is touched, off starts the arm, so that it is quite impossible for him to write even legibly. At the same time, curiously enough, he may be perfectly able to use his hand for other purposes ; but the moment he takes a pen in hand it becomes altogether unmanageable. Now as to the remedy. There is only one— that is, giving over the kind of work which has brought on the malady. The patient may do anything else he likes, but it is useless to attempt to carry that one on, and though it appears late in life, the patient must seek another vocation. The sooner this is fairly faced the better. If it comes on in one who can afford complete relaxation, who can go off and travel for a time, he may come back perfectly well and able to take pen in hand again for a moderate time only. Not so one who can only take partial rest ; that is simply useless. Many poor clerks, compelled to give over work, put themselves under treatment, and just when they begin to get well they get sold out of all they possess. Better for them had they at once taken to

PARAPLEGIA PARAPLEGIA 361

something which would have permitted them to use the other hand until the affected one got well. The great remedy is the continuous galvanic current and rest. Cod-liver oil, good food, and strychnine, with change of air, should be had if possible.

Paralysis agitans, in common language called *Shaking Palsy*, is a malady in certain respects resembling that just discussed. The disease consists in a want of power in coordinating the muscles, and also, it may be said, of keeping them at rest. The shaking commonly begins by affecting the hands and arms, but later it may affect any part of the trunk or limbs. Very often the head is early affected; later, even the jaws may be affected. This agitation is increased by any mental effort, especially to call into play the muscles affected. The disease is progressive, and by and by the whole body becomes affected, so that the patient can hardly walk, being always induced to run. All this time the senses are unimpaired, and the patient is acutely sensible of his misfortune, which often sadly interferes with his occupation. His bodily powers by and by become impaired, for he is often unable to sleep at night, and even unable to take his food in comfort from the unceasing agitation. These cases commonly occur in men advanced in life, and in them little benefit is to be hoped for; but it may occur earlier, and then we may hope to alleviate if not to cure the condition by the use of strychnine, iron and galvanism, especially of the continuous current. Even in the most favorable cases the prognosis is bad; but with care a long life may be possible, and as the intellect is quite clear, much good work may be done in it.

There are still certain forms of paralysis to be attended to. These are due to mineral poisons slowly imbibed until the system become impregnated. The mineral poisons thus giving rise to nervous symptoms are *lead* and *mercury*. Lead gives rise to paralysis through wasting of the muscles, affecting first of all the muscles which raise the forearm and arm, afterward those of other portions of the body. The malady is always symmetrical. Mercury gives rise to tremors rather than paralysis. Its effects are best seen in miners, or in those who gild looking-glasses by the old process. (See LEAD and MERCURY.)

PARAPLEGIA denotes paralysis, or loss of power, over the lower extremities and lower half of the body, and it is always dependent upon some change in the nervous system, and generally upon some disease in the spinal cord. The spinal cord may be looked upon as a long prolongation of the brain; like the latter, it is made up of nerve fibres and nerve cells, and it also sends forth a vast number of nerves of motion and sensation to the trunk and extremities. When, however, these nerves or their nerve centres are destroyed, the power of motion or sensation, or of both, is lost, and paralysis ensues.

Causes.—Inflammation of the spinal cord or its membranes, cancer of the cord, or any other tumor pressing upon it, or growing into its substance; fracture or dislocation of the vertebræ or bones forming the spinal column; a stab or gun-shot wound of the spine; hemorrhage into or softening of the spinal cord, are among the chief causes of paralysis. It may come on in cases of hysteria without there being any true paralysis at all; and, finally, a person, for various reasons, may simulate paraplegia.

Symptoms.—In most cases there is tingling and numbness of the legs and feet, occasional twitchings, followed by loss of the power of moving them; sensation is generally interfered with, but not absolutely gone. If the cause be due to an accident, the paralysis may appear at once; if to cancer or any tumor, the symptoms may come on gradually; if to inflammation, as after exposure to cold and wet or from syphilis, the paralysis may come in a very few days, and often terminate fatally from its extension upward and involving most important parts. In most cases the bladder is also paralyzed, so that there is retention of urine; and as the patient has no control over that organ, it becomes full and distended; the urine remains there and decomposes, becoming thick and purulent, and having a strong ammoniacal odor. It often happens that the patient's urine is constantly dribbling away, but this is due to the bladder being too full, and the overflow, as one may term it, comes away. The patient generally loses power over his bowels, and the stools may pass away unconsciously. If the affection spread upward, the abdominal and intercostal muscles become involved, and there is great distress in breathing; presently the arms are paralyzed, and the patient dies of suffocation, as he cannot expand his chest. This happens chiefly in the inflammatory cases, while, if the paralysis be due to other causes, the parts affected will be below the seat of injury, and the sufferer may go on for many months and even years; but then, in most cases, he is an invalid, and can hardly help himself about at all. Those cases which depend upon a syphilitic state of the constitution may generally be much benefited, if not cured, by iodide of potassium. In many cases a certain amount of power is regained; but, as a rule, even in cases of so-called recovery, there is an impaired gait, and the use of the legs is never fully regained. The danger depends much upon the cause. Any fracture or dislocation of the spine is always serious, but even then life may be prolonged for many weeks; as a rule, the higher the injury the greater the danger. A tumor of the cord will gradually make its progress onward and finally kill. Inflammation of the cord is generally fatal within a week or a fortnight ; otherwise a slow recovery may be looked for. When the cause depends on syphilis or rheumatic fever, great improvement may sometimes proceed from treatment.

Treatment.—In all cases of paraplegia the patient should lie on a water-bed, if possible, so as to prevent the formation of bed-sores, which are very liable to form in this disease. Great cleanliness must be observed, and any excreta removed when passed. A draw-sheet must be placed beneath the patient, and removed when required. The urine must be drawn off by a catheter at least twice a day, if the patient cannot pass it, or if it dribbles away ; often, too, it is a good thing to wash the bladder out night and morning with warm water. The feet should be kept warm in hot flannels, but the heat must not be too great, for the feet are very liable to blister in this affection. The diet must be light and nourishing, and modified to suit the patient's palate in long-standing cases. No bleeding must be used ; no mercury is to be given, except in cases of a syphilitic origin, and not always then ; no blisters need be applied, as they do no good. For acute cases, an ice-bag may be laid along the spine, and this gives relief sometimes. In chronic cases, when the paralysis is made out clearly to be incurable, the only thing one can do is to make the rest of life as easy as possible for the patient. In hysterical cases treatment is of much avail. It is too common for people to look upon a case of hysteria as synonymous with a case of shamming ; but this is a totally wrong view to take of the subject. It is very common among both sexes, and especially among young women, to find cases in which the emotional faculties seem developed out of proportion to the intellectual ones ; such people are what are ordinarily called of a nervous and excitable temperament ; they often indulge in emotional excitement, and this generally assumes a religious aspect, varying in its development according to the people associated with the patient in ordinary life. Intense mental worry, great grief, loss of a relation, and numerous other causes tend to produce an excitement of the emotional faculties, while at the same time they are not duly balanced by a well-taught intellect ; this is what is meant by " giving way to the feelings." In some cases this goes on to such a degree that the will is not exerted by the patient, and cannot be exerted unless some strong stimulant, as electricity, etc., is given to the nervous system. At one time the voice is lost or an arm is palsied ; in another case the leg is paralyzed, and it is put down often as hip-joint disease. And these cases do not occur because the patient won't use the limb ; it is because she cannot unless you apply a shock. Such cases are nearly always worse in young people, and their state, which, when once recognized, is easily cured, is often pronounced incurable, and the poor creature falls into the hands of quacks and wise women ; while, if her tone of thought has taken a religious turn, she is looked upon by the sect as one specially afflicted by Divine Providence. And it often happens that hysterical paraplegia may be cured suddenly ; and then, to the surprise of surrounding friends, she, who had been confined to bed for months, is suddenly able to walk. Of this nature are the reputed cures and miracles which now and then are heard of ; and undoubtedly recovery does occur, because a mental shock is really given in those cases. Intense faith, a sudden fire, and a shock of electricity, all act in a similar way, and most cases of recovery are brought about by one or other of these means. Telling a patient she will recover on a certain day has sometimes a similar effect, and the more especially if it is said by some one who harmonizes with her train of thought. The best treatment for such cases is not to oppose their views, nor, on the other hand, sympathize with them too much. Daily reading some sensible book, removing all trashy novels, trying to engage the mind on some amusing topic, avoiding all excitement, and some light occupation, as sewing, knitting, or wool-work, will be most likely to do good. Cold bathing, electricity for a short time every morning, and firm but kind discipline will promote a cure.

PARASITES are animals or vegetables which live upon other organisms. The mistletoe which grows upon the oak tree is a familiar example. Those found in man are of two kinds—1, animal ; 2, vegetable. The animal parasites may affect the skin, hair, intestinal canal, or almost any internal organ. Those which attack the skin are the *Acarus scabiei*, or itch insect, the pediculus, or ordinary louse, the flea, and the bug. The tapeworms (*Tænia solium* and *Tænia mediocanellata*, and *Bothriocephalus latus*), the roundworms (*Ascaris lumbricoïdes*), and the threadworms (*Oxyuris vermicularis*) are met with in the intestinal canal. The hydatids are animals like bags or bladders of water which occur in the internal organs of the body, and more especially in the liver ; they are produced by ova, which, escaping from the tapeworm of the dog (*Tænia Echinococcus*), are swallowed in drinking water, and, passing into the system through the alimentary canal, become developed into cysts or bags containing fluid. Another worm, the *Trichina spiralis*, is sometimes, but in this country very rarely, met with in the muscles of the body. It is caused by eating diseased pork or sausages, and much excitement was produced in Germany some years ago by a number of persons eating half-cooked sausages and becoming the victims of this disease.

Since all these parasites are produced from pre-existing living organisms, and enter the human body from without, it is most important to note in what way they enter, and to take proper precautions. Drinking contaminated water or eating food which is affected with the parasite are two most common causes of this disease. Hydatids, for instance, are descendants of the tape-worm of the dog, wolf, or other animal ; this tape-worm infests the alimentary canal of those animals, and the ova or eggs which are met with in the last segments of the worm pass at different times out of the intestines with the excreta, then they are probably washed by the rain into a dike or running stream, and a person drinking the water may become affected ; or else he may eat some water-cresses or some plant which has been growing in the water and on which the ova have settled, and so they enter the system.

It is remarkable that these ova do not produce in man a tape-worm, but are developed into bags or cysts in some internal organ, and these, when swallowed by a dog, will reproduce the entire worm. It is necessary, therefore, for uncooked vegetables to be most carefully washed in clean water before being eaten, and whenever a dog is seen to pass a worm the parasite should at once be burned or thoroughly destroyed so as to remove all source of danger.

Tape-worms may be produced by eating the raw meat of the pig, ox, or cow ; all such meat should be well cooked, but neither that nor salting are quite preventative, though the danger is much lessened. Smoking appears to kill the worm, and also a temperature equal to that of boiling water. But it is not proper to eat of such meat at all except dire necessity compels. The presence of tape-worms may be known by persons passing small pieces consisting of one or more segments, of a white color, longer than they are broad, and not unlike pieces of tape. There is generally more or less hunger, unsatisfied appetite, and a feeling of discomfort in the stomach. The worm affects the adult more commonly than children ; various remedies have been recommended, but it may generally be easily got rid of by a dose of a preparation of male fern taken early in the morning, while the individual has been fasting a few hours previously. (See FERN.) It should be remembered that the worm diminishes in breadth near the head, and the neck is, therefore, long and slender, and so when the parasite escapes, this part should be carefully sought for, as then the patient may be sure that the white worm has been expelled and will trouble him no more. Pieces several feet in length may come away, while the head still remains to form a fresh worm afterward. The parasite should be burned when it is expelled.

The *Round-worm*, in shape, size and general appearance, is very much like the common earth-worm, but the latter is redder and not so pointed at its two extremities ; the earth-worm also has little projections on its under surface, which probably aid it in locomotion, while they are absent in the parasite. It is found in the small intestines, or that portion of the alimentary canal which is next to the stomach ; they may occur singly or several together, and are either vomited up or passed by the bowel ; it is more common in children than in adults. A purgative or a dose of rhubarb or aloes will generally suffice to get rid of the worm. When the worm is present the patient generally has colicky pains in the stomach, fœtid breadth with nausea, or vomiting, and bad appetite. Santonin is perhaps the medicine most certain to expel this worm ; it may be combined with a purgative ; it forms the chief ingredient in the so-called "worm powders."

The origin of parasites which affect the skin is of course well known. In the case of itch, there are two kinds of insect, male and female ; these burrow a short way into the skin, and there the female lays her eggs, which in a short time become developed into their mature form ; they are chiefly met with between the fingers and on the arms and trunk ; they are not common on the head or face ; they produce an intolerable itching, and the patient vigorously scratches the affected parts. They are produced by dirt and direct contact with clothes or people similarly affected. (See ITCH.)

There are several varieties of the pediculus or louse ; they affect the hair of the head or genitals, or they may attack the whole surface of the body or only the armpits. (See ECTOZOA.)

PAREGORIC ELIXIR is the name commonly given to compound tincture of camphor, a preparation which owes its activity to opium. This elixir is sometimes ordered for children, under the belief or pretence that it contains no opium, but the modern paregoric elixir contains one grain of opium in half an ounce of the elixir, so that a teaspoonful or two of it is a tolerably full dose. Formerly it was stronger than it now is, and often gave rise to mishaps.

PAREIRA is the wood of the root of the *Cissampelos Pareira*, a climbing plant growing in Brazil. It occurs in cylindrical pieces, sometimes split, and covered with a most peculiar bark, if such it can be called ; its substance is vascular and porous. It has a sweetish odor and taste, and contains a principle which has been called cissampeline and pelosine. The preparations, as contained in the pharmacopœia, are a decoction, an extract, and a liquid extract, or concentrated decoction ; of these the first and the last are mainly used. Pareira contains a bitter substance, and may act like columba, but is chiefly used in maladies affecting the bladder, especially chronic catarrh of that organ. Sometimes it is combined with an acid, especially nitric acid ; sometimes with an alkali, especially liquor potassæ, seldom alone. Opinions consequently differ vastly as to its efficacy in such complaints as those of which we have spoken (which, be it said, are commonly very intractable), some thinking it of vast and others maintaining that it is of little or no value. Altogether, in the mean time, we can hardly be said to know if it has any influence at all beyond serving as a vehicle to the potent remedies mentioned. The dose of the decoction is about two fluid ounces, of the liquid extract two fluid drachms.

PARONYCHIA. (See WHITLOW.)

PAROTID GLANDS are two in number, one on each side, just below and in front of the ear. They secrete a great portion of the saliva, and are most active when mastication is going on. When inflamed they become painful, and form the disease known as "mumps ;" the patient is then unable to open the mouth for a few days, but the disorder will soon pass away. (See MUMPS.)

PAROTITIS. (See MUMPS.)

PARSLEY is *Petroselinum sativum*, and belongs to the order *Umbelliferæ*. It is a well-known seasoning herb, and was at one time included in the Materia Medica. It possesses certain medicinal properties which might be turned to account when other remedies are not at hand. Parsley acts on the system as a diuretic, emmenagogue, and as a carminative. In all affections of the bladder and kidneys, gravel, or stone, this plant was formerly very largely used, while the seeds, taken two or three times a day, are said to exert a powerful effect on the uterine secretions. At one time a poultice made of the bruised leaves and stems, with vinegar and water, was considered a specific for the bites of all venomous reptiles. Parsley chewed has the property of destroying any fœtor in the breath, or the smell imparted to it from spirits, onions, or other articles. If dried and preserved in bottles excluded from the air, it retains its flavor for a long time, and is very useful in flavoring omelets and similar dishes.

PARSNIP is the cultivated variety of the wild parsnip—*Pastinaca sativa*. On account of the woody fibre it contains it is less digestible than potatoes. It contains also less sugar, starch, and flesh-forming matters. Hence on the introduction of the potato into the Old World it became very much less consumed. It has a peculiar flavor, which is much liked by some people, and it still continues to be used as an article of diet, especially with boiled salt fish and beef. In Holland parsnips are much used in soups, while in Ireland cottagers make a sort of beer by mashing the roots and boiling them with water and hops, and afterward fermenting

the liquor. A kind of marmalade preserve has also been made from parsnips, and even wine, which in quality has been considered to approach the far-famed Malmsey of Madeira.

PARTURITION is the process of giving birth to a child, and is synonymous with the term labor or confinement. (See LABOR.)

PATELLA is the anatomical name for the knee-cap. (See KNEE-JOINT.)

PEAS are the product of various species of the genus *Pisum*, belonging to the leguminous order of plants. Both in a fresh state and when dried they are a valuable article of diet, and contain a large quantity of caseine in a digestible form. The green pea contains more sugar and less caseine than when dried. Dried peas are a wholesome and nutritious addition to other kinds of food. When added to soup they are agreeable and economical, and in the form of flour when ground they may be advantageously made into puddings or bread with wheaten flour.

PECTORILOQUY. (See STETHOSCOPE.)

PELLITORY, or **PYRETHRUM,** also known as Pellitory of Spain, is the root of a plant, *Anthemis Pyrethrum*, growing along the coast of the Mediterranean, and imported from its eastern shores. The root is spindle-shaped, and is ordinarily cut into pieces two or three inches long. It has a thick brown bark and breaks with a resinous fracture, dark brown in color with black shining parts. Its tincture is the preparation used, but even that is seldom employed, and then only as a gargle. Chewed, it gives rise to a feeling of pricking in the mouth and a flow of saliva. Hence it is called a masticatory. It is only used as a local stimulant.

PELVIC CELLULITIS is an inflammatory symptom of the cellular tissue which surrounds the bladder and womb. It may come on after a difficult labor, or after an abortion, or after an attempt to procure premature labor.

Symptoms.—It is characterized by great pain in the lower part of the abdomen, with loss of appetite and strength. There is always a good deal of fever attending it, and the patient is generally worse at night. Shivering and rigors cause much distress ; the patient feels weak and prostrate and loses color, so that the face has a sallow, earthy look. This disease is very apt to go on for several weeks, during which time the patient becomes emaciated, and of course is obliged to keep her bed. Very often the inflammation goes on to form an abscess, and in time this points and discharges a great deal of matter. It may burst inwardly, or sometimes in the skin over the groin, and the escape of pus is generally attended with relief. The convalescence is, however, very prolonged, and it may be weeks or months before the woman finally recovers her former strength.

Treatment.—The patient must be placed

in bed in a cheerful and well-ventilated room. Her diet must consist of milk, beef-tea, broth, eggs, jellies, etc., and while it is light and nourishing it should be given in abundance so as to keep up her vital powers ; a small chop can be taken if desired by the patient, but on no account must the woman be kept too low. Three or four glasses of port wine a day will do much good, or some bottled stout may be given if preferred. To relieve the pain hot fomentations or hot lin-seed-meal poultices must be applied, but they should be made as light as possible, as their pressure may cause pain. It is useful also to remove the weight of the bed-clothes from the abdomen by placing an ordinary fire-screen or some similar contrivance over the body. Great care should be taken to prevent the formation of a bed-sore ; for this purpose great cleanliness must be observed, and the patient should lie on a water-bed or water-cushion. (See BED-SORES.) At first saline medicines and anodynes may have to be given, afterward quinine, or some other tonic. Convalescence will be much aided by change of air, and especially by residence at the sea-side if not too cold. If an abscess form it will either burst or have to be opened, and be treated in the usual manner.

PELVIS is a space formed by the haunch bones and lower part of the spinal column which communicates freely with the abdom-inal cavity above, but is elsewhere closed. It contains the bladder in front, and the rectum behind, besides several important vessels and nerves ; in the female it contains, in addition, the uterus or womb, which is situated between the bladder and rectum. The pelvis is wider and shallower in the female than in the male, and in this way the process of parturition is facilitated ; from this cause, also, the hips are wider apart than in man, and this makes the difference in the gait between the two sexes when walking.

PEMPHIGUS, or POMPHOLYX, is a skin disease characterized by the presence of large blisters called bullæ on the surface of the skin ; often the arms and hands or the lower extremities are more affected than the trunk ; the disease may be either acute or chronic, and sometimes very troublesome to heal. The bullæ or blisters are surrounded by a narrow red raised aureole, or ring ; when the blisters burst a scab may form, and that spot may heal while a fresh one will form else-where. Zinc ointment must be applied ex-ternally, and small doses of arsenic may be given internally ; the general health must be improved by a light and nourishing diet. The bullæ should be pricked with a needle to let the fluid escape.

PENNYROYAL, the *Mentha Pulegium*, is only employed in medicine in the form of oil ; even that is not now officinal. The oil, which is yellowish and of a peculiar odor, is obtained by distillation from the plant, which grows in marshy places. The oil is used in the same cases as peppermint and spearmint. The herb itself has obtained a false reputation as an abortive.

PEPPER is the fruit of the *Piper nigrum*, a shrub belonging to the family *Piperaceæ*. It grows both in the East and West Indies. There are two sorts sold in the markets— " white" and " black." The white is pro-duced by the same plant as the black, and consists of the berry from which the skin or bark has been removed. The pepper berries contain an active principle called *piperine*, and an acrid resin as well as a volatile oil.

PEPPERMINT—the *Mentha piperita*—is only used nowadays as oil obtained by dis-tillation from the well-known plant. This oil is colorless, turns pale yellow by keeping, has a warm aromatic taste, and the odor of peppermint. Peppermint water is now made by distilling together a large quantity of water and a small quantity of the oil. The essence of peppermint consists of the oil dis-solved in a small quantity of rectified spirit. Peppermint water is largely used as a vehicle, but to some people the odor and flavor are both very disagreeable and should be avoid-ed. The essence may be given dropped on sugar to infants for spasms or gripes, and many adults find the same prescription grate-ful. The oil is used as an adjunct to purga-tives to prevent them griping.

PEPSIN is a modern remedy, if, indeed, it can be called a remedy at all, seeing that it consists as far as possible of the digestive principle secreted by the stomach, made use of to aid digestion. The best pepsin is made from pigs' stomachs ; it is known as *Pepsina porci ;* it is free from acid and starch, and has an odor by no means disagreeable. The pepsin most frequently used is of French or-igin—the Pepsine Boudault—and is obtained from the stomach of the calf. It is also large-ly mixed with starch, so that it is, compara-tively speaking, inert. It is grayish-white in color, always acid, and very often its smell is quite disagreeable. Pepsin has this pecu-liarity, that when acidulated with hydrochloric acid greatly diluted and kept at a tempera-ture of 100° F., it speedily dissolves all albu-minous substances, reducing all to a com-pound of nearly uniform character called peptone. This differs altogether in its char-acter from any other albuminous substance, in that it is easily miscible with water and readily diffusible, resembling in this a saline substance. Other acids, especially lactic, have the same power as hydrochloric, but in a much less degree.

The value of pepsin in a certain number of cases is undoubted ; but the exact cases for its employment are not yet quite manifest. It has been mainly used in cases of dyspepsia where there was reason to believe that the secretion of gastric juice was imperfect. It should in these cases be given during or after a meal. Sometimes it is made up in the form of a lozenge, to be swallowed just after food ; but this is not, perhaps, the best way. Pepsin has this peculiarity, that when it acts' on albuminous substances it does not itself become destroyed, but is quite capable of acting on quantity after quantity of the albuminous or fibrinous material, provided only a sufficiency of dilute acid be supplied. Apparently, therefore, the best plan would be to give this substance in a dose of three or four grains along with food, and from time to time thereafter to swallow small quantities of diluted hydrochloric acid, a few drops at a time, until digestion is complete. In cases of great debility of stomach, especially in old people, the habitual use of pepsin may render life easy and pleasant where formerly it was unendurable. It will be of great value where there is regurgitation of half-digested or half-putrid food, especially in cases of obstruction at the pylorus ; for it has been tolerably clearly proved that substances which are digested in the stomach are absorbed there, so that if albumen be converted into peptone, and there absorbed, there will be no necessity for it to pass the pylorus, nor will there be time for it to undergo decomposition. After a time a few drops of weak carbolic acid might be given to prevent putrefaction of what remains.

PERICARDITIS is an inflammation of the pericardium or membrane which envelopes the heart. Involving so important an organ as the heart, and interfering so gravely with its action, it is therefore a disease of serious significance. Pericarditis very rarely comes on as a primary disease ; it generally comes on as a secondary complication in the course of rheumatic fever or Bright's disease. More rarely it occurs by the spread of inflammation from adjacent parts, as in cases of pleurisy and pneumonia ; sometimes from the presence of an aneurism or other tumor in the chest, and still more rarely from external injuries, as gun-shot wounds, stabs, etc., for these generally prove fatal before there is time for the inflammation to appear.

Symptoms.—It will be most convenient to give the signs of pericarditis as they occur in a case of rheumatic fever. In such cases the patient will have had pain and swelling of the joints some days previously ; the skin will be hot and perspiring freely, while the temperature may have risen to 102° or 103°, or even higher. There will be loss of appetite, thirst, headache, and a white, furred tongue. If now pericarditis should come on, the temperature will rise higher by a degree or two ; the pulse will be quick and excited ; there will be pain across the chest, which is worse when the patient takes a deep breath ; the face will be pale, the breathing shallow and hurried, as the pain is worse on inspiring deeply ;

the patient will lie flat on his back, and feel faint if any attempt is made to raise him. If now one listen over the region of the heart, a rough sound will be heard, because the opposed surfaces of the pericardium are no longer smooth but rough. In a day or two there will also be signs that the cavity of the pericardium is full of fluid, and as this, pressing upward and laterally, displaces the lungs, the breathing is still more interfered with. There may also be heard numerous new sounds, called râles, over the chest, for the air-passages contain more mucus, and the air, in passing over them, causes these rattling noises. These symptoms are generally attended with an abatement or even cessation of all pain in the joints. So the patient goes on for a few days, when, if recovery takes place, the temperature begins to decline, the breathing is easier, the pulse lessens in frequency, and a general improvement may be noticed. If, however, the case goes on from bad to worse, there is generally low, muttering delirium ; weak and muffled sounds are heard over the heart ; there is greater prostration, and the patient dies from exhaustion or syncope. In all cases of recovery the absorption of the fluid is a slow process, and convalescence is retarded. In many cases, as the pericarditis abates, there may be signs that the endocardium, or lining membrane of the heart, is also affected, and so, while they recover for a time, there is left behind much serious mischief, which will sooner or later shorten the life of the patient. It is a remarkable fact that in some years it is much more common for inflammation of the heart to accompany rheumatic fever than at other times, nor is there at present any satisfactory explanation to be given. In some cases the muscular wall of the heart is inflamed as well as the outer covering, and this seems to be the case in most of those which are fatal, so that it adds to the danger of a case. In Bright's disease, when pericarditis comes on, it is generally found that the patient has suffered from the chronic form of this disease for a long time, and it generally supervenes in those who are the subjects of the gouty and contracted kidney. There is great pain over the front of the chest, and the rubbing sounds may be heard on placing the ear to the chest ; in other respects the symptoms are very similar to those mentioned above. Nor in those for whom there is some other cause for the inflammation are there any other marked symptoms than those described ; but in addition there will be the evidence of an aneurism, pleurisy, pneumonia, stab, wound, etc., according to the nature of the cause.

Treatment.—In all cases perfect rest must be enjoined in bed, and the horizontal position kept ; in some cases, however, of old standing heart disease the patients cannot lie down, and then they must be propped up with pillows. No exertion should be made by the sufferer, nor should he attempt to rise for any natural purpose ; nor should he be moved from bed to bed in the early days of the illness, for whenever movement is made the heart is called upon to do more work, and in its weak, inflamed state it may cause a fatal faint, or syncope. The diet must consist of milk, an egg beaten up in the milk, beef-tea, broth, etc., but no solid food should be given, nor anything which distends the stomach. Stimulants must be given according to the requirement of each case, and according to the age of the patient ; but no rules can be laid down here as to the quantity to be given daily. The nurse and medical man should look out carefully for any signs of bed-sores, and try and prevent their forming by perfect cleanliness and a smooth bed ; they are very apt to form in these cases. (See BED-SORES.) When the pain is great across the chest, and the breathing very quick, four or six leeches will give a great deal of relief by abstracting the blood ; then a hot linseed-meal poultice, not too heavy, should be laid across the chest and changed every three hours ; but care should be taken that the leech-bites have stopped bleeding, or otherwise the moist heat will make them bleed more, and exhaust the patient too much. Cotton-wool may be applied to the chest in some cases instead of a poultice. As such patients sweat profusely, the chest should be wiped dry and gently just after the poultice is removed, or the moisture soon cools and makes an uncomfortable feeling. Opium may be given at bed-time to procure sleep, but with great care in such as have affection of the lungs at the same time. Now and then, in the course of rheumatic fever, the pulse and temperature suddenly rise, and there is intense headache and delirium. In these cases of hyperpyrexia, or excessive fever, the thermometer may rise to 107′, 109°, or even higher. Most of these prove fatal, but some cases have lately been cured by the use of the cold-bath, or by packing in sheets dipped in cold water and often renewed. The great principles of treatment in pericarditis are—1. To give the patient light but very nutrient diet. 2. To avoid all exertion. 3. To try and abate any pain, cough, or other troublesome symptoms. 4. To sustain the patient's strength until such time as the acute inflammatory symptoms have abated. The period of convalescence follows, and those tonics may be given, and the rules may be adopted which are laid down in the article on FEVERS.

PERICARDIUM is the name given to the membrane which surrounds the heart and incloses it in a bag. Thus there are two opposed surfaces, each of which, in health, is perfectly smooth, and enables the heart to move freely with the least amount of friction. The sac or bag secretes a small quantity of fluid so as to lubricate the parts, and this is much increased in quantity in cases of hydropericardium, when there is a dropsical accumulation of the fluid ; this condition arises in cases of disease of the heart, kidneys, and lungs, and it is generally associated with a dropsical condition of other organs.

PERCUSSION is the art of ascertaining the state of internal organs by means of the sound produced by striking the part over them with the fingers, or by means of striking on a little ivory plate placed over the part. It requires a delicate ear and much practice to be able to use percussion skilfully. (See STETHOSCOPE.)

PERICRANIUM is the tissue which overlies the skull, and aids in nourishing the bone which lies beneath it.

PERINÆUM is the name applied by anatomists to the region at the lower part of the body which is perforated at its centre by the anus. It is bounded in front by the genitals, behind by the buttocks, and at the sides by the inner surfaces of the thighs. It corresponds to the outlet of the pelvis, or bony gristle at the lower part of the trunk, this outlet being bounded in front by the pubic bones, behind by the coccyx or terminal bone of the spinal column, and at the sides by portions of the pubic and ischial bones, and by a fibrous structure called the great sacro-sciatic ligament. In the deep parts of this region are contained the lower portions of the rectum and the generative and urinary organs, with their appendages. At the surface it measures four inches from before backward, and from two to three and a half inches in breadth at its widest part. The skin is loose, dark in color, and covered by short hairs ; around the anus it is puckered by radiated folds, and it is traversed from before backward, exactly in the median line of the body, by a prominent line called the *raphé*. This region is one of great importance and interest to the surgeon, as it is here that the incisions are made through which a stone can be removed from the bladder. It is occasionally the seat of abscess, connected in most instances with perforation of the walls of the rectum or urethra, and terminating in the one case in fistula, and in the other in urethral or perineal fistula. Injuries, especially heavy falls upon some hard body, are especially dangerous in this region, as they often cause laceration of the urethra and retention of urine.

PERIOSTEUM is a delicate covering to the bones ; it is richly supplied with blood, and takes an important part in the proper nourishment of bone.

PERITONEUM is a serous and smooth membrane which forms the outer coat of the stomach, liver, intestines, and some other organs of the abdomen and pelvis ; it also lines the wall of the abdominal cavity ; in this way a cavity is formed between its two surfaces, and this only contains a little fluid, just enough to moisten the opposed surfaces and to enable the intestines, etc., to glide over each other with the least possible amount of friction. When this membrane is inflamed it gives rise to great pain and sets up *perito-nitis.* (See PERITONITIS.) When fluid accumulates in the cavity formed by the peritoneum, as in cases of dropsy, ascites is then produced. Cancerous growths may be found in this membrane, causing a form of peritonitis, and also deposits of tubercle may be met with. Any injury to the membrane, as stabs or wounds from without, may set up a serious inflammation.

PERITONITIS is an inflammation of the peritoneum. Its danger will vary with the cause; it may be produced by external injuries, as a stab or gun-shot wound in the abdomen. These cases are often fatal. Cancer and tubercle may bring it on, but this is generally only a part of the malady, and the chief seat of disease is elsewhere. Any tumor of any organ in the pelvis or abdomen may cause it, as hydatid cysts, ovarian tumors, etc. Stricture, herniæ, and ulceration of the intestinal canal will bring it on, and it is often associated with typhoid fever.

Symptoms.—Pain over the abdomen, vomiting, and a raised temperature are the chief symptoms; the pulse is quick and small, the countenance anxious and sunken, the legs drawn up, so as to relieve the pain. The pain is worse on any movement, and is very wearing to the patient. In some cases of blood-poisoning, as in pyæmia and puerperal fever, etc., there may be peritonitis and yet no pain. Colic, which often comes on in lead poisoning, must not be mistaken for this disease. There will be no fever then, while there will be the occupation of the patient to guide one, and the individual will have a blue line on the gums. Some cases of hysteria may simulate peritonitis; here, again, the temperature is normal, and there are the usual signs of hysteria.

Treatment.—Perfect rest in bed must be enjoined; hot fomentations, made as light as possible, must be applied over the abdomen; the weight of the bed-clothes should be taken off from the patient as far as possible. Opium must be given to relieve the pain. Cooling, saline drinks and iced water may allay the thirst and sickness, while milk and beef-tea must be given every three hours or oftener, so as to keep up the patient's strength. But each case will vary so with the cause as that any given case might require a somewhat different line of treatment. The only thing that a good nurse can do is to see that the room is cool and well-ventilated, to secure perfect cleanliness, to give light and nourishing diet, and to enjoin perfect rest. Peritonitis often forms a serious complication in cases of typhoid fever. (See TYPHOID FEVER.)

PERITYPHLITIS signifies inflammation around the cæcum. It may come on after taking an indigestible meal, as partaking freely of nuts, etc., or a foreign body, as a pin, cherry-stone, etc., may lodge in the vermiform appendix, and set up inflammation around. The symptoms are great pain on the right side of the abdomen and fulness there; great tenderness on pressure; sickness, faintness, furred tongue, and fever. The treatment will be the same as that recommended in cases of peritonitis. (See PERITONITIS.)

PERRY. (See CIDER.)

PERSPIRATION, or **SWEAT,** is the secretion which is always taking place from the skin. Under ordinary circumstances, no liquid water appears on the surface of the skin unless very active exertion be made. The quantity secreted varies very much according to the temperature and moisture of the air, and the state of the blood and nervous system. A small quantity of carbonic acid and urea are excreted, but the greater part by far is water; the secretion is acid, and contains fatty matters derived from the sebaceous glands of the skin. (See SKIN.)

PERTUSSIS. (See WHOOPING COUGH.)

PERU, BALSAM OF. (See BALSAM OF PERU.)

PERUVIAN BARK. (See CINCHONA.)

PESSARIES are instruments intended for introduction into the vagina for maintaining the womb in its proper place and attitude. These are of various shapes and characters, but it is bootless to describe them here, since almost every practitioner has got some notion of his own on the subject of which are best for each particular case, and such mechanical appliances must always be selected and adjusted by a practitioner. The pessaries we here refer to are what are called medicated pessaries, and consist of some drug made up into a conical mass, and introduced into the vagina in cases of disease of the womb. These are sometimes capable of giving great relief, and are never to be neglected, especially where it is found necessary to employ sedatives locally. The best material for making these pessaries is the hard oil obtained from the fruit of the cacao-tree, which is separated in preparing some of the varieties of cocoa in ordinary use. This oil is hard as frozen butter, but is easily melted; mixed with a little olive oil or other similar substance it easily takes up the remedy when stirred with it, and when introduced melts slowly so as to allow of the remedy being gradually absorbed. Perhaps the best sedatives to be mixed with cacao butter are extract of belladonna and extract of opium. Acetate of lead and sulphate of zinc may also be added if it is necessary to arrest discharges. Tannic acid, too, is very usefully combined with it; as is alum or similar substances. In

most cases injections do better than pessaries, but when sedative applications are desired
the pessary allows of their reaching the parts
more gradually and in more divided doses.

PESTILENCE. (See FEVER; PLAGUE.)

PETECHIÆ are the spots which make
their appearance in certain diseases wherein
is much disorder of the blood. Chief among
these are typhus fever, purpura, and scurvy.
In typhus the spots begin as irregular patches
of a dusky hue, which may be elevated above
the skin so as to be perceptible to the finger.
If there is much blood derangement, these
spots are very dark in color, and it is these
dark-colored spots indicative of blood effusion
and alteration, which are called petechial.
In scurvy and purpura the patches are very
much larger than in typhus, but in patients
who have been exposed to the conditions
which gives rise to scurvy or purpura, any
acute disease may produce petechial spots in
the skin and mucus membranes. Petechiæ
on the surface of the pleuræ are indicative
after death of violent suffocation.

PHAGEDÆNA, derived from the Greek
word φάγω, to gnaw, is used to express a
variety of ulceration which destroys the tissues more rapidly and to a greater extent than
ordinary forms of ulcer. The subjects of this
local affection are usually individuals who
have been debilitated by some severe febrile
disorder of a typhoid character, or who have
been subjected to the influence of cold and
wet, foul air, bad and insufficient food,
fatigue, and excessive indulgence in spirits.
It is generally preceded by some sore or
wound, and its local causes are irritation of
the open surface and gross neglect of cleanliness. A very superficial sore, such as that
formed by the application of a blister, may,
under the above-mentioned constitutional
and local influences, rapidly become phagedænic and produce much destruction of the
soft parts. It has been most frequently met
with in connection with venereal ulcers, especially in those cases in which the patients
have been submitted to a prolonged and excessive use of mercury. Phagedænia varies
in intensity in different cases ; it is sometimes
so mild as to be scarcely distinguishable from
ordinary ulceration, and in other instances it
spreads with so much rapidity and destroys
so great an extent of the surface of the body
that there seems to be very little difference
between it and the affection known as *hospital
gangrene.* This latter form of phagedæna is
met with in noma, cancrum oris, and the
sloughing throat of scarlatina. It is believed
by many surgeons that phagedænic ulceration is caused by poisoning of the blood, in
consequence of the absorption of putrid matter.

In phagedæna there is a large and rapidly
spreading ulcer, the edges of which are
formed of sharply-cut, indented and undermined skin. The surface of this ulcer is un
even and of grayish color, and is covered by
a dark colored, thin and very fœtid discharge,
which is often marked by streaks of blood.
The integument surrounding the ulcer is
swollen, and of a dusky-red color. The ulcerative process is attended with severe gnawing pain.

Treatment.—In the treatment of this affection it is necessary that the patient be supplied with good nourishment, and that alcoholic stimulants be given freely, but at regular intervals. Opium is generally administered for the purpose of relieving the severe
pain and of allaying nervous irritation. The
diet should consist chiefly of fluid food, such
as milk, beef-tea, and strong broths. The
patient should be kept in bed in a large and
well-ventilated room. The bowels should be
kept open by mild purgatives, but great care
must be taken to avoid diarrhœa, as the subjects of phagedæna may rapidly sink under
any excessive drain upon the system. The
local treatment consists in cleansing the surface of the ulcer by frequently syringing it
with some disinfectant lotion, as a solution
of carbolic acid, of permanganate of potash,
or chloralum, and in relieving the pain by the
application of poultices or poppy fomentations. In severe cases, where the ulceration,
in spite of this treatment, is spreading with
rapidity and attacking important parts of the
body, the surgeon often finds it necessary to
apply the actual cautery, or some strong
caustic. Of caustic applications, fuming
nitric acid seems to be the one most in favor.

PHALANX is a technical term for the
small bone of a finger or toe ; each finger
and toe, therefore, has three phalanges.

PHARYNX is the upper part of the gullet,
and may be seen at the back part of the mouth
behind the tonsils. It communicates in front
with the mouth so as to allow the food to
pass down into the stomach ; above, with
the nose, as is shown in cases of vomiting ;
and below and in front of it lies the larynx
or windpipe, from which food is prevented
from entering by means of the epiglottis.
The Eustachian tubes, one on each side,
also enter the pharynx, and connect it
with the ear ; it is along these tubes that
inflammation may extend when the throat
is affected, as in cases of common cold or
scarlet fever, and so give rise to deafness.

PHENIC ACID. (See CARBOLIC ACID.)

PHLEBITIS means strictly an inflammation of a vein ; in cases of pyæmia, a clot may
form in a vein and give rise to some of the
symptoms ; but it is doubtful if actual inflammation of the walls of a vein ever occurs.
(See PYÆMIA.)

PHLEBOLITHES are small concretions
or particles, made up chiefly of calcareous
matter, which now and then form in the small
veins. They are seldom recognized during

life, and rarely give rise to any symptoms. No treatment is required, and the only result of their presence is that the circulation through that vessel is obstructed, and the blood has to go round another way.

PHLEBOTOMY means cutting a vein, as in the ordinary operation of bleeding or venesection. (See BLEEDING.)

PHLEGM is a common name for the fluid coughed up from the air-passages. (See EXPECTORATION.)

PHLEGMASIA DOLENS, also known as *white leg* or *milk leg*, is a malady affecting women, in childbed especially. It is most likely to occur in women who have been weakened by flooding or other causes. It sometimes occurs toward the latter stages of cancer of the womb. One or both extremities may become affected, and it is said that the left is more frequently its subject than the right. The limbs become brawny, but do not pit on pressure, and from altered conditions of the circulation the parts become quite white. It has been assumed, rather than proved, that it depends on coagulation of the blood in the veins, and it has been called obstructive phlebitis. Others, again, have said, that the obstruction in the veins is quite a secondary matter. Be that as it may, the disease commonly occurs in a first pregnancy, mostly within a month after labor. The *symptoms* of phlegmasia may be thus enumerated. They begin with fever, headache and pain, not infrequently preceded by a rigor. In about four-and-twenty hours the limb begins to swell ; the two limbs are seldom affected together. This swelling commences in the foot or ankle, and gradually creeps upward. Sometimes, however, it begins in the thigh, and passes downward, or all parts may be nearly simultaneously affected. The limb is hot and tender, swollen to nearly twice its size, pale white in color, tense, shining, and elastic, but not bagging on pressure. This acute stage may remain three weeks or longer, but the limb remains useless a very much longer time, only gradually recovering its normal size, power, and pliability.

Treatment.—The treatment may be said to be in the main prophylactic ; that is to say, it should be the aim to prevent such an accident, if there be any likelihood of its occurring ; and if it threatens, the aim should be to arrest the malady as speedily as possible. The great thing is absolute rest ; the patient must not be moved. Bran poultices should be applied the whole length of the limb, or hot-water fomentations freely employed ; but the former are best, as necessitating less inconvenience to the patient. Opium should be given internally to soothe pain, or chloral may be used for a like purpose. At a later period, good food, wine, bark, and iron are to be given, and bandages and hot fomentations, with gentle rubbings applied to the stiffened limb. The malady rarely if ever proves fatal.

PHOSPHORIC ACID, in its diluted form, is used in medicine, but not very extensively. It may be prepared directly from phosphorus by oxidation, by means of nitric acid. The liquid thus obtained, when diluted, is colorless, and has an agreeable taste. Its reaction, of course, is acid, and generally it may be said to act much in the same way as sulphuric acid, but is not nearly so astringent. It may be given in good large doses, and it has been suggested that this would render it of value when combating the tendency of urine to become alkaline in the urinary passages. It is a good medicine when given freely diluted with water, the pleasant acid taste rendering it agreeable to many parched invalids. To many it may be given freely as a drink, sufficient only being added to the water to render it pleasantly acid. The ordinary dose is about half a drachm diluted. The chief compounds of phosphoric acid employed in medicine are phosphate of ammonia, chiefly used for aiding the solution of uric acid in the urine, where there is a tendency to the formation of uric acid calculi. Phosphate of soda or tasteless purging salts is a capital laxative for children, and may be given in their food with ease. Phosphate of iron is a remedy of undoubted value as a ferruginous tonic, while it wants the binding and irritating qualities of some other iron salts. *Hosford's Acid Phosphate* is an excellent form in which to administer phosphoric acid.

PHOSPHORUS is a waxy-looking substance of very peculiar properties, obtained from bones. The bones are first of all acted on by oil of vitriol, and the solution which is thus formed is subsequently distilled with charcoal. The phosphorus, which is volatile, passes over, is collected in a cool receiver, and subsequently moulded into sticks. These it is necessary to keep under water, otherwise they would take fire. When freshly prepared, phosphorus is colorless, and semi-transparent, but after a time a film forms on its surface. When set on fire, it burns with a bright flame, giving rise to fumes of phosphoric acid. It may be converted into a peculiar form, known as red or allotropic phosphorus, which possesses properties quite distinct from those of the ordinary variety. Ordinary phosphorus readily takes fire ; the other form does not till it has been heated to a high temperature, and it is said not to be poisonous, whereas the ordinary variety is intensely so. Phosphorus is not often given internally ; if given in large doses, it acts energetically as a poison, and seems almost invariably to give rise to a form of degeneration of the liver. In smaller doses it has been given in nervous diseases, and as a stimulant in low nervous fevers. It has also been given in other diseases, with no very marked benefit. Indeed, it may be said that up to the present time no malady has been treated with any peculiar benefit by phosphorus. Moreover, it is dangerous and unpleasant, and except some other use be found for it, perhaps it had better be left out of the list of available remedies. The dose is about one fifteenth of a grain, and should be given dissolved in olive oil.

PHOSPHORUS DISEASE. (See LUCI-FER-MATCH-MAKER'S DISEASE.)

PHOTOPHOBIA, or dread of light, is a symptom common to a good many diseases of the eye. The patient its subject, shuns light in every way, and if introduced into a room with a bright light, obstinately shuts the eyes, and cannot be persuaded to open them. It is characteristic of no one disease. (See OPHTHALMIA.)

PHRENOLOGY is a system of mental philosophy designed and taught by Drs. Gall and Spurzheim on the continent, and by Mr. George Combe in England. The believers in this system maintain that it is an infallible index to the mind, and that the elevations or depressions in the head exhibit to the student of phrenology the whole nature of the mind, as on a map or chart. The brain in phre-nology is divided into three parts—the an-terior, middle, and posterior. The anterior, or front portion, is supposed to contain all the intellectual and perceptive faculties. The moral sentiments or emotions are situ-ated in the middle region, and the animal propensities are confined to the back of the head. Each of these portions of the brain is divided again into individual organs, hav-ing special functions assigned to each. Phrenologists enumerate about forty differ-ent organs or bumps, as they are familiarly called, each independent of the other, and capable of development or repression as the character is governed and controlled by the educated will. These dispositions are said to be affected by constitutional temperament, of which three varieties are recognized. The vital or vigorous and powerfully physical temperament ; the motive or bilious, charac-terized by dark hair and eyes, bony structure, and muscular development ; and the mental or nervous temperament, distinguished by delicacy of body, great susceptibility, and a light elastic frame, with fair hair and eyes. Phrenologists divide the mental organs into five groups, and there are many drawings and casts of the head illustrating this division and the situation of each faculty. Phrenol-ogy was at one time more studied than it is at present, but associated with mesmerism and electro-biology, as it is called, it has yet many votaries.

PHTHISIS is the technical name for pul-monary consumption. (See CONSUMPTION.)

PHYSIOLOGY is the science which treats of the history and functions of the human body and its several parts. It teaches the function and nature of every texture and por-tion of the body, and enlightens us as to the economy and use of the organs of the human system. Physiology is divided into animal and vegetable, and again into human and comparative physiology. All the principal portions of human physiology are treated under their several heads in the present work.

PHYSOSTYGMA, better known as the Ordeal bean of Calabar, in West Africa, is a substance recently introduced into practice, and though it has had the advantage of an unusually careful and accurate physiological investigation, has not yet come fairly into general use. The beans are the fruit of a tree belonging to the leguminous order, and each contains two lobes easily reduced to powder, and having a taste somewhat similar to that of other seeds of the order, which are wholesome enough. The active principle has been called physostygmine. It is mostly contained in the kernel, and may be extracted by alcohol. This extract is officinal, and is the only preparation which is so. When taken internally these beans, or even a por-tion of one of them, give rise to serious symp-toms. The bean was used in Old Calabar and neighboring countries as a test for witches, hence the title Ordeal beans. If any one accused of this, to the savages there, horrid crime, could devour a bean and still live, the charge was repelled and without hurt ; but if, as most frequently happened, the trial was attended with fatal results, then general suspicion was supposed to be con-firmed. In point of fact, if an excessive dose was taken, the patient sometimes recovered by vomiting , but if only a moderate dose was swallowed, then death most surely fol-lowed. It seems to act especially on the heart and spinal cord, paralyzing the former, and arresting the functions of the latter. The spinal cord, under its influence, seems to lose the power of fulfilling its functions, the affer-ent nerves ceasing to act before the efferent. The most apparent action of the Calabar bean is however, to cause contraction of the pupil of the eye. It has accordingly been used in certain forms of long sight to remedy that by applying it locally to the eye. It has also been given with some apparent success, real or fancied, in tetanus, even in the trau-matic variety ; but so many remedies have been vaunted for this—none having been found really efficacious—that men are scepti-cal. The dose of the extract is from one-sixteenth to one-fourth grain internally given as pill.

PIA MATER is one of the coverings of the brain and spinal cord. (See BRAIN and SPINAL CORD.)

PICROTOXINE is an active principle contained in *Cocculus Indicus*, which, as al-ready pointed out, is mainly used for nefarious purposes. Picrotoxine is not used in medicine. Even cocculus has been used mainly for destroying vermin.

PIGEON-BREAST is a condition of the chest due to malformation, generally indicative of deficient respiratory space. In pigeon-breast, the chest, instead of being more or less rounded or flattened from before backward, is flattened from side to side, and projects in front. This renders the chest exceedingly narrow from side to side, and causes the breastbone to project as it does in birds, so as to form a ridge in front. This arrangement sadly cramps the space available for the lungs, for the size of the heart does not greatly vary. Moreover, the shape is unfavorable to chest movements, and is accompanied by other imperfections of build, which indicate weakness of constitution. The shape of the chest is mainly of importance as indicating a tendency to consumption, which is very unfavorable to the subject of lung disease.

PIGMENT is the coloring matter which is found in the blood, bile, urine, and in nearly all the fluids of the body; it gives the color to the skin, and is, of course, most abundant in the negro, being deposited in the *rete mucosum* of the skin; it is due to the action of sunlight upon pigment that people are liable to become tanned. It is present in excess in cases of Addison's disease, and sometimes in pregnant women. It is in excess, also, in cases of melanotic degeneration. (See DEGENERATION.) It is more abundant in old age than in youth or childhood, and it gives the iron-gray color to the lungs in advanced age.

PILES. (See HÆMORRHOIDS.)

PIMENTO is the unripe fruit of the *Eugenia Pimento* or Allspice tree, which grows in the West Indies. An oil is distilled from the fruit in England. The pimento is a small, round berry, brown and rough on the surface, having the remains of the calyx manifest on it. The oil is yellow, and heavier than water. Like the oil of cloves, it appears to consist of two ingredients, similar in composition. The odor is aromatic, and the taste is hot and pleasant. It is mainly used as a flavoring ingredient in cooking, but the oil may be added to purgatives to prevent griping. The only officinal preparation is pimento water, mainly used as a vehicle.

PINS AND NEEDLES is a popular phrase applied to that peculiar numbness and pricking of the arm, hand, foot, or leg, which is so commonly felt after pressure or a long-continued constrained attitude. It is caused by some interruption to the circulation, and is generally removed by rubbing or exercise. If it should continue, it may be the precursor of some serious attack, and medical advice should be sought.

PITCH, or as the officinal variety is more

strictly termed, Burgundy Pitch, is the resinous exudation from the spruce fir, imported from Switzerland. It is hard and brittle, yet gradually takes the form of the containing vessel. Generally it is of a dull, reddish-brown, of a peculiar odor and taste. It is not bitter. It consists almost entirely of a resin, but a little volatile oil is also contained in it. This gives it its perfume. The resin is similar to that obtained from other plants of the turpentine group. The only preparation is the well-known pitch plaster, which is intended as a slight stimulant to the skin. The pitch plasters ordinarily sold often contain no pitch whatever.

PITYRIASIS is a trifling redness of limited portions of the skin, with a scaly or brawny condition of the cuticle covering the part. It is most common on the head, and is known to nurses as scurf and dandruff. The head itches, and directly it is rubbed quantities of little scales, formed of epithelium, are detached. A similar affection is sometimes found at the bends of the joints and on the trunk. When it occurs in a severe form the hair grows thin and comes out in considerable quantity, but not sufficiently to cause baldness. A solution of borax and mild astringent ointments are often found enough to cure this condition; hard friction to the skin must be avoided. It must not be confounded with *pityriasis versicolor* or *chloasma*, which is a parasitic disease. (See EPIPHYTA.)

PLACENTA is a flat, rounded body which is formed in the womb during pregnancy and which serves to connect the circulation of the mother and child, and so enable the latter to carry on the function of nutrition and respiration until birth. This structure comes away a few minutes after delivery, and it is commonly known as the after-birth. The placenta is often diseased in cases of abortion and premature birth.

PLAGUE is a form of low fever associated with swellings of the glands, carbuncles, and petechiæ, or hemorrhage in the substance of the skin. It has been known for many years under different names, as the black death, Levant plague, pestilential fever, and glandular pestilence.

History.—Before the end of the seventeenth century this disorder seems to have prevailed in many countries of Europe, and to have been endemic, occasionally bursting out over a wider area. In London, for the first seventy years of the century, not a year passed by without a few deaths being recorded as due to this disease, while it appeared in an epidemic form in 1603, 1625, 1636, and 1665. The last epidemic was so terrible in its consequence as to be known as the Great Plague, but it was followed by a marked decline, and the deaths afterward became fewer and fewer, and after 1679 none have been recorded in the metropolis. During the eighteenth century, although there was a marked diminution in the frequency and extent of the epidemic, yet there were several outburts in Europe. In Poland and in some parts of the Baltic, in 1710 ; in Provence, Marseilles, and other parts of Southern France, in 1720-71 ; at Rochefort in 1741, in Sicily in 1743, in several districts of Portugal in 1757, in Wallachia, Podolia, etc. in 1770, and at Moscow in 1771. In the present century the disease has chiefly broken out in Egypt, Syria, Asia Minor, and the coast of Barbary. The "Pali Plague" was first noticed in India, in Cutch, in the summer of 1815, after a period of great scarcity and distress. From that time until 1826, it prevailed in different places in Guzerat, spreading to Scinde in a north-westerly direction, and also toward Ahmedabad and other places in the British possessions eastward. After the beginning of 1821, there was no recurrence of the fever until 1836, when it was observed in the town of Pali, then the principal depot of traffic between the coast and the north-west provinces of India. It spread to numerous places in Marwar in that year and in 1837 and for the first half of 1838. In 1849 there was a similar fever in Grumah and Kumaon, on the southern slopes of the Himalayas, and in 1853 in Rohilcund.

Causes.—Certain conditions have always been found to favor the development and spread of this disease : Residence upon marshy alluvial soils along the Mediterranean, or near certain rivers, as the Nile, Euphrates, and Danube ; low, over-crowded, or badly-ventilated dwellings ; a warm, moist atmosphere ; decomposing animal and vegetable matter ; insufficient and unwholesome food, and physical and moral wretchedness. Those who have lived in an elevated situation have escaped the disease when it has appeared in the district. The plague, like typhus, has often followed in the wake of famine and other calamities. In the neighborhood of Bengazi, on the African coast, between Alexandria and Tripoli, an outbreak occurred in 1858 ; but for two or three years previously there had been an unusual drought, and the cattle had perished in an unusual degree. In 1857, the destitution of the Bedouin tribes became extreme, and then the pestilence commenced. The plague has generally been preceded by a great prevalence of the ordinary fevers, bowel complaints, pulmonary affection, and catarrh. In the spring of 1665, catarrhs and lung affections were very common in London, and in the middle of the summer the plague appeared. In the third week of September, 1665, no less than 8000 deaths occurred in the course of the week, although two thirds of the inhabitants had fled by that time from the city. The influence of season on the plague seems to be very marked. In England the pestilence was

most severe, in the four epidemic years above mentioned, from the middle of July to the first or second week in October. The plague at Marseilles was most fatal in the autumn months, and this was the case at Moscow in 1771, when more than 1200 deaths took place for several days out of a population of 150,-ooo. Cold weather seems to put a stop to the progress of this epidemic disorder. The terrible mortality that ensues from this pestilence renders it very important that quarantine should be strictly enforced, and that the most stringent rules should be made and carried out for the protection of other countries from its introduction by shipping and other channels of intercourse.

Symptoms.—This disease seems to vary in its character somewhat in each epidemic, and even in the same outbreak, but the older records are not sufficiently clear as to be thoroughly trustworthy, and the absence of it from England for so many years has prevented medical men in modern times from observing its course. Swelling of glands comes on in the groin, arm-pits, and neck. The carbuncles are generally on the upper or lower limbs ; less frequently on the chest, back, or cheek. They may vary in number from one or two up to a dozen ; they vary also in size and in their tendency to become gangrenous. The petechiæ are small hemorrhages into the skin, and may be found scattered all over the body. The fever symptoms are chiefly those of shivering, nausea, vomiting, lassitude, headache, and giddiness ; the countenance is heavy and stupid, and the eyes suffused and watery. There is then heat of the skin, great thirst, frequent vomiting, a coated tongue, fetid breath, a weak pulse, and great prostration. In some there is excitement and delirium, in others heaviness and stupor. The bowels are generally relaxed, and the stools dark and offensive. The urine is passed in less quantity than usual, and may even be bloody. Bleeding may also occur from the mouth, stomach, bowels, and air passages. In some the intellect is clear to the end, and in other cases convulsions and coma may come on.

Treatment.—Not much can be done when an outbreak occurs with respect to giving any medicine wih a beneficial effect. But much may be done by avoiding any over-active measures, and allowing the patient a pure air and an equable temperature. Bleeding, active purgation, and the use of mercury must be carefully avoided. The diet must be light and nourishing, and the patient's strength supported as far as possible. The treatment, in fact, is the same as that prescribed for typhus. (See TYPHUS FEVER.) The great object, however, should, be to carry out such sanitary measures as may prevent, and to avoid those unhealthy districts which favor, an outbreak. By these means it seems probable that the plague will never visit our shores.

PLASTERS are of many kinds. We have a recognized formula for at least a dozen different sorts. They consist of an adhesive mixture spread when warm on leather, linen or paper according to requirements. The common plaster known as diachylon is made from litharge or oxide of lead. The objects for which plasters are used are to promote absorption, to support a part, or to keep the two edges of a wound together. (See DIA-CHYLON.) A plaster of resin is called adhesive plaster, and another with soap is called soap plaster.

PLASTER OF PARIS is sulphate of lime, known as gypsum, and is found in large quantities in the neighborhood of Paris. It is used when mixed with water, after being reduced to a powder by that, for making casts of any object, and it is a most useful material in the hands of the anatomist, to retain casts of interesting anatomical or surgical cases for reference and for study. It has been also recommended for the purpose of acting as a splint for fractures.

PLASMA is the name given to the colorless fluid of the blood in which the corpuscles are suspended. (See BLOOD.)

PLETHORA is a fulness of blood, which may arise from many causes—excess in the amount or quality of food and drink partaken of. Persons most subject to plethora are those of a corpulent and florid nature, and such as are inactive and not willing to take out door exercise. Judicious diet, abstinence from malt liquors, wine, and spirits, except under certain restrictions, sponge baths and friction of the skin, and daily exercise, are the best remedies.

PLEURA is a serous membrane which covers each lung, and also lines the inside of the chest or thorax. It is a thin fibrous tissue, covered with a very smooth layer of epithelium which in health secretes a small quantity of serous fluid, and by moistening the opposed surfaces, causes the lungs to expand and glide over the chest-wall with the least amount of friction. Sometimes this membrane is inflamed, and the patient is then said to have *pleurisy* ; or there is an accumulation of serous fluid in the pleura, as in some cases of dropsy, from disease of the heart, or kidneys, causing *hydrothorax* ; in these cases the lung is compressed from the presence of the fluid in the pleural bag or cavity. When blood is effused into the pleura, it is called hæmothorax, and this is a very fatal disease ; it may be caused by a large vessel in the chest giving way and rupturing, or by a stab or wound from the outside through the chest-wall. Air is found in the pleura in cases of *pneumothorax*, and this condition, although of rare occurrence, may come on in the course of consumption, when a cavity in the lung becomes ruptured. *Empyema* is the name given to the disease in which pus is found in the pleural cavity.

PLEURISY is an inflammation of the pleura or serous membrane, which covers the lungs and lines the greater part of the cavity of the chest. In health this membrane is quite smooth, and lubricated by a small quantity of fluid, so that the lungs can move upon it with the least possible amount of friction. When this membrane is inflamed, however, it becomes roughened, and in most cases a large quantity of fluid is secreted, in consequence of which the lung on that side is compressed against the spine, and there is much distress in breathing, as the patient has only one lung available for the purpose of respiration. In those cases in which both sides are affected with pleurisy it follows that there is imminent danger of suffocation, as the lungs are unable to aerate the blood properly, and so, unless relief be afforded, or the inflammation subside quickly, death is very likely to ensue ; but, fortunately, double pleurisy is of very rare occurrence.

Causes.—Exposure to wet and cold is the most common cause ; but it may come on after an accident in which the ribs are broken, or in cases of stabbing in the chest or from a gun-shot wound, and other external injuries. In nearly all cases of pneumonia, or inflammation of the lung itself, there is more or less pleurisy ; but then very little fluid is effused. If the patient be in bad health previously, the effused fluid is apt to be purulent, and then the case is called empyema. (See EMPYEMA.) This is not uncommon in children after scarlet fever. In cases of pyæmia, pericarditis, phthisis, and many blood diseases, pleurisy is apt to supervene and to add to the danger.

Symptoms.—The patient first complains of a severe catching pain in the affected side, and this is made worse on taking a deep inspiration, or on coughing ; the pain is usually confined to one spot, and on listening there one may hear a rubbing sound, due to the roughened surfaces moving on each other. There is also a feeling of weakness and lassitude, the pulse quickens, the tongue is coated white ; there may be headache, thirst, and loss of appetite ; the temperature is raised, and the usual febrile symptoms appear. In a day or two the breathing becomes worse, because effusion of fluid is now going on ; the sufferer keeps to his bed, and lies on his back in a diagonal position, so as to enable the healthy lung to expand, while the affected is too sore to rest on. These symptoms go on for several days without much change being observable ; but they vary in intensity according to the amount of the effusion ; in some very bad cases there is much distress and anxiety of countenance, the respirations are quick and shallow, the face pale, while the lips are livid ; any exertion, as moving in bed or talking, increases the discomfort. In less severe cases the distress lessens as the fever abates, and the breathing becomes more regular. Then comes the time when the fluid begins to be absorbed, and when the lung commences to expand again ; but this takes up a very variable time, so that no rule can be laid down as to the duration of a pleurisy, some being of a very slight nature, while others may take weeks or even months, before they are really cured ; but long before this the severe symptoms have abated, and the chief trouble is shortness of breath on any exertion being made. In most cases the patient is liable to pain in the chest afterward, and to a recurrence of pleurisy on being overheated, or on exposure to cold and wet. If the inflammation be due to cancer of the pleura, or to an aneurism of the aorta, or to phthisis, the symptoms special to those diseases will also be present, and tend to aggravate the complaint and increase the danger. There are, doubtless, many cases in which people have dry pleurisy or inflammation without effusion, and in these the chief symptom will be pain in the side. But it must not be imagined that pain in the side always means pleurisy, as it may arise from many causes—as shingles, muscular exertion, indigestion, etc. (See PLEURODYNIA.)

Treatment.—The patient must at once be placed in bed in an atmosphere of about 60° to 65° F., and the air should be tolerably moist. This can be effected by boiling some water in a kettle on the fire, and letting the steam occasionally escape into the room. The great thing is to avoid any great variations of temperature, and especially any chills to the surface of the body. Nor should the patient be moved about from one room to another, if such movement cause any distress, nor should he be allowed to talk more than he can help. Three or four leeches applied to the spot where the pain is greatest will give much relief, and then when the bites have finished bleeding, a large hot linseed-meal poultice should be applied to the chest ; but care must be taken that the bleeding does not recommence on applying the heat, as too much blood may in that way be drawn, and tend to exhaust the patient and make him feel faint. Cotton-wool may also be applied for a similar purpose. It is best to lie on a mattress rather than on a feather bed ; as the body is then kept cooler, and it is easier to get at the patient. Light food must be given, and milk is generally borne the easiest in the early or febrile stage of the complaint ; a light pudding, eggs, beef-tea, broth, jelly, and fish may be given when the appetite returns, and the tongue begins to clean. Stimulants should be given in much moderation, as in the majority of cases they are not much needed, and if given in excess, tend to oppress the patient and hurry the breathing. Restlessness at night is a common symptom, but this must be borne as well as possible, for any anodyne, etc., only tends to increase the difficulty of breathing, and adds to the distress. Blisters should not be applied in the early stage when there is any fever, but they must be used later on, so as to hurry the absorption of the fluid ; or tincture of iodine may be painted over the affected side for the same purpose. In cases of double pleurisy, it may be required to tap the chest and let the fluid out, and in some cases bleeding from the arm may then be attended with benefit. During recovery patients should be careful about not going out too soon, and the more especially if the weather is cold, foggy, or wet. They should sit up at first in the afternoon, and may go from one warm room to another ; but they should not exert themselves too much if the breathing becomes hurried in doing so. Tonics may then be given to improve the general health, and the patient may return to his ordinary diet. All cases should be careful for some weeks to avoid exposure to bad weather, should not go out after sunset if possible, avoid getting over-heated, and always wear flannel next the skin.

PLEURODYNIA signifies pain in the sides. It is a symptom produced by several conditions. It may be caused by an attack of pleurisy or inflammation of the lining membrane of the chest-wall ; the pain is then of a shooting character, increases on taking a deep inspiration, and accompanied by a furred tongue, quick pulse, high temperature, and the usual symptoms of a febrile condition. The application of four or six leeches, followed by hot fomentations, or hot linseed-meal poultices, will relieve, if not remove, the pain in many cases. (See PLEURISY.) Pleurodynia is often met with in women who suffer from over-lactation or leucorrhœa, or who have borne children fast, or who, from any cause, are in a nervous and debilitated condition ; such women generally have the pain on the left side, or under the left breast ; they often have a headache, pain on pressing on each side of the spine, pain across the loins, and sometimes a choking feeling in the throat ; they feel weak and nervous and low-spirited. This pain is not inflammatory in its origin, and seems to depend upon an altered nutrition of the nervous centres ; its treatment will consist in rest, fresh air, removal of any mental worry, good and light diet and tonic or strengthening medicines ; often a pill of asafœtida, taken three or four times a day, will relieve the distressing nervous symptoms. Pleurodynia may come on when shingles or herpes are present. In this disease the pain generally precedes the vesicular rash; and may be persistent for some time after its disappearance. The internal administration of quinine will do good, while locally an anodyne liniment may be rubbed in night and morning. Pain in the side is also an accompaniment of a fractured rib or a blow on the side. A wide flannel bandage or strips of plaster passed half-way round the chest, so as to prevent the affected side from moving more than possible, will give relief. Pleurodynia is a symptom also in those who have a troublesome cough, and their distress is increased by the pain which each paroxysm of coughing brings on. The pain seems due to the intercostal muscles becoming tired with their undue exertions. It is similar to the stiffness which one feels after a ride on horseback when not accustomed to the exercise. The treatment must consist in relieving the cough, and applying a flat, wide flannel belt or bandage round the chest, which will support the chest-wall, and give great comfort to the patient. Sometimes pleurodynia is caused by a neuralgia of the intercostal nerves, and is worse at certain points when touched. The hypodermic injection of morphia will give relief, or the side many be rubbed with a liniment containing opium. Lastly, pleurodynia may be caused in some cases of disease of the heart and stomach. When the heart is affected, the treatment will consist in rest, in quieting the heart's action, and in applying a belladonna plaster to the side over the seat of pain. If the stomach is the seat of disease, the case has generally to be treated as one arising from indigestion. Whenever the pain is inflammatory, leeches, mustard poultices, hot fomentations, turpentine, stupes, and other counter-irritants are useful; when it is non-inflammatory, and depends upon an anæmic and weakened state of the constitution, tonics, and especially quinine, are useful. (See PAIN.)

PLEUROSTHOTONOS is a term applied to the lateral convulsions sometimes seen in cases of tetanus, when the patient throws himself from side to side.

PLUMBER'S GOUT is so called because men working with lead in any shape are more liable to attacks of gout than those in other occupations; it is not a separate disease from gout, but it signifies the mode in which the disease has its origin. (See also LEAD-POISONING.)

PLUMBISM is a condition in which the individual having been exposed to the action of lead has brought his system under the influence of that poison; plumbers are more liable to gout than painters; the latter are more liable to colic. (See LEAD-POISONING.)

PLUMMER'S PILL. (See MERCURY.)

PNEUMONIA, or INFLAMMATION OF THE LUNGS, may come on of itself, or it may follow in the course of some other disease, and the symptoms may then differ somewhat. Pneumonia is often associated with many of the fevers as typhus, typhoid, and measles, also with pyæmia and some other blood disorders, and in these cases it adds to the gravity of the complaint; but the main symptoms of pneumonia are then either masked or modified by the associated disease under which the patient is suffering. Simple uncomplicated pneumonia of one lung, or part of a lung, is not a formidable affection, and about nine-tenths recover with proper treatment.

Symptoms.—The first symptoms that appear are shivering or severe headache, pain on one side of the chest, furred tongue, and a high temperature; in the course of a day or two the skin will be hot and burning, the lips dry, the tongue covered with a white, moist fur; the patient breathes quickly, and is glad to be quiet in bed, and not be disturbed by talking; he feels a sense of pain and tightness on the affected side of the chest; he has a troublesome cough, and spits up frothy, viscid phlegm, tinged with blood; the urine is high-colored and diminished in quantity. In children the wings of the nostril are dilated at each inspiration, and they breathe very rapidly. In four or five days the symptoms are at about their height; on the seventh or eighth day, in most cases of recovery the temperature falls rather rapidly, the febrile symptoms abate, and the patient feels much better; his tongue cleans, the appetite returns, and the breathing is easier. For some time, however, he feels short of breath, and some weeks may elapse before the lung clears up and becomes sound again. In some severe cases, such a favorable termination must not be looked for; the inflammation may spread to the other lung, and cause great distress of breathing, and bring on a livid condition of the lips; there may be much delirium, and more so in those of intemperate habits, and at times the inflammation does not clear up, but passes into one of the forms of consumption.

Treatment.—The patient must at once be put to bed in a room with a temperature of from 60° to 65° F., and the air should not be too dry. Hot linseed-meal poultices or hot stupes must be applied to the chest, and changed when they become cool. If there is much pain, a few leeches to the side will give relief. For these cases the same rules as to diet, medicines, and precautions during convalescence may be laid down as are described in the article on PLEURISY. In very severe cases the treatment must vary with the special requirements of the case.

PNEUMOTHORAX is a condition in which there is air in the cavity of the pleura. In consequence of this the lung collapses, and the patient is unable to use the lung on the affected side. The most common cause of pneumothorax is when a cavity in the lung in a case of consumption bursts into the pleura;

less frequently, an abscess of the lung may bring on a similar result. In addition to the symptoms of the disease under which the patient has been previously suffering, there will be sudden pain on the affected side of the chest, great pallor of the face, much difficulty and distress in breathing, and a general collapse of the vital powers. In most cases such an accident puts an end to the life of the sufferer. Pneumothorax may also be produced by external conditions, as when a man is stabbed between the ribs, or when the chest is perforated by a pistol-shot, etc. In both cases air enters the cavity, the lung collapses, and the patient breathes with the greatest difficulty. Such cases are of a very serious nature, and generally prove fatal in the course of a day or two. Very little can be done, except keeping the patient quite quiet in bed, and covering over the wound. In such cases surgical aid must at once be sought. (See LUNGS, WOUNDS OF.)

PODOPHYLLIN is the dried underground stem of the *Podophyllum peltatum*, the American May apple, also known as Mandrake. From it is extracted a resin much more extensively used than the native plant. This resin, known as podophylline resin, or better as podophyllin, has also been called vegetable mercury, from its influence on the liver. Its powder is grayish-yellow, with a sweet odor, and a taste first sweet, then bitter. From the powder of the root podophyllin is obtained by exhaustion by means of alcohol. This strong tincture is then distilled so as to recover the spirit, and the remaining fluid added to water containing hydrochloric acid. This effectually throws down all the resin, which is then collected, washed, and dried. Besides this resin, the root contains a substance called Berberin. The resin is used as a powerful purgative, resembling the resins of jalap and scammony. It seems, however, to act more on the liver than these do ; at all events, it usually empties the gall bladder, and so makes a show of acting on the secretions of the liver. It commonly gives rise to griping, and for this reason is seldom used alone, being commonly combined with other purgatives, or such substances as belladonna. It is best, however, to give a small quantity of this remedy, say the eighth of a grain, along with another, say the compound extract of colocynth, so as to increase the efficiency of the former, or to continue its action over a longer period, for podophyllin is always long in acting. It is, too, by itself, very uncertain ; at one time causing hardly any motion of the bowels, at another an excessive flow. It is a bad remedy to take habitually, as it very soon seems to lose its influence. It is commonly used when it is desired to empty the bowels thoroughly, and so relieve the portal system, and to empty the gall bladder. It may be given

along with bitartrate of potass in dropsy ; a useful adjunct is an alkali of some kind, such as Rochelle salts.

POISONING might be defined as the untoward results of any substance given internally, or absorbed from the external surface. More briefly still, it might be defined as the evil effect resulting from the administration of a poison ; but this necessitates a definition of the word poison which it is hard to give. Almost any substance we can mention, if given in too great quantity, or if the quality has deteriorated by keeping, may produce poisonous effects ; and as these are usually unmistakable enough, it may be best to confine our attention to these. It is, however, generally agreed to limit the term poison to such substances as give rise to symptoms of illness by virtue of their own inherent properties, and not to qualities merely superadded for the time being. Thus, boiling water, if swallowed, would be quite as fatal as any poson could be; but this being no inherent quality in the water, but simply dependent on the heat added to it, we could not speak of the evil results as poisoning. Usually, too, it is customary to limit the idea of a poison to such substances as give rise to injurious effects when taken in small doses ; but there are substances which, given in considerable quantity, are exceedingly fatal, while given in small quantity they produce no symptoms of importance.

The symptoms, then, which constitute those of poisoning, i.e., the product of a substance which we agree to call poisonous, vary greatly in kind and character. Some substances act almost entirely on the parts with which they are brought in contact, and the symptoms which arise from such we designate *local ;* but others have little or no influence on the part to which they are applied ; they only act when they have passed into the blood, and reached some remote organ, usually the brain. Such action would be described as *remote.*

The *local effects* of poisons of course vary exceedingly ; some, like strong sulphuric acid and caustic potass, act on the internal organs as they would on the external—they burn or corrode them, and thus cause their destruction or death. But there are weaker substances, chemically speaking, which yet cause death, though not exactly in the same way. Thus arsenic, when introduced into the stomach, sets up much inflammation, and the result of this inflammation may be death of the part and of the individual. In both cases, death is due to injury to the stomach, but in the case of the corrosive substance, this is from chemical action ; in that of the irritant, like arsenic, it is due to the inflammation set up from irritation by the substance. There are, however, certain local effects of a totally different kind. Thus aconite produces a numbness and tingling in the parts with which it is brought in contact, while opium undoubtedly allays pain and irritation when it is applied.

The *remote effects* of poisons also differ among themselves. Thus, the effect produced by an injury to such an organ as the stomach is similar to the shock produced by a railway injury to a limb, and sometimes kills in exactly the same way ; but in the case of arsenic there is something more, for not only does arsenic set up inflammation of the stomach—vomiting, purging, and the like —it is capable of acting through the nervous system so as to produce convulsions or paralysis. Here, it might be said, that we have three different kinds of effects—first, the inflammation of the stomach and bowels, giving rise to symptoms similar to what might be produced in many other ways ; then the constitutional depression produced by these —the quick pulse, symptoms of fever, etc. ; and finally, the peculiar nervous effects already alluded to. It was for a long time disputed whether a poison acted on remote parts by means of the nerves or by means of the blood, and many experiments were performed to settle the question. This was at length fairly done, and now we hold that for a poison to exert its dangerous effects upon a remote part, it must be absorbed and carried thither. It is quite true that a poison may kill without this, as in the case of strong sulphuric acid, where the injury done to the stomach causes death ; but to enable opium, say, to produce more than a merely local effect, it must pass into the blood and reach the brain through that medium. The view that some poisons do act directly upon and through the nerves is mainly borne out by the rapidity with which some poisons, as prussic acid, prove fatal. This is so great as almost to preclude the possiblity of its reaching the brain through the blood, but the rate of circulation is great, and even in poisoning with prussic acid, there is time for the poison to reach the central nervous system. There are many things which influence the action of a poison, perhaps the most important being the quantity or dose ; for many poisons are in small quantities valuable remedies, though in larger they kill. Habit, too, has a most important influence. Thus, opium-eaters consume with impunity quantities of the drug which would kill one not habituated to its use. There is, however, a notion that certain drugs prove cumulative, as it is called—that is, when given for any length of time, they produce suddenly a poisonous effect. Digitalis is said to act in this way, but our experience does not incline us to take that view. As a rule, mineral substances cannot be taken in constantly-increasing doses. Lead, for instance, being introduced into the system, tends to accumulate

there, and ultimately to give rise to symptoms of poisoning ; arsenic is eaten in quantities by the Styrian peasantry without evil consequences ; while the stomach becomes more and more impatient of antimony as it is given, until at last, worn out by constant sickness and vomiting, the patient succumbs.

Symptoms.—To the public, however, it is of far more consequence to be able to form some idea of the symptoms of poisoning than to be acquainted with details of this kind. It may be surmised that an individual is suffering from the effects of some poisonous substance if, shortly after food or drink, he be seized with violent pain in the stomach, with vomiting and purging, especially if convulsions or paralysis is present, or if the individual suffer from great giddiness or delirium, or if there be a great tendency to sleep. It is chiefly on a combination of these three that we must rely for a certain diagnosis, and sometimes any diagnosis is impossible during life, however strong suspicion may be. Should suspicion of poisoning enter into the mind of any one, it would be his duty promptly to arm himself with further aid and support, by having recourse to the services of the best medical practitioner within reach, and, if necessary, to communicate his fears or suspicious to the gentleman so called in. This is of the utmost importance for the safety of the patient, and, of course, secures the individual from a charge of officiousness, to say nothing worse. If any one is suspected, his conduct is to be narrowly watched, for that often gives the clue desired ; very few guilty people are able to so dissemble as to give rise to no suspicion—generally their part is overdone, whether that be pretended interest or callousness. This, of course, refers to criminal poisoning, for in cases of accidental poisoning, as soon as suspicion is aroused, aid is most eagerly sought and information and assistance eagerly proffered. Accidental poisoning may occur at any time, so many poisonous substances being now employed in arts and manufactures, and a knowledge even of general principles may be of exceeding great value when life and death are hanging in the balance.

Treatment.—Taking it for granted that we have to do with a case of poisoning, we shall proceed to lay down certain rules which apply, more or less, to all cases, and which may be called into play by any one. Suppose an individual has swallowed poison, accidentally or purposely, and it becomes necessary for the bystanders to interfere for his safety, three things have to be done : 1st. *To get rid of the poison ;* 2d. *To stop its effects ; and*, 3d. *To remedy the evil it has done.* One or other must come first, but it does not greatly matter which of the first two really is first attended to : the first always comes best. When the question arises whether are we to get rid of the poison before stopping its action or after, one rule enforces itself, that is to save time. "Whatever is readiest is best," is emphatically the rule in dealing with poisoning. Better the poorest remedy given at the moment, than the very best given an hour later. A considerable number of all poisons are what might be styled self-evacuating ; having been taken, they set up vomiting and purging, and are thereby eliminated. In such cases, all that is necessary to be done is to aid the self-evacuating process, especially to aid the vomiting, and so perhaps get rid of the poison altogether. Ordinarily, two kinds of means are employed to get rid of the poisonous substance in the stomach ; these are stomach-pump and vomiting. It requires considerable skill to use the stomach-pump, and usually where one is to be obtained, there is also to be obtained the skilled aid necessary for its employment. In passing the tube down into the stomach, the grand rule is to use as little force as possible, and to make the point of the tube slide along the posterior wall of the gullet. Occasionally, grievous accidents have arisen from unskilful use of this instrument ; and so any one not acquainted with it and attempting to use it, should attend implicitly to this rule. The great advantage of the stomach-pump is that it allows you to wash the stomach out. With a properly constructed instrument, it is possible to reverse the current, and so to effectually wash out the stomach. There are, however, certain cases—as where violent corrosives have been swallowed—where the tissues are so much softened that an attempt to pass the stomach-pump would very likely end in driving it through the tissues, and so such attempts must be strictly avoided. The advantage of the stomach-pump is that it requires no action on the part of the stomach to empty that organ. In cases where the stomach is paralyzed, as it sometimes is in opium poisoning, this is of very great importance.

In cases where, from whatever cause, the stomach-pump cannot be employed, we have left to us the self-evacuation known as *vomiting*. This sometimes is one of the results of the poison itself ; in others it must be excited. If, as most irritants do, the poison have given rise to vomiting, it may only be necessary to encourage it. This is best done by tickling the fauces with a feather, and by copious draughts of lukewarm water. This process, though exhausting, must be continued until everything seems expelled from the stomach.

Sometimes, however, there is no vomiting, and then something must be given to cause the stomach to get rid of its contents. Here the same rule that the readiest is best prevails. It is useless, or worse than useless, to wait till an emetic is brought from the druggist's shop ; if that be far away, the resources of the locality must suffice. Three things may be made use of as emetics, which are to be found almost everywhere. These are *mustard, salt,* and *smelling-salts*, besides the stimulation of the fauces with the finger, and the use of lukewarm water. Smelling-salts are not suitable for all cases, but are good in a certain number of cases of poisoning, especially by vegetable substances, which give rise to narcotic symptoms. The dose of this is a teaspoonful given in a pint of lukewarm water, and followed up by copious draughts of the same. Mustard is a better emetic, and is generally to be had ; its use is limited to those cases where there is no violent irritating effect produced by the poison. Usually, it suits best where there is a sedative effect produced by the poison, and the stomach requires a stimulant to call its action into full play. The dose of mustard is a tablespoonful mixed up with a pint of lukewarm water, and followed by copious draughts of the same. Salt can always be had, and a handful of this dissolved in water will usually suffice to produce copious vomiting, and so the evacuation of the stomach contents. Ipecacuanha is a most useful emetic in cases where the stomach has been already irritated, and it is desirable to effectually get rid of any irritant substance which may remain. It is best given as ipecacuanha wine, half an ounce for a dose.

Frequently, however, the simple plan of getting rid of the poison will not suffice. Its effects have to be neutralized or remedied. That means, practically, that some antidote must be given. Now, no one antidote is suited to all emergencies—the antidote must be suited to the poison, and accordingly we must consider each poison or group of poisons separately with this view. The object of most antidotes is to render the active poison an inert substance, after which treatment may be employed to remedy the mischief already done. Most antidotes, therefore, are chemical agents which attack the poison, and render it insoluble, and so inert ; but some are of a kind whose virtues seem to be opposed to those of the poison—in short, what used to be called a counter-poison.

The ultimate end of all treatment is to keep the patient alive ; much may therefore require to be done to obviate the tendency to death which we cannot here recapitulate or even include in the treatment of poisoning, being common rules in the treatment of all diseases. To allay sickness and vomiting, if excessive ; to preserve strength ; to procure rest in one set of cases ; to keep the patient awake and from yielding to the sedative influence of the drug in another ; in all, to carry him through the period of danger, which varies in length in the case of many poisons, but which may be said to be distinctly limited, and to constitute one of the elements of safety in all.

Such are, generally speaking, the ends to be kept in view in dealing with a case of poisoning.

Kinds of Poisons.—The classification of poisons into certain groups has long been of the crudest description, and is still exceedingly imperfect. Long ago they were spoken of as mineral, vegetable, and animal. Even now some adhere to that grouping. It is, however, desirable to arrange them in some fashion, however imperfect, according to the effects they produce, and so the old crude classification into irritants, narcotics, and narcotico-irritants, is better than none. We have incidentally pointed out certain broad distinctons which enable us to give some better idea than these, however. Some poisons we pointed out, like sulphuric acid, when strong, act chemically by destroying the vitality of the parts to which they are applied. Such we may call *corrosives ;* others act as irritants, especially to the stomach, and may be called *irritants ;* but of these irritants there are at least two groups : those which irritate the stomach, but do not produce any other symptom than would an acute inflammation of that organ, however produced ; others, like arsenic, not only give rise to inflammation of the stomach and its consequences, but also produce certain specific effects characteristic of their action. In the case of arsenic these are mainly nervous ; in the case of mercury, they are salivation or sloughing about the mouth ; in the case of antimony, intense prostration of strength, and so on. These we may called *specific irritants.* After these come a great group which affect the nervous system in various ways, some producing sleep, others delirium, some calming nervous action, some exciting it, and giving rise to convulsions. All these had better be classed together, in the first instance, as *neurotics*—substances, that is, affecting the nerves. Each of these has its appropriate symptoms, and often an appropriate antidote.

The *Corrosives* are poisons which act by virtue of their chemical properties. They, when swallowed, destroy the surface and sometimes the deeper parts of all the organs with which they are brought in contact. The consequences of such injury, in short, are as violent as may follow destruction of a pair of limbs. They speedily bring about death. The chief corrosives are the three strong mineral acids *sulphuric acid, nitric acid,* and *hydrochloric acid.* The three chief alkalies act in a somewhat similar fashion ; these are caustic *potass, soda,* and *ammonia.* Moreover, these same substances, if diluted so as not to give rise to softening and perforation of the stomach, may yet be sufficiently powerful to give rise to fatal inflammation. These poisons give rise to tolerably characteristic symptoms. The symptoms begin immediately after swallowing ; the taste and feel are characteristic. Whatever they touch is altered, and they commonly occasion a vomiting of bloody matter. The remedies to be applied differ in the case of the acids and alkalies, the one being in point of fact a kind of antidote to the other. The strong mineral acids give rise to vomiting, not so often to purging, and the lips and tongue are commonly marked. First of all they are white with sulphuric and hydrochloric acid, and afterward they become black. Nitric acid always gives rise to a yellow mark.

The best remedy for these acids is some weak alkali, not caustic potass, or soda, nor even their carbonates, but some such substance as magnesia, beaten up with water or milk, and given in considerable quantity. The carbonate of magnesia is not so good as the calcined magnesia for this purpose, as it sets free a large quantity of carbonic acid, which might prove troublesome by distending the stomach.

The alkalies must be dealt with in exactly the opposite fashion—they must be neutralized by some weak acid : vinegar is perhaps the best thing to give, but any weak acid like acetic acid, citric acid, or tartaric acid may be given. Oxalic acid must not be given, being itself a deadly poison. For alkalies and acids, too, oils may be given with advantage. With neither acids nor alkalies must the stomach-pump be used. The vomiting which commonly follows the exhibition of such substances should be fostered by diluent drinks, as linseed-tea, gruel, exceedingly thin arrow-root, etc.

Sometimes these poisons attack the larynx where it joins the gullet, and may even cause suffocation. Should such a fate impend, it is quite proper to open the wind-pipe by *laryngotomy* or *tracheotomy.* Finally, we must note that these substances frequently cause death long after they have been swallowed. They destroy the tissues with which they come in contact ; the consequence is that, if the patient recovers and these sores heal in the gullet, the coats, as is usual, contract. This goes on, the gullet becoming narrow and narrower, till at last the patient may perish of actual starvation. This is a danger not to be overlooked, and so the medical attendant will do well in a case of this kind to pass a probang from time to time, to make sure that the gullet is not contracting.

Some vegetable acids must not be overlooked. Chief among these is *oxalic acid,* which is one of our most deadly poisons if given in quantity. This acid is frequently used to remove iron stains from linen, etc., or to clean brass vessels, so that accidents may result from it at any time. Vomiting commonly follows ; if not, a little warm water should be given ; but neither alkalies nor the stomach-pump should be used. The best thing to give is lime—even common plaster knocked down from the wall and ground up with milk or water suffices. For all the acids hitherto mentioned whiting in water is a capital remedy—perhaps the best.

Together, after this, we may group nearly all vegetable irritants and a good many mineral ones, including the *salts of zinc, tin, silver, chrome, and iron*. These act by giving rise to vomiting and purging, the common irritant symptoms, and the best way of dealing with them is to promote the vomiting in the first instance, and afterward giving demulcent drinks or eggs beaten up with milk. Sometimes a substance containing tannin, as oak bark, catechu, kino, etc., had better be prescribed, especially for zinc and silver ; but for the last common salt furnishes the best antidote. As regards simple vegetable irritants, including gamboge, scammony, elaterium, croton-oil, castor-oil seeds, euphorbium, etc., the grand rule is to favor vomiting till everything seems got rid of, and then to treat the case like one of inflammation of the stomach and bowels.

There are some peculiar substances to which a word more is due. *Phosphorus*, for instance, seems a most extraordinary kind of poison. It gives rise to symptoms of a most peculiar kind, especially affecting the liver. For it, unfortunately, there is no true antidote ; the great thing is to get rid of the substance, and that is best done by making use of the stomach-pump ; chalk and water and magnesia had also better be given. The vomited matters in this form of poisoning gleam in the dark.

Arsenic gives rise, as we have seen, to mixed symptoms, some dependent on irritation of the stomach, some on its peculiar influence on the nervous system. It is not possible here to lay down the marks diagnostic of arsenical poisoning, but its treatment consists in aiding the escape of the poison from the stomach by giving diluents and favoring vomiting. Raw eggs, beaten up with milk, are also useful. Animal charcoal, calcined magnesia, and a variety of other substances have also been recommended. The best antidote of all is the hydrated oxide of iron. This may be prepared by taking a druggist's stock bottle of tincture of the perchloride of iron, and adding to that liquor potassæ or caustic ammonia. The whole should then be run through a tow filter (made by sticking a morsel of tow or hemp in a funnel), and washing the filtrate. The solid part is to be used. The same may be given for poisoning by prussic acid.

Antimony is peculiar, in that it produces extreme depression. The best remedy for poisoning by most antimonial preparations is some substance which contains tannin. Black tea does so to a large extent when boiled ; in case of difficulty, therefore, a few ounces of tea should be thrown into boiling water, or better, a small quantity of boiling water added to it, the whole boiled for a few minutes, strained, reduced with cold water till fit to drink, and swallowed. Magnesia should be also given, if chloride of antimony has been the substance used.

Mercury is poisonous mainly in one form —*corrosive sublimate*. This gives rise to symptoms a good deal resembling those of ordinary corrosive poisoning ; but speedily the mouth becomes affected, and salivation or even sloughing follows. For corrosive sublimate, albumen is the best antidote. To that end white of egg should be beaten up with milk and freely administered.

Lead differs from most substances of its class, in that it gives rise to constipation rather than purging. This rule is not, however, by any means absolute. Should the bowels be confined, castor-oil must be given ; but acute poisoning from lead is, comparatively speaking, rare—chronic poisoning is that which we most frequently encounter. (See LEAD POISONING.)

Of the poisons called *neurotic, opium* occupies the chief place. The symptoms it gives rise to are totally different from those alluded to in the foregoing section. Soon after it is swallowed the patient becomes drowsy, and gradually deep sleep creeps on. till he can hardly be aroused ; if not roused in time, he sleeps the sleep of death. It is in such cases that the use of the stomach-pump and emetics is most beneficial. Without them the patient would almost certainly sink ; but when the poison is removed from the stomach, provided the individual can be kept alive for a few hours, he will recover perfectly. If any emetic is given, it should be mustard or sulphate of zinc. The patient must be kept moving about, for if he sleeps, he dies. Strong coffee and the galvanic battery are also useful adjuncts.

The treatment of opium poisoning is also the treatment of the great class of substances which it represents, but some of these admit of special treatment ; and even opium poisoning itself may sometimes be managed by the use of a special remedy—*belladonna*. In almost every instance neurotics have got a certain period within which they prove fatal ; that being passed, the patient gradually recovers. Now as regards opium, this is certainly within twenty-four hours ; if, therefore, the individual can be kept alive during that time he is safe. Very often a full dose of belladonna assists greatly in this.

There is a group of poisons often held to be allied to opium, but in reality widely different ; in point of fact, they produce delirium rather than sleep, and have hence been called deliriants. These include *hyoscyamus, belladonna, stramonium*, and *datura*. All of these substances are poisonous, though no death is recorded as the result of hyoscyamus. The treatment of these is in many respects similar to that to be adopted for opium. Emetics are to be promptly given ; if the substance has been given for some length of time, and debility has set in, so that the stomach does not readily react, then the stomach-pump should be used. The emetics, too, should be stimulant, as sulphate of zinc or copper, mustard or common salt, *never* ipecacuanha or tartar emetic. Moreover, it may be necessary to give something to stop the action of the poison ; some substance containing tannin, as tea or coffee, is best ; not prepared as for ordinary use, but boiled as hard as may be.

In some respects the actions of *aconite* and *prussic acid* are alike—both produce speedy and deadly results ; both seem specially to influence the heart. The treatment for aconite is similar in all respects to that just recommended for belladonna and its allies, but that for prussic acid is different. In point of fact, prussic acid proves so speedily fatal that there is seldom time to do more than give the patient some ammonia. Were there more time, and were any hydrated peroxide of iron, such as is used in arsenical poisoning, at hand, it ought to be given ; but such is the deadly power of the poison that there is seldom time to treat it ; not that prussic acid is so speedily fatal as is supposed, for a man has had time to run up and down two flights of stairs, and even then a considerable time did elapse before death, after a very large dose. Cold effusion has been recommended, but we fear would be like most other remedies for the poison—a remedy too late to be of any use. Iron and a free dose of ammonia should be the remedies.

There is another very fatal group of poisons, of which *nux vomica*, with its alkaloid *strychnine*, is the type. These poisons give rise to violent convulsions, similar to those produced by tetanus ; hence we conclude that this poison acts mainly, if not entirely, on the spinal cord. Here, too, we must try to get rid of the poison as speedily as possible if that be in our power ; but if the patient is already fully under the influence of the drug, any attempt to use the stomach-pump is sure to bring on a fit of spasms, which will alike prevent its passage and exhaust the patient. An emetic might be used, or, at all events, tried ; but the patient should be kept as quiet as possible. There is no antidote. Animal charcoal may be used, so may tannin, and a hundred other things ; but the only hope we can have is in enabling the patient to weather the storm by giving him, from time to time, a whiff of chloroform and ether mixed, so as to allay the spasm, to prevent suffocation in it and to avoid the danger of fatal collapse from exhaustion in the interval.

All these poisons are of a kind with which we have not much to do ; but they introduce us to a fresh group, from which not one of us is safe some time or other. *Anæsthetics* and noxious gases, at least of one kind, are closely allied. The anæsthetics in common use are chloroform and ether, or a mixture of the two. Now, no case whatever of anæsthesia from these agents is absolutely without risk, and so men ought to be chary of recommending their use, save in serious cases. When it is merely to remove a tooth or to cut off a bit of skin it is far better to suffer the pain than to risk the dangers of anæsthesia. Recently another anæsthetic, nitrous oxide, has been reintroduced. This, no doubt, is very useful for short operations—as removing teeth ; but it is unsuited for more prolonged ones. Anæsthesia depends on reducing the oxidation of the blood to a minimum, or, at all events that enters into the phenomena of its production, and so it is akin to poisoning with gases, like carbonic acid. Carbonic oxide, which is the main element in causing death from charcoal fumes, is a different kind of agent. Carbonic acid, inhaled simply seems to prevent the entire evolution of carbonic acid by the lungs ; carbonic oxide, on the other hand, seizes upon the blood corpuscles or coloring matter, fixes them, and renders them unable to take up or give off oxygen. In all cases except a few where the heart is paralyzed, as in some chloroform cases, the danger is an arrest of the respiratory process. That may depend on paralysis of the muscular power necessary to effect the movements of the chest, and to the change of the air, or on some other cause. Be that as it may, in any danger from anæsthesia, we have mainly to direct our attention to this—to see that the air passages are clear, that the tongue does not occlude them, and to persist in those movements which we know as artificial respiration (see DROWNING). Stimulants, too, should be used. If the heart has been brought to a standstill, galvanism may be used ; but the grand remedy is merely artificial respiration, which we must endeavor to effect as thoroughly as possible.

Most of the poisons here treated of are dealt with under separate headings, and fuller instructions are there given as to how we can obviate their tendency to produce death.

POLYDIPSIA is a technical term for excessive thirst ; in some cases of diabetes an immense quantity of fluid is taken in the course of twenty-four hours. This symptom, however, is occasionally observed in other cases in which there is no sugar in the urine.

POLYPUS is generally a pear-shaped tumor, attached by its thin end or stalk to some mucous membrane. Polypi are covered by mucous membrane, and within have a kind of semi-gelatinous contents. They may be detached by pulling them down with a pair of forceps, and strangling the stem with a piece of wire. They occur most frequently in the nose and in the womb, and in both situations give rise to troublesome symptoms. Removal is the best kind of treatment when it is possible. (See NOSE.)

POLYURIA is a term to signify an excessive flow of urine. (See DIABETES.)

POMADES. (See HAIR.)

POMEGRANATE-ROOT BARK is a remedy said to be of very great value against worms when it can be obtained fresh. The fruit of the pomegranate is tolerably well known. The rind of the fruit was at one time used as an astringent, and may yet be so where no better is to be had ; but as it owes its efficacy to the tannin which it contains, we in this country prefer to use that substance. There is an officinal decoction of the bark of the root ; but it is rarely, if ever employed. Indeed it is mainly of use medicinally, as already said, for its astringent properties.

POMPHOLYX. (See PEMPHIGUS.)

POPPY CAPSULES are the capsules of the opium poppy, gathered before they are quite ripe. They, therefore, contain a little opium, and the numerous seeds in their interior, called maw-seeds, contain a bland oil ; consequently a decoction of these capsules contains a doubly soothing property from the opium and from the oil. This decoction is the mode in which poppy capsules are mainly employed. A warm fomentation is prepared by boiling some of these capsules in water, and any injured part bathed with the fluid while yet warm. Sometimes a poultice is made from this fluid, and applied to bruises and other injured parts where the skin is whole. The quantity of opium contained in these capsules is very small and very variable. No preparation of the capsules ought to be used internally, though two officinal preparations of them still remain which are intended for this purpose, viz., an extract and syrup. This syrup used to be given to children, but ought to be entirely given up—it is dangerous. Far better use laudanum or morphine, and then we know what we are dealing with.

PORRIGO, or SCALD HEAD, was commonly applied in former times to any affection of the head where there were scabs, and a moist discharging surface. The term is now applied to cases in which impetigo has been irritated, and the small pustules have run together, and made an angry, red, and raw surface, which scabs over and discharges a watery fluid. It is a disease common in childhood, and often occurs on the chin, or round the corners of the mouth, or on the head. It looks like an eczema eruption. The part should be washed often with oatmeal and hot water, so as to remove the scabs ; then olive oil or zinc ointment should be freely smeared over the raw surface ; soap ought not to be used. The bowels should be opened by Gregory's powder, or a mixture of rhubarb and magnesia ; the diet should be light and nourishing, and since such children are generally pale and unhealthy, a little iron wine should be given two or three times a day. (See IMPETIGO and ECZEMA.)

PORT. (See WINES.)

PORTAL VEIN is the important vessel which, receiving the venous blood on its way from the stomach, spleen, and intestines, carries it on to the liver, to be distributed through that organ. (See LIVER.)

PORTER. (See BEER.)

POTASS is the hydrated oxide of the metal potassium, or kalium, which itself is only a chemical curiosity, but whose salts are of unspeakable value to mankind.

Liquor Potassæ, or solution of potass, is prepared from carbonate of potass, by adding to its solution quicklime ; when heated, after a time carbonate of lime forms, and the clear fluid is caustic potass in solution ; this diluted to the proper strength is liquor potassæ ; evaporated to dryness, and cast into moulds, it constitutes solid caustic potass, which is used for a variety of purposes. Liquor potassæ is colorless, very acrid, and has a soapy feel. If kept in glass bottles containing lead it attacks them, hence it is usually kept in bottles of green glass. Large doses of this substance may do much injury, and even a small dose, if concentrated, may prove fatal. The dilute solution used in medicine is a very valuable antacid, not only as neutralizing any free acid, but as tending to bring the stomach to a normal condition, being a sedative to its lining membrane. Sometimes a weak solution of caustic potass is used as a wash in certain forms of skin disease.

Caustic potass, in the form of stick, is usually met in little pieces about the size of a pencil. These should be quite white, but usually have a grayish tinge ; they speedily melt when exposed to the atmosphere, and therefore require to be kept in closely stoppered bottles. This substance speedily dissolves animal tissues, forming a kind of soap with them. Another preparation, formerly a good deal used for a similar purpose, was potassa cum calce—potass and lime : this does not melt so readily. It is sometimes used for making openings over abscesses when the patient dreads the knife, but always leaves an indelible scar. Caustic soda is now a good deal used where caustic potass used to be, seeing that it is a good deal cheaper.

Carbonate of Potash is made from pearl ashes ; these again from the ashes of wood. It occurs in small white grains, somewhat crystalline in appearance, and strongly alkaline. It attracts moisture from the atmosphere sufficient after a time to melt it; it must therefore be kept in carefully stoppered bottles. Carbonate of potash is less alkaline than caustic potass, but is too much so to be freely used internally ; in point of fact, it is chiefly used as alkaline lotions, which are applied to the skin in certain forms of disease of that part of the body, and in rheumatism and gout.

Bicarbonate of Potash is made by passing a stream of carbonic acid through a solution of the former salt. It occurs in large crystals, which do not absorb water from the atmosphere, and which have a mildly alkaline taste ; the crystals are readily soluble in water. Bicarbonate of potash is largely given internally, and may be taken in large doses, which speedily make the urine alkaline, and frequently increase that secretion. This is, perhaps, the favorite preparation of an alkali for internal use, and some use it very largely.

Alkalies have, indeed, a most extensive and most important application in medicine. Thus, applied to the stomach they induce a copious flow of the digestive fluid ; and thus, though alkaline themselves, give rise to a powerful acid secretion in abundance, thereby materially aiding digestion. The strong caustic alkalies, as already said, may be used, and often are used, for the destruction of warty growths, the hard edges of sores which will not heal, and so on. It must, however, be borne in mind that these substances readily permeate the tissues, soak into them and destroy them, sometimes to a much larger extent than is desired. The fluid, too, formed in rubbing into the skin or other parts is apt to run, and precautions must be taken to avoid that. It is, perhaps, best to use a piece of blotting paper to surround the part which we desire to destroy ; and as soon as we think the destructive action has gone far enough it is best to wash the surface with vinegar and water. It must also be borne in mind that the parts are destroyed to a much greater extent by using this caustic than would be imagined. Sometimes a large extent of surface sloughs after its being applied a little too vigorously. The carbonate of potass, in the proportion of a dram to a pint of water, is an admirable application in certain forms of skin diseases. In nettle-rash and prickly heat there is nothing nearly so good, and even in the malady called eczema there is no application so valuable. It may be tried in all cases where there is much itching, and if crusts be present, it will speedily remove them. It is in the stage where the whole skin seems to weep that this application is most beneficial in eczema ; later when there is rawness only, it ceases to be of use.

In that peculiar skin disease common in young people about puberty, where the face is covered over with little red dots with yellow tops, this plan will be found most useful. Strong yellow soap should be plentifully used, or the pustules should be touched with rather a strong solution of carbonate of potass ; at all events, the yellow tops ought to be kept from forming. Alkaline lotions, similar to those mentioned, are employed to remove the chalk stones of gout. The swellings should be kept enveloped in cotton wool or lint soaked in such a lotion, and kept moist by a covering of oiled silk.

Internally, besides being used to help digestion, alkalies may be given to neutralize acidity ; but as a rule, such treatment is a mistake. The cause of the acidity should be dealt with, not the acidity itself. They are also employed, mainly as carbonate or citrate, internally, to increase the alkalinity of the blood and urine. In the system they seem to favor the conversion and oxidation of various substances, and so seem useful in various ways ; partly by helping oxidation, partly in rendering the products of oxidation more soluble.

When there is excess of uric acid in the urine, whatever be its origin, bicarbonate of potass may be, as a rule, given with benefit ; some prefer to give the citrate of potass. In any case the object will be best attained by giving the alkali in the form of an effervescing draught, say 20 grains of the bicarbonate of potass with a little sugar, to which, when dissolved, a tablespoonful of lemon-juice may be added, and drank when effervescing. In this case the citrate is swallowed, but in the blood it is converted into a carbonate, and as such appears in the urine. The alkaline is, in the meantime, the favorite mode of dealing with rheumatic fever. There is considerable doubt as to whether this plan enables the individual to recover more promptly, or saves him from the risk of heart disease ; certainly it, as a rule, diminishes the pain and adds to his comfort ; 20 or 30 grains should be given every three hours, so as to keep the urine alkaline. As to the influence of alkalies on the urine in increasing or diminishing the amount of its products we can say little. It is usual to give acetate of potass and citrate of potass as diuretics ; but it is usual to give many things the efficacy of which rests on no solid foundation.

Acetate of Potass is prepared by adding acetic acid to bicarbonate of potass, or rather *vice versa*. It appears as beautiful white foliated satiny masses, neutral quite in reaction, and very readily absorbing water from the atmosphere. When taken internally, it is absorbed, and appears in the blood as carbonate of potass. Its action is commonly reckoned to be diuretic ; in very large doses it is slightly purgative. It is most frequently used as a diuretic, but sometimes also to render the urine alkaline, which it does, though itself neutral through being converted into a carbonate. The dose is from 20 to 60 grains.

Citrate of Potass is prepared much as the acetate by neutralizing carbonate of potass by citric acid. It is a white crystalline powder, which tends to deliquesce, and is slightly acid to the taste. Citrate of potass is pleasant to the taste, and agrees better with the stomach than most other preparations of the alkali. It is given in fevers as a cooling drink, and being, like the acetate, converted in the blood into carbonate, renders the urine alkaline. It is used, therefore, in various maladies when this is desired, especially in the form of an effervescing drink.

Tartrate of Potash, which is not often used, is made by neutralizing the acid tartrate by means of carbonate of potash. It exists as small crystals, without any distinguishing shape. In small doses it is diuretic, being, like the other vegetable salts of potash, converted into the carbonate. In large doses it is purgative, and is added to vegetable purgatives to increase their action. To this end it is usually given in doses of from two drams to half an ounce.

Acid Tartrate of Potash is better known as cream of tartar. (See CREAM OF TARTAR.)

Sulphate of Potass is a waste product in the manufacture of nitric acid. The residue is the acid sulphate, so some further steps have to be taken to render it neutral and pure. It is of no great value in medicine ; from its excessive hardness it is sometimes employed to aid in the trituration of vegetable substances. In this way it is employed in compound ipecacuanha powder and compound colocynth pill. The old compound ipecacuanha powder (Dover's powder) contained nitrate of potass (saltpetre) instead of this salt, and there is some reason to believe that it was more efficacious than the more modern preparation.

Nitrate of Potash, or saltpetre, is better perhaps known from its commercial uses than its medicinal properties ; nevertheless these are considerable. It is procured mainly from India by treating the washings of the soil with wood-ashes, after which the saltpetre is crystallized out. It occurs in crystalline masses or cakes, or in broken six-sided prisms, striated lengthwise. It is tolerably soluble in water, and has a peculiar cooling taste. It seems, when given in large doses, to act on the heart ; but in smaller doses it is of some value as a cooling remedy. Some authorities value it highly in acute rheumatism ; but its efficacy, like those of most other remedies in this malady, is more than doubtful. It has also been given in dropsies with a view to act upon the kidneys. The dose is ordinarily about 20 grains, and, if intended as a refrigerant, ought to be given while dissolving, not after being made into a solution.

Chlorate of Potass, is made by passing a stream of chlorine gas through a mixture of carbonate of potass and slaked lime. In this way chlorate of potass, chloride of calcium, and carbonate of lime are formed. The salt occurs as flat transparent crystals, has a cooling taste, and is not very soluble in water. There is an officinal lozenge made of the substance ; but if it is to be used as a lozenge— and that perhaps is the best way of using it —it is best to suck the crystals, allowing them to melt gradually in the mouth. Even this way they melt slowly. Chlorate of potass acts as a refrigerant, as does nitre, and it undoubtedly does much good used in the fashion we have just pointed out in certain forms of ulceration of the tongue and mouth, especially of a syphilitic taint, and due to mercurial impregnation. In all diseases of the throat of a malignant nature, where there is usually a tendency to the formation of a deep fur, and of this fur to decompose, a mixture of this salt with hydrochloric acid is of great value. The notions prevalent as to its action on the blood, etc., are simply absurd. Ten or twenty grains may be given internally for a dose ; but it is best given in the way we have indicated—that is, by sucking, or combined with hydrochloric acid.

Permanganate of Potass is made by heating chlorate of potass with peroxide of manganese and caustic potass. The product has subsequently to be boiled, to convert it into the purple maganate. It occurs in dark-looking needle-shaped crystals, readily soluble in water, which takes from them a magnificent purple hue. Its officinal preparation is a solution called liquor potassæ permanganatis, one of the most valuable substances employed in medicine, not internally perhaps, but by oxidizing and decomposing all the semi-putrid substances with which it comes in contact. Hence, as a lotion, it may be applied to foul ulcers, gangrenous parts, foul mouths, etc. It is, however, chiefly used as a disinfectant for the hands, etc., after touching foul sores or dead bodies before touching others. Its strength as an irritant, too, is considerable ; hence it may be used as an injection well diluted, for gleet, leucorrhœa, etc. It is not worth while giving it internally, but diluted, so as to be transparent or nearly so, and of a fine bluish-purple hue, it may be used as a wash, gargle, etc , with advantage. Ten grains to the ounce is about the proper strength.

Bichromate of Potass occurs in large red crystals of a tabular form. It is mainly employed as a dye stuff, and in the preparation of some drugs, as valerianate of soda.

Bromide of Potassium is made by adding bromine to caustic potash. It occurs as white cubical crystals, and owes its activity entirely to the bromine it contains. It is given instead of bromine, especially in epileptic and epileptiform seizures, and often with much success. The dose is 20 or 30 grains, beginning with five, and going upward. (See BROMINE.)

Iodide of Potassium, like the former, owes its efficacy to the iodine it contains, and not to the potass. It is made by mixing iodine and caustic potass. It is given where iodine should be given, being less irritating. Its dose varies from 2 to 30 grains or more, according to the purpose to be fulfilled. (See IODINE.)

Sulphuretted Potass, or liver of sulphur, is made by heating sulphur and carbonate of potass together. The salt has a strong smell of sulphuretted hydrogen. It is almost entirely used as a local remedy in skin diseases, parasitic, or otherwise. Internally it is readily absorbed, but its influence is not quite clear. Baths of it are of great use in chronic skin diseases and chronic rheumatism. There is an officinal ointment, which should be prepared just before use. If required internally, it is best administered as a natural mineral water.

Ferrocyanide of Potassium and *Ferridcyanide of Potassium* are only used as tests, or in the preparation of other remedies. By themselves they are not administered.

POTATO is the name given to the underground stem of the *Solanum tuberosum*, a plant belonging to the natural order Solanaceæ. The native country of this plant is South America. It was first grown in the British islands by Sir Walter Raleigh, in his garden at Youghal in Ireland. The part of the plant used as an article of food is the tuber or underground stem. The potato contains 75 per cent of water, and weight for weight contains less alimentary matter than most vegetable productions. It contains, however, starch, fibrine, and albumen, and mineral matters, which render it a very important article of diet. They should never be depended on alone, but as an addition to a diet with fat or flesh formers they are invaluable. During the potato famine in Ireland no substitute was found equal to them, and scurvy was the frequent result of their absence from the diet of the poor.

Potatoes are cooked in various ways, but the best methods of cooking are those where the saline matters are prevented from being lost in the medium (as water) in which they are cooked. They may be eaten raw as a salad with vinegar, and this has been found especially valuable in cases of scurvy, where uncooked vegetable food has not been procured for a length of time. The starch is often separated from the potato, and used to adulterate corn-flour, arrow-root, and other amylaceous foods. The scrapings of a potato may be used as a cold cataplasm with advantage in small burns. The potato contains a certain quantity of an alkaloid which is dissipated by heat, but it possesses no poisonous properties.

POULTICES are ordered for the purpose of soothing pain, or promoting by their warmth the formation of matter, and it is of the utmost importance that they should be well made and properly applied, and before being put on the skin they should be smeared with sweet-oil or glycerine, to prevent any particle sticking. As regards bread and linseed-meal poultices, no better authority can be quoted than Abernethy, who was singularly minute, and properly so, in his directions.

Bread and-water poultice he directs to be made as follows : " Put half a pint of hot water into a pint basin ; add to this as much of the crumb of bread as the water will cover ; then place a plate over the basin, and let it remain about ten minutes ; stir the bread about in the water, or, if necessary, chop it a little with the edge of the knife, and drain off the water by holding the knife on the top of the basin, but do not press the bread as is usually done ; then take it out lightly, spread it about one third of an inch thick on some soft linen, and lay it upon the part."

Linseed-meal poultices, says the same authority, should be made as follows : " Scald your basin, by pouring a little hot water into it, then put a small quantity of finely ground linseed-meal into the basin, pour a little hot water on it, and stir it round briskly until you have well incorporated them ; add a little more meal and a little more water, then stir it again. Do not let any lumps remain in the basin,but stir the poultice well, and do not be sparing of your trouble. If properly made, it is so well worked together that you might throw it up to the ceiling, and it would come down again without falling to pieces ; it is in fact like a pancake. What you do next is to take as much of it out of the basin as you may require, lay it on a piece of soft linen, let it be about a quarter of an inch thick, and so wide that it may cover the whole of the inflamed part."

Bran Poultices are frequently required, being useful as fomentations. A linen or flannel bag should be made of the size required, and loosely filled with bran, then boiling water should be poured upon it until it is thoroughly moist ; next it is to be wrung out in a coarse towel, and applied as directed.

Yeast poultices are made by taking one pound of flour and one ounce of yeast, boiling them together, laying on linen, and applying.

PRECIPITATE, WHITE. (See MERCURY.)

PRECORDIUM is the region of the chest which lies in front of the heart ; it corresponds to the lower sternal and left infra-mammary regions. (See CHEST.)

PREGNANCY may be reckoned to include within its meaning all the changes which take place in the ovum after its fertilization, whether these relate to the embryo or to the mother. In a treatise of this nature it is not possible to consider all the bearings of the subject, so we shall confine our observations to a few of the most important and practical. To this end we shall confine ourselves to the subject in two of its aspects only : viz., as regards the signs of pregnancy, and the diseases of pregnancy.

The signs of pregnancy are derivable from various sources, more or less accurate, some being of comparatively little value, others being absolutely certain. Probably the first thing to excite suspicion that she is pregnant on the part of a female is the cessation of the menstrual flow. The time arrives when this should make its appearance, and it fails to do so. Other circumstances may have arisen which lead her to suppose that she is pregnant, and this confirms the fact. Or yet, again, it may be that no such suspicion enters the mind, but a second period comes round, and still no sign of the ordinary flow. By this time other signs have appeared, and to one who is willing to be convinced, these will probably be quite sufficient to satisfy the mind. But this cessation by itself is far from a certain sign. We have already pointed out that such an occurrence is frequent where a female becomes the subject of ill-health, whatever the nature of it may be. We have pointed out especially that such an event is common among women who are anæmic and pallid in their complexion, so that by itself alone this indication is, comparatively speaking, worthless ; its value is, however, much greater in a strong healthy woman than in a weak and delicate one. There are, however, many instances where this indication is entirely absent. Thus, it not unfrequently happens that for a good number of years a married woman may never see the flow at all ; for no sooner does one pregnancy terminate, and the child is reared to the stage appointed by nature, than a new one begins. Moreover, in a considerable proportion of cases the flow continues for perhaps a month or two after the commencement of pregnancy.

Another indication of considerable importance is *morning sickness*, especially taken in conjunction with the foregoing. Generally this symptom sets in about the fifth or sixth week of pregnancy, so that the omission of the second menstrual period and the appearance of this sickness, may, taken together, be considered fair evidence of the existence of pregnancy. This sickness is peculiar. It usually commences immediately on getting up, may be severe for the time, but usually it does not last long ; in most cases it passes away in about half an hour, and the patient is well for the rest of the day. Occasionally it lasts longer, and may even persist for the whole period from morning to morning ; when this is the case it becomes dangerous, interfering with nutrition, for the patient can take no food, or, if it is taken, is immediately

brought up again. This sickness generally disappears about the third or fourth month, but may persist longer, while in many women it does not appear at all. This morning sickness is not, however, an invariable sign of pregnancy : notably it may be produced by irritability of the stomach, and extra indulgence in food or drink the previous evening ; it may likewise be produced by disease or misplacement of the uterus when there is no pregnancy. Its great value is as confirmatory evidence. Sometimes there is in pregnant women an extraordinary flow of saliva, but that of course, taken by itself, is worthless.

The changes which take place in the female *breast* are important. These begin about six weeks or two months after the commencement of pregnancy. The breasts feel fuller and tenser than usual, seem heavy, and sometimes throb and tingle, especially about the nipples. They increase in size and firmness, and in their interior may be felt a kind of knotted mass. This is highly important and characteristic, as this indicates that the enlargement is due to the milk-secreting apparatus, and not to the mass of fleshy tissue alone, which is common enough from various causes if a woman increases in *embonpoint*. After a time these contain milk. Another characteristic feature in the breasts is, that round about the nipples they become very dark, the darkness increasing and extending as pregnancy advances. Round about the nipples, too, appear little dark prominences about the size of millet-seed. These are characteristic ; but the breasts may enlarge from other enlargements of the uterus than that due to pregnancy, and some women, especially those of dark complexion, have naturally a dark ring round about the nipple. Milk, too, may appear in the virgin breast, so that none of these signs are by themselves conclusive ; like the others, they must be taken in conjunction. Then their value is great, especially the enlargement in thin women, and the darkening if in a fair-complexioned woman. The sign most commonly relied on is the most fallacious of all—that is, enlargement of the abdomen. The abdomen may be enlarged from a score of causes, so that except there be a strict and accurate investigation by a skilled individual of the cause of the enlargement, such an observation may be merely misleading. Once, however, the uterus begins fairly to enlarge, there is no great difficulty to the skilled practitioner, and up to that time, of course, the enlargement of the abdomen will be but slight.

Between the fourth and fifth month of pregnancy there occurs an incident which is usually convincing to the woman—that is, *quickening*. This term is given to the mother's perception of the first motions of the fœtus. It has been likened to various things—to a slight pulsation among others. Slight as the motion may be, however, it not unfrequently gives rise to faintness in the mother ; gradually these movements become stronger, until sometimes they prove exceedingly troublesome. The first perception of these movements is by the female commonly laid down as the half-term of pregnancy, but this is not strictly accurate. Not long after these arrives a period when a skilled practitioner is able to diagnose with absolute certainty the presence of a fœtus in the uterus ; this he does by means of the stethoscope. By applying this instrument over the abdomen of the female, just between the umbilicus and the nearest point of the pelvis, one may hear the beating of the heart of the fœtus. This sound, which is called the fœtal tic-tac, from the resemblance to the sound of a watch, is an absolute sign of pregnancy, nothing can produce it except the heart of a living fœtus, and so its detection implies pregnancy. There is, too, no danger of mistaking it for anything else, as the only sound likely to be confounded with it would be the sound derivable from the mother's circulation ; but this fœtal sound is double, and the heart beats just about twice as fast as does that of the mother.

All these are signs, more or less valuable, of pregnancy, and there are others which, however, we need not touch upon here. Not unfrequently, however, the pregnancy is concealed ; these signs have either not been observed or overlooked, and the first sign of pregnancy is the setting in of labor pains at the end of the ninth month. In no case can a practitioner say with absolute certainty that a female is pregnant until about the fifth month of pregnancy. He can of course pronounce it highly probable, but nothing more. Accordingly one should be careful in bringing accusations of pregnancy, which it might be difficult to prove or disprove.

The *Diseases of Pregnancy* can hardly be discussed here ; one or two more may, however, be alluded to. The salivation spoken of may be excessive, and may require remedy ; if so, the usual remedy for salivation may be given ; perhaps the best is to give the patient some pieces of alum and chlorate of potass, telling her to suck these from time to time. This often has the desired result. The sickness and vomiting of pregnancy are frequently much more serious matters ; they may, indeed, go so far as to endanger the life of the female from inanition, for sometimes it becomes impossible for her to keep anything in her stomach. Various plans have to be tried if the sickness is troublesome. The patient ought to keep the recumbent posture as much as possible ; the food should be as light and as easy of digestion as possible ; the time should carefully be watched so as to give it at any moment the sickness may go, and then a quantity should

never be given which will endanger the repetition of the vomiting. Still that may not suffice. Ice should therefore be given, and sometimes champagne will be kept down when nothing else will. All kinds of remedies have been tried, chief among them being prussic acid and oxalate of cerium, and sometimes all will prove vain. At the last it may be absolutely necessary to bring on premature labor to save the life of the mother. Ordinarily the appearance of this premature labor, or abortion as it is commonly called, requires to be carefully guarded against. Certain broad rules have been laid down already on this subject. (See ABORTION.)

Albuminuria is an accident which sometimes arises in pregnancy, and when it does may prove a most formidable complication. The exact cause of this albuminuria is not quite clear : ordinarily it is set down to pressure of the enlarged womb on the renal veins. Frequently it is not discovered until labor sets in, when the first indication of its presence may be a violent convulsion. If these convulsions set in during pregnancy before labor begins it may be necessary to empty the womb to save the mother. (See PUERPERAL CONVULSIONS.) There are many other maladies incident to the state of pregnancy, which we cannot, however, discuss. There is one simple enough apparently, which, however, frequently gives rise to a good deal of trouble—that is, constipation. Confined bowels should be carefully guarded against, and the best antidote is an occasional teaspoonful of castor-oil the first thing in the morning if sickness will permit.

Premature birth is said to take place when a child is born between the sixth and ninth month of fœtal life. It may come on of itself in some cases where there is a constitutional taint, or from fright, or injury, or habit, and then the case is like a labor, only less severe. If a premature birth is induced for the sake of killing the offspring, as in the case of an unmarried person, the offence is a criminal one, and renders the guilty parties liable to severe punishment. There are, however, some rare cases in which there is such deformity on the part of the mother that premature birth must take place to save the life of mother and child ; but this procedure is only justifiable after careful inquiry into the nature of the case, and a consultation of skilled and independent medical men.

PREPARED CHALK. (See CHALK.)

PRESBYOPIA is the name given to a defect in the eyesight, produced generally in advanced life. (See EYE and VISION.)

PRESERVED MEAT.—The practice of preserving animal food was observed by the nations of antiquity, and the feasts of the Romans were remarkable for animal products brought from all parts of the world. The arts of smoking pork and salting beef were known to our ancestors in Europe ; it is only in modern times that the idea has been conceived of bringing animal food from distant parts of the world, so as to supply the demand in the increasing populations of Europe for a large supply of animal food. We may date the attempt at preserving animal food first, for use at a future time, to Polar expeditions. At the International Exhibition in 1851 this subject attracted considerable attention, and at that time meat preserved in tin cases for use in ships and for exportation was exhibited. From that time the subject has attracted more or less attention.

The process by which the meat is prepared is a very simple one, and consists simply of exposing meat, from which the bone has been separated, in a tin case to a heat above that of boiling water. To do this, the tin with the meat is placed in a tank containing water holding in solution some salt, which will allow the water to be heated up to 250° or 260° Fahr. The tin is exposed to this temperature for some time ; the tin case is covered with a lid in which a little hole is made, and when the process is supposed to be completed the pinhole is soldered down, and the tin is air-tight.

Experience has shown that meat preserved in this way can be kept for any length of time. The effect of the cooking seems to act in one of two ways. According to one theory, the exposure of the meat to heat drives off all the free oxygen from the tissues of the meat, which is the active agent in putrefaction. According to another theory, all putrefaction is produced by living germs in the air, which by the process of steaming are destroyed.

Besides steaming in tins, many other processes have been adopted for carrying meat into Europe from America and the Antipodes. Dipping the meat in boiling fat, enveloping it in paraffine, covering it with ice, and merely salting it, have all been tried with more or less success. The " tinned " process is, however, the most successful.

The question has arisen whether the meat thus preserved retains its digestive and nutritive qualities, so as to render it a fit substitute for fresh meat. There is no doubt that the " tinned" meat, from its exposure to a high temperature, possesses qualities different from those of meat cooked at a lower temperature, but this has nothing to do with its digestible power or its nourishing properties. It does not appear from any chemical analysis or experiment on its use that the tinned differs at all from fresh meat. In prisons in England, where it has been tried on a large scale, no difference has been found in the health of prisoners after having taken it for months. In work-houses where it has been tried, the old people prefer it to the inferior fresh meat often served up to them.

The same reports come from lunatic asylums, ships, and institutions where it is employed. Every now and then a case is found in which the occlusion of air or germs has not been perfectly effected, and in which decomposition has set in, but under no other circumstances has any objection to its use been substantiated.

The tinned meat requires little or no cooking. It may be taken cold with hot potatoes or any other form of vegetable food. It may be heated and served up as a stew, but the ingenuity of an ordinary cook will suggest a hundred ways in which it may be placed upon the table.

The price of this meat will be seen to be much less than fresh meat when it is considered that it contains no bones and no water. It is calculated that one pound of this meat in the dieting a family will go as far as two pounds of fresh meat.

PROBANG is an instrument formed of a slender piece of whalebone with a piece of ivory or sponge at its extremity, for pushing bodies down the gullet or œsophagus into the stomach.

PROBE is an instrument for trying the depth and extent of wounds.

PROCIDENTIA is another term for prolapse of the womb.

PROGRESSIVE LOCOMOTOR ATAXY is a disease which is characterized mainly by the peculiar gait the patient assumes when walking, very much resembling the walk of a drunken man. The disease depends upon a grave disease of the spinal cord, by which its functions are more or less impaired, and in consequence of which the individual loses in a great measure control over his movements. In the ordinary movements that we make in locomotion there is a certain harmony of action between the muscles of our extremities. It is quite true that we can move an arm or leg of one side quite independently of the other side, but it is also true that as soon as we learn to walk we use our muscles in a certain order ; infants acquire this by experience, as every mother knows how awkward their first motions are ; animals seem to have this faculty very early, for most of them can walk the first day of their existence. Now this faculty of co-ordination—this faculty of harmonizing the movements of independent parts—is lost in cases of this disease.

Causes.—These at present are not clearly made out. It seems that exposure to cold and wet is the most frequent cause ; it seems most common in those who are engaged in draining, and in those who work for days together in water with large leathern boots on, as those who are making docks, etc. Such men get hot and perspire at their work, while they are very cold, if not wet, in their feet. It comes on, as a rule, in middle life, and seldom occurs among the young or the aged. As its name implies, it is essentially chronic in its course, and when once begun, it progresses gradually, and goes on for many years.

Symptoms.—The three most marked symptoms are so-called rheumatic pains in the limbs and chiefly in the legs, the want of harmony of movement, and more or less loss of sensibility in the lower extremities. At first the patient feels a sense of numbness and tingling in his legs, and he has "flying pains" about him; he is able to work, but does not feel so steady as before. By degrees he finds that when walking he loses partial control over his movements; he can walk several miles a day, but finds he is awkward in starting; in time his legs move incoherently, and in making a step the foot does not go directly forward, but is projected irregularly; his gait is so awkward that passers-by fancy he must have been drinking. If now he is asked to walk with his eyes shut he is much worse, and would fall without assistance; in turning round he is awkward, and also in starting; but once set going, he can walk a long distance without fatigue. Yet there is no paralysis; for if he sit down and bend his leg, he can resist well all efforts to straighten it, and this is not the case where paraplegia is present. He cannot always be certain of the nature of the ground on which he stands, nor, unless he look, can he be always certain whether he is on a wooden floor or on a stone pavement. His spirits are usually bad, and at times he bursts into tears without any apparent cause. He may have a desire to pass water frequently, but he has perfect control over his bladder and motions. In some cases there is loss of hearing or dimness of vision, in consequence of some nerve tracts in the brain becoming involved in the disease, but this is not a marked feature. The intellect is unimpaired, and he can read, eat, drink, and sleep well.

Treatment.—Very little can be done, if anything, in curing this disorder. Various medicines have been tried, but without any marked benefit; tonics, and especially iron and quinine, seem to do most good by improving the general health. The patient should be warmly clothed and live well; he need not keep in the house, but should walk out every day. No local treatment of the spinal cord does any good. As these cases generally occur among the poor, they are, in consequence, prevented from obtaining a livelihood, and, being driven to the workhouse, are not always able to live on the best of food. They should be encouraged to learn some simple occupation in which they can use their heads, and which does not involve any exercise or much manual labor.

PROGRESSIVE MUSCULAR ATROPHY involves, as its name implies, a grad-

ual wasting of the involuntary muscles. It is well known that in cases of lead-poisoning, wrist-drop is apt to occur from atrophy of some of the muscles of the fore-arm. In this affection, however, the atrophy or wasting is much more general. As a rule it begins in the arms, and is often most noticeable in the upper arm and shoulder, so that the patient is prevented from raising his hand to his head. It affects both sides, and extends pretty equally on each side.

Causes.—These, at present, are not understood; it may occur in children as well as in adults; it has been known to be hereditary in families, so that several children of the same parents have been carried off in turn. It is though a very rare disease, nor is it yet determined whether it depends on an alteration of the nervous system or of the muscular system, although most authors are in favor of the former view. It is a disease which is very chronic; it goes on gradually from worse to worse, until finally the muscles of respiration become involved, and death may ensue from suffocation.

Symptoms.—The first symptom generally noticed is a wasting of the muscles of the arm or leg, but more commonly the former; the fingers are used awkwardly in picking up anything; there is a numbness and tingling in the extremities, and occasional twitchings of the muscles of the part. In time the wasting is more marked, and the loss of power is proportionate to the amount of wasting. The patient cannot raise his arm nor flex it properly; if bent, he cannot resist any one trying to unbend it; he cannot make his hand reach his head without assistance. On examining the parts chiefly affected great wasting will be noticed, and the bones can be felt through the emaciated tissues; when the muscles of the shoulders waste the head droops forward slightly, and the patient has a high-shouldered appearance. In a similar way the legs waste, so that walking is performed with difficulty, and finally the sufferer has to keep his bed. Yet all the while the general health is not much impaired; he can eat, drink, and sleep well; the mental faculties are not affected, and his chief distress is the progressive weakness. But in time other parts get affected; as long as only the extremities are atrophied, loss of power alone ensues, and locomotion is rendered difficult; after a while the muscles of the chest will begin to waste, and the patient becomes short of breath. As this goes on the expansion of the chest is interfered with, and the sufferer is liable to bronchitis and congestion of the lungs; he has not strength to spit up the accumulated phlegm in his air-passages; exposure to cold or damp air makes him have a distressing cough and aggravates his symptoms. Hence it is always a bad thing for them to have catarrh or bronchitis, as it generally carries them off suddenly ; in very severe cases hardly any expansion of the chest-walls occurs, and death really occurs by suffocation.

Treatment.—For this disease, when once developed, but little good can be obtained by any drug. Iron, quinine, strychnine, and various tonics have been tried, but none of them seem to have any influence in checking the onward progress of the disease. Shampooing, electricity, and friction should be tried, and for a time benefit seems to result. The general health should be kept up by a nourishing diet ; the body should be kept warm. and flannel must be worn next the chest. All exposure to cold and wet should be avoided, and although out-door exercise should be taken when the weather is fine and dry, yet such people should not go out after sunset nor risk an exposure to the night air, as they might catch cold or obtain some lung complication. In the later stages the patients have to be propped up in bed, as they are too weak to support themselves ; in such cases all that can be done is to adopt any means that may please or give comfort to the patient, and so render more easy the inevitable end.

PROLAPSE OF THE WOMB is said to occur when that organ descends lower than usual ; it may come on after confinements, and is chiefly met with in those who stand a great deal, as washerwomen, etc., and in those who get up too soon after a labor. Mechanical treatment is best for this condition, and the patient should wear a pessary for the purpose. Much discomfort and distress is caused by a prolapse ; often there is difficulty in passing water and in defæcation. (See PESSARIES and UTERUS.)

PROLAPSUS is a term applied to the falling down or protrusion of any of the soft organs of the body, through their natural passages. Thus, *prolapsus ani* is the falling down of the rectum through the anus. *Prolapsus uteri* is the protrusion of the womb at the vulva. *Prolapsus irridis* is applied to the protrusion of the iris through a wound in the cornea.

PROOF SPIRIT is made by adding three pints of distilled water to five pints of rectified spirit. It contains 49 per cent of alcohol, and its specific gravity is .920. It is employed in making some of the tinctures of the Pharmacopœia.

PROPTOSIS is the name given to that peculiar condition of the eye-ball in Grave's Disease. It is often seen in a much milder degree, and may constitute a kind of deformity without at all interfering with vision. The cause seems to be a form of swelling of the cushion on which the eye rests.

PROSTRATION. (See SHOCK.)

PROUD FLESH is a term applied to the granulations of a wound when healing by

suppuration. (See INFLAMMATION and GRANULATIONS.)

PRUNES OR **DRIED PLUMS** are seldom employed by the physician, though sometimes in domestic practice. The smaller and more acid specimens ought to be used if the substance is to be used at all. These, however, though tending to relax the bowels, are of little use. It they do give rise to relaxation they generally, too, produce griping, and very likely flatulence. The common prunes may be used *ad libitum.*

PRURIGO is a form of skin disease characterized by the appearance of small clear blebs or pimples, which may alter their character so as to become scabs. These scabs are, however, more frequently produced by the scratching of the sufferers, who tear the skin with their nails till it bleeds, and so these minute crusts are formed. The itching which accompanies prurigo is almost intolerable, and is always aggravated by heat, so that the subjects of it dare hardly approach a fire or go to bed. This malady is peculiarly prevalent among old people, and one variety of it is accordingly characterized as *Prurigo senilis.* The malady is most obstinate, sometimes refusing all relief till the patients are weary of their lives.

Many assert that this malady is invariably due to those horrid insects, body-lice, and we are not prepared to deny that in a great number of cases, especially in elderly people, this is so. Neither are we, however, prepared to deny that some forms of prurigo may arise from nervous irritation, without the agency of parasites. In all cases this rule is imperative : let the under linen be well searched, especially in the morning, when these parasites are torpid. And let not the rank or position of the sufferer stand in the way of this. In these days one never knows with whom they are brushing elbows, and these vermin may and do get hold of persons in a higher sphere of life in a wonderful manner. It is, however, among the lower orders one sees prurigo and vermin most commonly associated, and in them there is but one remedy. They must be stripped, put in a warm bath, and well washed with carbolic-acid soap, much stronger than that commonly used, and their clothes must all be baked—boiling often does not suffice. The skin is best anointed with carbolic-acid ointment,or washed with a weak solution of the same. A weak solution of corrosive sublimate is frequently of great use in relieving the itching, and in destroying the cause of it. If the malady is due to other than parasitic causation other remedies must be used, such as are employed for what is techincally known as pruritus, or itching. (See PRURITUS.)

PRURITUS is the name given to the main symptom of the disease prurigo, as well as to others of skin diseases, *i.e.*, itching.

It sometimes gives rise to intolerable torments far worse than actual pain. Pruritus may affect the whole body, but much more commonly it affects certain special tracts. Among these the neighborhood of the organs of generation is a somewhat frequent site, and to this the term *pruritus pudendi* has been applied. It may be due, especially in children, to the presence of worms, and these should be carefully looked for ; but in a considerable number of instances nothing can be seen on the skin beyond the effects of scratching. For such itching various remedies may be tried, lead and opium lotion being one of the best. Lime water may be also tried, and if there is any discharge, magnesia. Prussic acid is sometimes used as a lotion, but requires great caution. In females it not unfrequently depends on uterine disease, and all remedies will prove useless until that is alleviated. Great cleanliness is, of course, essential, and stimulating food and hot fiery drinks must be avoided. In some cases the wet pack locally gives more relief than anything.

PRUSSIC ACID, also known as HYDRO-CYANIC ACID, is one of the most potent poisons known, but used aright it is also a valuable medicine. It is made by distilling yellow prussiate of potash (ferrocyanide of potassium) with sulphuric acid. The ordinary acid, which is very dilute, is colorless, and has a peculiar odor and tastes very slightly acid, the marks of its acidity passing readily away, it being very volatile. The ordinary acid contains only 2 per cent of the anhydrous acid, and that known as Scheele's only 4 ; nevertheless both are powerful poisons. There is an officinal preparation in the Pharmacopœia called Acidum Hydrocyanicum Dilutum. The acid is now a good deal used in the form of cyanide of potassium. This is largely employed by photographers, and has been the cause of several accidents, being almost as dangerous a poison as the acid itself. The anhydrous acid is probably the most intense poison known, destroying life with the greatest rapidity, not appearing to affect any one organ, but apparently arresting the functions of all. If the acid be strong, death may follow a dose in a few seconds, but under ordinary circumstances a fatal result, though speedy, does not follow with the same rapidity. The final act in destroying life seems to be paralysis of the heart.

As to any *antidote*, usually death occurs so speedily that there is time for no remedies ; sometimes ammonia is tried with a view to overcome the prostration, but as a rule with little avail. If the dose were small, and time permitted, it would be well to employ the hydrated oxide of iron, as in arsenical poisoning.

Greatly diluted prussic acid, applied to the skin, diminishes sensibility, and so if there is

much pain or itching in the part, such an application often does good. In skin diseases, where there is much itching and the skin is not broken, there can be no better application than a very weak solution of cyanide of potassium or of hydrocyanic acid. For this purpose thirty grains of cyanide of potassium may be added to a pint of water, or half a dram (fluid) of the acid may be added to six ounces of water. *This only if the skin is unbroken.* Moderate doses allay irritability of the stomach, and are frequently used in all painful affections of that organ, in ulcer, cancer, and especially neuralgia. Sometimes, too, it is employed with benefit in vomiting, but the exact cases in which it is beneficial are not quite clear.

So, too, in some chest affections, prussic acid is used with advantage. A certain number of cases of asthma seem connected with disease or irritability of the stomach, in these prussic acid may be tried with advantage. In whooping-cough, too, this remedy is often successful in allaying the violence of the paroxysm, though not in shortening the duration of the disease. Some recommend it in functional or other diseases of the heart, when palpitation is most violent.

The vapor, which consists of 10 or 15 drops of hydrocyanic acid added to an ounce or so of water at the ordinary temperature—the vapor being inhaled—is an admirable remedy for some forms of irritation of the lung, especially such as induce violent cough in consumptive individuals. Indeed, save that it is desirable to give it by the mouth when the stomach is concerned, this seems the best way of administering it. The dose by the mouth is from 2 to 5 drops freely diluted. It is apt to lose strength by keeping.

PSOAS ABSCESS is the abscess formed by the side of the spine in consequence of various diseases of the bodies of one or more lumbar vertebræ ; it gravitates along the muscles of the pelvis and points under the skin at the upper and inner parts of the thigh. A soft fluctuating swelling is produced, which increases rather rapidly in size and extends inward and downward reaching in some instances as far as the knee. This constitutes the lower portion of a large abscess extending as high as the spinal column in the loins, and which, as it passes from the abdomen into the thigh under the structure known as Poupart's ligament, is constricted to a narrow neck. This affection, like lumbar abscess, is serious in consequence of its almost invariable connection with advanced ulcerative disease of the spine. When the swelling in the thigh is large and painful, and when deep-seated fulness can be made out along the lower part of the abdomen on the corresponding side, and when one finds angular curvature and remote symptoms of disease of the spine, there can be very little doubt as to the presence of a psoas abscess. But where the abscess is small and the symptoms of suppuration extending from the thigh to the spine are not well marked, the diagnosis is not so easy, and the psoas abscess may be readily mistaken for a rupture or for an aneurism. Very little can be done for the *treatment* of psoas abscess. So long as the swelling does not cause much pain and grows slowly it should be left alone. Should, however, the abscess attain a large size, and the distended skin become red and inflamed and threaten to give way, the surgeon will find it necessary to let out the contained pus, either by repeated tapping or by making a free incision, under a veil dipped in a mixture of carbolic acid and olive oil.

PSORIASIS is a dry scaly disease of the skin ; it is chronic in its course and characterized by slightly raised red patches covered by white, shining, opaque scales ; these scales often come off in great numbers, so that on waking in a morning the patient finds his bed full of little branny particles. Sometimes the spots are circular, small and numerous, and scattered over the skin ; sometimes they are ring-shaped and the centre is healthy, while the disease spreads at the circumference ; sometimes large patches of irregular shape occur, and most often they are seen at the knees and elbows ; at other times the patches assume a figure-of-eight form. The edges are always well defined and with a tendency to be circular ; when the scales are rubbed off a dry and red surface is left. The name *lepra* was formerly given to the ring-shaped variety of psoriasis, but the term has now fallen into disuse. Psoriasis in all its forms runs a very chronic course, lasting not unfrequently for many years. When cured it is prone to come back again. Some persons have an attack of psoriasis every year ; spring and autumn are the seasons when it most frequently appears. The red patches of psoriasis are due to inflammation of the skin ; the scales are due to excessive formation of epithelium on the inflamed surface. The rash is often accompanied by much itching ; it occurs on the coarse and dry parts of the skin, and not where the sweat-glands are abundant. The disease is never communicated from one person to another, although a tendency to it is certainly hereditary ; it may come on as a consequence of syphilis. On the palms of the hands and soles of the feet it may be mistaken for eczema.

Treatment.—The treatment consists in paying attention to the state of the stomach, and regulating the diet carefully; arsenic is the best remedy, and it may be given in small doses two or three times in the day, but its action must be carefully watched. In many cases the local application of tar or pitch will suffice for a cure.

PTOSIS is the term applied to paralysis of the upper eyelids, so that it falls and covers the eye, the patient being unable to open that eye save by means of his fingers. The condition is mainly of importance as an indication of brain mischief, for this more frequently follows hemorrhage, or other damage to the cerebrum, than any other symptom. The muscle which raises the eyelids is governed by the same nerve which guides the movement of most of the muscles of the eyeball. Consequently, drooping of the upper eyelid is very often accompanied by squinting, the remaining muscles of the eyeball dragging it out of its accustomed situation.

PTYALIN is the active principle of saliva. (See Saliva.)

PTYALISM means an increased and involuntary flow of saliva. It attends the action of some medicines, especially the preparations of mercury, also iodide of potassium. (See Iodine, Mercury.)

PUERPERAL CONVULSIONS are commonly held to include the convulsions which occur both before, during, and after labor. Here, however, we must confine ourselves mainly to the two former, sometimes termed the *Eclampsia* of the pregnant and puerperal states. The convulsions generally occur quite suddenly. The spasms are violent and intensified, *i.e.*, of the kind called clonic, and they are attended by complete unconsciousness. Most frequently the whole body is affected, though sometimes only half of it is so ; and as they pass away the consciousness does not perfectly return, but stupor, more or less complete, continues. When the convulsions are partial, consciousness may not be lost. Such convulsions are most common in the later months of pregnancy, and just before labor, and occur more frequently in those in childbed for the first time. The fits usually follow each other in rapid succession, and each one lasts from half an hour to two hours or more, including the comatose period after each. By and by consciousness returns, but there is no knowledge of what has occurred in the interval. These convulsions are not unattended with danger, and as they not infrequently occur when no previous appearance of illness has threatened, they are greatly dreaded by pregnant females.

Nevertheless, they may be, so to speak, predicted, and so far avoided. Their cause is now known, at least approximately, and being known can be avoided. The convulsions are generally admitted to be due to renal mischief, most probably setting up albuminuria and urænia, and subsequently convulsions. What the kidney mischief may be is not quite clear. Sometimes beyond the albuminuria which ordinarily has lasted some time, there may have been little sign, though sometimes of course it has been known that the woman has been the subject of kidney disease before she became pregnant. Curiously enough, these last are not the most unfavorable cases, although the kidney mischief may be greatly aggravated thereby, and ultimately prove fatal. As far as the convulsions are concerned, those are most fatal where no previous mischief was known.

In patients who become the subjects of puerperal convulsions there may be signs of kidney disease beforehand, as swelling about the face, and especially below the eyes. But something more than this is required to account for the convulsions. It has been supposed by some that the pressure of the enlarged womb on the veins coming from the kidneys has been enough to give rise to the uræmia and the convulsions. Were that so, puerperal convulsions ought to be much more common than they are ; but this, too, is peculiar, that very often the removal of the fœtus is sufficient to cause their arrest, which it would not were they due to uræmia entirely. If convulsions come on during pregnancy, labor commonly begins too, and the child is expelled. If not, especially toward the end of pregnancy, it is highly desirable to evacuate the contents of the uterus, knowing that this often arrests them should the woman be attacked at a period at which the child could live. If labor has set in, sometimes the rupturing of the membranes and discharge of the waters will procure cessation of the convulsions. If they do not cease promptly, however, no time is to be lost ; the uterus must be emptied, by turning or forceps, as the case may be, and so both lives may be saved. Inhalation of choloroform, or a mixture of chloroform and ether, is strongly recommended by some.

PUERPERAL FEVER is a continued and contagious fever occurring in connection with childbirth. It comes on within a week or ten days after confinement, and must not be mistaken for *weed* or ephemera, which is a harmless kind of milk-fever. Puerperal fever is a very dangerous disorder, and it is one far easier to prevent than to cure. In some respects it is allied to erysipelas, and those who have been attending such cases have at times given puerperal fever to their patients. It is very important that women should not go into a large general hospital to be confined, for it has been shown over and over again that in that way many only go in to die ; and whereas in the surrounding districts no cases may have occurred, yet in a hospital some are sure to occur, and when once it has broken out, it is very difficult to get rid of it. It is far better for a woman to be confined at home in a dirty alley than to go into the most comfortable ward of a general hospital. Nor are special hospitals for women much better in this respect, for the mere herding of the women together when in

that state is injurious, and if an epidemic of fever happens to break out it is attended with dangerous results. No one who has been lately near a case of scarlet fever, or, in fact, any fever, or a case of erysipelas, either as doctor or nurse, should go near a woman in her confinement, and any one attending a puerperal-fever case should not, of course, go near another woman in labor. It is only by such strict rules that you can prevent the spread of this fatal disorder.

Symptoms.—There is headache, with shivering and rigors ; there is a diminution of the supply of milk, and the usual discharge lessens in quantity and even ceases. The temperature rises ; the tongue is dry and coated, and there is much thirst and prostration. No spots are, as a rule, observable on the skin, but in some cases there may be small petechiæ. The bowels are generally loose, and the urine is turbid and contains blood or albumen. The mind at first clear, soon becomes clouded, and the mother will take no notice of her child ; delirium, of a low, muttering character, comes on, and death takes place generally from exhaustion or syncope.

Treatment.—Very little can be done when once the fever is well developed. The woman will, of course, be in bed in a cool, well-ventilated room, but without draughts. The diet must be light and nourishing, and consist of milk, beef-tea, eggs, and stimulants as required for each case. The bed-hangings must be removed, and also any carpet, etc. ; these should be heated in an oven, so as to become disinfected. Carbolic acid should be used freely, or chloride of lime may be placed in saucers about the room. The patient's strength must be supported as well as possible until the crisis is passed ; the treatment will be the same as that described under typhus. (See TYPHUS FEVER.)

PUERPERAL MANIA is a form of insanity which comes on after a confinement, and may sometimes cause the patient to commit suicide. It is most common in unmarried women, and it is probably dependent on the mental distress and anxiety they undergo in concealing their state. Most cases had better be removed to an asylum, as, among the poor especially, proper nursing cannot be obtained in these cases. The mania comes on within a week or ten days after the labor ; the milk ceases, and there is generally an aversion for their offspring. They talk very wildly, and often use bad language. They may often hurt themselves unless watched. Under proper treatment they generally recover. No lowering measures must be adopted : a nourishing diet, rest, a quiet room, and moderate stimulation, are the best for this disorder. (See INSANITY.)

PULMONARY CONSUMPTION. (See CONSUMPTION.)

PULSE.—If the finger be placed upon an artery, such as that at the wrist, what is known as the pulse may be felt ; this is because the elastic artery dilates with each beat of the heart at regular intervals ; the pulse does not quite correspond to the beat of the heart in time, but occurs just after it, and the farther the artery is from the heart the longer is the interval. It follows that the pulse will be quick or slow, regular or irregular, according to the action of the heart at the time, and therefore it is useful as a guide in many diseases. Further, the wall of the artery may lose its elasticity in old age, or in some cases of fibrous or atheromatous degeneration, and then it becomes more rigid and it is harder work for the heart to send the blood through such vessels ; or, as in some cases of fever, etc., the elastic walls may lose their tone and become relaxed, so that the blood flows faster through such vessels, and the capillaries in front become fuller of blood : this change takes place chiefly through the influence of the nervous system. The average rate of a pulse in a healthy man is about 75 beats in a minute.

PUPIL is the name given to the central aperture in the iris of the eye, by which light can enter and act upon the retina. The pupil dilates in the dark, and contracts by a strong light. In man it is circular ; in the cat it is oval ; in most animals the iris is made up of muscular fibres. Belladonna dilates the pupil, opium contracts it ; in most cases of debility the pupil is large, while in nervous people it is generally small. (See EYE.)

PURGATIVES are remedies whose special function it is to cause an unloading of the alimentary canal. The group contains very various members, some gentle, some violent, in their action ; some acting mainly by increasing the motion of the bowel, some again by increasing its secretion. Some act on one part, some on another ; thus aloes seem to act almost entirely on the great gut, castor-oil little, if at all, on it. Laxatives are commonly included in the list ; but purgatives, strictly speaking, have a wider action. Rhubarb, senna, aloes, and jalap are comprehended among ordinary simple purgatives. More powerful purgatives, also called cathartics, are colocynth, scammony, castor oil, and podophyllin. Some seem to increase the liquid flow from the bowels. Most saline substances are in this group, such as tartrate and bitartrate of potass, Rochelle salts, phosphate of soda, Glauber's salts (sulphate of soda), Epsom salts (sulphate of magnesia), etc. Many of these are combined with advantage ; thus jalap and cream of tartar, scammony and the same, rhubarb and magnesia, senna and sulphate of magnesia, are all frequently given together, the one acting as an adjunct to the other's efficacy. But as many purgatives, especially the more violent, gripe

severely, or otherwise give rise to unpleasant effects, it is common to add to them some substance to prevent griping. These adjuncts are either hot substances, like ginger, red pepper ; or aromatic oils, like peppermint, etc. ; or they may be sedatives, like belladonna, hyoscyamus, etc.

A few purgatives act specially on the liver, and are called cholagogues ; chief among these are the preparations of mercury, podophyllin, and perhaps taraxacum. For particulars with regard to each, see under the appropriate heading.

PURPURA is characterized by an eruption of spots called petechiæ, or patches called ecchymoses, which are caused by hemorrhage into the skin, varying in tint from bright red to violet. The small spots are round, the larger more irregular in shape ; they do not disappear when the finger is pressed upon them. For the first few hours of their appearance the spots are of a pale pink, and slightly raised ; they then become level and deepen in color, finally becoming orange-colored and yellowish as they fade away ; while the old ones disappear, fresh ones keep coming. In some severe cases the hemorrhage takes place not only into the skin, but from the nose and alimentary canal ; blood may also appear in the fæces and urine. In mild cases there is little or no disturbance of the general health ; in severe cases there may be febrile symptoms, lassitude, and pains in the limbs. It occurs in those who live well, and it is not produced, like scurvy, from want of vegetable food ; its cause is not known. Tannin and gallic acid, iron and turpentine, have been given in this disease, and generally with good effect. Petechial spots may occur in the course of typhus fever and some other diseases, but these must not be confounded with purpura.

PURPURIC FEVER is a term used to designate those fevers in which the eruption assumes a deep purple color, and which does not disappear on pressure ; it used to be a synonym for typhus fever, but it also may be used in cases of malignant scarlet fever or small-pox when the rash is petechial ; such cases are nearly always rapidly fatal, and marked by great prostration and bleeding from the organs. (See MALIGNANT DISEASES.)

PUS is a term applied to the fluid contained in abscesses, and discharged from the surfaces of ulcers and granulating wounds. Healthy pus is of a white or pale yellow color, of creamy consistence, free from smell, and chemically neutral, being neither acid nor alkaline. Its usual specific gravity is about 1.030. If left standing for some time in a tall vessel it separates into a thin upper layer of transparent fluid and a thick and opaque yellow deposit. The fluid layer is composed of water, albumen, salts, chiefly chloride of sodium, fatty acids, and extractive matter. The deposit is almost entirely composed of *pus-corpuscles* or *globules*, which are minute spherical vesicles measuring from 1-500oth to 1-2000th of an inch in diameter. Within these globules are contained three or four small dark bodies, called nuclei, and a number of very minute granules. If a drop of acetic acid be added to a specimen of pus placed under the microscope, the granules will be seen to disappear and the nuclei to become more distinct. If water be added instead of acetic acid, the globules swell and become less opaque. These bodies are identical with bodies existing in the blood which are called white corpuscles, in contradistinction to the more numerous and darker bodies known as the red-blood corpuscles. There is also a close resemblance between pus-globules and the corpuscular bodies existing in chyle and lymph. In foul and unhealthy ulcers the discharge does not consist of inodorous pus, but of a thin fœtid and dirty fluid, called *ichor* or *ichorous pus*. When mixed with blood, pus is said to be *sanguineous pus*, or *muco-purulent* discharge; it is a mixture of healthy pus and of an increased secretion from a mucous membrane. This fluid is observed especially in diseases of the air-passages and of the lining membrane of the bladder.

PUSTULES are prominences formed on the surface of the skin in some diseases; the epithelium of the skin is raised, and beneath is some pus or matter. It is found in cases of small-pox, ecthyma, impetigo, and in some kinds of acne. When pricked, the matter exudes, and the pustule may dry up and heal, forming a scab, which falls off in a few days.

PUTRID FEVER is a term formerly used for cases of typhus fever; it is now occasionally used to designate very bad forms of scarlet or typhus fever or small-pox, when those diseases have assumed a very malignant form, and are accompanied by purple spots on the body which do not disappear on pressure.

PYÆMIA is a disease with well marked constitutional and local symptoms which is supposed to be due to the introduction into the system of pus or the constituents of pus. It is closely allied to septicæmia, puerperal fever, and erysipelas, and is often connected with inflammation of one or more veins. Some few cases of this disease have been reported in which the patients were quite free at the time of the attack from wound or sore; but usually the pyæmic symptoms follow a severe injury or a surgical operation, or occur in the course of some chronic suppurative affection. Pyæmia often results from compound fractures and operations on the bones, especially amputations, and is one if not the chief of the causes of death in the surgical wards of large city hospitals. The constitutional or general symptoms of pyæmia

resemble those of typhoid fever; the local symptoms consist in the formation of abscesses in the liver, lungs, and joints, and occasionally in hæmorrhagic and pustular affections of the skin. The following is the usual history of a case of pyæmia following a severe compound fracture: About the seventh or tenth day after the injury the patient has a severe attack of shivering, which is followed by intense fever, headache, and, perhaps, vomiting. The pulse then becomes frequent, and the bodily temperature increases from 99° to 103° or 104°. These symptoms persist, and the patient becomes restless and at times delirious. The attacks of shivering are frequently renewed, and in the intervals there is profuse perspiration. The tongue is dry and brown, and there is a peculiar sallow or tawny appearance of the skin over the whole body. The patient becomes much emaciated, very prostrate, and finally sinks from extreme exhaustion on about the fifteenth or twentieth day after the injury. The local symptoms presented in connection with this typhoid condition are painful swelling of one or more joints, pain in the region of the loin, shortness of breath, cough, and persistent expectoration, irregular patches of a bright red color scattered over the skin, especially near the joints, small pustules on the skin somewhat resembling those of small-pox. The first symptom indicates suppuration within the affected joints, and the chest symptoms, inflammation of pleuræ and lower portion of the lungs, with formation of small pulmonary abscesses. Suppuration in the liver is indicated by pain and swelling on right side of abdomen, and by jaundice.

The following are the *predisposing causes* of this disease: exhaustion from a long previous illness, as dysentery or fever, and from deprivation of food; organic disease of kidneys; profuse hemorrhage during or after an operation; unhealthy employment and residence in foul and badly ventilated quarters; chronic alcoholism, and intemperance both in eating and drinking. The most frequent predisposing causes are impure air, such as is contained in overcrowded surgical wards of a large hospital, and neglect of the patient's wounds leading to the accumulation of decomposing and putrid material about the raw surfaces. The duration of an attack of pyæmia varies much in different cases. Death may occur on the third or fourth day, or the symptoms may continue for a month or six weeks, and then terminate fatally. In acute pyæmia death most commonly takes place between the seventh and tenth days. Cases are sometimes met with in which the pyæmic symptoms are slight, and are prolonged for a period of three or four months, or even longer. To this form has been given the name of chronic pyæmia. Pyæmia is a very grave affection, and when acute and associated with frequently repeated chills and mischief in the lungs, is, in almost all, if not all cases, rapidly fatal. Indeed, recovery from any form of pyæmia is a very rare occurrence.

Treatment.—The treatment of pyæmia, like that of other acute and exhausting diseases of a typhoid character, usually consists in the free administration of alcoholic stimulants and concentrated fluid nutriment. Quinine in large doses seems in some cases to do good. Strict attention should be paid to the nursing of the patient, who is generally quite helpless. The room should be well ventilated, and freed of carpet and all unnecessary articles of furniture. Some disinfectant solution should frequently be sprinkled over the floor, and care be taken to remove and disinfect at once soiled sheets and bed-clothing, and to burn the dressings at every change.

PYELITIS is a technical term applied to a disease of the kidney, in which there is a formation of pus in that organ or in the ureter.

PYLORUS is the name given to the end of the stomach which is directed to the right side. (See STOMACH.)

PYRETHRUM. (See PELLITORY.)

PYREXIA is a technical term for fever, or the febrile condition ; it is present whenever the temperature of the body is above 99° Fahr. ; it is generally associated with headache, furred tongue, quick pulse, hot skin, and high-colored urine, with a feeling of fatigue and lassitude. Besides being met with in what are ordinarily called fevers, pyrexia is found in all cases of inflammation, in acute rheumatism, catarrh, etc.

PYROLIGNEOUS ACID. (See ACETIC ACID.)

PYROSIS, also known as water-brash, has been by some employed to signify heartburn with or without eructations of sour burning fluid. The term gastralgia is, however, commonly employed to signify heartburn pure and simple ; pyrosis, the acid eructations which commonly accompany it. As to the causation of pyrosis, see INDIGESTION. Its remedies must be considered with reference to the various sources of its production. In a goodly number of cases it is due to fermentation of the food ; if so, sulphurous acid will be found to give relief. In some cases an alkaline stimulant, like aromatic spirit of ammonia, is the best thing for momentary relief.

PYTHOGENIC FEVER is a term synonymous with typhoid fever ; it is not, however, a word in common use. (See TYPHOID FEVER.)

Q.

QUARTAN AGUE is said to occur when the fever comes on every third day, as is described in the article on INTERMITTENT FEVER.

QUASSIA is the wood of a tree growing in the West Indies, termed *Picræna excelsa*. It arrives in logs or billets, is grayish-brown externally, and light yellow internally. The wood is tough, but not very heavy, and is usually sold as chips. Sometimes drinking-vessels, carved out of the wood, are sold. These are to be filled with water at night, allowed to stand till morning, when the contents are consumed. The quassia wood is intensely bitter, and yields its bitterness very readily to water. Its preparations are an extract, a tincture, and an infusion. Of these, the infusion is mainly used, chiefly as a vehicle, for which it commends itself, being one of the very few bitters which contain no tannin, and so does not blacken with iron. Quassia is a pure bitter, but not an agreeable one. It is used sometimes in indigestion, but calumba has there mostly superseded it. In indigestion, with loss of power and irritability of the stomach, it may well be given along with either an acid or an alkali, according to the period of digestion. Sometimes, but rarely, it has been used as an antiperiodic ; in this way it seems to be devoid of efficacy. It is probably most useful combined with a preparation of iron and an acid in recovery from prostrating illness.

QUICKLIME. (See DISINFECTANTS.)

QUICKSILVER. (See MERCURY.)

QUININE is the most important constituent of cinchona bark, and has now, to a very great extent, superseded the crude substance as a remedy. Pure quinine is not employed in medicine, being quite insoluble in water ; but the sulphate takes its place. There are other alkaloids contained in these barks, especially cinchonine, and the relative proportions of these vary in certain kinds, cinchonine being more abundant than quinine. The bark known as calisaya bark is that which contains most quinine. Sulphate of quinine is snow-white and crystalline, the crystals being feathery. It possesses the curious property of fluorescence, *i. e.*, certain rays of light falling in a solution of quinine, though themselves invisible, cause the solution to yield light. The salt is neutral, and requires an acid to dissolve it in water if a solution of any strength is to be made.

The effects of quinine are manifold. Applied to the white corpuscles of the blood, and all bodies resembling them, it arrests their motion, and apparently kills them ; it also, within certain limits, arrests putrefaction even more powerfully than creasote. On the digestive tract quinine acts as do most other bitters ; it gives rise to an increased flow of mucus, and to a small extent also that of the gastric juice. Especially will it be serviceable to arrest the putrefactive changes in food which has been retained in the stomach without being digested, and so gives rise to flatulence, acidity, etc. Quinine, after being swallowed, passes into the blood, and in great measure is evacuated by the kidneys, almost unchanged.

The effect of quinine on the sense of hearing is peculiar. If taken in large doses, it speedily gives rise to noises in the head, singing in the ears, and sometimes deafness ; sight, too, may become dim, or even blindness for a time ensue ; headache is also produced, frontal in site and severe in character ; generally the pain is of a dull heavy kind, the face is flushed and hot, and the eyes suffused. These effects of large doses of quinine go by the name of cinchonism. Moreover, in large doses, quinine has the power of markedly reducing temperature ; for this reason it has been largely given in acute rheumatism, pyæmia, and some forms of fever. Sometimes quinine in these cases has been given in enormous doses, 20 grains, frequently repeated, being not uncommon. It is true that in these cases the temperature has sometimes been reduced, but the patient has died all the same. The most important use of quinine seems to be in malarious fevers, remittent or intermittent. The best plan of giving the remedy in these diseases seems to be to wait for a remission, then to give a full dose, at least 5 or 10 grains, and keep up the effect by an hourly administration of the remedy thereafter ; 2 grains will generally suffice as a dose for this purpose, but to arrest the paroxysm it is best to give a much larger quantity. Certain forms of neuralgia, of a distinctly remittent type, are best treated by quinine. A large dose should be given just before the expected attack ; 5 or 10 grains should suffice. Even ordinary neuralgias are frequently benefited by doses of quinine given during an intermission. Quinine is commonly prescribed in most forms of convalescence from acute disease. It is then ordinarily given in a dose of 1 or 2 grains dissolved in water or orange wine by a few drops of dilute sulphuric acid. In this way it is of undoubted service.

QUINSY is a common and troublesome affection, consisting of inflammation of the tonsils and adjacent parts of the fauces or back part of the mouth. It may occur at any age, but it is most common in young people ; and when once any one has been subject to it, it is very likely to recur on exposure to cold, so that some have an attack every year. Although painful at the time, no serious results may be anticipated, as it is a disease which is very amenable to treatment.

Symptoms.—The patient feels out of sorts after exposure to wet or cold ; he has a stiff and painful feeling in the throat ; the tongue becomes furred and white ; the appetite is bad ; there is often headache and pains in the limbs ; the temperature rises rapidly, and all the symptoms of a fever come on. The tonsils enlarge, so that the act of swallowing is made with difficulty, and the tonsils may be so large as almost to meet in the middle line, and quite prevent any solid food being taken ; at the same time there is swelling outside, just below the ear, which is painful on pressure. The enlarged tonsils may become full of pus, and when they burst they discharge much matter, and give at once much relief. The febrile symptoms last four or five days, and then pretty quickly subside ; in most cases the inflammation goes away without the formation of any matter ; generally, also, one side is more affected than the other.

Treatment.—The patient should at once go to bed, or at least keep in a room with a moist and warm atmosphere : any attempt to go out in the air only increases the malady, and makes the throat more sore than before. No solid food can be taken, and therefore beef-tea, hot milk, and soups must be given, and the thinner the fluid the more easily is it swallowed. Port wine is very valuable, and three or four glasses should be taken every day, and will be found to give great relief. Steam should be frequently inhaled by placing the mouth over a jug full of boiling water, but not over the mouth of a kettle, as the patient's mouth may be scalded. Gargles are of no use, as they do not go far enough back, and the effort of gargling is distressing to the patient. A hot bran or linseed-meal poultice should be placed round the throat at night, while during the day hot flannels should be worn. Sponging the outside of the throat with hot water will give great relief ; the inside of the throat may be sponged with some astringent lotion, as tannic acid or iron and glycerine, by which it may be kept constantly moist. A mixture containing chlorate of potass and bark is most useful in this affection, and it should be continued for some time until convalescence is established. Puncturing the tonsils with a small and narrow knife is very useful, even if it does not cause matter to escape. In some cases a leech or two behind or below the ear is useful, but blisters do no good. People who are liable to quinsy should be very careful to avoid, as far as possible, foggy and damp weather, as the disease is then very liable to recur. This affection might at first be mistaken for scarlet fever ; but the fever lasts for a shorter time, and there is no rash, nor is it followed by dropsy or swelling of the glands. In diphtheria there is less fever, but much more prostration, while a mem-

brane forms over the nostrils and a fatal re
sult often happens. (See SORE THROAT.)
QUOTIDIAN AGUE is said to occur

when the fever comes on every day, although
it may not come on exactly at the same hour.
(See INTERMITTENT FEVER.)

R.

RABIES. (See HYDROPHOBIA.)

RADIUS is a name given to one of the bones of the forearm.

RAILWAY INJURIES are of a somewhat peculiar nature, irrespectively of such forms of accidents as are mentioned elsewhere, such as fractures and dislocations. These injuries consist of concussions of the spine and spinal cord, and from the frequent absence of outward signs, and the obscurity of the early symptoms, are of a very insidious character, and their diagnosis is of the utmost importance to a medical man, as they so frequently are the sources of medico-legal inquiry. A well-known author, speaking of this class of injury, says : " That in no ordinary accident can the shock be so great as those that occur on railways. The rapidity of the movement, the momentum of the person injured, the suddenness of its arrest, the helplessness of the sufferers, and the natural perturbation of mind that must disturb the bravest, are all circumstances that of a necessity greatly increase the severity of the resulting injury to the nervous system, and that justly cause these cases to be considered as somewhat exceptional from ordinary accidents. This has actually led some surgeons to designate that peculiar affection of the spine that is met with in these cases as the ' RAILWAY SPINE.' Injuries of the spine and spinal cord have been already treated of generally, and it is hardly to the purpose to reconsider them specially in reference to the subject in hand, and we shall therefore pass on to such matters as relate to those cases where the fact of injury sustained on a railway has been the cause of litigation." Concussion of the spine from a direct and severe injury to the back may terminate, according to the same authority, in four ways : 1. In complete recovery, after a longer or shorter time. 2. In incomplete recovery. 3. In permanent disease of the spinal cord and its membranes. 4. In death. It is a very remarkable circumstance that, although the patient has apparently sustained in many cases a very trifling injury, the result is widely disproportionate, the reason of this being that the symptoms indicative of concussion of the spine and of the subsequent irritation and inflammation of the cord and its membranes are so slowly progressive. A patient is often quite unaware that anything serious has happened, feeling perhaps only violently jolted, and a little giddy or confused. After a while, however, when he has reached home, the effects of his apparently simple injury begin

to declare themselves. " A revulsion of feeling takes place ; he bursts into tears and becomes unusually talkative, and is excited ; he cannot sleep, or if he does, he wakes up suddenly with a vague sense of alarm. The next day he complains of feeling shaken or bruised all over, as if he had been beaten or had violently strained himself by exertion of an unusual kind. This stiff, strained feeling chiefly affects the muscles of the back and loins, sometimes extending to those of the shoulders and thighs. After a time, which varies much in different cases, from a day or two to a week or more, he finds that he is unfit for exertion and unable to attend to business." Such is generally the early history of a case of railway concussion. Sometimes the serious symptoms begin to develop immediately after the receipt of the injury, and in some cases not till long afterward, and most marked and distinct changes are visible in the countenance; the state of the memory, the thoughts become confused, all business aptitude is lost, the temper becomes irritable, the sleep disturbed, restless, and broken ; there are often loud and incessant noises in the head, the vision is frequently affected in various ways, the hearing, taste, smell, and the sense of touch become perverted ; the sense of speech is rarely affected, and usually the attitude of those afflicted is peculiar. There is a loss of freedom in the efforts of motion or movement, and the individual appears afraid to make such efforts ; the gait again is very characteristic : he walks unsteadily, and in a straddling manner ; the power of walking is very limited, and he is unable to ride ; the nervous power of the limbs will be found to , be affected ; sensation and motion, or both, may be impaired. Coldness of one of the extremities, owing to loss of nervous power and defective nutrition, is often noticed. The prognosis in these cases is very unfavorable, and patients have never been known to recover, completely and entirely, so as to be in the same state of health as before the accident.

Treatment.—With regard to the treatment of concussion of the spine brought on by such injuries, the first thing obviously is complete rest, and the patient should be compelled to lie on a prone couch, and the mind must be kept as much as possible at rest also ; ice-bags over the injured part of the spine ; internally the bichloride of mercury in quinine or bark ; nux vomica, strychnine, and iron are all of great value in certain cases. Salt-water douches to the spine, and galvanism, are recommended in some instances. The great thing to be done is to endeavor to improve the general health, and "prevent the development, if possible, of secondary diseases, such as phthisis, dependent on mal·nutrition, and a generally broken state of the health."

RAIN.—When, by the condensation of the aqueous vapor which forms the clouds, the individual vesicles unite, so as to become larger and heavier, they form regular drops, which come down as rain. The amount of rain which falls in any given place is measured by a rain-gauge or pluviometer. This consists of a cylindrical vessel, with a funnel-shaped lid, at the bottom of which is a small hole, through which the rain falls. A glass tube by the side of and in communication with the bottom of the vessel is marked with a scale, so as to show the quantity of rain which has fallen. When rain falls through moist air, the drops will, from their temperature, condense the vapor and increase in volume ; when, on the other hand, they traverse dry air, they tend to evaporate and lose in bulk. As a rule, most rain falls in hot climates, as there evaporation is most abundant. In London the rainfall is 23.5 inches in a year, at Bordeaux it is 25.8, at Madeira it is 27.7, at Havana it is 91.2, and at St. Domingo it is 107.6 inches. The quantity, however, varies with the seasons. An inch depth of rain on a square yard represents a fall of 46.74 pounds, or 4.67 gallons of water. On an acre it corresponds to 22,622 gallons, or rather more than 100 tons.

RAMOLLISSEMENT is a technical term for softening of the brain. It is described under CEREBRAL SOFTENING.

RANULA is a tumor situated below the tongue, bluish in color, translucent, and cystic in character. It sometimes attains such a size as to displace the tongue and impede its movements, causing serious inconvenience in mastication, deglutition, and articulation. It may be caused either by obstruction of a salivary duct or by the occlusion and dilatation of a mucous cyst, or dilatation of a bursa mucosa said to exist on the outer surface of the genio-hyoglossus muscle ; or it may be a new growth of itself, a myxomatous cyst. It may be healed by simple incision, or by cutting out a portion and evacuating the contents, and to prevent premature closing a strip of lint should be introduced ; or frequently the introduction of a seton suffices. In the case of cysts containing a thick putty-like material, the cyst wall must be dissected out entire. The injection of iodine is sometimes followed with good results. The contents of the cyst are gummy or albuminous in character, containing simple round mucous globules as their only structural element. Occasionally phosphatic concretions are met with.

RATTLE is a term applied to a noise in

the throat caused by the air passing through the mucus in the air-passages ; it often precedes death.

RECTIFIED SPIRIT is, in reality, alcohol with 16 per cent of water ; its specific gravity is .838 ; it burns with a blue flame, without smoke, producing carbonic acid and water. From this fluid, by removing the water by carbonate of potash, absolute alcohol may be obtained ; by adding water until the specific gravity of the fluid is .920, proof spirit is formed. (See PROOF SPIRIT.)

RECTUM is the lower portion of the large intestine, about eight inches in length, and largest in capacity just above the anus. It commences opposite the left sacro-iliac articulation, it descends obliquely toward the middle line as far as the lower end of the sacrum, then it bends downward toward the perinæum, and turning downward terminates in the anus. The chief diseases to which the rectum is liable are :

Abscess.—Some young weakly persons are subject to the formation of abscess in the areolar tissue, outside the rectum, and sometimes the collection of matter is situated at a considerable depth, as is shown by the tenseness and pain at the verge, with considerable difficulty in passing motions or making water ; the abscesses usually "point" at the margin of the anus, and should be opened at once, in order to prevent the formation of fistula. As the general health is very low in a constitution in which such collection of matter forms, the greatest attention must be paid to its improvement, and this state of things is not unfrequently associated with phthisis.

Fistula.—By the term fistula is meant a sinuous passage by the side of the rectum, the result of abscess, opening at the nates, sometimes not having any communication with the gut (*Blind External Fistula*), sometimes communicating with it, without, however, having any external orifice (*Blind Internal Fistula*). In some cases the internal and external apertures communicate, in which case the fistula is termed complete. These fistulæ are prevented from healing by the passage of feculent matters along them, and by the continuous state of motion of the part, caused by the action of the sphincter ani muscle, and, moreover, from their inability to contract. Of all the forms above mentioned, the complete is the most common and the most annoying, owing to the great pain and irritation it causes, and on account of the almost constant passing of fæcal mucus or flatus along it. The internal opening is usually about an inch and a half from the anus, and is to be felt as a small papilla, always within reach of the finger. Fistula is frequently co-existent with phthisis, probably due to tubercular inflammation of the rectum. The *treatment* of anal fistula is comparatively simple, and it consists in freely laying open the track or tracks, and dividing the sphincter ani, allowing it to heal up from below. Before the operation, the bowels must be well evacuated. In the case of either blind external or blind internal fiustulæ they must be reduced to the complete form, i.e., an external and internal opening must be made, by passing a probe throughout the tract. After the operation a good opiate is to be administered, to allay pain, and to prevent the action of the bowels. At the end of the third or fourth day a slight aperient should be given, and care taken subsequently to prevent the too early closing of the wound. The operation should not be undertaken in advanced cases of pulmonary phthisis, as the wound in all probability would not heal up, and the pulmonary symptoms will be aggravated from the fact of closing up the outlet of the purulent and other discharges. However, from the great irritation or pain with which such a condition is sometimes attended, as an obvious means of relief, the operation is justifiable.

Fissure.—Fissures are small chaps or cracks, forming, just inside the rectum in the mucous membrane, and usually situated at the posterior part. The affection is almost invariably connected with dyspepsia. The pain is most excruciating on passing a motion, or on any attempt to make an examination, the patient shrinking away, even at the mere approach of the finger. A slight operation, that of dividing the fissured track with a bistoury, is all that is necessary, and subsequent attention to diet and the general health.

Hæmorrhoids or *Piles* are fully treated of under HÆMORRHOIDS.

Prolapsus.—By prolapsus is meant the protrusion of the bowel at the anus. It may occur at any age, although *complete* prolapse is met with at the extremes of age. It is caused by a want of tone of the sphincter, constipation, or straining, ascarides, stricture, or stone in the bladder. In children, the bowel, after being replaced, should be prevented from falling again by applying a broad pad of lint, and drawing the buttocks firmly together with a broad strip of adhesive plaster, and children should be made to pass motions lying down. Bark, steel, quinine, and cod-liver oil to be given, and the parts douched after each motion with cold water. In all cases care must be taken to prevent constipation; and astringent injections, such as alum, rhatany, or perchloride of iron, with suppositories of tannic acid and cocoa butter, are very useful. Medical treatment failing, a surgical operation ought to be resorted to.

Stricture.—Stricture or contraction of the walls of the rectum is dependent upon organic or malignant disease, or in some cases there is a spasmodic stricture, which, however, is a concomitant of some pre-existing condition, such as piles or ulceration of the mucous coat. Stricture is rarely beyond the reach of the

finger, and the leading symptom of its existence is the fact of the motions, if solid, passing as pellets or in small pieces, and if fluid, being ejected as by a squirt. The treatment consists in keeping the bowels gently open, and in the careful use of a bougie as directed by a medical attendant. In severe cases operation is necessary. The stricture produced by malignant disease (scirrhous cancer, generally), is of course only to be treated by palliation. (See CANCER.)

Irritable Rectum. Pruritus.—Itching of the anus is a very common and troublesome affection, generally connected with some irregular state of the lining mucous membrane of the rectum, as for instance ascarides, but very often its cause may be urethral stricture (see STRICTURE), stone in the bladder, and in elderly men enlargement of the prostate or the existence of piles, and it is occasionally due to pediculi. The treatment most successful is the internal administration of tar in form of pill, arsenic, or confection of pepper; locally, tobacco-water enema, a mild mercurial ointment such as calomel and glycerine; cold bathing and plenty of exercise, and avoidance of spirits, coffee, and highly seasoned food.

Bleeding from Rectum. (See HÆMORRHAGE.)

Foreign Bodies in Rectum. (See FOREIGN BODIES.)

RECURRENT TUMORS is a term used when a tumor returns after removal. (See TUMORS.)

RED GUM, or STROPHULUS, is a simple form of skin eruption which occurs in infants. It may come very soon after birth, and generally within the first year of life; it occurs in those which have a delicate skin, and are of a scrofulous habit. This eruption is much allied to eczema, nor is it uncommon to find red gum on the body and eczema on the head. The disorder is a very harmless one, and may be brought out by any local irritation, as worms, improper diet, teething, etc. The rash is best marked on the back, where it appears in profusion as a number of minute red papules, attended by a trifling itching; sometimes the rash may come on all over the body. It is a disease of no practical importance, and may be treated by giving a very simple diet, consisting of milk or milk and lime-water, without thickening the food at all. A little rhubarb and magnesia, so as to act as a gentle purgative, should be given. Bathing the skin with tepid water is very useful, and if persisted in, zinc ointment should be applied to the affected part.

RED POPPY, *Papaver Rhœas*, though used in medicine, can hardly be spoken of as a substance of medicinal importance. The petals, which are well known, are of a bright red color, becoming dull red on drying. These petals are gradually added to boiling water, and the whole allowed to macerate for about twelve hours. After this the liquid is strained off and mixed with sugar, the whole forming 'a preparation known as syrup of *red* poppies. This preparation must not be confounded with the syrup of poppies, which is much more active. The syrup of red poppies is only used for its bright color, which makes the medicines with which it is mixed look agreeable, if they do not taste so.

RED PRECIPITATE. (See MERCURY.)

REFRIGERANTS are what we commonly call cooling medicines. They include saline and acid substances, some of them powerful, some of them weak. Refrigerants are at least of two kinds ; some have actually the power to diminish temperature, some only seem to allay thirst. Those which can diminish temperature are saline substances, given while dissolving. Some of these, as is well known, possess a power of diminishing temperature sufficient to freeze water. Others again, like acid fruits, seem only to possess refrigerant powers by allaying thirst, for a dry parched mouth is one of the most prominent indications of fever, and this being relieved there is often a belief that the bodily temperature is actually lessened. Of the salts which possess the power of diminishing temperature when dissolving, saltpetre may be taken as a type, though there are many others which are never given internally. The juice of grapes, oranges, and lemons, with the acids they yield—viz., tartaric and citric—are also useful, mainly, as already pointed out, in relieving thirst and moistening the parched mouth.

RELAPSING FEVER, or FAMINE FEVER, is a contagious disease which is chiefly met with in the form of an epidemic in periods of scarcity and famine. It is characterized by a very sudden attack of shivering or rigors ; a quick, full pulse ; white, moist tongue, afterward becoming yellow or brown ; pain at the pit of the stomach and vomiting ; an enlarged liver and spleen ; constipation of the bowels ; hot and dry skin ; no marked rash on the skin ; high-colored urine and pain in the limbs, with severe headache ; restlessness and often delirium ; then comes an abrupt cessation of the symptoms, on or about the seventh day, generally accompanied by copious perspiration. The febrile symptoms are then absent completely for a few days, the tongue becomes clean, the appetite returns, and the patient can often get up and walk about. Then comes the relapse, on or about the fourteenth day from the commencement of the fever, running a similar course to the first attack, but shorter in duration, and generally terminating about the third or fourth day of the relapse ; recovery generally ensues then, but there may be a second, third, or even a fourth or fifth relapse in some cases.

Causes.—Age seems to have very little influence in this disorder ; it attacks alike young and old, while typhus affects older people. *Sex* has very little influence in predisposing to this disease ; of 4917 collected cases, 2341 were males, while 2376 were females. The season of the year exerts very little influence on it ; typhoid fever is most prevalent in the autumn, and typhus fever during and toward the end of winter. No occupation seems to have a predisposing influence, but it is always most common among those who are idle and have a precarious living. Overcrowding and destitution favor the propagation of relapsing fever ; those who are exposed to cold and wet, to intemperance, mental and bodily fatigue, depression of spirits, etc., are thereby rendered more liable than others to catch the fever. Relapsing fever is certainly contagious : actual contact is not necessary, for the poison may be conveyed through the air from one person to another. Starvation and destitution are the two conditions which most of all tend to produce the disease ; from the records of the London Fever Hospital since 1847, 430 or 97.5 per cent of the patients admitted were paid for by the parish authorities, and were totally destitute : all were very poor, and not a single patient had been a servant in a private family. A large proportion of them had been literally starving for some time previous. The disease can hardly be mistaken for any other ; it is chiefly known by its sudden onset ; the severity of its symptoms during the first week ; the absence of a rash and the sudden cessation of all symptoms, followed in a few days by a second or even third relapse. Although most cases recover with only one or two relapses, yet they are left for some time in a very weak condition, and are liable to bronchitis and a peculiar form of ophthalmia. The mortality from this fever is but slight, not•being more than 1 in 21 cases. Most of the fatal cases die suddenly from syncope, or failure of the heart's action, or from suppression of the urine followed by coma.

Treatment.—The best treatment to be adopted is that which consists in placing the patient in a large well-ventilated room ; in promoting cleanliness, and in giving milk and other nourishing diet, such as has been described in the article on FEVERS. At the commencement of the attack the bowels should be opened, if required, by a purgative ; the skin should be frequently sponged with tepid water, but only one part of the body should be exposed at a time so as not to cause a chill to the surface. The vital powers must be kept up by milk, beef-tea, egg and milk, etc. Stimulants are not often wanted, but must be given with care when there is much prostration and failure of the heart's action. If there is any suppression of the urine, the bowels must be freely opened, the patient put into a hot-bath, and dry cupping applied to the loins. During convalescence tonics must be given, and the mineral acids with quinine are the best for this purpose. For some time after the fever has ceased the patient requires a liberal and generous diet before he will sufficiently regain his strength.

REMITTENT FEVER, or *Bilious Fever*, is a malarious fever, characterized by irregular repeated exacerbations, the remissions being less distinct in proportion to the intensity of the fever. It is accompanied by functional disturbance of the liver and frequently by yellowness of skin. There is a slight cold stage, which does not recur with every exacerbation ; an intense hot stage, with violent headache and gastric irritation ; and a slight sweating stage, which may be wanting. This disease is also known as jungle fever, bilious fever, bilious remittent, endemic fever, marsh remittent, and as gastric malarious remittent. In the article on Intermittent Fever the effect of malarious influences on the human system was described, and it was mentioned that whereas in temperate climates like our own intermittent fever of a mild character was met with, yet that the nearer we approached the tropics the fever assumed a greater severity, while in the tropics themselves the intermissions were less distinct, and the cases ran into the remittent form of fever ; so that there seems to be an intimate connection between these two forms of disease. Remittent fever is the gravest form of fever which arises from miasmatic or malarious influences, and it is most prevalent and fatal when high temperature and malaria act together. In intermittent fevers there is an absence of fever between the attacks, but in remittent fever, while the urgent symptoms abate in intensity, there is not any absence of fever, and the more severe the case the less marked are the remissions.

Symptoms.—The earliest symptom that comes on is oppression and discomfort at the pit of the stomach. The cold stage is not so well marked nor so long as in ague ; sometimes no rigors or shivering can be noticed, but merely a slight feeling of chilliness, alternating with flushes of heat. Although the patient feels cold, his temperature is at this time higher than usual, and may be above 100° Fahr., rising, as the hot stage advances, up to 106° F. or 107° F. When the hot stage comes on there is generally vomiting, and this continues throughout, causing much distress ; the tongue becomes furred and dry, and there is still a feeling of fulness and oppression in the pit of the stomach, the pulse rising up to 100 or even 120. The countenance is flushed, the eyes suffused, and the patient complains of violent headache and pains in the limbs and bones ; there is much restlessness and prostration. After these symptoms have lasted from six to twelve

hours they begin to abate ; a slight moisture may be perceived on the brow and neck, and then over the body ; the pulse and temperature go down ; the headache lessens ; the vomiting ceases, and some sleep may be obtained. This period of remission is not always observable in very severe cases. After an interval of from two to eight or ten hours the fever returns, and all the distressing symptoms are gone through again, to be followed soon by a second remission. The vomited matters at first consist of any food that may be in the stomach, and then a large quantity of watery fluid is brought up ; then bile becomes mixed with the vomited matters, giving them a greenish-yellow color, which may go on to be brown or even black from the presence of blood. The headache, at first throbbing, soon becomes a constant pain, with a feeling of tension across the forehead. Delirium of a violent form is rare, but there is often confusion of thought, and in some fatal cases a low, muttering delirium precedes the state of coma which ends in death. Sometimes petechial spots appear on the skin, and blood is found also in the vomit and in the stools. A slight degree of jaundice is common in many cases. The spleen is not often found enlarged.

The *duration* of the fever may be from five to fourteen days, but the length of time is much influenced by treatment. The fever may terminate in recovery, or may pass into an intermittent form, or death may ensue. Death rarely, if ever, occurs before the eighth day, and the patient sinks partly from exhaustion and partly from the action of the poison on the system. Sometimes an attack may come on in a few hours after exposure to the malarious influence, but generally it takes a week or ten days before the disease develops itself. For the causes of this fever further information may be obtained by reading the article on Intermittent Fever. Few could make any mistake in recognizing the difference between an intermittent and a remittent fever. Most cases of remittent fever recover, if properly treated ; the case is more favorable as the remissions become more marked, but less favorable when there is great prostration and exhaustion.

Treatment.—Place the patient in bed, and secure careful ventilation of the room. Iced water, soda-water, or lemonade may be given to check the thirst and vomiting ; an ice-bag may be applied to the head, and great relief may be afforded by packing the patient in a wet sheet, changing it when it becomes too warm, but in some cases it does not suit, and then bathing the surface with tepid water will do good. On the first sign of remission from 15 to 20 grains of quinine should be given, and this may be done though the tongue be foul and the headache continue. A free action of the bowels may be gained by giving a purgative at the onset, or by an injection. In the second remission another large dose of quinine should be given, and so on until cinchonism is produced. Mercury ought never to be given, and its use may prove very serious to the patient. Turpentine stupes and hot fomentations will relieve the pain at the pit of the stomach. During convalescence quinine and other tonics should be given, until the patient's health is re-established ; change of climate is then very desirable. (See INTERMITTENT FEVER.)

RENAL DISEASE. (See KIDNEY.)

RESIN. (See ROSIN.)

RESPIRATION is the process by which the air enters and emerges from the lungs, and in doing so causes the aeration of the blood, and converts the black venous blood into the scarlet arterial blood. (See LUNGS.)

RESPIRATOR is an instrument worn over the mouth by those who wish to avoid exposure to the night air in cases of consumption, winter-cough, etc. In this way warmer air is conveyed into the lungs, and this prevents any irritation of the windpipe and prevents a cough. There are several patent respirators, but a handkerchief held to the mouth will serve all practical purposes.

RETENTION OF URINE occurs in hysteria, and in some forms of paralysis, in some cases of drunkenness, after a confinement, etc. It is readily known by the patient not passing any water. The simple treatment required is to pass a catheter and draw the water off.

RETINA is a delicate membrane within the eye, chiefly made up of nerve fibres, which receive the impression of light from the external world. (See VISION and EYE.) It is liable to be affected in Bright's disease, in affections of the brain, and from external injury, when there is more or less loss of sight.

RETRO-PHARYNGEAL ABSCESS is said to occur when the abscess forms at the back part of the pharynx.

RHATANY, technically known as Krameria, is the root of a plant (*Krameria triandra*) growing in Chili. As imported there is a short root-stock, whence spring long red rootlets, which, were it not for the color, would look something like a rat's tail—whence, it is said, the name. The powdered root is also red. This has no smell, but has a sweetish taste, afterward very astringent. This is due to the quantity of tannin which it contains, and to which it mainly owes its properties. Its preparations are an extract, a tincture, and an infusion. These may be given whenever it is necessary to use tannin, but in practice rarely are so. Sometimes they are given for diarrhœa or dysentery, but rarely. The main use of the root is as an ingredient in tooth-powders. For this it is very useful; the powder being astringent, acts well on

spongy gums, and at the same time, mixed with chalk and a little alkali, promotes the removal of tartar from the teeth.

RHEUMATIC PARALYSIS. (See PARALYSIS.)

RHEUMATISM is one of those enigmatical disorders about which we know much and know little. We know what conditions give rise to it, or are most likely to do so, and we also know too well what its consequences may be ; but as to the immediate causation of the symptoms we know nothing. Besides, under the same title there are undoubtedly grouped diseases in many respects differing the one from the other ; all are known as rheumatism, yet among them are some of the most serious maladies to which flesh is heir, and yet again others of the slightest. The type of what we call rheumatism is *acute rheumatism*, also known as *rheumatic fever*. This malady is characterized by a high temperature, profuse sour sweats, and swelling and reddening of some of the larger joints of the body, most frequently the knee and ankle joints. These swollen joints are intensely painful, but as a rule the mischief passes away of its own accord. The great risk in a case of rheumatic fever is the danger of heart complication. Often in the course of the disease the pericardium, or the lining membrane of the heart itself, the endocardium, becomes inflamed. The pericarditis may not give rise to any very dangerous permanent mischief, but the inflammation of the other does. The portion of the endocardium most frequently attacked is that covering the valves, so that these become inflamed and thickened ; by and by, as time wears on, they contract, and so are rendered incompetent to efficiently fulfil their function of flood-gates, whence arise in time all the ills due to heart disease. Most cases of heart disease do, in point of fact, date their onset from a rheumatic attack. *Rheumatic fever* most frequently arises from cold and damp, especially if the individual has suffered from any cause of depression, as fatigue, improper food, or the like. It begins with restlessness and fever, with white or creamy tongue, and deranged bowels, constipated or relaxed. Presently the joints begin to ache, the pain increases till there is great swelling and tenderness all over one or more of the large joints of the body ; the hip joint is not, however, very often affected. There is by this time in most cases a very high temperature, 102° or 103° F., but it gradually increases, and in many cases becomes excessive ; this, indeed, constitutes one of the great dangers of the disease. Excessive bodily heat is apt to develop itself, and when the temperature rises above 105° F. there is always more or less danger to the patient, and every degree of increase adds to these in far more than geometrical progression, for by the time it reaches 108° recovery is as nearly as possible hopeless, and at 109° may be said to be quite so. In those cases where a high temperature develops itself, the sweat, which is ordinarily very profuse and of a strong acid odor, disappears, and its reappearance may be said to be the first sign of real amendment. The pain and tenderness in the joints, too, are very great. The patient can hardly bear the weight of the bed-clothes, still less can he bear the swollen limbs touched ; he himself dares not move, and even dreads the movements of others. The pulse is quick and of fair volume, and except the heart be affected it is regular. The thirst is extreme, while the tongue is coated with a thick white fur, which speedily renders the scanty saliva acid. The urine is high colored and full of urates— that is to say, it deposits on cooling a thick brick-dust-like sediment, which is re-dissolved on heating. It is difficult, too, to say when the patient has fairly seen the worst, for joint after joint may be affected, and even when the patient seems fairly on the road to recovery he may suffer a relapse. But the great feature of the disease is the tendency to implicate the heart. Curiously enough, the right or venous side of the heart is never affected, only the left, or arterial, and the pericardium, which is supplied with arterial blood. Sometimes the heart affection precedes the joint mischief, but this is not the rule. This process of heart implication has been described as *metastasis*. Now metastasis implies a change of site, but there is no change of site in rheumatism. It is true the heart becomes affected, but the joints do not improve on that account. The heart attack is as much a portion of the history of the disease as are the swollen joints. Another curious complication is chorea. These irregular and uncontrollable movements make their appearance generally with the heart complication, or shortly after. They ordinarily persist long after the rheumatic affection has ceased, and may even become permanent : they are most frequent in young people. As in the rheumatic affection of the heart, it is common to find fungi on the valves, deposited either by the blood, or formed on the valves by inflammatory changes—usually the former —these may be broken up by the blood current, and conveyed to some remote situation, as the brain, there to set up fresh mischief.

Now as to the *cause* of rheumatic fever. We shall not long dwell on such an unprofitable subject. It was long ago suggested that it was due to the presence of lactic acid in the blood, and that theory has been more or less upheld to the present time. It is worthless to speculate on the truth of this or no ; some, certainly, of the experiments carried out to support this view have been of the crudest possible description, altogether unworthy of attention. The excessive acidity of all the secretions has perhaps aided the lactic acid

view, and has given rise to a *mode of treatment* which is, with due modifications, perhaps the best—that is, the treatment by alkalis. To this end the bicarbonate of potass is given, either by itself, or effervescing with citric acid, in good large doses—30 grains or so, every four hours, continuing it till the pain begins to abate, and the urine is rendered alkaline. This plan is a good one undoubtedly if employed not rashly but with due vigilance, but there are doubts that in certain cases it has been overmuch employed, and has done harm. This, on the whole, is the plan we recommend. The joints ought at the same time to be wrapped in cotton-wool, put in some cases great good is derived by applying warm alkaline lotions next the skin, and placing cotton-wool over that. Then, too, the perspiration must be provided for—something must receive it, and if linen is next the skin this soon becomes cold and unpleasant with the patient's profuse sweating. Woollen cloths should therefore be placed next the skin, but should not be allowed to remain too long, or they act as a kind of poultice, giving rise to what are called *sudamina* on the surface ; hence there used to be a saying that the cure for rheumatism was six weeks in blankets. A favorite plan of treatment, began of late years, has been blistering. Large cantharides blisters have been placed round the limbs close to the affected joints, and allowed to remain there some hours ; after these have been removed large poultices are to be applied to favor the flow of fluid. Better, however, it would be to apply the blistering fluid with the poultices over that from the beginning, especially as this process is far less painful than the other. Dr. Herbert Davies, who introduced this plan, gave no medicine, but allowed the blisters to suffice for everything. There can, however, we think, be no doubt that the addition of effervescing alkaline draughts is an improvement ; this plan is mainly to be commended for the relief it gives to the pain in the joints. Many men like to give large doses of quinine. Now it is well known that quinine does materially reduce temperature, and it will do so in rheumatism as in other maladies, but the disease is not thereby remedied ; it pursues its course as before, and when the real time comes to use quinine with advantage, it has lost, by frequent repetition in large doses, its special virtues. By all means give quinine, but not till the pain has passed away, and the temperature is nearly natural. In a disease like rheumatic fever, where pain is one of the most prominent symptoms, it may readily be supposed that opium has been employed ; nevertheless, for some reason or another, it is not usually had recourse to, there being an idea that its exhibition, though useful for the time being, would tend to prolong rather than shorten the fever. This belief had its origin in the notion that the disease was due to some *materis morbi* which had to be eliminated, a view which is most probably not the true one. At all events, the current of opinion seems to set against the use of opium, and in favor of other sedatives. Nitrate of potass has been used and commended by some authorities ; it has been given both internally and externally—given, *ad libitum*, dissolved in water, internally to slake thirst, externally to cool the joints. Lemon-juice is another remedy which has not, like the last, received universal acceptation ; it may, however, be useful on account of the citric acid and potass which it contains. The diet during the brunt of the malady should be light ; beef tea and the like. By and by, when the patient becomes stronger, fish may be given, but meat must be reserved until a later period ; too early use of it may bring on a fresh attack. The patient should have plenty of drink supplied to him by the nurse, he himself not being allowed to move. Lemonade, made of lemons and water with sugar, is best, or soda or potass water may be given ; wine must, as a rule, be forbidden ; so must, above all things, beer. In convalescence, quinine or bark and ammonia, afterward iron and cod liver oil are to be prescribed.

Chronic Rheumatism is quite a different affection from rheumatic fever, since acute rheumatism may pass into the chronic stage, but most frequently the one is quite independent of the other. Most old people, especially if they have led a life of exposure and fatigue, are more or less affected with rheumatism, sometimes so far as to completely cripple them. The constitutional disturbance is slight, but the pain is sometimes great, both night and day, so as to wear out the patient by continual harassing. Some cases of this form of rheumatism do not suit well with heat, the pain being worst at night in bed ; but most of these cases are better for heat, and friction especially. If the patient has any syphilitic taint, as is sometimes the case in the worst instances, the pain at night may be terribly harassing. Some special forms of chronic rheumatism have acquired distinctive names. Thus there is a rheumatic affection of the loins we call *lumbago*. This is almost always aggravated by movements. Stiff neck, such as occurs after exposure to a cold draught of air, is another sample of chronic or sub-acute rheumatism, attacking a special part. *Myalgia* is the name given to pain in the muscles not due to rheumatism, and to be distinguished therefrom. Myalgia is not unfrequently found localized in the side, and there a kind of rheumatic pains are also ordinarily located. It is of importance to distinguish the one from the other, as the treatment of the two is different. Chronic rheumatism must not be treated by rote ; each case must be dealt with on its own merits.

We must try to improve the general health, and to procure the patient sound rest by sedatives if necessary. Multitudes of remedies have been tried, and some do better than others if the cures are properly selected. In most of them, iodide of potassium with bark and cod-liver oil are the most important remedies. If there is a gouty taint colchicum may be given, and some strongly recommend *Actæa racemosa ;* in most instances bicarbonate of potass is useless. Here local treatment is of great value, especially the local application of hot mineral waters, as douches and otherwise. Hot alkaline waters are, as a rule, preferable. Iodine paint to the affected joints sometimes does good, but not so much as hot alkaline lotions. Sulphur does good to many. All should wear flannel, and be careful in their diet. Beer, porter, and full-bodied wines should be prohibited.

Rheumatic Gout, as the malady is commonly called, is a most anomalous disease. It certainly has nothing to do with gout and nothing to do with rheumatism. The affection consists in an inflammation of the joints, chronic in character, and in some respects resembling gout, in others, rheumatism, but in all essential respects totally distinct. The affection is a most troublesome one, and not unfrequently cripples the patient, while defying the practitioner's art. It is said that more women suffer from this affection than men, but that is doubtful. There is often no constitutional predisposition, hereditary or otherwise, except some causes of general weakness. Rheumatic gout—or rheumatoid arthritis as it is also called—seizes on various joints, sometimes the large, sometimes the small ; but the hip, shoulder, knees, elbows, wrists, ankles, and hands are its favorite sites. Often it occurs in females at the turn of life, or about puberty. When acute, the disease comes on abruptly, something like rheumatic fever ; but more frequently it steals on gradually, the bowels being out of sorts, and the urine loaded and scanty ; the joints become stiff and painful, and are more or less swollen. This gives rise to lameness, and the joints may be felt to crackle as they move, something like a door on ungreased hinges. As the disease advances the stiffness grows greater and greater, while round about the joint are formed great masses of imperfect bone, while the cartilages covering the joints become absorbed : the joints are thus greatly deformed. The functions of the constitution are badly performed ; there is constant indigestion, the rest is disturbed, and the patient becomes painfully sensitive to the weather. No heart complication ever results from rheumatic gout. The *treatment* is very unsatisfactory. The best thing, we believe, short of a visit to a foreign spa, is rest and cod-liver oil. The bowels must be kept open by saline aperients, and sulphur waters generally suit well. Arsenical baths have been very highly commended, as have a hundred other substances. Iodide of potassium internally, and alkaline lotions applied so as to act like poultices externally, sometimes do good, but must not be continued too long for fear of impairing the constitution. So, too, any fixed appliance to the limb or joint may bring on fixation of the joint, an irreparable mischief.

RHEUMATOID ARTHRITIS. (See RHEUMATISM.)

RHINOPLASTIC OPERATION is a name given to an operation whereby a piece of skin can be taken from a healthy portion of the body and adjusted to a surface where the skin has been destroyed by injury or disease. By this method surgeons have been enabled to form a new nose by removing a flap of skin from the healthy forehead. In cases of severe burn on the arm, skin has been taken from the surface of the abdomen, and much deformity of the burned surface has been saved. Great skill is, however, required to bring about a successful result.

RHINOSCOPE is an instrument consisting of a small oval or circular mirror fixed on a handle about six inches in length ; when introduced into the mouth it should be passed into the back part of the fauces, with the mirror looking upward at an angle of about 45 degrees. The observer, sitting in front of the patient, wears on his forehead a large circular mirror, which reflects a strong light into the mouth from an adjacent lamp, in the same way as is described in the article on the Laryngoscope. In this way the back part of the nostrils may be carefully examined.

RHUBARB, as employed in medicine, consists of the roots of several plants, the species of which is not very accurately known, growing in Central Asia. The root, as imported, is always more or less perfectly deprived of its bark. The rhubarb makes its way from this district in two directions—•ne by way of Russia, the other by way of China. The Russian, commonly called Turkey rhubarb, occurs in irregular-shaped pieces, the rind of which has been removed. Its surface is smooth and yellow, its texture compact, its fracture uneven and gritty, marbled red and grey. The powder is bright yellow, and its smell rather pleasant, but the taste is bitter and disagreeable. These pieces usually have a hole drilled in them for slinging on a cord. The East Indian variety has the bark imperfectly removed, so that the surface is rounded instead of being angular. It is red and veined externally, not covered with a yellow powder like the Russian. The root is altogether more woody than the Russian variety, and the powder redder. Other specimens are in use, but are of inferior quality. Some are brought from India, and some are cultivated here. The peculiar purgative principle has never been separated from rhubarb, but it

contains a beautiful crystalline substance, chrysophanic acid. This occurs in needles of a golden lustre, not very readily soluble in water, but readily in alkalies. Oxalate of lime is found abundantly in some specimens, and gives the gritty character in Russian rhubarb. The best specimens of rhubarb are the grittiest, and so the quantity of oxalate of a lime, though in itself having nothing to do with its purgative action, have come to be a kind of test of the quality of the rhubarb.

The preparations of rhubarb are an extract, infusion, syrup, tincture, and wine, with a compound pill, and a compound powder. The compound pill, which is the pill in most general use as a laxative, contains rhubarb, aloes, myrrh, hard soap, and oil of peppermint. It is a most useful preparation. The compound powder, better known as Gregory's powder, consists of rhubarb, magnesia, and ginger. It is a great and deserved favorite in the nursery, and is used with benefit by older individuals.

Rhubarb, when taken into the mouth, turns the saliva yellow ; and the urine it turns reddish yellow ; if alkaline, purple red. In small doses it acts as a kind of tonic to the stomach and bowels ; in larger doses it is purgative, but it is apt to be followed by constipation, seeing that it exercises a kind of astringent effect, subsequent to its purgative action. On this account rhubarb, especially in the form of wine or tincture, is often prescribed in the early stage of diarrhœa, in order that it may carry off any irritant substances giving rise to the diarrhœa, and subsequently arrest the too violent action of the irritated bowel. Combined with alkalies, rhubarb is of infinite value in many forms of indigestion, depending especially on an irritable condition of the bowel. Habitual flatulence and distension are thus in many cases relieved. Females who are most subject to this, perhaps from sedentary habits, further reap the benefit of its action in habitually constipated bowels.

To children, rhubarb is commonly given as Gregory's powder, or along with a little gray powder. In either form it is exceedingly beneficial, as it also is along with a little bicarbonate of soda. Children are apt to eat things which disagree with them, giving rise to diarrhœa and the like. In such instances Gregory's powder is an invaluable remedy. (See GREGORY'S POWDER.)

A very good remedy for certain forms of indigestion, especially among females, is rhubarb and ginger tea. The two together are infused, and a wineglassful of the infusion taken every morning. Another very good plan is to chew the two together, swallowing the saliva. Ten grains suffice for this. This plan is highly recommended for piles. The purgative dose of rhubarb is 20 or 30 grains.

RHUS TOXICODENDRON, *Sumach,* is

not officinal, but is used a good deal in medicine, especially by homœopaths. The leaves of the plant, also known as the poison sumach, are employed. These contain a peculiar resin very acrid in character, insomuch that the juice blisters the part to which it is applied. If given in any quantity internally, it creates much irritation of stomach. It is said to act on the spinal cord something like strychnine, and has accordingly been given in paralysis. The results are, however, not quite positive. The powdered leaves may be given in one-grain doses.

RIBS are the bones which help to form the chest ; they are twelve in number on each side ; behind, they join the spine, while in front they are continued by cartilage to the breast-bone or sternum. The ribs are very liable to fracture from external injury. (See FRACTURES.)

RICE, a plant belonging to the natural order of grasses. The common rice of domestic use is *Oryza sativa*, and is a native of the East Indies. It is the principal article of diet of the Hindoos, Chinese, and other oriental nations. It is cultivated in other parts of the world, and is produced in abundance in the marshy grounds of North and South Carolina. Although this grain is largely consumed by the inhabitants of the world, it contains less flesh-forming matter than any other. It can only be the substantive article of diet of an indolent and feeble people. When eaten it should only be as an adjunct to other kinds of food. It cannot be healthily used as a substitute for potatoes or other fresh vegetables for any length of time. In cases of relaxation of the bowels in children or grown people it is a useful article of food. The Chinese and Mongolians distill a strong fiery spirit from rice, called "arrack." Some people mix the spirit called *toddy*, obtained from the cocoanut-tree, with the *arrack*, and thus increase its potency and improve its flavor.

RICKETS is the name given to a constitutional disease characterized by an unhealthy state of the system, which precedes for several weeks or months a peculiar disease of the bones, and of some other organs of the body ; there is curvature of the bones of the arms and legs and enlargement of their extremities.

Causes.—Some maintain that this disease is hereditary, while others are opposed to that view. There seem to be, however, predisposing causes either on the part of the parents or nurse, which have an unfavorable effect on the healthy development of the child. The state of the mother's health seems to affect the child more than that of the father ; there is no evidence to show that syphilis has any influence in producing a rickety child. Where the mother is pale and anæmic, and where without any actual disease there is a

state of general debility, and in cases where a family has been brought forth in rapid succession, the children are liable to become rickety. Deficient or improper diet, impure air constantly breathed, want of cleanliness and sunlight, cold, moisture, and deficient clothing seem to be the conditions which are generally found to precede the development of rickets ; of these causes, improper food is by far the most common. The children of the poor are fed badly, in many cases, from their birth. Brown sugar and butter, castor-oil and gruel are given them in the first few days of their existence, when their stomachs are far too tender to put up with such noxious things ; when awake, they are kept constantly at the mother's breast, and no time is allowed for the stomach to digest the food. Too often the food is thickened with corn flour, sweetened with coarse sugar and mixed with bread or biscuits, within a month after birth, while if constipation follow this mode of feeding, the unfortunate child is dosed with castor-oil. When it gets a little older and has existed four or five months, it is fed with the same food that the parents have : herrings, a piece of fried bacon, cheese, potatoes, cakes, and even beer are allowed to be swallowed by a child, when the delicate stomach should have nothing else but its mother's milk. So true is it that women know how to bear children, but not how to rear them. Nor is the child weaned when ten months old, but the mother goes on suckling it with the idea that she may retard the next pregnancy. And the succeeding offspring are brought up in a similar way when the mother's health has been impaired, and they suffer therefore to a greater degree. For the proper feeding of an infant, see DIET.

Symptoms.—No child is born rickety ; the impairment of the general health usually begins between the fourth and twelfth month. Most commonly the rickety condition is not noticed until the child begins to walk, or is affected by his first teething. At first the most ordinary symptoms are those which indicate irritation of the intestinal canal ; there may be diarrhœa alternately with constipation, enlargement of the abdomen, and more or less emaciation. The child is dull and languid, peevish and fretful ; the appetite is bad, and the sleep disturbed at night. If it tries to walk, it is "taken off its legs ;" it is thirsty and will drink plenty of water ; it has pain in the bones ; a pale face and flabby skin ; the hair on the head is thin, and blue veins marble the surface by their prominence ; the fontanelle remains open. In the next stage there are three symptoms to be chiefly noticed. I. A profuse perspiration of the head, neck, and upper part of chest. This sweating is worse at night ; beads of sweat may be noticed on the head, while the lower part of the body is dry and hot. 2.

There is a desire to kick the clothes off on the part of the child, as if with a wish to be cool ; so that the little patient lies with its naked legs on the counterpane. 3. There is general tenderness, so that the child cries when it is moved about. The urine is thick and deposits a pale sediment of phosphates on cooling. The next set of phenomena are those connected with the deformity of the skeleton. With the increasing paleness and flabbiness of the skin, the wrists and ankles enlarge, and the ends of the ribs are knuckled. The long bones of the extremities, and chiefly those of the legs, begin to yield, not being strong enough to bear the weight of the child. The deformity is very great in some cases, and such children are called knock-kneed or bow-legged. The spine is curved forward ; the head falls backward and the face looks upward ; lateral curvature of the spine is not so common, and with this curvature there is generally the deformity known as pigeon-breast. The back is flattened, there is a hollow under the armpit, the ribs are pressed in, and the breast-bone or sternum is more prominent than usual. At each inspiration the softened ribs are sucked in, and the space for the lungs and heart is much encroached upon. In this way also the bones of the arms become distorted, and the space so if the child tries to support itself by its arms and hands. The forehead is square and projecting. The head is generally unusually large, and the top flattened. The process of teething is generally delayed, and those that are through, decay, and soon fall out. The bones forming the pelvis are sometimes distorted, and add to the general mischief. Such children are generally of an inferior intellect, although sometimes thought by their mothers to be very precocious ; this seems due to the fact that such patients are more in the society of their elders, and have an old-fashioned way about them, because they cannot play with other children. After this the child may gradually get worse ; the emaciation goes on, the abdomen is more tumid, the softening of the bones and the deformity increase, and generally disease of the liver, kidneys, or spleen comes on. Death may occur from bronchitis or congestion of the lungs, or from diarrhœa, or from waxy degeneration of different internal organs, or from general dropsy. Children affected with rickets are liable to attacks of spasmodic croup (LARYNGISMUS STRIDULUS), convulsions, and chronic hydrocephalus. So death may end a life which to the unfortunate child has been one of unabated misery. Yet many cases do recover and grow up to adult life, but the deformity remains, and they are never so healthy as other people. The favorable symptoms will be, an increase in weight, an animated expression, and less pain in the limbs ; the pulse is less frequent, and the stools not so pale ;

the urine will return to its natural color, and the appetite is more natural. The growth of the limbs then goes on with great rapidity, and the muscles acquire a powerful development. Many of the dwarfs are examples of recovery from rickets ; they may possess plenty of strength in spite of their deformity ; they are are generally irritable and sulky, keeping aloof from their fellow-creatures in consequence of their misfortune being the subject of derision and mockery by their more fortunate brethren. The sooner the disease comes on after birth, the more likely is it to be fatal ; as a rule, if the disease be not far advanced, and if the deformity have not much affected the spine and chest, a favorable result may be looked for.

Treatment.—Improvement of the general health is the first thing to be sought after. The child should be placed in a warm and dry atmosphere, with due ventilation and pure air. The diet is most important, and should be given according to the rules laid down in the article on DIET, adapting it of course to the different ages of the patients. On a fine day the child should be wrapped up warm and carried out into the open air. The child should sleep alone ; the bed-clothes should be kept dry and clean. A warm salt-water bath should be given every morning if the child can bear it. All lowering remedies, as mercury and bleeding, must be carefully avoided. Steel wine or the syrup of the phosphate of iron, either alone or in conjunction with cod-liver oil, are very valuable remedies. Change of air and a visit to the sea-side may bring about excellent results if the parents can afford it. Lime-water may be mixed with the milk if the latter curdle on the stomach. Cod-liver oil should be taken after a meal, and with orange wine or as an emulsion. Raw meat, pounded in a mortar, is a good thing, but milk must form the principal article of food. (See DIET.)

RIGIDITY OF THE BODY (*Rigor mortis*) comes on naturally a few hours after death, and may last two or three days. Hence bodies should be laid out as soon as possible after the fatal event has occurred.

RIGORS, OR **SHIVERING**, come on after exposure to cold, in the commencement of an ague fit, and at the onset of many fevers ; during this period the temperature of the body is always raised.

RINGWORM is a skin eruption caused by the presence of a vegetable parasite. For the nature of this disease, and also its treatment, see EPIPHYTA.

ROCHELLE SALT is now technically known as tartarated soda, *i.e.*, cream of tartar neutralized by bicarbonate of soda. However known, the substance is valuable ; too little used perhaps. It exists as crystals, neutral in reaction, and readily dissolved in water. The taste is something like common salt, but not so bitter. It is most frequently administered effervescing as a seidlitz powder. Each of these powders contains two drachms of this salt with a sufficiency of bicarbonate of soda in the blue paper to cause effervescence when mixed in water with the contents of the white paper (tartaric acid). Fcr some people such a quantity of salt is quite sufficient to open the bowels, easily and freely ; others require more. This can easily be managed by telling the druggist to add a drachm or two drachms of the Rochelle salt, according as it is required. to the blue paper of a seidlitz powder. The effervescing material requires no addition.

RODENT ULCER, or NOLI ME TANGERE. (See LUPUS.)

ROSE COLD. (See HAY FEVER.)

ROSE PETALS AND ROSE HIPS, the produce of the *Rosa gallica* and *canina,* are introduced into the pharmacopœia in the form of confection, as a basis for pills. There is also an acid infusion of red-rose petals containing sulphuric acid, which is useful as an astringent, or as a vehicle for more powerful remedies. In themselves rose petals are slightly astringent, but by themselves are seldom used as such. Most commonly these preparations serve as agreeable vehicles for more powerful remedies.

ROSEMARY is only used in the form of oil. This oil is distilled from the flowering tops of the *Rosmarius officinalis,* which mainly grows in Southern Europe. The oil has the fragrance of the plant. It is colorless and soluble in spirit. The spirit is the only officinal preparation, but the oil itself is most frequently used. It is a powerful stimulant, and may be given in hysteria and nervousness in females ; some forms of headache are greatly improved by it. It is also used as a rubefacient. The oil is contained in soap liniment, and compound tincture of lavender. The oil may be given in doses of a drop or two on sugar, or the spirit may be added to various kinds of medicines as a stimulant and aromatic.

ROSEOLA, or rose-rash, sometimes also known as false measles, in a good many respects resembles the eruption of measles, but is not infectious nor contagious, and there is no watering at the eyes and nostrils, and no cough. The skin is mottled of a rose color, the patches being of no great size and of irregular shape ; sometimes the eruption appears as a cross of small slightly raised rose-colored spots. At first the eruption is bright red, but gradually it fades and finally disappears. The constitutional symptoms are slight. The rash fades in from three to six days. Sometimes the throat is affected slightly, as in scarlatina, which has led some to believe that the malady consists of a mixture of scarlatina and measles, but of that there is no proof whatever.

Various maladies give rise to a roseola ; the most important of these is syphilis. Usually syphilitic roseola is the earliest of the constitutional symptoms ; it commonly makes its appearance within six weeks of the primary attack, but it may be so slight as to give rise to no inconvenience, and so frequently escapes observation. Syphilitic roseola ordinarily consists of a number of rose-colored spots completely isolated and even with the surface, but sometimes they are fused together so as to give rise to patches which are above the surface, and so merge imperceptibly into the papules which commonly follow in order of secondary symptoms. At the same time the fauces present a rim of redness corresponding to the external rash. Belladonna sometimes produces a roseolar rash, but not very often. In infancy, stomachic derangement or dentition often gives rise to such an eruption, and it occasionally precedes the eruption of small-pox.

Treatment.—Little treatment is necessary. The bowels had better be opened by a saline purgative (Rochelle or Epsom salts), and the diet restricted ; after that a few doses of any alterative tonic will suffice to restore wonted health.

ROSIN is the hardened exudation of pine trees, of which there are three kinds in commerce—black, white, and yellow. It is only used in pharmacy to give consistence to plasters.

ROUND WORMS. (See ENTOZOA and PARASITES.)

ROYAL OINTMENT. (See BASILICON.)

RUBBING. (See INUNCTION.)

RUBEFACIENTS are irritant substances, which when applied to the skin give rise to heat, redness, and other signs of slight local inflammation. After removal it may or may not happen that the cuticle peels off. If applied for a longer period, the cuticle is raised and blood or serum forms underneath ; *i.e.,* we find a blister. A great number of substances are included in the list of rubefacients, but not many are used on account of the intractable character of most. About the simplest is the compound camphor liniment. and weak ammonia, in the favorite form of hartshorn and oil (freshly prepared) is also useful. The best is perhaps a mustard poultice, or Rigollot's mustard leaves ; volatile oil of mustard and oil of turpentine are more powerful, and require some skill in application. Corrosive sublimate and iodine are still more irritant ; they may be made to give rise to severe inflammation. Rubefacients are especially useful for getting rid of slight local pains or dissipating slight local inflammation.

RUBEOLA is the Latin name for measles. (See MEASLES.)

RUE is chiefly employed in the form of oil. which is distilled from the leaves and unripe fruit of the *Ruta graveolens,* a plant which grows throughout Europe. This oil is greenish yellow in color. It has a very disagreeable odor and an acrid taste ; it turns brown by keeping. Rue, or its oil, is a powerful stimulant to the part to which it is applied, and hence it has been given as a stimulant in flatulence. In hysterical affections especially, where the menstrual functions have been in abeyance, it is sometimes given with benefit. In these it may be employed as an enema. It is not, however, largely employed. It has falsely obtained a reputation as an abortive. The oil may be used as a local rubefacient, but is seldom employed for this purpose ; the dose of it internally is from two to three drops.

RUM is a spirit distilled from fermented sugar and molasses in the West Indies. Its peculiar odor depends on butyric ether, and a flavor given by the addition of pine apples. It is often used, mixed with honey and milk, for colds and hoarsenesses. (See DISTILLED SPIRITS.)

RUMINATION is the term applied to the action in a section of the animal kingdom of devouring the food rapidly, and then casting it up to chew at leisure. The action is familiar in many domestic animals, and is called "chewing the cud." Sometimes a similar kind of process in miniature is seen in man as the consequence of disease. Ordinarily, if the food be chewed slowly and swallowed leisurely, it is speedily attacked by the gastric juice, and as speedily dissolved, but if bolted and indigestion follow, it may return into the mouth in a condition which admits of being chewed and swallowed again. It is a sign, therefore, of an abnormal condition, and sometimes of serious disease.

RUPIA is the term given to the latest and most disagreeable form of syphilitic eruption. These eruptions, ordinarily, go through various stages, beginning with roseola, passing into papules or little hard masses which scale off, and leave no mark. Next they seem to have a yellowish top and ulcerate, and finally they appear as blebs, which disappearing leave a kind of dark-greenish crust on the surface of a wound. This wound of the surface ulcerates and forms pus ; this is added to the crust, which goes on drawing from the ulcerat-

ing surface beneath until they acquire a considerable height. These crusts go on growing to a considerable size, and may then break off, leaving a raw unhealthy ulcerating surface. From this surface a new and larger scab is formed, and so the process goes on, except it be restrained.

Treatment.—In dealing with these sores we must not temporize ; if not got rid of they may spread further and further, and even one leaves behind an indelible white sunk cicatrix which is unmistakable. It is therefore of prime importance to get rid of them early, especially if, as they often do, they affect the face. Local measures are essential, but before doing any good these hard and unsightly crusts must be got rid of. That may be done, but not effectually, by poulticing ; the process is tedious and unsatisfactory, so something better should be employed. Solution of potass and glycerine mixed and applied to the crusts, the whole being kept moist by a supply of the same material on lint, will speedily cause them to soften and fall off. When they have thus fallen off and left the ulcerating surface bare, these had better be destroyed, and the best thing for so doing is a strong solution of corrosive sublimate. This kept applied to it for a time will ordinarily suffice to make the part take on healthy action, after which citrine ointment is the best dressing. Cod-liver oil and iodide of potassia with mercury should be given internally at the same time.

RUPTURE is a word commonly applied to cases of hernia. (See HERNIA.)

RYE is a grass whose botanical name is *Secale cereale*. It yields a very nutritious flour, and when made into bread assumes a dark appearance, hence it is called "black bread." Although rye contains more starch and sugar than barley, it is not used for fermentation, on account of the rapidity with which it passes into an acid condition. Rye bread is sour to the taste on this account. The grain is subject to a disease which gives it a spined or horned appearance. This is the result of a fungus, which is injurious when eaten ; but under the name of ergot of rye is valuable medicinally in uterine cases. (See ERGOT.)

S.

SABADILLA, also known as cevadilla, is the dried fruit of the *Asagræa officinalis* of Mexico. From it is obtained the alkaloid veratria. The fruit is light brown in color, about half an inch long, and contains a few seeds. These seeds are blackish brown and shining. They have an intensely bitter taste, which is also acid. The seeds are only used as a source of veratria, not being themselves employed.

SAFFRON is the stigma of the flower of the *Crocus zativus*, growing in Greenland and Asia Minor, but cultivated in Southern Europe. This portion is orange-red in color, and when dried as collected constitutes hay-saffron. Sometimes it is packed and pressed into a parcel ; this constitutes cake-saffron. Saffron readily yields its coloring matter, so that when moistened and pressed against a piece of paper it stains it of an orange color.

This coloring matter is readily soluble in water and in alcohol, and sulphuric acid turns it from orange-red to blue. It is hence called polycroite. The only preparation of saffron is a tincture, which is little used in medicine save for its coloring properties. Saffron is contained, however, in decoction of aloes, aloes and myrrh pills, compound tincture of cinchona, ammoniated tincture of opium, and tincture of rhubarb. It is somewhat surprising that it is included in so many preparations, for it is almost useless, and is very expensive. It has indeed some reputation abroad for favoring the menstrual flow, but in this country such a belief hardly prevails. It is very often adulterated, the so-called cake-saffron frequently containing no saffron at all, but only petals of marigold, or more probably safflowers.

SAGE is a well-known herb, botanically *Salvia officinalis.* It is in every garden, and is used in cookery and as a domestic remedy for sore throats as a gargle, with honey, alum, or any astringent. As a wash for ulcers about the mouth or lips it is safe and pleasant and often very efficacious.

SAGO is a form of starch obtained from several kinds of plants. That which is most commonly used is the produce of the sago palm (*Sagus lævis*), which grows in the islands of the Indian Archipelago. The sago is obtained from the cellular tissue in the interior of the trunk of the tree. It is a good and pleasant article of diet for the invalid.

SALADS contain other constituents besides mineral matters, but their value in diet is mainly due to these. It seems necessary, in order to preserve health, that the human body should frequently partake of the various constituents which compose it, and which naturally waste away with daily use. Thus fat, fibre, starch, and other things are daily partaken of, and it is not less necessary that mineral salts, which preserve the body in health, should also enter into food. All fresh vegetables and fruit contain these salts in large quantities, but in boiling and cooking, in any way, they are dissolved and thrown away, so that it is only by eating some form of uncooked food that we can obtain them. Salads are very desirable on this account, and it is very wise to allow all people in health, whether children or adults, to partake of fresh uncooked fruit or salads every day. Lettuces, water-cresses, endive, celery, beet-root, radishes, corn-salads, sorrel, and even dandelion, are valuable and pleasant as ingredients in a salad, and there are many other plants which might be eaten with advantage were it not for prejudice.

SALAD OIL. (See OLIVE OIL.)

SAL AMMONIAC, also known as hydrochlorate of ammonia or chloride of ammonia, is a salt all whose virtues are not yet exactly known. This we do know, that it does good in some forms of headache almost magically. It is also of value in some forms of liver disease and as a stimulant of the menstrual flow when that is in abeyance. It does not partake much of the properties of ammonia. Its taste is disagreeable, and it has to be given in large doses. Some recommend beer as the best vehicle. The dose is 20 or 30 grains.

SALT is a chloride of sodium, and exerts an extraordinary influence on animal as well as vegetable life. All marine animals seem to have their existence determined by this substance. It enters into the composition of the human body, and all over the world man uses it, when it can be obtained, as an addition to his food. Salt exists in large quantities in the bowels of the earth, and in many parts of Great Britain and the continent it is worked and brought up to supply the market. It is obtained in the form of rock salt and in brine springs, and when purified is sold as "bay salt," and "fine salt." Salt prevents the decomposition of animal and vegetable substances. It is used extensively for preserving meat in conjunction with saltpetre (nitrate of potash). Bay salt is often employed to make artificial sea-water baths, and the stimulating effect of the salt in the water is frequently beneficial to those who cannot obtain sea-bathing.

SALTPETRE. (See NITRE.)

SAL VOLATILE is really the carbonate of ammonia, but the name is most frequently given to its preparation—the aromatic spirit of ammonia. This spirit contains carbonate of ammonia, strong solution of ammonia, volatile oil of nutmeg, oil of lemon, spirit, and water. It is an agreeable and useful stimulant where it is desirable that the effects should not continue too long. It is of especial value in the depression which follows the use of alcohol in excess, or indeed in any form of temporary prostration from which it is necessary to rouse the patient. It is also of great value in the bronchitis of the aged.

SALICINE is an active principle of a bitter nature extracted from willow bark. The bark is stripped from the common willow and allowed to dry. It is very tough, and has a somewhat aromatic odor and a very bitter taste. The salicine when pure exists in white scaly crystals, and is soluble in water and alcohol. This is reddened by sulphuric acid, and is converted into an odoriferous principle similar to that obtainable from meadow-sweet. Salicine has been chiefly commended as an antiperiodic in intermittent fevers where quinine was not to be had. Undoubtedly it has some activity in this way, but nothing to compare with that of quinine. It is to be given in doses of from 12 to 20 grains, and in that quantity may be of use in certain cases when quinine does not suit. It has never come into general use, and is not likely to do so, though comparatively cheap.

SALIVA is the ordinary secretion which is met with in the mouth, and proceeds chiefly from the parotid submaxillary and sublingual glands, aided by the small glands in the mucous membrane lining the mouth. These glands are very active when stimulated, and pour a large quantity of fluid into the mouth, and this is chiefly so during the process of mastication. The saliva is a thin, watery fluid, and contains a small quantity of animal matter called ptyalin. This peculiar compound has the power of converting starchy foods into sugar, and this is important because the former body is insoluble in the stomach while the latter is very soluble. Ptyalin will not act upon fatty or proteid compounds. (See DIET.) An extreme flow of saliva, called salivation or ptyalism, occurs in some cases where mercury is taken internally ; many woods, as betel-wood, etc., also have this property of exciting an increased flow when chewed.

SALIVATION is a term applied to a condition in which there is increased flow of saliva with swelling of the mucous membrane of the mouth. In most cases it is caused by the action of mercury, but it has been found that many other agents may have the same effect. Iodide of potassium, antimony, croton-oil, castor-oil, opium, and foxglove have been known to produce the milder symptoms of salivation. An increased flow of saliva without swelling or ulceration of the gums and cheeks may be produced by irritation of the mucous membrane of the mouth and alimentary canal, and by mental influences.

The mercurial salivation commences with tenderness of the gums and inner surfaces of the cheeks, and pain when the teeth are brought sharply together. The patient experiences a metallic taste. The secretion of saliva is so much increased that it accumulates in the mouth, and necessitates frequent spitting, and during the night flows from the mouth and saturates the pillow. The daily amount of saliva, which in health is about ten ounces, increases to four or six pints. The tongue then swells, and the mucous membrane of the gums and cheeks becomes red and inflamed, and finally ulcerates. The breath has a very offensive and peculiar odor. In bad cases the ulceration extends, and by destroying the tissue of the gums exposes the bone of the upper and lower jaws. These symptoms are usually associated with those of gastric and intestinal irritation, and of nervous debility and excitement ; with the exception of the metallic taste in the mouth none of the above symptoms are peculiar to mercurial salivation, and a similar condition may be presented in cases of salivation due to constitutional causes. Cancrum oris, a

gangrenous affection of the mouth which is occasionally met with in children suffering from measles, is often attributed by the parents to the effects of mercury, supposed to have been administered for the treatment of the febrile disorder. Cases of severe salivation produced by the medicinal use of mercury are at the present extremely rare, as the effects of mercury are seldom allowed to proceed beyond slight redness and tenderness of the gums. In cases where intense salivation has been produced by the administration of large quantities of mercury, other symptoms of mercurial poisoning are generally present ; of these the most prominent are pallor, trembling, an eczematous eruption over the surface of the body, and general debility. In some peculiarly constituted patients mercury may produce all the latter symptoms, and not give rise to salivation or any affection of the mouth. The property of producing salivation is common to all the preparations of mercury used in medicine, the most active being calomel and blue pill. Mercury when introduced into the system in other ways than through the mouth and stomach, as by inunction, fumigation, and hypodermic injection, does not fail after a certain time to produce similar symptoms of salivation. In the treatment of local affections, especially venereal sores and cutaneous eruptions, the first appearance of the symptoms of mercurial salivation is generally presented by a decided improvement in these affections, the indurated bases of the sores have commenced to soften, and the rash on the skin is less distinct. Salivation may be produced either by a very large and poisonous dose of mercury, or by frequently repeated small doses. In some cases there is a long interval between the end of the course of mercury and the first appearance of symptoms of salivation, mercury being a cumulative poison which may be stored up in the body slowly and gradually until it is in sufficient force to give rise to salivation and other affections. Some individuals are extremely sensitive to the action of mercury, and become salivated after very small doses of calomel or blue pill. A case has been recorded in which two grains of calomel caused salivation, sloughing and ulceration of the throat, necrosis of the lower jaw, and death. Other individuals, on the contrary, are so constituted that they can resist for a long time the action of large and frequently repeated doses of mercury, or even remain quite invulnerable. A patient who has been subjected to two or more courses of mercurial treatment becomes much less susceptible on each occasion to the action of the medicinal agent, and is less liable to be salivated by the last than by any previous course. The early occurrence of salivation during a mercurial course is much favored by a want of attention to the cleanliness of the mouth, and by bad teeth, and soreness of the gums. Catching cold and even a slight exposure to cold and wet will often cause early salivation.

During a course of mercury great attention should be paid to the state of the mouth. The teeth should be frequently brushed, and the patient, in order to harden the mucous membrane of the gums and cheeks, should wash out the mouth occasionally with some astringent gargle and suck small pieces of alum. In cases of mercurial salivation the cause should at once be removed. The swollen and ulcerated mucous membrane of the mouth should then be frequently washed with a solution of chloride of lime, or of alum, or with brandy-and-water.

SALTPETRE. (See POTASS, NITRATE OF.)

SAMBUCUS, the botanical name of the elder tree, or dwarf elder, is *Sambucus ebulus,* while that of the black or common elder is *Sambucus nigra.* The berries are often made into wine ; and from the flowers is distilled a pleasant fragrant wash for the skin, called elder-flower water. (See ELDER FLOWERS.)

SANDAL-WOOD is the product of a tree growing in India and Ceylon, and also in the South Sea Islands. It occurs in billets of a dark-brown color externally ; internally the rings are well marked. The powder is blood red, and has a slightly astringent taste. It is mainly used for the coloring matter, which may be extracted by alcohol or ether, and by alkaline solutions. It is sandal-wood which gives the red color to the compound tincture of lavender and to Fowler's solution of arsenic. An oil of sandal-wood has recently come into use as a remedy for gonorrhœa. Fifteen or twenty drops are usually given for a dose ; but a good many people it does not suit ; many cases are not benefited by it. In all instances it gives rise to a good deal of pain. Often it is very effectual.

SANGUINEOUS APOPLEXY is said to occur when there is a rupture of a blood-vessel in the brain. (See APOPLEXY.)

SANITARY REGULATIONS. — The following simple rules have been drawn up from the sanitary papers of Dr. Lankester, Medical Officer of Health, St. James's, Westminster. They are intended to guide people in case of an epidemic breaking out in any town or village. It is only by the local authorities actively investigating the cause of an outbreak, and then isolating the cases as far as possible, that one can hope to check those ravages which disease makes at times over an infected area.

The following regulations, or some similar ones, should be printed in large type, and be posted on the walls in public places, and also be distributed among the people :

When Small-Pox is prevalent.

1. When this highly-contagious and fatal disease is prevalent in a district, the inhabitants should be made aware of the danger to which they are exposed, and the best means of preventing the attack of the disease, and of stopping it where it has already broken out.

2. In the first place, it cannot be too widely known that vaccination is one of the best means for preventing an attack of small-pox.

3. All persons should be revaccinated after twelve years of age.

4. When small-pox prevails in a family or in a neighborhood, every person should be immediately revaccinated under the direction of a legally qualified medical practitioner.

5. Every child should be vaccinated within three months after its birth.

6 When it has been ascertained that an individual has got small-pox, everything should be done to separate the person attacked from those around. Where it is deemed desirable to remove persons thus affected to a small-pox hospital, information can be obtained, and the means of conveying patients ascertained, by applying to the health officer of the district.

7. Where persons are found to be laboring under the disease, a medical man should be sent for immediately.

8. The following directions should in all cases be carried into effect : The room should be cleared of all needless woollen or other draperies which might possibly serve to harbor the poison. A basin, charged with chloride or carbolate of lime, or some other convenient disinfectant, should be kept constantly on the bed for the patient to spit into. A large vessel, containing water impregnated with chloride of lime or with carbolic acid, should always stand in the room for the reception of all bed and body linen immediately on its removal from the person of the patient. Pocket-handkerchiefs should not be used, and small pieces of rags employed instead for wiping the mouth and nose. Each piece, after being once used, should be at once burned. As of necessity the hands of nurses become frequently soiled by the secretions, a good supply of towels and two basins—one containing water with carbolic acid or chlorides, and another plain soap and water, should be always at hand for the immediate removal of the taint. All glasses, cups, or other vessels, used by or about the patient, should be scrupulously cleaned before being used by others. The discharges from the bowels and kidneys should be received on their very issue from the body into vessels charged with disinfectants, and immediately conveyed away. No persons should be allowed to enter the room except those who are attending upon the sick. Persons attending upon the sick should be scrupulous in

cleaning their hands and disinfecting their clothes before they go out of the sick-room, or communicate with those who have not got the disease.

9. When persons have had the small-pox, whether they recover or die, the room in which they have been ill should be disinfected. The floor should be washed with chloride of lime and water, or with carbolic acid and water. The paper should be removed by moistening with carbolic acid and water. The room should then be fumigated by burning sulphur in an iron dish, the fireplace and the crevices in windows and doors being closed by putting paper over them. The room should be exposed to the sulphur vapor for five or six hours ; or the room may be fumigated in the same way with chlorine vapor, which is procured by pouring oil of vitriol (sulphuric acid) on common salt and black oxide of manganese.

10. After the room has been fumigated, it should be whitewashed, and the doors and windows kept open for a week or a fortnight.

When Typhoid Fever is prevalent.

This fever is also called drain fever, from its constant association with bad and imperfect drainage in houses ; it is also called gastric or enteric fever, from its chief seat being in the stomach and bowels. At one time it was confounded with typhus fever, but it differs from that disease in its causes, history, and result ; hence the term typhoid (like typhus) was at first given to it. It is also called in some parts of the country low fever, from the great exhaustion and weakness which attends it.

The means by which this disease may be prevented from spreading are very simple, and depend upon the fact that the poison by which it spreads is almost entirely contained in the discharges from the bowels. Dr. W. Budd gives the following excellent directions for preventing the spread of this disease. He says the discharges from the bowels infect—

1. The air of the sick-room.
2. The bed and body-linen of the patient.
3. The privy and the cesspool, or the drains proceeding from them.

From the privy or the drain the poison often soaks into the well and infects the drinking water. This last, when it happens, is, of all forms of fever poisoning, the most deadly. In these various ways the infection proceeding from the bowel-discharges often spreads the fever far and wide. The one great thing to aim at, therefore, is to disinfect these discharges on their very escape from the body and before they are carried from the sick-room. This may be perfectly done by the use of disinfectants. One of the best is made of green copperas. This substance, which is used by all shoemakers, is very cheap, and may be had everywhere. A pound and a half

of green copperas to a gallon of water is the proper strength. A teacupful of this liquid put into the night-pan every time before it is used by the patient renders the bowel-discharge perfectly harmless. To disinfect the bed and body-linen, and bedding generally, chloride of lime is more convenient. All articles of bed and body-linen should be plunged, immediately on their removal from the bed, into a bucket of water containing a tablespoonful of chloride of lime, and should be boiled before being washed. The privy, or closet, and all drains communicating with it, should be flushed twice daily with the green copperas liquid or carbolic acid diluted with water. In the event of death the body should be placed, as soon as possible, into a coffin, surrounded with charcoal, sprinkled with disinfectants. Early burial is, on all accounts, desirable. In towns, and places where the fever is already prevalent, the last rule should be put in force for all houses ; the drains of all houses should be flushed daily with disinfectants, whether there be fever in them or not. As the hands of those attending on the sick become unavoidably soiled by the discharges from the bowels, they should be frequently washed. The sick-room should be kept well ventilated, day and night. The greatest possible care should be taken with regard to the drinking water. Where there is the slightest risk of its having become tainted with fever-poison water should be obtained from a pure source, or should at least be boiled before being drunk. Immediately after the illness is over, whether ending in death or recovery, the dresses worn by the nurses should be washed or destroyed, and the bed and room occupied by the sick should be thoroughly disinfected. These are most important rules. Where they are neglected the fever may become a deadly scourge. Where they are strictly carried out, it seldom spreads beyond the person first attacked. A yard of thin wide gutta-percha or a mackintosh sheet placed under the blanket, under the breech of the patient, is a great additional safeguard by effectually preventing the discharge from soaking into the bed.

When Scarlet Fever is prevalent.

Scarlet fever is a highly contagious disease, and spreads from one person to another, and is thus propagated in families, towns, and districts. It is, therefore, highly desirable that every one should understand the nature of this disease and the means of preventing its spreading. It is always attended with a scarlet eruption on the skin, and is mostly accompanied by a sore throat. Whenever children have sore throat, or an eruption on the skin, they should be separated from the rest of the family until a doctor has seen them, or these symptoms have disappeared.

There is every reason to believe that, during the progress of this disease, not only the eruption of the skin, but everything that is thrown off from the body of the infected person, is heavily laden with the germs or seeds which are capable of propagating the disease in another person. The discharges from the nose and throat are especially virulent. There is also reason to believe that the discharges from the bowels are the same. The kidneys are frequently dangerously diseased in scarlet fever, and the secretion from these organs is also highly contagious ; the power of spreading the poison by means of these secretions is not confined to their immediately leaving the body, but continues long after. It is on this account that when these secretions have found their way to the cesspool and sewer, they may still give off poison to the surrounding air, and persons breathing it may become infected. Taking these things into consideration, it will be seen that it is necessary, if possible, to destroy and annihilate this poison, before it leaves the room where the person is whose body has produced it. The following directions, drawn up by Dr. W. Budd, should in all cases be carried into effect :

1. The room should be cleared of all needless woollen or other draperies which might possibly serve to harbor the poison.

2. A basin, charged with chloride or carbolate of lime, or some other convenient disinfectant, should be kept constantly on the bed for the patient to spit into.

3. A large vessel containing water impregnated with chlorides, or with carbolic acid, should always stand in the room for the reception of all bed and body-linen immediately on its removal from the person of the patient.

4. Pocket-handkerchiefs should not be used, and small pieces of rag employed instead, for wiping the mouth and nose. Each piece, after being once used, should be immediately burned.

5. As the hands of nurses of necessity become frequently soiled by the secretions, a good supply of towels and two basins—one containing water with carbolic acid or chlorides, and another plain soap and water, should be always at hand for the immediate removal of the taint.

6. All glasses, cups, or other vessels used by or about the patient should be scrupulously cleaned before being used by others.

7. The discharges from the bowels and kidneys should be received on their very issue from the body into vessels charged with disinfectants.

By these measures the greater part of the germs which are thrown off by the internal surfaces may be robbed of their power to propagate the disease. The poisonous germs that are thrown off from the skin require a somewhat different treatment. The plan recommended by Dr. Budd for the purpose of preventing the poison from the skin being disseminated through the air, is to put oil all over the skin. This practice is to commence on the fourth day after the appearance of the eruption, and to be continued every day until the patient is well enough to take a warm bath, in which the whole person is well washed with disinfecting soap and warm water. These baths should be administered every other day, for four times, when the disinfection of the skin may be regarded as complete. This proceeding should not, however, be adopted without consulting the medical man who is in attendance on the patient. Speaking of the plans above recommended, Dr. Budd says : " The success of this method in my own hands has been very remarkable. For a period of nearly twenty years, during which I have employed it in a very wide field, I have never known the disease spread in a single instance beyond the sick-room, and in a very few instances within it. Time after time I have treated this fever in houses crowded from attic to basement with children and others, who have, nevertheless, escaped infection. The two elements in the method are, separation on the one hand, and disinfection on the other."

Summary of Facts in favor of Vaccination.

1. Persons who have once had the small-pox are not liable to take it a second time. In the last century, inoculation of small-pox was practised, because it was known that small-pox thus communicated was usually milder than when caught naturally.

2. Cow-pox is a modified form of small-pox, and it has been clearly proved that those who have had it are very much less liable to take small-pox than others, and are as effectually protected as those who have already had small-pox.

3. Cow-pox is communicated by vaccination. Jenner discovered that persons who had caught cow-pox from the cow escaped small-pox, and thus was led to advise vaccination.

4. Forty-five millions of the people of Europe died from small-pox in the hundred years preceding the introduction of vaccination at the beginning of this century ; while not more than two millions have died from this disease during the seventy years in which vaccination has been practised. In London, before the introduction of vaccination, every *tenth* death that occurred was due to small-pox ; now only one death in every *eighty-five* is due to this disease. Even greater difference has been observed in other towns and cities of Europe, as in Trieste, where the deaths from small-pox have been *seventy-five* times less since than before vaccination ; in Moravia *twenty-one* times less ; in Silesia *twenty-nine* times less ; in Westphalia *twenty-*

five times less ; and in Berlin *nineteen* times less. Where vaccination has been stringently enforced, death from small-pox has been still more diminished. In Ireland, where this disease was once a scourge, it has now become almost unknown, and a similar good result has been obtained in many districts in India. When small-pox has been rife among the inhabitants of a city or district, it has been repeatedly observed that the unvaccinated have perished, and the vaccinated survived or altogether escaped. Very carefully kept records in hundreds of places in Europe and elsewhere have proved this. Out of thirty vaccinated nurses constantly employed at the Small-pox Hospital, not one ever contracted the small-pox ; of the patients admitted to this hospital, from 1836 to 1851, thirty-seven in the hundred of those unvaccinated died, while only six in the hundred who had been vaccinated (well or badly) died.

5. In England there is a larger proportion of unvaccinated persons than in any other country of Europe, and consequently a greater number of deaths from small-pox are shown, by the public registration, to occur.

6. While small-pox kills so great a proportion of the unvaccinated whom it attacks, the inoculation of cow-pox, *i.e.*, vaccination, seldom or never produces any consequences of an injurious nature. The authenticated fatal cases of erysipelas so produced are not more numerous than those which follow the prick of a pin. There is no proof that those who have been vaccinated suffer from scrofula or any similar disease as a consequence of vaccination alone. According to the Registrar-General's returns these diseases have diminished in frequency since the introduction of vaccination. That which *follows* after a thing is not necessarily *caused* by it. It is a matter of necessity that persons, who have been vaccinated, should have diseases after this operation as well as before it, but these cannot rightly be attributed to vaccination.

7. Although very rarely the eruptions of other diseases have, through the carelessness of medical practitioners, been mistaken for cow-pox, yet it would be as reasonable to ask for the abolition of railways because of railway accidents, as to demand the abolition of vaccination on account of such accidental occurrences.

8. The great means whereby small-pox may be wholly exterminated is universal vaccination.

SANTONIN is a crystalline substance neutral in reaction, obtainable from the unexpanded flower-heads of certain species of artemisia. The flower-heads, which can, at first sight, hardly be distinguished from seeds, have a strong odor and bitter taste. To obtain santonin, these are bruised and boiled for a time with water and lime. To this fluid hydrochloric acid is added till the whole becomes curd, when it is set aside for the santonin to subside. The precipitate is well washed and otherwise purified till it is brilliantly white and crystalline. It must be kept away from the light. The crystals have but little taste and no smell, insoluble in cold water, but soluble and subliming with a moderate heat. These brilliantly white crystals become yellow by exposure to light. Santonin, if given in any quantity, colors the field of vision yellow, so that the patient sees everything of that color. Sometimes green takes its place. The substance is a capital remedy for worms, and being nearly tasteless is easily taken by children. It is useless against flat worms, but is valuable as a remedy for round worms, especially of the larger kinds. The dose is from 2 to 5 grains. It is best given in a little sugar or honey.

SARSAPARILLA is a remedy which has been extolled to the skies, and has sunk into almost complete neglect. Lauded at one time, it has been despised at another, both probably unjustly. The plant which yields it is a species of smilax, chiefly growing in Central America and the West Indies. The part employed is the underground stem or rhizome, whence numerous long rootlets are given off. The preparations of sarsaparilla are a simple and compound decoction, and a liquid extract. The compound decoction contains Jamaica sarsaparilla, guaiacum-wood turnings, fresh liquorice-root, and mezereon. This preparation is that most frequently used, in doses of from two ounces to a pint. Sarsaparilla contains, besides the ordinary root constituents, an oil, and a principle called smilacin. It has never been very carefully investigated.

All kinds of properties have been attributed to sarsaparilla—diaphoretic, diuretic, tonic, and alterative—but it has been mainly used as an anti-syphilitic. It was early introduced as a remedy in this complaint, and in the form of what was called the Lisbon Diet Drink, was largely used for a long time. Gradually it fell out of use, and is now seldom used in that complaint. It is, however, said, by some, that it has so fallen out of use because improperly used ; that only small quantities were given, and that in small quantities it is useless. It seems, according to some reputable authorities, to do much good in the skin eruptions of syphilis, if given in doses of not less than half a pint or pint of the decoction daily. It has been also used in some skin diseases, especially in those of a scrofulous origin, as a sudorific where the skin is dry and tending to disease, and in chronic rheumatism and gout. In all of these maladies, however, it has been customary to use the sarsaparilla merely as an adjunct to powerful remedies, or when other good has resulted, it has not been always quite clear that the benefit was traceable to the sarsapa-

rilla. If used at all, it should be used abundantly, and in the form of freshly prepared
decoctions, simple or compound.

SASSAFRAS is the dried root of the
sassafras-tree, growing in the United States
and Canada. It is most frequently met with
as chips, which have a peculiar pleasant odor
and a warm sweet aromatic taste. It is contained in the compound decoction of sarsaparilla. Its action is stimulant, and is supposed to be specially useful in chronic rheumatism and skin diseases. It is never given
by itself. Sometimes its oil is used.

SAVIN consists of the fresh and dried
tops of *Juniperus Sabina*, which is a native
of England. From it is distilled an oil, colorless or pale yellow, having the odor of the
tops. The tops themselves are covered with
minute leaves, pressed to the stem and arranged in four rows. They are dark green,
and have a disagreeable odor and taste. The
oil which is contained in the tops gives them
activity along with some resin. From the
tops are prepared a tincture and an ointment.

Savin acts as an irritant wherever applied,
externally or internally, and is reputed to
have special power over the womb, and so
is given to promote the menstrual flow. The
ointment is mainly used to keep a blistered
surface raw, when it is deemed desirable to
do this. It has been frequently given to procure abortion, and as the substance is highly
irritant, this practice is attended with great
danger. Its use is to be avoided in pregnancy as dangerous and liable to be misinter-
preted. The ordinary dose of the powder is
four or five grains, and of the tincture 20
drops to half a drachm.

SCABIES. (See ITCH; ECTOZOA.)
SCALD-HEAD. (See PORRIGO.)
SCALDS. (See BURNS.)
SCALP consists of those integuments
which cover the cranium or vault of the skull.
They are very firm and dense. The scalp is
covered with a delicate cuticle or scarf-skin,
and immediately beneath this is a thick
cutis or true skin; beneath the cutis is a
layer of fat and cellular tissue, containing the
bulbs of the hairs. This cellular layer adheres
very intimately to the subjacent tendinous
layer, which is the tendon of the occipito-
frontalis muscle; between this tendinous expansion and the bone is a delicate cellular
layer. The scalp is largely supplied with
blood-vessels and nerves.

Affections of Scalp. Tumors.—The most
frequently met with are the encysted, also
called *Wens.* (See WENS.) Fatty tumors
are sometimes met with, but are rare, and they
seldom grow to any great size; the treatment
consists in excising them. *Erectile tumors,*
i.e., masses composed of a congeries of dilated
vessels, mostly veins; these are best treated
by the ligature. *Malignant tumors* are met
with in the scalp, but they as frequently as

not originate in the bone. Medullary is the
most usual form; a malignant form of ulceration is not uncommon, frequently commencing with a degenerate wen; the only treatment
in either instance is early and free removal,
provided the glands are not implicated.

Injuries of Scalp. Bruises.—Owing to its
exposed condition, the scalp is naturally very
liable to external injury, and, owing to its
aforestated vascularity, the results may be
very serious. The ordinary result is the formation of a tumor full of blood, the result of
extravasation, and the condition is to be
treated on general principles. If, for instance, the swelling be over some largish
artery, such as the occipital or temporal,
steady and firm pressure must be maintained
on the vessel between the tumor and the
heart; the application of cold and pressure
will check further extravasation. Acute inflammation not unfrequently follows these injuries, and if suppuration occurs, free incisions must be made. Constitutional treatment must be attended to, and rest, antiphlogistics, and perhaps depletion. Absorption may be accelerated, after all inflammatory symptoms have passed off, by keeping
the part wet with a solution of muriate of
ammonia or tincture of arnica. It may be
mentioned that a blow upon the *back* of the
head, may produce a black *eye,* owing to the
extravasation of blood under the tendon of
the occipito frontalis, and its subsequent
gravitation forward.

Incised Wounds.—In the case of incised
wounds of the scalp, no matter how severe,
the treatment will consist in carefully cleansing both surfaces of the wound, and the parts
carefully adjusted and maintained in position
by strips of plaster, compresses and bandages. It is well *not* to put any sutures in;
two or three wisps of the hair growing on
the opposed edges of the wounds tied across
make an excellent method of obtaining union,
and act as a suture without the penetration.
If the scalp wound becomes "puffy," the
adhering lips of the wound must be separated
to let out the inclosed fluid, and a warm poultice and hot fomentations must be applied.
Tonics and ammonia should be administered
at an early period of any symptom of erysipelatous puffiness, and if the patient has been
in the habit of taking brandy, wine, or spirits,
he should be still allowed them in moderation. (See COMPRESSION, CONCUSSION, EC-
TOZOA, EPIPHYTA, FRACTURES, and PITYRI-
ASIS.)

SCAMMONY is a gum resin exuding
from the top of the root when the stem of
the living plant (*Convolvulus Scammonia*) has
been removed. The root itself is also now
officinal as well as the resin contained in
scammony, which is its active principle. The
plant grows in Asia Minor and is chiefly imported from Smyrna. The root somewhat

resembles a carrot. It is brown without and
white within, and is possessed of a peculiar
odor. The gum resin, the well-known scammony, is blackish green in color, and occurs
in irregular masses, covered with its own
powder, but breaks with a shining fracture.
From the gum which it contains this substance forms a lather if wetted and rubbed.
If spirit be added the resin is dissolved up,
leaving the gum behind. This resin is
brownish and brittle. If prepared from the
root instead of the gum resin it is fragrant.
Scammony used to be much adulterated,
especially with starch and chalk. The resin
forms no emulsion with water, as does scammony itself. Its composition is very similar
to that of jalap. The preparations of these
substances are confection of scammony, containing scammony, ginger, oil of caraway,
oil of cloves, syrup, and honey. The compound powder of scammony contains scammony, jalap, and ginger. Of scammony
resin is made a scammony mixture, by rubbing up the resin with unskimmed milk.
This resin is also contained in compound
extract of colocynth, and scammony itself
occurs in the compound colocynth pill, and
the colocynth and hydrocyanus pill. Scammony and its resin are powerful purgatives,
producing much watery discharge, and if not
guarded griping much. They are seldom
given by themselves, but are usually added
to other and less violent laxatives. Usually,
too, it is customary to give along with them
some aromatic and stimulant, or some sedative substance, to guard against griping. It
is sometimes used in dropsies, especially
among children, and the compound powder
is often used to get rid of worms. The dose
of scammony itself is about 5 grains, of the
resin 3, and of the compound powder 10
grains. The confection may be given in
doses of 20 grains.

SCAR. (See CICATRIX.)

SCARF-SKIN, or the **EPIDERMIS,** is
the upper layer of the skin ; small scales are
always being shed, but it is abundantly cast
off after scarlet fever, and some other febrile
disorders, also in cases of psoriasis, etc.

SCARIFICATION is a term used in surgery when the cuticle or external skin requires to be cut or lanced through only.
Sometimes in case of dropsy it is necessary
to do this in order to allow the fluid to escape,
and in cases of children's gums where the
tooth presses against the external skin, and
is ready to burst through, scarification will relieve the irritation and cause no pain.

SCARLET FEVER is an acute febrile
disease, producing a scarlet rash upon the
skin, attended by a sore throat, and often
swelling of various glands and sometimes followed by dropsy. Contagion is the main if
not the only cause of scarlet fever ; measles
and whooping-cough are more contagious ;

typhus fever and diphtheria less contagious.
The poison may be retained in clothes for a
year or more and then give rise to the fever.
Both sexes are equally liable to an attack ;
between eighteen months and five years
is the most common period to have the
fever ; no season has much influence upon
it, but in this country it is, perhaps, most
common between September and November.
Many people confuse the terms scarlatina and
scarlet fever, and imagine the former is a
milder and less dangerous affection ; this is
a great mistake, for scarlatina is only the
Latin name for scarlet fever and not a different form ; the term is too often adopted when
there is some doubt as to the nature of the
case, and then it is used to conceal ignorance. Scarlet fever may be very mild, or
malignant, or latent. The period of incubation is generally about a week, but may be
only twenty-four hours.

Symptoms.—1. *Mild Scarlet Fever.*—The
onset is sudden ; there is a sore throat, with
tenderness at the angles of the lower jaw
and stiffness at the back of the neck ; vomiting is very common, and chiefly so in children ; shivering and rigors come on, and occasionally convulsions in young children.
The temperature rapidly rises and will go up
to 104° or 105° ; the pulse is very quick ; the
tongue is covered with a thin, white fur ;
there is thirst and loss of appetite. This
stage lasts from twelve to thirty hours, and
then a rash comes out ; sometimes the earlier
symptoms are so slight that the rash is the
first thing noticed. The rash consists of
small scarlet dots, almost running together,
so as to give a flush all over the skin ; the
color disappears on pressure, but rapidly re-
appears when the pressure is removed. It
generally appears at first on the sides of the
neck and upper part of the chest, and in the
bends of the joints ; it then spreads downward and is found to come out last on the
legs ; it begins to fade on the fourth or fifth
day, and is generally quite gone within a
week. The sore throat is always present to
a degree ; there is redness and swelling of
the tonsils and soft palate, so that it is very
painful to swallow, while the glands beneath
the jaw also swell and are painful. The temperature is generally higher than in measles,
and much higher than in diphtheria, but it
rarely exceeds 105° ; the fall of the temperature is usually on the sixth or seventh day,
but it may be earlier or it may be prolonged.
In no fever is the pulse quicker than in this
disorder, and it may be 140 or 160 in a minute. Moderate delirium and headache are
often present in these cases. After the rash
has gone the epidermis is dry and harsh, and
about the ninth or tenth day it begins to peel
and is sometimes cast off in large flakes, and
this desquamation, or peeling, may last a few
days or occupy several weeks.

2. *Malignant Scarlet Fever* is characterized by an increased severity of the above symptoms ; there is great prostration, delirium, and sleeplessness ; the rash does not always come out well ; the face may be livid, and stupor and coma come on, and end in death ; the throat is ulcerated, and there is much difficulty in swallowing.

3. *Latent Scarlet Fever* is when the disease is so mild that until the sequelæ appear one is not aware of having had scarlet fever. There is no relation between the abundance of the rash and the danger to the patient. However mild the disease may be, the sequelæ may come on with great severity : one is just as liable to catch the fever from a mild case as from a severe one.

Sequelæ.—After the fever has passed there may follow a train of symptoms which are very inconstant in their character and of much danger to the patient. The throat may continue to be affected and the glands outside may be inflamed and swell, so that the child's head seems encased in a " collar of brawn ;" often these glands suppurate and a large ulcerated surface is then seen. Deafness may come on and a discharge from the ear. Bronchitis and pneumonia are not so common as in measles. Sometimes convalescence is retarded by abscesses forming in various parts of the body ; at other times there is a painful affection of the joints, which much resembles rheumatic fever. Renal dropsy is also one of the most usual sequelæ, but its frequency varies in different epidemics ; the face and loose parts of the skin are very pale and puffy, and this is best seen under the eyes and on the insteps ; the urine is scanty and dark from containing blood ; there is often headache, loss of appetite, and perhaps convulsions ; this complication often comes on two or three weeks after the first appearance of the rash. (See BRIGHT'S DISEASE and HÆMATURIA.)

Treatment.—Most cases recover in a week, except those which are malignant, and those where the woman is at the same time pregnant ; the latter condition much increases the danger, and hence women should then be extremely careful not to go near a case of scarlet fever. The mild cases must be nursed simply, and there is no remedy which will cut short an attack. The patient must be put to bed and have a milk diet, in the same way as has been fully described in the articles on FEVER and MEASLES, and need not therefore be repeated here. Hot flannels or cotton-wool should be wrapped round the throat, and steam may be inhaled into the mouth. Sometimes a compress of linen steeped in cold water and applied to the throat gives great relief. When dropsy comes on it shows the kidneys are affected, and the patient must be put to bed again, if he has been up previously. A hot bath and purgatives

must be given to remedy this state of things. Exposure to cold too soon after an attack of scarlet fever is often a cause of the dropsy, and so care should be taken to keep the child in the house for at least three weeks after the rash and until the peeling has finished. In this way also the child is less liable to give it to others. Malignant cases may be knocked down at once and die within forty-eight hours ; ammonia and brandy must be given when the state is one of great prostration. Gargles are not of much use to the throat : brushing the fauces over with tannin and glycerine, or with a solution of nitrate of silver, is the best remedy. In cases of discharge from the ear this must be syringed with warm water three or four times a day, and a little cotton-wool should then be pushed in. During convalescence tonics should be administered, and for this purpose iron and quinine are the best remedies. The reader is referred to the article on SANITARY REGULATIONS for an account of the disinfecting measures to be used.

SCIATICA is not a single disease, but a group of diseases of various kinds, but all affecting nearly the same region. That region is the lower portion of the hip and thigh, along which the sciatic nerve runs, whence the name. True sciatica is a neuralgic affection, but numerous other maladies, especially of a rheumatic origin, have been mistaken for it. The sciatica rarely occurs in youth, and rarely begins in old age—most frequently it commences between forty and fifty. One kind of sciatica—of the truly nervous kind—is associated, especially in females, with hysteria, or other signs of a nervous temperament. Frequently these suffer from neuralgia in other situations. The sciatica which occurs in older persons very often follows on cold, damp, and fatigue. It is especially troublesome in men who have broken down under their exertions, and show signs of premature age. Sciatica occurring in these individuals is exceedingly intractable, and there are very frequently spots in the neighborhood of the great nerve that are exquisitely tender to the touch. In this form of paralysis, too, the motion of the extremities is interfered with. There is loss of power and motion, or any attempt at it gives rise to great pain. Besides loss of motor power there may be loss of sensation of the ordinary kind. There may be greater sensibility to mere touch, but the power of discriminating possessed by the skin is diminished. As the nerve which supplies the lower extremities is concerned in this affection, not only is the motor power impaired and the sensory functions interfered with, but there is often a loss of governing power, so that any stimulus which ordinarily would have little effect while the central governing power had full control over the extremities, may give rise to spasmodic contractions or cramps of the muscles of the affected extremities.

In another group of cases where there is marked pain in the sciatic nerve, these seem due to inflammatory or other changes in its sheath subsequent to rheumatism or syphilis. These cases belong to a totally different category to the former, and the treatment applied to them must be as different. In these cases iodide of potassium and cod-liver oil are the great remedies ; not so with the neuralgic affections of the sciatic nerve. Bicarbonate of potass, which is often prescribed, is quite useless, and if persisted in for any length of time worse than useless. Iodide of potassium ought to be given in good large doses, larger in syphilitic than in rheumatic sciatica. Not less than 10 grains should be given three times a day to begin with, and the quantity should be gradually increased. Cod-liver oil is to be given as the patient can take it, and continued for a long time. Small doses are of little worth. Now, of the true neuralgias, it is especially important to give in the first variety—that is, the one which occurs in a decidedly nervous temperament, and is very likely the result of nervous exhaustion — tonics. Steel and strychnine should be given, and persevered in ; these may not suffice to wholly get rid of the pain, but they will strengthen the constitution, and so enable other remedies to be used with more advantage. The strychnine may be given either as liquor strychnine (solution of strychnine), from five to ten minims for a dose, or the tincture of nux vomica in like quantity may be prescribed. Liquor strychnine is best when given along with iron. Of iron the two best preparations are the saccharated carbonates and the neutral chloride. The carbonate may be given in doses of 20 or 30 grains, the chloride in 20 or 30 minim doses. The liquor ferri perchloride may be used if the other is not obtainable. Arsenic is a remedy not to be overlooked in dealing with sciatic neuralgia, especially if there is any likelihood of malarial complications. The preparation commonly employed is Fowler's solution, of which the dose is two or three minims, given immediately after food. It is true that arsenic is of more value in other forms of neuralgia than sciatica ; nevertheless its use in an obstinate case—and sciatica is very obstinate sometimes—should never be overlooked.

Of the local means of relieving sciatica, chief among them we would place the hypodermic injection of morphia, especially over the spot where the pain is most severe. If the spot be also tender, it may be necessary to use ether spray to alleviate the pain of the injection. The quantity injected should not in the first instance exceed one fifth of a grain ; but it may be shortly repeated if successful. The value of this injection lies as much in

the rest from pain it gives as in its action on the nerves. Very often, however, it is not possible for a sufferer to procure this injection of morphia at all times when his pain is severe, and it is hazardous to allow him to have the command of the injection, as he is apt to increase the dose unnecessarily and speedily. When this is the case a small blister over the pained spot, with some lead and morphia lotion to apply when the skin is removed, will do great good. Of course such a lotion must be very weak. An ointment may be made to produce similar effects. A good many physicians like to use atropine in small quantity along with morphia when given under the skin ; and some give it by itself in the same way, frequently with success. The dose to be given must not exceed the sixtieth part of a grain.

The local use of electricity in sciatica is a recent introduction, but already it has attained wide popularity. The kind of electricity is, however, important. That in ordinary use—induced electricity, whether the original current be magnetic or chemical, it matters not—is useless. The current must be continuous, and it is important that it should be as nearly as possible constant. (See ELECTRICITY.)

All forms of sciatica are apt to return, and so if a patient has once suffered from the malady he ought to take great care that it does not, or the consequences may be disastrous. To this end over-fatigue, bodily or mental, should be avoided, and flannel constantly worn next to the skin.

SCIRRHUS, a name applied to one variety of cancer. (See CANCER and TUMORS.)

SCORBUTUS. (See SCURVY.)

SCRIVENER'S PALSY, or **WRITER'S CRAMP.** (See PARALYSIS.)

SCROFULA is a constitutional condition generally inherited from one or both parents, and increased by bad feeding in early life. The most characteristic features of a scrofulous individual are : a heavy figure, dull pasty complexion, with a prominent upper lip and a coarse mould of countenance ; mind and body lazily disposed, nostrils expanded, and nose rather turned up. When children they are very liable to inflammation of the eyelids, giving a red, angry look to the part, while most of the eyelashes are absent ; often, too, the glands enlarge, and more especially those under the jaw and in the neck ; this swelling comes on from a common cold, or in the course of an illness, and sometimes the gland breaks up into an abscess, which points and leaves, after recovery, a nasty seamed scar ; such people generally have several of these scars, from abscesses having formed at different times. Eczema is another condition to which scrofulous people are very subject when young ; it appears on the head and behind the ears. Discharge from the ear, earache, and deafness are not uncommon symptoms. Bronchitis, inflammation of the lungs, and perhaps consumption, may ensue. Nor do the intestines escape, for on any slight irritation diarrhœa is apt to come on. Sometimes the mesenteric glands in the abdomen swell, and this may be associated with dropsy and chronic inflammation of the peritoneum. Nor are diseases of the joints uncommon, and these may go on for months or years, and be very distressing to the patient, being accompanied by discharge of matter and disease of the bone. Scrofulous people are, therefore, liable to a great many diseases in consequence of their constitutional malady. As a rule, persons subject to this affection ought not to marry, as their offspring, will be more or less affected; marriage between cousins thus affected should be strongly prohibited. The general health of such people may be much improved by careful feeding in childhood, cod-liver oil, sea-bathing, and an out-door country life. (See KNEE-JOINT.)

SCURF is a popular term applied to those cases in which the epithelial scales of the skin are shed. It is often so in the heads of children, where branny scales are shed. Washing, once or twice a week, the part with camphor water is a good thing. On a larger scale it is seen in cases of psoriasis and in some cases of eczema. Sometimes it is called scurvy, but this is quite a wrong designation.

SCURVY or **SCORBUTUS** depends upon a state of mal-nutrition, following the use of a diet which is deficient in fresh vegetable matter, and tending to death unless the causes producing it are removed. Scurvy has been known for many years. In the long sea voyages of the navigators of the fifteenth and sixteenth centuries the crews suffered most terribly from this disorder, and many lives were lost. It was looked upon in former times as an infliction of Providence, as a warning against those who presumptuously strove to seek after unknown lands; yet now we know that it is a disease which can be readily cured by adopting proper measures, and by means which every habitable country affords. Scurvy only occurs when fresh vegetable nutriment has been for some time completely or partially withheld. It is most common among sailors, because on long voyages they have so much salt food and no fresh vegetables. Yet it may occur among landsmen. In the Crimean war the allied armies suffered very severely in the winter of 1854 and in 1855. As soon as the supply of fresh vegetables and lime-juice became more plentiful the disease gradually disappeared.

Symptoms.—First, there is a change in the color of the skin, which is pale or sallow; then the mind becomes listless, and the patient is averse to taking exercise and seems apathetic. There are pains about the limbs, and so the sufferer is glad to lounge about and rest himself. Gradually purplish spots, or petechiæ, are observed, especially about the legs and thighs; they are not usually raised above the surface of the skin; then larger patches form, as if numbers of these small spots had run together; and often there is an appearance as if the patient had been bruised. The lips are pale, the face becomes bloated, the conjunctivæ of the eyes become swollen and red. The gums, at first pale, begin to swell at their free margins, so as to encroach upon and almost envelop the teeth : they then become spongy, dark-red, or livid, not painful, but disposed to bleed when irritated. Sometimes the teeth are loosened and fall out; there is also a sickening fetid odor from the breath. Chewing is now rendered impossible, and even fluid food is swallowed with difficulty. Often swellings occur in various parts of the body, and chiefly near the bend of a joint; the most common seat of this condition is the ham, and next the elbow, or beneath the jaw. There is often breathlessness and attacks of syncope or fainting, and this is dangerous; as sudden death may in this way take place; therefore, any one who is bad with scurvy should be kept in the recumbent posture, and not be allowed to sit erect. In bad cases ulceration of the skin often comes on, and may spread rapidly, and be attended with dangerous bleeding. Very little difficulty can occur in making out a case of scurvy, and especially if the antecedent conditions be known.

Treatment.—This must consist in supplying the patient with the material, by the deficiency of which his disorder has been produced. It is wonderful how, in a very bad case, an immense improvement will take place in a few hours by giving lime juice : among the vegetables which may be given are oranges, lemons, limes, cabbage, lettuce, potatoes, onions, mustard, and cress, dandelion, sorrel, scurvy-grass, and grapes. An ounce of lemon-juice should be issued daily when vegetables are short. The other articles of diet must be so arranged in a case of scurvy as to be easy of digestion. The following suggestions have been issued by the London Board of Trade for the information of shipowners and shipmasters :

"Every ship on a long voyage should be supplied with a proper quantity of lime or lemon-juice.

"The juice having been received in bulk from the vendors, should be examined and analyzed by a competent medical officer. All measures adopted for its preservation are worthless, unless it be clearly ascertained that a pure article has been supplied.

"Ten per cent of brandy (specific gravity, 930), or of rum (specific gravity, 890) should afterward be added to it.

"It should be packed in jars or bottles, each containing one gallon or less, covered with a layer of oil, and closely packed and sealed.

"Each man should have at least two ounces (four tablespoonfuls) twice a week, to be increased to an ounce daily if any symptoms of scurvy present themselves.

"The giving out of lime or lemon-juice should not be delayed longer than a fortnight after the vessel has put to sea."

SCYBALA is a term applied to the fæces, or contents of the bowels, when they are passed not in a natural and proper form, but in hard small masses, more like marbles or excretions of sheep than what is ordinarily considered healthy. This condition denotes a costive habit of body, and should be corrected by gentle purgatives or by diet.

SEA-BATHING. (See BATHING, SEA.)

SEA-SICKNESS is a condition well known as one of surpassing discomfort, and one, too, which seldom induces that sympathy which enables us to meet far more serious ills with greater equanimity. Much has been written on sea-sickness, but the exact mode of its causation is not yet quite manifest, though some of its causes are identical with those which produce nausea on shore. Some delicate people cannot ride with their back to the horses of a carriage or to the engine of a train. If they do they speedily become giddy and faint, with a tendency to sickness, though that is rarely induced. We have seen the same cause come into play more forcibly in a boat but a little way off land. The waves running past the boat and the course of the boat in the opposite direction tend to make an individual giddy, and so to favor the advent of sea-sickness in its aggravated form. Under such circumstances a fixed look on the shore at a distance may preserve the individual from being actually sea-sick, though he may be faint. Even on shore unpleasant sights and smells may cause nausea and sickness. It is no uncommon thing for a young student to get sick at his first operation, especially if his stomach is irritable, and, as is well known, evil smells are even more powerful than foul sights this way. To an individual with a tendency to nausea, the smell of bilge-water and tar, or of grease, oil, and the like on ship-board, still more the sight of others in the act of being sick, are powerful inducements to go and do likewise.

But these things, at least some of them, persist; sea-sickness does not. It either passes away on landing, or, if the voyage is one of some duration, it gradually leaves the traveller hungry as a hawk. It is quite plain, therefore, that the immediate cause of the feeling of sickness is the unaccustomed motion of the vessel; once the individual has become acclimatized to that, the feeling passes away, and the benefit of the sea air is felt.

Under ordinary circumstances an individual goes on board ship without any preparation; as soon as the vessel begins to feel a little the motion of the sea the passengers begin to feel queer, especially if they have been eating and drinking more than has been good for them before putting to sea. If to this the individual superadd giddiness, induced by looking at the sea rushing past, there is speedily an end to it: the sea claims her own. There can be no question of the fact that the motion is the main cause of the nausea, for it is much worse in a small boat dancing freely on the water than in a large vessel, which is comparatively steady, and it is worse in what is called a chopping sea than in a regular, even swell, especially if the vessel be small enough to respond to all the motions of the waves. The motion communicated to the vessel is communicated to the passengers; the crew have their sea legs on board. To them the motion is nothing; they balance themselves as easily as on land. But the freshly embarked passenger cannot do this; he cannot balance himself; he is in constant danger of falling, or seems to be so, and his body is agitated in endeavors to support himself. These violent efforts induce motion in the organs contained in the abdomen, and doubtless also the nerves which supply these. Of course, if these viscera are overloaded, the evil comes all the more speedily. A sudden feeling of nausea causes the entrance to the stomach to relax, the motion superadded to contraction of the abdominal walls speedily causes evacuation of its contents, and one act of vomiting begets another. Meanwhile the original cause of the mischief, the motion of the vessel, continues, and the stomach, now rendered irritable, responds more easily to this stimulus, and so the sickness is kept up. Just behind the stomach and liver lie an important group of nerves which partly control the heart's action, the motion of the organs in the abdomen much affect that, and so perhaps the feeling of nausea as well as the attempts at vomiting are kept up. But vomiting always ends in producing intense depression; it tries every muscle in the body —nothing exhausts like it, and when to the former nausea and retching are superadded this feeling of exhaustion, the full misery of sea-sickness is developed; but by and by, as the system becomes habituated to this motion, the new sensations pass away, perhaps to return no more.

Prevention is better than cure, and of nothing is this truer than sea-sickness. As, moreover, many of us cannot stay at home forever, it is better to try to understand the best method of avoiding the scourge if we can. Should we desire to avoid the terrible nausea and depression, it is best to have the bowels well opened the day before, so that they shall not be loaded. We should also take care that the stomach is not overloaded, but as retching on an empty stomach is not pleasant, it is desirable to take a little food an hour or two before embarkation. Drinking or smoking, especially in those not accustomed to either, are strenuously to be avoided, as tending to render the stomach irritable. When the individual goes on shipboard he should select a spot where the motion is likely to be least, that will be as near the centre as possible, and then as the motion of the body standing is greater than sitting, and sitting than lying, it is best for him to lie down flat on the deck if possible.

Treatment.—As to internal remedies, all kinds of things have been tried ; none do so well as spirit of chloroform, which used to be called chloric ether. Thirty drops, or even a teaspoonful, of this may be taken in a little water as soon as the traveller goes on board and has lain down. For it is not given with a view to cure, but with a view to prevent the nausea. Certain it is that in a goodly number of cases the spirit of chloroform either enables the stomach to meet the shocks better by stimulating it, or by soothing it ; at all events, in a short voyage there is a good chance of escaping. In a longer voyage, when the sickness is passing away, drachm doses of aromatic spirit of ammonia, with a little spirit of chloroform, should be given ; but a still better " pick-me-up" is iced champagne. Fortunately, the two are not incompatible.

SEBACEOUS GLANDS. (See SKIN.)

SEDATIVES are medicines which depress the vital powers without inducing any previous excitement. The only remedies of this class that can be trusted in the hands of non-professional persons are tobacco and diluted hydrocyanic acid. The effect of tobacco is attained by smoking, which even if injurious as a regular practice may sometimes be indulged in with advantage. Hydrocyanic acid is a deadly poison in large doses, but in very small doses (2 or 3 drops of the diluted preparation) it may be given without risk in t e sickness which accompanies pregnancy, or in other ordinary cases in which a sedative is required.

SENEGA is the root-stock and rootlets of the *Polygala Senega*, growing in North America. The rootlets have a peculiar heel on one side ; their color is grayish-yellow. The taste is sweetish and acrid, causing flow of saliva. The active principle, senegin, also causes sneezing when applied to the nostrils. Two preparations of senega are in use, viz., an infusion and tincture.

Senega seems to act mainly on mucous membranes, especially on those of the lungs. It also acts on the skin, and sometimes on the kidney. Some esteem it to possess a certain influence over the heart and womb. It is, however, used almost entirely in chest

disease, as a remedy in chronic bronchitis, whooping-cough, and the like. Here it is seldom prescribed alone, generally other substances, as paregoric and carbonate of ammonia, are combined with it. It has been used in renal dropsy and in painful menstruation. The dose of the tincture is a drachm, of the infusion half an ounce to an ounce.

SEIDLITZ POWDERS are cooling powders, which do not appear in the pharmacopœia, but are nevertheless useful as a gentle aperient. Each dose requires two powders to prepare it, the white paper containing an acid, the blue an alkaline powder. The latter consists of Rochelle salts, tartrate of potass and soda and bicarbonate of soda, the acid being usually tartaric acid. These are mixed together with water and drunk while effervescing. (See ROCHELLE SALT.)

SENNA, as used in medicine, is of two kinds, the so-called Alexandrian or Egyptian senna, and East Indian or Tinnevelly senna. The substance is the leaf of various species of cassia. They all have a peculiar odor, and all, if examined, will be seen to have one side shaped differently to the other at the base of the leaf where it joins the stalk. Senna readily yields its virtues to water. Its preparations are a confection, infusion, mixture, tincture, and syrup. The confection is a good, useful preparation, consisting of senna, coriander, tamarinds, cassia pulp, prunes, extract of liquorice, and sugar. It is useful in piles. The mixture, best known as *black draught*, contains sulphate of magnesia (Epsom salts), extract of liquorice, tincture of senna, tincture of cardamoms, and infusion of senna. The tincture contains, besides senna, raisins, caraway, and coriander ; the syrup, coriander and sugar. Senna is hardly ever given as powder ; the infusion is most commonly employed, except among children, where the tincture or syrup takes its place. About an ounce may be given of the infusion, the same of the mixture, and of the confection a drachm or more. The syrup is given to children in the dose of a drachm or more.

Senna, as is well known, is a purgative, stimulating the motion of the bowels, and also aiding slightly in promoting their flow, but a salt of some kind, Epsom or Rochelle, is generally added to increase its efficacy in this way. Senna is seldom given alone, as it is apt to gripe, and for this reason spices are usually administered along with it. Senna is more generally used than any other purgative when it is simply desired to have the bowels cleared out, as it is apt to leave no ill consequence behind. It should not, however, be given if there is any tendency to inflammation of the bowels.

SEPTICŒMIA is an acute disease which resembles pyæmia very much in its general characters, and which is supposed to be caused by the absorption into the blood of putrid material from the surface of a wound or ulcer. It generally occurs after phlegmonous erysipelas, sloughing, or other forms of unhealthy action about a wound which has been caused either by accidental injury or by the knife of the surgeon. The following are the most prominent symptoms of this disease : great prostration ; the patient lies helplessly upon his back, as in bad typhoid or typhus fever, and at last falls into a state of intense collapse ; this prostration is increased by profuse perspiration and obstinate diarrhœa ; the tongue is dry and brown, and there is often much irritability of the stomach indicated by nausea and frequent vomiting. The nervous symptoms are very characteristic ; the patient appears drowsy and apathetic, and sensibility seems to be lost. There is low muttering delirium with short intervals during which the patient seems quite conscious and discourses rationally. There is rarely much restlessness or violent delirium. The frequently repeated fits of rigors or chills so characteristic of pyæmia are generally absent in this disease. The patient sinks slowly, and during the last twenty-four or thirty-six hours of life is in a state of coma and collapse, during which the actions of the heart and lungs are carried on very feebly and almost imperceptibly.

SEROUS APOPLEXY is a term often used, but it is a wrong expression ; such cases are nearly always due to chronic Bright's disease, and should be called renal coma. (See APOPLEXY and COMA.)

SERPENTARY consists of the dried root of the *Aristolochia Serpentaria*, a native of the United States. It also goes by the name of Virginia snake-root, and must not be confounded with the black snake-root. The part employed is the root or root-stock with the rootlets attached. These are of a pure brown color, and have a peculiar odor and taste, something resembling camphor. It contains bitter matter, a volatile oil, and some resin. Its preparations are an infuson and a tincture. Serpentary acts as a stimulant, but some men think much of it, some little. It seems in the hands of some to do good in certain forms of indigestion and certain conditions of bowels. It also acts on the skin as a stimulant, though not very powerfully. It seems to do good in chronic rheumatism and subacute gout. The tincture is commonly employed in doses of a drachm. The powder and infusion are not often given.

SERPENTS, BITES OF. (See SNAKES, BITES OF.)

SERUM. (See BLOOD.)

SETON means a long wound artificially made under the skin, the walls of which wound are kept in a state of irritation and suppuration by the presence of some foreign body. It differs from an issue in being a tubular wound *under* the skin and not an *open*

ulcer. A seton may be established by transfixing a pinched-up fold of skin by a large flat needle armed with a strand of cotton or silk thread, or by passing a bistoury through the base of the fold and then carrying the thread through the canal thus made, by means of a small-eyed probe. After the thread has been allowed to remain at rest for two or three days and has set up irritation and some discharge, it is pulled a little further through the wound so that a fresh portion may be included and the soiled portion be cut away. This manœuvre is repeated every second or third day, and when the strand is almost used up a fresh strand is attached and substituted for it. Instead of cotton or silk thread many surgeons use a small flat band of india rubber, which is less liable to become clogged by dry and offensively smelling discharge. Setons are established for the purposes of setting up counter-irritation, and of causing a chronic discharge so as to produce a drain upon the system. With the former object in view they are often useful when applied to the temple in some affections of the eye, and to the back of the ear in cases of deafness. As a means of producing a constant drain upon the system a seton is often established in old people who are threatened with an attack of apoplexy, or who suffer from constitutional disturbance in consequence of the closing by cicatrization of a large chronic ulcer. Chronic abscesses and tumors with fluid contents are often treated by the introduction of a long strand of silk thread. As the fluid flows slowly away from the orifices of the seton irritation is set up in the walls of the sac, which contract, and are finally glued together by inflammatory conditions. (See ISSUES.)

SHAKING PALSY, also known as *Paralysis Agitans*, is a malady most common in advanced life. (See PARALYSIS.)

SHAMPOOING. (See TURKISH BATH.)

SHERRY. (See WINE.)

SHINGLES. (See HERPES.)

SHIVERING. (See RIGORS.)

SHOCK, which is also called *Collapse* and *Prostration*, is an immediate result of severe injury, and consists in general depression of bodily power, and in partial or complete arrest of the heart's action consequent upon an intense and violent impression upon the nervous system. A similar condition is produced by the action of certain poisons.

The following are the symptoms of well-marked shock : The surface of the body cold and very pallid ; the bloodless condition of the skin is most evident in the face and lips, presenting a strong contrast to the usually florid appearance of this portion of the body ; the forehead is covered by drops of cold clammy perspiration ; the breathing almost imperceptible ; the pulse weak, irregular, and in extreme cases imperceptible ; great muscular debility—*prostration*; the patient is in a state of stupor, and the sensibility is benumbed. The symptoms of shock vary very much in degree in different cases, according to the nature of the injury and to the bodily or mental condition of the patient. Their intensity is much increased in cases where there has been much hemorrhage, then there is more mental disturbance, and the patient presents all the symptoms of severe syncope. In some cases there are nausea, hiccup, and vomiting. In cases of injury to the head resulting in compression or laceration of the brain, the symptoms of shock may be associated with convulsions and palsy.

The duration of shock varies very much. In less extreme cases the symptoms subside in the course of one or two hours; in severe cases they may last for thirty-six hours or two days. In cases of recovery the patient passes from a state of shock to one of perfect or imperfect reaction. When the reaction is perfect the pulse becomes stronger and fuller, and the breathing deep and well marked. The most favorable signs are returning warmth of the surface of the body, and slight restlessness on the part of the patient, with an inclination to lie on his side. In the course of a few hours there may be some fever indicated by a hot skin, a flushed face, and bright eye, and a rapid pulse. These symptoms, however, in favorable cases soon pass off, and the complete recovery is established. With imperfect reaction, on the other hand, the febrile symptoms increase in intensity, and then after a time give way to symptoms of nervous excitement and general exhaustion. There are great mental excitement, with or without delirium, muscular trembling, and much restlessness. These symptoms are associated with others indicating rapid exhaustion, such as vomiting, a cold and moist skin, and a low fluttering pulse. In bad cases these symptoms increase in intensity, and finally the patient dies in a state resembling coma. In children convulsions often occur during the states of shock and of imperfect reaction. Patients who have been accustomed to take large quantities of beer, wine, or spirits, almost invariably present during this state of imperfect reaction all the symptoms of violent *delirium tremens*. In individuals who are naturally weak and delicate, reaction, though favorable in its course, and steadily progressive, may be very slow, so that complete recovery is not attained for several days after the occurrence of the injury.

By far the most frequent cause of shock is injury. *Cæteris paribus*, the more important and necessary to life the injured organ may be, the more intense are the symptoms of shock. Severe and even fatal shock may be caused by injuries which produce no morbid appearances in any part of the body. A blow over the pit of the stomach or compression of the testicle may often give rise to intense and alarming symptoms. In railway accidents shock is often produced without any visible injury or subsequent symptoms of injuries to internal organs. Sudden and violent injuries to limbs, with extensive crushing of the soft parts, and compound fractures, are always followed by shock. The most intense shock resulting from visible injuries is probably met with after burns and scalds, either when a considerable depth of soft structure has been destroyed, or when the injury though superficial has involved a considerable extent of the surface of the body. A very superficial scald is almost always fatal in children, when a considerable portion of the skin covering the front of the chest and abdomen has been thus injured. Intense pain in connection with any kind of injury generally causes much shock. With gun-shot wounds there is generally well-marked shock, which is favored or intensified by the circumstances under which the injury was received. Professor Longmore, of the Army Medical School at Netley, remarks on this point that panic may lead to severe symptoms of shock, although the wound is not of a very serious character. "A soldier," he says, "having his thoughts carried away from himself, his whole frame stimulated to the utmost height of excitement by the continued scenes and circumstance of the fight, when he feels himself wounded is suddenly recalled to a sense of personal danger, and if he be seized with doubt whether his wound is mortal, depression as low as his excitement was high may immediately follow." In all cases of shock following injury the symptoms are modified by the mental and bodily condition of the patient, by the nature of the accident, and the circumstances under which it has taken place, and by the amount of disturbance to the nervous system, and the organs of the chest. Much loss of blood increases to a considerable extent the intensity of shock. Symptoms resembling those of traumatic shock may be produced by the action of narcotic and corrosive poisons, and also by intense pain caused by disease.

Treatment.—The chief indications in the treatment of severe shock are to keep up the action of the heart and lungs, and to maintain the temperature of the body until the full effects of the sudden and violent impression upon the brain and nervous system have passed off. In a case where there has not been much hemorrhage, and where no large wound is present from which bleeding might be likely to occur before the arrival of a medical man, the patient should at once be placed in bed between warm blankets; a bottle of hot water should be placed near the feet, and one under each armpit; if suitable stone bottles are not at hand, bricks or any other bodies which will retain heat for a time should be used. Care must be taken to prevent burning of the patient's skin by wrapping the heated bottles or bricks in flannel. Brandy should be administered frequently, and in small quantities, the spirit being slightly diluted with hot water. If there be much nausea or vomiting, an injection should at once be made into the rectum of beef-tea and brandy, or milk and brandy. When the shock is so intense that the breathing ceases or becomes almost imperceptible, an attempt should be made at artificial respiration according to the methods described in the article on DROWNING.

In cases of shock associated with profuse hemorrhage, brandy should not be administered very freely, nor should much warmth be applied to the surface of the body until the bleeding vessels have been closed either by ligature or by pressure.

The treatment of the serious symptoms indicative of imperfect reaction should consist in supporting the system and preventing fatal exhaustion on the one hand, and in allaying nervous irritability and producing sleep by large doses of sedative drugs on the other hand.

SHORT-SIGHTEDNESS. (See EYE and VISION.)

SHORTNESS OF BREATH is a common symptom in many diseases of the heart and lungs. (See DYSPNŒA.)

SIALOGOGUES are substances which promote the flow of saliva; they are hardly used in medicine. Chief among sialogogues is horse-radish, which is used mainly as a condiment. Pellitory is also powerful this way, as is ordinary mustard. The mere motions of the gums in chewing acts this way, so that masticating a bullet or a piece of india-rubber frequently excites salivation.

SICK-HEADACHE. (See HEADACHE.)

SICK-ROOMS should be as capacious as possible, because then the patient has 'more air to breathe, and it does not require renewal so often as the air in a small room does, and thus a draught is prevented. A fair amount of ventilation is carried on by the door, windows, and fireplace, but at least twice a day the windows should be opened so as to cleanse the room. If the patient can leave the room for a short time, so as to allow of a free current of air, so much the better; if not, the patient should be lightly covered over, so as not to feel any draught. In cases of fevers and any contagious disorders, it is best to remove from the room all unnecessary articles, as curtains, hangings, carpets, etc., and let there be disinfectants about. (See DISINFECTANTS.) In chronic cases the sick-room should be made as cheerful as possible, and the amount of light should be regulated so as to please the patient; in cold weather the fire should be kept nice and bright; when possible, flowers should be placed in the room.

But there are a hundred little details and comforts which a practised nurse will look after. A thermometer should be in the room so as to have the temperature properly regulated. (See NURSING.)

SIGHT is a special function of the optic nerve, by which we become acquainted with the world around us. (See EYE, COLOR BLINDNESS, and VISION.)

SILK, OILED, is a very useful preparation of silk, which renders it impervious to water or grease, and is chiefly used in surgery to lay over dressings of wounds so as to keep the lint, saturated with water, from becoming dry by evaporation. Less expensive textures are prepared from india-rubber, but they have all a disagreeable smell, which does not exist in oiled silk.

SITZ-BATH is a form of the bath much used in the hydropathic practice, and occasionally prescribed by the regular physicians. It consists in the immersion of the hips in cold water, by sitting down in a tub containing a moderate quantity of the water. Tubs designed especially for the purpose may be had, but an ordinary tub will answer for the purpose, unless the patient is feeble. The sitz-bath is unquestionably a powerful remedy in certain diseases of the head and digestive organs, and usually braces the entire system; but it is not advisable to take the bath except under medical advice.

SKELETON.—The skeleton of a full-grown human being consists of 200 distinct bones, exclusive of the little bones in the internal ear. They are thus distributed:

The spine	26 bones.
Skull	8 "
Face	14 "
Ribs and breast-bone	26 "
Upper extremity	64 "
Lower extremity	62 "

These bones are divided into four classes, known as *Long*, *Short*, *Flat*, and *Irregular*. The *long bones* are those which exist in the limbs, and are employed in locomotion; these consist of a *shaft* and two *articular extremities*, these extremities being covered with what is termed articular cartilage, and being capable of mutual movement upon each other by one or another form of joint, the gliding movements of such joints being assisted by the presence of bags containing joint oil (*Synovia*), which is placed between these articular cartilages. The shaft of a long bone is cylindrical, or nearly so, and its extremities are expanded. The shaft consists of compact tissue, while the extremities are composed of spongy, having a thin layer of compact tissue coated over them. The long bones are the cubit, the two bones of the forearm, the thigh bone, the shin and splint bones, the bones of the fingers and toes, and the collar bone.

Short Bones are compact, strong bones,

having several articular surfaces for mutual adaptation, and are found in those parts of the body where strength and limited motion are required, such as in the wrist, bones of ankle and instep. They consist of spongy tissue, with a coating of compact structure.

Flat Bones afford broad flat surfaces for the attachment of muscles and for the protection of cavities ; they consist of two layers of compact tissue, containing a layer of spongy between them. They are the skull bones, blade bones, haunch bones, breast bones, and ribs.

Irregular Bones are those which, as their name would suggest, cannot be grouped with the previously named, such as the bones of the spine, jaw bones, and several of those bones which make up the skull.

The natural position of the human skeleton is erect, and this is in great measure due to combined muscular action ; moreover, the natural architecture of the skeleton adjusts its own centre of gravity, which tends greatly to this end : thus all those joints which transmit weight to the ground lie in one vertical plane, and such a line would be described as passing from the top of the head, through the joints between the head and first bone of the spine, through that between the last bone of the vertebræ and the sacrum, and through those between the sacrum and haunch bone, the hip, knee, and ankle. The spine, consisting of a great number of bones, peculiarly articulated together by interposed elastic cushions, increases in size from above downward, and moreover, possesses several well-marked curves. The object of these cushions and curves is to receive the shock of sudden blows and falls, and to disperse their effects ; again, the curves are arranged alternately, so as to distribute the weight with greatest advantage to the centre of gravity of the body, which passes through all the curves, and falls on the centre of the base of the column. It will be observed that all the bones of the limbs are slightly curved, thus assisting in the individual and mutual transmission of shock. The pelvis (sacrum and haunch bones) is very broad and strong in man, and the plane of its arch is in such a direction that the weight is transmitted vertically from the sacrum to the heads of the thigh bones. The thigh bone being curved inward, allows of the weight of the body being brought under the pelvis, and transmitted to the broad expanded ends of the bones forming the knee-joint. The foot, in its turn, consists of an arch, or rather a double arch, which receives the transmitted weight, at its crown directly through the leg bones. Thus it will be seen that the upper limbs take no part in the maintenance of this natural upright condition, the composition of the skeleton being so arranged as to be subservient to it.

SKIN forms the external covering of the body ; there is an upper layer called the epidermis, or cuticle, which is made up of flat, rounded cells, and which are being always shed off gradually and replaced by new ones ; beneath this is the tough cutis, or true skin, which is chiefly made of fibrous tissue ; in the skin are numerous hair-follicles and sebaceous or sweat glands. The skin not only serves as a coat to protect internal organs, but serves other useful purposes ; it eliminates a large quantity of water daily, which is called perspiration or sweat, and this is always going on, although, unless violent exercise be taken, it is insensibly carried on. Carbonic acid, urea and fatty matters are also excreted by the skin to a certain degree. When a blister is applied, it is the epidermis which is raised up, while a serous fluid is beneath ; so again, when the skin peels after scarlet fever, it is only the upper layer of the epidermis that is shed. To enable the skin to act properly, it must be kept quite clean, although this is very seldom done ; cold water bathing is not enough, but an occasional hot bath must be taken so as to thoroughly cleanse the pores. The skin is liable to many diseases, but the names only need be mentioned here, and the reader must refer to the articles on those subjects for further information.

1. Those diseases which begin or appear as pimples or papules : strophulus, lichen, and prurigo.

2. Those diseases which are vesicular at first, or begin with a little blister or watery head : eczema, herpes, miliaria, sudamina.

3. Those diseases which are pustular, or contain a little matter : impetigo, ecthyma, and small-pox.

4. Those depending on the presence of a parasite, either animal or vegetable : ringworm, itch, etc. (See Ectozoa, Epiphyta, and Parasites.)

5. Those diseases attended with tubercles, or raised lumps, larger than a pimple : acne, molluscum, keloid, lupus, cancer, yaws, elephantiasis.

6. Those attended by too much or too little coloring matter in the skin, and forming, therefore, white or dark-colored spots : leucoderma, freckles, vitiligo, xanthelasma, Addison's disease.

7. Those diseases where the skin is harsh and rough : ichthyosis, xeroderma.

8. Those diseases in which there are hemorrhages under the skin, and so purple spots or patches are seen which do not disappear on pressure : purpura, scurvy, bruises, malignant disorders.

9. Those diseases which are attended with fever, and where there is a rash, as in the exanthemata : scarlet fever, measles, erysipelas, etc.

Skin-grafting consists of removing some scales of epithelium from a healthy portion of skin, and applying them to an old ulcer which will not heal readily ; these scales thus grow and form new centres from which a healthy cicatrization will presently ensue.

SLEEPLESSNESS, technically known as *insomnia*, is one of the most troublesome conditions with which we are called upon to deal. Sleep is absolutely essential to all, for the repose and repair of the nervous centres, which during waking hours are constantly, though unknown to us, engaged in the fulfilment of certain important functions. In early life the greater part of time is occupied in eating and sleeping. In adult life, as a rule, about one third of our time is passed in this manner, but elderly people often suffer from sleeplessness.

Sleep is ordinarily preceded by a feeling of languor and heaviness, during which we see some of the unobserved functions of the nervous system making themselves apparent. Thus, first of all, the eyelids droop—we are not aware of any strain in keeping them open, yet the moment our attention, involuntary though it be, is taken off, they gradually sink. So too the head, ordinarily held erect, falls forward, and the limbs, fall into the easiest posture. Even if the individual lies down, it will be noticed that, as sleep comes on, a different posture is assumed, one which allows of the relaxation of all his limbs and all his muscles. The respirations too are slower, gentler, and more prolonged ; they are carried out with the least possible amount of work, and the same may be said of the heart. Sleep then might be defined as the condition of least action in the human body, mentally and physically.

It is quite plain that interfering with this repose, which is absolutely necessary for the perfect nutrition of the body, must be fearfully exhausting, especially if there is continuous bodily and mental exertion. Indeed death may result from continuous want of sleep, and this is sometimes had recourse to as a punishment in China. Continued sleeplessness is therefore a most serious thing ; it is often the first indication of insanity, and is one of the most troublesome symptoms of violently insane persons. Mental anxiety frequently banishes sleep, but when the body is worn out, sleep comes and relieves the sufferer. Violent passions, though for a time dispelling sleep, ultimately bring it on in the same way through bodily exhaustion.

To procure sleep, especially of a sound and refreshing kind, is often of the very highest importance. If any one is about to undergo severe mental or bodily exertion, a good sound sleep is of even greater importance than a good meal. Indeed sleep is of essential importance to enable all to perform their allotted tasks, and so a few words on the best means for procuring sleep may be of value.

Dyspeptics are seldom sound sleepers, and in many cases the first thing to be done is to get the digestion in good order. This little fact shows the importance of dealing with every case on its own merits, and not by mere routine. Thus, we have known an individual who passed sleepless or worse than sleepless nights, and was troubled with evil dreams and nightmare as soon as his eyes were closed, have chloral administered. As a consequence, the dreams and nightmare were worse than ever, but a blue pill and a black draught speedily secured sound and refreshing sleep. Constipation and interference with the functions of the liver are serious enemies to satisfactory repose.

To many a due amount of exercise in the open air is absolutely indispensable, if sleep is to be procured, and it is often observed that a buffeting with the wind causes sleep sooner than any other form of exertion. The diet too must be attended to, if sleep is to be refreshing ; here every man must be a law to himself, for what sometimes soothes and comforts one man may excite another, and altogether prevent sleep. There is, however, one great rule, and that is never to go to sleep with an undigested meal in the stomach, if it be at all a heavy one. To this end many do much better by making their chief meal early in the day, and only taking a light meal in the evening. Tea and coffee taken late in the day are particularly prejudicial to sleep. The influence of these, however, differs in different individuals ; to some tea is more stimulating than coffee, to others the reverse is the case.

Going to bed at a certain regular hour, be that what it may, is powerfully conducive to sleep ; habit here, as in other things, becomes all-powerful. It is, too, advisable to have thoroughly done with the work of the day, some time before going to bed. If that work has been of a bodily description, the rest will often be enough of itself ; if it has been headwork, a change is often best. For many individuals there is no preparation for sleep equal to a pipe and a novel, to others this would be poison.

For refreshing sleep, it is essential that the bedroom be well ventilated, and many who make it a regular practice to sleep with the windows open find it of great value. Undoubtedly the refreshing nature of the sleep is enhanced by fresh air. Then, too, the bed should be in the middle of the room, not in a corner of it ; no curtain of any kind should surround it. Feather beds are an abomination, a good firm mattress is best. The pillows should be adjusted to the height of the shoulders, so that when one lies in the natural position on one side, the head is in a line with the rest of the body—the neck straight, not to be bent either upward or downward.

To many a " night cap" is essential, be it

a glass of wine, a tankard of ale, or a tumbler of grog. If indulged in with discretion, there is nothing to be said against the practice, except that, should the individual be so situated as to have to go without his accustomed stimulant, he will most probably pass a sleepless night.

The great thing, in most cases to procure sleep, is to obtain absolute rest of mind. To men of active brain this is sometimes singularly difficult, and many plans have been proposed to overcome the difficulty. They all consist in this, in attending to something of absolutely no interest, and which is of a dull uniform nature.

Of course all these things fail, especially in the presence of pain, and then more powerful means must be tried. Chief among these are opium, morphia, and chloral hydrate, but no man ought to take either or any of these on his own responsibility, for thereby habits are readily acquired which may be hard or even impossible to get rid of. If ordinary means and ordinary remedies do not suffice, the sufferer ought to consult some one of skill, in whom he has full confidence ; that is important.

SLOUGH is the dead part of the tissue of the body which is separated and thrown off by the healthy part after inflammation. It often becomes necessary, when this process is taking place, to assist nature by removing this source of annoyance, and to prevent the foul odor that arises from it by the use of disinfectant lotions—such as weak solutions of carbolic acid. (See MORTIFICATION.)

SMALL-POX, or VARIOLA, is a febrile, eruptive, and contagious disorder, which in past times raged with much violence, but in recent periods has been vastly controlled by the discovery of vaccination. The most common varieties are : the *discrete*, in which the pustules are distinct ; the *confluent*, in which the pustules run together ; the *malignant*, which is often associated with purpura and an eruption resembling measles—a very dangerous form ; the *modified*, which comes on in those partially protected by vaccination, and a kind that runs a very mild course. In cases of small-pox there is : 1. The stage of incubation, which lasts twelve days, from the date of receiving the poison. 2. The stage of eruptive fever and invasion, lasting forty-eight hours. 3. The stage of maturation wherein the rash is fully developed, lasting about nine days. 4. The stage of secondary fever or decline, lasting a variable time, according to the severity of the disease. Discrete small-pox is the simplest form of the disease, and is rarely attended with danger to human life ; confluent small-pox destroys the greatest number of lives, and may prove fatal to as many as 50 per cent. In the distinct or discrete form, the primary fever is less intense than in the confluent form ; in the lat-

ter, there is often delirium, and more especially in those who are intemperate, such as draymen, potmen, tailors, compositors, etc. The malignant variety is terribly fatal ; the blood seems profoundly poisoned from the first, and is more fluid than usual ; bleeding from the mouth, nose, and bowels is not uncommon ; in women, there is also bleeding from the womb, and if they are pregnant, abortion will ensue. In modified small-pox, the patient is often able to go about the whole time, and the rash may suddenly decline on the fourth or fifth day, and recovery follow.

Symptoms.—The disease begins with shivering or rigors, pain in the back, vomiting, thirst, headache, and a general feeling of indisposition ; in children, convulsions may come on. In many cases the rash of small-pox in vaccinated cases is preceded by a more or less scarlet or roseolous rash which is mottled over the body. If the finger be pressed on the forehead, a shotty feeling may be noticed, for the rash of small-pox generally commences there ; at first a pimple forms, but afterward a pustule, and then it dries and scabs over, and leaves a pit or depression behind. When the rash comes out, the temperature falls, but rises again about the eighth or ninth day ; in mild cases, however, this secondary fever is hardly perceptible. The eruption usually appears first on the forehead, face, and wrists, and then on the rest of the body, coming out on the legs and feet about two days later. The eruption takes about eight days to arrive at its full development ; during this time there is much swelling of the face and eyelids, so that the patient cannot see for a few days ; in bad confluent cases the face seems covered with a mask, and a disagreeable odor proceeds from the body. Boils are apt to form in cases of confluent small-pox ; they are also very subject to pleurisy, pneumonia, and bronchitis ; sometimes the tongue is much swollen and dry, and the patient may be unable to close the mouth or to speak ; this is a very bad symptom. Inflammation of the ear, followed by an abscess, is not uncommon in this disorder. Erysipelas, gangrene, and pyæmia are now and then met with in the course of this disorder. Inflammation of the eye and ulceration of the cornea may add to the general mischief. For a pustule to form on the eye is very rare, but it is very common to see one in the soft or hard palate. Small-pox may be mistaken for measles, but in the latter disease there is little or no fever, the rash comes out in twenty-four hours, is vesicular and not shotty, and more abundant on the trunk, and not so much on the face and forehead.

Mortality.—The death-rate of confluent

small-pox is 50 per cent, and of discrete small-pox four per cent. Confluent small-pox is very rare in those who are vaccinated. Age has an influence on the disease, for it is most fatal in children and old people, but least fatal between 10 and 15 years of age. Small-pox is decidedly an infectious and contagious disorder ; riding in a cab or omnibus in which a patient has been recently, or even passing them in the street, will give the disease ; a mild case may give rise to a severe one, and *vice versâ*. It may attack an individual a second time, but this is a very rare occurrence.

Treatment.—There is no medicine which can check this disorder. The patient should be at once isolated, and it is best when an epidemic is about that small hospitals should be built away from other dwellings where these cases can be treated, and the spread of the disorder diminished. For diet, they may have milk, tea, gruel and beef-tea, chicken-broth, and, in fact, the treatment which has been laid down for fevers generally. There should be great pains taken to ventilate the room without having too much draught, and keeping it about a temperature of 60°. Great cleanliness must be observed, and all linen, clothes, etc., must be disinfected after being used. Bed-curtains, carpets, and hangings of any kind must be dispensed with. Flour, starch, or hair-powder may be abundantly peppered over the face and body to relieve the itching and discomfort, and to absorb any acrid discharge. It is doubtful if any good will arise from using anything to prevent pitting ; gutta-percha in chloroform does no good, but if the face be washed over in the early stage with nitrate of silver, it may lessen the marking ; olive oil, cold cream, and glycerine and water will relieve the patient when they are locally applied. After recovery, the stains are shallow and of a brownish color, becoming paler after a few months.

In the article on vaccination, the subject of the prevention of small-pox has been very fully entered into, and therefore need not be repeated here ; the latter article should be read with this one, if the reader is anxious to understand the relations of the two disorders. For information with regard to disinfection, etc, see SANITARY REGULATIONS.

SMELL is a special function of the olfactory nerves, which are two in number, and are distributed over the lining membrane of the nose.

SMELLING-SALTS. (See AMMONIA.)

SMOKING. (See TOBACCO.)

SNAKES, BITES OF.—The more rapidly the symptoms of poison appear after the bite of a snake, the more dangerous they are likely to be. The two fangs of the reptile commonly enter and produce two minute wounds, from which only one or two drops of blood may at first issue. A smarting, severe burning pain is immediately perceived, the part begins to swell, and a puffiness almost to the bursting of the skin spreads in a short time over the whole limb. There is fever, often with delirium, small pulse, pain in the region of the heart, and convulsions. These symptoms are attended with a feeling of anxiety and lassitude, laborious respiration, thirst, nausea, vomiting, and syncope. Death from the bite of a viper has been known to occur in thirty six hours. If the individual survive the first effects, the wounded part may become livid and gangrenous, and he may sink under the irritative fever set up. According to Fontaine, out of more than sixty cases of viper-bites only two were fatal, and in one of these gangrene commenced in the wound in three days, and the person died in twenty days.

Treatment.—The treatment in case of a bite from a poisonous snake should first be the application of a ligature between the part bitten and the heart, or of a cupping-glass, in order to prevent absorption. The wound should be enlarged, and well washed. If absorption has taken place, and the limb be swollen, the whole of the skin may be smeared with oil, and attention directed to the constitutional symptoms. Brandy and ammonia should be given to prevent depression. Strong acetic acid, which coagulates the poison, may be applied when the person is seen soon after the accident. There is no known antidote to the poison of the cobra. The serpent-charmers of the East appear to secure themselves from injury by extracting the poison-bags under the fangs, or by causing the snake to exhaust itself by biting other animals, before handling it.

SNAKE-ROOT. (See ACTEA and SERPENTARY.)

SNEEZING is a convulsive action of the respiratory muscles, caused by irritation of some part of the lining membrane of the nostrils, either by the presence of some particles of matter, such as dust or snuff, or owing to the congestion of the membrane induced by what is called a cold in the head.

SNUFF is usually composed of dried and powdered tobacco, but many herbs are used in the same way, and are sold under the name of cephalic snuffs, for headaches and the like. A pinch of snuff may sometimes be useful in relieving the irritation of the lining membrane of the nose and head by sneezing, but when taken to excess, snuff is extremely injurious.

SOAP, as used in medicine, is of two kinds, hard and soft ; both are made from olive oil, but into hard soap soda enters as an ingredient ; into soft soap, potash. Olive oil consists mainly of two substances, olein and palmetin ; these being made up of oleic acid and palmetic acid, combined with glycerine. If now to either of these an alkali be added, what is called saponification takes place, the acids combine with the alkalies, and glycerine is set free. The substance is no longer a fat, it is a soap. Hard soap is grayish white in color, but that commonly used, called castile soap, is veined and marbled. Soft soap again is a semi-fluid mass resembling honey. It is yellow and semi-transparent, often showing white points where crystals have begun to form. These soaps ought to be well neutralized by the alkali so as to have no greasiness about them.

Hard soap is employed in making soap cerate plaster and soap plaster, and in the preparation of a useful liniment commonly known as opodeldoc. This liniment contains hard soap, camphor, oil of rosemary, spirit, and water. Soft soap is used in turpentine liniment. The intention of the soap in this, and its addition to many other liniments, is to enable the part to which it is to be applied to be well rubbed without suffering from the results of friction on the skin. In these cases it is the rubbing which does the good, not the liniment.

Hard soap is often used as a basis for pill-making, but the other ingredients ought to be carefully selected. Thus, some substances, as resins of a purgative kind, are best given with an alkali; substances of an acid nature again should not be given with soap. Soap and water is a favorite injection with some practitioners, but the soap in that case is nearly useless; it is the mechanical result of the water which produces the desired motion, only the soap may soften the parts concerned.

SODA, as an alkali, is used as liquor sodæ, i.e., solution of caustic soda. This is made by heating carbonate of soda with slaked lime, when caustic soda is set free. This liquid is colorless, and has an intensely burning taste. The solution, when evaporated to dryness, constitutes caustic soda, or hydrate of soda. The solution is powerfully alkaline, and might be used in a good many cases instead of liquor potassæ, but this last substance seems to be preferred for internal use. The caustic soda in a solid form may be used in the same way as caustic potass for destroying the edges of ulcers, etc. It does not melt so readily, and is not apt to run on to places where it is not desired, but it is not so much used as the other.

Carbonate of Soda, or washing soda, is of great importance economically, not much medically, save for cleanliness. It is now made from common salt, but used to be made from sea-weed ashes. It occurs in large irregularly crystalline masses, which, when dried, yield up their water of crystallization and fall into powder. This is dried carbonate of soda. Neither is often used internally. The soda salt, mainly used for its alkalinity, is the

Bicarbonate of Soda, which only occurs in powder. It is only slightly alkaline,

and is not at all caustic. The preparations are an effervescing solution—medicinal soda-water, and a lozenge. Ordinary soda-water contains no soda, only carbonic acid; if it is desired to have soda in it, the specially prepared soda-water must be used. Bicarbonate of soda is much used as an antacid, and to render other substances alkaline. It sets better on the stomach than bicarbonate of potash. It does not act so much on the kidneys, and is not given in acute rheumatism. For ordinary antacid purposes, especially to allay heartburn, it is more used than the potash salt. The dose of bicarbonate of soda is from 5 to 30 grains, but more may be given, though seldom necessary.

Sulphate of Soda, better known as Glauber's salt, is a substance which most undeservedly has fallen out of repute. It is a waste process in making hydrochloric acid, is also found abundantly in certain mineral waters, as well as in sea-water. The salt occurs, when pure, in prisms, and is colorless, transparent, and neutral. Its taste is exceedingly bitter, and given internally acts as a purgative, producing copious watery motions. It is the most important purgative constituent of many mineral waters. It may be given in doses of 2 drams to half an ounce. It is best given mixed with some other purgative.

Acetate of Soda is only used in the preparation of arseniate and phosphate of iron. By itself it can hardly be said to be used, acetate of potash taking its place.

Sulphite of Soda is much more important, not for its soda, but for its sulphurous acid. In making it sulphurous acid is passed through carbonate of soda to saturation. It exists in prisms which have a slight odor of sulphurous acid, readily soluble in water. It is given internally in the same cases as sulphurous, especially to arrest vegetable growth in the form of sarcinæ, etc. Externally it may be used as a lotion, where the acid would not be desirable. The dose is 20 grains to a drachm.

Hyposulphite of Soda is frequently employed in the same way as the sulphate, but in the pharmacopœia it is only introduced for analytic purposes. It occurs in crystals readily soluble in water. Sometimes it is used as a mouth-wash.

Nitrate of Soda is a very deliquescent salt, and so cannot be made use of for gunpowder. It is only used in making arseniate of soda.

Phosphate of Soda, or tasteless purging salt, is got by adding to a solution of bone earth in sulphuric acid, carbonate of soda to neutralization, or more. The salt then formed appears in fine large crystals of a saline taste. In good large doses it purges, and having no disagreeable taste is very useful for children and delicate persons. It requires to be given in doses of half an ounce or so. It is best given in soup or broth, in which it is as nearly as possible tasteless. In smaller doses it acts on the kidneys, but is not much used this way. The dose is 20 or 30 grains. *Chlorinated Soda* owes its efficacy not to the soda it contains, but the chlorine. It is a bleaching solution constituted in the same way as bleaching powder, and used for similar purposes. It is alkaline in reaction, and is sometimes made into a poultice. Internally, it has been given to get rid of fœtid sloughs in the alimentary track, but is better used as a gargle, as in ulcerated mouths and sore throats. Externally it may be used much diluted as a wash to fœtid sores. It is not much given internally. The dose is 10 or 20 drops freely diluted.

Chloride of Sodium, or common salt, is more important as a food than a medicine. If not used ill-health follows, the bowels get disordered, and worms form. In large doses it is emetic, and it may even give rise to dangerous consequences. It is chiefly used as an emetic in cases of poisoning where no other remedy is at hand. Two or three tablespoonfuls may be given well stirred about in lukewarm water, followed by copious draughts of the same. Warm salt-water baths are frequently useful in chronic rheumatic pains. (See SALT.)

Citro-Tartarate of Soda is a salt in many respects similar to Rochelle salt, which contains tartaric acid only. It is this substance in the granulated form which is commonly called citrate of magnesia. If well prepared and well kept, it constitutes a good laxative and sets well on the stomach. If not kept in carefully stoppered bottles, the carbonic acid is gradually given off and it will not effervesce. The dose is about a drachm or two drachms. (See ROCHELLE SALT.)

SODA WATER is a well-known effervescing beverage, containing properly a weak solution of bicarbonate of soda with carbonic acid gas, which is pumped in till the water is well charged with it. It is then bottled, tightly corked, and wired. In many cases of fever and thirst this is a very pleasant and grateful beverage; and when mixed with a little brandy or wine it forms an exhilarating draught in periods of exhaustion and depression, often being preferable to champagne, as it contains no sugar.

SOFTENING OF THE BRAIN. (See CEREBRAL SOFTENING.)

SOLUTIONS in medicine are substances that have been dissolved in water, alcohol, or some other liquid. Every substance is soluble in something, and that something is called its *solvent* or *menstruum*. Thus limestone will not dissolve in water, but it will in strong acid; all the metals are insoluble in water, but soluble in some acid; resins are not soluble in water, but are in alcohol. Salt is soluble in water, but water can dissolve only a certain quantity of it, and then the solution is said to be *saturated*. Hot water dissolves more of some acids than cold, but when the hot water cools, it lets fall all the salt above the quantity that it could hold in solution when cold. Solutions are said to be *concentrated* or *inspissated* when some of the water is driven off by evaporation.

SORE EYES. (See OPHTHALMIA.)

SORE THROAT is a common symptom in many diseases. 1. It may accompany an attack of scarlet fever, when there will also be the usual rash on the second day. (See SCARLET FEVER.) 2. It may come on with an attack of diphtheria, in which case there will be an ashy gray membrane on the fauces and back part of the mouth, without much swelling. (See DIPHTHERIA.) 3. It may come on in the course of a common cold, and be slightly relaxed, or the throat may be inflamed and quinsy produced. The best plan is to wrap some warm flannel round the throat, inhale steam by putting the mouth over a jug of boiling water; keep in bed or in a warm room, so as not to breathe in a cold atmosphere, and have something warm at bed-time, so as to encourage a good perspiration. (See QUINSY.) 4. Relaxing and damp weather, or living badly and working hard, will in some people produce a relaxed condition of throat. For this two or three glasses of good port wine, and swabbing the throat with a solution of tannin and glycerine, or tincture of steel and glycerine, is the best remedy. 5. Sore throat is common with those who have to be exposed to all kinds of weathers: they should be treated as if they had quinsy. 6. Sore throat now and then comes on in clergymen, but it is very doubtful if it is caused by speaking too much. Cold bathing, out-door exercise and tonics, with regular living, will generally cure the case. They are generally at the time pale, thin, and out of health. 7. Sore throat is common in those who have had syphilis, and in them there is no swelling of the part, but generally ulceration of the tonsils. These ulcers have a grayish surface, are generally symmetrical, and have a rounded outline; there may be also other general symptoms of the disorder; but those who have once had a bad throat are very liable to another slight attack on taking cold. Iodide of potassium and mercury form the best remedy, while the throat should be brushed over with some astringent solution.

SPACE is important, because it allows a due amount of air for a person to breathe, nor is anything worse than to continuously work in a close atmosphere. Each man requires at least 800 cubic feet of air. (See VENTILATION.)

SPANISH FLY. (See CANTHARIDES.)

SPASMODIC CROUP. (See LARYNGISMUS STRIDULUS.)

SPASM means the violent and uncontrollable action of some particular set of muscles. Spasms are generally described as of two sorts, viz., *tonic* and *clonic*. In tonic spasms the muscles of a part contract violently, and remain rigid and immovable by the will of the patient for a greater or less length of time. Such contractions occur in tetanus and in ordinary cramp. Clonic spasms consist in sudden contractions and relaxations regularly alternating. The jumping of the legs and arms, which occur under certain conditions, are examples of this. Spasms, again, in the ordinary sense of the word as used by the vulgar, mean gripes, and commonly depend on indigestion and constipation. In most cases they are best relieved by a purgative, containing a good deal of stimulant substance, such as the essential oils. In children the so-called spasms depend almost invariably on imperfect digestion of food, which ferments in the bowels, and so give rise to diarrhœa and gripes. To do any permanent good in these cases, it is necessary to completely reform the diet, as they are perhaps most commonly due to giving starchy food too early, or the milk given turns sour. Lime-water given along with the milk is a good thing. One particular form of spasm, called *trismus nascentium*, is very fatal to children when newly born. It seems due to a foul atmosphere.

SPEARMINT, which grows naturally in marshy places in this country, is only officinal in the form of oil. This oil is colorless, or pale yellow, and is distilled from the fresh herb. There is an officinal preparation of it, viz., spearmint water, which may be used as a vehicle for other remedies. The oil is stimulant and carminative, and is given along with purgatives, to prevent them from griping. The dose of the oil is about one or two drops.

SPECULUM means a mirror or looking-glass. In surgery it is an instrument which is used for widening the natural passages and discovering the nature of disease which cannot be seen by the naked eye. It is chiefly used in cases of disease of the uterus.

SPEECH, as the main means of communicating our ideas one to another, must be looked upon as one of the most important of human faculties. The same faculty is possessed by some of the lower animals, especially parrots, but in them it is merely imitative. The mechanism of speech is peculiar—not confined to any one organ, though mainly depending on movements in those situated at the upper part of the windpipe, called the larynx. In it are situated two bodies, which unite the properties of cords and membranes. These move from before backward, and can be so adjusted by direct and indirect action of muscles that almost any part of them may be permitted to vibrate, or certain parts only. The cords commonly called the vocal cords are set in motion by means of air ejected from the lung, and according as a greater or less extent of each cord is allowed to vibrate, so a grave or a shrill note is produced. But this, though the origin of voice, is only a small part of speech ; most animals possess power of emitting sound so originating, but entirely want the faculty of speech. After the sound is produced by the vocal cord it has to be modulated in the upper portion of the throat and mouth, some sounds being produced in the throat, some by the tongue, some by the teeth, and some by the lips, the ultimate product being articulate speech. But speech also implies a language, if ideas are to be communicated, and here enters a totally new element.

It has been noted that in certain forms of brain disease the faculty of speech is lost. Sometimes this would seem to be due to a want of articulating power ; but in others it is a real want of language. This is known as aphasia, and is commonly associated with disease of one particular portion of the brain, and paralysis of one side. The individual is capable apparently of forming ideas ; but he cannot express them either by reading or writing. As far as the organ of voice is concerned, that is perfect as ever ; but the faculty of language is gone, and to articulate speech is impossible. (See APHASIA and APHONIA.)

Some forms of language cannot be spoken. Thus the emblematic language of the ancient Egyptians and Mexicans, commonly called hieroglyphics, *i.e.*, sacred carvings, was of this kind, while many savage languages are in an unwritten state. In this way we see that the faculty of speech is something very complex. Into its idea both the function of voice and the power of framing a language which shall contain a sufficient number of symbols to indicate daily wants, may enter. The part of the brain where the faculty of language seems to be centred is commonly assumed to be the left posterior frontal convolution. Injury or disease of this part gives rise in most cases to the condition spoken of as aphasia ; but any injury or disease which may intervene between this spot and the motor nerve centres which control the motions of the organ of voice may also interfere with the communication of the ideas, elsewhere framed, and commonly conveyed through speech. In this case, however, the individual would be able, if originally educated, to communicate his ideas in writing, which an aphasic individual cannot. There seems to be still another form of loss of speech where the individual forgets words and letters necessary to communicate ideas. This is commonly spoken of as *amnesia*, while loss of the power of written language is called *agraphia*. These different faculties have yet to be studied more carefully ; but the knowledge we even now possess enables us to understand the complexity and difficulty of the whole subject.

SPERMACETI, which is a nearly pure form of a fat called cetine, is obtained from the head of the sperm whale. The head of this animal is of enormous size, and in cavities in its upper jaw are lodged this substance, mixed with oil. When it cools the spermaceti crystallizes, and the oil is poured off. It occurs in white crystalline masses, and has little odor or taste. It consists of palmetic acid combined, not with glycerine, but a substance named ethal. Its only preparation is the well-known spermaceti ointment, consisting of spermaceti, white wax, and almond oil. This is largely used as an emollient, and applied to coverings of various kinds to keep them from adhering to sores.

SPIGELIA, the root of the *Spigelia Marilandica*, or Carolina pink, a native of North America. The root consists of a kind of head, whence are given off many rootlets of a brown color. It contains some oily and bitter matter, and used to be much employed for destroying worms. It is still used for that purpose in the United States. In large doses it purges considerably, and sometimes produces peculiar effects of a narcotic kind. Usually this substance is combined with a purgative when administered, which it may best be in the form of infusion. The dose ordinarily given is from a drachm to two drachms.

SPINA BIFIDA is a congenital swelling situated over some part of the spine. Its most frequent seat is in the region of the loins, but it is occasionally met with at the back of the neck, and less frequently on the back. It is due to arrested growth of the posterior arches of one or more vertebral bones ; the membranes which loosely envelop the spinal cord become distended with fluid, and are bulged out through the tissues in the walls of the canal, and form under the skin a soft and rounded tumor. When the malformation affects several of the vertebral bones the base of this tumor is broad, but when only one or two of the arches are deficient, or merely fissured, there is more or less of a pedicle or stalk. The size, conformation, and appearance of the tumor, and the symptoms caused by the malformation, differ very much in different cases. The state of things is usually as follows : in the lumbar region, just above the sacrum, and in the middle line of the back, is a large fluctuating and rounded tumor, evidently containing fluid, and the surface of which is covered by thin and distended skin. At the base of this tumor a fissure, or large hole, can generally be felt in the posterior part of the spinal column. When the child is placed upon its belly the tumor shrinks to a slight extent, and the skin becomes flaccid ; in the erect position of the child the tumor swells and the skin becomes stretched and smooth. As the child grows, serious nervous symptoms, such as convulsions and palsy of the lower extremities, make

their appearance. In most cases spina bifida terminates fatally, and the patient dies in convulsions, which in some instances are immediately preceded by giving way of the walls of the tumor. The affection, however, does not always cause death ; several cases have been recorded in which the patient attained an advanced age without suffering any ill effects from the tumor, which continued to grow, though not out of proportion to the rest of the body. A more favorable and occasional termination of cases of this kind is a closure, through adhesive inflammation, of the walls of the orifice between the spinal canal and the tumor. A closed and comparatively harmless cyst is thus formed, which is called a *false spina bifida*. The walls of the tumor formed in cases of spina bifida are composed of the skin and extended membranes of the cord, and sometimes a portion of the cord itself spreads out into a thin membrane. The contents of the tumor are a thin clear fluid, a portion of the cord and some of the spinal nerves.

In consequence of the close connection between the tumor in spina bifida and the contents of the spinal canal, all surgical attempts at a radical cure of this affection are extremely hazardous. The too frequent result of such interference is acute inflammation of the cord and its membranes, causing convulsions, palsy, and finally death.

In cases, however, where the tumor is increasing very rapidly, and is attended with severe symptoms of nervous irritation, which, if allowed to persist, would most certainly prove fatal, the surgeon generally feels disposed to give relief by puncturing the distended skin with a fine needle, so as to allow the fluid contained in the sac to flow away in drops. This proceeding has in some instances been attended with success. When the tumor grows slowly, and while the child remains in good health and free from acute nervous symptoms, the treatment should be limited to affording mechanical support by means of a bag-truss, air-pad, elastic bandage, or some suitable contrivance of the like kind, and to covering the surface of the tumor every evening with a layer of collodion.

SPINAL CORD may be looked upon as a prolongation of the brain downward. It lies within the spinal column in the vertebral canal, safe from any external violence, unless the injury be very severe. It sends off on each side numerous nerves which supply every part of the body. Like the brain, it is covered by three membranes, and it consists of two portions, a gray matter, where various nerve-cells are met with, and a white portion, which is formed of nerve-fibres, which convey motion and sensation. Any injury to the cord will cause more or less loss of motion and sensation in the parts below, and then paraplegia is said to occur. The cord is liable to inflammation, and the patient is said to have myelitis ; to chronic degeneration, causing progressive locomotor ataxy ; to cancerous and other tumors, causing paraplegia ; to destruction, through fracture or dislocation of the vertebræ surrounding it ; to concussion, as in a railway accident, and to inflammation of its membranes, or spinal meningitis. (See MENINGITIS, PARAPLEGIA, and PROGRESSIVE LOCOMOTOR ATAXY.)

SPINE, or spinal column, is composed of a number of strong pieces of bone called vertebræ ; they are twenty-four in number, and are divided into the cervical, dorsal, and lumbar vertebræ. Each piece is provided with a central hole or cavity, and when one fits over the other, a long canal is formed with bony walls, in which the spinal cord can lie with safety under ordinary conditions, and is preserved from harm. The spine, like other bones, is liable at times to fracture and dislocation, and such accidents are dangerous in proportion to the injury done to the delicate cord within. (See LATERAL CURVATURE and RAILWAY INJURIES.)

SPIRITS. (See ALCOHOL.)

SPITTING BLOOD. (See HÆMOPTYSIS.)

SPLEEN is an organ which lies on the left side of the abdominal cavity. It is connected with the lymphatic system, and plays an important part in the formation of the blood ; nothing certain, however, is yet known about its functions. It is much enlarged in some cases of leucocythæmia and in ague ; in the latter disease it is called the ague-cake. It is liable also to waxy degeneration. (See DEGENERATION.)

SPLINTS are requisite in cases where fracture or severe sprain necessitates the keeping of a limb or member in absolute rest, and in the present article such appliances will be treated somewhat in detail, both as regards their form and uses. In the article ACCIDENTS, a rough-and-ready method of maintaining rest and extension is described, such as improvised splints, made with walking-sticks, band-boxes, newspapers, etc., but when proper materials are at hand it will be much to the comfort of the patient if they be employed. Whatever material splints are made of, it is of the greatest importance that they be well *padded*, and such paddings may be made of cotton-wool, tow, strips of old blankets, lint, or soft linen. They should be maintained in position by strapping, bandages, or fillets, *i.e.*, broad tapes secured by buckles. The test of a fracture being in proper position, or of a sprained limb being in the best position of rest, is the feeling of ease on the part of the patient. In applying the retaining materials, care of course must be taken to make them firm, but not tight.

Fractures and severe sprains, moreover, may be treated by the application of "splints," which are applied in a plastic condition, allow-

ing of their subsequent setting firm, thus : pasteboard or gutta-percha, softened in boiling water, and accurately moulded to the limb ; these should be lined with wash-leather, and perforated with a gun-punch in several places to allow of the escape of perspiration. Again, a solid casing can be made with gum, starch, or dextrine, or plaster of Paris ; an ordinary flannel bandage being first applied, a jean roller bandage is to be evenly applied and its surface thoroughly smeared with these materials in solution during adaptation. Stiff shoe-leather soaked in boiling water is an excellent material. As these bandages or " splints" set very quickly and very hard, it is well to guard against œdema or swelling of the limb, so that before the application of the solidifying material a tape must be laid lengthwise on the limb, with its ends projecting above and below the bandage ; then, if the apparatus require removal, a pair of stout scissors or shears can be insinuated between it and the skin and thus avoid wounding the flesh. Splints may also be made of perforated sheet-iron, zinc, tin, or wire gauze, etc.

Special Splints for the Upper Extremity—Angled Splints. These consist of some light material, generally perforated metal having a movable joint, the various positions in which it can be fixed being obtained by an arrangement of slots and screws, such splints being necessary in fractures of the bones entering into the conformation of the elbow-joint, or after operations, such as resection, or the removal of dead bone or tumors.

The *Pistol-shaped Splint* is used for fracture of the lower end of the radius (Colles' Fracture), and consists of a piece of board, cut straight at first, and then bent downward in its own plane, being made of a suitable breadth to fit the forearm and hand, being in form somewhat like a pistol, its object being to fix the arm in such a position that the hand is bent toward the ulnar side. *Gordon's Splint* for this fracture is sometimes used, and consists of an anterior and posterior splint, having on the outside of the front splint a rounded, tapering, projecting margin, and the posterior splint tapering toward the hand.

Special Splints for the Lower Extremity— Liston's Splint. This consists of a narrow deal board, having at its upper end two holes through which a band passes, for the purpose of gaining extension from the perinæum, and at its lower extremity two deep notches, through which pass the turns of the bandage which bind it to the limb (foot), with a hollow on its side for the outer ankle ; its length should be from just below the axilla to four or five inches below the foot. Its use is for fractures of the thigh-bone. Its method of application is as follows : The splint must be thoroughly padded with wadding, tow, or old blanket, the ankle carefully bandaged, and the perinæal band adjusted, then the instep and ankle should be secured by means of the notches at the extremity. Then the requisite extension is made by tightening the perinæal band through the holes in the upper extremity of the splint.

Dr. Smith's Splint " consists of a couple of light iron rods, bent at such an angle as to suit the shape of the thigh and leg when slightly flexed. The rods are connected together at their lower end, and an interspace is left between them sufficient to receive the limb. From one rod to the other strips of bandage are fastened transversely, side by side, so as to form a shape, fitted to the shape of the leg and thigh ; upon this the limb is laid, and then the rods are attached to cords, which are suspended from a point above the bed, and which are regulated by pulleys." This form of splint is remarkably simple, cheap, and clean.

McIntyre's Splint is a convenient form of apparatus for fractures of the leg, or for use after operations on that limb. It consists of a concave iron splint, with a thigh-piece and foot-piece, and a joint at the knee regulated by a screw, so that it can be fixed at any angle and the limb kept perfectly at rest. The splint requires to be very carefully padded, as the sharp metal edges are liable to cause severe sores unless properly protected.

Dupuytren's Splint is in shape very like a long Liston's splint, only much shorter. It is useful in fractures or injuries of the lower limb, especially fractures of the fibula.

It would be impossible to mention every form of special splint which has been devised ; moreover, they have been adverted to and their method of application detailed in the article on fractures. (See FRACTURES.) Suffice it to say, that the general principles for their manufacture and adjustment, and some slight mechanical knowledge, and ordinary common-sense, will suggest a contrivance suited to an emergency until medical advice can be obtained.

SPONGE is organic porous marine substance, in reality the skeleton of a motozoon, found in the seas of the southern parts of Europe and America in large quantities. Though now generally used only for purposes of cleanliness, burned sponge was at one time employed largely as a remedy for goitre and other scrofulous tumors, its efficacy depending on the large quantity of iodine it contains. The subsequent discovery of other sources of iodine set it aside as an article of medicine. Sponges require great care and cleanliness if used in surgery, or in the cleansing and dressing of wounds and sores. They should be rinsed well and dried after each occasion for use, and great care should be taken that the same sponge be not used for two patients or for two purposes. On this account the use of lint, which can be thrown away after every dressing, seems preferable.

SPONGIOPILINE is an invention of Mr. Markwick, for which he obtained a prize at the Great Exhibition of 1851. It is intended to be used for fomentations and poultices, and consists of a mass of shreds of wool and sponge, backed by india-rubber, so that while the thick substance retains the moisture and heat, the water-proof back may prevent its escape. It is a very useful and cleanly substitute for a more elaborate poultice when such an application is quickly necessary.

SPONTANEOUS COMBUSTION. (See COMBUSTION, SPONTANEOUS.)

SPRAINS—Of the Back.—These are usually caused by a fall from a height, or from a weight coming down suddenly on the neck or shoulders. The structures suffering are the fibrous ones generally, such as muscular fascia, tendons, and ligaments. There is considerable swelling in the loins soon after the accident, and great pain on any attempt at motion. The inconvenience arising from a severe sprain in the back lasts a long time ; so that a person may be confined to his bed or sofa for a fortnight, and it may be many weeks or even months before he completely loses pain. There may be some transient effect produced on the kidneys, and blood may be found mixed with the urine for a few days, but rarely any bad effects ensue. The *treatment* consists in giving a mercurial purgative, followed by Dover's powder, poppy fomentation to the back, made with an old soft blanket covered with thin oil-cloth, and with dry blankets ; or the part may be covered with thick compresses of cotton-wool soaked in a solution of tincture of arnica, in the proportion of an ounce to a pint, and laying gutta-percha tissue or oil-skin over it. When the person can sit up, some stimulating liniment, or compound tincture of iodine, may be used, and a warm plaster applied to the loins.

Knee.—Sprains or ricks of the knee are very common and very painful, setting up great swelling in the articulation. The treatment of course depends upon the severity of the injury. If there be much pain and inflammation, leeches, hot fomentations, and poultices. In all cases perfect rest. Cold lotions, lint soaked in tincture of arnica, and well-applied bandages are the best methods of curing the results of the accident. The patient must not get about too soon.

Ankle.—The lower extremity is the most frequent seat of sprain of all the limbs, and particularly the ankle-joint, and the ridiculous fashion of wearing high-heeled boots, whereby the base of support for the body is diminished, is a frequent cause of the accident. In the slighter sprains of the ankle the ligaments are stretched, or, perhaps, a little lacerated ; but in the severe ones they are completely torn. Severe sprains are often mistaken for fractures, and should the case be one, when from swelling and pain there be any doubt, it should be treated as a fracture, bearing in mind that proper treatment of fracture is the best that could be adopted for a sprain.

SPRAY-PRODUCER. (See FLUIDS, ATOMIZED.)

SPRINGS. (See MINERAL SPRINGS.)

SPURIOUS CROUP. (See LARYNGISMUS STRIDULUS.)

SQUILL consists of the bulb of the sea-onion (*Urginea scilla* or *Scilla maritima*) sliced and dried. It grows along the shores of the Mediterranean, partly in the water. The bulb is pear-shaped, and often of considerable size. It is covered with brown scales overlapping like those of the lily. The outer ones are membranous, the inner white and fleshy, these being cut across. Squill is commonly seen n small white pieces, consisting of transversei sections of these scales. Squill has a bitter taste and is not easily powdered until well dried ; in that state they may easily be converted into powder, but if allowed the powder speedily absorbs moisture from the atmosphere, so that the powder becomes a solid adherent mass. Squill seems to owe its efficacy to a resinous substance, which is not, however, separated for use. Its preparations are, vinegar of squill, oxymel of squill, made by mixing squill vinegar with honey, a syrup, and a tincture. There is also a compound squill pill, a very useful preparation ; it consists of squill, ginger, ammoniacum, hard soap, and treacle. To this a little opium may be added. Squill is also contained in the ipecacuanha and squill pill. Given internally squill acts mainly, at least in ordinary doses, as an expectorant and diuretic. In larger doses it may produce vomiting and purging. It is chiefly given in lung diseases, to favor the secretion of a normal mucus and to render the secreted matters less viscid. This kind of secretion is mainly seen in advanced cases, so that squill is rarely given in acute cases. In these it seems probable that the irritant action, or stimulant action, might be a disadvantage, and so foster the malady we desire to cure. Squill is often given as a diuretic, but here too rarely by itself. Most frequently it is combined with mercury and digitalis. Sometimes it seems to be of special benefit in this form in dropsy from heart disease. but seems less likely to be of value where the kidneys are affected. The dose of powdered squill is about two grains, or the compound pill from 5 to 10 grains, of the tincture 20 drops, and of the oxymel half a drachm to a drachm.

SQUINT. (See EYE.)

STAPHYLOMA means an unnatural protrusion of the tunics of the eyeball.

Staphyloma of the Cornea.—Of this condition there are two varieties. In one the cornea, rendered soft and weak in consequence of a slow inflammatory process, yields to the pressure of the clear aqueous fluid collected in the anterior chamber of the eye, and forms a rounded or conical prominence in front of the globe which presses upon and, in some cases, protrudes between the eyelids. This condition is usually associated with more or less marked corneal opacity. In cases where the cornea remains clear the patient complains of impairment of vision and is often short-sighted. In the other variety of staphyloma a portion of the cornea has been destroyed by ulceration ; the gap thus formed is filled up by portions of protruded iris, which become adherent to its margins. The protruded and exposed iris is subsequently thickened by the formation of delicate scar tissue on its surface, but still yields to the pressure of the aqueous fluid and forms a projection in front of the globe. The most marked instances of this kind of staphyloma may be observed in patients who have had an attack of purulent ophthalmia, which has caused sloughing, and removal of nearly the whole of the cornea. Patients afflicted with the latter form of staphyloma usually suffer from frequent attacks of ophthalmia, and of pains and inflammation in the displaced iris. Distension of the staphyloma by accumulation causes much pain and irritation, which is generally relieved for a time by rupture of the protruded membrane. This, however, is always followed by closing of the orifice and reaccumulation of the aqeous humor. Sympathetic inflammation often attacks the opposite eye. The palliative treatment consists in guarding the eye against possible causes of irritation, and in applying the ordinary means of relief during the recurrent attacks of ophthalmia. When there is painful distension of the staphylomatous cornea and iris in consequence of a great accumulation of aqueous humor, considerable though temporary relief may be effected by making a small puncture into the thinnest and most prominent part of the projection. When the opposite eye is affected with sympathetic inflammation it becomes necessary to remove a part or the whole of the damaged globe.

Staphyloma of the Sclerotic.—This term is applied to protrusion of a portion of the sclerotic, due either to thinning of the membrane itself, or to thinning or rupture of the subjacent tunics—the choroid and retina. This condition may be caused by wounds of the sclerotic, blows on the eyeball, or slow inflammatory changes, resulting in a loss of firmness and diminished resistance in the tunics of the eye. Staphyloma may affect the anterior, lateral, or posterior portions of the globe of the eye. In cases of anterior or *ciliary* staphyloma may be perceived one or more bluish, small and irregular-shaped prominences, which contrast strongly with the surrounding portions of white and smooth sound sclerotic. The cornea and the walls of the anterior chamber generally remain healthy. The same changes occur in staphyloma of the lateral portions of the sclerotic. This affection, which is called equatorial staphyloma, is often associated with much impairment of vision and severe recurrent attacks of ophthalmia. Posterior staphyloma generally occurs at that part of the sclerotic which corresponds to the optic nerve and yellow spot. This is frequently a congenital condition, and is the cause of that defect of vision known as myopia, or short-sightedness.

STARCH is a substance found very abundantly in the vegetable kingdom. Its presence was at one time thought to be characteristic of plants, but it has recently been found in animals. It occurs in the form of irregularly shaped granules, which vary in size from $\frac{1}{1000}$ to $\frac{1}{7000}$ of an inch in diameter. These granules are simple or compound. They vary in size and shape, with every species of plant, and are insoluble in water, but are easily diffused through it. They are thus separated from the insoluble cellulose among which they are deposited in plants. In order to separate the starch, the plant is bruised or crushed, and put into a vessel of water, when the cellulose sinks, and the starch diffused through the water, which is decanted and set aside till the starch has deposited. On being mixed with water, and exposed to a temperature of 180°, the starch gelatinizes, and mixing with the water thickens it. This occurs in the cooking of starch, and lies at the foundation of pudding making. Starch is turned blue by iodine, which is the best test of its presence. It is composed of carbon, hydrogen, and oxygen, of which carbon constitutes one half by weight, and the hydrogen and oxygen are in the proportions to form water. When starch is taken as an article of diet, the carbon is burned in the system in contact with the oxygen of the air, and carbonic acid gas is formed and heat given out. Starch is readily converted into glucose, or grape sugar, by the action of nitrogenous substances, especially the salivine of the saliva, and it is in the form of glucose that it enters the blood of animals. All starch in food not converted into glucose is waste. Starch is therefore less readily convertible into aliment than sugar.

Starch is abundantly present in all the common forms of vegetable diet; it exists almost in absolute purity in arrowroot, tapioca, and sago. These substances are therefore not nutritious or flesh-forming, simply heat-giving to the human body. Potatoes and rice can never form the staple food of a vigorous people, because they consist chiefly of starch, and contain little or no flesh-forming matter. During the growth of plants, starch is converted into dextrine, gum, and sugar; it also assumes different properties in certain groups of plants; thus it exists in an amorphous form in seaweeds and lichens, and is then

called licnenine ; and there are other varieties, as inuline, found in the elecampane.

Starch is extensively used in the arts, and in surgery for making stiff bandages, which are put on wet, and dry hard and firm. It is also useful in a finely powdered state to dust over a delicate skin after washing, to dry it perfectly and prevent chapping.

STAVESACRE is the seed of the *Delphinium Staphisagria*, a plant growing in the south of Europe. The seeds have a curious cocked-hat shape, dark brown color, and are pitted on the surface ; they contain an alkaloid called delphinia. Stavesacre is no longer officinal. The seeds have considerable irritant properties, and give rise to vomiting and purging ; sometimes, also, they seem to have some stupefying effects. In ointments the powder has been a good deal employed for destroying vermin in the heads of dirty children. Internally it has been rarely used.

STERNUM. (See BREAST-BONE.)

STETHOSCOPE is a wooden instrument which conducts the sounds in the chest to the ear of the listener. It is very useful, and one can hear better with it than by placing the ear to the wall of the chest, and it is obviously more convenient in many cases.

STIFF NECK. (See NECK.)

STIMULANTS mean something having power to excite the organic action of an animal, or to increase the vital energy of an organ. A stimulant may be either local or general, as it is applied to a part or taken into the system.

STINGS. (See ACCIDENTS and INSECTS.)

STINKS. (See DEODORIZERS.)

STOMACH.—The stomach, from its important functions, controlling the whole system of nutrition, merits greater consideration than it is apt to receive at the hands of many. Any disease of such an organ implies so much interference with all other functions as to preclude, in great measure, their proper fulfilment. Even the functions of the brain are intimately dependent on those of this organ. Common acute inflammation, such as often affects other organs, is rare in the stomach, except when excited by some powerful irritant swallowed. On the other hand, the slighter form of inflammation, commonly called gastric catarrh, is much more common than is supposed, and is, indeed, the ordinary form in which the stomach resents ill-treatment : ordinarily, this form of malady is reckoned as indigestion merely. (See INDIGESTION.) The two most important maladies of the stomach are simple and malignant ulceration, the latter commonly going by the name of cancer. These two diseases are fully treated of under CANCER and GASTRIC ULCER.

STOMACH-ACHE. (See GASTRODYNIA.)

STOMACH-PUMP is an apparatus by means of which, in cases of poisoning, fluids can be introduced artificially into the stomach, or be withdrawn from this organ. It consists in a small pumping apparatus, to which is attached a long elastic tube of sufficient length to be passed down the gullet into the stomach. This tube, at the point where it passes into the mouth, is usually guarded from the action of the patient's teeth by a perforated gag of wood. The stomach-pump, though not used so frequently and indiscriminately as in former days, is, however, an invaluable and indispensable aid in the treatment of cases of poisoning by opium and other narcotics, and of extreme drunkenness caused by poisonous quantities of spirits. It may be laid down as a general rule that the stomach-pump ought always to be used when the patient, under the influence of a narcotic or alcoholic poison, is too much exhausted or too insensible to swallow emetics or antidotes, or where, as in cases of attempted suicide, he obstinately refuses to swallow. One or two pints of lukewarm water should first be pumped into the stomach, and then be withdrawn with part of the contents of the stomach and of the poison. This process should be repeated until the injected water, when pumped back again, is found to be clear and colorless. Very often, however, the simple introduction of the stomach-pump, or the presence of a small quantity of warm water, will cause vomiting; but in cases of intense narcotic poisoning the stomach is generally insensible to the presence of the tube, and requires to be well washed out. When in cases of poisoning the patient is able or willing to swallow, and vomiting can be produced by the frequent administration of warm drinks, the stomach-pump ought not to be used. This instrument is not always a harmless one, and when used by inexperienced hands, and in circumstances exciting haste and confusion, may do considerable mischief. The mucous membrane of the throat, gullet, or stomach may be wounded by the violent introduction of the tube, and some bleeding from the raw surfaces may be produced. A more serious accident is the introduction of the tube into the air-passages instead of the gullet and stomach. A case has been recorded in which, after death from sulphuric acid poisoning treated by the stomach-pump, the windpipe, bronchi, and large portions of the spongy tissue of the lungs were found choked and plugged with chalk mixture, which it had been intended to introduce into the stomach. Another danger attending the use of the stomach-pump is laceration of the mucous membrane of the stomach, strips of which are drawn into the orifices of the tube as the fluid contents of the stomach are being withdrawn. This occurs only in cases where the inner coat of the stomach has been softened by some corrosive agent, and on this account it has been laid down as a rule that the stomach-pump ought not, except under

special circumstances, to be used in cases of poisoning by the mineral acids.

STONE.—The solid precipitates of the urine give rise to the formation of concretions in the urinary passages, which are known under the names of *gravel*, *stone*, or *calculus*. The conditions of the constitution of individuals in whom they occur are termed *diatheses*, and the presence of gravelly or sedimentary deposits in the urine passed, together with any irregularity causing irritation in the urinary organs, should be most carefully attended to, with a view of preventing the formation, if possible, of calculus. In order to discover the condition of the urine, a microscope, a urinometer, test tubes, test papers, and reagents, to be afterward mentioned, are necessary. Urinary calculi are formed from the following salts : (1) uric acid, urate of ammonia, lime, magnesia, or soda ; (2) oxalate of lime ; (3) phosphates of lime, magnesia, or ammonia ; (4) cystine ; (5) uric or xanthic oxide. The existence of these several deposits may be detected as follows : (1) The lithic or uric acid deposit has, to the naked eye, a pink or reddish sandy appearance as sediment, the urine having been originally passed clear. The urine itself is acid, turning blue litmus paper red, and has a high specific gravity. The existence of urates in the urine denote a weak state of the system, and frequently some irregularity of the digestion, or error in diet will cause a deposit. It is most frequently met with either in childhood or between the ages of 40 and 60, and is hereditary. The symptoms of a fit of gravel are pain in the loins, spasmodic retraction of the testicle, frequent painful micturition, some fever, and derangement of the digestion. (2.) The oxalate of lime is deposited from urine which is highly acid, containing much lithate ; it appears under the microscope as minute octahedral crystals. (3.) The phosphates arise either from excessive mucous secretion in the bladder, or from an insufficiently acid condition of the urine. (4.) Cystine is rare, the urine being of a yellowish-green color, and having an aromatic or fœtid odor. (5.) The uric or xanthic oxide is the rarest of all the deposits, and has been chiefly discovered in children in the form of a calculus ; it appears to have much the same chemical character as cystine. Calculi are formed as follows : there being a nucleus in some part of the urinary passages, the prevailing deposit forms round it, generally concentrically ; this nucleus (see LITHOT-OMY) may either exist within the body or be introduced from without, but most frequently it is found to consist of uric acid or oxalate of lime. These small masses may enlarge and remain within the kidney (*renal calculi*), or they may pass by the ureter into the bladder, where they receive additions, constituting vesical calculi, frequently becoming fixed in some pouch in that viscus, or in the prostate gland. The stones when found have characteristic appearances, and can be readily enough recognized by their external aspect, or, of course, more thoroughly after section. Thus, (1) the uric or lithic acid, by far the most common, is generally oval, flattened, fawn or mahogany colored, and its section shows its formation in concentric laminæ. (2.) Phosphate of lime is rare as a stone ; it is pale brown, friable, and laminated. (3.) Triple phosphate forms white or pale gray stones, composed of small brilliant crystals. (4.) The fusible stone, formed of triple phosphate of lime, is a white friable mortar-like mass. (5.) The mulberry calculus is composed of oxalate of lime, and resembles the fruit of a mulberry, being dark red, rough, and covered with tubercles. Alternating calculi are composed of alternating layers of deposit.

Renal Calculus, or Stone in the Kidney, usually consists of uric acid or oxalate of lime. The symptoms of stone existing in the kidney are well marked ; there is a dull aching and feeling of weight in the loins, and a sharp pricking feeling in the region of the kidney. The urine is occasionally bloody, and there is frequent desire to pass water, great pain in the lumbar region generally, and a violent spasmodic retraction of the testicle of the side affected. The passage of such a stone down into the bladder should be expedited by diluents or diuretics, such as Vichy water, or solution of bicarbonate of potash, warm baths and fomentations, and cupping, and leeches to the loins. Calculi will frequently remain impacted in the kidney, causing abcess or wasting of the gland. The passage of a stone from the kidney into the bladder is very much the same as the preceding, only there is violent sickness and shivering, faintness, and often collapse. Warm baths, large doses of opium, and diluents are the remedies. (See GRAVEL.)

STORAX, or STYRAX, is a kind of liquid balsam obtained from the bark of a tree, *Liquidambar orientale*, growing in Asia Minor. This balsam is afterward purified. It occurs in two forms : a thick liquid of the consistence of honey, and brownish red, nearly solid masses, softening with heat. Storax if pure should be soluble in alcohol or ether, and is by chemical means capable of being broken up into a variety of products. Storax is not nowadays much used in medicine. It belongs to a group of substances which have fallen into disrepute. It is contained in compound tincture of benzoin, commonly called Friar's balsam. The whole group of balsams were at one time much employed as applications to cuts and wounds, and doubtless were of service, but with an improved system of dressing they went out, and are now little used in regular practice.

STOUT. (See BEER.)

STRABISMUS. (See EYE.)

STRAMONIUM commonly implies the leaves of the *Datura Stramonium*, or Thorn Apple, growing in this country, but the seeds of the same plant are also now officinal. The leaves are large and much indented at the edges, with a peculiar rank, disagreeable odor. These should be gathered when the plant is flowering. The seeds are very small, kidney-shaped, and rough on the surface, and have a peculiar taste. All parts of the plant contain an alkaloid identical with that contained in belladonna, but called daturia instead of atropia. This may be obtained in white crystals, which yield a peculiar odor on being moistened by sulphuric acid. The preparations are made from the seeds only, and are an extract and tincture. The leaves are mainly used for smoking.

The properties of stramonium are much like those of belladonna, as might be expected from their similarity of composition. Nevertheless, stramonium, more perhaps from habit than anything else, is most frequently given for maladies which are not usually treated by belladonna. Stramonium is in point of fact prescribed almost entirely for spasmodic lung affections, especially asthma. For this malady, whether merely spasmodic or partly dependent on disease of the organ itself, stramonium is usually prescribed in the form of tincture, or the leaves are given for smoking. These generally do well, and procure relief for a time at all events.

Stramonium is sometimes given with the intention of relieving pain. An ointment may be made of the leaves and spread over a painful part, but this plan is not often adopted. For smoking, twenty grains of the dried leaf may be made into a cigarette and smoked, taking care to inhale the smoke. This at first gives rise to cough, but by and by profuse expectoration follows, and then comes relief. Some mix stramonium with tobacco, but the smoke of this is more irritating, and cannot well be inhaled. In some cases stramonium fails altogether, and in all the dose must be increased. The *Datura tatula* has been used for smoking like the *Datura Stramonium*. A quarter of a grain to half a grain of this extract, and 20 minims of the tincture, are the ordinary doses.

STRANGLING. (See CHOKING.)

STRANGULATED HERNIA. (See HERNIA.)

STRICTURE, or contraction of any of the natural passages in the body, may occur as the effects of disease or injury ; but by the term *stricture*, in its general sense, is meant that affecting the urethra, or channel by which the urine passes from the body. Stricture of the urethra may be either spasmodic or permanent.

Spasmodic Stricture is of frequent occurrence in persons who have an irritable urethra, so rendered by repeated attacks of gonorrhœa, or who may have some slight organic stricture, and the symptoms are liable to come on after too much drink, irritation of the lower bowel from piles, etc., getting wet, horse exercise, or some unnatural condition of the urine. An inability to pass water after a too long voluntary retention of the urine in the bladder must be distinguished from permanent stricture, as it depends upon spasm of the neck of the bladder or urethra from some such cause as above. The symptoms are as follows : the individual has a great desire to pass water, and on straining finds himself unable to do so ; the bladder becomes distended, and appears as an increasing tumor above the pubes, and, if not relieved, the continued efforts at evacuation may terminate in rupture of the urethra and extravasation of the urine. (See EXTRAVASATION OF URINE.) In such cases of stricture, especially those arising from debauch of any sort, and when such symptoms have not *previously* existed, a hot hip-bath and a good dose of opium cause speedy relief. The tincture of iron, in 10 drop doses every 10 minutes, is often of use. If the symptoms still continue, the catheter must be passed, and a large one used ; for choice, a No. 8 or 9.

Permanent Stricture, or, as it is called, Organic Stricture, is a contraction of the urethral canal in one or more places, owing to the infiltration of plastic effusion, and fibroid degeneration of the tissues. A constriction is thus produced, varying in tightness, in some cases almost completely blocking up the canal, while in other and simpler ones it is very slight. Occasionally a fibrous band is found stretching across from one side of the canal to the other, forming what is termed a *bridle* stricture. Organic strictures are generally situated in that portion of the urethra just in front of its bulbous portion ; frequently they are found nearer the orifice. The most frequent cause of stricture is neglected gonorrhœa, and perhaps the ill effects of improper remedial agents : patients frequently treating themselves, or getting into the hands of the quacks. Stone in the bladder and injuries of the urethra, may also be cited as causes. The symptoms of an organic stricture are, difficulty in micturition, small stream of urine, generally forked or dribbling, pain during the act of making water, and frequent desire to do so. This form of stricture is frequently complicated with abscess, terminating in fistulæ or sinuses in the perinæum.

Treatment.—The treatment consists both of constitutional and mechanical means. As far as regards the constitutional, any stomach disorders, irritating urine, or inflammatory tendency must be removed, and temperance, rest, early hours, warm baths, and alkaline remedies will do much toward assisting such mechanical means as may from the

nature of the case be deemed necessary. The mechanical treatment of stricture is of such importance that experienced surgical advice must always be taken as early as possible ; and we can do little more in a work of this nature than refer to some of these methods. In the first place the stricture may be dilated by bougies, expanding instruments, a catheter retained in the bladder, caustics, incisions, or external division. The bougie is frequently advised to be used by the patient himself, after having been instructed in the method of using it ; it must be flexible and strongly made, to avoid its breaking in the passage. By the introduction of bougies of gradually increasing thickness, distension is combined with compression, and by this means the ring-shaped cicatricial constriction of the urethral canal is frequently overcome ; and, if applicable to the case, this constitutes by far the most satisfactory course of treatment.

The treatment by expanding instruments consists in the introduction of some appliance whereby mechanical distension is obtained, either sudden and forcible, or gentle and gradual. Many ingenious methods are in use, and are more or less effective in different cases and in different hands. Treatment by the retention of a catheter in the bladder is of value in cases of hard gristly cartilaginous strictures, and in cases of false passage. It consists in tying a small catheter in the bladder, and subsequently a larger one, until the stricture suppurates and becomes dilated. Caustics are occasionally applied to the canal by instruments specially adapted, called porte caustiques ; lunar caustic, or nitrate of silver, is the agent employed. Division of the constricting portion is effected in some instances by internal section, some contrivance being introduced carrying a cutting edge, such as the urethrotome. The urethra has in some cases of complication to be opened from without, in order that the urine may come away ; the operation by means of which this is effected is termed perineal section—an operation requiring great skill and considerable patience on the part of the surgeon.

STROKE is the popular name for one who is struck down in an apoplectic fit. (See APOPLEXY and HEMIPLEGIA.)

STROPHULUS. (See RED GUM.)

STRUMA. (See SCROFULA.)

STRYCHNINE is an alkaloid of a most potent character obtained from nux vomica or St. Ignatius's Bean. (See NUX VOMICA.)

STUMPS.—After amputation of a limb or of portion of a limb, the resulting stump is liable to several affections, and of these *neuralgia* is one of the most frequent ; it is most commonly met with after amputation below the knee and in the arm or forearm. In such cases the part must be carefully defended from pressure in the adaptation of an artificial limb. It depends on some change ·in the structure of the nerves in the stump, but if such change cannot be clearly detected, the treatment to be adopted is that used for neuralgia generally, such as iron internally, and the light application of lunar caustic to the part. In the case of the formation of neuromata, or nerve tumors, the course of treatment lies in their excision, or in a refashioning of the stump.

Exfoliation, or necrosis of the end of the bone or bones in a stump, occasionally occurs after an amputation, and the sequestrum may consist merely of a thin scale of bone, or in severe cases of a portion of bone involving the whole thickness of its extremity, tapering upward, of a cancellous texture. In some instances, when the stump has been badly formed, or the flaps have sloughed, the end of the bone projects, forming what is called a "*conical stump.*" The treatment of such cases is obviously a repetition of the original amputation higher up in the limb.

Bursæ sometimes form over the ends of bones in stumps, generally occurring after blows on them. The fluctuation and general character of these swellings closely resemble abscess. In the case of abscess early incision is necessary, and in the case of the bursæ rest and fomentation are generally sufficient.

Hemorrhage occurs usually a few hours after the stump has been formed, when the patient is warm in bed and has fully recovered from the state of shock. The treatment of such cases consists in the opening up of the flaps and applying ligatures or styptics, or both, to the bleeding points. Pressure in slight cases is often sufficient—at all events should be employed in the course of the main arterial trunk, until medical aid arrives.

STUPOR is that state of partial insensibility which often precedes coma ; it may be caused by a stroke, by drink, by opium, or carbonic acid poisoning, in cases of renal disease, etc. ; the treatment depends of course upon the cause. (See COMA.)

ST. VITUS'S DANCE. (See CHOREA.)

STYE. (See HORDEOLUM.)

STYPTICS are substances applied to a part to arrest bleeding. Most of these are astringents, and seem to act by causing the minute bleeding vessels to shrink, and so prevent further hemorrhage. Cold is the best and simplest styptic, especially if applied as ice. That will arrest most bleedings. Astringent substances, like galls in powder, catechu, etc., which contain tannin, matico in powder, alum, especially burnt, may all be employed. Perchloride of iron is also a powerful styptic ; but one of the most powerful of all is solid nitrate of silver, applied so as to touch the bleeding orifice. If a large vessel bleeds, it must be tied or twisted, or otherwise secured.

STYRAX. (See STORAX.)

SUBCUTANEOUS INJECTION. (See HYPODERMIC INJECTION.)

SUBINVOLUTION is said to occur when the womb does not return to its usual size after delivery, but is larger and heavier than it ought to be. Such women are liable to menorrhagia, pain in the back, and inability to walk far. Tonics must be given, and a liberal diet and rest in the horizontal position.

SUCKLING. (See LACTATION.)

SUDAMINA are minute vesicles, or little bladders, containing fluid, seen in profusion on the chest in cases of rheumatic fever and some other diseases ; they require no treatment.

SUDDEN DEATH is generally caused by disease of the heart and large vessels. It may be caused by an accident, as falling from a scaffold, or by being run over ; drowning generally takes at least five minutes to kill a person, and one may be resuscitated after having been in the water ten minutes, or even a little longer. Strangulation and hanging make a person insensible in a minute ; but death will not take place for three or four minutes if the person die by suffocation. If, however, the person breaks his neck in falling, he will die immediately. Poisoning very rarely causes sudden death, except where prussic acid is used, and then death may supervene in a minute or a minute and a half. Deaths by chloroform are also sudden. Cases of apoplexy generally die within twelve or twenty-four hours ; rarely, if ever, in less than three hours. Aortic disease and fatty heart are by far the most common causes of sudden death which occur in this country ; apoplexy or a stroke is never a cause ; syncope or fainting, rupture of an aneurism, ulceration of a vessel, profuse hæmoptysis are more rare causes of sudden death. In all cases an inquest must be held, and a post-mortem examination made.

SUDORIFICS are remedies which cause and promote perspiration. They are also called diaphoretics. Of course the simplest is heat ; but sometimes that alone does not answer well ; the skin does not open, and the heat becomes more disagreeable. If therefore heat alone be used, as in the Turkish bath, it is advisable to bathe the surface in water if the perspiration does not come freely. Of the sudorifics in common use only one or two deserve mention. These are the acetate of ammonia, which some esteem as a diaphoretic, others despise. But undoubtedly the two most important are the compound ipecacuanha powder, or Dover's powder, and the antimonial powder, or James's powder. Sometimes tartar emetic and laudanum are given. Sudorifics are very useful in certain stages of certain complaints. Thus, if an ordinary cold be caught, at the early stage, with shivering, dry skin, and discomfort, a good perspiration may completely dispel it.

SUFFOCATION means simply death for

want of air, and this may be produced by any cause which prevents the free access of atmospheric air to the lungs ; thus hanging, drowning, choking, and inhaling noxious gases, all induce suffocation. A frequent cause of suffocation in very young children is the anxiety of the mother to prevent cold air getting to them, and covering them up, head and all, to keep them warm. This often takes place in bed, when the infant sleeping with the mother slips down into the bed under the clothes, and breathes only the impure air which is confined there till it becomes asphyxiated and dies. These sudden deaths are often said to occur from fits or convulsions, when in reality they are simply cases of suffocation. (See APNŒA.)

SUGAR OF LEAD. (See LEAD.)

SULPHUR is employed in medicine in two forms—sublimed sulphur and precipitated sulphur, or milk of sulphur. Sublimed sulphur is commonly used. It is prepared by fusing virgin sulphur, and conducting the vapor into a cool chamber, where it consolidates into bright yellow powder without taste or smell. It burns with a blue flame, and produces the unpleasant fumes of sulphurous acid. The precipitated sulphur is pale yellow, and its powder is much finer. The preparations of sulphur are a confection and an ointment. The confection contains sulphur, cream of tartar, and syrup of orange-peel. It is a valuable laxative in piles, or where it is not desired to do more than gently open the bowels, as in fissure of the anus, or in strictures of the rectum. It is mainly, however, as an external application that sulphur is employed. Sulphur ointment still remains the great remedy for the itch, but it is useful in other forms of skin disease. (See ITCH.)

SULPHURIC ACID is the most powerful of all the acids. It is made by burning sulphur, and afterward oxidizing the sulphurous acid by the fumes of nitre. Sulphuric acid thus prepared is a heavy oily-looking fluid, commonly known as oil of vitriol. It is intensely acid, and speedily chars any vegetable substance added to it. Commercial oil of vitriol often contains arsenic from the use of impure sulphur. The diluted acid is used in two forms—as aromatic sulphuric acid, which is flavored by cinnamon and ginger ; and dilute sulphuric acid, in which water alone has been added. The strong sulphuric acid is rarely employed, even as a caustic ; it is unmanageable, and less powerful reagents are preferred. Internally the aromatic or dilute sulphuric acid is mainly used as an astringent. In this way it is of much service in the wasting sweats of consumption ; and it may be of service where there is a chronic mucous discharge from the bowels. It is also of importance as an astringent in diarrhœa, especially if combined with opium. The ordinary dose of dilute or aromatic sulphuric acid is

about 10 or 15 drops, well diluted with water or some such vehicle. In diarrhœa that quantity ought to be given with as much laudanum, if irritating substances have been expelled.

SULPHUROUS ACID is a remedy of some importance. It may be prepared in a variety of ways, but it is most commonly obtained by reducing sulphuric acids by means of charcoal. It is most easily prepared by burning sulphur,in the open air. It has the wellknown odor of burning sulphur. Sulphurous acid is a powerful deoxidizing reagent, and is powerfully destructive of vegetable life. Applied to the skin it causes some reddening ; and if any vegetable parasite is present, as is not unfrequently the case in skin disease, it is destroyed. Hence arises its value in such maladies. Internally, if there is any tendency to fermentation, and if fungi are present in the stomach, it does great good. Used as spray in certain forms of sore throat,sulphurous acid is also of great use. It may be freely applied, and subsequently used scmewhat diluted as a gargle. Sulphates and hydrosulphates, especially of soda, are frequently given internally in its stead. (See SODA.)

SUMACH. (See RHUS TOXICODENDRON.)

SUMBUL, or MUSK ROOT, is the root of a plant growing somewhere in Central Asia. It reaches us mostly by way of Russia, partly also by way of Bombay. (See MUSK.)

SUN-STROKE. (See HEAT-STROKE.)

SUPPOSITORIES are forms of remedies similar to medicated pessaries. They consist of some basis, most frequently called butter, which, while taking shape and possessing a certain consistence, shall yet melt gradually and so expose the medicated materials they contain to gradual absorption. They are generally introduced into the rectum before rest, and allowed to remain there. Most frequently they contain some sedative, as opium, morphia, or belladonna ; but occasionally also astringents.

SUPPRESSION OF THE URINE takes place when the kidneys do not secrete their proper amount of urine, and then the blood becomes poisoned, because those substances are retained in the blood which ought to be voided ; there is thus an important difference between these cases and those of retention of urine, which may arise from a stricture, or from paralysis of the bladder, and which are relieved by passing a catheter. In cases of suppression the loins must be cupped, and a sharp purge must be given. It often comes on at the end of old standing kidney disease, and hastens the termination of the illness.

SUSPENDED ANIMATION. (See DROWNING.)

SUTURES.—The edges of wounds or surgical incisions are approximated by what are termed *sutures*, and these sutures are applied by different modifications of needles and threads. The *needles* are various in shape and size, straight and cylindrical. straight and triangular, or curved and double-edged. Hare-lip pins are of great use in many forms of wound. The threads are either hempen, silken, catgut, horse-hair, or metal. It must be borne in mind that no suture should be used until all bleeding has ceased, and every foreign substance removed, and exact apposition attained. The needle should be passed through the integument so deeply that it does not give way on the natural tension of the parts, and the thread and its knot should not be drawn so tightly that they cut the pierced tissue, or strangulate it.

The different forms of suture are the *interrupted*, which consists in the approximation of the edges of a wound by entering a needle armed with a thread on one side of the wound or incision, and bringing it out through the other. The edges of the wound being held in apposition, either a double knot or a single one with a bow is tied, and the suture is fixed. In the *uninterrupted* suture the armed needle is passed continuously from one side to the other, until the whole length of the wound is traversed.

The *quill* suture is of use in cases where some degree of force is necessary to keep the edges of a wound together, and also for approximating the deeper parts ; it is applied by passing a double ligature and inclosing portions of quill, rolls of strapping, or pieces of bougie, as " points d'appui."

The *zigzag* suture is much on the above principle, is applied by thrusting the armed needle through the lips of a wound in the first place, and then entering it a short distance on, on the side of its emergence, then again passing it through the lips of the wound, and repeating the proceeding on the opposite side.

The *twisted* suture is applied with the assistance of hare-lip pins ; the pins transfix the lips of the wound at intervals, and the thread is twisted around each in succession as a figure of 8, passing from one to the next in order.

Metal sutures are preferable in some instances, as they give rise to less local irritation.

Sutures should be removed at an interval of a day or more, and those causing the greatest irritation should be removed first. As soon as irritation to any extent, in fact, is seen at their points of passage, they should be removed.

Adhesive plaster should be removed from a wound when it *gets black ;* it is then useless and irritating, and its place of course should be taken by a fresh piece of strapping.

SWEAT. (See PERSPIRATION.)

SWELLINGS. (See ANASARCA.)

SYMPATHY is an awkward sort of term applied to the evils which result from the influence of one kindred diseased organ on another. This same influence may be seen also in health. One of the best examples is the filling out of the breasts, which commonly takes place in women just before the monthly period. The breasts may also become swollen, hard and knotty in ovarian disease, as if the female were pregnant. The headache of indigestion is another familiar instance of sympathy. So too is the pain experienced in the right shoulder when the liver is diseased, the pain extending down the thigh when passing a stone from the kidney. The vomiting, which is one of the most troublesome things accompanying the passage of a gall-stone, is commonly spoken of as sympathetic, though it may not really be so. In short, the curious alliance between parts brought about either by an alliance of function or by a common origin of nerve-supply, might be illustrated by numerous examples, but by none more telling, save, perhaps, the aptness of one eye to become diseased when the other is.

SYNCOPE is a technical term for a faint produced by shock or excitement, or by the ailing power of a weak heart. It often is the proximate cause of death in heart disease. Brandy, ether, and other stimulants, should be used to rouse the heart to act more vigorously.

SYPHILIS. (See VENEREAL DISEASE and NODES.)

T.

TABES MESENTERICA is strictly a disease of childhood, and is a sure sign of a scrofulous constitution. It is in reality tubercular disease of the mesenteric glands, and is better recognized by general symptoms than by any discoverable enlargement of the glands, which seems to be the origin of the disease. Emaciation, loss of appetite, and relaxation of the bowels are among the earliest symptoms, and tenderness and distension of the abdomen suggest the existence of diseased glands. The condition of these glands can only be ascertained by very careful examination, and requires a practised hand to undertake. (See MESENTERY.) The course of the disease is slow, but its duration is difficult to estimate on account of the obscurity of the earlier symptoms. It seems to occur more frequently in boys than in girls, and is seldom found in children under three years of age, most commonly between the fifth and tenth years. The children do not necessarily die, they sometimes recover.

Treatment.—The treatment should consist

in relieving, if possible, the oppressed glands. An ointment of iodide of lead may be rubbed into the body twice a day, and the syrup of the iodide of iron given internally. The diarrhœa, which is so frequent a symptom of this disease, should be arrested as soon as possible by small injections of warm starch and opium. Unfortunately this diarrhœa is so often the result of tubercular ulceration of the bowels that all efforts prove unavailing to arrest it. A light farinaceous diet, with a little boiled mutton or fish for dinner, but no bread, salt, or solid food. Change of air, moderate exercise, and daily sponging of the body in tepid salt water contribute to the cure. The practice of thoroughly but gently rubbing the body, legs, and hips, and securing an amount of reaction after the bath is desirable ; and, as recovery proceeds, small tonic doses of quinine and tincture of iron may be given. Cod-liver oil is also a most valuable remedy in this disease from its earliest stages.

TÆNIA MEDIOCANELLATA, a tape-worm. (See ENTOZOA.)

TÆNIA SOLIUM, a tape-worm. (See ENTOZOA.)

TALIACOTIAN OPERATION is a name applied to the operation of forming a new nose, invented by Taliacotius, a celebrated Chinese surgeon, who lived about the beginning of the Christian era. He was the first who ever attempted to restore a lost nose, and his original idea was to cut a pear-shaped piece of cuticle or skin from the patient's arm, all but a small pedicle, or stalk, which remained attached to the original limb and supplied nutrition and life to the excised piece. This was spread over the framework of the nose, the edges of the cheeks being first scarified, and the arm bound up to the head and tip of the nose, where it remained until union had taken place between the new piece of skin and the surrounding edges, when the little point of union was severed, and the arm set free. When the patient objected to supply his own cuticle, Taliacotius was in the habit of obtaining the needed material from the arm, leg, or thigh of some one else. Mr. Liston, the great surgeon, revived this long-neglected operation, and formed new noses for his disfigured patients by cutting a piece of skin out of their foreheads, leaving it attached by a small kind of foot-stalk, and then inverting it on to the frame of the nose, when it was carefully plastered over and left to unite. In several cases this operation has been very successful.

TAMARINDS, though contained in the pharmacopœia, can hardly be said to be remedies of importance. The pulp of the fruit of the tamarind-tree, which grows both in the East and West Indies, is the part used. This pulp is sweetish and at the same time sour. The fruit as preserved and sent over to this country is used in confection of senna. The pulp is slightly laxative, and is rather pleasant.

TANNIC ACID, or TANNIN, is a powerfully astringent substance contained in oak-bark and a great variety of other vegetable products. It is obtained by exposing powdered galls to damp air for a short time ; next ether is added, and squeezed out of the mass ; the mass is again pulverized, and again ether is added ; this is squeezed out and added to the other, and the tannin is obtained by evaporation. Thus prepared, the acid is a yellowish white powder, of a very astringent taste. It turns all iron salts blue-black, and throws down gelatine. Tannin is a powerful astringent, as may be seen by applying it to the lips. It then causes the vessels to contract and the parts turn white. In the body it is converted into gallic acid, so that substance is more frequently given internally instead of tannin. Generally its effects are astringent and closely allied to those of gallic acid.

TANSY is an herb which grows wild in old fields in many parts of the United States, and is also cultivated to some extent in gardens. The whole plant has a stong and aromatic smell and a very bitter taste. The leaves were formerly much used for flavoring soups stews, etc., but other herbs have superseded it for this purpose, and it is seldom employed now except in domestic medicine. An infusion of the leaves will expel worms from children ; but the popular idea that a decoction of its leaves will act as an abortive is a delusion. Tansy is best in the spring months, when it is young and green ; it is sold also at the drug-stores in cakes dried and pressed.

TAPE-WORM.—There are three kinds of tape-worm which infest the intestinal canal ; these are fully described in the articles on ENTOZOA and PARASITES.

TAPIOCA is the starch obtained from the *Jatropha* or *Janipha Manihot*. The juice of the root is acrid and poisonous, but it is washed away, and the starch collected. From this starch is made in the country cassava bread ; after it has been perfectly purified it constitutes tapioca. Abroad, tapioca is sometimes used for a poultice ; here, only as an article of food.

TAPPING is the common or popular name for the operation known to surgeons as PARACENTESIS (which see).

TAR is obtained by the destructive distillation of various species of pine. It is a thick, black, treacly-looking substance, with a strong and peculiar odor. If water be shaken with it the water smokes up, some of its substance becomes brown, and has something of the smell of tar. This water was at one time much extolled as a medicine. The composition of tar is very complex ; its only preparation is an ointment consisting of tar and beeswax. From the various substances it contains, tar is a stimulant of value, especially for outward application. In some incorrigible forms of skin disease, especially in the hands and feet, tar has done good, especially if the disorder be of a scaly kind. In many of these cases it may be given internally as well as externally. Tar itself, or its vapor, has been used with great advantage in certain cases of lung disease, especially in chronic bronchitis and diseases complicated by it. The dose is about 30 grains made into a pill. An ounce or two of tar-water may be taken at a dose.

TARAXACUM. (See DANDELION.)

TARTAR is the deposit on the teeth which occurs in those who do not brush their teeth properly ; it may be scraped off ; cleanliness will prevent it from forming.

TARTAR EMETIC, or TARTARATED ANTIMONY, is the most important preparation of antimony. It is a powerful emetic and depressant, and in small doses it acts as a diaphoretic ; two grains often suffice to produce vomiting. It is mainly used for its depressant effects. It is not now so much employed as formerly.

TARTARIC ACID is procured from cream of tartar, a natural deposit from wines. First of all a tartarate of lime is formed, and from this the tartaric acid is set free by means of sulphuric acid. It exists in transparent, rather irregular crystals ; its taste is sour, but, on the whole, agreeable ; it is freely soluble in water. In the system, tartaric acid and the substance with which it is combined are converted into carbonates. Tartrates are nearly neutral or even acid salts, but this property of conversion enables us to give them where alkalies are required, and they sit on the stomach very much better. The acid may therefore be given as a cooling drink, and yet apear in an alkaline form in the urine. Most frequently this acid is used for the production of effervescing drinks : 10 grains or so is the quantity ordinarily used. (See EFFERVESCING DRAUGHTS.)

TAXIS signifies an attempt to return or *reduce* a rupture by simple manipulation. In cases of reducible and moderately-sized rupture, the contents of the hernial sac may under ordinary circumstances be readily replaced by slight pressure, or slip back spontaneously whenever the individual lies down. But when the rupture is strangulated, and the neck of the protrusion is tightly compressed by the opening in the abdominal walls, careful and delicate handling is required in order to overcome the resistance, and at the same time to avoid injury and rupture of the inflamed coats of intestine. The patient should then be placed in an easy recumbent position, with the hips and knees bent, and the thigh on the side of the rupture rolled inward in order to relax the muscular and tendinous structures about the neck of the sac. He should be charged to abstain as much as possible from moving the body and lower limbs, and to keep his head in one position and the mouth wide open. The surgeon, by gentle compression and kneading of the rupture, and by moving the parts at the neck of the sac, then endeavors to direct the distended intestine and other contents of the sac, through the canal leading to the abdominal cavity. The direction of the pressure is made to vary according to the anatomical nature of the rupture. In umbilical rupture, the attempt is made to pass the contents of the sac directly backward ; in inguinal hernia, outward and upward ; and in femoral hernia, first downward and backward and then upward and inward. In successful taxis, the rupture, when it contains intestine, generally first shrinks a little, and then suddenly disappears with a gurgling sound. When the sac contains much omentum, it is reduced slowly and gradually. The duration of the manipulation, in cases of obstinate rupture, should be adapted to the nature of the case, and the probable condition of the contents of the hernial sac. In cases where the rupture is indolent and free from pain and inflammation, and no remote symptoms of strangulation are present, the surgeon generally feels justified in continuing his manœuvres for twenty minutes or half an hour. In cases of strangulation, however, and especially after vomiting, the rupture ought to be handled with the utmost gentleness, lest the walls of the inflamed and probably gangrenous intestine be ruptured. If the strangulated intestine cannot be reduced by gentle taxis, the patient may be placed in a warm bath, and the attempt be repeated. This, however, is not in all instances a safe proceeding, as the patient may have been much exhausted by vomiting, etc. At the present day, after the failure of the first attempt at reduction, and in the presence of undoubted symptoms of strangulation, the surgeon places the patient under the influence of chloroform, again tries the taxis, and then, in case of a second failure, proceeds at once, while the patient is insensible, to perform a cutting operation. (See HERNIA)

TEA consists of the leaves of several varieties of a small shrub found in China and India. The leaves are gathered in the fourth year of the growth of the plant, which is generally dug up and renewed in its tenth or twelfth year. The leaves are cropped with care by gatherers, who wear gloves, wash frequently, and avoid eating things likely to affect the breath. The differences between teas result from the varieties of soil and growth, and also from the mode of curing and drying the leaves. Black tea consists of leaves slightly fermented, washed, and twisted. Genuine green tea is made of exactly the same leaves, washed and twisted, without fermentation ; but commercial "green" teas are often black teas colored with Prussian blue. Probably five hundred millions of men, or half the human race, now use tea. The chief action of tea depends firstly on its volatile oil (less in old than in new tea), which is narcotic and intoxicating ; and secondly, on a peculiar crystalline principle called *theine*. Theine excites the brain to increased activity, but soothes the vascular system by preventing rapid change or waste in the fleshy parts of the body, and so economizes food. Four grains of *theine*, contained in half an ounce of tea, act in this way ; but if one ounce of tea, containing eight grains of theine, be taken in a day by one person, then tremblings, irritation of temper, and wandering thoughts ensue. When the system is thus saturated with theine, it is useful to resort to cocoa as a substitute for a few days, when the symptoms subside, and the use of tea can be renewed ; but it is unadvisable ever to take it in such quantities as to occasion such symptoms. Tea contains also a quantity of tannic acid, which, being an astringent, is useful as a gargle in sore throat, and as an injection in some cases.

TEETH, CARE CF THE. (See TOOTHACHE.)

TEETHING. (See DENTITION.)

TEMPERAMENT is a term used by physiologists to distinguish a peculiar organization of the system in different individuals, and they are usually grouped into four classes. Physiologists recognize—

1. The *Sanguine* temperament, characterized by plumpness of body, fair or red hair, blue eyes, a soft, thin skin, active circulation, and a full, quick pulse.

2. The *Phlegmatic* temperament is distinguished by a round body, soft muscles, fair hair, pallid skin, and slow, languid circulation and pulse. All the functions, mental and bodily, are torpid.

3. The *Bilious* temperament, known by firmness of muscle and flesh, defined sharp features, black hair and dark complexion, a full, firm, and moderately quick pulse.

4. The *Nervous* temperament, characterized by a small, spare frame, quick, impulsive movements, and a delicate constitution ; the pulse is small and weak, and easily excited ; the whole nervous system is susceptible, the thoughts quick and imagination lively.

Some physicians place great reliance on the indications of temperament in the treatment of disease, and find that those who possess a sanguine temperament are most liable to acute inflammatory diseases ; the phlegmatic inclining to scrofulous complaints ; the bilious to affections of the liver and digestive organs ; and the nervous to mental disorders and diseases of the nervous system generally.

TEMPERATURE.—The temperature of an ordinary adult when a thermometer is

placed in the armpit is 98.4° ; in the mouth, 99.5° ; the blood is about 100°. In fevers this is much exceeded, and the heat of the patient may rise to 105° or 106° ; a higher temperature than this will generally prove fatal, unless it descend soon ; the highest temperatures recorded have been in some cases of rheumatic fever, when the body rose to 109°, and even to 111°. The temperature of a hot bath is about 98° ; of a tepid bath 70°—75°. In describing the fevers, the value of the temperature as a symptom is noticed in each case.

TENDO ACHILLIS is the longest tendon of the body, and the great leverage of the heel, being the extensor muscle of the leg. The ancients gave it the name from the fable that Thetis held the boy Achilles by the heel when she dipped him into the Styx, and made all the rest of his body invulnerable.

TESTICLES, the male secreting organs in the human body, two in number, situated in the scrotum, and containing the procreating fluid of the male.

TETANUS is an affection characterized by painful and rigid contraction of the voluntary muscles, which is persistent and aggravated from time to time by very severe spasms. The two chief forms of tetanus are the *traumatic*, when it occurs after wounds, and the *idiopathic*, which comes on in the absence of any manifest cause. In the former the spasms are usually severe and acute, in the latter they are milder and chronic. Traumatic tetanus, however, is sometimes a subacute or even a chronic affection. The following are the *symptoms* that may be presented in a severe attack of tetanus following a wound : After certain common symptoms, such as a feeling of general uneasiness, headache, and feverishness, have been experienced, the patient complains of stiffness of the jaws and at the back of the neck ; swallowing is difficult, the voice is low and husky, and there is a peculiar expression of the face due to con traction of the muscles which move the lips and eyelids ; the patient next suffers from painful cramp in the muscles of the face and neck, and, in consequence of permanent rigidity of the muscles of mastication and spasms of the gullet, is unable to take any food ; to this stage, in which the mouth is firmly closed, has been applied the name of *locked jaw ;* the spasms then attack the muscles of the abdominal walls, and violent pain is felt at intervals at the pit of the stomach ; the front of the abdomen is retracted, and the muscles during the severe paroxysms feel to the hand like a hard board ; the voluntary muscles of the back and limbs finally become affected and very painful, cramps are felt over the whole body, which as the affection progresses are divided by shorter and shorter intervals ; the bowels are generally bound, and there is often retention of urine ; the symptoms increase in intensity, and at last death occurs either from pain and exhaustion, or in consequence of spasms of the diaphragm and other muscles of respiration ; the mental faculties generally remain unimpaired, until very shortly before death. The usual duration of an attack of severe and fatal tetanus is from three to six days. Cases, however, have been recorded in which death occurred within a few hours after the commencement of the symptoms. The symptoms of *acute trau-matic tetanus* vary much in different cases ; the spasms may be restricted to a certain region or a certain set of muscles, or they may commence at the seat of the wound, and not, as is usually the case, in the muscles of the jaw. The ordinary tetanic symptoms may be complicated by epilepsy, delirium, and coma.

There is no injury to the surface of the body, however slight it may be, of which acute tetanus might not be a result, and there is no relation between the extent and degree of the injury and the intensity of the tetanic symptoms. It has been known to follow slight contusions and blows with a stick or cane. It rarely occurs after clean cuts, and is mostly connected with contused wounds involving nerves and the fibrous structures, as fasciæ, tendons, and ligaments. With regard to locality, it has been stated that tetanus occurs more frequently after wounds of the hands and feet and their respective digits. The interval between the receipt of the injury and the commencement of the tetanic symptoms, the so-called period of incubation, varies in different cases. In the majority, the symptoms come on between the fourth and the tenth day ; the period in many lasts from ten to twenty days, but is extended over the twenty-second day in only ten out of every hundred cases. It has never been known to exceed a month. The symptoms sometimes come after an interval of only a few hours, and one instance has been recorded in which a negro was attacked with tetanic spasms in a quarter of an hour after his hand had been punctured with a fragment of chinaware. The shorter the interval the more severe are the symptoms. Tetanus, when it occurs before the tenth day after the injury, is usually fatal ; in cases occurring after the tenth day the mortality is much reduced. Tetanus occurs much more frequently in males than in females, and in the latter its symptoms are less severe. Tetanus may occur at any period of life, but in more than half the number of recorded cases the patients were between ten and thirty years of age. It has been asserted that tetanus is most fatal in patients under ten years of age, and least fatal in patients between ten and twenty years of age. The accession of traumatic tetanus does not seem to be influenced in any way by morbid conditions of the body, or by previous states of

bad health. The healthy and the unhealthy, the strong and weak, are equally affected. Negroes and Asiatics are much more liable to attacks of tetanus than the white races. Europeans are not rendered more disposed to tetanus by residence in the tropics. It has been stated that tetanus is met with more frequently at periods of the year in which there are frequent and sudden changes of temperature.

The course and symptoms of an attack of idiopathic tetanus resemble very much those of the acute traumatic form, but are rarely so intense. The chief causes of the so-called idiopathic tetanus are exposure to cold and wet, and intestinal irritation. It is rarely met with in this country, but occurs frequently in the tropics.

The symptoms of tetanus may resemble very much at first sight those of hydrophobia, and in some cases the medical attendant experiences considerable difficulty in establishing a perfectly satisfactory diagnosis. The following are the chief points of difference in these two dangerous affections ; in tetanus the muscular spasm is persistent, and perfect relief never occurs for a single instant until a short time before death ; in hydrophobia spasms are always of brief duration, and alternate with periods of complete relaxation and relief ; the persistence of the muscular contraction in tetanus is most marked in the lower jaw, which in almost all cases remains fixed and immovable. In hydrophobia there is a constant flow of saliva, and the patient complains of great thirst ; in tetanus these two symptoms are usually absent ; the countenance in tetanus is generally expressive of intense suffering ; in hydrophobia, not so much of physical suffering as of excessive restlessness and mental excitement ; in the latter affection the mental faculties are always much disturbed, and the patient often falls into a state of violent delirium and maniacal excitement ; in tetanus, on the other hand, the mind usually remains undisturbed until the termination of the attack ; in hydrophobia there is an aversion to fluids, the very thought of which very much excites the patient ; in tetanus there is no mental aversion to fluids, but when an attempt is made to administer them the patient endeavors to express by action his inability to open the jaws and to swallow. Any reliable history as to the bite of a dog about six weeks or two months previously will at once establish the diagnosis in doubtful cases of hydrophobia. Tetanus, though a very dangerous affection, is not always fatal ; in acute cases, where the symptoms commence shortly after the receipt of a wound, recovery seldom occurs, but when the attack comes on after the tenth day from the receipt of the wound, and the tetanic symptoms last over fourteen days, recovery is the rule and death a rare exception. No case of recovery from hydrophobia has been hitherto recorded. Symptoms somewhat analogous to those met with in severe cases of tetanus are produced by poisonous doses of *strychnia* or *strychnine*. The symptoms of poisoning commence soon after the strychnine has been swallowed, and set in with shortness of breath, rigidity of the muscles of the neck and back, and painful tetanic spasms of the extremities ; the body is usually arched backward so as to rest on the head and heels. The muscles of the face are much convulsed, so as to produce a characteristic grinning expression called the risus sardonicus. All the voluntary muscles are attacked at about the same time, and there is no persistent contraction of the muscles of the jaw ; in these respects, and also from the prominence, among the symptoms, of backward arching of the body, and from the occurrence of intervals of complete intermission, the phenomena of strychnine poisoning differ from those of acute traumatic and idiopathic tetanus.

Treatment.—No continued success has yet attended the administration of any one of the numerous medicinal agents that have been tried in cases of severe tetanus Calomel, opium, chloroform, belladonna, aconite, quinine, Calabar bean, and Indian hemp have all been extensively used, in some cases with undoubtedly good results, in others with signal failure. No drug is yet known which has the power of arresting the course of the disease, and of controlling its severer symptoms. So long as tetanus is to be regarded as a disease which must run a certain course, the chief indications of treatment will be the support of the patient's strength and the relief of suffering and pain. Fluid and easily digested food, with wine or spirits, must be freely supplied, and when the patient is unable to open the mouth or to swallow, should be administered by injections or through an elastic tube passed through the nose into the gullet. Pain may be relieved by the internal administration of opium, by subcutaneous injections of morphine, or by inhalation of chloroform. In many cases painful and violent muscular spasm has been much allayed by the application, along the spine, of bladders of ice. Great care must be taken to guard the patient from all causes of excitement and irritation, and the room in which he is confined should be kept darkened and at a uniform temperature. It is very important that there should be a speedy and free evacuation of the bowels. In cases of traumatic tetanus following a wound, the injured part, if painful and inflamed, should be poulticed and kept as much as possible at rest.

TETTER, a disease of the skin which often appears on the face and the side of the mouth, and requires simple treatment, such as an alkali like bicarbonate of potash or soda, internally, and the application out-

wardly of powdered oxide of zinc occasionally.

THEINE. (See Tea.)

THERMOMETER is an instrument for measuring the temperature of a room. A tube of glass, with a bulb blown at one end but open at the other, is filled partly with mercury ; on heating the mercury, the bulb and tube become filled with mercury, and the vapor of mercury and all the air is driven out ; the open end is then hermetically sealed in the flame of the blow pipe. The freezing-point and the boiling-point are the two standards taken, because under ordinary conditions at the sea-level these are fixed points. The thermometer is immersed in melting ice, and then the point at which the mercury stands is scratched on the glass ; it is then placed in boiling water and the level of the mercury is noted. On the Fahrenheit scale this distance is divided into 180 degrees ; on the Centigrade scale into 100 degrees, and on the Réaumur scale into 180 degrees. The freezing-point is called zero on the last two scales, but 32 on the Fahrenheit scale.

THIGH, BROKEN. (See Fractures.)

THORACIC DUCT is a narrow tube, lying in front of the spine, which conveys the chyle and lymph from the receptaculum chyli into the blood by its communication with a vein at the root of the neck.

THORAX, an anatomical name for the chest. (See Chest.)

THORN-APPLE. (See Stramonium.)

THREAD-WORMS, or Oxyurides, are often found in children. (See Ascarides.)

THROAT. (See Sore Throat and Foreign Bodies.)

THROMBUS occurs when a plug is formed in a vessel during life ; it is generally met with in veins, but may occur in the heart or in an artery.

THRUSH is a common affection in children. It may be seen in the mouth as small white specks on the lining membrane, but this may be so also in various parts of the intestinal canal. For causes and treatment, see Aphthæ.

TIC, the common and short term for *tic douloureux*, is that form of neuralgia which specially affects the fifth nerve, the sensory nerve of the face. Either of its three branches may be affected, but in pure neuralgia it is most likely to be the uppermost. The other two, namely, the superior maxillary and inferior maxillary, are much more likely to be affected with a kind of counterfeit neuralgia or reflected pain, caused by bad teeth or gums. In all of these cases the jaws must be carefully examined ; and if any good is to be done, all bad stumps are to be removed and the gums, as far as possible, rendered free from tenderness. (See Neuralgia.)

TIN is a metal found in the form of various ores, of which tin pyrites and tinstone are the most important. It was first introduced as a medicine by Dr. Alston, the first professor of Materia Medica in the University of Edinburgh, and it has since been constantly employed as a vermifuge. The mode of administering it is to give at least half an ounce of the powder every morning for three successive days, while the stomach is empty, and then to carry it off with a brisk purgative. It is given in the form of electuary, made up with treacle or orange confection. It is undoubtedly effectual in cases of ascarides and lumbrici, but is less so in tænia. Its action is probably mechanical only, for there is no property in the intestinal secretions or other contents, in consequence of which tin could be dissolved ; and, besides, the worms are alive when discharged. A chloride of tin is a good disinfectant.

TINCTURES are solutions of any colored substance in spirits of wine ; when not colored the solution is called a spirit.

TOBACCO consists of the leaf of the *Nicotiana tabacum*, or tobacco-plant, growing in America. Another variety of the plant, the *N. Rustica*, is cultivated in Asia Minor for Turkish tobacco. The leaves are large and oblong, covered with short downy hairs and emit a heavy odor when they begin to dry. The dried leaves only are used.

Tobacco contains a peculiar and powerful alkaloid called *nicotine*. This, when freshly prepared, is colorless, but grows brown when older. It is exceedingly powerful and very poisonous. Tobacco is a powerful sedative, and causes, perhaps, through the faintness it induces, general relaxation of all parts of the body, especially of the muscles. It is hardly ever given internally, and the injection which used to be employed to procure relaxation in parts concerned in strangulated hernia has been displaced by chloroform.

Tobacco, in the form of snuff, may act as a powerful irritant, especially to the eyes and nose. But tobacco is almost invariably employed in the form of smoke, as from a pipe or cigar. Used thus there can be no doubt but that it produces a powerful sedative effect, calming and soothing, if used in the proper dose. This is easily known, as a dose too strong speedily turns the individual sick and faint. This may sometimes be useful in procuring relaxation of parts under the influence of muscular spasm, as in asthma. It is an important question whether tobacco used in moderation does good or harm. Used immoderately, like everything else, it is a great evil ; used in moderation it is often of good service. It has been said that it is apt to give rise to a certain form of blindness ; that would only arise from immoderate use. Its use by young people is not desirable ; they should not want soothing.

TOE-NAIL, INGROWING. (See NAILS.)

TOLU BALSAM is one of those sub-

stances allied to storax and balsam of Peru. They all contain cinnamonic acid, and possess very similar properties. Tolu is lighter in color and denser than is the Peruvian balsam. It is seldom or never used except as compound tincture of benzoin, or Friar's balsam, which contains the substance.

TONGUE in structure consists essentially of muscular tissue covered by mucous membrane. The muscular fibres, omitting those of muscles inserted into the organ, are arranged in two horizontal and several vertical layers, the former set lying immediately underneath the mucous membrane, and the latter passing vertically from between the horizontal layers, leaving intervals which are occupied by gland structure. The mucus membrane is furnished with papillæ. 1. The circumvallate, which are a dozen or so in number, and are arranged at the base of the tongue like an inverted V ; these papillæ are greatly concerned in taste, and are supplied by the glosso-pharyngeal nerve. 2. The fungiform ; these are scattered over the tongue, and are specially observed at the sides and tip. 3. The conical or filiform are distributed all over the tongue. The tongue is divided into two symmetrical halves by a fibrous septum, the existence of which is marked by a raphé in the median line.

Diseases.—*Tongue-tie* is a condition in which the frænum, or fold, seen on the under surface, extends to the tip, and appears to tie the organ down to the underlying structures ; its division, by means of a pair of blunt-pointed scissors, readily remedies the defect.

Inflammation of the tongue (glossitis) may be caused by wounds, or stings, or by the application of some acrid substance ; occasionally it comes on without any apparent cause. If the symptoms are not peculiarly urgent—*i.e.*, if there be no great pain or swelling, or threatening of occlusion of the fauces, a leech or two under the jaw and a smart purgative usually afford relief. If the inflammation be very sudden, its progress rapid, and suffocation threaten, then a few longitudinal incisions should be made on its surface to allow of the escape of fluids. In very severe cases, where these measures afford no relief, and the symptoms are very urgent, tracheotomy must be performed. Glossitis is sometimes brought on by the excessive use of mercury ; the treatment in such cases consists of purgatives, astringent lotions, and careful bandaging of the organ, and full doses of chlorate of potash internally.

Ulcers of the tongue may have their origin from several causes ; either from local irritation, such as decayed teeth, or from some derangement of the digestive organs, in fevers, or from syphilis, or from the prolonged and maladministration of mercury. In all cases there is a marked foulness of breath. The constitutional treatment of course varies with the case, and removal of all obvious irritation, attention to the bowels, and locally the application of a solid stick of lunar caustic, the sucking a few crystals of chlorate of potash, and in syphilitic ulceration, the application of a little calomel powder diluted with flour, are about the best remedies. Those connected with secondary or tertiary syphilis are the most intractable, and frequently defy all treatment. Malignant ulcers of the tongue are epithelial in their character, and their development is frequently ascribed to local irritation, such as a sharp stump of a tooth, the habit of smoking short clay pipes, etc., but such causes are very questionable. The margins of such ulcers are composed of hard granulating masses, implicating the substance of the tongue, and ultimately involving the glands at its base; under the jaw and in the neck they are attended with great pain, and are usually deeply excavated. The prognosis in these cases is very unfavorable. The treatment is very unsatisfactory, and consists in removal as the only chance for the sufferer.

Enlargement (hypertrophy) occurs in young persons, and is nearly always congenital. The tongue protrudes from the mouth, becomes ulcerated from contact with the lower teeth, and there is a constant dribbling of saliva. The treatment consists in attention to the state of the digestive system, bandaging the organ, and astringent lotions. In cases where this treatment is of no use, removal of a portion or the whole of the protruded part must be performed.

Tumors in connection with the tongue are sometimes met with. Of the most frequent occurrence is ranula, to which a special article is devoted (See RANULA), encysted tumors, closely resembling ranula, fatty tumors, and nævi.

The ducts of the salivary glands, the parotid, and submaxillary, are sometimes the seats of *concretions* composed of phosphate of lime and animal matter, oval in shape, of a brownish or yellowish color, and of variable size, sometimes being as large as a small egg. Occasionally they come away of their own accord by ulcerating through their confines, but the treatment consists in their removal by the knife and forceps.

Wounds of the tongue almost always bleed very freely ; in slight cases iced water or styptics will arrest the hemorrhage, or occasionally a vessel may be tied, or pressure may be kept up by a pair of common forceps, the blades of which are kept together by an elastic band. The edges of a severe cut or laceration should be approximated with sutures. All pain, swelling, etc., should be allayed by iced drinks and astringent and disinfecting gargles or washes.

TONICS are a class of remedies supposed to give strength and tone to the system, of which quinine and iron are examples.

TONSILS are two glandular structures, situated between the anterior and posterior pillars of the fauces, one on each side. (See FAUCES.) They are somewhat oval in shape, varying in size in different individuals. They consist of a congeries of mucous glands, and their internal surface is marked with small holes ; there are ducts leading from the cells in which the mucus is secreted. The use of the mucus is to lubricate the fauces during the passage of food, and it is expressed at the moment of deglutition. The tonsils lie in close proximity to some very important blood vessels — viz., posteriorly, the internal carotid artery and the jugular vein, while externally are the trunks of the temporal and external maxillary arteries, and between the vessels and the tonsil is the superior constrictor of the pharynx. Hence any operation upon the tonsils must be conducted with great care.

Diseases of the Tonsils. — *Tonsilitis,* or *quinsy,* is an inflammation of the tonsil and tissues immediately surrounding it, generally due to cold, exposure, or some peculiar condition of the body, sometimes to cutting the last molar teeth, the swallowing of some irritant, playing on wind instruments, etc. It commences with shiverings, feverish symptoms, redness, swelling, heat, and dryness of the fauces and tonsils. There is great pain in swallowing and attempts at articulation. It sometimes ends with a tardy ulceration, or suppurates, or by becoming erysipelatous spreads down the air-passages. Sometimes small pustules or follicular abscesses appear, forming a yellowish ulcerated surface ; again, a most formidable symptom is the formation of a thick, tough, whitish pellicle, resembling wash-leather, on the surface of the tonsil, pointing to diphtheria. (See DIPHTHERIA.) The *treatment* consists, in mild cases, of the internal administration of minderus spirit, the external application of linseed-meal or hemlock poultice, and a gargle of warm water. The application of lunar caustic is of great value. In the severe diphtheritic cases, swabbing with glycerine and perchloride of iron, tonic treatment, good living, fresh air, and stimulants are necessary. If abscess occur, it requires to be actively treated at once, as respiration and deglutition are impeded, and the matter should be evacuated. The left forefinger is to be introduced into the mouth, the tongue depressed, and a straight sharp-pointed knife, with its back resting on the tongue and its point directed backward, is plunged into the centre of the exposed tumor. The edge and point of the knife must be carried inward toward the middle line, and never outward, lest the vessels already mentioned be wounded. If bleeding be very severe, a strong solution of the perchloride of iron must be applied to arrest it.

Ulceration of the tonsils is frequently caused by the irritation of carious teeth, or cutting the wisdom teeth. Again, a frequent cause is syphilis, and in such cases constitutional treatment with iodide of potass, or some mercurial combined with a tonic, and the local application of lunar caustic, or in severe cases nitric acid, is indicated. A disinfecting gargle of carbolic acid or chlorinated soda is of great value in all such cases, as the odor of the breath is most offensive.

Enlargement of the Tonsil is often the result of quinsy, and is frequently met with in scrofulous children or adults of weak habit. The tonsils in such cases are greatly enlarged, projecting, as fleshy excrescences, into the back of the fauces, and sometimes almost entirely closing them, interfering greatly with deglutition, speech, and breathing, producing a peculiar guttural tone of voice, and causing snoring during sleep. In young children cod-liver oil and other antiscrofulous treatment may be of great service, but in adults nothing is of any permanent good but removal, or partial removal, of the hypertrophied gland tissue. The gland is seized with a pair of long-clawed forceps, dragged out, from its bed, toward the middle line of the fauces, and a blunt-pointed curved bistoury (with its edge, all but an inch or so, covered with lint or plaster) is passed behind it and made to cut its way toward the middle line of the fauces, and away from its pillars. The bleeding and pain are very slight generally, and any bleeding can be controlled with iced water or tincture of perchloride of iron. The enlarged gland tissue hardly ever returns. In removing tonsils in children it is as well to adminster chloroform.

TOOTHACHE is a most distressing ailment, too well known to need description, and is apt to attack any one, though some families and constitutions seem more subject to it than others. It is a sort of neuralgia, and frequently depends on the condition of the general health, which reacts on the nerves, and especially on any susceptible nerve which may be exposed to contact with the air in a decayed tooth. Decay in teeth is occasioned chiefly by the collection of particles of food, which set up a fermenting action and extend the process of decomposition gradually to the bone of the tooth, and so wear away the enamel, and form a little point of opening for the nerve to become affected. In order to avoid this, great care should be taken to keep the teeth well cleaned, and when possible to brush them after every meal. When any tiny speck of discoloration is perceived, it is wise at once to go to the dentist and have the decay removed, and the aperture filled with gold. Any accumulation of tartar on or around the teeth should be carefully removed. When, however, toothache has really seized hold of a victim, the only chance is either to summon up courage to have the offender extracted at once, or to try one of the many remedies which exist, and have in some cases been found beneficial. Creosote, chloroform, eau-de Cologne, or brandy on wool will often cause a cessation of pain for a time ; but these are simply stimulants, and often sedative measures are more effectual. When the gum is much inflamed, a leech applied to it and allowed to draw freely will often give relief, and a poultice of bread-and-milk held in the mouth is sometimes comforting. Experience, however, goes to prove that endurance is the only remedy to be relied on in toothache—that in time the pain will cease, and that if the sufferer cannot make up his mind to bear it, the only effectual cure is extraction.

TORMENTILLA is the root of an indigenous plant belonging to the rose tribe. Its properties are astringent, and the substance may be used as an astringent when no more powerful remedy is at hand.

TOURNIQUETS are instruments for the mechanical compression of a vessel in order to prevent hemorrhage. The first tourniquet was used in 1674, and was the invention of the French surgeon Morel, and was a rude contrivance, and consisted of a stick passed beneath a fillet or band, and twisted round so as to constrict the limb to the requisite degree of tightness. A great improvement was made upon this in the early part of the following century, by J. L. Petit, also a French surgeon, and his tourniquet consisted essentially of two metallic plates, which could be separated from one another by means of a screw, so as to tighten a strap which was connected with them, and also encircling the limb. The common tourniquet now in general use is based upon that just described, and it consists of a firm, narrow, flat pad to compress the artery, a strong piece of webbing or band to pass round the limb, and a bridge furnished with rollers over which the band passes, and a screw which raises the bridge and thus tightens the band. The pad must always be so arranged that it compresses the artery against the bone. On applying it the band should first of all be buckled tightly, when by turning the screw, great pressure is obtained. Care should be taken that the screw be opposite the buckel of the band.

Tourniquets are very useful in the absence of assistance, for, if properly adjusted, they at all events stop hemorrhage ; but they have this disadvantage, that when applied to an artery they also compress the vein or veins of the limbs, rendering the venous hemorrhage more profuse.

TRACHEOTOMY is the operation of opening the trachea, or windpipe, to save the patient's life. It is sometimes necessary in cases of croup, diphtheria, and when suffocation is imminent from the pressure of a foreign body in the air-passages. (See LARYN-GOTOMY.)

TRAGACANTH is a kind of gum obtained from a plant growing in Asia Minor. It is allied to gum acacia. The gum exists in flakes, not easy to powder till well heated. Part of it is soluble in water, and this suspends the rest, so that it forms a thick tenacious mucilage, much denser than that formed by gum arabic. This mucilage is used for suspending heavy powders. A compound powder consisting of tragacanth, gum acacia, starch, and sugar, mixed with hot water and allowed to cool, is useful in the same way.

TRAINING, that is to say a system of physical education, is too much neglected among us. It is quite true that for boating and the like training is looked to, but then the exercise is excessive, and likely in the long run to do harm rather than good. The systematic use of certain exercises and the regulation of food and diet are calculated to do good, but they must be sensibly conducted, or the reverse is the case. (See GYMNASTICS.)

TRANCE, a curious and interesting phenomenon, presented occasionally in what is understood by cataleptic conditions of the body. The symptoms are so peculiar, and so often connected with deceptive action on the part of the patient, that they are most difficult to investigate. (See CATALEPSY.)

TRICHINÆ are animal parasites which find their way into the muscles of the human body, and may prove fatal. (See ENTOZOA and PARASITES.)

TROCAR, a surgical instrument used for the purpose of perforating the abdomen or chest when fluid has to be drawn off by tapping, as it is called.

TRUSSES are mechanical contrivances for the support or for the prevention of the protrusion of any viscus, but most usually for the support of the parts concerned in abdominal rupture or hernia. If a hernial protrusion occurs in either sex, it should be advised that mechanical treatment be adopted at once, for no matter whether in infancy, youth, or middle age judiciously-applied trusses frequently effect a cure, without further surgical interference, and at all events cause but little trouble or annoyance. A physician should always be consulted as to the form of truss needed, and should himself take the necessary measurements, and himself apply the apparatus in the first instance. It is a great mistake, and one productive of the worst results, to leave the advice of a truss to an instrument-maker, and we often see instances, especially among the poor, of ill-fitting, ill-shaped trusses, which not only do no good at all, but in many cases do absolute harm, by increasing the mischief they are designed to alleviate. A truss should be firm, light, and elastic, and preserve its shape, and the strength of the spring should always be

equable, so that it may retain the rupture without irksomeness. A truss consists essentially of a pad attached to a metal spring having straps so arranged that it may be kept in the desired position in any of the various movements of the body. There are many different forms, whether single or double, named after their inventors, for the general principles of which see HERNIA. The following hints on trusses are of value, as the experience of an authority on the matter :

" In the majority of cases, the circular spring truss is the best form. The curve of the spring and the relative position of the pad with it should be appropriate to the configuration of the wearer. A single piece of metal should form the spring and foundation of the pad. As far as practicable the spring of the truss should pass around the bony rim of the pelvis, fitting closely to the figure, and should lie out of the region of the great muscles of the buttock (*glutæi*). The form of the spring may be designed after the French model or the German. The former resembles the coil of a watch-spring, and is very elastic and clinging ; the latter almost exactly fits the outline of the body in its state of repose : it is almost inelastic, and very hard. The French is always pressing inward, even when the wearer is at rest ; the German scarcely presses at all when the abdomen is soft, but resists with power when any expulsive force makes the abdomen swell. The best shape for the spring is one which forms a medium between the two. The pad should be of moderate dimensions. For the adult it should not exceed two and a half inches in length and two inches at the widest part. Its superior edge should follow the upper line of the spring, which falls a little from the shoulder or bend, where it lies in contact with the hip. The inner surface should be directed slightly upward. The proper shape for the pad, and the materials of which it should be constructed, may be varied to accommodate particular cases. The wearer generally discovers after a while which kind of pad is most free from annoyance ; that pad, however, is the best which maintains perfect and unintermitting retention of the hernia. Every pad should have attached to it two studs, one near its junction with the spring, and another at its lowest point. To the upper one the transverse strap, passing from the free end of the spring, is attached : the lower stud is used with the thigh strap, which should be always worn. It is loosely fastened on to the spring of the truss near its shoulder, and should fall along the hollow beneath the buttock. In the erect posture of the wearer this strap should be moderately tight ; it prevents the pad from shifting, and should never be discarded." The pad may be prevented from fretting the skin by covering it with fur, or by the interposition of some soft substance.

Trusses for ventral, umbilical, and femoral hernia are also constructed. In the case of crural or femoral hernia " the spring should fall somewhat suddenly from the point where it passes around the hip, and lie along the fold of the groin (Poupart's ligament). The pad should be rather small and convex. The cross-strap should fasten high up on to the shoulder of the spring, in order to keep the pad well down on the thigh. The thigh strap should start from near the pad, and return, after encircling the thigh, to the pad itself." In large hernia, or those which have become irreducible, a bag truss is indispensable. Trusses are also in use for the support of prolapse of the womb or rectum, and constructed of various forms by different makers. (See HERNIA.)

TUBERCULAR MENINGITIS. (See MENINGITIS.)

TUBERCULOSIS is a name applied to a form of fever which is accompanied by the formation of small bodies called tubercles in various tissues ; when they are deposited in the membranes of the brain they give rise to the disease known as tubercular meningitis, or acute hydrocephalus.

TUMORS, or new growths, are divided by pathologists into two main groups : one of innocent or benign growths, the other of malignant growths. The latter are distinguished by the following common characters : rapidity of growth, tendency to infiltrate and to replace the tissues of the affected part, tendency to recur after removal by operation, tendency to multiply locally and to infect other and remote parts of the body ; a tendency to destructive and progressive ulceration, inducing fatal exhaustion through pain, continuous discharges, and occasional loss of blood. To any tumor presenting these so-called characters of malignancy, the term cancer was applied by pathologists of a past generation, but at the present time, in consequence of the extensive use of the microscope in pathological research, there is a tendency to classify tumors with regard more to minute structure than to clinical characters. The tumors constituting the malignant differ much in consistency and in minute structure, but the great majority of them have been referred to one of the following two great divisions : that in which the growth is composed of some form of connective tissues, and that in which it is made up in great part of cells resembling in character those found in the epidermis, on mucous membranes, and in the ultimate lobules of secreting glands. To the former division belong tumors that are called sarcomata or fleshy growths, to the latter belong the true cancers or *carcinomata*. Structurally the two are distinct ; with regard to clinical characters and malignancy, their resemblance is very close, the chief distinctions in these respects being the facts that cancer almost in-

variably, and sarcoma seldom, affects secondarily the lymphatic glands, and that the latter usually appears at an earlier period of life. (See CANCER.)

Innocent or benign tumors may occur in almost any part of the body, and they may vary in character from so simple a growth as a wart up to formations which may endanger life or require some serious surgical operation for their removal. It would be useless to attempt in a work like this any useful classification of tumors, as any properly devised system would be unintelligible to the ordinary reader. The question with most people who find a tumor is forming is as to its being of a cancerous nature or not, and this can only be answered by obtaining the advice of a medical man. Much harm is done by the reckless way in which patients, to get rid of their malady, fall into the hands of those who pretend to cure them, while too often they only hasten on the fatal termination. The great majority of small tumors are harmless in character, and often cause inconvenience rather than any other distress, but in all cases proper medical advice must be taken before recourse is had to removal. (See FLOATING TUMORS.)

TURKISH BATH. (See BATH.)

TURMERIC is the underground stem of a plant (Curcuma) growing in Ceylon. It contains a substance bright yellow in color, which alkalies readily turn brown. A solution of this material in alcohol or paper smeared with it may be used as litmus, or as a test for alkalies. Turmeric itself is something of a stimulant, and is used as a condiment. It is turmeric which gives the bright yellow color to curry powder, into whose composition it enters.

TURPENTINE is a mixture of oil and resin obtained from various species of pine, and mainly produced in America. This substance is separated by distillation ; the oil of turpentine passes over, the resin is left behind. Turpentine as it flows from the tree is of a pale yellow color, about the consistence of honey ; but gradually, by exposure, becomes harder, the oil passing off and the resin remaining behind. Oil of turpentine, which is alone used internally, is a colorless fluid, with the peculiar odor and taste of the above liquid ; the resin is semi-transparent and yellow. The preparations of oil of turpentine are a confection, an enema, a liniment, an acetic liniment, and an ointment. The resin figures in a plaster and an ointment.

Applied to the skin, turpentine acts as a powerful stimulant : if used along with heat it may redden the part, or if its vapor be confined, even blister it. Its liniment is of value for stiff joints and chronic rheumatism. Turpentine stupes are valuable applications. A piece of flannel is wrung out of hot water as hot as the hands will bear, turpentine sprinkled on the surface, and so applied to the skin.

This application is of exceeding great value in inflammation of internal organs near the surface, as in slight peritonitis and pleurisy.

Internally, turpentine may be given either as a stimulant or for destroying worms. Often turpentine is given as a stimulant to the kidneys, but it may produce much irritation in the urinary tract. It is also valuable for arresting hemorrhage, especially if that is partly due to debility. It is frequently given as an injection, when it not only moves the bowels but also acts as a stimulant to the system at large. When swallowed, turpentine may likewise act as a purgative, but it is common to combine it with castor-oil. It is perhaps the most valuable remedy for tape-worms we possess, provided the patient is not made sick by the dose. If retained, it speedily causes the worm to be expelled dead. The dose of oil of turpentine as a stimulant and diuretic is half a drachm or a drachm ; to destroy worms half an ounce is given.

TYMPANITES is the term given to flat-ulent distension of the abdomen. The exact origin and nature of the gases which cause the bowels to swell up and resound like a drum has been often made the subject of speculation, but not very often of careful in-vestigation. Undoubtedly in certain diseases a period often comes when the bowels, from no very ostensible cause, swell up from wind, which apparently has been secreted by their walls. This commonly occurs only in very exhausting diseases, or in which there is great prostration. In typhoid fever, and in peri-tonitis of whatever origin, it is found and dreaded. It is as a rule of very evil omen. When tympanites does occur, the best appli-cation outside is turpentine in the form of stupe ; many, too, prefer to give turpentine internally, but that is a question to be settled in each individual case. Stimulants are as a rule are first given, and sometimes passing a long tube up the rectum carefully and gently may enable the gases to escape, and so afford unspeakable relief to the patient. In the last resource the bowels must be punctured with a fine hollow needle or trocar.

TYMPANUM. (See EAR.)

TYPHLITIS means an inflammation of the cæcum. (See INTESTINES.) It may be caused by eating nuts or some other indiges-tible food, which sets up an irritation when lodged there. There is much pain, a little fever, some vomiting, and constipation. Hot fomentations should be applied, an anodyne given, and the bowels should be opened by injections.

TYPHOID FEVER is a continued and infectious fever, caused chiefly by the influ-ence of bad drains and sewer-gas, lasting an uncertain period of from four to six weeks, and sometimes followed by a relapse. It is also known by the names low, enteric, gas-tric, pythogenic, drain, cesspool, bilious, in-fantile, remittent, and slow nervous fever, also as abdominal typhus fever.

Causes.—The exciting causes are conta-gion and spontaneous degeneration. Num-bers of cases go to prove that those nursing the sick from this disease very frequently catch it, and they probably do so from the emanations of the stools. Whenever any drain-age soaks from the surface into a well used for drinking purposes, or when sewer gases escape into a house by a leaky pipe, or when the traps are out of order, or when one drinks foul and stagnant water into which any drainage from manure can enter, then arise the conditions which excite the disease. Very few houses are properly drained, and whenever a storm occurs and the sewers are suddenly flushed, the gases escape upward into the waste-pipes of the houses along the route and overcome the resistance of the traps, so that a most noxious smell arises whenever the pan of a water-closet is raised. It is of the utmost importance that all water-closets should be outside the house ; that the waste-pipe should not communicate with the main sewer unless there be first a communi-cation with the open air, so that the back-ward pressure will never cause the gases to regurgitate into the house ; that just beneath the pan of a water-closet the waste-pipe should communicate with the open air and be carried up above the house-top ; that a cistern with a continuous supply of water should be supplied close to and above each water-closet, and that the cistern for the drinking water should be quite distinct from the other cisterns. In small places the dry-earth system should be adopted, and care must be taken that no leaking from an old cesspool can escape into the well for drinking purposes.

Symptoms.—The onset of typhoid fever is always very gradual and insidious ; it begins with feeling out of sorts, aching pains in the limbs, headache, loss of appetite, and chilli-ness ; for many days the sufferer is able to go about and think that there is not much the matter. Sometimes there is diarrhœa, or some intestinal disturbance ; then the pulse is quicker, the skin hot, and the tongue red and dry. The nights are disturbed and rest-less, and he does not care for any exertion. At the end of the first week, or often later, he takes to his bed, and it is found that he is feverish, has no appetite, is thirsty, and his bowels are generally relaxed. The urine is scanty and high-colored ; there is still more restlessness at night ; there is no stupid, heavy expression as in typhus, nor are the eyes suffused ; on the contrary, the face is often pale and the cheeks have a pink flush, and the eyes are clear and bright. Between the seventh and the twelfth day the peculiar eruption appears on the chest, abdomen, and back, and it consists of a few slightly raised rose-colored spots, which disappear on pressure under the finger and fade away in two or three days, but in the mean time others appear, so that several crops are noticed, and fresh ones may be seen every day; these spots are never petechial. If now the hand is pressed over the right side of the abdomen there may be a feeling or expression of pain, and one may also feel a gurgling under the fingers. About the middle of the second week delirium comes on, at first slight and only noticed at night, and then more constant, intense, and noisy. The tongue is dry, red, and glazed, and often cracked in various directions; in children, however, it may sometimes remain moist and white the whole time, and in very young cases also you do not always see any rash at all. As the disease advances the patient loses flesh and strength; he lies prostrate and perhaps unconscious of what is going on around, and if it end fatally he will become quite insensible, have a markedly high temperature, and fumble at the bed-clothes. If the disease progress favorably the amendment is very gradual, and for this the temperature is a pretty good guide. The temperature rises from the first, but not so suddenly as in typhus and relapsing fevers; at the end of the first week it may be 104° or 105°, being generally highest toward evening; it keeps high with slight oscillations for about twenty-one days, and then a fall may often be noticed in the morning, although it ascends again at night, and these daily variations are very marked and may cover three or four degrees; at about the thirtieth day, or a little later, the symptoms are decidedly less severe in ordinary cases; the tongue cleans; there is less prostration and delirium, and a general improvement is manifested. But then a relapse may ensue, and the temperature will again rise, and the patient go through a second attack, but this is much shorter than the first.

Complications.—Typhoid fever is a very dangerous disease, because there are so many accidents to which patients are liable. Diarrhœa may be very profuse and exhaust the patient, but as a rule diarrhœa is not a very bad symptom, and should be left alone, unless very profuse. Bleeding from the bowels, when it occurs in any large quantity, is a very dangerous sign; it is due to the ulceration of the intestines. Bleeding from the nose is not often a bad symptom. Perforation of the bowel is very likely to occur between the twenty-fifth and thirty-second day, and even later, and this may be brought on by any error of diet; it is attended by collapse, and is very fatal. Inflammation of the peritoneum, either with or without perforation, adds greatly to the danger. Bronchitis and pneumonia may supervene and increase the general mischief. Some cases are

very mild, others very severe, and there is perhaps no fever which varies more in its forms, nor about which so much anxiety and uncertainty must exist with regard to a successful issue, nor is one safe until recovery is fully established. In many cases it is most difficult to be certain of the nature of the case in the first week. It is most likely to be mistaken in children for acute tuberculosis; or it may be looked upon as the so-called gastric fever or gastric irritation; or it may resemble the symptoms of arsenical poisoning. It may be as well to say here that there is no such disease as *gastric fever;* it either means typhoid fever or it is a disturbance of the stomach and intestines from poisoning or eating unripe fruit. Whenever three or four cases occur together this fever may be suspected, and if any one die of similar symptoms within a week or two, and the cause is not clearly made out, an examination of the body must be made to settle the point, for many cases of arsenical poisoning have in this way been overlooked. In typhoid fever the main appearances after death are ulceration of the bowels, and chiefly near the cæcum and toward the end of the ileum, with enlargement of the spleen and mesenteric glands.

Treatment.—As regards ventilation, good nursing, cleanliness, and quiet, and with respect to disinfectants, etc., nothing more need here be said than is laid down in the article on Typhus Fever, and it need not be repeated. Yet there are some special points of importance. The diarrhœa need seldom be checked unless one is purged more than twelve or fifteen times a day, and then a little starch injection may be given; if there is much bleeding it may be requisite to give turpentine. It is a mistake to give medicines containing acids, as they often increase the purging and the bleeding. In fact, there is no medicine which can cure the fever. The diet must be very light, and no solid food should be taken under six weeks or two months, because, in consequence of the ulceration of the bowels, the coats are very thin and liable to burst. Eating an orange or a piece of potato, or drinking an effervescent draught will cause distension of the bowel and rupture it, just when the patient is otherwise doing well; the greatest precautions should be taken during the third and fourth weeks, as then it is most liable to occur. Milk must form the main article of diet, and then an egg or two may be beaten up in it, or a custard may be given and beef-tea; then a small piece of mutton and sole, and so on, gradually to more solid food. If there is much distension of the bowels, hot flannels on which is sprinkled a little turpentine will be found very useful. For information as to disinfection, see SANITARY REGULATIONS.

TYPHUS FEVER is a highly contagious

fever, attacking people of all ages, which occurs in an epidemic form, and generally in periods of famine and destitution. It has been known at different times under various names ; thus it has been called pestilential fever, petechial fever, brain fever, putrid continual fever, camp fever, jail fever, etc.

Causes.—In the individual, sex and age have no influence in determining an attack ; nearly equal numbers of both sexes catch it, and children, as well as adults and old people, are liable to it, but more cases have it after fifteen than before. Of 3456 cases admitted into the London Fever Hospital, nearly one half of the cases were thirty years of age or upward, one eighth were fifty or upward, while less than one sixth were under fifteen. Depressing mental influences, overwork, and anxiety render the system more liable to contagion ; those who are badly fed, and those who suffer from loss of a harvest ; people who have suffered the hardships of war, of civil strifes, and commercial distress, are often its chief victims. Overcrowding, dirt, and bad ventilation are important predisposing causes to this affection. Typhus is chiefly met with in cold and temperate climates, but not in the tropics. The chief cause of typhus is contagion, or the transmission of the disease from one person to another ; the other causes only render the system more liable to the action of the poison. Nearly all the evidence goes to show that typhus is essentially a disease that is caught by a healthy person coming in contact with one previously affected, and it is easily caught during convalescence. It is very rare for a person who has had the fever once to have it a second time.

Symptoms.—It is difficult to say how long the disease may be incubating in the system before it appears, but the period is certainly not constant, and seems to vary from a few hours to several days. The onset is marked by a severe headache, loss of appetite, and languor, and aching of the limbs ; the invasion of the symptoms is not so sudden as in relapsing fever, but much better marked than in typhoid fever. For three or four days the patient gets worse, being unable to go about, and feeling chilly and prostrate ; he then is worse at night and restless ; the skin is hot, the tongue coated ; there is thirst and sometimes vomiting ; by the third day of the disease most are obliged to take to their bed, while this is not the case in typhoid fever, which is a much more insidious disorder. There is a general aspect of a typhus case, which an experienced person will at once recognize ; the patient lies prostrate on his back with a dull and weary, if not stupid, expression ; the eyes are suffused and watery, and a dusky flush overspreads the face. As the disease progresses the eyes are half shut and the mouth open ; he lies moaning and

unable to move himself or answer questions ; the lips and teeth are dry and covered with sordes and look black ; the mouth is dry ; the tongue dry, brown, or black, and marked with cracks. The temperature rises from the first, and reaches 103° or 104° Fahr. by the middle of the first week ; the highest temperature reached in the fever is seldom less than 105°, although it may be higher, but the higher the point reached the greater is the danger ; the fever may slightly abate in favorable cases about the ninth or tenth day ; no marked fall, however, takes place until the end of the second week, and generally on the fourteenth day, when defervescence may take place suddenly, and the normal temperature (98.4°) be reached in twenty-four hours, but more commonly it takes two or three days for the descent to be accomplished. The temperature generally is highest of an evening ; when defervescence occurs, the temperature always goes below the normal line so as to mark 97°, or even 96°, and in a few days it becomes natural. This fall is a very good sign, and then the patient is generally out of danger. A very high temperature (106° or 107°) is a sign of serious gravity. In mild cases the fever begins to leave on the twelfth day in many cases. The pulse is generally 120 in a minute, but is very easily compressed under the finger ; the heart sounds in very severe cases are feeble, and the first sound may even be inaudible. A rash appears in nearly every case, and is very characteristic ; sometimes it looks as if there were a general mottling just beneath the skin, or distinct spots may appear of small size and purplish color ; they are irregularly rounded ; at first may disappear on pressure, but soon become petechial ; oftentimes the two kinds occur together, but sometimes separately. The rash appears on the fourth or fifth day, rarely later ; it comes on the back of the wrists first, in the armpits, and over the epigastrium ; then it more or less covers the trunk ; it seldom comes on the face and neck ; the rash has something of a measly look, but the other symptoms are much more severe than are seen in measles ; the rash lasts a variable time, but generally until the fourteenth or fifteenth day. No solid food can be taken, but the patient is always thirsty. The bowels in some cases are confined, in others they are open too much. There may often be heard rattling or wheezing noises in the chest, and the more so when the face is very dusky. The nervous symptoms are well marked ; restlessness, loss of sleep, and confusion of thought first come on ; then headache, giddiness, a buzzing in the ears, and deafness ; in most cases there is delirium, and the patient is beset with horrid fancies. In bad cases he lies picking the bed-clothes, twitching his hands, and muttering to himself or moaning ; or he may be quite unconscious with wide-open eyes, staring vacantly. Loss of the power of swallowing and insensibility are very bad signs, and generally precede death. The urine is passed involuntarily, as well as the motions in most cases, so that great cleanliness has to be observed.

The *duration* of typhus may be from three to twenty-one days, but about fourteen or fifteen days is the average time; if a case live more than this time it will generally recover. The termination in recovery is sometimes quite rapid, and the tongue will clean, the temperature fall, and the delirium cease in a day or two, but generally the improvement is more gradual and lasts over three or four days. Unlike typhoid fever, there is no relapse, so that when once the temperature has come down the best hopes may be entertained; nor is he liable to peritonitis or perforation of the bowel, as in typhoid fever.

The *death-rate* varies with the epidemic, being generally greatest at the commencement. Of children under ten years of age about 5 per cent die; of those over sixty years of age, 66 per cent die; the older the patient the greater is the danger; between thirty and forty, 21.5 per cent die; between twenty and thirty, 15.6 per cent die; between ten and twenty, 8.6 per cent die. Habits of intemperance increase the danger in those attacked; bulky people die more frequently than thin ones, black people more than white, and those who are overworked and have mental worry, etc., have the disease with the most severity.

Treatment.—The patient must be placed in a well-ventilated and large room, so that draughts will be avoided; he should have his bed so situated that the light from a window will not fall on his face, as this is annoying; all curtains, carpets, and bed-hangings should be at once removed; the bed should not be too soft, and a draw-sheet or mackintosh must be put under the patient. He should not be allowed to exert himself at all, but try and husband all his strength. The greatest cleanliness must be observed and all excreta removed at once, and carbolic acid or chloride of lime should be mixed with them, or any soiled linen may be put in a tub of water in which is some carbolic acid. Bed-sores are very liable to form on the back, and so the nurse must always be on the lookout, and try and prevent them coming by smoothing the sheets, drying the patient, and rubbing brandy or balsam of Peru over the part; better still, to have a water-cushion or a water-bed. The skin may be sponged down with tepid water,

the nurse drying and sponging one part at a time, so as to prevent any undue chill to the surface from exposure; this relieves the patient and partly removes that disagreeable smell so common from the skin in typhus cases. None but the nurse and doctor should see the patient; all noises must be stopped and great quiet enjoined; at night time there may be a small light in the room, but placed so as not to disturb the patient. Milk must be the chief article of diet, and is best given cold; an egg or two may be beaten up in it, and three or four pints of milk may be given in the twenty-four hours; this must be done regularly every two hours in equal quantities, and more especially must this be done at night or in the early morning when the prostration is the greatest. Beef-tea and broths, jellies, extract of beef, custards, etc., may be given if the patient can take them and wants so. For drinks in the early stage, lemonade, cold tea, soda-water, etc., may be given, but do not let them have too much effervescent drinks; when very bad the nurse will have plenty to do to get the milk down. Stimulants are very useful, but the quantity must vary with each case, and be left to the medical man's judgment; brandy is the best stimulant, and may be given with iced milk; too much must not be given at first, as it causes oppression and inability to take nutrient food; but afterward, in the stage of great prostration, its proper and careful administration may save the patient's life.

Albumen is often present in the urine in these cases, but calls for no special treatment. Much care must be taken that there is no retention of urine in the bladder in these cases, as that organ is very liable to be paralyzed. When the crisis has passed and the tongue cleans, some boiled mutton may be given; also jellies, light puddings, custards, etc. The stimulants may then be diminished and beer given if preferred. If, however, convalescence be retarded by bed-sores, or by the formation of abscesses, the stimulants must be continued and solid food given sparingly. In some cases the mind is childish for some time after recovery. A trip in the country, plenty of good food, and fresh air will complete a cure. For the prevention of typhus spreading, isolation must be adopted, and if a case occur in a crowded court, it should be removed to a fever hospital. For the measures to adopt with regard to disinfecting the clothes, room, etc., see SANITARY REGULATIONS.

U.

ULCERS consist in the gradual disintegration and separation of tissues, the healthy nutrition of which has been disturbed by local inflammatory changes, by impoverishment or poisoning of the blood, or by an injury to one or more of the nerves of the affected region. In this process the destroyed tissues break down into minute particles, or undergo

liquefaction ; in gangrene, to which ulcera- | CANCER, GASTRIC ULCER, DYSENTERY,
tion is closely connected, the open sore is | PHTHISIS, TYPHOID FEVER, etc.
formed by the separation of the dead tissues
in sloughs or large and visible masses. Ulceration may attack any organ or tissue ; it
is often met with in bone, and sometimes in
teeth ; the tissues most disposed to it are the
skin, mucous membrane, and connective or
areolar tissue. Nerves and blood-vessels
resist longer than other tissues the ulcerative
process, and may, in cases of rapidly-increas-
ing and sloughing ulcers, be seen isolated in
the midst of discharge and slough. The cornea
is a frequent seat of ulceration, which too often
causes blindness or serious impairment of vision, by resulting in opacity or perforation of
the membrane. Within the body ulceration
very frequently occurs in some part of the alimentary canal. Ulcer of the stomach, ulcer
of the duodenum, after severe burn, typhoid
and tubercular ulceration of the small intestine, syphilitic and dysenteric ulceration of
the colon and rectum, and fissure or painful
ulcer of the rectum are all well-known affections. The favorite seats of ulcers on the
surface of the body are the legs ; here the ulceration is generally due to local irritation
and obstruction in the circulation. Ulcers,
when present in parts of the body above the
knees are usually dependent upon some
constitutional affection, such as syphilis or
scrofula, or are connected with some form of
cancer. The face is often attacked with obstinate and spreading ulceration, of which
the most common examples are epithelioma,
lupus, and the rodent ulcer. Ulcers vary
much in their form and appearance, in their
rate of increase, and in the severity of the
symptoms to which they give rise. Some, as
the sloughing and phagedanic ulcers, spread
very rapidly, and are attended with bad general symptoms ; others, as the so-called callous ulcers, observed on the legs of old people, undergo very little change, and usually
cause very little pain, and rarely any constitutional reaction. Some are quite indolent,
and others give rise to excruciating pain ;
there is also much variety in their shape, the
general tendency, however, being to form
round or oval ulcers. These differences generally disappear when the ulcer takes on
healthy action and begins to cicatrize ; a
healthy granulating surface is then presented, which closes by contraction, and the
formation of a gradually extending zone of
delicate scar-tissue at its edges. Some
ulcers, and especially those formed in cancerous affections, obstinately resist local treatment, and continue to increase in size and
invade surrounding tissues until the patient
sinks from pain and exhaustion.

The following remarks apply only to ulcers
formed on the surface of the body. For
information concerning the ulcers of internal
organs, the reader is referred to articles on

The Inflammatory Ulcer is met with generally in front of and on the lower half of the
leg, and is usually due to slight injury, such
as a grazed or broken shin. As a rule, the
patient is either a plethoric individual, whose
health has been impaired by excesses in diet,
or one advanced in years, and exhausted in
consequence of hard work and insufficient
nourishment. The sore is small and circular
and usually single ; its base is covered by
small granulations of a brownish-red color,
from which there is a profuse discharge of
thin and acrid ichor ; the edges of the sore
are sharply cut, and the surrounding skin is
hot and red. The patient complains of
severe burning pain in the ulcer and over the
inflamed skin. The development of this
troublesome and painful affection is favored
and in many cases caused by negligence on
the part of the patient, or in consequence of
inability to discontinue active work. A slight
abrasion on the shin of an unhealthy individual suffering from congestion of the liver,
piles, and distension of the veins of the lower
extremities, will almost certainly degenerate
into a painful and inflamed ulcer if the part
affected be not kept at complete rest for a
few days. The prevalence of the inflammatory ulcer among the laboring classes is no
doubt due to the frequent occurrence in
individuals of this class of wounds and slight
injuries to the lower limbs, and to the necessity under which they lie of continuing work
and active exercise, even though suffering
from acute pain.

The *treatment* of inflammatory ulcer should
consist of complete rest of the affected limb.
The patient should remain in bed with the
limb elevated on a pillow. The ulcer should
be dressed with a light bread poultice, warm
fomentations, or a weak lead lotion. When
the pain has subsided, and the ulcer presents
the appearance of a healthy granulating sore,
water dressing or a weak solution of sulphate of zinc should be applied and the
limb be bandaged from the toes as far as the
midd'e of the thigh. Local applications alone
are quite useless. The patient must remain
in bed, in the recumbent position, until the
ulcer has changed into a rapidly-closing and
healthy sore.

Irritable Ulcer is the name applied to any
small sore which has an unhealthy appearance, obstinately resists treatment, and gives
great pain when touched at a certain point of
its surface. According to some, the acute
pain in this affection depends upon the exposure of a nerve on the surface of the ulcer.
This pain is always limited to one spot, the
rest of the raw surface being free from tenderness. This variety of ulcer may be much relieved by a poultice or poppy-head fomentations. The only effectual means of cure,

however, is division of the exposed nerve, by making a small incision across its track at a short distance above the painful spot.

Chronic, Callous, or Indolent Ulcer is of frequent occurrence among old and debilitated individuals, and in most instances affects the lower part of the leg. It is usually of considerable extent, and in some bad cases completely encircles the limb. The surface is smooth and glassy, is much depressed below the surface, and is surrounded by hard and white edges. The skin surrounding the ulcer is thick and callous ; the leg below the ulcer is hide-bound as it were, and the foot is often swollen. This ulcer, though large and formidable in appearance, is generally free from pain and remains indolent, except when much irritated. The hard edges then rapidly sink down, and a large and painful sloughing ulcer is formed.

The essential point in the *treatment* of chronic ulcers is to establish healthy and active granulations, and at the same time to reduce the thickening and induration of the parts around, so that these may yield to the contractile force of the scar-tissue formed over the granulating surface. The patient should keep in the recumbent position, and take good diet and a moderate amount of alcoholic stimulants. The ulcer should be poulticed, and afterward, when its surface is moist and bathed by a purulent discharge, should, together with the surrounding hard skin, be strapped and bandaged. By this treatment an indolent ulcer of an oval shape and not very large will generally, in the course of a month or six weeks, be completely closed. In cases, however, where the ulcer is very old and large, and involves a considerable extent of the circumference of the limb, although considerable improvement may be produced, and the raw surface be much reduced, it will seldom be possible to make the limb sound. There is always a tendency for the scar formed over a chronic ulcer to break down and to slough whenever the patient commences to walk about again and to take active exercise. The general idea that it is dangerous to close an old indolent ulcer is not an unfounded one, as the cicatrization of a chronic ulcer in a person of advanced age is often followed by symptoms of constitutional disorder and slight apoplectic strokes. In cases of this kind it is often thought necessary to establish a drain upon the system by making an issue or a seton wound.

Varicose Ulcer is met with in the lower extremity in connection with distended and varicose veins. It generally commences as a small simple or inflammatory sore, and then, in consequence of the congested state of the limb, persists and acquires the characters of an indolent ulcer. The varicose ulcer is seldom single ; usually one may observe about the ankle two or more torpid ulcers of oval

shape and varied size. The surface of each ulcer is smooth and a brownish-red color, and the edges are hard and somewhat elevated. The skin around and between the ulcers is generally swollen and red, and sometimes raw, in consequence of a chronic eczematous eruption ; the foot is swollen, and about the ankles are unnatural swellings caused by distension of the superficial veins ; the veins of the leg are much swollen and varicose, and the small subcutaneous veins form large purple patches of an arborescent appearance. A prominent cord, formed of one or more varicose veins, may generally be seen passing upward from the superior margin of each ulcer. A vein is sometimes laid open by the extension of a varicose ulcer, and bleeding takes place, which persists and causes much trouble and alarm, so long as the patient remains erect, and no means are applied locally in order to arrest the flow. As soon, however, as the patient is placed on his back and the lower limb is elevated, the bleeding is arrested. In cases of this kind a small pad of lint or linen rag should be placed over the bleeding point, and be retained there by plaster and a few turns of a bandage.

The varicose ulcer may usually be much reduced in size, or even completely closed, by rest in the recumbent position, by the application of mildly stimulating lotions to the raw surface, and by firm bandaging of the affected leg. The ulcer, however, will generally break out again if the patient becomes careless and takes active exercise without taking measures to afford support to the varicose veins. A bandage ought to be applied every morning, and be carried from the toes to the middle third of the thigh, or, what is a much better plan, the limb should be encased in an elastic stocking. In cases where several large and obstinate ulcers exist in connection with an extremely enlarged and varicose state of the veins of the leg and thigh, it will be necessary for the patient, before he can obtain any relief, to submit to a surgical operation, by which the larger superficial vessels may be obliterated.

Strumous or Scrofulous Ulcers usually result from the enlargement and suppuration of a lymphatic gland, or from the bursting of a small subcutaneous abscess. They are generally multiple and closely clustered, and most frequently affect the neck. They are very irregular in shape, and often run together. The individual ulcers are small, and show no tendency to increase much in size ; the edges are irregular and formed of thin and undermined skin of a pink or light purple color. The surface of each ulcer is composed of large pulpy granulations, from which there is constant discharge of thin yellow pus. In old and severe cases there is much thickening of the surrounding integument. These ulcers are usually associated with enlarged glands, pustular eruptions on the scalp and face, and with other manifestations of the morbid disposition known as scrofula. In these cases no relief can be obtained, except by a proper constitutional treatment (see SCROFULA). The local treatment should consist in the application of mildly stimulating lotions, such as a weak solution of tincture of iodine, or a lotion containing blue-stone (2 grains to 1 ounce of water). (See PHAGEDÆNA.)

UMBILICUS, the central spot of the abdomen, marked by a depression ; it is now and then protruded in infants, and may require a pad and bandage to keep it in its place ; it is commonly called the navel.

UNGUENTS. (See OINTMENTS.)

UNLEAVENED BREAD. (See BREAD.)

URATES, or LITHATES, form the common deposit in the urine known as sand or gravel. They are usually of a pink or drab color, and consist of uric acid in combination with potash, soda, and ammonia. They often appear after an ordinary cold, in many cases of fever, and from too much drinking. In such cases the urine is often more acid and more scanty than usual. (See URINE.)

UREA. (See URINE.)

URETER, a narrow tube passing down from each kidney into the bladder, and allowing of the passage of the urine.

URETHRA is the tube which allows of the passage of the urine from the bladder. It is liable to be hurt by accident or disease, and may be the seat of stricture, so that one cannot pass water readily. (See STRICTURE.)

URIC ACID is one of the constituents of healthy urine ; it now and then forms the nucleus of a stone or calculus. It is found in excess in the blood in cases of gout. (See URINE.)

URINALS should be thoroughly clean and amply supplied with running water. Night and morning they should also be well flushed down. It is usual to have some chloride of lime placed about, so as to remove any noxious odors. Great pains should be taken at all times to remove any accumulation of fluid. It is desirable that the walls should be made of glazed tiles rather than metal, wood, or slate, and the roof should freely communicate with the open air.

URINE is a secretion which is constantly going on from the kidney, and in this way a large quantity of water and various inorganic and organic matters are continually being taken from the blood ; and it is important that this should be the case, for if these materials were retained they would be productive of serious consequences. Urine is generally of a light amber color, of an acid reaction, turning blue litmus-paper red, of a peculiar odor, and saline taste. Its specific gravity on the average is 1.020, pure water being taken as 1.000 ; but this will vary with

the time of day and with the amount of liquid food absorbed into the system. The quantity of urine passed during the twenty-four hours varies a good deal, but on an average may be estimated at forty to fifty ounces. An average healthy man excretes about fifty ounces or 24,000 grains of water in a day. In this are dissolved 500 grains of urea and from ten to twelve grains of uric acid. The following table shows the composition of healthy urine :

	In 100 parts of solid matter.
Water................... 936.80	
Urea.............. .... 14.23	33.00
Uric acid.............. ..... 0.37	0.86
Alcoholic extract......... 12.53	29.03
Watery extract........... 2.50	5.80
Vesical mucus........... 0.16	0.37
Chloride of sodium....... 7.22	16.73
Phosphoric acid... ...... 2.12	4.91
Sulphuric acid........... 1.70	3.94
Lime... .... 0.21	0.49
Magnesia................ 0.12	0.35
Potash............. ...... 1.93	4.47
Soda................. .... 0.05	0.12

In addition the urine contains carbonic acid, oxygen, and nitrogen in a gaseous form, but very small quantities of the two latter substances. Urea is a nitrogenous product, and nearly all the nitrogen which daily enters the blood in such food as the proteid compounds passes off in the urine after being used up in the system. By taking an animal diet the urea is increased in amount, and diminished by living upon a vegetable diet. When uric acid is not properly eliminated by the kidneys it will be retained in the blood, and it is always present in that fluid in cases of gout. In many cases the constituents of the urine, instead of passing away in a soluble form, may become deposited in the solid state on their way from the kidney ; at first very small, they may increase in size, and form what is commonly known as a stone or calculus. Stones are chiefly formed in the kidney ; but they are met with in the bladder as well. (See OBSTRUCTIONS ; STONE.) Uric acid or urates and phosphates are the substances most commonly forming the greater part of a stone. In ordinary cases there is no sediment in the urine, but merely a faint cloud of mucus. Often, however, when one has a cold, or has had an excess of beer, or from various causes, there may be a pink or drab sediment when the urine cools. This is composed of urates, or uric acid in combination with the alkalies, potash, soda, and ammonia, which are daily passed in the urine. This sediment is of slight practical importance, and will disappear by drinking plenty of bland fluids, as tea, water, etc. The color of urine varies much : when a small quantity is passed, it is of a high color ; when plenty is voided, it is paler. This is seen in cases where a person has had two or three glasses of hot gin and water. This fluid rapidly runs through the kidneys, and the urine is then almost colorless and of low specific gravity. A large quantity of urine daily passed, associated with great thirst and dryness of skin, should make one suspect that diabetes is present. When a stone is present in the bladder, or when the patient suffers from stricture, the urine may be ammoniacal and smell disagreeably. There will most likely be also a deposit of phosphates. Retention of urine is said to occur when the urine is secreted by the kidneys but not voided from the bladder. It is at once relieved by passing a catheter. Suppression of urine is of much more serious import, and is due to mischief in the kidneys, and if not relieved soon will cause death by poisoning the blood. This often forms the last stage of Bright's disease, and commencing with drowsiness and often convulsions, passes through stupor into coma and death. The urine is of immense importance to the physician in finding out the state of a patient, and in making out clearly the nature of many disorders. Albumen, pus, blood, and sugar are the four impurities most commonly met with in the urine in cases of disease. Albumen may be known by heating a small quantity of urine in a test-tube, and adding nitric acid, when a white flocculent precipitate will come on. The patient is then said to be suffering from albuminuria. Albumen may be present in cases of Bright's disease, emphysema, chronic bronchitis, heart disease, most of the fevers, and chiefly in those which take on a malignant character, and in several other disorders. (See ALBUMINURIA.) Pus gives a greenish yellow deposit in urine, and it turns very tenacious when heated with a solution of potash. It may be caused by disease of the kidney or bladder, or by an abscess bursting into the bladder. Paralysis of the bladder in cases of paraplegia, a stone in the bladder, and old-standing strictures are the conditions most liable to favor the formation of pus in the urine. Blood is present in the urine in cases of acute Bright's disease, in heart disease, many fevers, blood poisoning by turpentine, cantharis, etc. (See HÆMATURIA.) Sugar is present in the urine in cases of diabetes; but the importance of this as a symptom varies much with the age of the patient, being far more dangerous in young than in old people. Elderly persons, and those who are consumptive, now and then pass slight amounts of sugar without its producing any symptoms. (See DIABETES). For pain or difficulty of passing urine, see DYSURIA.

URTICARIA. (See NETTLERASH.)

UTERUS is the technical name for womb, and is situated in the pelvis. It is chiefly composed of muscular fibres, which increase enormously in size in cases of pregnancy, and aid in bringing forth the child into the world. It contains a narrow cavity about two inches and a half in length, and it is lined by a mucous membrane. From this membrane is secreted the fluid which comes away at the ordinary monthly periods. On either side of the uterus is an ovary, which at certain times is connected with the womb by means of the Fallopian tube. The uterus is liable to many diseases—a polypus or a fibroid tumor may grow in its walls, or project into its cavity. The symptoms will probably be occasional hemorrhage, or bleeding in excess, pain in the back, and perhaps difficulty in passing water. The uterus may become bent upon itself, and sometimes cause distressing symptoms of pain, difficulty in micturition and defecation, and excessive menorrhagia. Sometimes there is ulceration of the womb, accompanied by leucorrhœa. After delivery the uterus does not always return to its normal size, but is larger and heavier than usual. Subinvolution is then said to have occurred; menorrhagia is then very likely to supervene, and the patient feels weak, and is unable to undergo much exertion. Prolapse or procidentia of the uterus comes on in those who have to stand about much, and who get about too soon after their confinement. The common symptoms of disease of the womb are pain in the back, and generally across the loins, of a bearing-down character, weight or discomfort in the pelvis, difficulty in micturition and defecation, with menorrhagia or leucorrhœa, and inability for walking or any exertion. Cancer of the womb is known chiefly by the excessive pain, a profuse and often fœtid discharge, occasionally menorrhagia and a marked cachexia.

Treatment.—Many diseases of the uterus occur in nervous women, and the constant pain and inability to go about much are apt to produce a sense of depression and melancholy, and to fix the patients' attention too much on their disorder. And this is perhaps intensified by their coming under the care of medical men who, devoting themselves to a special line of practice, are apt to unduly estimate the local malady, instead of looking at the constitution and general health of the patient. Most affections of the womb are to a great extent curable. Rest in the horizontal position, a moderate amount of out-door exercise without causing fatigue, wearing an abdominal belt, and the use of a cold or tepid hip-bath, will do much to alleviate any diseased condition that may be present. In cases of prolapse pessaries are very useful but nothing is more to be condemned than a mere mechanical treatment of these affections. The patients' attention should be diverted from their malady by having some light employment, as needlework, fancy work, etc., by cheerful companionship, and by reading useful books—by anything, in short, which prevents them thinking too much about their complaint. Most women improve much, and may quite recover, when the child-bearing period is past. Riding and driving seldom do much good, as they are attended with so much jolting, and often aggravate any pain. Tonics and astringents must now and then be given to improve the general health and alleviate any excessive hemorrhage. The chief thing to be done, however, is to improve the physical health of women during girlhood—in allowing them out-door exercise and more freedom in running about; in altering any absurd fashion of dress, as tight-lacing etc., so as to develop the chest and not compress any internal organs; in teaching them to swim and ride; in preventing them from keeping late hours at balls and parties, and from breathing impure air; in changing the artificial system of education of the present day;

in allowing them to read sensible books, and not inferior literature; and, finally, by letting them learn the elementary principles of health and diet, so as to enable them to become intelligent and efficient wives and mothers.

In cases of cancer of the womb, nothing can be done beyond alleviating any symptoms that may arise, and in rendering a painful disease as easy as possible to be borne.

UVULA is a muscular prominence covered with mucous membrane, which projects from the centre of the soft palate, hanging down like a tongue. It may be relaxed in cases of ordinary sore throat, or destroyed in some cases of syphilis, or it may be habitually too long, and cause a tickling cough. In such cases the end may be snipped off with advantage. (See SORE THROAT.)

V.

VACCINATION is a process by which a peculiar specific disease, known as the cow-pox or vaccinia, is introduced into the system with the view of protecting it against an attack of small-pox.

The cow-pox is a disease which never occurs spontaneously in man, but it may be readily communicated to him by inserting some of the matter into the system. Vaccinia, or the cow-pox, is not always prevalent in this country, but occurs casually and sometimes appears almost as an epidemic. When affecting the cow, the rash appears as a small vesicle or blister, and comes out on the teats and udder. The disease runs its course in a precise and definite manner, and lasts about three weeks. About four days after the invasion of the disease the animal may become slightly indisposed, and small red papules or pimples appear on the teats or near the udder. These soon become vesicular, and the top of the pimple becomes raised and pellucid, as if a little blister were present; this is due to the epithelial covering of the skin being raised by the effusion of some clear lymph, so that the spot has a pearly look. When these vesicles are well developed, the margin is raised and there is a central concavity, or cup-like depression; at first the skin around is of a natural color, but about the eighth or ninth day a pink blush or areola is seen around the vesicle, which extends gradually, so that in two or three days more the areola forms a zone half an inch wide, and there is some thickening and hardening of the skin around; at the same time the vesicles lose their pellucid, pearly look and become more opaque. By the twelfth day the lymph or fluid in the vesicle becomes more turbid, and the whole becomes drier and a crust begins to form; many vesicles burst, and as their contents escape they become dry and form scabs; the

scabbing is complete in six or eight days, and from the twentieth to the twenty-fourth day these crusts fall off spontaneously and leave slight depressions or pits behind, which remain permanent.

Such is a short account of the cow-pox, and it will be seen presently that the disease, when given to man, runs a precisely similar course. When cows affected in the above manner are milked, the vesicles burst, and the lymph which exudes from them is often found to produce sores of a definite and similar character on the hands of the milker, and they, in their turn, are the means of causing the disease to spread to other animals in the dairy. Among the dairy districts in the fertile vales of Gloucestershire there existed, more than a century ago, a popular notion that milkers who were thus infected with the cow-pox were incapable of taking the small-pox. This singular fact attracted the notice of Edward Jenner, in the year 1768, who at that time was an apprentice to a surgeon at Sodbury, near Bristol. At an early age he satisfied himself, by inoculating with small-pox several people who had had the cow-pox, that this notion had in it the elements of truth, and by dint of perseverance and accurate reasoning he at length disclosed to the world that discovery which has made the name of Jenner for all time illustrious, and saved thousands of his fellow-creatures from a painful death or a life-long disfigurement. He conceived the happy idea of giving man the cow-pox by inoculation, and then of transmitting it to others by inoculation from one human subject to another. Thirty years afterward (A.D. 1798), he published an "Inquiry into the Causes and Effects of the Variole Vaccinæ"; and in this work he established the following propositions: 1. That this disease (cow-pox), casually communicated to man, has the power of rendering him insusceptible of small-pox. 2. That the specific cow-pox alone, and not other eruptions affecting the cow which might be confounded with it, had this protective power. 3. That the cow-pox might be easily communicated to man whenever it was requisite to do so. 4. That the cow-pox, once engrafted on the human subject, might be continued from individual to individual by successive transmissions, conferring on each the same immunity from small-pox as was enjoyed by the one first infected direct from the cow. It is but seldom now that a child is vaccinated directly from the cow, although it is an open question whether it is not advisable every few years to obtain matter directly from that animal, as it is possible that lymph from the human subject may deteriorate by time.

Great as was the discovery of Jenner, it met with great prejudices in his day ; some looked on small-pox as a scourge to humanity which Providence occasionally sent as a beneficent warning for its sins, and they saw in an attempt to stop its ravages that man was impious enough to thwart the Divine will. Others objected that any matter from an inferior animal like the cow should be allowed to enter the human system. Many similar foolish statements were made to prejudice the people against vaccination, and years elapsed before it came into repute ; still more years elapsed before the Legislature enforced compulsory vaccination. As has been the case in every other scientific advance, ignorance and credulity have ever opposed what afterward has proved to be a vast benefit to humanity.

The symptoms observable in men after vaccination closely resemble those which have been above described as occurring in a cow affected with vaccinia. If a child be vaccinated with pure vaccine lymph taken from the arm of another child, nothing will be seen locally during the first two days, but at the end of the second or on the third day a small red pimple appears, which gradually increases in size, and on the fifth or sixth day it has become a vesicle or little blister of a pearly color, with well-defined raised edges, while the centre is depressed and concave. On the eighth day the vesicle has become perfect ; it is round and plump, and the edges are more defined and pellucid, while the centre is more concave. About this time also a red blush or areola is seen round the vesicle, and this continues to spread for a zone of from one to three inches ; the skin looks red and angry, and becomes hard and painful from an affection of the tissue of the skin. When this areola appears, the child generally presents constitutional symptoms ; sometimes they are very slight and pass by unnoticed ; others may be peevish and restless, and have some derangement of the bowels or enlargement and inflammation of the glands in the arm-pit.

On or about the tenth day the areola begins to fade, the vesicle dries in the centre, while the lymph gets opaque and turbid, so that by the fourteenth or fifteenth day a dark brown scab is formed, which dries, blackens, and falls off between the twentieth and twenty-fifth day ; a cicatrix or scar is left which becomes permanent, is generally circular, and marked with minute pits. Such are the stages through which the vesicle passes, but it is important to note that only on the eighth day is the vesicle in perfection, and it is then only that lymph should be taken. Thus, if the child be vaccinated on a Monday, matter should be taken on the following Monday, but not later, as after that time the lymph is not so protective. It happens occasionally that parents are much alarmed by skin eruptions occurring after vaccination, and they often lay it down heedlessly to the fault of the physician for introducing bad matter ; this is a great mistake, for in some children any constitutional disturbance will bring out an eczematous eruption, as is indeed often seen when they are teething. No alarm need be felt on this score, as the mischief is soon cured, and it depends upon some peculiarity in the child's constitution. At times a rose-colored rash may appear on the body, or a crop of papules or vesicles ; these are generally very transitory, and disappear when the scab falls off the arm. The shape of the scar, and also its size, will depend upon the way in which the vaccination is performed ; some make one puncture in three or four places on the arm, about an inch from each other ; others scratch or scarify the skin, and some make punctures very close together, so that when the vesicles form they coalesce or run together and form a large irregular scab. All these methods are equally efficacious, and are adopted according to the fancy of the operator. The course of the vesicle in the soft, smooth skin of a child is more characteristic than in an adult who has not before been vaccinated. If the lymph inserted be taken direct from the cow, the course may be accelerated or retarded ; sometimes the vesicle is delayed only for a day or two, and it has been said to lie dormant in the system for many weeks ; if now the child be again vaccinated, the original vesicles will resume their course, and the two will run on together. Should the child be incubating measles or scarlet fever, the areola may not form until these diseases have gone. Mere delay in the appearance of the symptoms will not hinder the protective influence, so long as the red areola appears before the child is exposed to small-pox. When acceleration of the symptoms occurs, the vaccination is generally useless and spurious ; if any doubt exist, the child should be done again after a short interval. The important rule to remember is this : " that if there is any deviation from the perfect character of the vesicle and the regular development of the areola, the vaccination is not to be relied on as protective against small-pox."

If the arm becomes much inflamed, a little cold cream spread over the red surface will give relief ; the child should not be allowed to rub the places, and any source of irritation should be avoided. As a rule, the regular phenomena of vaccination only occur once in a lifetime ; if lymph is introduced into the system of a person who has once been successfully vaccinated, spurious effects result ; a papule will at first form, to be followed by a little vesicle ; this is surrounded by an angry red areola, which may cause great irritation and itching. The symptoms begin early and arrive at their height on the fifth or sixth day, when they begin to decline ; on the eighth day the scab generally forms and soon falls off. Severe constitutional symptoms are more common in cases of revaccination than in primary cases, and in a very few exceptional cases erysipelas has supervened, while in others the lymph has acted as a poison and caused death by pyæmia.

Every child should be vaccinated in early infancy ; out of 20,590 deaths from small-pox in England during the six years 1856-61, no less than 5010 were in children under one year of age, so that there is great risk of catching small-pox if any delay occurs ; this observation more especially applies to those living in large towns, where the danger of infection is greatest, and it is still more needful in periods when an epidemic of small-pox is raging. Plump, healthy, well-fed children should therefore be vaccinated before they are three months old ; they are then free from the disturbances often caused in the system by teething. But there may be reasons why vaccination may be delayed ; the child may be suffering from acute disease, as measles or scarlet fever, or bronchitis, or from malnutrition and diarrhœa ; the general health should in such cases be first attended to, and then the operation may be performed. For similar reasons delay may take place if any skin-eruption be present ; herpes, eczema, and intertrigo, or the chafing which is often seen in the folds of the skin, have an injurious influence.

Should small-pox, however, be near at hand, discretion must be used, and it will be needful to perform vaccination in spite of these circumstances ; all such cases must therefore be left to the judgment of the medical man in attendance. No age is too early for vaccination if there is direct exposure to small-pox, and many infants have been saved who have been operated upon immediately after birth. The incubation of small-pox lasts twelve days, and the time needful for the development of the areola in vaccination is only nine days. Hence it is obvious that, although a

person has been exposed to and has actually imbibed the poison of small-pox, yet, if he be vaccinated within the first three days immediately following the reception of the infection, its protective influence will be felt in modifying the disease. Hence, then, we have this fact of great importance—that if the vaccination can be got to the stage of areola before the small-pox appears, life may be saved ; the loss in such cases of a single day may be most disastrous. Lymph should always be taken from healthy children, and from well-marked vesicles, just before the areola commences, or at any rate within a very few hours. After the eighth day the vesicle may yield more lymph, but it is weaker and not so protective ; if on the eighth day several small punctures are made in the pellucid, pearly vesicle all round the circumference, minute drops of clear lymph will readily exude. Care should be taken not to draw the slightest trace of blood, nor to use any pressure in squeezing out the lymph ; this simple operation is unattended with any pain to the child.

It is best for lymph to be inserted from the arm of one child to the arm of another, but as this cannot always be done, it is usual to adopt various means for preserving the lymph. For this purpose ivory points may be dipped into the lymph, and when the latter dries upon it, the point may be kept until required for use, or the fluid may be preserved in capillary glass tubes, from which air can be excluded, and this is a valuable and efficacious measure. In very rare cases children may be insusceptible to the influence of cow-pox ; a few cases fail to take the first time, but are successful on a second trial ; those incapable of taking cow-pox are probably incapable of catching small pox.

If people are successfully vaccinated they are, as a rule, forever protected against small-pox ; yet there are some cases in which some persons are liable to it, but even then they take it in a very mild and modified manner ; very rarely indeed does it leave any marked disfigurement or prove fatal. It was never maintained by Jenner that those who were successfully vaccinated were absolutely safe from an attack of small-pox ; but just as some who have had small-pox once may now and then have a second attack, so those who have once suffered from cow-pox may in like manner have the small-pox at some future time. These cases, however, occur so seldom, and when they do happen are so mild and harmless in their manifestations, that vaccination, when well performed, may be considered a most effectual safeguard against small-pox. Jenner himself saw cases of this kind, and in his own writings he has thus stated his opinion of the value of vaccination : "Duly and efficiently performed, it will protect the constitution from subsequent attacks of small-pox, as much as that disease itself will. I never expected it would do more, and it will not, I believe, do less." To have its due protective influence, the operation must be properly performed, and the phenomena must develop themselves in a due and regular manner. Experience has shown that, in order to thoroughly infect the constitution, a certain amount of local affection is as essential as a perfect character of the vaccine vesicle. The benefit derived from vaccination may be seen in the faces of the children of the poorer classes ; fifty years ago one child out of every three was marked with small-pox, while now hardly one in forty can be found to have any traces of that disease. From ignorance and carelessness there will always be a good many persons in a community who are unvaccinated, and this will be more especially the case when there has been no epidemic of small-pox for some time ; apathy and indifference are then felt for the operation, and when an epidemic again appears, these are the first to fall victims to their rashness. It is not uncommon to find that in a family of four or five some have been vaccinated and some have not and when small-pox appears, the death or disfigurement of those who are unvaccinated is a proof of the dangers of delay and a strong argument in favor of the simple operation. For a convincing summary of facts in favor of Vaccination, see SANITARY REGULATIONS.

Revaccination.—Whenever an epidemic of small-pox is prevalent, a panic takes place among the people, and every one is in a hurry to be revaccinated ; now, although it is no doubt of great importance that those who have been imperfectly vaccinated in infancy should again resort to this operation for further safety, it certainly is not wise to give way to panic, for at such times so great is the application for fresh lymph that the demand exceeds the supply, and thus persons are hastily and insufficiently done ; it follows from what has been said before that those who have four or more perfect cicatrices on the arm are free from danger for the rest of their lives ; those, on the other hand, who have one or two marks, and these perhaps not very perfect ones, should certainly be revaccinated when they attain the age of puberty, and the operation should be done with as much care as in the case of an infant. The matter should be taken from an infant's arm, and in no case whatever from the arm of a person who has been successfully revaccinated, as the lymph then is not sufficiently protective. A popular notion exists that the human constitution changes every seven years ; there is, however, no proof whatever of the truth of this assertion, and it seems pretty clear that a second vaccination about the age of puberty is all that is required. If, at the second operation, the arm rises, and all the usual phenomena appear, it is probable that the

effect of the first operation had worn off, and the patient was liable to a modified attack of small-pox ; if, on the other hand, no effects follow the second time, it is a sign that the original vaccination remains efficacious, and that no danger need be feared even when small-pox is prevalent. It is the custom at the London Small-pox Hospital to vaccinate every attendant and nurse when they first enter, and after an experience of thirty years it is important to note the fact that *not a single case* of small-pox has arisen among them. Although a certain small proportion of those who have been thoroughly vaccinated in infancy do take small-pox in a modified form after they are grown up, yet after effectual revaccination such a case hardly ever recurs. In Würtemberg, out of 14,384 soldiers who have been revaccinated, only *one* case of small-pox broke out during a period of five years. There can be no doubt that, on the outbreak of an epidemic of small-pox, all the persons in the house should be carefully examined, and those who have no cicatrices, or at least but imperfect ones, should at once be vaccinated, and not only those in the house, but others who live close to and have recently mixed with them, as is the case in a crowded court or alley. If this were sufficiently done, there would be no occasion for a panic, and an epidemic of small-pox would probably be at once stamped out.

VACCINIA. (See VACCINATION.)

VAGINA is the anatomical name for the female passage, and necessary to be explained because it sometimes occurs that in states of disease applications or injections are ordered to be applied to this part of the body, which is mentioned only by this term. Sometimes a slight membrane exists at the entrance to this passage, which prevents the proper and natural monthly discharge. When this is discovered it should be at once removed, and it is well if such an accident should be perceived and remedied in infancy, before any evil consequences can result from it.

VALERIAN is the root of a well-known plant, the *Valeriana officinalis.* The best plants grow in dry soils. The root consists of a kind of stock or head, whence numerous rootlets are given off. The color is light brown, the odor peculiar and characteristic. The roots contain valerianic acid and an oil. This oil contains two substances, valerole and valerianin, neither of very great importance. Valerole, by exposure, is slowly converted into valerianic acid. The preparations of valerian are an infusion, a tincture, and an ammoniated tincture, in which aromatic spirit of ammonia replaces the ordinary spirit.

Valerian acts as a powerful stimulant. It is mostly given in nervous diseases, especially in those of hysterical subjects, as well as in chorea and such like affections, as an antispasmodic. The value of valerian is not quite clear ; some esteem it highly, others rather scout its efficacy The ammoniated tincture is the best form of the remedy. The dose is a drachm.

Valerianic acid, though contained in valerian, is prepared from a totally different substance. Fusel oil, which is a waste product in the distillation of most forms of alcohol, tends, when kept, to pass by oxidation into valerianic acid. This may be done at once by chemical means, sulphuric acid and bichromate of potass being employed. The acid is then neutralized by carbonate of soda, and valerianate of soda is produced.

Valerianate of soda is hardly ever used itself in medicine, but is employed in the manufacture of another salt, *valerianate of zinc*. This salt occurs in fine scales, with the odor of valerianic acid. It is not readily soluble in water. Valerianate of zinc is commonly esteemed a valuable nervine tonic, though some prefer to give sulphate or oxide of zinc along with the ordinary tincture of valerian. It has been given in nervous affections, as chorea, epilepsy, and hysteria. It has also been given with advantage along with quinine in neuralgia. A valerianate of quinine is now made. The dose of valerianate of zinc in from 3 to 5 grains or more.

VALVES are usually folds of membrane which guard certain orifices ; they are met with in the course of the veins, and play an important part in the proper action of the heart. (See HEART.)

VANILLA, a delicious and fragrant orchidaceous plant growing in the West Indies, whose fruit, in the form of a long pod, is highly prized, on account of its delicate flavor, by confectioners, cooks, and chemists.

VARICELLA. (See CHICKEN-POX.)

VARICOSE VEINS. (See VARIX and VEINS.)

VARIOLA, a technical name for small-pox. (See SMALL-POX.)

VARIOLOID is the name given to the mild form which small-pox takes in persons who have been vaccinated or who have already once had the disease. It is always less virulent than small-pox itself, and is very rarely attended by serious results. (See VACCINATION.)

VARIX consists in dilatation and a convoluted state of the veins, due in most instances to an obstruction of the current of blood toward the heart. It occurs very often in the lower part of the rectum, where it constitutes hæmorrhoids ; and in the affliction known as varicocele the veins of the testicle are thus affected. The most frequent seats of varix, however, are the lower extremities, a condition being there established which is commonly termed that of " varicose veins."

In a well marked case of *varicose veins* the inner surface of the lower limb, from foot to groin, is studded with a number of soft, bluish swellings, varying in size and shape, and which are formed by a tortuous and dilated condition of the large saphena vein, which extends along the whole length of the limb. These swellings become more prominent when the patient stands up, or after constriction of the knee or thigh. The skin covering the tumors is generally thin and distended. This condition gives rise to stiffness and aching pain in the affected limbs, and even slight exercise is soon followed by a sense of fatigue. The skin about the ankles is puffy and is marked by purple patches of small veins, arranged in an arborescent form, etc. The feet are generally cold, and the toes of a bluish color. The skin of the leg is generally dry and itches very much ; it is very often red and inflamed, and the seat of an eczematous eruption. In old people, and in cases where the varicose condition is of long standing, large ulcers may form on the lower third of the leg, constituting the so-called *varicose ulcers.* An occasional serious result of varix is thinning and giving way of the skin over a distended vein and hemorrhage, which, so long as the patient remains in the erect position or allows the leg to hang down, continues, and may speedily become fatal, but which may be readily arrested by placing the patient on his back, elevating the limb, and applying slight pressure with a pad of lint and a bandage over the bleeding point.

The predisposing causes of varix are an inherited tendency and debility, due to old age, overwork, or long illness. It is believed by some that the distension of the veins is occasionally preceded by a gouty condition of the blood. The chief exciting cause is obstruction to the venous circulation applied either directly to the lower limb, as in the case of wearing tight garters, or indirectly, as in disease of the heart, congestion of the liver, or constipation with overloading and distension of the large intestine. Varix may be caused by the pressure upon the veins of the pelvis of tumors, or of the pregnant uterus. Pursuits necessitating much standing or walking very often give rise to the affection. It has been stated that cooks and soldiers are the people most especially prone to the formation of varicose veins.

Treatment.—The subjects of varix in the lower extremities, when old or debilitated, should be invigorated by good diet and medicinal tonics, such as quinine and preparations of iron. In cases where the patients are of middle age, and are full-blooded, it will be necessary to advise temperance and restriction of diet, and occasional free purgation, in order to relieve a congested liver and a distended rectum. In early stages of the affection much may be done to prevent the further development of the varix, and all its troublesome results of eczema, ulceration, etc., by removing all likely causes of obstruction to the venous circulation. The patient should avoid, as far as may be possible, standing or walking ; the limb for a period of six weeks or two months should be kept elevated ; the skin should be well rubbed with the hand night and morning, and during the day an elastic stocking should be worn, or the limb should be bandaged from toe to groin. Many operations have been devised for the purpose of producing permanent obliteration of the distended veins. Of these, the safest and the one most commonly practised consists in the application of needles and twisted sutures—tying the veins. These operations rarely produce a permanent cure, but are often of great service in cases where severe varix has resulted in painful eczema, or extensive and obstinate ulceration. In cases of old and severe varix, very little can be done to produce any abiding relief.

VEINS are vessels distributed through all parts of the body, and through which the blood returns to the right side of the heart after it has supplied the different organs and tissues. In addition to this widely distributed or systemic venous set, there are two special sets of veins, the pulmonary and the hepatic. The first serves to carry blood from the heart to the lungs, the second collects the blood from the minute or capillary vessels ramifying in the walls of the stomach and intestines, and carries it to the liver. The veins of the systemic set commence by minute branches, which, as they travel toward the heart, are found to increase in size and diminish in number until at last the blood is conveyed to the right auricular chamber of the heart by two large veins, the superior and inferior venæ cavæ. The veins of the heart open directly into the right auricular chamber of the systemic veins. There are two kinds, the deep and the superficial veins. The first accompany the arteries of the limbs and trunk, and are deeply situated ; the latter are placed immediately under the skin, and are prominent and readily visible, especially in old and thin persons. On tightly binding a handkerchief or bandage around the arm between the elbow and shoulder, the return of the venous blood to the heart from the parts below the constriction may be prevented. The superficial veins of the upper extremity become distended with blood, and their branches and large trunks may be seen extending from the fingers upward. From the results of this experiment, Harvey was led to the conclusion that the blood in the veins passes toward the heart, and that the veins commence in small branches, which gradually pass into large and less numerous vessels as they pass upward toward the heart. Most veins are provided with valves or small folds of the internal membrane which project into

the interior of the vessels. The usual arrangement of these valves is thus : two folds are placed at directly opposite points of the interior of a vein, their free margins are concave, and their attached margins convex ; where there is a free current of blood toward the heart, each fold is driven outward and is applied closely to the inner surface of the walls of the vein, so that the channel remains free. If, however, the current of blood is obstructed, or has a tendency to flow backward from the heart, these folds are driven inward and come together in the centre of the calibre of the vein, and thus prevent any further regurgitation. These valves are most numerous in the veins of the lower extremities ; in the largest and smallest veins they are absent.

An *incised wound of a vein* is much less serious, even when the vessel is large, than a similar injury of an artery. In the former injury there is an even stream of dark-colored blood, the flow of which can be readily arrested by pressure below the wound. From a wounded artery, on the other hand, there is a profuse flow in jets of bright red blood, which, when the vessel is large, cannot be arrested save by completely closing the vessel *above and below* the wound. Even where a large vein has been completely divided, as in amputation of a limb, the flow of blood usually soon ceases, although the vessel does not contract. Arteries, both large and small, to contract when cut through, but still not sufficiently to obstruct the bleeding. An artery never becomes permeable again after it has been wounded, and the blood is carried along the enlarged collateral vessels. A wound in the walls of a vein, when properly treated, closes by adhesion, and the calibre of the vessel remains free. In former times, when the operation of bleeding was often performed, and persons were bled habitually at certain seasons of the year, the incision at the front of the elbow was repeatedly made at the same spot and into the same veins.

Inflammation of one or more veins, or *phlebitis*, may be acute or chronic. The most frequent causes of acute inflammation are incised or punctured wounds of veins, irritation of the surrounding tissues, or the application of a ligature. In some rare cases inflammation attacks veins without any appreciable cause. Acute phlebitis occurring in a healthy individual, who is submitted to suitable and careful treatment under good hygienic conditions, usually runs a favorable course, and terminates in complete and speedy recovery ; but when it attacks a " bad patient," and one who has been debilitated by previous disease or insufficient nourishment, it constitutes a very serious affection, in consequence of a tendency to the formation of diffused abscesses in the affected part, of the intense constitutional reaction, and of the very probable occurrence of pyæmia. The following are the symptoms of ordinary acute phlebitis attacking the superficial veins of a limb. Intense pain over the starting point of the inflammation, and tenderness and redness of the skin along the course of the inflamed vessels ; the limb is swollen and œdematous, and its surface is marked by a pale diffused blush ; the patient suffers from more or less inflammatory fever, and complains of general uneasiness, headache, and nausea ; the pulse is high and strong, the skin hot, and the tongue dry at its centre and of a bright red color at its tip and edges. As the acute symptoms subside the course of the inflamed veins is marked by hard cords, which remain for a long time after convalescence. In bad cases of phlebitis, abscesses form about the affected veins, and the whole limb becomes red and much swollen ; the constitutional symptoms take on a typhoid character, and there is much prostration ; finally, pyæmic symptoms come on, as pain and swelling of one or more joints, jaundice, a cough, and shortness of breath, to which the patient succumbs. In chronic phlebitis the symptoms both local and constitutional are much less severe, although they last for a long time and give rise to much trouble and anxiety. The affected limb is stiff and painful, and its movements are interfered with. There is much œdema of the parts from which the inflamed vein proceeds, and considerable induration along the branches and trunk of the affected vessel.

The treatment of acute phlebitis should consist in the administration of tonics, and occasionally sedatives in order to relieve pain. The part affected must be kept at perfect rest, and if the veins of the lower limb be inflamed the patient should be kept in bed. Warm fomentations should be applied over the inflamed region, and care be taken to guard the patient from draught. Cooling lotions or cold applications of any kind ought to be avoided. The patient should be allowed a generous diet. When suppuration takes place the surgeon generally considers it necessary to make early and free incisions, not only to relieve suffering but also to prevent diffusion of the suppurative process. In chronic inflammation, with much thickening of veins, the affected limb should be kept at rest as much as possible, and be supported by a firmly applied flannel bandage.

Air, entrance of into Veins.—In the records of practical surgery have been repoited several instances of sudden death or of alarming prostration during the course of cutting operations on the neck and armpit. The history of these cases may be briefly summed up : during the removal of a tumor in either of the above regions, a distinct gurgling or hissing sound is suddenly heard, and the blood at the bottom of the wound becomes frothy from admixture with bubbles of air. The patient either dies at once, or falls into a deadly faint, complaining of great oppression over the heart and of a sensation of impending sudden death. There are some few instances of tardy recovery from this condition, but death either immediate or delayed for two or three hours is the usual result. The phenomena observed before death in these cases, and the post-mortem examination, together with the results of physiological experiments on animals, indicate very clearly that the alarming symptoms just described are due to the entrance of air into veins, and its transmission to the right side of the heart along the vessels whose proper function it is to return back venous blood to this central organ. It is easy to account for this introduction of air into a wounded vein. When the chest is expanded in the movement of inspiration, a vacuum is formed within this large cavity, which is filled up by a downward rush of air through the windpipe, and by a rush of blood in a similar direction, through the jugulars and other veins of the neck. If during an operation, as for instance the removal of a tumor, one of these veins be wounded, and the gap thus formed be kept stretched for a short time, a deep inspiration, as it suddenly withdraws the blood from the lower part of the opened vessel, might cause a sucking in of external air in considerable quantity through the wound. The immediate cause of death is a failure in the supply of arterial blood to the brain, from the arrest of the circulation, but whether this arrest be due to distension of the heart with air, to inaction of the valves of the heart in consequence of the presence of air, or to a stoppage of the flow of blood through the lungs in consequence of the admixture of air-bubbles, is still a matter of speculation.

VENESECTION, the operation of bleeding from a vein, is one nowadays of comparative rarity, although in some cases it is of undoubted value. The veins selected are generally those at the bend of the elbow, and of these the *median cephalic* is to be preferred, as there is less danger of wounding the brachial artery than in operating on the *median basilic*, which lies immediately over that vessel, although most blood can be drawn from it. The operation is thus performed : The patient lying down or sitting, a tape or narrow bandage is fastened firmly round the arm above the bend of the elbow, sufficiently tight to arrest the flow of blood in the veins, but not to stop the pulse at the wrist. The surgeon next takes the patient's arm and extends it, and fixes the hand under his left armpit, if he is operating on the right arm, and, *vice versâ*, if on the left ; next by gently rubbing the part, he causes as great an afflux of blood as possible to the vein, which he keeps up with his left thumb, at the same time that the four other fingers seize the limb, and by being placed behind it make tense the skin. Then the surgeon takes the heel of the lancet between his thumb and forefinger, and steadies his hand by making a *point d'appui* on the surface of the limb with his other fingers. The opening of the vein is to be made by a simple puncture, and obliquely to the direction of the vein. Immediately after the puncture the blood spirts out with greater or less rapidity, and its flow is facilitated by making the patient grasp a stick or a roller bandage. When the necessary quantity of blood has been withdrawn, the bleeding at the point is arrested by placing the left thumb upon it, and at the same time removing the bandage from the arm. After cleaning the part, a compress of lint, maintained in place by the application of a figure of 8 bandage (see BANDAGES), and fixed by a pin, complete the proceeding.

If the external jugular vein in the neck is selected as the point to bleed from, the operation is as follows : The patient lying down, the vessel (which is directed obliquely from before backward, across the middle of the sterno-cleido-mastoid muscle), is compressed a little above the clavicle. The puncture is made in the middle of the neck, ought to be large enough, and made in a direction across the fibres of the superficial skin-muscle, the platysma, else the wound would close immediately the puncture was made, and the blood would escape into the sub-muscular tissue. The bleeding is arrested by a well-adapted compress.

VENEREAL DISEASES are divided into *gonorrhœa, chancres,* and *syphilis*. By some writers it is maintained that all three forms of this disease are produced by one poison, while others maintain that syphilis alone is due to a specific poison.

Gonorrhœa is an acute inflammatory process going on in the lining membrane of the urethra. This is at first attended with pain in making water, and the passing of a clear watery mucus from the urethra. This heals up at first, but the discharge returns, having a turbid and puriform character. The attack is attended with more or less feverishness ; other symptoms may follow—excoriations, swellings, or even abscess may occur.

Treatment.—Gonorrhœa may get well of itself. The discharge subsides, the inflammatory symptoms disappear, and the discharge becomes of a mucous character, and is then called a *gleet*. It is usual, however, to have recourse to remedies. It is recommended in the early stages to try and arrest the disease by what is called the ectrotic or abortive treatment. This consists in applying nitrate of silver in solution by means of a syringe to the inflamed urethra. This should not, however, be attempted by the patient himself. If no opportunity exists of apply-

ing this treatment, the patient must be treated according to general principles. Rest must be enjoined, the diet must be low, the parts should be supported, antimony in small doses should be given, as also mild aperients ; leeches may sometimes be necessary, and a hip-bath administered. The inflammatory stage being over, less stringent measures may be adopted, and various stimulating agents—such as copaiba and cubebs, may be given. As the patient recovers, should the discharge continue, injections of sulphate of zinc should be employed.

Sometimes the gonorrhœal virus affects the eyes. The treatment must be the same as for an ordinary case of ophthalmia. (See OPHTHALMIA.) Warts sometimes accompany gonorrhœa. The best method of removing them is by knife or scissors, and touching the exposed parts with nitrate of silver or some other escharotic.

One of the evil consequences of gonorrhœa is stricture of the urethra. This disease may come on independent of gonorrhœa in the form of what is called spasm of the urethra. In this condition the muscular coats of the urethra contract, and, rendering the passage narrower, the urine either flows slowly or not at all. It is brought on by exposure to cold, the effects of a debauch, the presence of irritation about the anus or the action of cantharides. The symptoms often retire as quickly as they come on, by the use of chloroform, a hip-bath, fomentation, some form of sedative by the mouth, or a gentle purgative. Tincture of muriate of iron may be subsequently given with advantage. (See STRICTURE.)

The inflammatory action of gonorrhœa may produce stricture. The same symptoms may occur, and there may be difficulty of making water, or suppression altogether. In such cases the symptoms are alleviated by the ordinary treatment. Should an abscess outside the urethra cause the stricture, the abscess should be opened. In all cases where the urine is not passed the catheter must be employed, and for this purpose medical aid should be sought. The third and most difficult form of stricture is that when the canal is narrowed by chronic structural change. This constitutes true or organic stricture. The symptoms of this form of stricture may for some time escape notice. The urine is passed in an attenuated stream, sometimes twisted, or scattered, or dribbly. Sometimes there is pain and uneasiness in the bladder when it is full. There is an increased tendency to micturition, and the water passes away frequently involuntarily after micturition. There is also often a gleety discharge.

The treatment of such cases is simple, but must always be conducted under surgical superintendence. It consists mainly in inducing the absorption of the enlarged tissue by the introduction of instruments called bougies. At first a small-sized bougie is introduced, and subsequently larger ones, till at last the urethra acquires its natural size, and allows the urine to flow naturally.

Chancres, the second form of venereal diseases, are those sores which form on the organs of generation, either after impure intercourse, or independent of it. They never assume the appearance of the true chancre with indurated edges, and are easily cured by external applications. Oxide of zinc ointment, the black wash, and other simple dressings may be applied externally, while cooling medicine may be given, and rest enjoined.

Syphilis, the third form of venereal disease, is produced by a specific virus whereby a sore is formed which is called a chancre. Such sores exhibit various characters, but they are reduced to two heads : first, those which are not followed by subsequent effects, and those which are. The treatment of the first class of sores should always consist of an attempt to cure them at once by the ectrotic treatment. This should be effected by means of fusing nitric acid, the acid penitrate of mercury, potassa fusa, or chloride of zinc. The surface may be then dressed with lint. Water dressings may be afterward applied.

The second form of sore, which is really the symptom of a formidable disease, is characterized as " a superficial erosion situated upon an indurated base." The sore has a regularly oval or elongated form, sometimes not bigger than a millet-seed, and rarely attaining the size of a shilling. The surface presents a pearly gray aspect without granulations. The general surface is usually cup shaped, as if scooped out by means of a gouge. The induration of the base and margins of the sore are its great characteristics. The time of the incubation of this sore has been put down at from ten to forty days. No treatment of this sore can prevent the development of constitutional symptoms. The *treatment* consists in the same methods as those recommended for the simple sore. The great mass of medical men, however, recommend some form of mercurial treatment. Mercury is given internally, and the sore itself is treated with mercurial ointment. These sores may proceed in their course and produce sloughing and phagedenic sores demanding in their treatment the most special care. Such sores are frequently attended with *bubo*, the result of irritation of the absorbents. The treatment of a syphilitic bubo requires the same general treatment as an ordinary abscess. (See Abscess.)

Although the primary sore may be healed, the true specific chancre is followed by certain general symptoms which are called *secondary* and *tertiary*. The syphilitic poison in these cases pervades the whole system, and certain well-known symptoms follow. In the

course of a few days or weeks a state of the system comes on in which there is sallowness of the countenance, more or less emaciation, a sense of lassitude and muscular debility, headache, with palpitation, and other signs of disturbance of the heart's action, œdema of the lower extremities, and a tendency to bleeding at the nose. Following these symptoms are a variety of affections of the skin known as syphilitic eruptions. These eruptions have a copper color, a rounded form, a tendency to desquamate, and have no irritative quality.

The treatment of such cases should consist of an attempt to relieve the system of the accumulated syphilitic virus. For many years the only general remedy for this state was considered to be some of the preparations of mercury to the extent of salivation. Although it has been shown that the constitutional symptoms may be cured without mercury, a large number of physicians still recommend this treatment. Others have recommended chlorate of potass, iodide of potassium, sarsaparilla, and a hygienic or tonic treatment. No one, however, should presume, when suffering under the various phases of these diseases, to attempt to treat themselves, but apply to the nearest intelligent practitioner who does not advertise his powers of curing these diseases by some secret treatment or vaunted remedy.

VENTILATION is a subject of much importance to public health, but one which is often much neglected. By ventilation one must understand a due supply of fresh air in the twenty-four hours, so as to allow plenty of oxygen to enter the lungs and properly aerate the blood. But it is a bad thing if there is too much draught with the ventilation, as in this way the patient is often cold, and this may do harm. For this reason the poor, who huddle together in winter in a small room, prefer a stifling atmosphere with a warm temperature to a ventilation which in their case is generally accompanied by a draught. Again, in the wards of a hospital it is usual to have the windows open to let fresh air in, but this is sometimes bad for cases of bronchitis and Bright's disease, as the cold air blowing upon the patients increases their malady. To obviate this, there should be a corridor running parallel with each ward, and supplied with plenty of windows on each side ; the one set, communicating with the open air, should be opened, while those communicating with the ward are closed, and *vice versâ ;* in this way, by repeating the process several times a day, enough fresh air will be brought in the room without too much draught being felt. In workshops, factories, etc., this entry of fresh air is very important, as a room soon becomes contaminated when many are working in it. An ordinary fire and the usual crevices of a window or door are agents in promoting a current of air through a room ; the smaller a room the oftener has the air to pass through it to sustain a proper amount of purity. Each person should have at least 800 cubic feet of space, with a due current of air. A bedroom should have the windows open for the greater part of the day. so as to thoroughly aerate it. (See AIR, RESPIRATION, and LUNGS.)

VENTRICLE is the name applied to a cavity ; thus there are ventricles in the brain and in the heart. (See BRAIN and HEART.)

VERATRIA is a powerful alkaloid which exists in several vegetable substances, but is itself mainly obtained from sabadilla seeds. The process for obtaining it is complicated, but consists in the separation of the alkaloid by means of alcohol, and afterward purifying the product thus obtained. The pure veratria is crystalline and almost insoluble in water, but freely so in alcohol. Its taste is powerfully acrid, and it is excessively irritating to mucous membranes, especially the nose. From this cause it gives rise to violent sneezing when applied to the nostrils. There is an officinal preparation of veratria not much used, viz., the ointment of veratria.

Veratria acts very powerfully on the skin, and still more so on mucous membranes, producing irritation on these, but afterward sedative effects. The true use of veratria is not yet quite clear.

VERATRUM ALBUM, or WHITE HELLEBORE, contains veratria, and owes its influence to that substance. The root-stock of the plant is employed, but it is no longer officinal. It is generally seen in sections an inch or two long, with the rootlets projecting from it. The color is yellowish brown, lighter within, and it has an exceedingly acrid and bitter taste. White hellebore used to be employed mainly in the form of vinum veratriæ—white hellebore wine. This substance when swallowed gives rise to much vomiting and purging, and was at one time used, as was black hellebore, largely in the treatment of mania. It gave rise to much disturbance of the alimentary canal, and greatly reduced the patient's strength, but did no good. It has fallen completely into disuse, as has the black hellebore ; but another veratrum, which, however, is said to contain no veratria, has come into general use, especially in America, where it is a native. (See VERRATRUM VIRIDE.)

VERATRUM VIRIDE, or GREEN HELLEBORE, is known as Indian poke, and is said to have been long in use among the aborigines of North America. The root-stock, which is thick and fleshy, is the part used. It gives rise to tingling, and has a peculiar acrid taste like others of the class. Its only preparation is the tincture. This is an excessively powerful preparation, given even in small doses ; repeated at moderate intervals it produces

much sickness and great prostration of strength. Even doses of a few drops of the tincture will in a short time give rise to these sensations. It acts apparently specially on the heart, which it controls ; not long after it has begun to take effect, the pulse gets small and the strength is greatly diminished. It does not purge when given as tincture, but produces vomiting. Its effects have been compared to those of colchicum, but though analogous they do not coincide. In America the drug has been largely given in inflammations, especially of the lungs. It seems to do good in rheumatism, but does not cut short the attack.

In pneumonia, green hellebore has been employed with benefit. It lowers the temperature, and seems to favor the local changes necessary to recovery. The remedy requires careful handling.

VERDIGRIS is an acetate of copper, a product which results from the action of some fermenting substance on copper. It is more dangerous as a poison than valuable as a remedy. Not unfrequently its presence in badly-cleaned copper cooking utensils has given rise to somewhat serious results. Sometimes, but very rarely, it is applied externally ; internally it is never used. It has been mixed with honey and applied, by means of a camel's-hair pencil, to some half-vitalized body, as warts, for the purpose of destroying them, and so getting rid of them.

VERMILION. (See MERCURY.)

VERTEBRA. (See SPINE.)

VERTIGO, or GIDDINESS, is that peculiar sensation wherein we seem to be standing quite still, and objects running round us. This commonly causes loss of balance, and the individual may fall down. In a good many cases he is able to recover himself without falling, especially if he can lay hold of anything to steady himself with for a moment. In most cases giddiness depends on an insufficient or improper supply of blood to the brain. Thus, in giddiness after a severe illness, in attempting to stand upright, we see imperfect blood supply. In other instances the blood supply is impure from containing too much alcohol, or the products of imperfect food-metamorphosis. In old people, when the vessels become hardened and unyielding, as well as incapable of due resilience, we often find giddiness a permanent symptom.

Thus it is seen that vertigo is rather a symptom than a malady, and a symptom, too, of very varying significance, for sometimes apparently over-fulness of the vessels gives rise to a kind of giddiness. If, for instance, the face is flushed and the head hot, it may be desirable to give some purgative medicine, whereas the kind referred to first of all as occurring in convalescence, is best remedied by a glass of wine. The subsequent management depends on the same principle. Where there is weakness, good food and exercise are the best remedies ; in the other, saline purgatives, with some diuretic.

Not unfrequently vertigo depends upon or foreruns brain disease, and such brain disease may be very intractable in character. Headache is commonly associated with such vertigo. Thus the symptom of giddiness, taken by itself, may teach us nothing beyond directing attention to the case which, if carefully studied, will gradually reveal itself to the skilful practitioner.

VESICA, an anatomical name for the Bladder.

VESICAL CALCULUS. (See STONE.)

VESICANT is any remedy, as Spanish fly, acetic acid, etc., which can raise a blister on the skin. (See BLISTERS.)

VESICLE is a small blister on the skin containing a little clear fluid within. It is seen in cases of eczema, herpes, and erysipelas ; also when a blister is applied.

VIBICES are patches of hemorrhage which occur in the skin in cases of purpura ; they are also known as ecchymoses ; when very small they are called petechiæ. (See PETECHIÆ.)

VILLI are small prominences on the inner or mucous lining of the intestinal canal, which take an active part in the absorption of the food. (See INTESTINES.)

VINEGAR, as employed in medicine, is only a dilute form of acetic acid. It is used as a refrigerant and as a solvent for some medicinal substances yielding their properties more readily to vinegar than to alcohol. The officinal vinegar is made from malt ; that in most common use is made from wine. It really matters little which is used. Vinegar and water is a favorite local application for cooling, but it is not nearly so efficacious as spirit and water.

VIOLETS, the plant of the *viola odorata*, is no longer officinal. The root was at one time used, but recently only the petals have been. These were used to prepare a syrup, which was a beautiful preparation, but rather useless. The root had more active properties, especially emetic.

VIRGINIA SNAKE-ROOT. (See SERPENTARY.)

VISCERA is a name applied generally to any of the internal organs ; thus the liver, kidneys, spleen, etc., are spoken of as the abdominal viscera.

VISION.—The eye is the organ of vision, and by it we perceive those phenomena which reveal to us objects in the world around us. The eye is a nearly circular body, placed safely in a bony cavity called the orbit, and acted upon by certain muscles, so that we can direct our gaze in any direction we please. In front of the eye is a circular window called the cornea, which enables light to enter ; behind this is a circular ring of muscular fibres, called the iris, which is variously colored in different people ; in the centre of this ring is an opening called the pupil, and this corresponds to the centre of the eye, and allows of the transmission of light. Now the iris, being muscular, can contract or dilate, and thus the size of the pupil will necessarily vary. When the light is very strong the pupil contracts, so as to let in less light, and conversely in the dark, the pupil expands, so as to allow as much as possible to enter. Placed behind the iris, but close to it, is a transparent double convex lens, and further behind is the retina, a delicate membrane made up chiefly of nerve-fibres, and when the rays of light fall on this membrane they excite in the brain the sensation of vision. The retina is concave, and at the back of the eyeball ; the greater part of the eye is filled up in the centre with a transparent gelatinous substance known as the vitreous body, while between the cornea and iris is a similar clear substance known as the aqueous humor. Thus it will be seen that in health there is a clear, transparent passage for the rays of light to pass into the eyeball, so as to fall upon the sensitive surface of the retina ; and further, this passage is guarded by the iris, which, by contracting or dilating, may diminish or increase the quantity of light entering. Now the use of the double convex lens is to bring the rays of light to a focus on the retina. Parallel rays, or those proceeding from a distant object, are in this way brought to focus, while non-parallel rays are more or less deflected from their previous direction. The position of the lens can be slightly altered at will, and this is of much use in determining the size and clearness of an object ; the eye, in fact, accommodates itself to various distances, as otherwise anything afar off will appear dull and blurred. This is the case in some diseases. If, for example, we look at two objects, one of which is at the distance of a yard from the eye, and the other at two yards when we fix the first the other becomes dim ; while if we fix the second, the other in turn becomes indistinct. Errors in the apparatus for this accommodation of vision give rise to the affections known as myopia and presbyopia, or short-sight and long-sight. The usual cause of short-sight is too great a convexity of the lens, so that the focus is formed in front of the retina, and not upon it. It may be remedied by slightly concave glasses, which, by their diverging power, correct the want of accommodation and throwing the focus further back cause the image to be formed on the retina. Working at a microscope, and constantly looking at small objects, as reading and writing, tend to produce myopia. It is common in the case of young people, but diminishes with age. Presbyopia, or long-sight, is the opposite condition. The eye can see distant objects very well, but those which are very near badly. This is because the lens is not convex enough, and does not allow a proper convergence of the rays of light upon the retina. It is corrected by using convex glasses according to the degree to which the patient is affected. It generally occurs in old people.

People are *color-blind* when their retina will not perceive some of the rays of light. (See COLOR-BLINDNESS.) The bluish tinge seen in solutions of quinine and horse-chestnut is called fluorescence, and is due to the fact that extra rays of the spectrum are then made visible to the human eye. The special sense of sight is carried on by the optic nerve, and if this is diseased, more or less blindness will ensue. A cataract, or disease of the lens and opacities in the cornea, cause blindness by preventing the light entering the eye. The mischief may in some of these cases be removed. We may see flashes of light without any rays of light entering from the outer world. Any sensation which will stimulate the retina in a way similar to what ordinary light does will cause a flash to appear before the eyes. A gentle current of electricity passed through the temples will give the appearance of summer lightning. In dreams, and in some diseases of the brain, flashes of light may also appear. When an eye is exhausted by looking too long at a bright color, another color, called the complementary color, will appear on removing the gaze. This is due to the retina being tired for a short time, and a false impression is therefore conveyed to the brain. Thus, on looking at a scarlet object with a fixed gaze, a green one will appear on looking away, and every one who has gazed at the sun a short time will have observed a black disk on the pavement when he looks down again. There are an immense number of optical delusions which in sensible people are corrected by the experience of the other senses, but which in foolish and emotional people may lead to extravagances and erroneous impressions. (See EYE.)

VISION, DOUBLE, may be produced at will in perfectly healthy eyes : " If a person hold the two forefingers in a line from his eyes, so that one may be more distant than the other, by then looking at the nearest, the more distant will appear double" (Dr. Arnott, " Elements of Physics," vol. ii.). It occurs, moreover, as a frequent disorder of vision. either dependent upon changes in the nervous, transparent, or muscular structures of the eye, or arising without any intrinsic morbid changes, and in sympathy, as it were, with some disease affecting a near or remote organ of the body. Thus affections of the stomach, worms, toothache, headache, and chronic affections of the brain are often associated with double vision, which ceases after the subsidence of the primary malady. There

are two kinds of double vision : In the first, the patient sees double, treble, etc., with one eye alone. This is called *polyopia*. In the second kind the patient sees double with both eyes open—*diplopia*, or double vision. Double vision may be produced through long-continued exercise of the eye in reading or writing by a bad light, and by straining the sight in reading small type. In cases of this kind the disorder is usually transitory, and may be speedily relieved by resting the eyes or closing them for a short time.

In recent squinting, due to irritation in the stomach or intestines, or to some severe affection of the brain, the patient sees double. In ordinary congenital or long-acquired squinting of one eye, double vision does not generally occur, as the patient, by habit, has learned to use only the sound eye, and to give all his attention to this.

VITILIGO is a technical name for white patches on the skin, caused by loss of pigment, or the usual coloring matter, at that spot.

VITILIGOIDEA is a name given to certain yellow patches which now and then are met with round the eyelids and elsewhere on the skin. (See XANTHELASMA.)

VITREOUS BODY is the name given to a gelatinous semi-fluid substance which fills up the central portion of the eye. It is quite transparent, and allows of the transmission of light. (See EYE AND VISION.)

VITRIOL. (See SULPHURIC ACID.)

VOICE is the product of the vocal cords, situated in the larynx, at the upper portion of the windpipe. The structure is fully explained in the article on LARYNX. Voice may be lost from various causes. The condition is termed aphonia. Thus voice is often lost in hysterical people, in whom a smart electric shock will generally bring it back speedily enough. It may be lost in disease, as in ulceration of the larynx, or in malignant disease of the same. In these cases the nature of the malady is decided by examination with the laryngoscope, and the remedy will of course depend on the nature of the malady. (See APHONIA.)

VOMITING means the ejection of the contents of the stomach upward, instead of into the bowel. The act is a complex one, and seems due to two factors, viz., contraction of the walls of the stomach itself, and contraction of the abdominal walls, the contents of the abdomen thereby in their turn pressing on the stomach itself. The causes of vomiting are very various—irritation of the stomach itself, whatever be its cause, will give rise to ejection of its contents ; but vomiting occurs in many other maladies. When gall-stones or small urinary calculi are passing there is usually sickness and vomiting ; in Bright's disease there is vomiting too, and in the maladies of the brain among children vomiting is an invariable symptom. To arrest vomiting, ice swallowed in small lumps is a capital remedy. Bismuth is good, especially with small doses of opium. In all cases the quantity of the remedy used should be small. Bulky preparations will most probably be rejected.

VOMITING BLOOD. (See HÆMATEMESIS.)

VOYAGES are frequently undertaken as a means of health, especially by persons with delicate chests, and often are attended with most beneficial results. We shall only lay down one or two rules as to their selection, but these are important : Voyages should never be undertaken by any too delicate to stand a little knocking about. Sometimes people are sent away this way who die on their voyage, and whom it was positive cruelty to send abroad. A short voyage is useless. One to Australia and back is good. The time for sailing should be late autumn, after the equinoxes. Some few private stores should be taken, but these will suggest themselves. The first rule is the guide to everything else.

W.

WAKEFULNESS. (See SLEEPLESSNESS.)

WALKING as a means of exercise is invaluable to people in sufficiently robust health to undertake it. It should never be allowed to be carried so far, however, as to produce more than a sufficient degree of fatigue to make it pleasant, and should never be indulged in to the foolish extent it is by some young men. The great thing in walking is the boot. That should not be too heavy, strong, with good thick broad soles and low heels. Walking boots should always be made to lace ; what are called side-springs are an abomination. A few nails are a decided improvement.

WARD'S PASTE was a patent medicine long used for piles. The confection of black pepper was intended to take its place, and is a good deal used in that troublesome affection. (See PEPPER.)

WARMING in most large institutions and churches, etc., is accomplished by means of hot water, which circulates in pipes through the building and radiates heat. A furnace in the basement heats the water, which then rises in the pipes, while the cooler water descends back to the boiler to be in turn reheated, and so a continual current is set up. An ordinary room may be heated by a stove or fire ; the latter is much to be preferred. A room should not be more than 65° Fahr., for beyond this point the heat is relaxing.

WARTS are papillary tumors, the varieties of which depend upon their locality. The most common are those situated about tne hands or fingers, or sometimes on the face, and more rarely on other parts of the body ; they chiefly affect young persons, and their structure is hypertrophied papillæ, closely adherent to each other, and covered with thick cuticle. A somewhat scarce variety occurs upon the scalp occasionally, and almost invariably in women after adult age, although it has been met with in males, and from its presence and form gives great pain and inconvenience in brushing the hair. A third variety is occasionally met with beneath or at the side of the finger or toe nails. These originate beneath the skin and protrude beyond the free margin of the nail. They are generally very painful and troublesome. Warts of a peculiar nature, arising from venereal causes, are met with under the foreskin and between the labia, and are liable to rapid propagation from their close contact with neighboring parts. Such warts are undoubtedly contagious.

In some persons warts appear to be hereditary, and the period of life up to the time of puberty seems to be that in which these growths flourish. They sometimes appear curiously, suddenly, and as suddenly disappear. It is very probable that the poison of decomposing animal matter is, under certain conditions, capable of exciting these warts. This is partly inferred from the fact that those engaged in the manipulation of dead and morbid tissues are frequently affected by them. As a rule, warts do not materially increase in size, supposing that they do not entirely disappear, although occasionally, owing to some permanent source of lcoal irritation, they may "take on" a semi-malignant (see TUMORS) character, especially those occurring about the face.

Treatment. — The treatment of simple warts, such as occur on the fingers or scalp, is very simple. The best method consists of their destruction by the glacial acetic acid, which may be either dropped upon them or painted thickly over them with a brush, care being taken to apply a little oil or glycerine to the tissue contiguous to the wart, so as to avoid blistering it. Lunar-caustic, tincture of the perchloride of iron, a drop of pure nitric acid, or the acid nitrate of mercury, are all good and frequently efficient remedies. The scalp warts are often most effectually treated by ligaturing their bases with a loop of silk or thin silver wire, and allowing them to drop off. Venereal warts of a not very general character are best treated by snipping them off with scissors, or by the application of powdered oxide of zinc, or equal parts of powdered savine and diacetate of copper. In the case of rapidly growing warts, and those which are evidently degenerating in their appearance, excision of the growth and of the integument from which it grows is the advisable treatment.

The variety noticed as growing from under the nails of the fingers or toes is best treated by pulling out the papillæ constituting its bulk by forceps, separately.

WASTING DISEASES.—Wasting is a very common symptom in many disorders. In the adult it is observed in all severe cases of fever, but then they recover their weight during convalescence. In cancer and consumption it is a most marked symptom, also in many cases of disease of any internal organ; it is chiefly noticed also in those who are liable to degeneration of their tissues. (See DEGENERATION.) It is clearly impossible to enumerate all the causes of wasting, or to give any account of the symptoms associated with it, as it is common to so many varieties of disease. In children, in whom very few symptoms can be observed, wasting is a valuable indication of mischief. It may arise from insufficient nourishment, or over-feeding, or from unsuitable foods (see DIET), or from chronic diarrhœa, which more or less exhausts the child (see DIARRHŒA); or from chronic vomiting, generally depending on some gastric disturbance; or from the child having rickets (see RICKETS), or from the presence of worms in the intestinal canal (see ENTOZOA). Inherited syphilis, consumption, and tuberculosis will also cause much wasting. A part may waste from want of use, and so it is common to see a wasting of an arm or leg in cases of paralysis. (See TABES MESENTERICA).

WASTING PALSY. (See PROGRESSIVE MUSCULAR ATROPHY.)

WATER is a compound of hydrogen and oxygen, in the proportion by weight of two parts of the former and sixteen of the latter. Ordinary water is too well known to require a full description. It should be clear, colorless, and deposit no sediment on standing, nor on evaporating a drop on a glass slide. Covering a large portion of the earth's surface, it is invaluable to man in many respects; in the form of ice in the Arctic regions it forms a bridge of communication between distant places; in its liquid state it supplies animal and vegetable matter with the means of sustenance; it supplies man also, by means of the seas and rivers, with means for communication with foreign lands; in the ocean and in lakes and rivers are also contained myriads of fish and other products which are useful as food; as steam it is useful as a motor power in economizing labor-advancing civilization, and improving the condition of man. It occurs in various forms—ice, hail, dew, hoar-frost, rain; snow and hail are but various kinds of water. On evapora-

tion from the surface of the earth, it forms clouds, which, when they condense, give back the water to the thirsty earth.

Pure water is only known to the chemist, for all the ordinary kinds of water contain either gaseous, saline, or organic matters. It freezes at 32° F., or at zero on the Centigrade scale, into a number of crystalline forms. It evaporates at all temperatures and boils at 212° F. or 100° C., under the ordinary atmospheric pressure at the sea level. Above 39° F. water expands by heat; below this point it expands gradually, thus differing from most fluids, which contract by cold; it is due to this fact that pipes burst in a house in the winter, and that a jug may become broken when ice forms. Sea water, however, contracts regularly on lowering the temperature. The density of water at 60° F. is taken as unity (1.000), and it is the standard by which the specific gravities of all solids and liquids are compared in this country; the barometric pressure at the time should be noted to insure accuracy; the mercury should stand at 30 inches or 760 millimetres. A cubic inch of pure water weighs at 60° F. in air 252.456 grains. Nearly all ordinary compounds contain water, and therefore shrink on exposure to heat, for then the water is driven off. Nearly all crystallizable bodies contain water, and when heated become powdery. There are various kinds of natural waters, viz., rain water, spring, mineral, river, and sea water.

Rain water is never really absolutely water, as it contains gases which it absorbs in passing through the air; melted ice and melted snow are perhaps the purest forms of water which can be obtained naturally. The water of our lakes contains various inorganic and organic impurities from the rivers which flow into them or the springs which supply them. The beautiful color of the Swiss lakes seems due to the floating about of innumerable fine particles brought down by the swift mountain torrents.

River water contains less saline matter than spring water, but it also contains various organic impurities according to the district through which it passes. Near large towns it may contain a good deal of sewage, or refuse from manufactories; it contains also fish-spawn, leaves, silt or mud, according to the rapidity of the current. Before, therefore, it can be used for drinking purposes, it must be filtered through beds of sand, gravel, etc., so as to remove the impurities; any running stream has a self-purifying power, because it continually exposes fresh portions of the water to the air, and so the organic matters get oxidized. It is thus very important that for the due supply of a large town there should be a rapid current, absence of sewage matter from the towns above, and proper filtration, so as to separate inorganic impurities. The presence of organic matter may be roughly estimated by putting two or three drops of permanganate of potash into half a gallon of water ; if pure, there will be a pink tinge ; if impure, it will be colorless, or a faintly brown precipitate is produced. River and rain water are ordinarily known as *soft* waters, because they contain little or no lime ; hence they are more useful for washing and other domestic purposes.

Spring water, although it may look transparent, always contains saline matters, and chiefly the lime salts ; hence such water, although very agreeable to drink and quite wholesome, is known as *hard water*, and soap curdles in it and does not produce a good lather. Carbonate of lime, common salt, sulphate of lime, and carbonate and sulphate of magnesia, are the salts most usually present in spring water. They are held in solution partly by the carbonic acid which all such waters contain ; this is seen on boiling the water, when the carbonic acid is driven off, and on cooling the water looks turbid and deposits a small amount of sediment consisting of the above salts ; to this cause is due the fur or incrustation on the inside of kettles and boilers. Pure water is very insipid, and it is to the gases and saline impurities of ordinary spring water that its refreshing properties are mostly due. The danger from impure water is due to the organic matters and those derived from drains, sewers, etc. An excess of lime in spring water is said to cause the Derbyshire neck, known also as bronchocele or goitre.

Mineral waters contain iron, sulphur, and various salts, according to the nature of the soil through which the water has percolated. Those which contain iron are called chalybeate waters ; those which contain carbonic acid are pungent and effervescent ; artificially prepared they are known as soda and seltzer waters. Some have sulphur or sulphuretted hydrogen in solution, and are very nauseous. Others contain saline matters, as the springs at Epsom, and hence the well-known medicine commonly called Epsom salts. (See MINERAL WATERS.)

Sea water is largely impregnated with common salt and with chloride of magnesium, to which it owes its bitter taste. From the vast surface of the seas pure water is constantly evaporating to form the clouds ; into it run the contents of myriads of rivers, while the sea itself constantly returns to the earth marine plants, fish, guano, kelp, etc., which are useful to man. The mean specific gravity of sea water is 1.027, and the quantity of salt it contains varies from 3.5 to 4 per cent. Hence it is easier to keep afloat in salt than in fresh water. The following table shows the composition of the sea water of the British Channel :

Water	961·74372
Chloride of sodium	28·05948
Chloride of potassium	0·76552
Chloride of magnesium	3·66658
Bromide of magnesium	0·02929
Sulphate of magnesia	2·29578
Sulphate of lime	1·40662
Carbonate of lime	0·03301
Iodine	traces
Ammonia	traces
	1000·00000

For chemical purposes water is obtained by distillation ; this may be done on a small scale by heating water in a glass retort, and allowing the vapor to pass over into a receiver which is kept cool ; the vapor condenses and pure water is obtained ; however, the first and the last portions distilled should be thrown away, as the first may contain volatile waters, and the last saline impurities, when the contents of the retort get too low. The specific gravity of steam is 0.662, of ice 0.94 ; hence ice is lighter, and therefore floats in water. Water may be produced when an electric spark is passed through a vessel containing a mixture of hydrogen and oxygen, in the proportion of two volumes of the former and one of the latter ; a slight explosion occurs and a few drops of moisture are produced. Water dissolves many substances, and therefore cisterns are best made of slate and not of lead, as that metal is acted on by the water, and may give rise to colic and lead poisoning if swallowed. Iron pipes, and not lead pipes, should be used for the conveyance of water for the same reason. (See FILTERS.)

WATER BEDS, or WATER-CUSHIONS, are very useful in many cases of fever, paralysis, and long-standing disease ; they aid in preventing any undue pressure on a part, and so prevent the formation of bed-sures ; they are also of great comfort to a patient, and enable one to rest much better than on an ordinary bed.

WATER-BRASH is a common symptom in indigestion. It is caused by the rising up in the œsophagus, or gullet, of a watery fluid secreted by the glands of the stomach. (See INDIGESTION and PYROSIS.)

WATER-CLOSETS are inventions of which no one has any particular reason to be proud. With a show of cleanliness they combine essential nastiness and a good deal of real danger. In large towns their use can hardly be dispensed with ; in the country, with imperfect drainage and water supply, they are an intolerable nuisance. Earth should always be used instead of water in country places ; it is preferable in every way. To keep water-closets moderately safe, the first thing is ventilation of the sewers ; if not, they ventilate themselves into the house by means of the water-closets. A pipe should therefore lead from the house sewer to the highest point of the building. From time to time the pipes and traps ought to be examined, to see that the former are entire and not leak ng, and that the traps are in working order. A portion of disinfectant, fluid or solid, should be used at least once every day. The best is carbolic acid, in powder or in solution. With these precautions the water-closet system may be worked with no very great risk, but the best thing is to get rid of them wherever possible.

WATER-CURE. (See HYDROPATHY.)

WATER IN THE CHEST. (See HYDROTHORAX.)

WATER ON THE BRAIN. (See HYDROCEPHALUS.)

WAX is the *Cera* of the Roman physicians, and is a peculiar substance occurring in the textures of plants, and sometimes on their surface. It is also formed by animal organisms, and in the largest quantity by the common bee. It is obtained for use after the honey which the waxen cell incloses has been expressed, and is fused in boiling water and strained. This is the yellow wax of commerce. The white wax of the shops is made by bleaching the common yellow wax and exposing it to the air. It is largely used in the preparation of plasters, ointments, and cerates, and also by dentists and mechanical surgeons in taking moulds or models of any part of the body.

WAXY DEGENERATION of the liver, or kidney, or spleen. (See DEGENERATION.)

WEAK ANKLES. (See ANKLES.)

WEAKNESS. (See DEBILITY.)

WEANING. (See LACTATION.)

WEIGHT.--The average weight of the human body may be taken at 154 lbs. Such a body would be made up of muscles and their appurtenances, 68 lbs., skeleton, 24 lbs. ; skin, 10½ lbs. ; fat, 28 lbs. ; brain, 3 lbs. ; thoracic viscera, 2½ lbs. ; abdominal viscera, 11 lbs. ; blood, 7 lbs. About five pounds more blood will remain in the tissues and will not drain away, and therefore it is reckoned with them. The female weighs less than the male. The following is the weight of the chief internal organs :

	Male.	Female.
Brain	48-53 oz.	40 –45 oz.
Heart	10 "	9 "
Lungs	18-20 "	15 –18 "
Liver	50-60 "	45 –50 "
Kidney	5- 5½ "	4½- 5 "
Spleen	4- 6 "	4 – 6 "

These are but averages ; a great variation in the weight of people is met with, depending on their age, stoutness, sex, height, and mode of living.

WEIGHT AND HEIGHT.—Within the last few years public attention has been drawn to the fact that weight is as important an indication of the general condition of the human body as any other evidence, and many physicians make a practice of weighing their patients periodically at each consultation. The habit of being weighed has almost become an amusement, and in railway stations, shops, and many places of recreation, weighing-machines are to be seen in constant request. It will easily be seen, however, that to know the correct weight of an individual without reference to height is of little advantage, but if a standard be ascertained as to the proper proportion which weight should bear to height, then we know how much a person ought to weigh, and can treat him accordingly. One of the earliest efforts made to obtain anything like a fixed relation between weight and height was that of Dr. Boyd, who weighed a certain number of inmates of the Marylebone workhouse. He took the weight and height of 108 persons suffering from consumption, and found they measured 5 ft. 7 in., and weighed 90 lbs. He then measured and weighed 141 paupers not in ill-health, and found their average height was 5 ft. 3 in., and they weighed 134 lbs. This subject attracted the attention of the late Dr. John Hutchinson, and he determined to take the height and weight of persons of all classes of the community. In this way he collected the height and weight of upward of 5000 persons. This list, however, included persons who exhibited themselves as giants and dwarfs, and other exceptional cases. He therefore reduced his instances to 2650 persons, all of whom were men in the prime and vigor of life, and included sailors, soldiers, firemen, policemen, draymen, gentlemen, paupers, and pugilists. This group of cases was intended to make one class as a set-off against another, so as to get a fair average. The following is the result of Dr. Hutchinson's osbervations :

Height. Ft. In.	Weight. lbs.
5 1	120
5 2	126
5 3	133
5 4	139
5 5	142
5 6	145
5 7	148
5 9	155
5 9	162
5 11	169
5 11	174
6 0	178

Of course the result of these observations can only be considered as approximate, but they are sufficient to show that among a set of healthy men there is a healthy standard of height and weight. In examining this table, Dr. Lankester found that for every inch increased in height we have five pounds more in weight, and this rule holds good for all practical purposes. Starting with a person 5 ft. in height, who, according to the assumed law, should weigh 115 lbs., we obtain the following results :

Height. Ft. In.	Weight. lbs.
5 0	115
5 1	120
5 2	125
5 3	130
5 4	135
5 5	140
5 6	145
5 7	150
5 8	155
5 9	160
5 10	165
5 11	170
6 0	175
6 1	180
6 2	185
6 3	190
6 4	195

Although this law is approximately good for a certain number of cases, even above and below this table, it is practically found, and especially in the case of children and growing persons, that there is a wide difference of weight at heights below 5 ft. Attention may also be drawn to the fact, that there will constantly occur in the community instances of persons where either the muscular or bony systems are excessively developed, and who, consequently, weigh more or less than their height.

The conclusion we come to with regard to all these weighings and measurings is, that all ordinary departures from the average height and weight of the body deduced from Dr. Hutchinson's tables are due to an increase or decrease of the fat or adipose tissue of the body. Thus, taking the composition of a human body weighing 154 lbs. and measuring 5 ft. 8 in., it will be found to contain 12 lbs. of fat. It is then mainly due to the diminution or increase of this substance that human beings vary in weight, and it is important to find out whether this fat be of any use or value in the system, and whether the indications afforded by the weighing-scales should not afford some suggestions for caution in diet and regimen. Besides exerting a primary influence on the growth of the body, fat subserves many other purposes and is essential to animal life. When there is too little deposited for the purposes of life, then serious disease has already commenced, or may set in ; while, on the other hand, a redundancy of this deposit may seriously interfere with the functions necessary to life. It is from this point of view that the value practically of a knowledge of the height and weight of individuals becomes apparent. When the weight of a person is much below his height, then it may be suspected that some disease has set in, which may go on to the destruction of life. One of the earliest symptoms of consumption is a tendency to loss of weight. Long before any symptoms are present of tuberculous deposits in the lungs, this loss of weight is observable in persons afflicted with consumption. At this stage of the disease, a large amount of evidence renders it probable that the fatal advance of this disease may be prevented. This fact has been admitted by the practice introduced during the last thirty years of administering cod-liver oil and fatty substances to those who are threatened with consumption. In fact, it may be stated generally that, wherever the weight is much below the height, suspicion should be aroused and the indication regarded. The other side of the question should not be forgotten ; in certain families and individuals there is a tendency to develop adipose tissue unduly. However free from fat may be the food, what little it contains is arrested in the tissues of these individuals and they become "fat"—that is, they weigh more than their height. Sometimes this is deposited all over the system so as not to be an obvious obstruction to the functions of life ; but it can be well understood that, when two men of equal stature, say 5 ft. 8 in., one having to carry 154 pounds and the other 168 pounds, the latter will be at a disadvantage. This arises from two causes. The heavier man, carries, in the first place, greater weight, and in the second place his heart has to project into the tissues of the body a larger amount of blood in order to keep him alive. For every pound a man weighs above his height his system is at a disadvantage, and he suffers in various ways. When fat is equally distributed about the body, then no immediate disadvantage is felt ; but when fat is accumulated in particular parts of the body, interfering with the functions of particular organs, then its evil influences become speedily apparent. When persons weigh much above their height it is obviously a matter of importance that they should, as much as possible, relieve the tax on their muscular and circulating system by diminishing their weight ; but this must be done with caution. The sudden withdrawal of accustomed articles of food is unwise, and it is better gradually to lessen the fatty portions of diet than to go to extremes. When looked at carefully, there can be no doubt that the relation of height and weight are very important as regards health and the chances of life. Whenever the weight is below the height there is a fair suspicion of scrofulous or tuberculous disease, and when the weight is greatly in excess of the height there is a tendency to those sudden impairments of muscular and circulating powers which may lead to premature and sudden death. (See BANTINGISM.)

WENS are encysted tumors, most frequently met with on the scalp or eyebrows. The origin of these tumors is in obstruction or imperfect congenital development of the sebaceous follicles dilated by the accumulation of their contents. If existing on the scalp they are generally multiple, and the cyst wall strong and tough, and at first but loosely adherent to the surrounding tissues. If irritation be set up by continuous pressure or friction, the cyst wall becomes intimately adherent to these tissues. The contents vary from being merely an accumulation of the natural sebaceous secretion to several forms of its perversion. Sometimes they are semifluid or honey-like, sometimes atheromatous, sometimes steatomatous or fibrinous, occasionally purulent. Hairs or eyelashes are frequently met with in their cavity, and in encysted tumors, which exist in the ovaries, hair, skin, teeth, or bones are met with.

Treatment.—The treatment of such tumors consists in their removal : if very small, evacuation by pressure is sufficient ; but if large and unattached, a simple incision through the integument and down upon the cyst wall, with the subsequent enucleation of the entire cyst and its contents. Supposing the tumor to be very large, and its cyst wall thin and adherent, removal must be effected by regular dissection. It must be borne in mind that, unless the entire cyst or bag is removed, there is every probability of the tumor returning. These tumors may occur in the neck, and a somewhat favorite locale is just under the angle of the lower jaw. Those occurring on the eyelids (tarsal tumors) have extremely thin walls, and it is rarely necessary to dissect them out, as by eversion of the eyelid and rupture of the sac from the under surface all deformity from cicatrix is avoided, and the result is all that is required. (See EYELIDS.) It is well to remark that these wens should be removed by a surgeon as soon as they are noticed, as the scars increase in size, and are horribly unsightly if situated in any prominent place, and their removal is safe and generally unattended with any great pain.

WET-NURSING may be required in those cases where on the death of the mother, or because she is incapable of suckling her infant, another woman who has been recently confined is employed for the purpose of giving the child sustenance. It is better to have a wet-nurse than to bring the child up by hand or by bottle. Precautions should be taken to see that the wet-nurse is in good health at the time.

WETTING THE BED, so frequent in children and so well known to nurses, requires careful attention, and should not always be treated as an avoidable habit and punished accordingly—though it is often necessary when a child is of sufficient age to understand to create habits of self-control by severe measures. In the first place, however, pains should be taken to ascertain whether the irritable condition of the bladder be not produced by the too alkaline condition of the water, or by the presence of worms in the rectum, which cause great irritation in the surrounding nerves, and so involuntarily lead

to the discharge of the water. If after all precautions have been used to discover a local cause for the habit, and none seems to exist, the occurrence of it must be treated as a fault, especially if it occur in the daytime, when indolence and indifference alone can account for such an uncleanly practice. Frequently, however, in young children, it will be found to cease altogether after a few doses of suitable medicine to allay one or other of the causes of irritation.

WHEALS are red and white marks on the skin, which are seen in cases of nettle-rash, and in some forms of indigestion ; tepid water will generally relieve the tingling, but the treatment must be directed to the cause.

WHEY is the watery part of milk, the part which separates when curds are made. It is a wholesome and pleasant drink, and in cases of cholera or fever is very often beneficial.

WHISKEY. (See DISTILLED SPIRITS.)

WHITE BLOOD CORPUSCLES are rounded, often granular cells, which are seen in the blood with the aid of a good microscope. (See BLOOD.) They are in excess in cases of leucocythæmia and lymphoma.

WHITE LEG. (See PHLEGMASIA DOLENS.)

WHITE PRECIPITATE. (See MERCURY).

WHITE HELLEBORE. (See VERATRUM ALBUM.)

WHITE SOFTENING OF THE BRAIN. (See CEREBRAL SOFTENING.)

WHITE SWELLING. (See KNEE-JOINT.)

WHITES. (See LEUCORRHŒA.)

WHITLOW, called also paronychia, is a very common and, if neglected, serious affection. It signifies an abscess of the fingers, and it may arise from various causes, and has various localities and intensities. The simplest form is one which is limited to the surface. The finger is swollen, inflamed, and intensely painful, and the integument generally vesicates. This form frequently begins by inflammation of the matrix of the nail, which nail may be eventually shed. The *treatment* consists of poultices, fomentation, and attention to the state of the bowels. A more serious form of the disease is one which affects the deeper structures, and attacks the subcutaneous areolar tissue, and this much resembles a boil, and the swelling, tension, and pain are very considerable. The affected parts should be freely incised to evacuate pus, and this proceeding must be followed by fomentations and poultices. In the case of a painful tip to a finger, which is very painful and does not seem inclined to suppurate, it should be well rubbed with lunar caustic.

The worst form of whitlow is the tendinous whitlow or *bone-felon ;* and the disease originates in the deep fibrous tissue of the finger, or in the periosteum or bone. It is characterized by the most excruciating pain from the very outset of the disease. Pus forms early, and the constitution is affected with frequently severe inflammatory fever. If this state of things be permitted to continue, there is no relief for the symptoms until nature has evacuated the pus herself : but then the joints are disorganized, the tendons have sloughed, the bones become carious or necrosed, and if recovery takes place, it is with stiff, useless digits, requiring amputation. The *treatment* of such cases consists in the early and free evacuation of the pent-up matter by a deep, vigorous incision *down to the bone*. The pain at the time is most acute, but the relief is instantaneous, and in all probability a useful finger is retained. These bone-felons not unfrequently spread into the palm of the hand, forming palmar abscesses, or may extend underneath the annular ligament, and the matter burrow up into the tendinous sheaths of the muscles of the forearm. In opening such abscesses in the palm, the incision should be made over and down upon the metacarpal bone, taking care to avoid wounding the digital artery or palmar arch.

The subsequent treatment of bone-felons consists in poulticing, fomentation, and the administration of tonics, and in taking care that stiffness of the fingers be avoided after the free incisions, and when the tissues have become healthy, by early passive motion, and inunction of ointment, such as creosote or resin. All dead skin is to be carefully removed. Their most frequent causes are the inoculation of decaying animal or vegetable matter, and the effect of such on a somewhat low state of health. It must be borne in mind that the discharge from them is contagious.

WHOOPING-COUGH is a disease of great frequency in childhood, and a large proportion of infant mortality is due to this cause. It belongs to that class of disorders which is called zymotic. It is contagious, but differs in this respect from other contagious diseases, that whereas they are communicable by a third person who may all the time be unaffected, this is not so with whooping-cough, and this fact has some practical importance in a children's hospital. A ward for measles and scarlet fever should be kept quite separate from the main building, and all the nurses, etc., should also be distinct ; but for cases of whooping-cough it is enough if other children are kept from going into the ward, while the nurses, etc., may go about without spreading the disease. Whooping-cough is known in different parts as hooping-cough, chin-cough, kink-cough, and pertussis. It may be defined as a disorder in which a convulsive cough consists of a long series of short and forcible expirations, and then a deep and loud inspiration, and repeated more or less frequently during each paroxysm ; it lasts several weeks, occurs once in a lifetime, and is most common in childhood. Whooping-cough has been known since the middle of the seventh century, and has always of late years been prevalent in this country ; it seems to be most fatal in those years in which measles are also prevalent. No disease kills more children under one year of age than whooping-cough ; nearly 70 per cent of all the cases occur under two years of age, and not more than 5 per cent of the deaths are recorded as above five years of age.

Symptoms. —The earliest is a common cold or catarrh, accompanied by a cough : there is also a slight amount of fever, restlessness, and sometimes running at the eyes and nose. The cough in a few days becomes more troublesome, and some glairy fluid may be brought up from the chest ; in a week or ten days, but oftener later, the child will begin to have the characteristic whoop ; the cough comes on in paroxysms, and is more frequent by night than by day ; each paroxysm begins with a deep and loud inspiration, followed by a succession of short and sharp expirations, again followed by a deep inspiration, and the repeated expiration ; this may go on several times, and last one or two minutes, according to the severity of the case. Just before each attack comes on, the child clings to its nurse or mother ; it sits in an erect position ; during the paroxysm the face is flushed, the veins in the head and face prominent, the eyes suffused and watery, and generally there is some glairy fluid expelled from the mouth, or vomiting may come on. After the paroxysm the child will rest for a time and appear pretty well until the next attack comes on. In bad cases there may be twenty and thirty paroxysms a day, and several fits of coughing besides, without the whoop being heard. In ordinary cases there are from four to ten spasmodic attacks in the twenty-four hours. These symptoms last for three or four weeks, and then the cough abates in severity and frequency, and finally ceases altogether ; even when there is no whooping, the child may continue to have a troublesome cough for some time. In most cases there is some bronchitis attending this complaint, and this is shown by the hurried breathing, rise of temperature, and by hearing rattling noises over the chest. The more mischief there is in the lungs, the greater is the danger to the child. Convulsions are a sign of bad import, and this is generally the way in which such cases die. Whooping-cough cannot be made out until the characteristic whoop appears, and then there can be no difficulty in recognizing the disease.

Treatment.—In all cases it is best for the

child to keep in the house as soon as the mal-ady has declared itself ; in a very mild case it need not be kept in bed, but it should be in a room of a warm and even temperature, and protected from draught ; it can then be allowed to play about as it likes. If there is any lung affection it must be put to bed, and hot linseed-meal poultices placed round the chest. Other children must not be allowed to come near it unless they have had an attack previously, for in this way its spreading may be prevented. The child must be fed in the usual way, but solid food should be given sparingly. Where the infant is emaciated, and has some other disease, as rickets, etc., the treatment proper for that disease may be continued. Steel wine is very valuable in cases of whooping-cough, and more especially when there is no fever, and during convalescence ; it may also stop the diarrhœa, which is now and then present. If there is any prolapse of the bowel, the part should be sponged lightly with a solution of sulphate of iron, and at once returned. This is often due to the excessive diarrhœa, and steel wine must be given internally. Numberless remedies have been tried to cure whooping-cough, but none have succeeded. Iron, alum, zinc, sulphuric acid, etc., have all failed to do much. The most hopeful remedy is belladonna if given in large doses, and the symptoms watched ; children can bear more of this drug the younger they are, but it is a dangerous remedy, and can only be given with the greatest care. Warm clothing must be worn, and during convalescence a nourishing diet, moderate exercise in the air when fine, a tepid bath in the morning, and a tonic, as steel wine or cod-liver oil, must be enjoined.

WILLOW.—The bark of the willow is sometimes used in medicine, but only to produce its active principle, salicine. (See SALICINE.)

WIND IN THE STOMACH. (See FLATULENCE, INDIGESTION, and INWARD FITS).

WINDPIPE is the main tube, or trachea, which allows of the passage of air from the mouth and nostrils into the lungs. It can be felt in the throat, and when pressed gives an uncomfortable feeling of impending suffocation. (See LUNGS and FOREIGN BODIES.)

WINDS enter most essentially into the climate of any region, the prevalent wind giving it a character of its own. This has been alluded to in dealing with house-building and climate, and need not be further alluded to here. Extreme cold, if dry and still, may be borne much more easily than a higher temperature if windy. The continual renewal of the air next the body abstracts the heat more rapidly in the one case than the other. (See CLIMATE and HOUSE.)

WINE is the name generally given to fermented liquors when no foreign ingredient is added to flavor them. Thus the fermented wort of malt is called malt wine when hops are not added. Some wines are made from the juice of various fruits fermented, as currants, gooseberries, elderberries, and others. The term wine, however, is more especially applied to the fermented juice of the grape. Of all fruits the grape is best adapted for making wine. The reason is that the juice of the grape contains tartaric acid, and this acid forms an insoluble salt with potash. Thus the acid of the wine is in the form of an insoluble supertartrate of potash, which is called tartar, and when purified is sold under the name of cream of tartar, and when burned is converted into carbonate of potash or salt of tartar. The acids contained in other fruits, as the citric acid in the orange, the malic acid in the apple and pear, form soluble supersalts with potash, are retained in the fermented juice, and render the wine so sour that sugar has to be added to cover their acidity. This is the case with all the fruits from which home-made wines are prepared, and, in fact, the source of the objection to them as ordinary beverages.

Wines generally contain more alcohol than beers (see BEER) and less than distilled spirits. (See DISTILLED SPIRITS and ALCOHOL.) The quantity of alcohol varies very much in different kinds of wines ; and, in fact, the quantity of alcohol is the first element which determines the price of wines. An import duty is levied on all wines coming into this country, and those containing below a certain percentage of alcohol pay less than those above the same point. Wines, however, are not consumed for their alcohol alone. They contain other ingredients which they derive from the grape-juice, which give them taste and flavor. Thus, when the fermentation of the grape-juice is not complete, a certain quantity of sugar is left, and according to the quantity of sugar wines are said to be "sweet" or dry. While hocks, clarets, and other light wines contain little or no sugar, port, sherry, and champagne always contain a large amount. In the case of port and sherry this sugar is added during the manufacture, in order to enable them to keep and bear exportation. At the same time that a large quantity of the tartaric acid contained in the juice of the grape is thrown down while the "must" is being fermented and the wine is in the cask, the whole of the tartaric acid is not got rid of, and a certain quantity is retained. In order to get rid of this, the wines of Spain are exposed to a process called "plastering," which consists in mixing with the grapes a certain quantity of gypsum, or plaster of Paris. The quantities of alcohol, sugar, and acid found in one imperial pint of certain of the wines commonly consumed will be found in the following table:

Wine.	Water.	Alcohol.	Sugar.		Tartaric Acid.
			oz.	grs.	grs.
Port	16	4	1	2	80
Brown Sherry	15½	4½	0	360	90
Pale Sherry	16	4	0	80	170
Claret	18	2	0	0	161
Burgundy	17½	2½	0	0	160
Hock	17¾	2¼	0	0	127
Moselle	18¼	1¾	0	0	140
Champagne	17	3	1	133	90
Madeira	16	4	0	400	100
St. Elie (Greek)	16	4	0	22	44
Santorin (Greek red)	16⅓	3½	0	40	60

From this table it will be seen that it is erroneous to suppose that ports, sherries, and madeiras are free from acidity. They do not contain so much tartaric acid as the lighter French and German wines, but the taste of the acid is covered by the sugar they contain. The sugar in wine is often a very prejudicial agent. As it exists in most wines, it is in a state in which it more readily ferments than when in the form of common sugar. Hence patients are recommended to take "dry" wines. The fact is, with regard to ports, sherries, and madeiras, they can hardly be said to be wines at all. They are all made on the same principle—that of adding to the genuine wine certain quantities of sugar and brandy. In short, it may be stated that all these wines are manufactured by the taking the wine of one brewing and adding to it the "must" or unfermented juice of a second quantity, and adding the pure brandy distilled from a third portion. It has been a great misfortune that these wines have been thrust into circulation in our country, and a taste has been acquired for them which it seems now impossible to annihilate.

When the stimulus of alcohol is required in disease, it is no doubt better to secure it through pure wines, such as those of France, Germany, or our own country, than in the saccharine compounds presented to us from Spain and Portugal. If larger quantities of alcohol are required in disease, it is better presented in the form of brandy or whiskey. The latter spirit is now sold so pure that it may without hesitation be used in the sick-room as a substitute for strong wine. A theoretical objection has been urged against the use of spirits and water. It is said that the stomach, through the action of endosmosis, absorbs the water, and leaves the spirit to act as an irritant on the stomach and surrounding organs. This is said not to be the case with the alcohol and water in wines, when the two are held in a much closer chemical union.

There are three other qualities in wines which demand some consideration. The first is what is called the *bouquet* and *flavor* of wines. These things are sometimes con founded, but they are really different. The vinous flavor is common to all wines, but the bouquet is peculiar to certain wines. The substance which gives flavor to all wines is œnanthic ether, and is formed during the fermentation of the grape-juice. When separated from the wine, this substance is anything but pleasant to the taste and smell. It is composed of an acid—œnanthic acid—which forms an ether with the alcohol. The bouquets of wines are formed in the same way by some of the acids found in the grape-juice after fermentation, combining with the ethyl of the alcohol, and forming ethers. Many of the bouquets thus formed are well known, and they consist of ethers formed by ethyl with acetic, proprionic, pelargonic, butyric, caprioc, and caprylic acids. As far as we know at present, these ethers do not in any way exert medicinal effects on the system. All we know is, they are, many of them, most agreeable to the taste and act upon the tongue as delicious odors of flowers upon the nose. These are the things which make one wine more pleasant to drink than the other, and which give the highest price to the best of wines. They are not detectable by chemical agency, and it is the taste of these bouquets, and nothing else, which gives to one wine the value of 25*s.* a bottle, and another 2*s.* 6*d.*, when all other qualities are precisely the same.

Another point in the nature of wines is their coloring matter. Some wines are what are called "red," and others are "white." Ports, clarets, burgundies, are all red; while some of the wines of Greece, Germany, Hungary, and other parts of the world are red also. The red colors of these wines have been analyzed with some care, but they do not seem to exert any influence upon the system. The most important agent in them is tannic acid, or tannin, which exists in some wines to a very large extent. It is especially present in ports and clarets, and less in burgundy. It gives an astringency to red wines which is not found in white. The large quantity of tannin in port gives it a tendency to deposit a sediment, which is known by the name of "crust," and which is found on the lower side of the bottle after keeping. This crust consists of oxidized tannic acid, which becomes insoluble, and carries down with it a blue coloring matter, and the saline matter contained in the wine. The longer port is kept the more of the crust it throws down. By this process port wine loses its color and density, and acquires a purer flavor, and its price is proportionately enhanced. Port wines kept twenty, or thirty, or forty years command, when originally good wines, almost

fabulous prices in the market. This, however, is a mere matter of taste, and such
wines have no dietetical or medicinal qualities
to recommend them. Even the assertion that
they may be taken with more impunity
than new wines is problematical. They do
not seem to contain so much alcohol as wines
not kept, and may be taken in larger quantities on that account.

The other coloring matters described by
chemists are a *blue* and *brown* coloring matter. The latter is found in dark white wines
as well as in red wines. The brown coloring
matter is found in port wine, when all the
tannic acid and blue coloring matters are
thrown down. The blue coloring matter is
derived from the skins of the red grapes from
which red wines are made. These skins are
also the source of the tannin. The brown
coloring matter is more or less present in the
skins of red and white grapes.

The other matters which give a character
to wines are the saline compounds. These
substances, which constitute the "ashes" of
all burned vegetable tissues, exist in very
varying quantity in all fruits, and are found
dissolved in the juices of fruits ; hence we find
them remaining in the wine after fermentation
of the juice. The most abundant of these
salts is the bitartrate of potash (cream of tartar), of which we have already spoken. In
addition to this, wines contain tartrate of
lime, tartrate of alumina, tartrate of iron,
chloride of sodium, chloride of potassium,
sulphate of potash, phosphate of alumina.
These salts occur in the proportion of from
one to four parts in the one thousand of wine.
They do not make much difference in the
flavor or action of wines ; but their presence
or absence is one of the surest indications of
the genuineness of a wine. Those who manufacture wines with alcohol and water, and
add a certain quantity of good wine to give a
flavor, do not usually add these mineral constituents, which are always the best test of a
genuine wine.

In conclusion, we may say, with regard to
the medicinal and dietetical use of wines—

1. That, where they are employed for the
sake of the stimulating effects of alcohol, it is
a matter of indifference which may be administered, remembering that some wines are
twice the strength of others.

2. The bouquet and flavor of wines render
them more agreeable than to drink any form
of mixed spirits or beer.

3. Wines are less likely, especially when
administered on an empty stomach, to do
harm to the coats of the stomach than any
mixture of brandy.

4. All sugared wines, as port and sherry,
shoud be interdicted in gouty states of the
system, and in diabetes, and in dyspepsia
attended with wind in the stomach.

5. For all dietetical purposes, clarets, hocks,

and the dry wines of Greece, especially the
latter, are to be preferred before all others.

6. Where it is desirable to secure an astringent effect, the red light wines are to be pre-
fered to the white.

7. The tartaric acid of wines is not injurious, and does not increase acidity in the stomach or the blood. It is an error to suppose,
on this account, that unsugared wines may
not be given where there is a tendency to
form lactic acid in the stomach or lithic acid
in the blood.

8. Where powerful stimulants are required,
it is better to give brandy, gin, whiskey, or
robur, than even the stronger wines.

9. Pure spirits with water are better than
wines manufactured from impurely distilled
alcohol, which contains fusel oil, and acts in-
juriously on the nervous system.

WINTER-COUGH is a very common
symptom in cases of emphysema and chronic
bronchitis. It is generally worse every winter, and may go away in the summer altogether. Those who are exposed to the weather are very liable to it ; also those who are
intemperate, and those who have heart and
kidney disease. The best thing for those who
can afford it is to keep in the house in bad
weather, or seek some milder climate, but
this can seldom be done. Wearing a respirator, not talking in the open air, and avoiding
fogs and night air, are useful measures. (See
BRONCHITIS and EMPHYSEMA.)

WINTER-GREEN. (See CHIMAPHILA.)

WISDOM-TEETH are generally cut between twenty and twenty-five years of age ;
they are four in number, two in each jaw,
and are placed at the back part of the mouth.
Sometimes a little discomfort attends their
coming through the gum.

WOMB. (See UTERUS, INVERSION OF
WOMB, INVOLUTION, PREGNANCY, and PRO-
LAPSUS.)

WOOL may either be bought at the chemist's, under the name of cotton-wool, when it
is very fine and white and soft, or at the
draper's, under the name of wadding, when
it has a sort of glaze or thin skin over it,
which enables it to be cut into lengths, and
when opened so that the skin is outside it
forms a valuable dressing and protection from
the air for burns and scalds, and is also
largely used as a warm covering for rheumatic limbs and joints, and in cases where flannel appears to be too harsh and unyielding a
material.

WORMS include all those parasites which
infest the intestinal canal ; they are commonly divided into three classes—the tape-
worms, the round worms, and the thread-
worms. (See ANTHELMINTICS, ASCARIDES,
ENTOZOA, and PARASITES.)

WORMWOOD is the flowering herb of
the *Artemisia Absinthium*, and is the flavoring ingredient in the liqueur *absinthe*. The

odor is disagreeable and the taste very bitter. The substance contains a bitter principle abstracted by alcohol, called absinthine. The plant itself, or an infusion of it, is a powerful bitter tonic, and it is said also anthelmintic. The liqueur is said to give rise to peculiar affections of the nervous system, different from those of ordinary alcoholism.

WOUNDS. (See ACCIDENTS, DISSECTION WOUNDS, GUN-SHOT WOUNDS.)

WRIST-DROP is a symptom occasionally met with in cases of lead-poisoning; the patient is then more or less unable to raise the wrist, as the extensor muscles of the arms are wasted and paralyzed. (See LEAD POISONING).

WRITER'S CRAMP is a wasting of the muscles of the ball of the thumb, caused, as the name implies, by too much using of them; rest and electricity are the best remedies. (See PARALYSIS.)

WRY-NECK consists in a remarkable but not very uncommon distortion of the head and neck, which in the majority of cases is congenital. In a well-marked instance of this affection the following appearances are presented: the entire head is bent forward and downward, and is approximated to the tip of the shoulder, usually on the right side; the face is directed forward, slightly upward, and to the opposite or left side; the right side of the neck is traversed in a direction from above downward and forward by a tense and hard subcutaneous band which is formed by the persistent and unnatural contraction of the sterno-mastoid muscle—the muscle which in the healthy state may be distinctly seen passing from the back of the ear on each side downward along the side of the neck to the upper margin of the breast-bone; the side of the face which corresponds to the contracted sterno-mastoid muscle is usually smaller than the opposite half; any attempt made to restore the head to its normal erect position will always cause severe pain. This distortion, when congenital, may be due to some disorder or deficiency in development of the fœtal nervous system, to uterine pressure associated with an irregular position of the child in the womb, or to violence produced during delivery, as forcible twisting of the neck by the rough usage of forceps. The congenital wry-neck is usually slight and almost inappreciable for some months after birth, but as the child grows up and begins to take active exercise, the distortion rapidly increases, and soon gives rise to uneasiness and even suffering.

The most frequent cause of non-congenital forms of wry-neck is rigidity of the muscles on one side of the neck in consequence of rheumatic or inflammatory affections. The distortion in some few instances is due to paralysis of one sterno-mastoid muscle, the head being drawn to the opposite shoulder by the unopposed contraction of the fellow muscle. A condition resembling genuine wry-neck may be produced by the following causes: disease of the bones or joints of the cervical portion of the spine, the retractile action of an extensive scar on one side of the neck, extensive scrofulous ulceration along the neck, swelling and induration of cervical glands.

A variety of wry-neck is occasionally met with in which there are incessant spasmodic contractions of one sterno-mastoid muscle, approximating the head to the shoulder by violent, jerking movements. This troublesome affection is always acquired, and seldom comes on before the age of twenty-five or thirty years. It occurs more frequently in females than in males. After it has lasted for a long time great pain is left by the patient, in consequence of the violent and repeated movement of the head and the vertebral bones in the neck, and much debility results from want of sleep. The convulsive movements in most cases persist during the life of the patient, although they may be relieved from time to time by galvanism and by a change of air and scene. Subcutaneous section of the lower part of the affected sterno-mastoid muscle arrests the spasmodic movements, but these in the course of a month or six weeks usually return again with the same activity. Occasionally spasmodic wry-neck is a temporary affection excited by gastric and intestinal irritation or congestion of the liver.

Treatment.—In cases of genuine wry-neck, whether congenital or acquired, cure may sometimes be effected by the use of a collar or machine contrived in order to keep up prolonged and gradually increasing extension of the contracted sterno-mastoid muscle. This kind of treatment is usually associated with frequently repeated shampooing of the affected side of the neck. In old and advanced cases, however, this treatment will prove ineffectual, and then the last resource of the surgeon will be subcutaneous division of the contracted muscle. This operation, followed by gradual elevation of the head by means of a collar, generally results in permanent cure.

X.

XANTHELASMA is a disease of the skin in which yellow, slightly raised patches occur on various parts of the body. It is most common around the eyelids, but is seen also on the elbows, knuckles, and other parts of the body. It is sometimes associated with jaundice, but it is a condition of no practical importance and requires no treatment. It is of very rare occurrence, and gives rise to no troublesome symptoms.

XANTHIC OXIDE, or **XANTHINE**, is sometimes met with in the form of a calculus, or stone in the bladder : such stones are usually small, but are of such rare occurrence as to be looked upon more as curiosities than as possessing any practical interest.

XANTHINE. (See XANTHIC OXIDE.)

XERODERMA, or **ICHTHYOSIS**, is a form of dry skin sometimes met with in children and adults. It is usually congenital, and may occur in many members of the same family. The skin is dry, harsh, and rough. On the face the epidermis is usually comparatively smooth ; on the neck it is rough and has a branny appearance ; on the rest of the body, cracks are seen on the skin. Such patients do not generally enjoy good health, and are liable to palpitation of the heart.

Treatment.—This is usually of but little avail. Olive oil will remove the scales and improve the general appearance, while cod-liver oil and steel may be taken internally for the benefit of the health.

Y.

YEAST as used in medicine is chiefly employed in making poultices. These are applied to old sores, but are not so useful as charcoal poultices or those of chlorinated soda.

YELLOW FEVER is an infectious, continued fever, beginning with languor, chilliness, headache, and pain in the back ; the countenance is flushed and the eyes moist and suffused ; the skin gradually acquires a lemon or greenish-yellow color ; there is generally a wandering of the mind, and often delirium ; the patient is restless and watchful, or he may pass into a state of drowsiness and then coma ; there is an uneasy feeling at the pit of the stomach, and vomiting, at first of a clear, glairy fluid, and afterward of a coffee-ground appearance ; there may also be irrepressible hiccough, and shrieking or melancholy wailing. Sometimes the disease progresses with fearful rapidity, running through all the stages and putting an end to the patient's life within twenty-four hours. The vomiting of the dark-colored fluid is indicative of a fatal termination.

Yellow fever was first recorded in the West Indies in 1647, and since then it has been more or less present up to the present time. In St. Thomas and San Domingo the disease seems to be permanently located. It appears to affect those who live in the low country more than those on the hills ; it does not spread, as a rule, to parts more than 3000 feet above the sea. A certain amount of heat is essential to the development of this fever ; few cases are observed where the temperature is less than 72° Fahr. ; and the first frost puts an end to its spread. Dr. Maclean, who has had much experience of disease in the tropics, thus lays down the differences between yellow and remittent fevers : Yellow fever is specifically distinct from remittent fever. Yellow fever is unknown in India, where true malarial fevers abound. There is in yellow fever an absence, for the most part, of that periodicity which is so characteristic of true malarial fevers—*i.e.*, the remissions and exacerbations. Men do not pass from recovery to health, as is the case in such a marked degree in yellow fever, after which there is no, or very little, evidence of the existence of any cachexy. Malarial fevers exist and are destructive at a temperature at which yellow fever is at once destroyed. Albuminous urine is almost invariable in yellow fever, only occasional in remittent. There is in yellow fever a great deal of bleeding from various parts of the body ; in remittent fever, this is generally absent. Quinine has a power over the malarial fevers, but not over yellow fever. Men suffer from malarial fevers again and again ; second attacks of yellow fever are very rare.

Treatment.—The patient should have a hot bath in the first stage, and then, going to bed, he should have warm drinks so as to encourage sweating ; this may be followed by a purgative so as to have the bowels well open. Mercury need not be given ; nor is quinine of any use. The sickness is very distressing, but may be relieved by lime-water or by a few drops of chlorodyne or chloroform ; creosote and hydrocyanic acid do not seem to be of any use for this purpose. Stimulants must be given according to the needs of each case. The great objects in treatment are to sustain the vital powers, to moderate the febrile excitement, and to check any distressing symptoms that may arise. Of course, competent medical advice should be obtained at the earliest possible moment.

YELLOW GUM, or the jaundice of new-born children, comes on two or three days after birth, and then the child's skin is of a yellow color, the urine very dark, and staining the cloths a deep yellow, while the motions are light. It is a simple disorder, which will soon pass away. It is due to the liver, engorged from the lungs, not acting properly at first. The child should be put to the breast, and the mother's milk is generally sufficiently aperient at first to open the bowels ; if not, a little grey powder may be given at bed-time. It may be some days before the yellow tinge has quite gone from the ski

YELLOW-WASH is a lotion made by dissolving corrosive sublimate in lime-water.

Z.

ZINC is introduced into the Pharmacopœia in the metallic form for the preparation of its chloride.

The *oxide of zinc* is made by heating the carbonate. It is a white powder, without taste or smell, and turns yellow by heating. Its only preparation is an ointment, which is very useful. If given internally in large doses it causes vomiting, but is never used with that intention. It is chiefly given as a nervine tonic and astringent. It is used, as are all the other preparations of zinc given internally, in chorea, epilepsy, hysteria, and neuralgia. Externally, the ointment is very useful as an application to raw, weeping surfaces. The dose is 5 or 10 grains.

Calamine is a form of oxide no longer officinal. Its ointment, known as Turner's cerate, long had a reputation where now the oxide is used.

Chloride of zinc is made by dissolving zinc in hydrochloric acid. A solution of it is officinal. When made into a paste with flour or any similar substance the chloride acts as a powerful escharotic. In weaker solutions it is a useful astringent. Chloride of zinc paste is sometimes used to destroy cancerous masses and malignant ulcers, so as to obtain a healthy fresh surface. A solution of this was long used for disinfectant purposes, under the title of Sir W. Burnett's solution.

Sulphate of zinc is the most important salt of the metal. It is got by dissolving the metal in sulphuric acid, as when hydrogen is prepared. The salt occurs in crystals, very much alike to those of Epsom salts, but gives off water instead of abstracting it from the atmosphere. Sulphate of zinc given internally in fair doses, gives rise to vomiting, speedily and surely. It is thus one of our best emetics in suitable cases, but must not be administered where there is already irritation. It is given in smaller doses like the oxide, as a tonic, in nervous complaints, chorea, epilepsy, hysteria, and the like. It is well combined with valerian. Externally, sulphate of zinc is very largely used in various forms of discharge, and is a most valuable astringent. The dose as an emetic is from 15 to 20 grains, as a tonic 3 to 5. As a lotion 3 grains may be dissolved in an ounce of water.

Carbonate of zinc and *acetate of zinc* are as yet little employed ; their effects are intermediate between those of the oxide and of the sulphate.

ZYMOSIS is a technical term, applied to actions of a peculiar and not much-understood nature, and allied to fermentation. There are various fevers which seem to have their origin in some poison which enters the system, and there for a time the poisonous germs seem to multiply and increase ; thus if a person be inoculated with a most minute quantity of small-pox matter he will catch the disease, if unprotected, and in the course of a few days hundreds of pustules will appear, and from each of these pustules a little fluid may be taken, and thousands of persons might in this way have the disorder. Now although we are ignorant as to the exact nature of the poisons in the different fevers, there are fair grounds for assuming that, when a small dose of any poison of this class enters the blood, it there goes through a process of multiplication just as yeast does during fermentation. All these poisons are contagious. There are seven principal diseases of the zymotic class, according to the nomenclature of the Registrar-General, and eleven others less common : 1, small-pox ; 2, measles ; 3, scarlet-fever ; 4, diphtheria ; 5, croup ; 6, whooping-cough ; 7, continued fevers, including typhus, typhoid, and simple continued fever ; 8, quinsy ; 9, erysipelas ; 10, puerperal fever ; 11, carbuncle ; 12, influenza ; 13, dysentery ; 14, diarrhœa ; 15, cholera ; 16, ague ; 17, remittent fever ; 18, rheumatism. It is most important to remember that all zymotic diseases are in a great measure preventible, and if proper precautions were observed and sanitary measures regularly carried out, an immense number of lives might be annually saved to the country.

THE END.

INDEX.

www.ingramcontent.com/pod-product-compliance
Lightning Source LLC
Chambersburg PA
CBHW020857210326

41598CB00018B/1700